Lecture Notes in Computer Sci

Edited by G. Goos, J. Hartmanis, and J. va

T0238299

Springer
Berlin
Heidelberg
New York
Hong Kong
London
Milan
Paris
Tokyo

Tandy Warnow   Binhai Zhu (Eds.)

Computing and Combinatorics

9th Annual International Conference, COCOON 2003
Big Sky, MT, USA, July 25-28, 2003
Proceedings

 Springer

Series Editors

Gerhard Goos, Karlsruhe University, Germany
Juris Hartmanis, Cornell University, NY, USA
Jan van Leeuwen, Utrecht University, The Netherlands

Volume Editors

Tandy Warnow
University of Texas at Austin
Department of Computer Science
One University Station, C0500
Austin, TX 78712, USA
E-mail: tandy@cs.utexas.edu

Binhai Zhu
Montana State University
Department of Computer Science
EPS 357
Bozeman, MT 59717, USA
E-mail: bhz@cs.montana.edu

Cataloging-in-Publication Data applied for

A catalog record for this book is available from the Library of Congress.

Bibliographic information published by Die Deutsche Bibliothek
Die Deutsche Bibliothek lists this publication in the Deutsche Nationalbibliografie;
detailed bibliographic data is available in the Internet at <http://dnb.ddb.de>.

CR Subject Classification (1998): F.2, G.2.1-2, I.3.5, C.2.3-4, E.1, E.5, E.4

ISSN 0302-9743
ISBN 3-540-40534-8 Springer-Verlag Berlin Heidelberg New York

This work is subject to copyright. All rights are reserved, whether the whole or part of the material is
concerned, specifically the rights of translation, reprinting, re-use of illustrations, recitation, broadcasting,
reproduction on microfilms or in any other way, and storage in data banks. Duplication of this publication
or parts thereof is permitted only under the provisions of the German Copyright Law of September 9, 1965,
in its current version, and permission for use must always be obtained from Springer-Verlag. Violations are
liable for prosecution under the German Copyright Law.

Springer-Verlag Berlin Heidelberg New York
a member of BertelsmannSpringer Science+Business Media GmbH

http://www.springer.de

© Springer-Verlag Berlin Heidelberg 2003
Printed in Germany

Typesetting: Camera-ready by author, data conversion by PTP-Berlin GmbH
Printed on acid-free paper   SPIN: 10927991   06/3142   5 4 3 2 1 0

Preface

The papers in this volume were presented at the 9th Annual International Computing and Combinatorics Conference (COCOON 2003), held July 25–28, 2003, in Big Sky, MT, USA. The topics cover most aspects of theoretical computer science and combinatorics related to computing.

Submissions to the conference this year were conducted electronically. A total of 114 papers were submitted, of which 52 were accepted. The papers were evaluated by an international program committee consisting of Nina Amenta, Tetsuo Asano, Bernard Chazelle, Zhixiang Chen, Francis Chin, Kyung-Yong Chwa, Robert Cimikowski, Anne Condon, Michael Fellows, Anna Gal, Michael Hallett, Daniel Huson, Naoki Katoh, D.T. Lee, Bernard Moret, Brendan Mumey, Gene Myers, Hung Quang Ngo, Takao Nishizeki, Cindy Phillips, David Sankoff, Denbigh Starkey, Jie Wang, Lusheng Wang, Tandy Warnow and Binhai Zhu. It is expected that most of the accepted papers will appear in a more complete form in scientific journals.

The submitted papers were from Canada (6), China (7), Estonia (1), Finland (1), France (1), Germany (8), Israel (4), Italy (1), Japan (11), Korea (22), Kuwait (1), New Zealand (1), Singapore (2), Spain (1), Sweden (2), Switzerland (3), Taiwan (7), the UK (1) and the USA (34). Each paper was evaluated by at least three Program Committee members, assisted in some cases by subreferees. In addition to selected papers, the conference also included three invited presentations by Jon Bentley, Dan Gusfield and Joel Spencer.

We thank all the people who made this meeting possible: the authors for submitting papers, the program committee members and external referees (listed in the proceedings) for their excellent work, and the three invited speakers. Finally, we thank the Computer Science Department of Montana State University for the support and the local organizers and colleagues for their assistance.

July 2003   Tandy Warnow, Binhai Zhu

Program Committee Chairs

Tandy Warnow, University of Texas at Austin, USA
Binhai Zhu, Montana State University, USA

Program Committee Members

Nina Amenta, UC Davis, USA
Tetsuo Asano, JAIST, Japan
Bernard Chazelle, Princeton University, USA
Zhixiang Chen, University of Texas at Pan American, USA
Francis Chin, University of Hong Kong, China
Kyung-Yong Chwa, KAIST, Korea
Robert Cimikowski, Montana State University, USA
Anne Condon, UBC, Canada
Michael Fellows, University of Newcastle, Australia
Anna Gal, University of Texas at Austin, USA
Michael Hallett, McGill University, Canada
Daniel Huson, Tuebingen University, Germany
Naoki Katoh, Kyoto, Japan
D.T. Lee, Academia Sinica, Taiwan
Bernard Moret, University of New Mexico, USA
Brendan Mumey, Montana State University, USA
Gene Myers, UC Berkeley, USA
Hung Quang Ngo, SUNY at Buffalo, USA
Takao Nishizeki, Tohoku, Japan
Cindy Phillips, Sandia National Labs, USA
David Sankoff, University of Montreal, Canada
Denbigh Starkey, Montana State University, USA
Jie Wang, University of Massachusetts at Lowell, USA
Lusheng Wang, City University of Hong Kong, China

Organizing Committee

Gary Harkin, Montana State University, USA
Michael Oudshoorn, Montana State University, USA
John Paxton, Montana State University, USA
Jeannette Radcliffe, Montana State University, USA
Rocky Ross, Montana State University, USA
Denbigh Starkey (Chair), Montana State University, USA
Year-Back Yoo, Montana State University, USA

Referees

Eric Bach
David Bryant
Prosenjit Bose
David Bunde
Jin-Yi Cai
Bob Carr
To Yat Cheung
Zhe Dang
Olaf Delgado-Friedrichs
Allyn Dimock
Nadia El-Mabrouk
Stanley P.Y. Fung
Frederic Green
Minghui Jiang
Tao Jiang

Anna Johnston
Mikio Kano
Neal Koblitz
Francis C.M. Lau
Hendrik Lenstra
Chi Ming Leung
Xiang-Yang Li
Kazuyuki Miura
Cris Moore
Subhas C. Nandy
William D. Neumann
Kay Nieselt-Struwe
Ojas Parekh
Krzysztof Pietrzak
Chung Keung Poon

Md. Saidur Rahman
Romeo Rizzi
Yaoyun Shi
Shang-Hua Teng
Cao An Wang
Guohua Wu
Xiaodong Wu
Hongwei Xi
Jinhui Xu
Siu-Ming Yiu
Xizhong Zheng
Xiao Zhou
Dengping Zhu
David Zuckerman

Table of Contents

Graph Theory/Algorithms I

Automata/Petri Net Theory

Graph Theory/Algorithms II

Complexity Theory II

Distributed Computing

Web-Based Computing

Complexity Theory III

Graph Theory/Algorithms III

Computational Geometry II

Graph Theory/Algorithms IV

Scheduling

Computational Geometry III

Graph Drawing

LIAR!

Joel Spencer

Courant Institute
New York University
spencer@cs.nyu.edu

Paul is trying to ascertain an unknown x, out of n possibilities, from an adversary Carole by asking a series of q queries. In the standard "Twenty Questions" Paul wins if and only if $n \leq 2^q$. In Liar Games, Carole is allowed, under certain restrictions, to give an incorrect response. Throughout this study Carole will be restricted to give at most k incorrect responses, or lies. Asymptotic analysis here will be for k arbitrary but fixed, $k = 1$ being a natural and interesting case. The game with ten queries, one hundred possibilities, and (at most) one lie is amusing to play.

The basic liar game, in which Paul can make arbitrary binary queries, has been well studied. The basic bound is that if

$$n \left[\binom{q}{0} + \binom{q}{1} + \ldots + \binom{q}{k} \right] > 2^q$$

then Carole (who is allowed an adversary strategy) will win. We shall indicate a two proofs of this basic bound and arguments for why the converse almost (but not quite!) holds.

There is a natural connection between Liar Games and Coding Theory. The protocol must allow Paul to determine x despite (at most) k "errors" in the "transmission." The major distinction is that Paul's queries are sequential and can depend on previous responses. Some have described Liar Games as Coding Theory with Feedback.

In the Half-Lie Game Carole has a further restriction. If the correct answer is yes then Carole *must* say yes. In medical terminology, there can be false positives (though at most k of them) but no false negatives. Fixing k we define $A_k(q)$ to be the largest n for which Paul wins the Half-Lie Game with q queries. At COCOON 2000 Cicalese and Mundici found the asymptotics when only one lie is permitted:

$$A_1(q) \sim \frac{2^q}{2q}$$

We shall give a different proof of this result and a generalization to any fixed number k of lies.

The Half-Lie Game has two useful interpretations. I take $k = 1$ for simplicity. One is as a vector game. In the middle of the game the state is a position vector $\boldsymbol{p} = (a, b)$ where a is the number of x for which Carole has not yet lied and b is the number of x for which Carole has lied once. Now Paul's query can also be given in vector form and the two potential new states have simple descriptions.

T. Warnow and B. Zhu (Eds.): COCOON 2003, LNCS 2697, pp. 1–2, 2003.
© Springer-Verlag Berlin Heidelberg 2003

The second, more combinatorial, view is in terms of 1-trees. A 1-tree is a subset of $\{Y, N\}^q$ consisting of one designated root $\boldsymbol{r} = (r_1, \ldots, r_q)$ and, for each i with $r_i = N$, a "child" $\boldsymbol{r}' = (r'_1, \ldots, r'_q)$ that agrees with $\boldsymbol{r}$ in the first $i-1$ coordinates and has $r'_i = Y$. These 1-trees represent possible sets of response sequences of Carole with a particular x. We argue that Paul wins if and only if it is possible to pack n of these 1-trees (more generally, k-trees) into $\{Y, N\}^q$. This leads to some intriguing combinatorial questions.

At a meeting in Dagstuhl in Combinatorial Games last year Elwyn Berlekamp noted that the Half-Lie games corresponds to Coding Theory over the Z-channel. We have found that the analysis readily extends, indeed is more natural, over arbitrary channels. By a channel we here mean a finite set of possible lie types. For example: let Paul ask ternary queries with answers A, B, C. We allow Carole to "lie" with B when the answer is A and with C when the answer is B, but otherwise she cannot lie. Fixing the channel Ch and the number of lies k let $A_{k,Ch}(q)$ be the largest n for which Paul wins. The methods developed for the Half-Lie problem allow us to find the asymptotics of $A_{k,Ch}(q)$. To our surprise, the asymptotics depend only on the number E of lie types. When the queries are t-ary we show

$$A_{k,Ch}(q) \sim \frac{t^k}{E^k} \frac{t^q}{\binom{q}{k}}$$

This is joint work with Ioana Dumitriu (M.I.T.) and Catherine Yan (Texas A&M).

Experiments for Algorithm Engineering

Jon Bentley

Avaya Labs Research
Basking Ridge, New Jersey, USA
jbentley@avaya.com

Hoare introduced and analyzed the Quicksort algorithm in the early 1960s [6]. By the early 1970s, a highly tuned version of that algorithm was implemented in the Unix system's main memory sort function, qsort. For two decades, that code performed admirably.

In the early 1990s, Alan Wilks and Rick Becker once again used that old, reliable program, and were stunned by the results. A run that should have taken a few minutes was cancelled after hours. They studied their input data and eventually produced an eight-line program that showed that their run would have taken weeks. They enclosed that program in a superb bug report that they e-mailed to me (details are described by Bentley [1]). They had stumbled across a problem that Hoare had foreseen: selection of the partitioning element. An implementation cleverness that had stood the test of two decades worth of time had finally failed catastrophically.

Doug McIlroy and I set out to solve their problem. Experiments showed that the venerable implementation of Quicksort failed on a disturbing variety of inputs. We looked at other production-quality implementations and found that they all exhibited such behavior. We thus faced the task of building our own implementation of the Quicksort algorithm to meet the interface specifications of the qsort utility. Bentley and McIlroy describe the resulting algorithm and sketch its history [4].

This talk describes many of the experiments that took place behind the scenes in that process, and the tools used to perform them. In addition to using profilers to time particular implementations, we also used a program to generate a cost model that would give us insight into the relative costs of critical operations (details are given by Bentley [2]). Simple experiments allowed us to measure the effectiveness of a large family of sampling schemes for selecting the partitioning element. Caching was not critical in the early 1990s but now plays a key role in the performance of sorting implementations; simple experiments accurately predict the performance of various sorting algorithms under caching. Many of these issues are addressed by Bentley [3].

We set out to build a production-quality sorting algorithm, and we were successful at the task. The algorithm has been widely used since it was introduced, and we have yet to hear of it catastrophically failing. The effort also led to some pleasant theory, including the ternary search trees described by Bentley and Sedgewick [5] and McIlroys [7] killer adversary for any implementation of Quicksort. I will describe some of my experiences in experiments leading to new theory.

T. Warnow and B. Zhu (Eds.): COCOON 2003, LNCS 2697, pp. 3–4, 2003.
© Springer-Verlag Berlin Heidelberg 2003

References

1. J. L. Bentley. Software Exploratorium: The Trouble With Qsort, *UNIX Review*, Vol. 10, 2, pp. 85–93, February 1992.
2. J. L. Bentley. Software Explorations: Cost Models for Sorting, *UNIX Review*, Vol. 15, 4, pp. 65–72, April 1997.
3. J. L. Bentley. *Programming Pearls*, Second Edition, Addison-Wesley, Reading, MA, 2000.
4. J. L. Bentley and M. D. McIlroy. Engineering a sort function, *Software–Practice and Experience*, Vol. 23, 1, pp. 1249–1265, 1993.
5. J. L. Bentley and R. Sedgewick. Fast Algorithms for Sorting and Searching Strings, *Proceedings of the 8th Annual ACM-SIAM Symposium on Discrete Algorithms*, pp. 360–369, January 1997.
6. C. A. R. Hoare. Quicksort, *Computer Journal*, Vol. 5, 1, 1962.
7. M. D. McIlroy. A killer adversary for quicksort, *Software–Practice and Experience*, Vol. 29, pp. 341–344, 1999.

Empirical Exploration of Perfect Phylogeny Haplotyping and Haplotypers

Ren Hua Chung and Dan Gusfield*

Computer Science Department,
University of California,
Davis, Davis CA 95616, USA
gusfield@cs.ucdavis.edu

Abstract. The next high-priority phase of human genomics will involve the development of a full *Haplotype Map* of the human genome [15]. It will be used in large-scale screens of populations to associate specific haplotypes with specific complex genetic-influenced diseases. A key, perhaps bottleneck, problem is to computationally determine haplotype pairs from genotype data. An approach to this problem based on viewing it in the context of perfect phylogeny was introduced in [14] along with an efficient solution. A slower (in worst case) variation of that method was implemented [3]. Two simpler methods for the perfect phylogeny approach that are also slower (in worst case) than the first algorithm were later developed [1,7]. We have implemented and tested all three of these approachs in order to compare and explain the practical efficiencies of the three methods. We discuss two other empirical observations: a strong phase-transition in the frequency of obtaining a unique solution as a function of the number of individuals in the input; and results of using the method to find non-overlapping intervals where the haplotyping solution is highly reliable, as a function of the level of recombination in the data. Finally, we discuss the biological basis for the size of these tests.

1 Introduction to SNP's, Genotypes, and Haplotypes

In diploid organisms (such as humans) there are two (not completely identical) "copies" of each chromosome, and hence of each region of interest. A description of the data from a single copy is called a *haplotype*, while a description of the conflated (mixed) data on the two copies is called a *genotype*. In complex diseases (those affected by more than a single gene) it is often much more informative to have haplotype data (identifying a set of gene alleles inherited together) than to have only genotype data.

The underlying data that forms a haplotype is either the full DNA sequence in the region, or more commonly the values of *single nucleotide polymorphisms (SNP's)* in that region. A SNP is a single nucleotide site where exactly two (of four) different nucleotides occur in a large percentage of the population. The

---

* Research Supported by NSF grants DBI-9723346 and EIA-0220154

T. Warnow and B. Zhu (Eds.): COCOON 2003, LNCS 2697, pp. 5–19, 2003.
© Springer-Verlag Berlin Heidelberg 2003

SNP-based approach is the dominant one, and high density SNP maps have been constructed across the human genome with a density of about one SNP per thousand nucleotides.

1.1   The Biological Problem

In general, it is not feasible to examine the two copies of a chromosome separately, and *genotype* data rather than haplotype data will be obtained, even though it is the haplotype data that will be of greatest use.

Data from m sites (SNP's) in n individuals is collected, where each site can have one of two states (alleles), which we denote by 0 and 1. For each individual, we would ideally like to describe the states of the m sites on each of the two chromosome copies separately, i.e., the haplotype. However, experimentally determining the haplotype pair is technically difficult or expensive. Instead, the screen will learn the $2m$ states (the genotype) possessed by the individual, without learning the two desired haplotypes for that individual. One then uses computation to extract haplotype information from the given genotype information. Several methods have been explored and some are intensively used for this task [4,5,8,23,13,22,20,21]. None of these methods are presently fully satisfactory, although many give impressively accurate results.

1.2   The Computational Problem

Abstractly, input to the haplotyping problem consists of n *genotype* vectors, each of length m, where each value in the vector is either 0,1, or 2. Each position in a vector is associated with a site of interest on the chromosome. The position in the genotype vector has a value of 0 or 1 if the associated chromosome site has that state on both copies (it is a *homozygous* site), and has a value of 2 otherwise (the chromosome site is *heterozygous*).

Given an input set of n genotype vectors, a *solution* to the *Haplotype Inference (HI) Problem* is a set of n pairs of binary vectors, one pair for each genotype vector. For any genotype vector g, the associated binary vectors v_1, v_2 must both have value 0 (or 1) at any position where g has value 0 (or 1); but for any position where g has value 2, exactly one of v_1, v_2 must have value 0, while the other has value 1. That is, v_1, v_2 must be a feasible "explanation" for the true (but unknown) haplotype pair that gave rise to the observed genotype g. Hence, for an individual with h heterozygous sites there are 2^{h-1} haplotype pairs that could appear in a solution to the HI problem.

For example, if the observed genotype g is 0212, then the pair of vectors 0110, 0011 is one feasible explanation, out of two feasible explanations. Of course, we want to find the explanation that actually gave rise to g, and a solution for the HI problem for the genotype data of all the n individuals. However, without additional biological insight, one cannot know which of the exponential number of solutions is the "correct one".

2   Perfect Phylogeny

Algorithm-based haplotype inference would be impossible without the implicit or explicit use of some genetic model, either to asses the biological fidelity of any proposed solution, or to guide the algorithm in constructing a solution. Most of the models use statistical or probabilistic aspects of population genetics. We will take a more deterministic or combinatorial approach.

The most powerful such genetic model is the population-genetic concept of a *coalescent*, i.e., a rooted tree that describes the evolutionary history of a set of sequences (or haplotypes) in sampled individuals [24,16]. The key observation is that "In the absence of recombination, each sequence has a single ancestor in the previous generation." [16].

There is one additional element of the basic coalescent model: the *infinite-sites* assumption. That is, the m sites in the sequence (SNP sites in our case) are so sparse relative to the mutation rate, that in the time frame of interest at most one mutation (change of state) will have occurred at any site. This assumption is usually made without contention in SNP data.

Hence the coalescent model of haplotype evolution says that without recombination, the true evolutionary history of $2n$ haplotypes, one from each of $2n$ individuals, can be displayed as a tree with $2n$ leaves, and where each of the m sites labels exactly one edge of the tree, i.e., at a point in history where a mutation occurred at that site. This is the underlying genetic model that we assume from here on. See [24] for another explanation of the relationship between sequence evolution and coalescents.

In more computer science terminology, the no-recombination and infinite-sites model says that the $2n$ haplotype (binary) sequences can be explained by an (unrooted) *perfect phylogeny* [11,12]:

Definition. Let B be an $2n$ by m 0-1 (binary) matrix, and V be an m-length binary string. A *rooted perfect phylogeny for B* is a rooted tree T with exactly $2n$ leaves that obeys the following properties:

1) Each of the $2n$ rows labels exactly one leaf of T.

2) Each of the m columns labels *exactly one* edge of T.

3) Every interior edge (one not touching a leaf) of T is labeled by *at least* one column.

4) For any row i, the columns that label the edges along the unique path from the root to leaf i specify the columns of B where row i has a value that is different from the value of V in that column. In other words, knowing V, that path is a compact representation of row i.

If we don't know V at input, the *unrooted perfect phylogeny* problem is to determine a V so that B has perfect phylogeny rooted at V.

The classic Theorem of Perfect Phylogeny is that a binary matrix B has a perfect phylogeny if and only if for each pair of columns, there are no four rows in those columns with values (0,0), (0,1), (1,0) and (1,1). Moreover, if the columns of B are distinct, then there is only one perfect phylogeny for B. Note that an edge to a leaf need not have a column label.

The above condition is known as the "Four-Gametes Test" in the population genetics literature, and is known as the "Compatibility Test" in the phylogenetic literature.

2.1   The Perfect Phylogeny Haplotype (PPH) Problem

Under the coalescent model of haplotype evolution, the HI problem now has precisely the following combinatorial interpretation:

Given a set of n genotypes G, find an HI solution consisting of at most $2n$ distinct haplotype vectors B, such that the vectors in B fit a perfect phylogeny. This is the unrooted version of the PPH problem. See Figure 1 for a trivial example.

If a root sequence V is specified, then the perfect phylogeny is required to have the root sequence of V. The two versions of the problem are actually equivalent in the sense that an instance of one variation can be reduced to the other variation, so that an algorithm for one variation can be used for both variations.

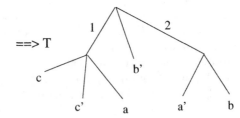

Fig. 1. A simple example where there are two solutions to the HI problem, but only the one shown solves the PPH problem. Matrix Q is created from matrix M, doubling the rows to prepare for the PPH solution B.

3   Three Algorithms and Programs for the PPH Problem

In this section we briefly describe three algorithms and three programs (GPPH, HPPH and DPPH) for the PPH problem. The first algorithm for the PPH problem was developed in [14]. After that publication, two additional methods were developed and presented [1] and [7]. Program GPPH follows the basic approach presented in [14], with some modifications as detailed below. Programs DPPH and HPPH follow the algorithms developed in [1] and [7], respectively.

The three programs are available at: cs.ucdavis.edu/~gusfield

3.1 Solution by Graph Realization: Program GPPH

The first algorithm for the PPH problem, given in [14], is based on reducing the problem to a well-studied problem in graph theory.

Let E_r a set of r distinct integers. A "path set" is an *unordered* subset P of E. A path set is "realized" in a undirected, edge-labeled tree T consisting of r edges, if each edge of T is labeled by a distinct integer from E_r, and there is a contiguous path in T whose labels consist of the integers in P. For simplicity, we refer to each integer in E_r as an "edge", since in the tree T that we seek, each edge will be uniquely labeled with an integer in E_r. Note that since P is unordered, its presentation does not specify or constrain the order that those edges appear in T. In quite different terms, from the 1930's to the 1960's Whitney and Tutte and others studied and solved the following problems:

The Graph Realization Problem. Given E_r and a family $\Pi = P_1, P_2, ..., P_k$ of path sets, find an undirected tree T in which each path set is realized, or determine that no such tree exists. Further, determine if there is only one such T, and if there is more than one, characterize the relationship between the realizing trees.

In [14] we reduce the PPH problem to the graph realization problem, and have implemented this approach in a program called here GPPH[1]. Program GPPH implements a slightly different reduction than is described in [14]. The actual reduction is detailed at:

wwwcsif.cs.ucdavis.edu/~gusfield/recomberrata.pdf.

After reducing a problem instance, program GPPH solves the graph realization instance, using a variation [10] of Tutte's classic algorithm for graph realization [25]. In GPPH, we implement the reduction and the graph realization solution separately. The reduction takes $O(nm)$ time to create an instance of the graph realization problem, and the graph realization module runs in $O(nm^2)$ time, and is completely general, not incorporating any particular features of the PPH problem. Having a general solution to the graph realization problem gives us a program that can be used for other applications of graph realization besides the PPH problem. It is of interest to see how this general approach performs relative to methods that incorporate particular insights about the PPH problem.

Tutte's graph realization algorithm is a method that recursively solves graph realization problems for a subset E of E_r, and a family of path sets $\Pi(E)$, formed from the family Π by restricting each path set in Π to E. Given E and $\Pi(E)$, and knowledge of what decisions have already been made, the algorithm chooses an edge e (called here the "pivot" edge) from E and determines, by examining

---

[1] This program was previously called PPH, but now that multiple programs exist for the PPH *problem*, each is given a distinct name.

the path sets in $\Pi(E)$ and using general insights about paths in trees, whether there are edges in E that must be on one particular side of e or the other side of e. It also determines whether there are pairs of edges in E that must be separated by e, and if so, whether these pairs form a bipartition. To do that, it constructs a graph $G(E)$ containing one node for each edge in E, and one undirected arc between every pair of edges (e', e'') in E that must be separated by e. If $G(E)$ is bipartite, Tutte's algorithm arbitrarily puts the edges from one side of $G(E)$ on one side of e, and the edges from the other side of $G(E)$ on the other side of e. It then recursively solves the graph realization problem for the two sets of edges on either side of e, and when those two subproblems have been solved (returning two subtress), the algorithm determines how to attach those two subtrees to e. Hence, the recursion is central to the method. If $G(E)$ were determined not to be bipartite, then the algorithm correctly determines that there is no tree realizing all the path sets in E_r.

There are other solutions to the graph realization problem not based on Tutte's algorithm. The method in [2] is based on a general algorithm due to Lofgren, and runs in $O(nm\alpha(nm))$ time, where α is the inverse-Ackerman function. That algorithm is the basis for the worst-case time bound established in [14], but we found it to be too complex to implement. In [14] it was explained that after one PPH solution is obtained, by whatever method, one can get an implicit representation of the set of all PPH solutions in $O(m)$ time.

In program GPPH, about 1000 lines of C code were written to implement the reduction part of the method and about 4000 lines of C code were written to implement the graph realization part of the method.

3.2   Algorithm and Program HPPH

Although obtained independently of Tutte's method, the method of Eskin, Halperin and Karp [7] can be viewed as a specialization, to the PPH problem, of Tutte's general graph realization method. The method is developed under the assumption that the root V is known, and it exploits the rooted nature of the PPH problem (even when V is not given, a solution to the PHH problem is a rooted tree) to find simpler, specific rules to use when picking a pivot edge e, and to determine the placement of the other edges to e. In particular, it builds a tree top-down from the root, by choosing for a pivot edge e, a "maximal" edge, i.e., one that is guaranteed not to be below any of the edges that are not yet in the tree. Finding such a maximal pivot edge is a simple matter in the context of the PPH problem using the "leaf-count" idea in [14]. Because of the maximality, the general operation in Tutte's algorithm of determining which edges are on which side of e becomes simpler. One determines for each other edge e', whether e' must be placed below e (on the same path from the root as e), or must not be placed below e, or whether both placements are permitted. As in Tutte's method, edges in the last category are considered pairwise to find pairs of edges where exactly one of the pair must be placed below e, but either one of the pair can be the edge chosen to be below e. These pairs are represented by a graph, and the algorithm checks that the graph is bipartite. If so, one set of the bipartition is arbitrarily chosen to be placed below e and the other chosen to not be placed below e. After the algorithm makes that decision, it recurses until a tree is built or no further progress can be made. The algorithm can also implicitly represent the set of all PPH solutions by noting where it could have made alternative choices.

We call the program implementing this method the HPPH program. Because the HPPH program is a specialization of Tutte's general method using simpler rules to determine and implement the pivot, it is expected to run faster in practice than the GPPH program. Moreover, the top-down structure of the method has the consequence that the recursion used to describe the algorithm in [7] is actually tail-recursion, so that the method can be implemented iteratively without any explicit recursion. This can speed up the program, particularly for large data sets, and we have implemented HPPH without recursion. Viewed as a specialization of Tutte's general method, it is of interest to see how much of a speed-up the simpler pivot rules provide. About 1500 lines of C code were written to implement HPPH.

Segue to DPPH. One of the central details in HPPH (and also in the DPPH method to be discussed) is that for certain pairs of columns (c, c') in M, we can determine from looking at those columns alone (without *any* other information from M) whether the edge labeled with c must be placed below the edge labeled with c', or the c' edge must be placed below the c edge, or neither edge can be placed below the other. This is called a "forced relationship". The HPPH program spends $O(nm^2)$ time at the start to examine the columns in pairs to find any forced relationships (actually it determines another weaker relationship as well). After this first stage, the algorithm does operations whose best time-bound is $O(nm^2)$.

3.3   Algorithm and Program DPPH

The method in DPPH [1] is not based (explicitly or implicitly) on a graph realization algorithm, but is based on deeper insights into the combinatorial structure of the PPH problem and its solution. These insights are exploited to avoid any recursion or iteration, in contrast to the GPPH and HPPH programs. Rather, after the initial $O(nm^2)$-time stage, where all the forced relationships are found, the algorithm finds a PPH solution in $O(q^2)$ time, where q is the minimum of n and m, by a simple depth-first search in a graph that encodes the forced relationships and the additional decisions that remain to be made. Moreover, the graph represents the set of all solutions in a simple way. We will not fully detail that here, but we can explain how the graph determines the number of solutions, assuming there is a solution.

We use the column indices in M for vertex labels, and let GF be a graph with a node c for each column c in M, and an edge between two nodes c and c' if there is a row in M with a 2 in both columns c and c'. Let k be the number of connected components of GF. Now mark any edge (c, c') where c and c' are in a forced-relationship, and let z be the number of connected components of the subgraph of GF induced by the marked edges. Then the number of solutions to the PPH problem, assuming there is one, is exactly $2^{(z-k)}$. Hence it is actually easier to count the number of solutions than to find one.

Because of the deeper insights encoded in DPPH, we expect it will be the fastest method in practice. About 1500 lines of C code were written to implement DPPH.

4   Generating the Test Data

We used the program created by Richard Hudson [17] to generate the haplotypes. That program is the widely-used standard for generating sequences that reflect the coalescent model of SNP sequence evolution. The program allows one to control the level of recombination through a parameter r. When r is set to zero, there is no recombination, and hence the haplotypes produced are guaranteed to fit a perfect phylogeny.

After obtaining $2n$ haplotypes from Hudson's program, the haplotypes are randomly paired to create the n genotype vectors given to the three programs. The three PPH programs were tested with thousands of datasets and all of the outputs were verified to see that perfect phylogenetic trees were created and that the output haplotypes explain the input genotypes. In the output file, the input genotypes and their corresponding haplotypes are reported. The output file also reports the perfect phylogenetic tree and whether the tree is unique or not. If it is not, the number of different trees which fit the input data is reported. Program DPPH also produces the implicit representation of the set of all solutions in a simple format.

5   Performance Comparison

For genotype data with 50 individuals and 50 sites, the three PPH programs typically solve the problem in less than one second on a computer equipped with AMD K6 1.33GHz CPU and 256 MB RAM. However, GPPH spends notably longer than the others when the data is large. The relative speed of program DPPH compared to the other two methods, increases as the data size increases. It typically spends under two minutes to handle genotype data with 500 individuals and 1000 sites. This makes it possible to handle long sequences with large samples.

Our basic expectation that DPPH would be the fastest, followed by HPPH and GPPH in that order, was confirmed, as is shown in Figure 2. We should note that our implementation of HPPH included a number of minor ideas not made explicit in the original paper [7], that considerably sped up the execution. Also, Shibu Yooseph [27] has implemented DPPH and reports running times that are two to three times faster than our implementation. We implemented the graph realization program used in GPPH literally as described in [10] and did not attempt any optimizations, although many are clearly possible.

		Average Running Times (seconds)		
sites	individuals	GPPH	DPPH	HPPH
30	50	0.6574	0.0206	0.0215
100	100	1.5697	0.3216	0.37345
300	150	9.31835	3.041	4.497
500	250	36.12485	11.5275	21.5901
1000	500	256.1847	75.5935	189.4267
2000	1000	2331.167	639.93	1866.569

Fig. 2. The comparison of the running times of three methods. Each number is the average of 20 datasets.

6 Related Topics

6.1 Uniqueness of the Solution: A Strong Phase Transition

For any given input of genotypes, it is possible that there will be more than one PPH solution. When designing a population screen and interpreting the results, a unique PPH solution is very important. So the question arises: for a given number of sites, how many individuals should be in the sample (screen) so that the solution is very likely to be unique? The general issue of the number of individuals needed in a study was raised in [14]. Theoretical and empirical results on this question, addressing the number of individuals needed in a study as a function of the number of distinct haplotypes in the population, appear in [9]. Here, we report on several experiments which determine the frequency of a unique PPH solution when the number of sites and genotypes changes. Intuitively, as the ratio of genotypes to sites increases, the probability of uniqueness increases.

The input datasets are generated by Hudson's program [17] discussed earlier. While generating the data, we added an essential, biologically relevant restriction that every mutation in the tree generating the haplotypes must be on an edge that has at least 5 percent of total leaves beneath it. That is, at that site, the least frequent allele must appear in at least 5 percent of individuals in the sample. Without adding this restriction, the data are not biologically realistic, and the frequency of a unique PPH solution remains low even when the ratio of genotypes to sites is high. The table in Figure 3 shows the frequency, for datasets with 50 and 100 sites respectively, that a unique PPH solution is observed in 5000 datasets, as the number of individuals in the sample varies. Note that the number of individuals is the number of input genotype vectors, so that the underlying tree generating the haplotypes has twice that many leaves.

The table shown in Figure 3 shows a phase transition between 19 and 20 individuals. This is almost certainly related to the 5% rule. When the number of individuals (genotypes) reaches 20, the number of leaves in the underlying tree is 40, where the 5% rule requires that each mutation (site) be placed on an edge with at least two leaves below it. Hence a mutation on an edge attached to a leaf is prohibited. Before then, a mutation could be placed on such an edge, and when that happens, the solution is very unlikely be unique. A more

sites	individuals	frequency	sites	individuals	frequency
50	10	0.0000	100	10	0.0000
50	19	0.0002	100	19	0.0000
50	20	0.7030	100	20	0.7260
50	28	0.7050	100	28	0.7236
50	30	0.9520	100	30	0.9774
50	40	0.9926	100	40	0.9966
50	50	0.9974	100	50	0.9994
50	60	0.9990	100	60	0.9996
50	70	0.9994	100	70	0.9998
50	80	0.9998	100	80	1.0000
50	90	0.9998			
50	100	1			

Fig. 3. The frequency of a unique PPH solution increases when the number of individuals increases. 5000 datasets were simulated for each entry.

interesting phase transition occurs between 20 and 30 individuals. Notice that the frequency of uniqueness is close to one before the number of individuals reaches the number of sites. This may seem unexpected, and is explained by the fact that there are edges in the tree that receive more than one mutation (site) in Hudson's program. Hence the ratio of the number of individuals to the number of edges in the tree that receive one or more mutations, is higher than suggested by the results shown here. However, the biologically relevant comparison is of the number of individuals to the number of sites of interest, and so the good news is that the number of individuals needed in a sample, in order to have a high probability of a unique PPH solution, is relatively low compared to the number of sites.

6.2   Handling Haplotypes Generated with Recombinations

The PPH problem is motivated by the coalescent model without recombination. However, the programs can be useful for solving the HI problem when the underlying haplotypes were generated by a history involving some amount of recombination. In that case, it is not expected that the entire data will have a PPH solution, but some intervals in the data might have one. We can use one of the PPH programs to find maximal intervals in the input genotype sequences which have unique PPH solutions. We first find the longest interval in the genotype data, starting from position 1, which has a unique PPH solution. We do this using binary search, running a PPH program on each interval specified by the binary search. Let us say that the first maximal interval extends from position 1 to position i. We output that interval, and then move to position 2 to determine if there is an interval that extends past i containing a unique PPH solution. If so, we find the maximal interval starting at position 2, and output it. Otherwise, we move to position 3, etc. We continue in this way to output a set of maximal intervals, each of which contains a unique PPH solution. This also implicitly finds, for each starting position, the longest interval starting at that position that contains a unique PPH solution.

In principle, the intervals that are output could overlap in irregular, messy ways. However, we have observed that this is rarely the case. Generally, the output intervals do not overlap, or two intervals overlap at one site, i.e., the right end of one interval may overlap in one position with the left end of the next interval. This provides a clean decomposition of the data into a few intervals where in each, the data has a unique PPH solution.

A program called PPHS, which is available at the website mentioned earlier, was implemented to find the maximal intervals. We performed several experiments for the genotype data with different recombination rates, using 100 genotypes and 100 sites. The input data for the experiments was again generated by Hudson's program [17]. Therefore, we were able to check if the output haplotypes differed from the original haplotypes, in any interval. The results are shown in Figure 4. The most striking result is that when the recombination rate is moderate, the accuracy of the PPH solutions inside each interval, compared to the original haplotypes, is very high. In fact, when $r = 4$, in each of the fifteen runs described in Figure 4, the unique PPH solution found by the algorithm precisely recreated the input haplotypes in each interval. That is, the PPH program found the correct haplotype pairs perfectly in each interval. One might think that this must always be true, since the PPH solution is unique in each interval, but genotypes that can be explained with haplotypes that fit a perfect phylogeny need not have been generated that way.

There are many ways that such a decomposition can be used. The most obvious is to reduce the amount of laboratory work that is needed to fully determine the correct haplotypes. For example, in a problem with 100 sites and 10 intervals, we can form new shorter genotype vectors with one site per interval, hence 10 sites. If the correct haplotype pairs for these shorter genotype sequences are determined, we can combine that information with the (assumed correct) haplotype pairs determined in each interval by a PPH program. The laboratory effort is reduced to one tenth of what it would be to determine the haplotypes from 100 sites. Another approach is to input the shorter genotype sequences to one of the statistical-based methods for haplotype inference. These methods can handle a moderate level of recombination and are generally quite accurate, but their running times increase greatly with an increasing number of sites.

7   How Large Should the Test Data Be?

In this paper we tested programs and data with up to 2000 sites. This is a larger number of SNPs than has so far been observed to fit the perfect phylogeny model. Here we discuss the question of how many SNP sites should be included in tests of PPH programs.

We define a set of binary sequences as "tree-compatible" if the sequences pass the four-gametes test. That is, the data do not have two positions (sites) where four sequences in the set contain all four combinations 0,0; 0,1; 1,0, and 1,1. We

	r = 4		r = 16		r = 40	
Experiments	Errors	Intervals	Errors	Intervals	Errors	Intervals
No. 1	0	7	0	17	9	18
No. 2	0	7	1	17	26	17
No. 3	0	5	3	20	0	20
No. 4	0	5	1	16	0	22
No. 5	0	4	2	14	5	24
No. 6	0	3	1	12	1	22
No. 7	0	7	0	18	7	20
No. 8	0	4	0	14	0	25
No. 9	0	7	0	18	12	20
No. 10	0	10	0	15	1	24
No. 11	0	9	1	12	5	19
No. 12	0	5	0	18	1	18
No. 13	0	7	0	16	27	22
No. 14	0	8	0	14	1	25
No. 15	0	10	0	19	0	24
Average	0	6.5	0.6	16	6.3	21.3

Fig. 4. Fifteen experiments with 100 individuals and 100 sites, performed with three different recombination rates. r is the recombination rate used in Hudson's program [17]. When r is high, the probability of recombination is high. The Interval count is the number of intervals output by program PPHS. The error count is the total number of intervals where the given unique PPH solution is not correct for that interval. Hence the number of errors reported can be larger than the number of intervals.

define a "tree-compatible interval" as an interval where the subsequences are tree-compatible. Sequences that are tree-compatible can be derived on a perfect phylogeny. Therefore, in considering the size of relevant input to PPH programs, the key question is: What is the longest tree-compatible interval that one could reasonably expect to find in a set M of binary-encoded biological sequences (SNPs, or other binary sequences) of current or future interest? We define that length as m^*.

The correct value of m^* is certainly unknown at present, and the existing literature related to this issue (mainly from studies of haplotype structure in humans, and studies of linkage disequilibrium in a few other organisms) represents a minuscule fraction of the molecular diversity studies that are desired and that are expected to be conducted in the future. However, there is already good evidence establishing that m^* is much larger than 30 (a number suggested by some of the studies in humans, for example [6]).

Relevant data can come from two sources: actual sequence and SNP data that can be directly examined for tree-compatible intervals, and less direct studies of linkage disequilibrium (LD). LD causes (or is defined by) a high correlation between the occurrences of alleles at two sites. If f_A is the frequency of allele A at one site, and f_B is the frequency of allele B at a second site, and f_{AB} is the joint frequency of those alleles at the two sites, LD at the two sites can be measured by the deviation of f_{AB} from $f_A \times f_B$. Assuming infinite sites, long intervals of high LD suggest long tree-compatible intervals, since high LD is generally caused by little recombination between the two sites. LD can be measured without directly determining whether the interval is tree-compatible, and so LD is a more indirect, but much more available, indicator of m^* than is provided by the full SNP sequences in the sample. Presently, very few studies have determined full SNP sequences in large populations, so data on LD is very central in estimating m^*.

In published data on humans, there is considerable variance in the level of LD observed. For example, surveys of LD suggest that among Nigerians, high levels of linkage disequilibrium extend to intervals of only 5 Kb, but in Northern Europeans, high levels extend to intervals of 60 to 80 Kb. Depending on the number of individuals in the survey, that could translate to tree-compatible intervals of between 5 and 80 SNPs. In even more homogeneous populations, for example in Finland or Iceland or Pennsylvania Dutch, even longer tree-compatible intervals of SNPs are expected. Generally, structured subpopulations which are the result of recent migration, historical bottlenecks (reduction of the population size), or admixture (the recent mixing of two distinct populations) are expected to have even longer tree-compatible intervals, and the exact lengths are not yet known. Smaller, more local populations should have even longer tree-compatible intervals. Moreover, structured, local, subpopulations, are very important in human genetics research, so that even if 30 were the human-average number of tree-compatible SNP sites in an interval, there are important studies where one expects, and will search for, longer tree-compatible intervals. Domesticated plants and animals are expected to have longer tree-compatible intervals than humans, and the few studies that have been done are consistent with this expectation [19].

It is also important to understand how the data reported in [6] were obtained. Those studies were looking for long intervals in humans of high LD containing a small number of haplotypes in the given sample. SNP sites were sampled in the genome at a density that was sufficient to identify these long intervals, but once found, the researchers were not interested in determining the total number of SNP sites in the intervals. Therefore, typically under 30 SNPs were found in intervals up to 200,000 nucleotides long. But a human interval of that length is expected to have between 200 and 400 SNPs, and so it is incorrect to interpret the number of SNPs in an interval reported in [6] as an estimate of m^*. Moreover, the number of SNPs found will also be a function of the size of the sample. Fine scale linkage and association mapping aimed at identification of specific genes and variants will often lead to the determination of many more SNPs in candidate regions. And large scale resequencing on a genomic scale will also produce high densities of SNPs throughout the genome [18].

Moreover, there is variation in the density of SNPs, independent of the amount of recombination in a region. For example, recent investigation of the laboratory mouse genome [26] found long regions of the genome that contained 45 SNPs per 10Kb, and other long regions that contained only 1 SNP per 10Kb. Hence, it is expected that in the mouse genome we could encounter tree-compatible regions with the same number of nucleotides, where one region has many more SNPs than the other.

In general, the value of m^* depends both on the diversity of biological sequences (from different organisms, regions of the genome, populations and subpopulations) that are of importance and that will be, or have been, studied, and the number of sequences in those studies. More fundamentally, for any given sample, the length of the longest tree-compatible interval is a function of the ratio of the (site) mutation rate, and the recombination rate in that sample. Among all the organisms, genomic regions, populations and subpopulations that are of interest, that ratio is expected, and has already been seen, to vary over a large range [18]. Hence it is not correct to conclude that m^* has been established from the limited, specific studies recently conducted in humans.

In short, the correct value of m^* is unknown, and there is very little that is known, empirically or theoretically, that establishes good bounds for m^*. There is enormous diversity in biology, with a vast number of unexplored organisms, populations, subpopulations and genomic regions of interest. Any suggestion that m^* is already known ignores this. More specifically, the suggestion that m^* is bounded by 30 does not reflect the little data that is already known, and the understanding that compared to the human studies recently done, much longer tree-compatible regions should exist in sequences from subpopulations and organisms whose diversity has not yet been studied.

Acknowledgements. Thanks to population geneticists Chuck Langley, Peter Morrell and Steven Orzack for advice on the discussion in Section 7.

References

1. V. Bafna, D. Gusfield, G. Lancia, and S. Yooseph. Haplotyping as perfect phylogeny: A direct approach. Technical report, UC Davis, Department of Computer Science. July 17, 2002.
2. R. E. Bixby and D. K. Wagner. An almost linear-time algorithm for graph realization. *Mathematics of Operations Research*, 13:99–123, 1988.
3. R.H. Chung and D. Gusfield. Perfect phylogeny haplotyper: Haplotype inferral using a tree model. *Bioinformatics*, 19(6):780–781, 2003.
4. A. Clark. Inference of haplotypes from PCR-amplified samples of diploid populations. *Mol. Biol. Evol*, 7:111–122, 1990.
5. A. Clark, K. Weiss, and D. Nickerson et. al. Haplotype structure and population genetic inferences from nucleotide-sequence variation in human lipoprotein lipase. *Am. J. Human Genetics*, 63:595–612, 1998.
6. M. Daly, J. Rioux, S. Schaffner, T. Hudson, and E. Lander. High-resolution haplotype structure in the human genome. *Nature Genetics*, 29:229–232, 2001.
7. E. Eskin, E. Halperin, and R. Karp. Efficient reconstruction of haplotype structure via perfect phylogeny. Technical report, UC Berkeley, Computer Science Division (EECS), August, 2002.
8. M. Fullerton, A. Clark, Charles Sing, and et. al. Apolipoprotein E variation at the sequence haplotype level: implications for the origin and maintenance of a major human polymorphism. *Am. J. of Human Genetics*, pages 881–900, 2000.

9. S. Cleary and K. St. John. Analysis of Haplotype Inference Data Requirements. Preprint, 2003.
10. F. Gavril and R. Tamari. An algorithm for constructing edge-trees from hypergraphs. *Networks*, 13:377–388, 1983.
11. D. Gusfield. Efficient algorithms for inferring evolutionary history. *Networks*, 21:19–28, 1991.
12. D. Gusfield. *Algorithms on Strings, Trees and Sequences: Computer Science and Computational Biology*. Cambridge University Press, 1997.
13. D. Gusfield. Inference of haplotypes from samples of diploid populations: complexity and algorithms. *Journal of computational biology*, 8(3), 2001.
14. D. Gusfield. Haplotyping as Perfect Phylogeny: Conceptual Framework and Efficient Solutions (Extended Abstract). In *Proceedings of RECOMB 2002: The Sixth Annual International Conference on Computational Biology*, pages 166–175, 2002.
15. L. Helmuth. Genome research: Map of the human genome 3.0. *Science*, 293(5530):583–585, 2001.
16. R. Hudson. Gene genealogies and the coalescent process. *Oxford Survey of Evolutionary Biology*, 7:1–44, 1990.
17. R. Hudson. Generating samples under the Wright-Fisher neutral model of genetic variation. *Bioinformatics*, 18(2):337–338, 2002.
18. C. Langley. U.C. Davis Dept. of Evolution and Ecology. Personal Communication, 2003.
19. J.Z. Lin, A. Brown, and M. T. Clegg. Heterogeneous geographic patterns of nucleotide sequence diversity between two alcohol dehydrogenase genes in wild barley (Hordeum vulgare subspecies spontaneum). *PNAS*, 98:531–536, 2001.
20. S. Lin, D. Cutler, M. Zwick, and A. Cahkravarti. Haplotype inference in random population samples. *Am. J. of Hum. Genet.*, 71:1129–1137, 2003.
21. T. Niu, Z. Qin, X. Xu, and J.S. Liu. Bayesian haplotype inference for multiple linked single-nucleotide polymorphisms. *Am. J. Hum. Genet*, 70:157–169, 2002.
22. S. Orzack, D. Gusfield, and V. Stanton. The absolute and relative accuracy of haplotype inferral methods and a consensus approach to haplotype inferral. Abstract Nr 115 in Am. Society of Human Genetics, Supplement 2001.
23. M. Stephens, N. Smith, and P. Donnelly. A new statistical method for haplotype reconstruction from population data. *Am. J. Human Genetics*, 68:978–989, 2001.
24. S. Tavare. Calibrating the clock: Using stochastic processes to measure the rate of evolution. In E. Lander and M. Waterman, editors, *Calculating the Secretes of Life*. National Academy Press, 1995.
25. W.T. Tutte. An algorithm for determining whether a given binary matroid is graphic. *Proc. of Amer. Math. Soc*, 11:905–917, 1960.
26. C. Wade and M. Daly et al. The mosaic structure of variation in the laboratory mouse genome. *Nature*, 420:574–578, 2002.
27. Shibu Yooseph. Personal Communication, 2003.

Cylindrical Hierarchy for Deforming Necklaces

Sergei Bespamyatnikh

Department of Computer Science, University of Texas at Dallas,
Box 830688, Richardson, TX 75083, USA
besp@utdallas.edu

Abstract. Recently, Guibas *et al.* [7] studied deformable necklaces –
flexible chains of balls, called beads, in which only adjacent balls can
intersect. In this paper, we investigate a problem of covering a necklace
by cylinders. We consider several problems under different optimization
criteria. We show that optimal cylindrical cover of a necklace with n
beads in $\mathbb{R}^3$ by k cylinders can be computed in polynomial time. We also
study a bounding volume hierarchy based on cylinders.

1 Introduction

Our study is motivated by the representation and manipulation of molecular
configurations, modeled by a collection of spheres. The representation of three-
dimensional geometric structure of a molecule with spheres where each atom is
viewed as rigid sphere is a common approach. The sizes of spheres depend on the
atom types. There are recommended values for the radius of each atom sphere
called van der Waals radius. The distance between the centers of every pair of
spheres is also known. In this model, the spheres of atoms in a chemical bond
interpenetrate. The fused spheres have been studied in computational geometry.
Halperin and Overmars [8] proved useful properties of the sphere model. For
example, the combinatorial complexity of the boundary of a molecule is linear.

Recently Guibas *et al.* [7] studied the sphere model under motion. They call
the sphere model a *necklace* and the spheres *beads* and we borrow this terminology. Deforming necklaces and efficient algorithms for this problem are needed
in many computational fields including computer graphics, computer vision,
robotics, geographic information systems, spatial databases, molecular biology,
and scientific computing. There is a literature on using spheres in engineering
modeling [4]. Computational problems involving motion can be stated in the
Kinetic Data Structure Framework (KDS for short) [6,5]. A motion of necklaces
models (i) the Brownian motion of molecules where necklaces move as rigid
bodies, and (ii) molecular dynamics where necklaces undergo local changes only.
Applied to deforming necklaces an efficient KDS can be design using bounding
volume hierarchies. Guibas *et al.* [7] focused on a simple variant of hierarchy
that uses spheres. They analyzed two ways of defining the spheres in hierarchy. A *sphere hierarchy* of a necklace is defined to be a balanced tree whose
leaves correspond to the beads. To each internal node is assigned a *cage* that is a
bounding sphere. A *wrapped hierarchy* is a sphere hierarchy of a necklace where

T. Warnow and B. Zhu (Eds.): COCOON 2003, LNCS 2697, pp. 20–29, 2003.
© Springer-Verlag Berlin Heidelberg 2003

the cage corresponding to each internal node is the minimum enclosing sphere of the beads in the canonical sub-necklace associated with that node. Another hierarchy called a *layered hierarchy* is defined by making the cage of an internal node to be the minimum enclosing sphere of the cages of its two children [12].

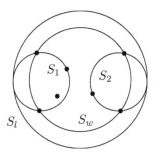

Fig. 1. S_w is the wrapped sphere of 7 points and S_l is the layered sphere constructed for the spheres S_1 and S_2.

The wrapped hierarchy is slightly more difficult to compute than the layered hierarchy although it is tighter fitting, see Fig. 1 for an example and [4] for detailed comparison of two hierarchies. The idea of using cylinders comes from this example. The spheres S_1 and S_2 can be covered by a cylinder of volume smaller than the volume of S_l. The use of cylinders for molecular representation is not a new idea. The cylinders are widely used in software packages visualizing proteins. Structural components of proteins – helixes – are depicted by cylinders. For example, the human deoxyhaemoglobin 4HHB is a protein with high concentration of helices; it contains 32 helices. The helix structure is shown on Fig. 2. Packing helices into cylinders and related problems can be found in [13].

We want to represent a molecule using cylinders. Problems of covering spheres in $\mathbb{R}^3$ by cylinders are computationally difficult. Recently, Zhu [15] considered the problem of covering points by cylinders. The problem is NP-hard even for points (spheres of radius 0). In this paper we address the following problem.

Necklace packing into cylinders. Let $\mathcal{N}$ be a necklace consisting of n beads $B_1, B_2, \ldots, B_n$ in $\mathbb{R}^3$. Let k be an integer $1 \leq k \leq n$. Find k cylinders $C_1, C_2, \ldots, C_k$ and a partition of the necklace into k subnecklaces $\mathcal{N}_1, \mathcal{N}_2, \ldots, \mathcal{N}_k$ such that, for each $1 \leq i \leq k$, the cylinder C_i contains the beads of the necklace $\mathcal{N}_i$ and a function $F(C_1, C_2, \ldots, C_k)$ is minimized. Examples of function $F()$ can be (i) the sum of volumes of the cylinders, or (ii) the maximum radius of a cylinder, or (iii) the maximum volume of a cylinder, etc.

We show that, for reasonable functions F, the problem of necklace packing can be solved in polynomial time. The algorithm exploits the sequence property of necklaces and is in contrast to NP-hardness result of packing a distributed points [15].

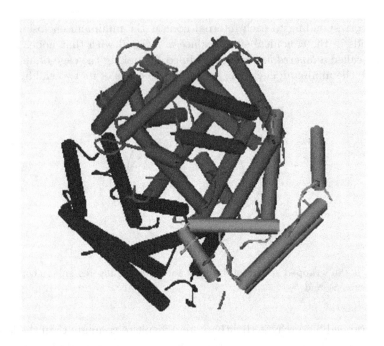

Fig. 2. The helices of human deoxyhaemoglobin 4HHB.

We use the necklace packing to generate a cylinder hierarchy. As in the sphere hierarchy [7], there are two options here: wrapped and layered hierarchy. We define a cylinder cage to be the *smallest enclosing cylinder* of underlying cylinders where a cylinder is measured using the function $F()$.

We also mention that the smallest radius cylinders find applications in projective clustering [2,9].

2   Smallest Enclosing Cylinder

Given a necklace $\mathcal{N}$, we want to find the smallest cylinder containing all the spheres of $\mathcal{N}$. Unfortunately, it seems that the property of spheres of being in a necklace does not help in computing the smallest enclosing cylinder. There are several results on computing a cylinder with the smallest radius [1,3,14]. Agarwal *et al.* [1] gave a $O(n^{3+\delta})$ algorithm for any $\delta > 0$. Schömer *et al.* [14] found a $O(n\varepsilon^{-2}\log\varepsilon^{-1})$ time algorithm that computes $(1 + \varepsilon)$-approximation of the smallest radius for any $\varepsilon > 0$. Chan [3] improved the running time to $O(n/\varepsilon)$ using convex programming. Zhu [15] obtained a practical algorithm with running time $O(n\log n + n/\varepsilon^4)$. We give a practical algorithm that computes $(1+\varepsilon)$-approximation of the smallest cylinder under various objective functions $F()$. We denote i-th bead in the necklace by $B_i(o_i, r_i)$ where o_i is the center of B_i and r_i is its radius. The coordinates of a point $p \in \mathbb{R}^3$ are denoted as $(x(p), y(p), z(p))$ or (p_x, p_y, p_z).

Algorithm 1

Step 1. Compute

$$\text{Box} = [c_x - a_x, c_x + a_x] \times [c_y - a_y, c_y + a_y] \times [c_z - a_z, c_z + a_z],$$

the bounding box of the spheres in the necklace $\mathcal{N}$. For example,

$$c_x - a_x = \min_{1 \le i \le n} \{x(o_i) - r_i\}.$$

Step 2. Let $r = 2\max(a_x, a_y, a_z)$ and let

$$D = [c_x - r, c_x + r] \times [c_y - r, c_y + r] \times [c_z - r, c_z + r].$$

Step 3. Generate a grid of size $2/\varepsilon \times 2/\varepsilon$ on each face of the cube D. Let G be the set of all grid points.

Step 4. For each line $p_1 p_2$ defined by two grid points $p_1, p_2 \in G$, find the smallest cylinder C with center line $p_1 p_2$ that contains all the beads of the necklace. Compute $F(C)$.

Step 5. Compute the smallest value of $F(C)$ obtained in Step 4.

Theorem 1. *Let $F(C)$ be one of the following functions: (1) the radius of the cylinder C, or (2) the volume of the cylinder C. Algorithm 1 computes $(1 + \varepsilon)$-approximation of the smallest enclosing cylinder in $O(n/\varepsilon^4)$ time.*

The exact algorithm by Agarwal *et al.* [1] can be applied to the minimization of the radius. What if one want to minimize the volume of a cylinder?

Theorem 2. *The smallest volume cylinder enclosing a necklace of n beads can be computed in $O(n^6)$ time.*

Proof. First, we show $O(n^6)$ bound. The volume of a cylinder C of radius r and height h is $\text{vol}(C) = \pi r^2 h$. Let C^* be the smallest volume cylinder enclosing a necklace $\mathcal{N}$. We assume that the beads of $\mathcal{N}$ are in general position, i.e. there is no unbounded cylinder whose surface touches to more than four beads. The surface of C^* has three components: two disks D_1 and D_2 and the cylindrical surface S that can be unwrapped to a rectangle. Note that each disk D_1 and D_2 touches a bead since C^* has the smallest volume. The cylinder C^* has a property that either

   (i) the surface S touches four beads, or
   (ii) the surface S touches only three beads.

   In the first case the center line of C^* is determined by four touching beads and two disks D_1 and D_2 are tangent to two extreme beads in the direction of the center line. There are $O(n^4)$ possible ways to choose these beads. Each combination of beads generate $O(1)$ center lines of a cylinder touching them simultaneously. We compute each of these lines and find two corresponding disks D_1 and D_2. This gives us the volume of the cylinder. We check in linear time if the cylinder contains all the beads of $\mathcal{N}$. Thus, the first case can be processed in $O(n^5)$ time.

In the second case the location of the cylinder in the space is determined by five beads: three beads B_1, B_2, B_3 touching S and two beads B_4, B_5 touching the disks D_1 and D_2. Note that beads B_1, B_2 and B_3 are distinct but two sets $\{B_1, B_2, B_3\}$ and $\{B_4, B_5\}$ can intersect. There are $O(n^5)$ choices to select the beads. Each tuple generates $O(1)$ cylinders minimizing the volume. Each cylinder can be checked if it contains all the beads of $\mathcal{N}$. The running time in the second case is $O(n^6)$.

Next, we describe a more complicated algorithm with better asymptotic run-time.

Theorem 3. *Using advanced techniques the smallest volume cylinder enclosing a necklace of n beads can be computed in $O(n^{5+\delta})$ time where $\delta > 0$ is arbitrary small constant.*

Proof. We apply the parametric search technique of Megiddo [11]. We consider the following decision problem.

> **Decision Problem.** Given a necklace $\mathcal{N}$ and a parameter V, decide if there is a cylinder of volume V that contains all the beads of $\mathcal{N}$.

Let L be a line in $\mathbb{R}^3$ and let C be the smallest cylinder with centerline L containing the necklace. Let $(\xi_1, \xi_2, 1)$ be the *direction* of L, i.e. the line L is parallel to the line $L(\xi_1, \xi_2)$ passing through the origin $O(0,0,0)$ and the point $(\xi_1, \xi_2, 1)$. Consider i-th bead B_i. To simplify notation, let $o_i = (x_i, y_i, z_i)$. There are two planes, denoted by π_i^- and π_i^+, orthogonal to the direction $(\xi_1, \xi_2, 1)$ and tangent the bead B_i. Their equations can be written as $\xi_1 x + \xi_2 y + z = h_i^-(\xi_1, \xi_2)$ and $\xi_1 x + \xi_2 y + z = h_i^+(\xi_1, \xi_2)$.

The values of $h_i^-(\xi_1, \xi_2)$ and $h_i^+(\xi_1, \xi_2)$ can be computed as follows. Let o_i' be the projection of the point o_i to the line $L(\xi_1, \xi_2)$, see Fig. 3. Then the coordinates of o_i' satisfy $x(o_i') = \lambda_i \xi_1, y(o_i') = \lambda_i \xi_2$ and $z(o_i') = \lambda_i$ for some real number λ_i. The segment $o_i o_i'$ is orthogonal to the segment Oo_i'. Thus

$$\lambda_i \xi_1 (\lambda_i \xi_1 - x_i) + \lambda_i \xi_2 (\lambda_i \xi_2 - y_i) + \lambda_i (\lambda_i - z_i) = 0$$
$$\lambda_i (\xi_1^2 + \xi_2^2 + 1) = \xi_1 x_i + \xi_2 y_i + z_i. \tag{1}$$

Let $\alpha = \xi_1^2 + \xi_2^2 + 1$. Clearly, $\alpha > 0$. The planes tangent to B_i intersect the line $L(\xi_1, \xi_2)$ at the points $H_i^+(\lambda_i^+ \xi_1, \lambda_i^+ \xi_2, \lambda_i^+)$ and $H_i^-(\lambda_i^- \xi_1, \lambda_i^- \xi_2, \lambda_i^-)$ where $\lambda_i^+ = \lambda_i + r_i/\alpha$ and $\lambda_i^- = \lambda_i - r_i/\alpha$, see Fig. 3. The values of $h_i^-(\xi_1, \xi_2)$ and $h_i^+(\xi_1, \xi_2)$ can be obtained by substituting the points H_i^- and H_i^+ to the equations of the planes π_i^- and π_i^+

$$h_i^-(\xi_1, \xi_2) = \xi_1 x_i + \xi_2 y_i + z_i - r_i$$
$$h_i^+(\xi_1, \xi_2) = \xi_1 x_i + \xi_2 y_i + z_i + r_i.$$

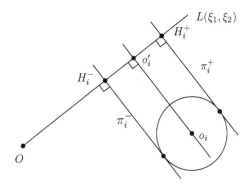

Fig. 3. The points $H_i^\bullet$ and H_i^* .

Suppose that H_i^+ and H_j^- are the extreme points on the line $L(\xi_i, \xi_2)$ among all the points $H_l^-, H_l^+, l = 1, \ldots, n$. Then the points H_i^+ and H_j^- determine two disks of the cylinder C. The height of C is

$$h = |H_i^+ H_j^-| = \sqrt{(\lambda_i^+ \xi_1 - \lambda_j^- \xi_1)^2 + (\lambda_i^+ \xi_2 - \lambda_j^- \xi_2)^2 + (\lambda_i^+ - \lambda_j^-)^2}$$
$$= |(\lambda_i^+ - \lambda_j^-)|\sqrt{\alpha}.$$

Note that $\lambda_i^+ \leq \lambda_j^-$. Then

$$h = (\lambda_i^+ - \lambda_j^-)\sqrt{\alpha} = (\lambda_i + r_i/\alpha - \lambda_j - r_j/\alpha)\sqrt{\alpha} \tag{2}$$
$$= (\xi_1(x_i - x_j) + \xi_2(y_i - y_j) + (z_i - z_j) + (r_i - r_j))/\sqrt{\alpha}. \tag{3}$$

Since the volume of the cylinder is $V = \pi r^2 h$, the radius of the cylinder is $r = \sqrt{V/(\pi h)}$. The existence of the cylinder of radius r containing the beads is related to to transversals [1]. A line L is called a *(line) transversal* of the necklace $\mathcal{N}$ if it intersects every bead of $\mathcal{N}$. Let $\mathcal{T}(\mathcal{N})$ denote the set of all transversals of $\mathcal{N}$. The reduction to transversals is as follows. The set of balls $B_i(o_i, r_i)$ can be wrapped by a cylinder of radius r if and only if (i) $r_i \leq r$ for all i, and (ii) the balls $B_i'(o_i, r_i - r)$ admit a transversal. For our purposes we need a constrained reduction: the set of balls $B_i(o_i, r_i)$ can be wrapped by a cylinder whose center-line is parallel to $L(\xi_1, \xi_2)$ and radius r if and only if (i) $r_i \leq r$ for all i, and (ii) the balls $B_i'(o_i, r_i - r)$ admit a transversal parallel to $L(\xi_1, \xi_2)$. Agarwal *et al.* [1] solves the transversal detection problem using minimization/maximization diagrams of bivariate functions. We apply their approach to other functions since we use different parametrization.

A line L' parallel to the line $L(\xi_1, \xi_2)$ can be parametrized using two more variables ξ_3 and ξ_4 as follows:

$$L'(\xi_1, \xi_2, \xi_3, \xi_4) = \{(\xi_3 + t\xi_1, \xi_4 + t\xi_2, t) \mid t \in \mathbb{R}\}.$$

Let $\pi'(\xi_1, \xi_2, \xi_3)$ be the plane containing all the lines $L'(\xi_1, \xi_2, \xi_3, \xi_4), \xi_4 \in \mathbb{R}$. Consider l-th bead B'_l. We define two functions $f_l(\xi_1, \xi_2, \xi_3)$ and $g_l(\xi_1, \xi_2, \xi_3)$ as follows. Intersect B'_l with the plane τ_l passing through its center and parallel to the xz-plane. The equation of the plane τ_l is $y = y_l$. Let σ_l^+ and σ_l^- be two hemispheres defined by cutting B'_l by τ_l. We assume that σ_l^+ lies in the halfspace $y \geq y_l$ and σ_l^- lies in the halfspace $y \leq y_l$. If the plane $\pi'(\xi_1, \xi_2, \xi_3)$ intersects the bead B'_l, then there are two lines $L'(\xi_1, \xi_2, \xi_3, a)$ and $L'(\xi_1, \xi_2, \xi_3, b), a \leq b$ tangent to the hemispheres σ_l^- and σ_l^+, respectively. We define $g_l(\xi_1, \xi_2, \xi_3) = a$ and $f_l(\xi_1, \xi_2, \xi_3) = b$. In the second case $\pi'(\xi_1, \xi_2, \xi_3) \cap B_l = \emptyset$ we define $f_l(\xi_1, \xi_2, \xi_3) = +\infty$ and $g_l(\xi_1, \xi_2, \xi_3) = -\infty$. The functions f_l and g_l have property that the line $L'(\xi_1, \xi_2, \xi_3, \xi_4)$ intersects the bead B'_l if and only if $f_l(\xi_1, \xi_2, \xi_3) \leq \xi_4 \leq g_l(\xi_1, \xi_2, \xi_3)$. The line $L'(\xi_1, \xi_2, \xi_3, \xi_4)$ is transversal of the beads if and only if

$$\max_{1 \leq l \leq n} f_l(\xi_1, \xi_2, \xi_3) \leq \xi_4 \leq \min_{1 \leq l \leq n} g_l(\xi_1, \xi_2, \xi_3).$$

We show that the functions $f_l(\xi_1, \xi_2, \xi_3)$ and $g_l(\xi_1, \xi_2, \xi_3)$ have *constant description complexity* [1], that is, the graph of each function is a semi-algebraic set in $\mathbb{R}^4$ defined by a constant number of polynomial equalities and inequalities of constant degree. We assume that i and j are fixed. By Equation (3) the radius of cylinder satisfies

$$r^2(\xi_1^2 + \xi_2^2 + 1) = (\xi_1(x_i - x_j) + \xi_2(y_i - y_j) + (z_i - z_j) + (r_i - r_j))^2. \quad (4)$$

The lines tangent to the hemispheres σ_l^- and σ_l^+ are at distance $r - r_l$ from the center o_l. This is equivalent to the condition that the line $L(\xi_1, \xi_2)$ is at distance $r - r_l$ from the point $p_l(x'_l, y'_l, z'_l)$ where $x'_l = x_l - \xi_3, y'_l = y_l - \xi_4, z'_l = z_l$. Substituting i by l and o_i by p_l in Equation (1) the nearest point on $L(\xi_1, \xi_2)$ to p_l has coordinates $(\lambda\xi_1, \lambda\xi_2, \lambda)$ where

$$\lambda = (\xi_1 x'_l + \xi_2 y'_l + z'_l)/(\xi_1^2 + \xi_2^2 + 1). \quad (5)$$

Therefore the tangent lines satisfy

$$\begin{aligned}(r - r_l)^2 &= (\lambda\xi_1 - x'_l)^2 + (\lambda\xi_2 - y'_l)^2 + (\lambda - z'_l)^2 \\ &= \lambda^2\alpha - 2\lambda(\xi_1 x'_l + \xi_2 y'_l + z'_l) + (x'_l)^2 + (y'_l)^2 + (z'_l)^2 \\ &= (x'_l)^2 + (y'_l)^2 + (z'_l)^2 - \lambda^2\alpha. \end{aligned} \quad (6)$$

Plugging λ from (5) and x'_l, y'_l, z' we obtain a polynomial of constant degree.

Recently Koltun and Sharir [10] proved that the overlay of two trivariate diagrams has $O(n^{3+\delta})$ complexity. Applied to the maximization diagram of $f_l(\xi_1, \xi_2, \xi_3)$ and the minimization diagram of $g_l(\xi_1, \xi_2, \xi_3)$ we obtain $O(n^{3+\delta})$ bound for the fixed pair (i, j). The decision problem can be solved in $O(n^{5+\delta})$ time since there $O(n^2)$ pairs (i, j). The parametric search technique allows to solve the optimization problem within the same bound $O(n^{5+\delta})$.

3   General k

In this Section we show how to find an optimal necklace packing into cylinders. Our algorithm is based on a dynamic programming approach. Essentially, a polynomial time algorithm is possible since the problem is decomposable into polynomially many subproblems.

Theorem 4. *Let $F(C)$ be one of the following functions: (1) the radius of the cylinder C, or (2) the volume of the cylinder C. Let $F(C_1, \ldots, C_k)$ is defined to be either minmax or minsum of the values $F(C_1), \ldots, F(C_k)$. The problem of necklace packing into cylinders minimizing $F()$ can be solved in $O(n^{5+\delta})$ time for the case (1) and in $O(n^{7+\delta})$ time for the case (2).*

Proof. Let i and j be two integers such that $1 \leq i \leq j \leq n$. Let $\mathcal{N}(i, j)$ denote the sub-necklace of $\mathcal{N}$ with beads $B_i, B_{i+1}, \ldots, B_j$. Let $C(i, j)$ be the optimal cylinder covering the necklace $\mathcal{N}(i, j)$. By Theorem 3 the cylinder $C(i, j)$ can be found in $O(n^{5+\delta})$ time if $F(C)$ is the volume function. Let $\Xi(i, j)$ denote the value $F(C(i, j))$. Therefore the values $\Xi(i, j)$ for all i and j can be computed in $O(n^{7+\delta})$ time. The bound for the first case follows if we apply $O(n^{3+\delta})$-algorithm by Agarwal *et al.* [1].

Let m be an integer $1 \leq m \leq k$. Let $\mathcal{C}(j, m) = \{C_1, C_2, \ldots, C_m\}$ be the set of m cylinders in the optimal packing of the necklace $\mathcal{N}(1, j)$ with m cylinders. Let $\Phi(j, m)$ denote the value $F(C_1, C_2, \ldots, C_m)$. Then

$$\Phi(j, 1) = \Xi(1, j) \quad \text{for all } 1 \leq j \leq n$$
$$\Phi(j, m) = \min_{1 \leq i < j} \{\Phi(i, m-1) + \Xi(i+1, j)\} \quad \text{for all } 1 \leq j \leq n \text{ and } 1 \leq m \leq k$$

The dynamic program computes the values $\Phi(j, m)$ using the above equations. This takes $O(n^2)$ time. Clearly, the optimal value of the necklace packing is $\Phi(n, k)$. The theorem follows.

We remark that the approach based on dynamic programming can be applied to many other objective functions that are decomposable.

4   Cylindrical Hierarchy

We consider two hierarchies wrapped and layered. The cage of a node in the wrapped hierarchy is defined as the optimal cylinder covering all beads in the corresponding subtree. In the layered hierarchy the cage is the smallest cylinder containing the cages of its children. At first glance, computation of the wrapped hierarchy is more difficult than computation of one cylinder. We show that the computing time is the same for exact problem and slightly bigger for the approximate one.

4.1   Wrapped Hierarchy

Lemma 1. *The wrapped hierarchy can be constructed in*
   (a) $O(n^{3+\delta})$ time exactly if the objective function is based on radius only,
   (b) $O(n^{5+\delta})$ time exactly if the objective function is based on volume,
   (c) $O((n \log n)/\varepsilon^4)$ time approximately for either radius based function or volume based function.

Proof. The main computational task is to construct all the cages of the hierarchy. We construct each cage independently using the algorithms from Theorems 1 and 2. The algorithm is recursive. Let $t(n)$ be the running time for computing the optimal cylinder (or its approximation) for n spheres. Let $T(n)$ be the running time to construct the hierarchy for n beads. Then

$$T(1) = const, \quad \text{and} \quad T(n) = t(n) + 2T(n/2) \ \text{ if } n \geq 2.$$

The lemma follows since (a) $t = n^{3+\delta}$, (b) $t = n^{5+\delta}$, and (c) $t(n) = O(n/\varepsilon^4)$.

4.2   Layered Hierarchy

In the layered hierarchy we have a new problem.

   Cage Problem. Let C_1 and C_2 be two cylinders in $\mathbb{R}^3$. Find an optimal cylinder C that contains C_1 and C_2 where the quality of a cylinder is measured by a function $F()$ as in the necklace packing problem.

Lemma 2. *The optimal cage can be found in $O(1)$ time. Thus the layered hierarchy can be constructed in $O(n)$ time.*

Proof. We note that it is necessary and sufficient to substitute the cylinders C_1 and C_2 by four circles in their boundaries, see Fig. 4.

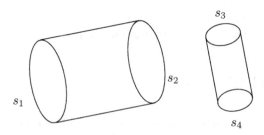

Fig. 4. The four circles s_1, s_2, s_3 and s_4 on the boundary of two cylinders.

The problem of finding an optimal cylinder has a constant complexity. We parameterize the cylinder using six variables. Let $a = (a_x, a_y, 0)$ and $b = (a_x +$

$b_x, a_y + b_y, 1)$ be two points that are intersection of the center line of the cylinder and the planes $z = 0$ and $z = 1$. We can assume that the center line of an optimal cylinder is not parallel to the plane OXY by perturbation argument. Two disks on the boundary of the cylinder can be parameterized as planes $b_x x + b_y y + z = h_1$ and $b_x x + b_y y + z = h_1$. The volume of the cylinder can be expressed using the variables $a_x, a_y, b_x, b_y, h_1, h_2$. The optimal value of volume can be found in $O(1)$ time.

The total running time is linear since the recurrence for the running time is $T(n) = O(1) + 2T(n/2)$.

References

1. P. K. Agarwal, B. Aronov, and M. Sharir. Line traversals of balls and smallest enclosing cylinders in three dimensions. *Discrete Comput. Geom.*, 21:373–388, 1999.
2. P. K. Agarwal and C. M. Procopiuc. Approximation algorithms for projective clustering. In *Proc. 11th ACM-SIAM Sympos. Discrete Algorithms*, pp. 538–547, 2000.
3. T. Chan. Approximating the diameter, width, smallest enclosing cylinder and minimum-width annulus. In *Proc. 16th Annu. ACM Sympos. Comput. Geom.*, pp. 300–309, 2000.
4. S. De and K. J. Bathe. The method of finite spheres. *Computational Mechanics*, 25:329–345, 2000.
5. L. Guibas, F. Xie, and L. Zhang. Kinetic data structures for efficient simulation. In *Proc. IEEE Intern. Conf. on Robotics and Automation*, 3:2903–2910, 2001.
6. L. J. Guibas. Kinetic data structures — a state of the art report. In P. K. Agarwal, L. E. Kavraki, and M. Mason, editors, *Proc. Workshop Algorithmic Found. Robot.*, pp. 191–209. A. K. Peters, Wellesley, MA, 1998.
7. L. J. Guibas, A. Nguyen, D. Russel, and L. Zhang. Collision detection for deforming necklaces. In *Proc. 18th Annu. ACM Sympos. Comput. Geom.*, pp. 33–42, 2002.
8. D. Halperin and M. Overmars. Spheres, molecules, and hidden surface removal. *Comput. Geom. Theory Appl.*, 11(2):83–102, 1998.
9. S. Har-Peled and K. Varadarajan. Projective clustering in high dimensions using core-sets. In *Proc. 18th Annu. ACM Sympos. Comput. Geom.*, pp. 312–318.
10. V. Koltun and M. Sharir. The partition technique for overlays of envelopes. In *Proc. 43nd Annu. IEEE Sympos. Found. Comput. Sci.*, pp. 637–646.
11. N. Megiddo. Applying parallel computation algorithms in the design of serial algorithms. *J. ACM*, 30(4):852–865, 1983.
12. S. Quinlan. Efficient distance computation between non-convex objects. pp. 3324–3329, 1994.
13. J. Sadoc and N. Rivier. Boerdijk-coxeter helix and biological helices. *The European Physical Journal B*, 12(2):309–318, 1999.
14. E. Schömer, J. Sellen, M. Teichmann, and C. K. Yap. Smallest enclosing cylinders. *Algorithmica*, 27(2):170–186, 2000.
15. B. Zhu. Approximating 3D points with cylindrical segments. In *Proc. 8th Ann. Internat. Conf. Computing and Combinatorics*, pp. 420–429, 2002.

Geometric Algorithms for Agglomerative Hierarchical Clustering*

Danny Z. Chen and Bin Xu

Department of Computer Science and Engineering,
University of Notre Dame, Notre Dame, IN 46556, USA
{chen,bxu}@cse.nd.edu

Abstract. Agglomerative Hierarchical Clustering (AHC) methods play an important role in the science of classification. In this paper, We present exact centroid and median AHC algorithms and an approximate single-link AHC algorithm for clustering data objects in the d-D space for any constant integer $d \geq 2$; the time and space bounds of all our algorithms are $O(n \log n)$ and $O(n)$, where n is the size of the input data set. Previously best algorithmic approaches for these three methods take at least quadratic time in the worst case. We implemented these AHC algorithms, and the experimental results show that our algorithms are quite efficient and practical.

1 Introduction

AHC is especially important in chemical information research (e.g., for molecule structure analysis and compound selection). For example, recent research has suggested that hierarchical clustering methods perform better than the more commonly-used non-hierarchical clustering methods in property-prediction studies and in separating active molecules and inactives [4]. In fact, AHC algorithms have been implemented by many commercial softwares, e.g., SAS (SAS Institute Inc), S-Plus (Bell Labs), Oracle (Oracle Corporation), BCI (Barnard Chemical Information Ltd), and Xplore (Medical Device Technologies, Inc).

More precisely, AHC methods can be described as performing the following general process:

1. Begin with n clusters, each consisting of exactly one input object.
2. Search for the most "similar" pair of clusters. Let the two such clusters be labeled as A and B.
3. Merge clusters A and B. Label the cluster resulted from the merge as C, and update the data structure for the clusters to reflect the deletion of A and B and the insertion of cluster C.
4. Repeat steps 2 and 3 a total of $n - 1$ times (by then, the n input objects are all in one cluster).

* This work was supported in part by Lockheed Martin Corporation and by the National Science Foundation under Grant CCR-9988468.

T. Warnow and B. Zhu (Eds.): COCOON 2003, LNCS 2697, pp. 30–39, 2003.
© Springer-Verlag Berlin Heidelberg 2003

In this paper, we consider three kinds of AHC methods for spatial data objects in the d-D space $\mathbb{R}^d$ ($d \geq 2$): centroid, median, and single-link. The criterion that we use for merging objects/clusters is their similarity or dissimilarity, which is based on the Euclidean metric (i.e., the L_2 metric).

The centroid and median methods are intended for spatial data consisting of interval-scaled measurements. Suppose that we are given a set S of n points $p_1, p_2, \ldots, p_n$ in the d-D space $\mathbb{R}^d$ (for any constant integer $d \geq 2$). We denote by x_{ij} the j-th coordinate of the point p_i, $i = 1, 2, \ldots, n$ and $j = 1, 2, \ldots, d$. We first consider the centroid method. The *centroid* of a cluster R is the point $\bar{x}(R)$:

$$\bar{x}(R) = (\bar{x}_1(R), \bar{x}_2(R), \ldots, \bar{x}_d(R))$$

such that its j-th coordinate $\bar{x}_j(R)$ is

$$\bar{x}_j(R) = \frac{1}{|R|} \sum_{p_i \in R} x_{ij}$$

Note that $\bar{x}(R)$ need not be an input point of S. The dissimilarity between two clusters A and B is defined as the L_2 distance between the two centroids of A and B. When merging clusters A and B to form a new cluster R (i.e., $R = A \cup B$), the centroid of R can be obtained as follows:

$$\bar{x}(R) = \frac{|A|}{|R|}\bar{x}(A) + \frac{|B|}{|R|}\bar{x}(B).$$

Just like the centroid method in which we use the centroid of each cluster as a representative for the cluster, in the median method, each cluster also has a representative point which we simply call the *representative*, and the dissimilarity between two clusters A and B is the L_2 distance between the two representatives of A and B. For a cluster of only one point, its representative is the point itself. When merging two clusters A and B to form a new cluster R, with their representatives being $\bar{x}(A)$ and $\bar{x}(B)$, the representative $\bar{x}(R)$ of R is

$$\bar{x}(R) = \frac{1}{2}(\bar{x}(A) + \bar{x}(B))$$

Due to the similarity between the centroid and median methods, we can use a unified algorithm for handling both methods, with the difference being only at the calculation of the centroids (in the centroid method) and the representatives (in the median method).

The single-link algorithm was first introduced by Florek *et al.* and Sneath. The dissimilarity between any two clusters A and B is defined as $d(A, B) = \min_{p_i \in A, p_j \in B} d(p_i, p_j)$, i.e., the minimum of all the pairwise distances between the points in the two sets A and B. All commonly-used AHC algorithms of both types have at least a quadratic running time. For example, the AHC algorithms in BCI's chemical information software package all run in quadratic time [4] and could be impractical.

The single-link method is a common AHC method. It can be naturally based on computing the minimum spanning tree (MST) (which can be easily transformed to a desired cluster hierarchy. Different single-link algorithms have been proposed. For example, Murtagh used Bentley and Friedman's MST algorithms in coordinate spaces, which run in $O(n \log n)$ time on average; thus the single-link algorithm in [7] takes on average $O(n \log n)$ time. However, a bad case occurs when the data points are equally distributed among a few very widely separated clusters, for which those MST algorithms take $O(n^2)$ time. Agarwal *et al.* [1] presented an efficient algorithm for computing the Euclidean MST in $\mathbb{R}^d$, with an $O(n^{2-2/(\lceil d/2 \rceil+1)+\epsilon})$ time bound for any value $\epsilon > 0$. Hence a single-link algorithm based on the MST solution [1] takes nearly $O(n^2)$ time for a large d.

In this paper, we present faster exact algorithms for the centroid and median AHC methods and a $(1+\epsilon)$-approximate algorithm for the single-link AHC method, using the exact or approximate Euclidean measure of dissimilarity between clusters. Our main contributions are summarized below.

1. All our AHC algorithms run in an optimal $O(n \log n)$ time and $O(n)$ space for n input points in $\mathbb{R}^d$ ($d \geq 2$). Our efficiency does not depend on any distributions of the input data. Our approximation algorithm produces provably good solutions (which are often sufficient in applications).
2. We implemented and experimented with all our AHC algorithms. In fact, some of our key geometric components, such as dynamic closest pairs and well-separated pair decompositions, are not only theoretically complicated but also substantially non-trivial to implement. Our experimental results, especially on general data sets of large sizes and high dimensions, show that our AHC algorithms actually perform much faster than their theoretical predictions and are quite efficient. We believe that our fast AHC algorithms are of considerable practical importance.

2 Exact Centroid and Median AHC Methods

We use a unified algorithm to handle both the centroid and median methods, with the only difference being at calculating their respective representatives for the clusters (see Section 1 on this). In each method, a cluster is represented by a representative point, and the dissimilarity between two clusters is the Euclidean distance between their representative points. Two clusters with the smallest dissimilarity are merged to form a new cluster (with its own representative point). Hence, a key operation is to identify among a set of points under consideration two points with the smallest Euclidean distance. A difficulty is: The point set is dynamically changing in the algorithm, i.e., we need to delete points which are the representatives of the merged clusters and insert points which are the representatives of the newly formed clusters. Note that dealing with dynamic points in higher dimensions is also difficult.

2.1   The Main Algorithm

We make use of the dynamic closest pair technique [3] in our centroid and median AHC algorithms. For the dynamic closest pair problem, one would like to design a data structure that can efficiently update the closest pair as points are inserted into and deleted from S in an on-line fashion.

Bespamyatnikh [3] gave an optimal algorithm for solving the dynamic closest pair problem, using a data structure of size $O(n)$ and maintaining the closest pair in $O(\log n)$ worst-case time under the deletion/insertion of arbitrary points. However, The algorithm [3] is very complicated theoretically (and is actually even harder to implement efficiently). Our exact centroid and median algorithms are based on Bespamyatnikh's dynamic closest pair approach, and we implemented these algorithms effectively.

Bespamyatnikh's approach is based on the fair-split tree data structure [3, 6], which represents a hierarchical subdivision of the space $\mathbb{R}^d$ into a set of axis-parallel boxes. Our fair-split tree T is for a set of clusters (i.e., storing their representative points). More precisely, each leaf node of T stores the representative of exactly one cluster. To calculate the representative of a new cluster after merging two clusters in the centroid method, we also need to record the number of points in each cluster.

Our main algorithm works as follows.

1. Construct the fair-split tree T for the input point set S by inserting the points of S into T one by one, maintaining the fair-split tree structure and the closest pair in T.
2. While there are at least two points in T, do
   a) Let (p, q) be the closest pair of the representative points stored in T.
   b) Merge the clusters for p and q, and compute the new representative z for the new cluster.
   c) Delete p and q from T and insert z into T, maintaining the fair-split tree and closest pair.

The tree T is constructed from a hierarchical subdivision of $\mathbb{R}^d$ into axis-parallel boxes. It is easy to see that our above main algorithm is carried out essentially by a sequence of on-line operations $op_1, op_2, \ldots, op_N$, where $N = O(n)$ and each op_i is one of the three kinds of operations: (1) inserting an arbitrary point in T, (2) deleting a point from T, and (3) maintaining the closest pair of the points in T. We omit the maintenance of the fair-split tree and closest pair under deletions and insertions of points (see [3] for more details) due to the space limit.

Our maintenance algorithms need to keep performing several operations on T such as point location (finding the leaf node in which a point is stored), point deletion, and point insertion. However, a difficulty is: The height $h(T)$ of T thus maintained need not be $O(\log n)$. Actually, $h(T)$ can be as large as $O(n)$. To enable each operation on T to run in $O(\log n)$ time, we further apply the sophisticated dynamic tree techniques [2,3]. Due to the space limit, we also omit the details of this part from this version.

2.2   The Final Result

The fair-split tree T for the input point set S can be built in $O(n \log n)$ time and $O(n)$ space. There are $n - 1$ iterations in our main algorithm, and in each iteration we delete two points from and insert one point into T, while maintaining the closest pair in T. Since each operation (deleting or inserting a point) takes $O(\log n)$ time, the $n - 1$ iterations altogether take $O(n \log n)$ time. Hence, we obtain the following result for our exact centroid and median algorithms.

Theorem 1. *Let S be a set of n points in $\mathbb{R}^d$, with $d \geq 2$ being a constant integer. The exact centroid and median AHC problems on S can be solved in $O((24d + 1)^d \cdot (36d + 19)^d \cdot d \log d \cdot n \log n)$ time and $O(dn)$ space based on the Euclidean distance metric.*

3   Approximate Single-Link AHC Method

As discussed in Section 1, a single-link AHC solution can be based on computing the minimum spanning tree (MST). By using the well-separated pair decomposition technique [5,6], one can compute a sparse graph G_s for S, such that the MST of the graph G_s is a good approximation of the exact Euclidean MST (EMST) on S.

3.1   Well-Separated Pair Decomposition

Roughly speaking, two point sets are said to be *well-separated* if each set represents a cluster of points such that the distance between these two clusters is significant with respect to the larger radius of two balls each containing one of the two clusters. Below is a more precise definition of well-separation.

Definition 1. *[5,6] Given a real number $s > 0$ (called the* separation value*) and two finite point sets A and B in $\mathbb{R}^d$, A and B are said to be well-separated with respect to s if there are two disjoint d-D balls C_A and C_B of the same radius, such that C_A contains A, C_B contains B, and the distance between C_A and C_B is at least s times the radius of C_A.*

Definition 2. *[5,6] Let S be a set of n points in $\mathbb{R}^d$, and $s > 0$ be a given separation value. A well-separated pair decomposition (WSPD) for S (with respect to s) is a sequence of pairs of non-empty subsets of S, denoted by $\{A_1, B_1\}$, $\{A_2, B_2\}$, ..., $\{A_m, B_m\}$, of size m, such that*

1. *$A_i \cap B_i = \emptyset$ for each $i = 1, 2, \ldots, m$.*
2. *For every unordered pair $\{p, q\}$ of distinct points of S, there is exactly one pair $\{A_i, B_i\}$ in the sequence, such that either (a) $p \in A_i$ and $q \in B_i$, or (b) $p \in B_i$ and $q \in A_i$.*
3. *A_i and B_i are well-separated with respect to s, for each $i = 1, 2, \ldots, m$.*

Callahan and Kosaraju [5,6] showed how such a WSPD of size $m = O(n)$ on S can be computed using a binary tree T, called *split tree*. Here we briefly describe their main idea. The split tree is similar to a kd-tree. The algorithm in [5,6] computes the axis-parallel bounding box of all points in S, which is then successively split by d-D hyperplanes, each of which is orthogonal to an axis. As a box is split, each of the two resulting boxes contains at least one point of S. When a box contains exactly one point of S, the splitting process stops (for this box). The resulting binary tree T of such boxes stores the points of S at its leaves, one leaf per point. Each node u of T is associated with a subset of S, denoted by S_u, which is the set of all points of S that are stored in the subtree of T rooted at u.

Callahan and Kosaraju [5,6] showed that the split tree T for S can be computed in $O(n \log n)$ time, and, from T, a WSPD of size $m = O(n)$ can be obtained in an additional $O(n)$ time. Each pair $\{A_i, B_i\}$ in this WSPD is represented by two nodes u_i and v_i of T, with $A_i = S_{u_i}$ and $B_i = S_{v_i}$.

3.2   The Final Result

An approximate EMST is a spanning tree of S whose total weight is no more than $1 + \epsilon$ times that of the exact EMST of S, where $\epsilon > 0$ is related to the approximation factor. To compute an approximate EMST, we perform two steps once a WSPD for S based on a separation value $s > 0$ is given:

1. Construct a sparse graph $G_s = (V, E)$, where V consists of all points of S, and E consists of the edges such that for every pair $\{A_i, B_i\}$ in the given WSPD for S, G_s contains exactly one edge (a, b) connecting two points $a, b \in S$ such that $a \in A_i$ and $b \in B_i$.
2. Compute an MST in G_s using the standard graph techniques for MST.

Then based on [5,6], we have the following result on our approximate single-link AHC algorithm.

Theorem 2. *Let S be a set of n points in $\mathbb{R}^d$ for a constant integer $d \geq 2$, $s > 2$ be a constant separation value, and $\epsilon = s^{-2}$. Then the $(1 + \epsilon)$-approximate single-link AHC problem on S can be solved in $O(n \log n + (\epsilon^{-d/2} \log \frac{1}{\epsilon}) \cdot n)$ time and $O(\epsilon^{-d/2} n)$ space based on the Euclidean metric.*

4   Experimental Results

To study the practical performance of our AHC algorithms presented in Sections 2 and 3, we implemented all of them using C++ on a Sun Sparc 20 workstation running Solaris. Our experimental study has several goals. The first goal is to find out, in various AHC settings, how the execution times of our exact and approximate AHC algorithms vary as functions of two key parameters: the input data size $n = |S|$ and the input data dimension $d \geq 2$. Our second goal is to

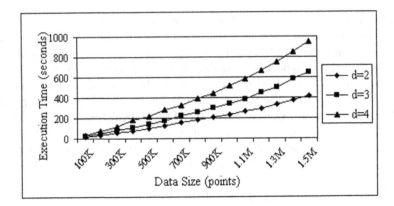

Fig. 1. The relation between the execution times of our exact centroid AHC algorithm and the data sizes, for data sets in 2-D, 3-D, and 4-D.

develop efficient software for our exact and approximate AHC algorithms, and make it available to real AHC applications. Besides, we were quite concerned that the theoretical upper time bounds of our exact centroid and median AHC algorithms (i.e., $O((24d + 1)^d \cdot (36d + 19)^d \cdot d \log d \cdot n \log n)$) and approximate single-link AHC algorithm (i.e., $O(n \log n + (s^d \log s)n)$ for a constant $s > 2$) appear to be quite high for various practical settings in which d and s must be rather "big". Hence, our third goal is to compare the experimental results with these theoretical time bounds to determine the practical efficiency of our AHC algorithms.

All input data sets used in our experimental study were generated randomly. To mimic the input data sets for real AHC applications, we conducted a large number of experiments with various data distributions such as uniform, Gaussian, Laplace, and other correlated distributions. Our experiments have shown that the effects of different data distributions on the efficiency of our exact and approximate AHC algorithms are not significant, and hence we may use input data sets of any distributions to illustrate the performance of all our AHC algorithms. Due to the space limit, we omit the results of many experiments in this version.

Each point for the curves in Figures 1, 2, 3, and 4 represents the average of 30 experiments whose input data sets were generated with different seeds for the random number generator that we used.

4.1   Experimental Results of Our Exact AHC Approaches

Since the only (minor) difference between our exact centroid and median AHC algorithms is at their ways of calculating the new representative point after merging two clusters, there is no essential difference between these two algorithms. Our experimental results that are omitted from this version also proved that.

We hence illustrate only the centroid AHC algorithm with the experimental data below.

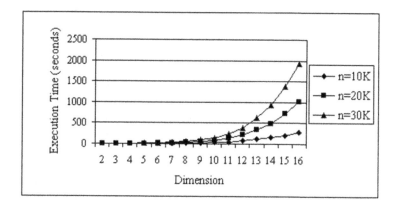

Fig. 2. The relation between the execution times of our exact centroid AHC algorithm and the dimensions, for data sets of sizes 10K, 20K, and 30K.

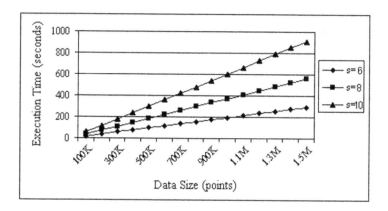

Fig. 3. The relation between the execution times of our approximate single-link AHC algorithm and the data sizes, for 2-D data sets.

We considered the data sizes n of input point sets varying from 100K to 1500K = 1.5M, in fixed dimensions $d \geq 2$. Figure 1 shows the execution times of our exact centroid AHC algorithm as functions of n, in dimensions 2, 3, and 4.

In Figure 1, the three curves for $d = 2$, 3, and 4 all indicate that the execution times of our exact centroid AHC algorithm increase almost linearly with respect to the increase of the data sizes. More precisely, the execution times increase very slowly as the data sizes increase (with a small positive slope). For example, the slopes are about 28.3 seconds/100K, 43.9 seconds/100K, and 65.5 seconds/100K for dimensions 2, 3, and 4, respectively. Comparing to the theoretical upper time bound, the increase of the execution times is substantially slower with respect to the increase of the data sizes.

A key issue is how our execution times depend on the dimensionality of the input data sets. Since the dimension value $d \geq 2$ acts as the power parameter in the theoretical time bound, d is certainly a crucial factor to the execution times as well.

To determine the relation between the execution times of our exact centroid AHC algorithm and the dimensions of the input data sets, we considered dimensions varying from 2 to 16, with data sizes 10K, 20K, and 30K. As shown in Figure 2, the increase of the execution times is much slower than what the theoretical time bound predicts. For example, for data sizes 10K, 20K, and 30K, when the dimension d increases from 2 to 10, the execution times increase 26.9, 38.1, and 38.6 times, respectively; when the dimension d increases from 10 to 16, the execution times increase 7.8, 10.2, and 11.9 times, respectively. These numbers are all significantly smaller than the increase of the upper time bound.

4.2   Experimental Results of Our Approximate AHC Approach

To find out how the execution time of our approximate single-link AHC algorithm varies as a function of the input data sizes, we considered data sets of sizes varying from 100K to 1.5M, and the separation values s varying from 6 to 10. Figure 3 shows some of the experimental results for 2-D data sets.

In Figure 3, the three curves for different values of s all indicate that the execution times increase almost linearly with respect to the increase of the data sizes. Further, the execution times increase very slowly as the data sizes increase (with a small positive slope). For example, the slopes are about 19.8 seconds/100K, 36.7 seconds/100K, and 61.5 seconds/100K for the s values 6, 8, and 10, respectively. Comparing to the theoretical upper time bound, the increase of the execution times is clearly significantly slower with respect to the increase of the data sizes.

We considered dimensions d varying from 2 to 16, with the separation values $s = 6, 8$, and 10, and data sets of sizes 20K. As shown in Figure 4, the increase of the execution times is much slower than what the theoretical time bound predicts. For example, for the values of $s = 6, 8$, and 10, as the dimension d changes from 2 to 4, the execution times increase 77.4, 71.8, and 55.6 times, respectively; as the dimension d changes from 4 to 16, the execution times increase 4.6, 2.3, and 1.7 times, respectively. All these numbers are obviously much smaller than the increase of the theoretical upper time bound.

4.3   Concluding Remarks

From the above experimental results on all our exact and approximate AHC algorithms, one can see that comparing to their theoretical time bounds, the

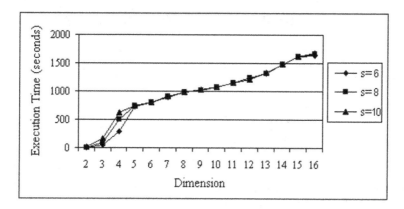

Fig. 4. The relation between the execution times of our approximate single-link AHC algorithm and the dimensions, for data sets of sizes 20K.

increase of the execution times of all our AHC algorithms is quite slow with respect to the increase of the data sizes and the dimensions. Therefore, these AHC algorithms are expected to be quite efficient and practical in real applications.

References

1. P.K. Agarwal, H. Edelsbrunner, and O. Schwarzkopf, Euclidean Minimum Spanning Trees and Bichromatic Closet Pairs, *Discrete Comput. Geom.*, 6 (1991), 407–422.
2. S.W. Bent, D.D. Sleator, and R.E. Tarjan, Biased Search Tress, *SIAM J. Comput.*, 14 (1985), 545–568.
3. S.N. Bespamyatnikh, An Optimal Algorithm for Closest Pair Maintenance, *Discrete Comput. Geom.*, 19 (1998), 175–195.
4. R.D. Brown and Y.C. Martin, Use of Structure-Activity Data to Compare Structure-Based Clustering Methods and Descriptors for Use in Compound Selection, *J. Chem. Inf. Comput. Sci.*, 36 (3) (1996), 572–584.
5. P.B. Callahan, Dealing with Higher Dimensions: The Well-Separated Pair Decomposition and Its Applications, Ph.D. thesis, Dept. Comput. Sci., Johns Hopkins University, Baltimore, Maryland, 1995.
6. P.B. Callahan and S.R. Kosaraju, A Decomposition of Multidimensional Point Sets with Applications to k-Nearest-Neighbors and n-Body Potential Fields, *J. ACM*, 42 (1995), 67–90.
7. F. Murtagh, A Survey of Recent Advances in Hierarchical Clustering Algorithms, *The Computer Journal*, 26 (4) (1983), 354–359.

Traveling Salesman Problem of Segments*

Jinhui Xu, Yang Yang, and Zhiyong Lin

Department of Computer Science and Engineering
State University of New York at Buffalo
Buffalo, NY 14260, USA
{jinhui,yyang6,zlin}@cse.buffalo.edu

Abstract. In this paper, we present a polynomial time approximation scheme (PTAS) for a variant of the traveling salesman problem (called *segment TSP*) in which a traveling salesman tour is sought to traverse a set of n ϵ-separated segments in two dimensional space. Our results are based on a number of geometric observations and an interesting generalization of Arora's technique [5] for Euclidean TSP (of a set of points). The randomized version of our algorithm takes $O(n^2(\log n)^{O(1/\epsilon^4)})$ time to compute a $(1 + \epsilon)$-approximation with probability $\geq 1/2$, and can be derandomized with an additional factor of $O(n^2)$. Our technique is likely applicable to TSP problems of certain Jordan arcs and related problems.

1 Introduction

In this paper, we study the following generalization of the Euclidean traveling salesman problem (TSP), called *segment TSP* problem: Given a set S of n ϵ-separated segments in 2-dimensional space, find a tour to traverse all the segments such that the total Euclidean distance of this tour is minimized. We assume that all the endpoints of segments have integral coordinates, the minimum distance between any pair of segments is at least ϵ, and the size of the smallest bounding square of S is $O(n)$ (we call such segments as ϵ-separated segments). The tour traverses each input segment completely. However, it could enter and leave a segment multiple times, with each time visiting a portion of this segment. Since the Euclidean TSP can be reduced to this problem, the segment TSP is clearly NP-hard.

Our study on this problem is motivated by applications in radiosurgery [9, 29] and laser engraving. Radiosurgery (e.g., stereotaxic radiosurgery [29]) is a minimal-invasive surgical procedure of using a set (as many as 400 [29]) of radiation beams to destroy tumors. The beams are often originated from grid points in a discretized beam space and appear consecutively on a set of ϵ-separated segments (i.e. portions of the grid lines).

The traveling salesman problem (normally defined on a set of nodes) is a classical problem in combinatorial optimization, and has been studied extensively

---

* This research was supported in part by an IBM faculty partnership award, and an award from NYSTAR (New York state office of science, technology, and academic research) through MDC (Microelectronics Design Center).

in many different forms [4,5,11,13,19,21,23,25,28]. For non-metric TSP, most of the previously known algorithms are heuristic, and none of them guarantees a fixed approximation ratio for the general non-metric TSP problem. For metric TSP, an early approximation algorithm by Christofides [12] showed that a 1.5-approximation algorithm exists for any instance of the metric TSP problem. In [27], Papadimitriou and Yannakakis proved that the general metric TSP problem is MAX SNP-hard, indicating that a polynomial time approximation scheme (PTAS) is unlikely.

Better approximation algorithms have been obtained for special metric TSP problems. In [16], Grigni, Koutsoupias and Papadimitriou showed that the un-weighted planar graph (i.e., every edge has weight one) TSP problem with the shortest-path metric has a PTAS. Later, Arora *et al.* [4] gave a PTAS for the weighted planar graph TSP problem. A breakthrough in the metric TSP comes from the discovery of PTAS for the Euclidean TSP. Arora [5] and Mitchell [23] each independently obtained a PTAS for this problem. Based on a spanner of the set of points, Rao and Smith [26] later gave an improved algorithm to Arora's algorithm, which runs in $O(n \log n)$ time.

Although each of the above results provides powerful techniques for a number of problems, it seems that they all have difficulty to solve the segment TSP. The major obstacle is the follows. In segment TSP, when performing a recursive partition on the set of segments, there could be many segments or subsegments inside each (even the smallest) subproblem. Furthermore, the optimal tour may enter or leave each segment (or subsegment) multiple times and any point on the segment could be a candidate of such a point (called *entry point*). Thus a dynamic programming on such a subproblem might take exponential time.

Based on a number of interesting observations, together with a generalization of Arora's technique, we present in this paper a PTAS for the segment TSP problem.

Due to space limit, we omit many details.

2   Preliminaries

Let $S = \{s_1, s_2, \cdots, s_n\}$ be a set of n input segments with each segment $s_i = \overline{p_i q_i}$ represented by its two endpoints p_i and q_i. A traveling salesman tour π of S traverses each segment in S and thus contains two types of line segments: input segments (or subsegments) and segments connecting two input segments. To distinguish them, we call the former *segments* and the latter *bridges*. The intersections of segments and bridges are called *entry points* of π on the corresponding segments.

The bounding box B of S is the smallest square box which covers all segments in S. A dissection of the bounding box B is a recursive quadtree decomposition T of B such that each non-empty leaf square has unit size. A leaf square may contain no endpoint at all but only the interior of some input segments. A leaf square is said non-empty if it contains some portion of input segments.

A *portal* of a square is a prespecified point on the boundary of the square such that a traveling salesman tour can deviate itself to pass through it. Such deviated tour will be called a salesman path. An m-regular set of portals is the set of points evenly-spaced on each edge and the corner of the boundary of a square.

A traveling salesman path is (m, r)-light with respect to the dissection T if it crosses each edge of any square at most r times and always at portals.

Similar to the definition of portals on the boundary of squares, we also define a set of evenly-spaced portals on the input segments. A traveling salesman path is (m, r)-connected with respect to the dissection T if the portion of each input segment inside a leaf square has m portals and the traveling salesman path has at most r entry points on that portion of segment with each entry point coincident with a portal.

3   Main Difficulties and Ideas

Since our algorithm for the segment TSP problem follows the main steps of Arora's scheme, we first briefly introduce Arora's algorithm, and then point out the main difficulties encountered in the segment TSP and our main ideas for overcoming them.

For the Euclidean TSP problem, Arora's algorithm performs the following main steps.

1. Perturb the set of input points such that they are well rounded to integral points and well separated.
2. Recursively partition the set of rounded points and build a randomly shifted quadtree such that each leaf contains at most one rounded point.
3. From bottom to top, use dynamic programming on every node of the quadtree to compute the shortest bent (m, r)-light path visiting all the points in that node, where $r = O(1/\epsilon)$ and $m = O(\frac{\log n}{\epsilon})$.

It was shown in [5] that the bent (m, r)-light path is a $(1+\epsilon)$-approximation of the optimal path with probability $\geq 1/2$, and can be computed in $O(n(\log n)^{O(1/\epsilon)})$ time.

The success of the above approach on the Euclidean TSP relies on several key facts.

1. The set of points can be well separated such that each leaf node of the quadtree contains at most one rounded point.
2. An (m, r)-light path can be *efficiently* computed by dynamic programming in each node of the quadtree.
3. An optimal path can be "uncrossed" (or *patched*) at the boundary of each square of the quadtree and transformed to an (m, r)-light path.

Unfortunately, when the objects change from points to segments, none of the above facts seems to be true. The main reason is because each leaf square may contain many segments, and each segment may have an unbounded number of entry points, seemingly suggesting that the dynamic programming may take exponential time.

To overcome these difficulties, our main ideas are to introduce portals on each segment and use the portals as the candidates for entry points. To reduce the total number of entry points inside each leaf square, we require the computed traveling salesman path to be (m, r)-connected, that is, the path can have at most r entry points on the portion of each segment inside a square. Locally, the (m, r)-connected path could be much longer than the optimal solution. Thus a major challenge is to bound the total increased length by such a restricted path. Unlike the Euclidean TSP problem where the increased length by an (m, r)-light path can be relatively easily bounded by counting the number of crossings of the traveling salesman tour with the dissection lines (e.g., the total increased length is $\epsilon \times \#of crossings$) which is no more than $O(OPT)$, the increased length in the (m, r)-connected path however depends on the total number of entry points of an optimal TSP tour which seemingly has no connection with the value of OPT. By making use of the properties of the segments and observations on the behavior of bridges, we prove an interesting upper bound on the total number of entry points in an optimal solution which enables us to bound the total increased distance by the (m, r)-connected path.

Our algorithm performs the following main steps.

1. Perform a dissection on the bounding box B of the segments such that each non-empty leaf square has a unit size, and then make a random (a, b)-shift.
2. Use dynamic programming to compute an (m, r)-light and (m, r)-connected TSP path for each non-empty square in the dissection in a bottom-up manner.

Based on a detailed analysis, we show the following main theorem.

Theorem 1. *Given a set of n ϵ-separated segments on plane for some constant $\epsilon > 0$, one can compute a $(1 + \epsilon)$-approximation of the segment TSP with probability at least $1/2$ in $O(n^2 (\log n)^{O(1/\epsilon^4)})$ time.*

4   Algorithm and Analysis

As mentioned in last section, to solve the segment TSP problem efficiently, we need to reduce the complexity of each subproblem associated with a node in the quadtree dissection. Since the high complexity may be due to either the number of segments or the number of entry points. Below we discuss our ideas on each of them.

The following lemma shows that the number of segments in a leaf square is bounded.

Lemma 1. *In any leaf square Q, the number of segments in S intersected by Q is bounded by a constant (i.e., $O(1/\epsilon)$), and each segment intersects two edges of Q.*

Thus, our focus is on how to reduce the number of entry points. The following lemma shows some property of an optimal tour.

Lemma 2. *An optimal TSP tour does not intersect itself except at the entry points.*

To reduce the number of entry points, we introduce $m = O(\frac{\log n}{\epsilon})$ portals on the portion of each segment inside a square of the dissection, and restrict the TSP tour to enter and leave a segment only at its portals. Based on the set of portals, we compute an (m, r)-connected path to approximate the optimal tour. The following key lemma bounds the quality of such an approximation.

Lemma 3. *For a set of segments S, there exists an (m, r)-connected traveling salesman path which is a $(1 + 3C_0\sqrt{\epsilon} + (12 + C)\epsilon)$-approximation of the optimal solution.*

To prove this lemma, we need several other lemmas on the optimal TSP tour. We first study the behavior of bridges in an optimal tour.

We call two bridges which share the same entry point e on a segment s as a *two-part-bridge*. If the two parts are in different sides of s, then the two-part-bridge, say b, is called a *bi-bridge*, e a bi-entry, and s a bi-segment of b. If the two parts of b are in the same side of s, then we call b as a *V-bridge* (see Figure 1), e as a *V-entry*, and s as a *V-segment*. The two parts of a *V*-bridge b may also intersect with some other segment, say s', and form an A-like shape (see Figure 1). We call s' as an A-segment of b. In a square Q, if a *V*-bridge b has an A-segment in Q, then b is called a *shared V*-bridge in Q, otherwise an *unshared V*-bridge in Q.

Lemma 4. *In an optimal TSP tour π, a bi-bridge b either has its entry point coincident with an endpoint of the segment s or its two parts are collinear (i.e., on a straight line).*

Corollary 1. *In an optimal tour, all bridges are aligned along a set of interior-disjoint line segments whose endpoints are either V-entries or endpoints of input segments.*

Lemma 5. *Let s_1 and s_2 be any pair of segments in S which are connected by a bridge in an optimal tour π. Then there is no V-bridge in π which intersects s_1 and s_2 consecutively and has one of them as its V-segment.*

To bound the number of entry points inside a leaf square Q, we classify all the segments in Q into classes based on the pair of intersected edges (of Q). (Note that by Lemma 1, each segment in Q intersects two edges.) Since all segments are disjoint, there are no more than 5 classes (see Figure 1). In each class C, we call the first and last segments (along the intersected edges) as bounding segments.

Lemma 6. *In an optimal tour π, any segment s in a leaf square Q intersects at most 6 shared V-bridges in Q.*

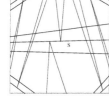

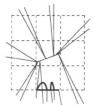

Fig. 1. Illustrating the proof of Lemma 6.

Fig. 2. Illustrating the proof of Lemma 7.

Corollary 2. *Let Q be a leaf square intersecting k segments of S, and π be an optimal tour of S. Then the total number entry points produced by all shared V-bridges in Q is at most $12k$.*

Lemma 7. *Let Q be a leaf square containing k segments. In an optimal tour π, all unshared V-bridges in Q produce in total $O(k)$ entry points in Q.*

The above lemmas about V-bridges suggest that to reduce the total number of entry points inside a leaf square, we should focus on reducing those entry points generated by bi-bridges.

Lemma 8. *Let Q be a leaf square containing k segments. Then the total length of all bridges in an optimal TSP tour π inside Q is bounded by a constant C_0.*

Proof. Sketch of the proof: Our main idea for proving this lemma is to first remove Q and all its contained segments and bridges. We then build an Euler graph along the outside boundary of Q such that the structure of π outside Q is preserved and all segments contained in Q are attached to the outside structure. The total length of the added edges for building up the Euler graph is no more than a constant (e.g., 24). Since the Euler graph provide an TSP tour to the set of segments, thus should have larger total length than π, which implies that the total length of all bridges in Q is less than 24.

In a leaf square, a bi-bridge either touches the opposite sides or the neighboring sides. The following two lemmas help us to bound the entry points generated by bi-bridges.

Lemma 9. *In a leaf square Q containing k segments, there are at most C_0 bi-bridges touching the opposite edges of Q (See Figure 3).*

For a bi-bridge b touching the neighboring sides e and e', we say segment s and s' intersected by b are the surface segments if they are the farthest pair of segments (along b) in Q. We define the *surface distance* of b as the distance between its corresponding surface segments along b. Let δ be the minimum surface distance among all bi-bridges intersecting e and e'.

Note that to bound the entry points generated by bi-bridges, we can ignore those bi-bridges which intersect only one segment in Q as their generated entry

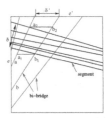

Fig. 3. The number of bi-bridges touching opposite sides of Q is a constant.

Fig. 4. The maximum distance (along e or $e^{\bullet}$) between two neighboring bi-bridges is $\geq \delta$.

points can be bounded by the total number of crossings between grid lines and the bridges, thus can be handled in a way similar to the one used for (m, r)-light path in the Euclidean TSP problem [5]. Thus we can assume that $\delta \geq \epsilon$.

Lemma 10. *Let Q, e, e' and δ be defined as above. For two neighboring bi-bridges a and b intersecting e and e', their maximum distance (along e or e') δ' is $\geq \delta$.*

Proof. Sketch of the proof: Consider Figure 4. If $\delta' < \delta$, we can modify the optimal tour in the following way to obtain a shorter tour. Draw $\overline{a_2 b_2}$ parallel to e'. Similarly, draw segment $\overline{b_1 a_1}$ parallel to e', and delete bridge $\overline{a_1 a_2}$ and $\overline{b_1 b_2}$. Since $\delta' \geq \overline{a_2 a b_2} \geq \overline{b_1 a_1}$, and $\overline{a_1 a_2} \geq \delta$, $\overline{b_1 b_2} \geq \delta$, the new tour induced by such a modification is shorter than optimal tour. A contradiction.

Lemma 11. *In a leaf square Q, the number of bi-bridges intersecting two neighbouring edges of Q is $O(\sqrt{1/\epsilon})$.*

Lemma 12. *Given a set S of ϵ-separated segments, there exists an optimal TSP tour π such that $E(\pi, S) = O((\sqrt{1/\epsilon} + C')|OPT|)$, where $E(\pi, S)$ is the number of entry points of π and C' is a constant.*

Now, we are ready to prove Lemma 3.

Proof. If an (m, r)-connected TSP path is used to approximate the optimal TSP tour π, the increased cost comes from either moving the entry points to their closest portals or reducing the number of entry points on segments which contain more than r entry points. By the proof of Lemma 12, the number of entry points on each segment in a leaf square is $O(\sqrt{1/\epsilon} + C')$ which is bounded by $O(r)$.

Thus the total increased cost is mainly due to changing locations from entry points to portals, which is $\epsilon E(\pi, S) = \epsilon(\sqrt{\frac{4C_0}{\epsilon}} + C')|OPT| \leq (3C_0\sqrt{\epsilon} + (12 + C)\epsilon)|OPT|$.

Therefore this (m, r)-connected TSP path is a $(1 + 3C_0\sqrt{\epsilon} + (12 + C)\epsilon)$-approximation of the optimal TSP tour.

For the bottom-up dynamic programming, we are able to prove an interesting "patching" lemma for entry points and establish the main result in Theorem 1.

Lemma 13. *For a segment s of length l and containing more than 3 entry points from a TSP tour π, there exists a constant g and a set of segments on s whose total length is at most $g \times l$ and whose addition to π changes it into another TSP tour which has two entry points on s.*

Based on Lemma 13, we can show that there exists a procedure to change an optimal TSP tour π and make each segment to be (m, r)-connected. Further, if we charge the increased cost to the corresponding segment, the following lemma holds.

Lemma 14. *The expected cost for dynamic programming charged to a segment s can be bounded by $\epsilon \times E(\pi, s)$, where $E(\pi, s)$ is the number of entry points on S in the unchanged π.*

After proving all the lemmas, we are ready to prove our theorem.

Proof. First, let us analyze the running time of our algorithm. Our algorithm begins with randomly (a, b)-shifted dissection which takes $O(n^2 \log n)$ time. The time complexity of this step is actually equal to the number of nodes in the dissection tree T. Since in the dissection we may have n^2 non-empty leaf squares, $|T|$ is thus $O(n^2 \log n)$.

Next, our algorithm performs a dynamic programming to computer the TSP tour. It begins from leaf squares, and then bottoms up to the entire bounding box of the set of segments. The running time required depends on the number of lookup table entries. Observe that for each square in the quadtree, on each boundary edge we have m portals and use at most r of them. Moreover we have m portals on each segment in a leaf square and use at most r entry points on this segment. Thus we define the (m, r)-*multipath problem as follows*. The input of this problem is:

1. A square in the shifted quadtree.
2. A multiset of $\leq r$ portals on each of the four edges of this square such that the sum of the sizes of these multisets is an even number $2p \leq 4r$.
3. A multiset of $\leq r$ entry points on each of the k segments in this square such that each entry point is on one of the p paths.
4. A pairing $\{a_1, a_2\}, \{a_3, a_4\}, \ldots$ between the $2p$ portals.

By this observation, we know that the number of entries in the lookup table is the number of different instances of the (m, r)-multipath problem in the shifted quadtree. Let k be the number of bridges in a leaf square. It is easy to see that in a quadtree with T squares, this number is $O(T(m+4)^{4r}(m+2)^{kr}(4r+2kr)!)$.

In the bottom up procedure, suppose all the nodes with depth $> i$ have been solved. Now we have a square S at depth i which is the parent of four squares S_1, S_2, S_3 and S_4 at depth $i + 1$. The algorithm enumerates all the possible combinations of portals on the inner edges of the children squares. The number of combinations is $O((m + 4)^{4r}(4r)^{4r}(4r)!)$.

Thus the running time of our algorithm is $O(T(m+4)^{8r}(m+2)^{kr}(4r)^{4r}((4r)!)^2 (4r + 2kr)!)$, which is $O(n^2(\log n)^{O(1/\epsilon^2)})$ if we take $r = 18g/\epsilon + 4$ and $m = 36g \log L/\epsilon$ $(L = O(n))$.

Suppose the optimal tour has length $|\text{OPT}|$. While we force the tour to be (m, r)-connected, this will increase a cost of $(3C_0\sqrt{\epsilon}+(12+C)\epsilon)|\text{OPT}|$ by Lemma 3. The MODIFY procedure for segments and square edges will increase a cost of $\epsilon \cdot \sum E(\pi, S) \leq 2C_0\sqrt{\epsilon} + (C_0 + 12 + C)\epsilon|\text{OPT}|$ by Lemma 14 and Lemma 12. To force the tour to be (m, r)-light, we introduce a cost of $\epsilon|\text{OPT}|/2$ (refer to [5]). So totally the increased cost is $\leq (5C_0\sqrt{\epsilon}+(C_0+25+2C)\epsilon)|\text{OPT}| \leq C''\sqrt{\epsilon}|\text{OPT}|$ where $C'' = 6C_0 + 25 + 2C$. Let $\epsilon_0 = C''\sqrt{\epsilon}$, equally we get $\epsilon = \frac{\epsilon_0^2}{(C'')^2}$.

Therefore we got a $(1 + \epsilon_0)$-approximation tour with running time of $O(n^2(\log n)^{O((C'')^4/\epsilon_0^4)}) = O(n^2(\log n)^{O(1/\epsilon_0^4)})$.

References

1. E. M. Arkin, Y.-J. Chiang, J. S. B. Mitchell, S. S. Skiena, and T. Yang. On the maximum scatter TSP, *Proc. 8th ACM-SIAM Sympos. Discrete Algorithms*, pp. 211–220, 1997.
2. E. M. Arkin and R. Hassin, Approximation algorithms for the geometric covering salesman problem, *Discrete Appl. Math.*, 55:197–218, 1994.
3. E. M. Arkin, J. S. B. Mitchell, and G. Narasimhan, Resource-constrained geometric network optimization, *Proc. 14th Annu. ACM Sympos. Comput. Geom.*, pp. 419–428, 1998.
4. S. Arora, M. Grigni, D. Karger, P. Klein, and A. Woloszyn, Polynomial Time Approximation Scheme for Weighted Planar Graph TSP, *Proc. 9th Annual ACM-SIAM Symposium on Discrete Algorithms (SODA)*, pp. 33–41, 1998.
5. S. Arora, Polynomial-time Approximation Schemes for Euclidean TSP and other Geometric Problems, *Journal of the ACM*, 45(5) pp. 753–782, 1998.
6. B. Awerbuch, Y. Azar, A. Blum, and S. Vempala, Improved approximation guarantees for minimum-weight k-trees and prize-collecting salesman, *Proc. 27th Annu. ACM Sympos. Theory Comput.*, pp. 277–283,1995.
7. A. Blum, P.Chalasani, D. Coppersmith, B. Pulleyblank, P. Raghavan, and M. Sudan, The minimum latency problem, *Proc. 26th Annu. ACM Sympos. Theory Comput. (STOC 94)*, pp. 16
8. S. Carlsson, H. Jonsson, and B. J. Nilsson, Approximating the shortest watchman route in a simple polygon. Manuscript, Dept. Comput. Sci., Lund Univ., Lund, Sweden, 1997.
9. D.Z. Chen, O. Daescu, X. Hu, X. Wu, and J. Xu, Determining an Optimal Penetration among Weighted Regions in 2 and 3 Dimensions, SCG'99, pp. 322–331.
10. X. Cheng, J. Kim, and B. Lu, A Polynomial Time Approx. Scheme for the problem of Interconnecting Highways, *J. of Combin. Optim.*, Vol.5(3), pp. 327–343, 2001.
11. J.L. Bentley, Fast Algorithms for Geometric Traveling Salesman Problems, *ORSA J. Comput.*, 4(4):387–411, 1992.
12. N. Christofides, Worst-case analysis of a new heuristic for the traveling salesman problem, In J.F. Traub, editor, *Symposium on New Directions and Recent Results in Algorithms and Complexity*, Academic Press, NY, pp. 441, 1976.
13. A. Dumitrescu and J.S.B. Mitchell, Approximation algorithms for TSP with neighborhoods in the plane, *Proc. 12th ACM-SIAM Sympos. Discrete Algorithms (SODA'2001)*, pp. 38–46, 2001.
14. N. Garg, A 3-approximation for the minimum tree spanning k vertices, *37th Annual Symposium on Foundations of Computer Science*, pp. 302–309, Burlington, Vermont, October 14–16 1996.

15. M. X. Goemans and J. M. Kleinberg, An improved approximation ratio for the minimum latency problem, *Proc. 7th ACM-SIAM Sympos. Discrete Algorithms (SODA '96)*, pp. 152–158, 1996.

16. M. Grigni, E. Koutsoupias, and C.H. Papadimitriou, An Approximation Scheme for Planar Graph TSP, *Proc. IEEE Symposium on Foundations of Computer Science*, pp. 640–645, 1995.

17. J. Gudmundsson and C. Levcopoulos, A fast approxmation algorithm for TSP with neighborhoods. Technical Report LU-CS-TR:97–195, Dept. of Comp. Sci., Lund University, 1997.

18. R. Hassin and S. Rubinstein, An approximation algorithm for the maximum traveling salesman problem. Manuscript, submitted, Tel Aviv University, Tel Aviv, Israel, 1997.

19. M. Jünger, G. Reinelt, adn G. Rinaldi, The Traveling Salesman Problem. In M. O. Ball, T.L. Magnanti, C. L. Monma, and G. L. Nemhauser, editors, *Network Models,* Handbook of Operations Research/Management Science, Elsevier Science, Amsterdam, pp. 225–330, 1995.

20. S. Kosaraju, J. Park, and C. Stein, Long tours and short superstrings, *Proc. 35th Annu. IEEE Sympos. Found. Comput. Sci. (FOCS 94)*, 1994.

21. E.L. Lawler, J.K. Lenstra, A.H.G. Rinnooy Kan, and D.B. Shmoys, editors. *The Traveling Salesman Problem,* Wiley, New York, NY, 1985.

22. C. Mata and J. S. B. Mitchell, Approximation algorithms for geometric tour and network design problems, *Proc. 11th Annu. ACM Sympos. Comput. Geom.*, pages 360–369, 1995.

23. J.S.B. Mitchell, Guillotine Subdivisions Approximate Polygonal Subdivisions: Part II – A simple polynomial-time approximation scheme for geometric TSP, k-MST, and related problems, *SIAM J. Computing*, Vol.28, No. 4, pp. 1298–1309, 1999.

24. J.S.B. Mitchell, Guillotine Subdivisions Approximate Polygonal Subdivisions: Part III – Faster Polynomial-time Approximation Schemes for Geometric Network Optimization, Manuscript, April 1997.

25. J.S.B. Mitchell, Geometric Shortest Paths and Network Optimization, chapter in *Handbook of Computational Geometry*, Elsevier Science, (J.-R. Sack and J. Urrutia, eds.), 2000.

26. S.B. Rao, and W.D. Smith, Improved Approximation Schemes for Geometrical Graphs via "spanners" and "banyans" *Proc. 30th Annu. ACM Sympos. Theory Comput.,* May 1998.

27. C.H. Papadimitriou and M. Yannakakis, The Traveling Salesman Problem with Distances One and Two, *Mathematics of Operations Research*, Vol. 18, pp. 1–11, 1993.

28. G. Reinelt, Fast Heuristics for Large Geometric Traveling Salesman Problems, *ORSA J. Comput.*, 4:206–217, 1992.

29. A. Schweikard, J.R. Adler, and J.C. Latombe, Motion Planning in Stereotaxic Radiosurgery, *IEEE Trans. on Robotics and Automation*, Vol. 9, No. 6, pp. 764–774, 1993.

30. X. Tan, T. Hirata, and Y. Inagaki, Corrigendum to 'An incremental algorithm for constructing shortest watchman routes'. Manuscript (submitted to internat.j.comput.geom.appl.), Tokai University, Japan, 1998.

Subexponential-Time Algorithms for Maximum Independent Set and Related Problems on Box Graphs[*]

Andrzej Lingas and Martin Wahlen

Department of Computer Science, Lund University,
Box 118, S-22100 Lund, Sweden
{Andrzej.Lingas,martin}@cs.lth.se

Abstract. A box graph is the intersection graph of orthogonal rectangles in the plane. We consider such basic combinatorial problems on box graphs as maximum independent set, minimum vertex cover and maximum induced subgraph with polynomial-time testable hereditary property Π. We show that they can be exactly solved in subexponential time, more precisely, in time $2^{O(\sqrt{n}\log n)}$, by applying Miller's simple cycle planar separator theorem [6] (in spite of the fact that the input box graph might be strongly non-planar).

Furthermore we extend our idea to include the intersection graphs of orthogonal d-cubes of bounded aspect ratio and dimension. We present an algorithm that solves maximum independent set and the other aforementioned problems in time $2^{O(d2^d b n^{1-1/d}\log n)}$ on such box graphs in d-dimensions. We do this by applying a separator theorem by Smith and Wormald [7].

Finally, we show that in general graph case substantially subexponential algorithms for maximum independent set and the maximum induced subgraph with polynomial-time testable hereditary property Π problems can yield non-trivial upper bounds on approximation factors achievable in polynomial time.

1 Introduction

In [5] it was shown that the existence of a subexponential algorithm for Independent Set would imply the existence of subexponential algorithms for SAT, Colorability, Set Cover, Clique, Vertex Cover and any other search problems expressible by second order existential formulas whose first order part is universal.

Moreover, as we observe in Section 3 of this paper, proving substantially subexponential upper time bounds for maximum independent set, or equivalently maximum clique or minimum vertex cover[1] and generally for maximum induced subgraph with polynomial-time testable hereditary property Π problems could

---

[*] This work is in part supported by VR grant 621-2002-4049
[1] Recall that minimum vertex cover is always complement of maximum independent set.

© Springer-Verlag Berlin Heidelberg 2003

lead to better approximation factors achievable for maximum independent set and the maximum induced subgraph with hereditary property Π problems in polynomial time. This however seems very difficult or unlikely. For instance, the $O(n/\log^2 n)$ approximation factor for maximum independent set has not been improved for more than decade [3]. This kind of evidence of the difficulty of deriving subexponential algorithms, e.g., for maximum independent set, does not necessarily exist in case of special graph classes for which relatively good approximation algorithms for maximum independent set running in polynomial time are known.

In this paper we study maximum independent set, minimum vertex cover and maximum induced subgraph with polynomial-time testable hereditary property Π problems on the so called box graphs which are the intersection graphs of orthogonal rectangles in the plane (e.g., the problem of maximum independent set on box graphs is known to admit an $O(\log n)$-approximation polynomial-time algorithm [1]). We show that the aforementioned problems can be exactly solved in subexponential time, more precisely, in time $2^{O(\sqrt{n}\log n)}$, by applying Miller's simple cycle planar separator theorem [6].

Our application of the planar separator theorem seems highly non-standard since it relies rather on the structure of unknown optimal solutions, i.e., maximum cardinality subsets of disjoint rectangles, than the structure of the input box graph which might be strongly non-planar!

Furthermore we extend our idea to include the intersection graphs of orthogonal d-cubes of bounded aspect ratio and dimension. We present an algorithm that solves maximum independent set and the related problems in time $2^{O(d2^d bn^{1-1/d}\log n)}$ on such box graphs in d-dimensions. We do this by applying a separator theorem by Smith and Wormald [7] in a similar fashion.

The structure of our paper is as follows. In Section 2, we present and analyze the aforementioned subexponential algorithms for maximum independent set and the related problems on box graphs in two and d dimensions. In Section 3, we observe that for general graphs substantially subexponential algorithms for maximum independent set and the maximum induced subgraph with polynomial-time testable hereditary property Π problems can yield non-trivial upper bounds on approximation factors achievable in polynomial time

2   Subexponential Algorithms

To begin with, we need the following lemma showing that we may assume w.l.o.g the rectangles defining a box graph to be placed on a small integer grid.

Lemma 1. *Let S be a set of n orthogonal rectangles in the plane. In time $O(n\log n)$, S can be one-to-one transformed into a set U of orthogonal rectangles on an integer $O(n) \times O(n)$ grid such that two rectangles in S overlap iff their images in U overlap.*

Proof. Draw through each edge of each rectangle in S co-linear line. Sort the resulting vertical lines by their X-coordinates as well as the resulting horizontal lines by their Y-coordinates. Move the vertical lines horizontally so the distance between two consecutive lines becomes 1. Analogously, move the horizontal lines vertically. in order to obtain an $O(n) \times O(n)$ integer grid. The aforementioned line movements transform rectangles in S into rectangles on the grid preserving overlaps.

The following lemma is crucial in the design of our subexponential algorithm for maximum independent set and related problems. It makes use of Miller's simple cycle planar separator theorem [6].

Lemma 2. *Let R be a set of $k \leq n$ non-overlapping orthogonal rectangles on an integer $O(n) \times O(n)$ grid. There is a straight-line cycle, composed of at most $O(\sqrt{k})$ segments lying on the grid which does not properly intersect any rectangle in R and splits it into two subsets, each having at most $\frac{2}{3}k + O(\sqrt{k})$ rectangles.*

Proof. We construct a biconnected plane straight-line graph G on the grid including the vertices of the rectangles in R as follows. Let b be the minimum size rectangle on the grid that includes all rectangles in R. The corners of b become also vertices of G. For each vertex v of a rectangle in R we draw a vertical line up to the first intersection with another rectangle in R or b. At the point of the intersection, we create a new vertex of G. Symmetrically, we draw a vertical line down to the first intersection with a rectangle in R or b, creating a new vertex of G at the point of the intersection. Next, from each left vertex of a rectangle in R we draw a horizontal line to the left up to the first intersection with another rectangle in R, b, or a drawn vertical line, creating a new vertex of G at the intersection. Symmetrically we proceed with right vertices of rectangles in R. Two vertices in G are adjacent iff they are connected by a fragment of the perimeter of a rectangle in $R \cup \{b\}$ or one of the drawn lines free from other vertices. It is clear that G is biconnected, has $O(k)$ vertices and all its faces are rectangular.

Let us assign the weight $\frac{1}{4k}$ to each vertex of rectangle in R and the weight 0 to all remaining vertices in G. By Miller's simple cycle planar separator theorem [6], there is a simple cycle in G on $O(\sqrt{k})$ vertices which separates the vertices of rectangles in R into two parts, each containing at most $\frac{2}{3}4k$ vertices. Since the simple cycle is composed of edges in G, it does not cross any rectangle in R. Hence, the simple cycle separates the rectangles in R, possibly touching some of them, into two parts, each containing at most $\frac{2}{3}k + O(\sqrt{k})$ rectangles.

Suppose that R is a maximum cardinality set of k non-overlapping rectangles in S and that we can guess the separating cycle for R specified in Lemma 2. Then, we can determine the rectangles in S within the cycle and those outside, and recursively find maximum independent set for each of these two sets to output their union. In the recursive calls, the size of the grid can be easily trimmed down to $O(m) \times O(m)$ where m is the number of input rectangles.

Instead of guessing such a separating cycle, we can enumerate such simple cycles by considering all permutations of $O(\sqrt{k})$-element subsets of the grid points and testing whether they yield simple cycles. It takes time $2^{O(\sqrt{k}(\log n + \log k))}$ as the number of the subsets and their permutations is $\binom{O(n)^2}{O(\sqrt{k})}O(\sqrt{k})!$ and the tests can be done easily in polynomial time. Next, we run recursively our procedure for all of the enumerated cycles. Among the pairs of optimal solutions for the two subproblems induced by each of the cycles, we choose that yielding the largest cardinality union.

The recursion depth is $O(\log k)$. At its bottom, we have subproblems of picking a constant number of pairwise disjoint rectangles that can be solved in polynomial time. Note that $\frac{2}{3}k + O(\sqrt{k}) \leq \frac{3}{4}k$ for sufficiently large k. Therefore, a subproblem can be generally described by $O(\log k)$ fragments of bounding cycles of total length $O\left(\sum_{i=1} \sqrt{(\frac{3}{4})^i k}\right) = O(\sqrt{k})$. Hence, the total number of subproblems is $\binom{k}{O(\sqrt{k})} O(\sqrt{k})! \binom{O(\sqrt{k})}{O(\log k)}$, i.e., $2^{O(\sqrt{k} \log k)}$ and a single subproblem can be solved on the basis of the smaller ones in time $2^{O(\sqrt{k}(\log n + \log k))}$.

Theorem 1. *Let S be a set of n orthogonal rectangles in the plane. A maximum cardinality subset of S composed of non-overlapping rectangles can be found in time $2^{O(\sqrt{n} \log n)}$.*

Corollary 1. *Let G be a box graph on n vertices. Given a box representation of G, a maximum independent set and thus minimum vertex cover of G can be found in time $2^{O(\sqrt{n} \log n)}$.*

To find an f-approximation of a maximum independent set in the box graph G, we can run our method starting from a hypothetical set of disjoint rectangles of size at most $\lceil n/f \rceil$. Hence, we easily obtain the following generalization of Corollary 1.

Corollary 2. *Let G be a box graph on n vertices. Given a box representation of G, an f-approximation of a maximum independent set of G can be found in time $2^{O(\sqrt{n/f} \log n)}$.*

We can partly generalize Theorem 1 and Corollary 1 to higher dimensions by utilizing the following fact proved in [7].

Fact 1. *Given n interior-disjoint orthogonal d-cubes, of aspect ratio bounded by b, in R^d, there exists an orthogonal d-cube such that $\leq 2n/3$ d-cubes' interiors are entirely inside it, $\leq 2n/3$ d-cubes' interiors are entirely outside, and $\leq (4 + o(1))bn^{1-1/d}$ are partly inside and partly outside.*

By enlarging the boundary of the separating d-cube by the boundaries of the cubes intersected by it on its outside, we immediately obtain the following corollary from Fact 1.

Corollary 3. *Given n interior-disjoint orthogonal d-cubes, of aspect ratio bounded by b, in R^d, there exists an orthogonal d-solid with $(4 + o(1))2^d n^{1-1/d}$ vertices such that $\leq 2n/3 + (4+o(1))bn^{1-1/d}$ d-cubes' interiors are entirely inside it and $\leq 2n/3$ d-cubes' interiors are entirely outside it, and each of the d-cubes has its interior either inside or outside the solid.*

The boundary of the separating, orthogonal d-solid can be described by $O(2^d bn^{1-1/d})$ vertices and $O(d2^d bn^{1-1/d})$ edges. By a straightforward generalization of Lemma 1.1 to include d-dimensional boxes, we may assume without loss of generality that the d-cubes and the solid are placed on an integer grid of size $O(n)^d$. Thus, we can enumerate the boundaries of such separating solids in time $\left(O(2^d bn^{1-1/d})^{n^d}\right) \times \left(O(2^d bn^{1-1/d})^d\right) O(2^d bn^{1-1/d})^d$. The second term upper bounds the number of ways of choosing vertex neighbors for the $O(2^d n^{1-1/d})^d$ vertices. By straightforward calculations, the enumeration time and thus the number of the boundaries of such separating solids are bounded from above by $O(2^{O(d2^d bn^{1-1/d}\log n)})$. Now, we can partly generalize the procedure from the proof of Theorem 1 to d-dimensions, plugging in Corollary 3 instead of Lemma 2, to obtain the following theorem by straightforward calculations.

Theorem 2. *Let S be a set of n orthogonal d-cubes of aspect ratio bounded by b, in R^d. A maximum cardinality subset of S composed of non-overlapping d-cubes can be found in time $2^{O(d2^d bn^{1-1/d}\log n)}$.*

Corollary 4. *Let G be a box graph on n vertices in d-dimensions. Given a d-dimensional box representation of G, where each d-cube has an aspect ratio bounded by b, a maximum independent set and thus a minimum vertex cover of G can be found in time $2^{O(d2^d bn^{1-1/d}\log n)}$.*

Corollaries 1, 2, 4 can be immediately generalized to include the problem of *maximum induced subgraph with polynomial-time testable hereditary property Π.* The latter problem is to find a maximum cardinality subset of vertices of the input graph which induces a subgraph having the property Π. If the property Π holds for all induced subgraphs of a graph whenever it holds for the graph then we call it *hereditary*. Next, we call Π *polynomial-time testable* if one can determine in polynomial time if it holds for a graph.

If Π holds for arbitrarily large graphs, does not hold for all graphs, and is hereditary then the problem is NP-complete (cp. GT21 in [4]). Examples of such properties Π are "being an independent set", "being bipartite", "being a forest", "being a planar graph".

The following theorem is a generalization of not only Corollary 2 but also of Corollary 1 since an f-approximation for $f = 1$ means an exact solution. We leave its proof which is a straightforward generalization of that of Corollaries 1, 2 to the reader.

Theorem 3. *Let G be a box graph on n vertices. Given a box representation of G, an f-approximation of a maximum induced subgraph of G with polynomial-time testable hereditary property Π can be found in time $2^{O(\sqrt{n/f}\log n)}$.*

The proof the consecutive theorem is in turn a straightforward generalization of that of Corollary 4.

Theorem 4. *Let G be a box graph on n vertices in d-dimensions. Given a d-dimensional box representation of G, where each d-cube has an aspect ratio bounded by b, an exact solution to a maximum induced subgraph with polynomial-time testable hereditary property Π problem can be found in time $2^{O(d2^d bn^{1-1/d}\log n)}$.*

3 Subexponential Algorithms versus Approximation Algorithms

Algorithms of substantially subexponential time complexity can yield non-trivial approximation algorithms. We shall exemplify this statement with the problem of maximum independent set.

Theorem 5. *Suppose that the maximum independent set admits an algorithm running in time $O(2^{\phi(n)}n^l)$ where $\phi \leq n$ and l is a constant. If the function $\gamma(n) = y$ is well defined by the equation $\phi(y) = \log n$, it is monotone and $\gamma(2n) = O(\gamma(n))$ then the maximum independent set problem admits an $n/\gamma(n)$ approximation algorithm.*

Proof. Divide the set of vertices of the input graph into $\lfloor n/\gamma(n) \rfloor$ disjoint subsets, each of size in the interval $[\gamma(n), \, 2\gamma(n))$. For each of the intervals, run the subexponential algorithm in order to find a maximum independent set in the subgraph induced by this subset. It takes polynomial time by the properties of the function $\gamma(n)$. By pigeon hole principle, at least one of the algorithm runs will return an independent set of size $k\gamma(n)/n$ where k is the maximum size of an independent set in G.

Theorem 5 can be immediately generalized to include the problem of maximum induced subgraph with polynomial-time testable hereditary property Π.

Theorem 6. *Suppose that for some polynomial-time testable hereditary property Π the problem of maximum induced subgraph with the hereditary property Π admits an algorithm running in time $O(2^{\phi(n)}n^l)$ where $\phi \leq n$ and l is a constant. If the function $\gamma(n) = y$ is well defined by the equation $\phi(y) = \log n$, it is monotone and $\gamma(2n) = O(\gamma(n))$ then this problem admits an $n/\gamma(n)$ approximation algorithm.*

Corollary 5. *If for some polynomial-time testable hereditary property Π the problem of maximum induced subgraph with the property Π admits an algorithm running in time $O(2^{n^{1/l}})$, where l is a constant, then this problem admits an $n/\log^l n$ approximation algorithm.*

References

1. P.K. Agarwal, M. van Kreveld and S. Suri, Label Placement by Maximum Independent Set in Rectangles, *Computational Geometry: Theory and Applications*, 11(3-4), pp. 209–218, 1998.
2. M. Bern and D. Eppstein. Approximation Algorithms for Geometric Problems. In D. S. Hochbaum, editor, *Approximation Algorithms for $\mathcal{NP}$-Hard Problems*, chapter 8, pages 296–339. PWS Publishing Company, Boston, MA, 1996.
3. R. Boppana and M. M. Halldórsson. Approximating maximum independent sets by excluding subgraphs. *BIT* 32(2): 180–196, 1992.

4. M. R. Garey and D. S. Johnson. *Computers and Intractability: A Guide to the Theory of NP-completeness.* Freeman, New York, NY, 1979.

5. R. Impagliazzo, R. Paturi and F. Zane. Which Problems Have Strongly Exponential Complexity? Proceedings 1998 Annual IEEE Symposium on Foundations of Computer Science, pp 653–663, 1998.

6. G. Miller. *Finding small simple cycle separators for 2-connected planar graphs.* Proc. 16th Ann. ACM Symp. on Theory of Computing (1984) pp. 376–382.

7. W.D. Smith and N.C. Wormald. Geometric separator theorems. In proc. ACM STOC'98.

A Space Efficient Algorithm for Sequence Alignment with Inversions

Yong Gao[1], Junfeng Wu[1], Robert Niewiadomski[1],
Yang Wang[1], Zhi-Zhong Chen[2]*, and Guohui Lin[1]**

[1] Computing Science, University of Alberta, Edmonton, Alberta T6G 2E8, Canada
{ygao,jeffwu,niewiado,ywang,ghlin}@cs.ualberta.ca
[2] Mathematical Sciences, Tokyo Denki Univ., Hatoyama, Saitama 350-0394, Japan
chen@r.dendai.ac.jp

Abstract. A dynamic programming algorithm to find an optimal alignment for a pair of DNA sequences has been described by Schöniger and Waterman. The alignments use not only substitutions, insertions, and deletions of single nucleotides, but also inversions, which are the reversed complements, of substrings of the sequences. With the restriction that the inversions are pairwise non-intersecting, their proposed algorithm runs in $O(n^2m^2)$ time and consumes $O(n^2m^2)$ space, where n and m are the lengths of the input sequences respectively. We develop a space efficient algorithm to compute such an optimal alignment which consumes only $O(nm)$ space within the same amount of time. Our algorithm enables the computation for a pair of DNA sequences of length up to 10,000 to be carried out on an ordinary desktop computer. Simulation study is conducted to verify some biological facts about gene shuffling across species.

1 Introduction

Sequence alignment has been well studied in the 80's and 90's, where a bunch of algorithms have been designed for various purposes. The interested reader may refer to [2] and the references therein. Normally, the input sequence is considered as a linear series of symbols taken from a fixed alphabet, which consists of 4 nucleotides in the case of DNA/RNA sequences or 20 amino acids in the case of proteins. An alignment of two sequences is obtained by first inserting spaces into or at the ends of the sequences and then placing the two resulting sequences one above the other, so that every symbol or space in either sequence is opposite to a unique symbol or space in the other sequence. An alignment, which is associated with some objective function, can be naturally partitioned into columns and the column order is required to agree with the symbol orders in the input sequences.

* Supported in part by the Grant-in-Aid for Scientific Research of the Ministry of Education, Science, Sports and Culture of Japan, under Grant No. 14580390. Part of the work done while visiting at University of Alberta.
** Corresponding author. Supported in part by NSERC grants RGPIN249633 and A008599, and a REE Startup Grant from the University of Alberta.

In other words, these alignments use substitutions, insertions, and deletions of symbols only and are called normal alignments. For our purpose, we limit the input sequences to DNA sequences. In the literature, a number of algorithms [2] have been designed to compute an optimal normal alignment between two DNA sequences, under various pre-determined score schemes. A score scheme basically specifies how to calculate the score associated with an alignment. In this paper, the score schemes are symbol-independent. The simplest such score scheme would be the notion of *Longest Common Subsequence*, where every match column gets a score of 1 and every other column gets a score of 0 and the objective is to maximize the sum of the column scores. In the more general linear gap penalty score scheme specified by a triple (w_m, w_s, w_i), every match column gets a score w_m, every substitution column gets a score w_s, and every insertion or deletion (an *indel*) gets a score w_i. The linear gap penalty score schemes assume every single nucleotide mutation is independent of the others, which appears to be not biologically significant. The most commonly used score scheme nowadays is the so-called *affine gap penalty score scheme*. Given an alignment, a maximal segment of consecutive spaces in one sequence is called a *gap*, whose length is measured as the number of spaces inside. An affine gap penalty score scheme is defined by a quadruple (w_m, w_s, w_o, w_e), where w_m is the score for a match, w_s is the score for a replacement/substitution, and $w_o + w_e \times \ell$ is the penalty for a gap of length ℓ. Intuitively, affine gap penalty score schemes assume single nucleotide mutations might depend on its neighboring nucleotide mutations. As an example, under the affine gap penalty score scheme $(10, -11, -15, -5)$, the following alignment has a score of 4:

```
12345678901234567890012345
-CCAATCTAC----TACTGCTTGCA
 ||| ||| |    ||||| ||
GCCACTCT-CGCTGTACTG--TG--
```

In fact, one can easily verify that the above alignment is optimal for DNA sequences CCAATCTACTACTGCTTGCA and GCCACTCTCGCTGTACTGTG under the specific score scheme.

Alignments are used to reveal the information content of genes in DNA sequences and to make inferences about the relatedness of genes or more general inferences about the relatedness of long-range-segments of DNA sequences. With many genomic projects been carried out, emphasis has been put on the study on the genetic linkage among species and/or organisms. The sequences thus under consideration could come from different species and inferences can be made about the relatedness of species. In this aspect, the normal sequence alignment has the limitation on using evolutionary transformations to substitutions and indels. It has been widely accepted that duplications and inversions are common events in molecular evolution [3,7]. Some pioneer work on studying sequence alignment/comparison with inversions is done by Wagner [6] and Schöniger and Waterman [5]. In particular, in [5], a dynamic programming algorithm is developed to compute an optimal (local) alignment with inversions. The algorithm runs in $O(n^2 m^2)$ time and consumes $O(n^2 m^2)$ spaces, where n and m are the lengths of two input sequences respectively.

The high order of running time seems computationally impractical, nonetheless, it is the huge space requirement that actually makes the computation infeasible. For instance, suppose the algorithm needs $\frac{1}{4} n^2 m^2$ byte memory, then an optimal alignment between two sequences of length 250 won't be carried out in a normal desktop of 1Gb memory. Schöniger and Waterman [5] designed an algorithm to compute a *suboptimal* alignment with inversions restricted to a constant number of *highest scoring inversions*. This latter algorithm runs in $O(nm + \sum_{i=1}^{k} \ell_i)$ time and requires an order of $O(nm + \sum_{i=1}^{k} \ell_i)$ space, where k is the number of restricted highest scoring inversions and ℓ_i's are their lengths respectively. In this paper, we develop a space efficient algorithm to compute an *optimal* alignment with non-restricted inversions, in terms of both the number and their scores, within $O(m^2 n^2)$ time. The algorithm is non-trivial in the sense that deep computation relationships are carefully characterized.

Problem Description. An inversion of a DNA substring is defined to be the reverse complement of the substring. In this paper, we won't put a limit on the number of inversions in the alignments, but we do want to put a lower bound on the lengths of inversions, for the reason that we believe inversions correspond to some conserved coding regions and a conserved region must have a least number of nucleotides inside. We let L denote this lower bound. In the extreme case where $L = 0$, it becomes the problem considered in [5]. The inversions are restricted to be on one of the two input sequences (by symmetry) and they are not allowed to intersect one another. Moreover, the substrings from the other sequence against which those inversions are aligned are not allowed to intersect one another either.

The score schemes used in this paper are all affine gap penalty ones. For the sake of comparison, the parameters in the schemes may be different in different simulations. We associate each inversion with a (constant) penalty of C [5], to compensate for the fact that the inversion is an evolutionary event.

Now it is ready to formulate the computational problem we will be considering in the paper. The input is a pair of DNA sequences S_1 and S_2 over the nucleotide alphabet $\Sigma = \{A, C, G, T\}$, together with an affine gap penalty score scheme (w_m, w_s, w_o, w_e), inversion lower bound L, and inversion penalty C. The goal is to compute an optimal alignment between S_1 and S_2 with inversions. For simplicity, we associate *standard similarity* with a normal optimal alignment, which takes substitutions and indels only; and we use *inversion similarity* to associate with an optimal alignment taking inversions in addition (called an *optimal inversion alignment*). Both similarities (also called sequence identities hereafter) are defined to be the ratio of the number of match columns in the alignment over the length of the shorter sequence.

Organization. The paper is organized as follows: In the next section we will detail our algorithm to compute an optimal alignment between a pair of DNA sequences S_1 and S_2. The algorithm runs in $O(m^2 n^2)$ time and requires only

$O(mn)$ space, where m and n are the lengths of S_1 and S_2, respectively. In Section 3, we show the simulation results of our algorithm applied to the instances tested in [5]. We conclude the paper with some remarks in Section 4.

2   The Algorithms

Let $S_1[1..m]$ and $S_2[1..n]$ denote the two input sequences of length m and n, respectively. For simplicity of presentation, we simplify the problem a little by requiring that gaps in non-inversion regions and inversion regions are counted separately and independently. We let $\overline{S[i]}$ denote the complement nucleotide of $S[i]$ (the Watson-Crick rule); let $\overline{S_1[i_1..i_2]}$ denote the inversion of $S_1[i_1..i_2]$, that is, $\overline{S_1[i_2]}\ \overline{S_1[i_2-1]}\ldots\overline{S_1[i_1]}$.

2.1   A Less Space Efficient Algorithm

In this subsection, we introduce a less space efficient dynamic programming algorithm which is conceptually simple to understand. This algorithm is for linear gap penalty score scheme (that is, we won't use w_o and w_e here, but use w_i instead which is the score for an indel). The more complicated yet more space efficient algorithm in Section 2.2 for the affine gap penalty score scheme will be a refinement of this simple algorithm.

Inversion Table Computation. Suppose that $S_1[i_1, i_2]$ and $S_1[i_3, i_4]$ are two inversion substrings in sequence S_1, then by the length constraint and non-intersecting requirement we have: $i_2 - i_1 \geq L - 1$, $i_4 - i_3 \geq L - 1$, and either $i_2 < i_3$ or $i_4 < i_1$. For every quartet (i_1, i_2, j_1, j_2), where $0 \leq i_1 \leq i_2 \leq m$ and $0 \leq j_1 \leq j_2 \leq n$, let $I[i_1, i_2; j_1, j_2]$ denote the standard sequence similarity between two substrings $\overline{S_1[i_1..i_2]}$ and $S_2[j_1..j_2]$ (that is, without inversions). As an easy example, $S_1 = \text{ACGT}$ and $S_2 = \text{ACGA}$, then $S_1[2..4] = \text{CGT}$ and thus $\overline{S_1[2..4]} = \text{ACG}$, and $S_2[1..3] = \text{ACG}$. Therefore, $I[2, 4; 1, 3] = 3w_m$.

Let $w(a, b)$ denote the score of matching nucleotide a against nucleotide b, where a and b are both in the alphabet (which includes nucleotides A, C, G, T). Therefore,

$$w(a, b) = \begin{cases} w_m, \text{ if } a = b, \\ w_s, \text{ if } a \neq b \end{cases}$$

Let $S_1[0] = S_2[0] = -$. To assist the *inversion table I* computation, I is enlarged to include the following boundary entries:

- $I[i + 1, i; j_1, j_2] = (j_2 - j_1 + 1)w_i$, where $0 \leq i \leq m$ and $0 \leq j_1 \leq j_2 \leq n$;
- $I[i_1, i_2; j + 1, j] = (i_2 - i_1 + 1)w_i$, where $0 \leq i_1 \leq i_2 \leq m$ and $0 \leq j \leq n$.

The recurrence relation for computing a general entry $I[i_1, i_2; j_1, j_2]$ is as follows:

$$I[i_1, i_2; j_1, j_2] = \max \begin{cases} I[i_1 + 1, i_2; j_1, j_2 - 1] + w(\overline{S_1[i_1]}, S_2[j_2]), \\ I[i_1 + 1, i_2; j_1, j_2] + w_i, \\ I[i_1, i_2; j_1, j_2 - 1] + w_i, \end{cases}$$

where $1 \leq i_1 \leq i_2 \leq m$ and $1 \leq j_1 \leq j_2 \leq n$.

Notice that table I contains in total $\frac{1}{4}(m+1)^2(n+1)^2$ entries and filling each of them takes a constant amount of time.

Dynamic Programming Table Computation. Let $DP[i,j]$ denote the score of an optimal inversion alignment between prefixes $S_1[1..i]$ and $S_2[1..j]$, where $0 \leq i \leq m$ and $0 \leq j \leq n$. The boundary entries of the *dynamic programming table* DP are:

- $DP[0,j] = j \times w_i$, where $0 \leq j \leq n$;
- $DP[i,0] = i \times w_i$, where $0 \leq i \leq m$.

Recall that every inversion in S_1, as well as its aligned segment in S_2, must have length at least L. The recurrence relation for computing a general entry $DP[i,j]$ is as follows:

- When $i \geq L$ and $j \geq L$,

$$DP[i,j] = \max \begin{cases} DP[i-1,j-1] + w(S_1[i], S_2[j]), \\ DP[i-1,j] + w_i, \\ DP[i,j-1] + w_i, \\ \max_{\substack{1 \leq i' \leq i-L+1 \\ 1 \leq j' \leq j-L+1}} \left\{ DP[i'-1,j'-1] + I[i',i;j',j] \right\} - C. \end{cases}$$

- In the other case,

$$DP[i,j] = \max \begin{cases} DP[i-1,j-1] + w(S_1[i], S_2[j]), \\ DP[i-1,j] + w_i, \\ DP[i,j-1] + w_i. \end{cases}$$

Since the dynamic programming table contains $(m+1)(n+1)$ entries and entry $DP[i,j]$ takes up to $O(ij)$ time to fill, the running time of the overall algorithm is $O(m^2n^2)$.

The correctness of the algorithm follows directly from the recurrences and therefore we have the following conclusion.

Theorem 1. *The inversion similarity between $S_1[1..m]$ and $S_2[1..n]$ can be computed in $O(m^2n^2)$ time using $O(m^2n^2)$ space.*

2.2 A Space Efficient Algorithm for Affine Gap Penalty Score Schemes

In the algorithm in Section 2.1 we divide the computation into two phases. In the first phase we prepare all the possible inversions together with their all possible aligned substrings from the other sequence and the associated scores. Direct extension can lead to an algorithm for affine gap penalty score schemes, running in the same amount of time and requiring the same amount of space. In the following, we present a single-phase computation. We fill the dynamic programming table row-wise. Furthermore, when filling the ith row, we compute all possible inversions ending at position i. We will prove that there are $O(mn)$ such possible inversions and each can be calculated in constant time based on the intermediate results of computation for the $(i-1)^{\text{th}}$ row. This single-phase computation still takes $O(m^2n^2)$ time, nonetheless requires $O(mn)$ space only.

Let $DP_M[i,j]$, $DP_I[i,j]$, and $DP_D[i,j]$ denote the scores of an optimal inversion alignment between prefixes $S_1[1..i]$ and $S_2[1..j]$, where $0 \le i \le m$ and $0 \le j \le n$, such that the last operation is either a match or a mismatch, an insertion, and a deletion, respectively. The boundary entries of these dynamic programming tables are:

- $DP_M[0,0] = DP_I[0,0] = DP_D[0,0] = 0$;
- $DP_I[0,j] = w_o + j \times w_e$, where $1 \le j \le n$;
- $DP_D[i,0] = w_o + i \times w_e$, where $1 \le i \le m$.

Let $I_M^{i,j}[i',j']$, $I_I^{i,j}[i',j']$, and $I_D^{i,j}[i',j']$ denote the sequence similarities between inversion $\overline{S_1[i'..i]}$ and $S_2[j'..j]$, where $1 \le i' \le i \le m$ and $1 \le j' \le j \le n$, such that the last operation is either a match or a mismatch, an insertion, and a deletion, respectively. Again, to assist the computation for these 3 inversion tables, they are enlarged to include the following boundary entries:

- $I_M^{i,j}[i+1,j+1] = I_I^{i,j}[i+1,j+1] = I_D^{i,j}[i+1,j+1] = 0$;
- $I_I^{i,j}[i+1,j'] = w_o + (j-j'+1)w_e$, where $1 \le j' \le j$;
- $I_D^{i,j}[i',j+1] = w_o + (i-i'+1)w_e$, where $1 \le i' \le i$.

The recurrence relations for dynamic programming tables computation are:

$$DP_M[i,j] = \max \begin{cases} w(S_1[i],S_2[j]) + \max \begin{cases} DP_M[i-1,j-1], \\ DP_I[i-1,j-1], \\ DP_D[i-1,j-1]; \end{cases} \\ \max_{\substack{1 \le i' \le i-L+1 \\ 1 \le j' \le j-L+1}} \left\{ I_M^{i,j}[i',j'] + \max \begin{cases} DP_M[i'-1,j'-1], \\ DP_I[i'-1,j'-1], \\ DP_D[i'-1,j'-1] \end{cases} \right\} - C \end{cases}$$

$$DP_I[i,j] = \max \begin{cases} DP_M[i,j-1] + w_o + w_e, \\ DP_I[i,j-1] + w_e, \\ DP_D[i,j-1] + w_o + w_e, \\ \max_{\substack{1 \le i' \le i-L+1 \\ 1 \le j' \le j-L+1}} \left\{ I_I^{i,j}[i',j'] + \max \begin{cases} DP_M[i'-1,j'-1], \\ DP_I[i'-1,j'-1], \\ DP_D[i'-1,j'-1] \end{cases} \right\} - C \end{cases}$$

$$DP_D[i,j] = \max \begin{cases} DP_M[i-1,j] + w_o + w_e, \\ DP_I[i-1,j] + w_o + w_e, \\ DP_D[i-1,j] + w_e, \\ \max_{\substack{1 \le i' \le i-L+1 \\ 1 \le j' \le j-L+1}} \left\{ I_D^{i,j}[i',j'] + \max \begin{cases} DP_M[i'-1,j'-1], \\ DP_I[i'-1,j'-1], \\ DP_D[i'-1,j'-1] \end{cases} \right\} - C \end{cases}$$

Before computing these i^{th} row entries, $I_M^{i,j}$, $I_I^{i,j}$, and $I_D^{i,j}$ tables are pre-computed using the following recurrence relations.

$$I_M^{i,j}[i',j'] = w(\overline{S_1[i']}, S_2[j]) + \max \begin{cases} I_M^{i,j-1}[i'+1,j'], \\ I_I^{i,j-1}[i'+1,j'], \\ I_D^{i,j-1}[i'+1,j'] \end{cases}$$

$$I_I^{i,j}[i',j'] = \max \begin{cases} I_M^{i,j-1}[i',j'] + w_o + w_e, \\ I_I^{i,j-1}[i',j'] + w_e, \\ I_D^{i,j-1}[i',j'] + w_o + w_e \end{cases}$$

$$I_D^{i,j}[i',j'] = \max \begin{cases} I_M^{i,j}[i'+1,j'] + w_o + w_e, \\ I_I^{i,j}[i'+1,j'] + w_o + w_e, \\ I_D^{i,j}[i'+1,j'] + w_e \end{cases}$$

Notice that the computation of $I_M^{i,j}$, $I_I^{i,j}$, and $I_D^{i,j}$ tables needs the values in $I_M^{i,j-1}$, $I_I^{i,j-1}$, and $I_D^{i,j-1}$ tables only. It follows that we may just keep 3 inversion tables $I_M^{i,j-1}$, $I_I^{i,j-1}$, and $I_D^{i,j-1}$ after computing entries $DP_M[i,j-1]$, $DP_I[i,j-1]$, and $DP_D[i,j-1]$. These 3 tables are then used in computing entries $DP_M[i,j]$, $DP_I[i,j]$, and $DP_D[i,j]$, where we create 3 new inversion tables $I_M^{i,j}$, $I_I^{i,j}$, and $I_D^{i,j}$. After that, those 3 inversion tables $I_M^{i,j-1}$, $I_I^{i,j-1}$, and $I_D^{i,j-1}$ will no longer be used and thus can be deallocated.

In other words, we need only in total 9 2-dimensional tables during the overall computation, which consume $O(mn)$ space. The overall running time $O(m^2n^2)$ is obviously seen from the recurrences, where trying all possible combinations of i' and j' for pair (i,j) dominates the computation.

Theorem 2. *The inversion similarity between $S_1[1..m]$ and $S_2[1..n]$, using affine gap penalty score schemes, can be computed in $O(m^2n^2)$ time using $O(mn)$ space.*

3   Simulation Results

In the example instance in the introduction, $S_1 = $ CCAATCTACTACTGCTTGCA and $S_2 = $ GCCACTCTCGCTGTACTGTG. Under the affine gap penalty score scheme specified by $(10, -11, -15, -5)$, we showed there an optimal normal alignment. Using the lower bound $L = 5$ and inversion penalty $C = 2$, the following shows an optimal inversion alignment, associated with a score of 43:

```
1234567890 123456 789012345
-CCAATCTAC gcagta TTGCA
 ||| ||| | || |||  ||
GCCACTCT-C GCTGTA CTGTG
```

In which the lower case substring "gcagta" is an inversion from $S_1[10..15] = $ TACTGC.

In [5], it has been calculated under the same score scheme that $S_1[10..15]$ *vs.* $S_2[10..15]$ is the highest scoring inversion and $S_1[7..9]$ *vs.* $S_2[13..15]$ is the second highest. And using these 2 highest scoring inversions, the output inversion alignment in their paper is exactly the same as ours as shown. Therefore, our work confirms that the inversion alignment computed using 2 highest scoring inversions for the above instance in [5] is in fact an optimal one.

A biological instance used for simulation in [5] consists of a DNA sequence from *D. yakuba* mitochondria genome using nucleotides 9,987–11,651 and a DNA sequence from mouse mitochondria genome using nucleotides 13,546–15,282. Under the affine gap penalty score scheme specified by $(10, -9, -15, -5)$ and inversion penalty $C = 20$, by pre-computing a list of 400 highest scoring inversions, the alignment output in [5] found an inversion substring in *D. yakuba* consisting of nucleotides 7–480 which aligns to nucleotides 58–542 from mouse. The putative organization of genes in the two DNA sequences is described in Table 1:

Table 1. Putative organization of genes in the two DNA sequences.

	D. yakuba	mouse
URF6	1–525	519–1 (inverted)
tRNA Glu		588–520 (inverted)
cytochrome b	529–1,665	594–1737

So the identified inversion to some extent detects the biologically correct inversion. In our simulation, we use the same score scheme and again add the lower bound on the inversion lengths $L = 5$.

Unfortunately, the algorithm didn't detect any *good* inversions. What it found are three short inversions $S_1[344..349]$ which aligns to $S_2[326..331]$, $S_1[387..391]$ which aligns to $S_2[357..361]$, and $S_1[456..463]$ which aligns to $S_2[417..424]$. With these three inversions, the detected inversion identity is 0.6853, contrast to the standard identity 0.6847. We did another experiment by cutting out the two URF6 genes from both sequences and calculating their inversion identity, namely, the first 525 nucleotides from *D. yakuba* and the first 519 nucleotides from mouse. It turned out that the standard identity between these two substrings is 0.5780 and the inversion identity remains the same as the standard one without any inversion detected.

Since the inversion algorithm didn't detect any meaningful inversions, we modified the algorithm to detect reversals, which only reverse a substring but not take the complement. We define the reversal identity similarly to be the ratio in an optimal reversal alignment. For the two URF6 genes, by setting the length lower bound $L = 5$, we found a lot of reversals which are listed in the Table 2. The reversal identity is 0.6301 as detected.

By setting $L = 300$, we found a reversal substring $S_1[128..513]$ which is aligned to $S_2[121..507]$. The alignment score is improved from 152 to 167 with a little bit identity sacrifice down to 0.5742. The standard identity between these two segments is 0.5622 with alignment score 55 (an optimal standard alignment

Table 2. Fragmental reversal segments and their aligned partner.

D. yakuba	mouse	D. yakuba	mouse
15–23	16–25	266–276	277–287
37–41	37–44	283–302	294–315
44–73	47–70	316–322	329–335
78–82	75–79	324–342	337–350
100–114	105–119	350–378	358–383
130–165	134–172	383–387	388–392
180–192	187–199	394–436	399–438
209–227	216–238	437–447	439–449
244–249	255–260	459–512	461–504
254–261	265–272	518–525	510–519

is shown in the left side of Figure 1); The reversal identity between them is
0.5596 with alignment score 110 (an optimal reversal alignment is shown in the
right side of Figure 1).

```
Alignment Score:55
Matches:    217
Identity:    0.562176165803109
          1234567890123456789012345678901234567890
          ----------------------------------------
   0  TAATAACTAAAAGTTTTTGATACTCATACATTTTATTTTTAAT-TTTTTT
   0  ||||  ||  | | |  |||||||||  | |    ||| |
   0  ----AACTCCAACATCATCAACCTCATACA--TCA--ACCAATCTCCCAA

   1  AGGAGGAATACTTGTTTTATTTA--TTTATGTTACATCATT-AGC-TTCT
   1  | |  || || ||  || |  |  || || ||  || ||| |
   1  ACCATCAAGA-TTAATTACTCCAACTTCATCATA-ATAATTAAGCACACA

   2  AATGAAATATTTAA--TTTATCAATTA-----AATTAACTTTATTTTCCA
   2  ||| ||| |   || | | |||| ||   ||| ||| |   ||
   2  AATTAAAAA---AACCTCTAT-AATCACCCCCAAT--ACTAAAAAACCCA

   3  TATTTATTTTATTTTTTATAT---TTATTTTATCAATAATTCTTGA-TAA
   3  ||| |  ||  ||    |  ||   | ||| |    || ||  |     ||
   3  AAATTA----ATCAGTTAGATCCCCAAGTCTCTGGAT-ATTCCTCAGT--

   4  AACTTCTATTACT-TTATTTTTAATAAATAACGAA-ATACAATCT--ATT
   4  | | ||  |  | ||   ||| | || ||| || || |   || ||
   4  AGC-TATAGCAGTCGTATATCCAA-ACACAACCAACATCCCCCCTAAATA

   5  ATTGAAATAAATTCTTATTTTAC--AGAAA--ATTC-TTTATCTTTAAAT
   5  | | ||| ||| | |||| || | ||| ||| |  | |  |||
   5  AATTAAAAAAA--C-TATTAAACCTAAAAACGATCCACCAAACCCTAAAA

   6  AAATTATATAATTTTCCAACAAATTTTGTAACAATTTTA-TTAA---TAA
   6  |||| ||  |  |||||||   |||||||  | |    ||        ||
   6  CCATTAAACAA----CCAACAAACCCACTAACAATTAAACCTAAACCTCC

   7  ATTATTTATTAATTACTTTAATTGTTGTAGTAAAAATTACTAAAC-TATT
   7  ||| |   || ||||||| ||  |   ||  |  | || || || |
   7  ATAAATAGGTGAAGGCTTTAA-TGCT-AACCCAAGACAACCAACCAAAAA

   8  TAAAGGTCCT-ATCCGA----
   8  ||| |  | || | | |
   8  TAATGAACTTAAAACAAAAAT
```

```
Alignment Score:110
Matches:    216
Identity:    0.559585492227979
          1234567890123456789012345678901234567890
          ----------------------------------------
   0  AGCCTATCCTGGAAATTTATCAA----AT-CATTAAAAATGATGTTGTTA
   0  | |   ||| || |||||| || |||| ||  | | |
   0  AAC---TCC---AACATCATCAACCTCATACATCAACCA--ATCTCCCAA

   1  ATTTCATTA--ATTATTTATTAAATAATT-ATTTTAACAATGTTTTAAAC
   1  |  ||| |  |||| ||| | | || ||  || ||     |||| |
   1  A--CCATCAAGATTAATTACTCCA-ACTTCATCATAATAA----TTAAGC

   2  A-ACCTTTTAATATATTAAATAAATTTCTATTTCTTAAAAGACATTTTAT
   2  | |||    ||  |||||| ||| ||||  ||       ||
   2  ACAC-----AA---ATTAAAAAAACCTCTAT------AATCACCCCCAAT

   3  TCTTAAATAAAGTTATTATCTAACATAAAGCAATAAATAATTTTTATTTC
   3  | |||        ||| ||  || ||   || |         ||    ||
   3  ACTAAAA-------------AACCCAA---AATTAATCAGTTAGATCCC

   4  --ATTATCTTCAAAATAGTTCTTAATAACTATTTTATTTATATTTTTAT
   4  |  ||||      ||| || || || ||||   ||   | ||   |
   4  CAAGTCTCT---GGATATTCCTCAGTAGCTATAGCAGTCGTATATCCAAA

   5  TTTA---TTTATACCTTTTATTTCAATTAAAATTAACTATTTAATTTATAA
   5  |        || ||  ||  | |||||||| ||||||| || ||  |
   5  CACAACCAACATCCCCCCTAAATAAATTAAAAAAACTATTAAACCTAAAA

   6  A-G-T------AATCTT----CGATTA------CTACA----TTGTATTT
   6  | | |      ||||| | ||| |||  |  ||  |     ||
   6  ACGATCCACCAAACCCTAAAACCATTAAACAACCAACCAACAAACCCACTAACA

   7  ATTTATTTT----GTTCATAAGGAG--GATTTTTTTAATTTTTA--TTTT
   7  ||| |   |     ||||| ||||||||  | |  ||  | || ||
   7  ATTAAACCTAAACCTCCATAAATAGGTGAAGGCTTTAATGCTAACCCAAG

   8  ACATAC---TCATAGTTTTTGA----AAATCAATAAT
   8  ||| ||   || | | |||  ||   ||| ||| |||
   8  ACA-ACCAACCAAAAATAATGAACTTAAAACAAAAAT
```

Fig. 1. An optimal standard alignment (left) and an optimal reversal alignment (right)
between $S_1[128..513]$ and $S_2[121..507]$.

4 Conclusions

The space efficient algorithm developed in this paper enables the computation of an optimal inversion alignment between two DNA sequences of length up to 10,000bp on a normal desktop with 1Gb memory. Previous algorithms either fail on long sequences or only produce a suboptimal inversion alignment restricted to a number of pre-computed highest scoring inversions. The simulation conducted shows a disagreement with previous simulation. Further investigation is necessary, typically on the selection of suitable score schemes.

The recurrences for computing DP_M, DP_I, and DP_D tables are written for the case where gaps inside inversion segments and gaps inside non-inversion segments are treated separately and independently. If two gaps from different categories are adjacent to each, then they might be counted as one gap. The recurrences can be slightly modified, where one copy of inversion penalty C should be merged to $DP_M[i'-1,j'-1]$, $DP_I[i'-1,j'-1]$, and $DP_D[i'-1,j'-1]$ during the computation, to take care of this case.

Our algorithm can be easily modified to compute an optimal reversal alignment between sequences. Some simulation has been run which shows something different from inversions. We have also done some preliminary simulation study on applying the inversion and reversal algorithms to detect similar secondary structure units for RNA sequences, which correspond to reversed substrings, in the RNase P Database (http://www.mbio.ncsu.edu/RNaseP/home.html) [1]. The result will be reported elsewhere.

Upon acceptance, we were informed of a recent work [4], where Muthukrishnan and Sahinalp consider the problem of minimizing the number of character replacements (no insertions and deletions) and reversals and propose an $O(n \log^2 n)$ time deterministic algorithm, where n is the length of either sequence.

Acknowledgments. We thank Dr. Patricia Evans (Computer Science, University of New Brunswick) and Bin Li (Biological Science, University of Alberta) for helpful discussions.

References

1. J. W. Brown. The Ribonuclease P Database. *Nucleic Acids Research*, 27:314, 1999.
2. D. Gusfield. *Algorithms on Strings, Trees, and Sequences.* Cambridge, 1997.
3. C. J. Howe, R. F. Barker, C. M. Bowman, and T. A. Dyer. Common features of three inversions in wheat chloroplast dna. *Current Genetics*, 13:343–349, 1988.
4. S. Muthukrishnan and S. C. Sahinalp. An improved algorithm for sequence comparison with block reversals. In *Proceedings of The 5th Latin American Theoretical Informatics Symposium (LATIN'02)*, LNCS 2286, pages 319–325, 2002.
5. M. Schöniger and M. S. Waterman. A local algorithm for DNA sequence alignment with inversions. *Bulletin of Mathematical Biology*, 54:521–536, 1992.

6. R. A. Wagner. On the complexity of the extended string-to-string correction problem. In D. Sankoff and J. B. Kruskal, editors, *Time Warps, Strings Edits, and Macromolecules: the Theory and Practice of Sequence Comparison*, pages 215–235. Addison-Wesley, 1983.
7. D. X. Zhou, O. Massenet, F. Quigley, M. J. Marion, F. Monéger, P. Huber, and R. Mache. Characterization of a large inversion in the spinach chloroplast genome relative to *marchantia*: a possible transposon-mediated origin. *Current Genetics*, 13:433–439, 1988.

On the Similarity of Sets of Permutations and Its Applications to Genome Comparison

Anne Bergeron[1] and Jens Stoye[2]

[1] LaCIM, Université du Québec à Montréal, Canada,
anne@lacim.uqam.ca
[2] Technische Fakultät, Universität Bielefeld, Germany,
stoye@techfak.uni-bielefeld.de

Abstract. The comparison of genomes with the same gene content relies on our ability to compare permutations, either by measuring how much they differ, or by measuring how much they are alike. With the notable exception of the breakpoint distance, which is based on the concept of conserved adjacencies, measures of distance do not generalize easily to sets of more than two permutations. In this paper, we present a basic unifying notion, *conserved intervals*, as a powerful generalization of adjacencies, and as a key feature of genome rearrangement theories. We also show that sets of conserved intervals have elegant nesting and chaining properties that allow the development of compact graphic representations, and linear time algorithms to manipulate them.

1 Introduction

Gene order analysis in a set of organisms is a powerful technique for phylogenetic inference. Current methods are based on notions of *distances* between genomes, which are usually defined as the minimum number of such and such operations needed to transform one genome into the other one. Distance matrices can either be used directly as data for phylogenetic reconstruction, or in more qualitative attempts to reconstruct ancestral genomes [9]. All these methods, with the notable exception of the *breakpoint distance* [6], are closely tied to initial choices of allowable rearrangement operations. They are also *pure* distances, in the sense that similarities between genomes are purposefully ignored.

The breakpoint distance is based on the notion of conserved adjacencies. Compared to other distances, it is easy to compute, but it often fails to capture more global relations between genomes [17]. Nevertheless, conserved adjacencies have two highly desirable properties:

1. They can be defined on a set of more than two genomes, allowing for the identification of similar features in a family of organisms.
2. They are invariant under optimal rearrangement scenarios, in the sense that it is not necessary to break adjacencies to explain how a genome evolved from another one [10,15,21].

T. Warnow and B. Zhu (Eds.): COCOON 2003, LNCS 2697, pp. 68–79, 2003.
© Springer-Verlag Berlin Heidelberg 2003

Table 1. Condensed mitochondrial genomes of six Arthropoda

Fruit Fly	1	2	3	4	5	6	7	8	9	10	11	12	13	14	15	16	17
Mosquito	1	2	3	4	5	6	8	7	9	−10	11	12	13	14	15	16	17
Silkworm	1	2	3	4	5	6	7	8	9	10	11	12	14	13	15	16	17
Locust	1	2	3	5	4	6	7	8	9	10	11	12	13	14	15	16	17
Tick	1	3	4	5	6	7	8	9	10	11	−2	12	13	14	15	16	17
Centipede	1	3	4	5	6	7	8	9	10	11	−2	12	16	13	14	15	17

A first generalization of adjacencies is the notion of *common intervals* that identify subsets of genes that appear consecutively in two or more genomes [13,22]. Common intervals identify more global relations between genomes, but often lose the invariant property of adjacencies with respect to optimal rearrangement scenarios. For example, all optimal sortings by reversals of the permutation (1 3 2 5 − 4 6) break, in some of the intermediate permutations, the common interval (2 3).

Are adjacencies the only structures that are invariant under biologically meaningful rearrangement operations? No. There exists a class of common intervals, called *conserved intervals*, that may be the best of two worlds. We will show that these structures capture both local and global properties of genomes; are invariant under most rearrangement scenarios; and their number and nature can be computed in linear time.

2   Permutations, Gene Order, and Rearrangements

In the following we will take for granted the simplifying hypothesis that the genes of an organism are ordered and oriented along linear or circular DNA molecules. For example the 37 mitochondrial genes of the Fruit Fly are listed in [7], with minus signs to reflect orientation, as:

```
cox1, L2, cox2, K, D, atp8, atp6, cox3, G, nad3, A, R, N, S1, E, -F,
-nad5, -H, -nad4, -nad4L, T, -P, nad6, cob, S2, -nad1, -L1, -rrnL, -V,
-rrnS, UNK, I, -Q, M, nad2, W, -C, -Y
```

The first gene is arbitrary, since mitochondrial genomes are circular molecules. When organisms with the same gene content are compared, one of them is chosen as a base organism, and all identical strips of genes are converted to integers. By extension, these are also called "genes". Table 1 presents the result of this transformation applied to the mitochondrial genomes of six *Arthropoda*, with Fruit Fly as base organism. The original 37 genes have been divided in 17 blocks: some represent isolated genes, and others represent longer strips. For example, 10 stands for S1, and 11 for E, -F, -nad5, -H, -nad4, -nad4L, T, -P, nad6, cob, S2, -nad1.

Various techniques are then used to compare the resulting permutations. The *distance* approaches focus on the differences between two particular genomes. For example, Fruit Fly differs from Mosquito by the *reversal* of gene 10, and the *transposition* of genes 7 and 8. One can count the minimal number of reversals and/or transpositions necessary to transform each genome into any other, yielding a distance matrix for the set of species. Explicit rearrangement scenarios, that is, sequences of operations that transform optimally one genome into another, are also used to reconstruct ancestral genomes.

Another approach, the *breakpoint distance*, counts the lost adjacencies between genomes. It does not rely on particular rearrangement operations or an evolutionary model, and it has an associated measure of similarity: the number of *conserved* adjacencies. For example, given the circularity of the genomes, Fruit Fly and Mosquito have 12 conserved adjacencies, and their breakpoint distance is 5.

Such a similarity measure extends easily to sets of species. For example, the first four species of Table 1 share 6 adjacencies: $[1, 2]$, $[2, 3]$, $[11, 12]$, $[15, 16]$, $[16, 17]$, and $[17, 1]$. When comparing all six species, the only left adjacency is $[17, 1]$: this lack of conserved adjacencies is a direct consequence of how the data was transformed. Does this mean that losing common adjacencies amounts to losing all common structures?

A quick glance at Table 1 reveals that the six permutations are very "similar". For example, the genes in the interval $[1, 12]$ are all the same, with small variations in their ordering. This is also true for the genes in the intervals $[3, 6]$, $[6, 9]$, $[9, 11]$, and $[12, 17]$. It turns out that such intervals, together with conserved adjacencies, play a fundamental role in rearrangement and distance theories, ancestral genome reconstructions, and phylogeny.

The following *family portrait* gives a representation of the conserved intervals of the permutations of Table 1:

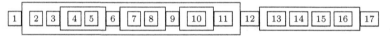

This representation boxes the elements in rectangles, which can be glued together to form larger objects. It takes its roots in *PQ*-trees [8] that are used to represent sets of permutations. All permutations of Table 1 fit the representation with the following conventions: (1) free objects within a rectangle can be reordered, or can change sign, (2) connections between rectangles are fixed. This representation also captures the features that should be invariant in biologically plausible rearrangement scenarios within the family.

In order to illustrate this last point, consider the two following rearrangement scenarios that transform Silkworm into Locust using a minimal number of *reversals* (operations that reverse the elements of a consecutive block while changing their signs).

1 2 3 4 5 6 7 8 9 10 11 12 14 13 15 16 17	1 2 3 4 5 6 7 8 9 10 11 12 14 13 15 16 17
1 2 3 -4 5 6 7 8 9 10 11 12 14 13 15 16 17	1 2 3 4 -14 -12 -11 -10 -9 -8 -7 -6 -5 13 15 16 17
1 2 3 -4 -5 6 7 8 9 10 11 12 14 13 15 16 17	1 2 3 4 -14 5 6 7 8 9 10 11 12 13 15 16 17
1 2 3 5 4 6 7 8 9 10 11 12 14 13 15 16 17	1 2 3 4 -13 -12 -11 -10 -9 -8 -7 -6 -5 14 15 16 17
1 2 3 5 4 6 7 8 9 10 11 12 -14 13 15 16 17	1 2 3 5 6 7 8 9 10 11 12 13 -4 14 15 16 17
1 2 3 5 4 6 7 8 9 10 11 12 -14 -13 15 16 17	1 2 3 5 4 -13 -12 -11 -10 -9 -8 -7 -6 14 15 16 17
1 2 3 5 4 6 7 8 9 10 11 12 13 14 15 16 17	1 2 3 5 4 6 7 8 9 10 11 12 13 14 15 16 17

Those two scenarios are fundamentally different, even if they both use six reversals. The right one uses much longer reversals than the left one, and the right one *breaks* conserved intervals between Silkworm and Locust in intermediate permutations, namely $[3, 6]$, $[1, 12]$, and $[12, 17]$. If a rearrangement scenario is expected to reflect the various intermediate species between Silkworm and Locust, the right one looks highly suspicious. Recent papers address these problems in various ways, for example by assigning weights to operations [1], or with probabilistic studies of the possible scenarios [16].

The two main flaws of the second scenario – long reversals and breaking conserved intervals – are closely tied: breaking conserved intervals, as we will show in Sect. 6, often involves long range operations that radically disturb a genome. In this sense, conserved intervals can be used as an intrinsic measure that allows to screen out rearrangement scenarios, or phylogenetic hypotheses, without the need of arbitrary weights or probability measures.

3   Conserved Intervals

This section presents a formalization of the notion of conserved intervals, together with properties that allow the development of linear time algorithms to manipulate them.

Definition 1. *Let G be a set of signed permutations on n elements. An interval $[a, b]$ is a* conserved interval *of the set G if:*

1) either a precedes b, or $-b$ precedes $-a$, in each permutation, and

2) the set of unsigned elements that appear between a and b is the same for all permutations in G.

An elementary consequence of this definition is the fact that if $[a, b]$ is a conserved interval, so is $[-b, -a]$. We will consider these intervals as equivalent.

Table 1 contains several examples of conserved intervals. Their description is eased by the fact that the identity permutation belongs to the set G. When this is the case, all conserved intervals can be identified with their positive endpoints $a < b$, and the set of elements that appear between a and b is $\{a + 1, \ldots, b - 1\}$. The following example illustrates a more general case. Consider the two permutations:

$$P = \quad 1 \; 2 \quad 3 \quad 7 \; 5 \quad 6 \; -4 \; 8$$
$$Q = \quad 1 \; 7 \; -3 \; -2 \; 5 \; -6 \; -4 \; 8$$

In this example, $[1, 5]$ and $[2, 3]$ are conserved intervals, but not $[1, 6]$. The other conserved intervals of P and Q are $[1, -4]$, $[1, 8]$, $[5, -4]$, $[5, 8]$, and $[-4, 8]$. The diagram representation of these intervals, with respect to the permutation P, is:

When the identity permutation is not in G, it is always possible to *rename* the elements of G such that conserved intervals will be intervals of consecutive elements. For example, if one composes[1] the permutations P and Q of the above example with the inverse permutation P^{-1}, the first permutation becomes the

---

[1] Here, composition is understood as the standard composition of functions. Dealing with signed permutations requires the additional axiom that $P(-a) = -P(a)$.

identity permutation $Id = P^{-1} \circ P$. In general, it is elementary to transform a set of conserved intervals to its equivalent up to renaming. It is a consequence of the following proposition:

Proposition 1. *Let R be a permutation and G a set of permutations, denote by $R \circ G$ the set of permutations obtained by composing each permutation in G with R. The interval $[a, b]$ is conserved in G if and only if the interval $[R(a), R(b)]$ is conserved in $R \circ G$.*

Some intervals, such as $[1, -4]$ for the set $\{P, Q\}$ in the above example, are the union of smaller intervals: $[1, -4] = [1, 5] \cup [5, -4]$. Intervals that are not unions are specially useful:

Definition 2. *Conserved intervals that are not the union of shorter conserved intervals are called* irreducible.

Sets of conserved intervals can be simply characterized by the corresponding set of irreducible intervals. Indeed, disjoint irreducible intervals, as highlighted in the diagram representation, are either *chained* or *nested*. The following proposition captures the basic properties of these structures.

Proposition 2 ([5]). *Two different irreducible conserved intervals $[a, b]$ and $[c, d]$ of a set G of permutations, are either:*
   1) disjoint,
   2) nested with different endpoints, or
   3) overlapping on one element.

Overlapping irreducible intervals form chains linked by their successive common elements. A chain of $k - 1$ intervals $[a_1, a_2][a_2, a_3] \ldots [a_{k-1}, a_k]$ will be denoted simply by its k links $[a_1, a_2, a_3, \ldots, a_k]$. For example, $[1, 5, -4, 8]$ is a chain of the set of conserved intervals of P and Q. A *maximal* chain is a chain that cannot be extended. We have:

Proposition 3. *Every irreducible conserved interval belongs to a unique maximal chain.*

One consequence of Proposition 3 is that maximal chains, as sets of links, together with isolated genes, form a partition of the set of genes. This will reveal useful to construct data structures to keep track of conserved intervals.

A set of permutations on n elements can have as many as $n(n-1)/2$ conserved intervals, but at most $n - 1$ irreducible intervals. These bounds are achieved with sets containing only one permutation. A key observation, that will eventually lead to linear time algorithms to compute the number of conserved intervals, is the following:

Proposition 4. *Each maximal chain of k links contributes $k(k - 1)/2$ to the total number of conserved intervals.*

Finally, we will want to construct sets of conserved intervals for the union of two sets of permutations. Definition 1 implies that the set of conserved intervals of a union of two sets of permutations is the intersection of their sets of conserved intervals. The following proposition, shown in [5], relates these sets to their respective irreducible intervals when both sets of permutations have at least one permutation in common.

Proposition 5. *Let P be a permutation that is contained in both sets of permutations G_1 and G_2. The interval $[a, b]$ is a conserved interval of $G = G_1 \cup G_2$ if and only if there exist two chains of irreducible conserved intervals, with respect to P, with $k \geq 0$, $m \geq 0$:*

$$[a, x_1, \ldots, x_k, b] \quad in \; G_1,$$
$$[a, y_1, \ldots, y_m, b] \quad in \; G_2.$$

The interval $[a, b]$ is irreducible if and only if $\{x_1, \ldots, x_k\}$ and $\{y_1, \ldots, y_m\}$ are disjoint.

Variable Geometry Genomes. Although the definition of conserved intervals was given for permutations that model genomes composed of single linear chromosomes, they can be adapted to other types of genomes. For details, see [5].

4   Algorithms

This section discusses three algorithms. The first one is an adaptation of an existing algorithm that computes the conserved intervals of two permutations. The second one computes the conserved intervals of a set of permutations. The third one, finally, computes the conserved intervals of two sets of permutations, directly from their two individual sets of conserved intervals.

Conserved Intervals of Two Permutations. Conserved intervals between two permutations are strongly related to the notion of connected components of the *overlap graph* of a signed permutation. This graph plays a fundamental role in the sorting by reversals problem [11], and the sorting by reversals and translocations problem [12]. In the last few years, linear algorithms to identify these components have been devised [2]. The following algorithm is adapted from [4], and identifies all irreducible conserved intervals[2] $[a, b]$ of a permutation π with the identity permutation such that both a and b have positive sign in π. The case of negative endpoints is treated by reversing π.
    For example, for the permutation

$$P = \quad 0 \;\; -4 \;\; -3 \;\; -2 \;\; 5 \;\; 8 \;\; 6 \;\; 7 \;\; 9 \;\; -1 \;\; 10,$$

Algorithm 1 identifies the positive irreducible conserved intervals $[6, 7]$, $[5, 9]$, and $[0, 10]$. It will identify $[2, 3]$ and $[3, 4]$ on the reversed permutation.

---

[2] In the original paper, these were called *framed common intervals*.

Algorithm 1 (Positive irreducible intervals with the identity permutation)

1: stack 0 on S
2: stack n on M
3: $M_0 \leftarrow n$
4: **for** $i = 1, \ldots, n$ **do**
5:      // Computation of M_i
6:      unstack from M all elements m smaller than $|\pi_i|$
7:      $M_i \leftarrow m$
8:      stack the element $|\pi_i|$ on M
9:      // Identification of irreducible intervals
10:      unstack from S all indices s such that $(|\pi_i| < \pi_s$ or $|\pi_i| > M_s)$
11:      **if** $i - s = \pi_i - \pi_s$ and $M_i = M_s$ **then**
12:         output $[\pi_s, \pi_i]$
13:      **end if**
14:      **if** π_i is positive **then**
15:         stack the index i on S
16:      **end if**
17: **end for**

The algorithm assumes that the input permutation is in the form $\pi = (0, \pi_1, \ldots, \pi_{n-1}, n)$. Define M_i to be the nearest unsigned element of the permutation that precedes π_i and is greater than $|\pi_i|$. (Set M_i to n, if such an element does not exist). The following lemma relates the values of M_i to conserved intervals.

Lemma 1. *If $[\pi_s, \pi_e]$ is a positive conserved interval of π and the identity permutation, then $M_s = M_e$.*

The algorithm uses two stacks: S contains the possible start positions of conserved intervals; M contains possible candidates for M_i. The top of S is always denoted by s. The top of M is always denoted by m.

Proposition 6 ([4,5]). *Algorithm 1 outputs the positive irreducible conserved intervals of a permutation π with the identity permutation in $O(n)$ time.*

Corollary 1. *By applying Algorithm 1 both to $\pi = P^{-1} \circ Q$ and to the reverse of π, the irreducible conserved intervals of two permutations P and Q can be found in $O(n)$ time.*

Conserved Intervals of a Set of Permutations. In order to find the irreducible conserved intervals of a set of permutations, the first step is to compute the irreducible intervals of each permutation with one particular permutation from the set, say π_1, using Algorithm 1, and then merge together the resulting sets of irreducible intervals. For example, computing the irreducible intervals of the set:

$$
\begin{array}{rrrrrrrrrrrr}
Id = & 0 & 1 & 2 & 3 & 4 & 5 & 6 & 7 & 8 & 9 & 10 \\
P = & 0 & -4 & -3 & -2 & 5 & 8 & 6 & 7 & 9 & -1 & 10 \\
Q = & 0 & 5 & -7 & -6 & 8 & 9 & 1 & 2 & 3 & -4 & 10
\end{array}
$$

Algorithm 2 (Irreducible intervals of $G_1 \cup G_2$, both containing the identity permutation)

1: stack 0 on S
2: **for** $i = 1, \ldots, n$ **do**
3:     **if** there is an interval $[x, i]$ in I_1 **then**
4:         unstack from S all elements larger than x
5:     **end if**
6:     **if** there is an interval $[x, i]$ in I_2 **then**
7:         unstack from S all elements larger than x
8:     **end if**
9:     **if** s and i belong to the same chain both in I_1 and I_2 **then**
10:         unstack s from S and output $[s, i]$
11:     **end if**
12:     **if** there is an interval that starts at i in I_1, and one in I_2 **then**
13:         stack i on S
14:     **end if**
15: **end for**

---

would first yield the two sets of maximal chains $\{[0, 10], [2, 3, 4][5, 9], [6, 7]\}$ (of P and the identity) and $\{[0, 10], [1, 2, 3], [5, 8, 9], [6, 7]\}$ (of Q and the identity), respectively, in graphic representation:

Assume that each set of irreducible conserved intervals is given by its maximal chains. Since these form partitions of the genes that are endpoints of conserved intervals, there exists a data structure with the following properties: (1) For each index from 1 to n, it is possible to determine in constant time the interval, if any, that starts and/or ends at this index. (2) It is possible to determine in constant time if two intervals belong to the same chain.

Let I_1 and I_2 be two sets of irreducible conserved intervals of sets of permutations G_1 and G_2 that have one permutation π_1 in common. For the moment we will assume that π_1 is the identity permutation. Then Algorithm 2 finds all irreducible conserved intervals of $G_1 \cup G_2$. It uses a stack S that contains possible start positions – or, equivalently, elements of the identity permutation. The top of the stack S is always denoted by s.

The correctness and time complexity of Algorithm 2 are established by the following theorem, whose proof can be found in [5].

Theorem 1. *Algorithm 2 outputs the irreducible intervals of $G = G_1 \cup G_2$ in $O(n)$ time, given I_1 and I_2, the irreducible intervals of two sets of permutations G_1 and G_2 that both contain the identity permutation.*

Corollary 2. *Let I_1 and I_2 be the irreducible intervals of two sets of permutations G_1 and G_2 that both contain a permutation P. The irreducible intervals of $G = G_1 \cup G_2$ can be found in $O(n)$ time by applying Algorithm 2 to $I'_1 = \{[P^{-1}(a), P^{-1}(b)] \mid [a, b] \in I_1\}$ and $I'_2 = \{[P^{-1}(a), P^{-1}(b)] \mid [a, b] \in I_2\}$.*

Corollary 3. *The set of irreducible conserved intervals of a set of permutations G can be computed in $O(|G|n)$ time and $O(n)$ additional space.*

Conserved Intervals of Disjoint Sets. Finally we are interested in computing the conserved intervals of two sets of permutations $G_1 = \{P_1, \ldots, P_k\}$ and $G_2 = \{Q_1, \ldots, Q_m\}$ that not necessarily have a permutation in common, given their sets of irreducible conserved intervals I_1 and I_2, respectively.

This can be done in linear time by properly combining Algorithms 1 and 2. The idea is to select one permutation from each set, say P_1 from G_1 and Q_1 from G_2, and compute the conserved intervals of these two by Algorithm 1. Then observe that the two sets $\{P_1, Q_1\}$ and $G_1 = \{P_1, \ldots, P_k\}$ have a joint permutation P_1, and hence their common irreducible intervals can be computed by Algorithm 2. Similarly, $\{Q_1, P_1, \ldots, P_k\}$ and $G_2 = \{Q_1, \ldots, Q_m\}$ contain a joint permutation Q_1, so their common irreducible intervals can also be computed by Algorithm 2.

5 Similarity and Distance

The number of conserved intervals of a set of permutations is a measure of similarity, but it can easily be transformed into a distance between two permutations, or two sets of permutations. The basic idea is that two sets of conserved intervals can be compared with the cardinality of their symmetric difference.

Definition 3. *Let G_1 and G_2 be two sets of permutations on n elements, with respectively N_1 and N_2 conserved intervals. Let N be the number of conserved intervals in $G_1 \cup G_2$. The* interval distance *between G_1 and G_2 is defined by $d(G_1, G_2) = N_1 + N_2 - 2N$.*

Note: The interval distance satisfies the fundamental properties of a mathematical distance since one can prove that the relation is symmetric, reflexive, and satisfies the *triangle inequality:* $d(G_1, G') + d(G', G_2) \geq d(G_1, G_2)$.

A detailed comparison of the interval distance with other rearrangement distances can be found in [5]. The behavior of the interval distance is a consequence of the fact that it is affected be the length – or number of genes – involved in a rearrangement operation: short reversals, for example, are less disturbing than long ones. In particular, the amount of disruption due to a single rearrangement operation can readily be computed. For example, we have the following:

Proposition 7. *Suppose that P and Q have n elements, then:*
1) if P is obtained from Q by reversing k elements, then the interval distance between P and Q is $k(n - k)$;
2) if P is obtained from Q by transposing two consecutive blocks of a and b elements, then the interval distance between P and Q is $(a+b)(n-(a+b))+ab$.

Since the interval distance is affected by length, the practice of collapsing identical strips of genes should be questioned. Indeed, as we saw in the example of Sect. 2, the integers resulting from such a transformation stand for strips of genes that vary greatly in length. We believe that whole genome comparison should use all available information, and that length of segments is relevant to the study of rearrangement scenarios, as advocated in [19].

6   Links With Rearrangement Theories

In Sect. 2, we gave an example of how conserved intervals could be used to evaluate optimal reversal scenarios between two genomes. Reversals are one of the many operations that are currently used to model genome evolution: the main other ones – among those that do not need to model duplication of genes – are transpositions, reverse transpositions, translocations, fusions, and fissions.

In this section, we want to characterize the rearrangement operations, or scenarios, that *preserve* conserved intervals:

Definition 4. *Let P and Q be two permutations, and ρ a rearrangement operation applied to P yielding P'. We say that ρ preserves the conserved intervals of P and Q if the conserved intervals of $\{P, Q\}$ are contained in those of $\{P', Q\}$.*

Keeping in mind the graphical representation of the conserved intervals, it is easy to identify the operations that preserve conserved intervals: only rearrangements within blocks are preserving. To be more formal, note that all operations, except fusions, destroy some adjacencies that existed in the original permutation: the number and nature of these adjacencies is a key concept.

Definition 5. *Let ρ be a rearrangement operation that transforms P into P'. A breakpoint of ρ is a pair of elements that are adjacent in P but not in P'.*

In other words, breakpoints are where one has to cut P in order to apply ρ. Reversals and translocations have 2 breakpoints, transpositions have 3, and fissions have 1.

Consider the irreducible intervals of P and P' with respect to P. Adjacencies in P either belong to a (smallest) irreducible interval, or are *free*. For example, in the diagram

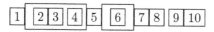

the adjacency $(3, 4)$ belongs to the interval $[1, 5]$, $(2, 3)$ belongs to $[2, 3]$, and $(8, 9)$ is free. Note that when two or more adjacencies belong to the same irreducible interval, then none of these adjacencies is conserved between P and P'.

Theorem 2 ([5]). *Reversals, transpositions, and reverse transpositions are preserving if and only if all their breakpoints belong to the same irreducible interval, or are free. Translocations and fissions are preserving if and only if all their breakpoints are free.*

It turns out that most rearrangement operations used in optimal scenarios are indeed preserving. It is outside the scope of this paper to discuss these results in detail: they involve the *cycle* structure of a permutation, which are special subsets of the breakpoints of a permutation P with respect to a permutation P'. The following result has been proved in various disguises in recent years [4,11, 14]:

Theorem 3. *All the breakpoints of a cycle belong to the same irreducible interval.*

In the sorting by reversals theory, a *sorting* reversal is defined as a reversal that decreases the reversal distance by 1. It is shown [11,20] that the breakpoints of sorting reversals, except for one type called *component merging*, belong to a single cycle, thus we have:

Corollary 4. *All sorting reversals, except component merging, are preserving.*

Component mergings are a rare type of reversals in optimal scenarios: they break at least two irreducible intervals, thus they often involve long reversals.

The theory of translocations, fusions, and fissions [12,18] relies on the properties of sorting by reversals, thus most sorting reversals are preserving. Finally, transpositions are a more delicate matter since sorting transpositions are not (yet) characterized. Nevertheless, it is known that transpositions that increase the number of cycles – a desirable property when sorting permutations – have all their breakpoints in the same cycle [3]. Thus we have:

Corollary 5. *All transpositions that create two adjacencies are preserving.*

7   Conclusion

We have introduced a new similarity measure for permutations, based on the concept of conserved intervals. Conserved intervals have very interesting properties with respect to preserving the usual genome rearrangement operations. We believe that conserved intervals are a fundamental concept of rearrangement theory: they provide the unifying grounds to understand the variety of operations that are used to model genome evolution. Supported by recent results on the expected size of rearranged genome segments, one could go as far and claim that any rearrangement scenario that breaks conserved intervals is mathematical rambling without connection to evolutionary reality.

References

1. Y. Ajana, J.-F. Lefebvre, E. R. M. Tillier, and N. El-Mabrouk. Exploring the set of all minimal sequences of reversals – an application to test the replication-directed reversal hypothesis. In *Proc. WABI 2002*, volume 2452 of *LNCS*, pages 300–315. Springer Verlag, 2002.
2. D. A. Bader, B. M. E. Moret, and M. Yan. A linear-time algorithm for computing inversion distance between signed permutations with an experimental study. *J. Comp. Biol.*, 8(5):483–492, 2001.

3. V. Bafna and P. A. Pevzner. Sorting by transpositions. *SIAM J. Disc. Math.*, 11(2):224–240, 1998.
4. A. Bergeron, S. Heber, and J. Stoye. Common intervals and sorting by reversals: A marriage of necessity. *Bioinformatics*, 18(Suppl. 2):S54–S63, 2002. (Proc. ECCB 2002).
5. A. Bergeron and J. Stoye. On the similarity of sets of permutations and its application to genome comparison. Report 2003-01, Technische Fakultät der Universität Bielefeld, 2003. (Available at www.techfak.uni-bielefeld.de/ stoye/rpublications/report2003-01.pdf).
6. M. Blanchette, T. Kunisawa, and D. Sankoff. Gene order breakpoint evidence in animal mitochondrial phylogeny. *J. Mol. Evol.*, 49(2):193–203, 1999.
7. J. L. Boore. Mitochondrial gene arrangement source guide. www.jgi.doe.gov/programs/comparative/Mito_top_level.html.
8. K. S. Booth and G. S. Lueker. Testing for the consecutive ones property, interval graphs and graph planarity using *PQ*-tree algorithms. *J. Comput. Syst. Sci.*, 13(3):335–379, 1976.
9. G. Bourque and P. A. Pevzner. Genome-scale evolution: Reconstructing gene orders in the ancestral species. *Genome Res.*, 12(1):26–36, 2002.
10. D. A. Christie. *Genome Rearrangement Problems*. PhD thesis, The University of Glasgow, 1998.
11. S. Hannenhalli and P. A. Pevzner. Transforming men into mice (polynomial algorithm for genomic distance problem). In *Proc. FOCS 1995*, pages 581–592. IEEE Press, 1995.
12. S. Hannenhalli and P. A. Pevzner. Transforming cabbage into turnip: Polynomial algorithm for sorting signed permutations by reversals. *J. ACM*, 46(1):1–27, 1999.
13. S. Heber and J. Stoye. Finding all common intervals of *k* permutations. In *Proc. CPM 2001*, volume 2089 of *LNCS*, pages 207–218. Springer Verlag, 2001.
14. H. Kaplan, R. Shamir, and R. E. Tarjan. A faster and simpler algorithm for sorting signed permutations by reversals. *SIAM J. Computing*, 29(3):880–892, 1999.
15. J. D. Kececioglu and D. Sankoff. Efficient bounds for oriented chromosome inversion distance. In *Proc. CPM 1994*, volume 807 of *LNCS*, pages 307–325. Springer Verlag, 1994.
16. B. Larget, J. Kadane, and D. Simon. A Markov chain Monte Carlo approach to reconstructing ancestral genome rearrangements. Technical report, Carnegie Mellon University, Pittsburgh, 2002.
17. B. M. E. Moret, A. C. Siepel, J. Tang, and T. Liu. Inversion medians outperform breakpoint medians in phylogeny reconstruction from gene-order data. In *Proc. WABI 2002*, volume 2452 of *LNCS*, pages 521–536. Springer Verlag, 2002.
18. M. Ozery-Flato and R. Shamir. Two notes on genome rearrangements. *J. Bioinf. Comput. Biol.*, to appear.
19. D. Sankoff. Short inversions and conserved gene clusters. *Bioinformatics*, 18(10):1305–1308, 2002.
20. A. Siepel. An algorithm to find all sorting reversals. In *Proc. RECOMB 2002*, pages 281–290. ACM Press, 2002.
21. G. Tesler. Efficient algorithms for multichromosomal genome rearrangement. *J. Comput. Syst. Sci.*, 65(3):587–609, 2002.
22. T. Uno and M. Yagiura. Fast algorithms to enumerate all common intervals of two permutations. *Algorithmica*, 26(2):290–309, 2000.

On All-Substrings Alignment Problems

Wei Fu[1], Wing-Kai Hon[2], and Wing-Kin Sung[1]*

[1] School of Computing, National University of Singapore,
Singapore,
{fuwei,ksung}@comp.nus.edu.sg,
[2] Department of Computer Science and Information Systems,
The University of Hong Kong, Hong Kong,
wkhon@csis.hku.hk

Abstract. Consider two strings A and B of lengths n and m respectively, with $n \ll m$. The problem of computing global and local alignments between A and all m^2 substrings of B can be solved by the classical Needleman-Wunsch and Smith-Waterman algorithms, respectively, which takes $O(m^2 n)$ time and $O(m^2)$ space. This paper proposes faster algorithms that take $O(mn^2)$ time and $O(mn)$ space. The improvement stems from a compact way to represent all the alignment scores.

1 Introduction

Sequence comparisons have been studied extensively for the last few decades [4, 5]. Many useful metrics are defined, which find applications in various areas. *Alignment score* is perhaps the most popular metric in the literature. Examples of its usages can be found in text processing, such as syntax error correction [1] and spelling correction [3], or in comparing graphs [6].

Due to the breakthrough in bio-technology in recent decades, biological sequences such as DNA, RNA or protein, are being rapidly produced in the laboratories everyday. Sequence comparison now plays an important role in extracting useful information from these raw biological data. Scientists are often interested in comparing a short sequence (usually, a known gene or a motif) with many substrings in a long sequence (usually, a genome). In such scenario, it is good for us to perform a preprocessing so that the above kind of comparison between the two strings can be done efficiently afterwards. This motivates the All-Substrings Alignment Problem: Given two strings A and B of lengths n and m respectively, with $n \ll m$, the problem is to find the alignment scores between A and all substrings $B[i..j]$. The alignment can be global alignment, suffix alignment, local alignment, or semi-global alignment.

A naive solution to this problem is to compute the alignment scores between A and all substrings of B explicitly using the classical Needleman-Wunsch and Smith-Waterman algorithms [7,9]. Such approach requires $O(m^2 n)$ time and $O(m^2)$ space. This paper solves these problems by devising an $O(mn^2)$-time algorithm. This is significant since n is much smaller than m.

---

* Research supported in part by NUS Academic Research Grant R-252-000-119-112.

T. Warnow and B. Zhu (Eds.): COCOON 2003, LNCS 2697, pp. 80–89, 2003.
© Springer-Verlag Berlin Heidelberg 2003

For global alignment, our improvement is from the observation that, although the alignments between A and different substrings of B vary, their scores can actually be 'compressed' under suitable transformation. Precisely, for any fixed prefix $A[1..k]$, the global alignment scores between $A[1..k]$ and all substrings of B can be stored in $O(mn)$ space instead of the trivial $O(m^2)$ space. We also show that this compressed structure for $A[1..k]$ can be computed directly from that for $A[1..k-1]$. These ideas lead to our algorithm, which is incremental in nature. Moreover, the compressed structures themselves can serve as a compact index for storing all the global alignment scores between A and any substring of B, so that retrieving a particular value takes $O(\log n)$ time.

For local, semi-global, and suffix alignments, we observe that among the $m^2 n$ alignments between $A[1..k]$ and $B[i..j]$ for all $1 \le k \le n$ and $1 \le i < j \le m$, many of them are the same. More precisely, there are at most mn^2 useful values, which can be found in $O(mn^2)$ time. To retrieve a particular alignment score between $A[1..k]$ and $B[i..j]$, we only need to do a range maximum query over these $O(mn^2)$ value.

The organization of the paper is as follows. Section 2 gives basic definitions of the problems. Section 3 describes the $O(mn^2)$-time algorithm for solving the All-Substrings Global Alignment Problem, while the other alignment problems are discussed in Section 4. Finally, Section 5 concludes with some open problems.

2   Preliminaries

We formally define some notations which are used throughout the paper.

Definition 1. [8] An alignment *between two sequences is formed by the insertion of spaces in arbitrary locations along the sequences so that they end up with the same length.*

Having the same length, the augmented sequences of an alignment can be placed one over the other, creating a mapping between characters or spaces in the first sequence and characters or space in the second sequence. In general, we assume that no space in one sequence is aligned with a space in another sequence.

To measure the similarity of an alignment, we need the concept of *score scheme*.

Definition 2. *Let Σ denote a set of characters and let Σ' denote $\Sigma \cup \{\sqcup\}$, where the symbol $\sqcup$ represents a space. A* score scheme *over Σ is a function $\delta : \Sigma' \times \Sigma' \to \mathbb{Q}$, where $\mathbb{Q}$ denotes the set of rational numbers.*

Given a score scheme, the *score* of an alignment is defined as follows.

Definition 3. *The* score *of an alignment is the summation of the δ function over every pair of the corresponding characters in the given alignment.*

We now define the problems that we are interested in this paper. Let $A[1..n]$ and $B[1..m]$ be two sequences of characters over an alphabet Σ.

Definition 4. *The* global alignment score *between two sequences is defined as the score of the highest scoring alignment between the two sequences. The* All-Substrings Global Alignment Problem *on (A,B) is to find the global alignment scores between A and any substring of B.*

Definition 5. *The* suffix alignment score *between two sequences is defined as the score of the highest scoring alignment between a suffix of one sequence and a suffix of the other one. The* All-Substrings Suffix Alignment Problem *on (A,B) is to find the suffix alignment scores between A and any substring of B.*

Definition 6. *The* semi-global alignment score *between two sequences S and T is defined as the score of the highest scoring alignment between S and a substring of T. The* All-Substrings Semi-global Alignment Problem *on (A,B) is to find the semi-global alignment scores between A and any substring of B.*

Definition 7. *The* local alignment score *between two sequences is defined as the score of the highest scoring alignment between a substring of one sequence and a substring of the other one. The* All-Substrings Local Alignment Problem *on (A,B) is to find the local alignment scores between A and any substring of B.*

3   All-Substrings Global Alignment

In this section, we consider the All-Substrings Global Alignment Problem on (A, B) between two sequences $A[1..n]$ and $B[1..m]$. We first assume the followng score scheme.

1. $\delta(x, \sqcup) = \delta(\sqcup, x) = -1$.
2. $\delta(x, y) = \begin{cases} 1, & \text{if } x = y \text{ and } x, y \neq \sqcup \\ -1, & \text{if } x \neq y \text{ and } x, y \neq \sqcup \end{cases}$

Let $H_k[i, j]$ denote the global alignment score between a prefix $A[1..k]$ of A and a substring $B[i..j]$ of B. Our problem is thus equivalent to finding all the values $H_n[i, j]$ for every $1 \leq i \leq j \leq m$. This can be solved based on the following lemma.

Lemma 1.

$$H_k[i, j] = \max \begin{cases} H_k[i, j - 1] & + \delta(\sqcup, B[j]) \\ H_{k-1}[i, j] & + \delta(A[k], \sqcup) \\ H_{k-1}[i, j - 1] + \delta(A[k], B[j]) \end{cases}$$

The framework of computing H_n is simple: First compute H_0, then complete the matrices H_k for $k = 1, 2, \ldots, n$. Note that each H_k has $O(m^2)$ entries, and each entry can be filled in $O(1)$ time by Lemma 1. Thus, we get H_n in $O(m^2 n)$ time. For the space, it takes $O(m^2)$, since to compute any matrix H_k, we only need to keep H_{k-1}.

When m is very large (which is common in the problem we are modeling), the above time and space complexities are not practical. We propose an alternative way to solve this problem in $O(mn^2)$ time and $O(mn)$ space, making use of a simple modification on the matrices. The next section shows such modification, while the details of the new algorithm is discussed afterwards.

3.1   Transformation of H_k's

Instead of computing H_k's, we compute another set of matrices F_k's as follows.

Definition 8. $F_k[i, j] = H_k[i, j] + j - i + 1 + k.$

Similar to Lemma 1, we have a corresponding formula that relates the adjacent matrices F_{k-1} and F_k.

Fact 1.
$$F_k[i, j] = \max \begin{cases} F_k[i, j-1] \\ F_{k-1}[i, j] \\ F_{k-1}[i, j-1] + 2 + \delta(A[k], B[j]) \end{cases}$$

Note that $H_k[i, j]$ can be retrieved in constant time given $F_k[i, j]$. Thus, our problem can be solved by finding all the values of $F_n[i, j]$. Although the definition of F_k looks similar to H_k, the following properties make the computation of F_k a better choice over H_k.

Fact 2. For $1 \le k \le n$ and $1 \le i \le j \le m$,

1. $F_k[i, j] \le F_k[i, j+1]$ if $j + 1 \le m$
2. $F_k[i, j] \le F_k[i-1, j]$ if $i \ge 2$
3. $0 \le F_k[i, j] \le 3k$.

Proof. (sketch.) Statement 1 follows directly from Fact 1, and Statement 2 can be proved by induction on j. For Statement 3, it can be proved as follows. In any highest-scoring global alignment of $A[1..k]$ and $B[i..j]$, each character in A or B contribute at least -1 in the alignment score, so $H_k[i, j] \ge -(j-i+1+k)$. This implies $F_k[i, j] \ge 0$. On the other hand, there are at least $|j - i + 1 - k|$ character-space alignment, and at most k matching-character alignment. Thus $H_k[i, j] \le k - |j - i + 1 - k|$. This implies that $F_k[i, j] \le k + (k + j - i + 1) - |j - i + i - k| = k + 2\min\{k, j - i + 1\} \le 3k$.

The above facts imply that the values of F_k are bounded and are monotonic increasing in each row (and decreasing in each column). We next give a related definition, and then show that F_k can be stored in a compact manner.

Definition 9. *For each row i of the matrix F_k, a row interval point is the value j such that $F_k[i, j-1] < F_k[i, j]$. Note that we assume $i \le j \le m$. Similarly, for each column j of the matrix, a column interval points is the value i such that $F_k[i, j] < F_k[i-1, j]$.*

Lemma 2. F_k *can be stored compactly in $O(km)$ space. Any value of $F_k[i, j]$ can be queried in $O(\log k)$ time.*

Proof. For each row i of F_k, we use an array of size $3k + 1$ such that the x-th entry stores the smallest j with $F_k[i, j] \ge x$. Note that these values are in essence the row interval points. As there are m rows in F_k, the total space is $O(km)$.

Given such a representation, each value of $F_k[i, j]$ can be found by a binary search in the array corresponding to the i-th row. The query time is $O(\log k)$.

3.2   Computation of F_n

We aim at computing F_n by first computing F_1, and then F_k based on F_{k-1} efficiently for $k = 2, 3, \ldots, n$. For each F_k, instead of computing all the entries, we compute its compact representation as stated in Lemma 2.

We first discuss the computation of the compact form of F_1. For any i, j such that $1 \le i \le j \le m$, under the current score scheme, we have

$$H_1[i, j] = \begin{cases} -(j - i + 2) & \text{if } B[i..j] \text{ does not contain } A[1] \\ -(j - i - 1) & \text{if } B[i..j] \text{ contains } A[1] \end{cases}$$

Accordingly, we have

$$F_1[i, j] = \begin{cases} 0 & \text{if } B[i..j] \text{ does not contain } A[1] \\ 3 & \text{if } B[i..j] \text{ contains } A[1] \end{cases}$$

Then, to store F_1, we just need to keep an array $L[1..m]$ such that $L[i]$ stores the smallest j such that $B[i..j]$ contains $A[1]$. In fact, $L[i]$ stores the only row interval point of row i of F_1 (Recall that there is one row interval point in each row of F_1). Afterwards, we can determine whether the value of $F_1[i, j']$ is 0 or 3 by comparing j' with $L[i]$, for any j'.

The array L can be computed easily by scanning B once as follows:

1. Initialize all entries of L to 0.
2. Traverse B from $B[1], B[2], \ldots, B[m]$, and set $L[i] = i$ if $B[i] = A[1]$.
3. If $L[m] = 0$, set $L[m] = m + 1$.
4. Examine L backwardly from $L[m - 1], L[m - 2], \ldots, L[1]$. If $L[k - 1] = 0$, set $L[k - 1] = L[k]$.

This gives the following lemma.

Lemma 3. *The compact form of F_1 can be computed in $O(m)$ time and $O(m)$ space.*

Next, we show how to compute the compact form of F_k from that of F_{k-1}. Firstly, based on Fact 1 and recursively expand ths F_k terms on the right side, we have an alternative definition of F_k as follows:

$$F_k[i, j] = \max \begin{cases} F_k[i, j - 1] \\ F_{k-1}[i, j] \\ F_{k-1}[i, j - 1] + 2 + \delta(A[k], B[j]) \\ F_k[i, j - 2] \\ F_{k-1}[i, j - 1] \\ F_{k-1}[i, j - 2] + 2 + \delta(A[k], B[j - 1]) \\ F_{k-1}[i, j] \\ F_{k-1}[i, j - 1] + 2 + \delta(A[k], B[j]) \end{cases}$$

$$\vdots$$

$$= \max \begin{cases} \max_{i \le j' \le j} F_{k-1}[i, j'] \\ \max_{i+1 \le j' \le j} F_{k-1}[i, j' - 1] + 2 + \delta(A[k], B[j']) \end{cases}$$

$$= \max \begin{cases} F_{k-1}[i, j] \\ \max_{i+1 \le j' \le j} F_{k-1}[i, j' - 1] + 2 + \delta(A[k], B[j']) \end{cases}$$

The above definition is intuitively simpler than that in Fact 1, as each entry of F_k depends only on entries in F_{k-1}. In contrast, the definition in Fact 1 relies on some other entries in F_k itself.

Now, let $G_k[i,j]$ denote $\max_{i+1 \leq j' \leq j} F_{k-1}[i, j'-1] + 2 + \delta(A[k], B[j'])$, where we assume $G_k[i,j] = 0$ if $i = j$. Then, we have

$$F_k[i,j] = \max\{F_{k-1}[i,j], G_k[i,j]\}.$$

The critical part to compute F_k is to compute G_k. For G_k, we observe that

Observation 1. *For any $1 \leq k \leq n$ and $1 \leq i \leq j \leq m$,*

1. $G_k[i,j] \leq G_k[i, j+1]$ *if $j \leq m-1$*
2. $G_k[i,j] \geq G_k[i+1, j]$ *if $i \geq 2$*
3. $0 \leq G_k[i,j] \leq 3k$.

Proof. (sketch) Statement 1 follows from definition of G_k, while Statement 2 can be proved by induction on j. For Statement 3, since we have $G_k[i,j] \leq F_k[i,j]$ and $G_k[i,j] \geq F_{k-1}[i, j-1] + 2 + \delta(A[k], B[j])$, it thus follows immediately from Fact 2(3) that $0 \leq F_k[i,j] \leq 3k$.

These observations suggest that G_k can be stored using row interval points. If we have collected all the possible row interval points of G_k, it is easy to see that they can be merged with those of F_{k-1} to get the compact form of F_k, using time linear to the total number of row interval points we consider.

We now claim that in $O(km)$ time, we can find all the possible row interval points of G_k, and the number of which is at most $O(km)$. Suppose that this is true, we have the following results.

Lemma 4. *Given the compact form of F_{k-1}, we can compute the compact form of F_k in $O(km)$ time and $O(km)$ space.*

Theorem 1. *The compact form of F_n can be computed in $O(mn^2)$ time and $O(mn)$ space.*

Proof. Follows directly from Lemmas 3 and 4.

The remaining part focuses on proving the previous claim. Firstly, we show a new property of G_k.

Definition 10. *A corner point of the matrix G_k is a tuple (i,j) satisfying*

$$G_k[i, j-1] < G_k[i,j] \text{ and } G_k[i+1, j] < G_k[i,j].$$

Observation 2. *For any corner point (i,j) of G_k, either*

1. $i = j$, *or*
2. $i + 1$ *is a column interval point of column j in F_{k-1}.*

Proof. (sketch) All (i, i) are defined to be corner points, as $G_k[i, i-1]$ or $G_k[i+1, i]$ are undefined. For the other corner points (i, j), we have $G_k[i, j] > G_k[i, j-1]$. Then from definitions of $G_k[i, j]$ and $G_k[i, j-1]$, we can derive that $G_k[i, j] = F_{k-1}[i, j-1]+2+\delta(A[k], B[j])$. Now, with $G_k[i+1, j] < G_k[i, j]$, we get $F_{k-1}[i, j-1] + 2 + \delta(A[k], B[j]) > G_k[i+1, j] \geq F_{k-1}[i+1, j-1] + 2 + \delta(A[k], B[j])$. This implies $i + 1$ is a column interval point of column j in F_{k-1}.

The above shows that the number of possible row interval points for G_k is at most $O(km)$, as there are at most $O(km)$ column interval points in F_{k-1}. The following algorithm completes our claim by showing that the column interval points of F_{k-1} can be found from the row interval points of F_{k-1} in $O(km)$ time and $O(km)$ space.

The algorithm for finding the possible row interval points of G_k is as follows.

1. Examine the compact form, or the row interval points, of F_{k-1}. Compute the *corner points* of F_{k-1}.
2. Sort the corner points (i, j) in the *descending* order of j first, and then in *descending* order of i.
3. Compute the column interval points of F_{k-1} from the corner points of F_{k-1}.
4. Compute the possible column interval points of G_k.
5. Compute the possible row interval points of G_k.

Step 1 can be completed by processing the row interval points of F_{k-1} row by row. Step 2 can be completed by bucket sort. Step 3 can be completed by processing the corner points in the sorted order of Step 2. We use $L[j][x]$ stores the smallest i such that $F_{k-1}[i, j] = x$. Initially, we set all the $O(km)$ entries of L to be 0. When we process (i, j), we let $x = F_{k-1}[i, j]$ and fill $L[j'][x] = i$ for $j' = j, j+1, \ldots$ until it is not equal to 0. Then the array L, and thus the column interval points, will be set correctly. (The correctness is postponed in the full paper.) Step 4 can be completed based on Observation 2. For Step 5, it converts the possible column interval points of G_k to the possible row interval points of G_k, which can be done by a similar approach as Step 1 through Step 3, which converts the row interval points of F_{k-1} to the column interval points. Finally, all the above steps take $O(km)$ time and $O(km)$ space. This completes the proof of the claim. $\blacksquare$

3.3   Generalizing the Score Scheme

Finally, we consider the general score scheme as follows.

1. $\delta(x, \sqcup) = \delta(\sqcup, x) = -\beta$.
2. $\delta(x, y) = \begin{cases} 1, & \text{if } x = y \text{ and } x, y \neq \sqcup \\ -\alpha, & \text{if } x \neq y \text{ and } x, y \neq \sqcup \end{cases}$

where we consider α and β to be rational numbers.

By setting $F_k[i, j] = H_k[i, j] + (j - i + 1 + k)\beta + k\alpha$, and using similar tricks as before, we have the following theorem.

Theorem 2. *Under the general score scheme, the All-Substrings Global Alignment Problem can be solved in $O(mn^2)$ time and $O(mn)$ space.*

4   Other Alignment Problems

Given two strings $A[1..n]$ and $B[1..m]$, this section discusses the All-Substrings Alignment Problem for suffix, semi-global, and local alignment. Denote $C_k[i,j]$, $D_k[i,j]$, and $E_k[i,j]$ be the suffix alignment score, the semi-global alignment score and the local alignment score between $A[1..k]$ and $B[i..j]$, respectively. Our problem is to compute $C_k[i,j]$, $D_k[i,j]$, and $E_k[i,j]$ for all $1 \le k \le n$ and $1 \le i \le j \le m$. The three lemmas below state the recursive equations for $C_k[i,j]$, $D_k[i,j]$, and $E_k[i,j]$.

Lemma 5.

$$C_k[i,j] = \max \begin{cases} C_k[i,j-1] & + \delta(\sqcup, B[j]) \\ C_{k-1}[i,j] & + \delta(A[k], \sqcup) \\ C_{k-1}[i,j-1] + \delta(A[k], B[j]) \end{cases}$$

Lemma 6.

$$D_k[i,j] = \max \begin{cases} D_k[i,j-1] \\ D_{k-1}[i,j] & + \delta(A[k], \sqcup) \\ C_{k-1}[i,j-1] + \delta(A[k], B[j]) & \text{if } A[k] = B[j] \end{cases}$$

Lemma 7.

$$E_k[i,j] = \max \begin{cases} E_k[i,j-1] \\ E_{k-1}[i,j] \\ C_{k-1}[i,j-1] + \delta(A[k], B[j]) & \text{if } A[k] = B[j] \end{cases}$$

Using the generalized score scheme as in Section 3.3, C_k, D_k, and E_k satisfy the following lemma.

Lemma 8. *For any i, j, k $(1 \le k \le n, 1 \le i \le j \le m)$ such that $j-i \ge k+\lfloor k/\beta \rfloor$,*

1. $C_k[i,j] = C_k[j - \lfloor k/\beta \rfloor - k, j]$
2. $D_k[i,j] = \max_{i \le i' \le j-(k+\lfloor k/\beta \rfloor)} D_k[i', i' + (k + \lfloor k/\beta \rfloor)]$
3. $E_k[i,j] = \max_{i \le i' \le j-(k+\lfloor k/\beta \rfloor)} E_k[i', i' + (k + \lfloor k/\beta \rfloor)]$

Proof. For Statement 1, consider the suffix alignment between $A[1..k]$ and $B[i..j]$ which achieves the highest score (say, $A[i_1..k]$ with $B[i_2..j]$). Since any suffix of $A[1..k]$ is of length at most k, the suffix alignment contains at most k pairs of matching characters, which contributes at most score k to the alignment. On the other hand, since the alignment score must be at least 0, the suffix alignment can contain at most $\lfloor k/\beta \rfloor$ pairs of character-space match. Thus, $j-i_2+1 \le k+\lfloor k/\beta \rfloor$, and if $j - i \ge k + \lfloor k/\beta \rfloor$, we have $C_k[i,j] = C_k[j - k - \lfloor k/\beta \rfloor, j]$. Statements 2 and 3 can be proved similarly.

4.1 Computing C_k's

By Lemma 8(1), we observe that many $C_k[i,j]$ values are redundant. Precisely, we only need to compute $\mathcal{C} = \{C_k[i,j] \mid 1 \leq k \leq n \text{ and } j - i \leq k + \lfloor k/\beta \rfloor\}$, while the other entries can be derived by Lemma 8(1). Thus, we have

Lemma 9. *We can compute the values in $\mathcal{C}$ in $O(mn^2)$ time and $O(mn)$ space.*

Proof. For each $k = 1, 2, \ldots, n$, we can apply Lemma 5 to compute those $C_k[i,j]$ with $j - i \leq k + \lfloor k/\beta \rfloor$ in $O(km)$ time and $O(km)$ space. The lemma thus follows. $\square$

Lemma 10. *Given the values in $\mathcal{C}$, for any $1 \leq k \leq n$ and $1 \leq i \leq j \leq m$, $C_k[i,j]$ can be retrieved in $O(1)$ time.*

Proof. If $j - i \leq k + \lfloor k/\beta \rfloor$, $C_k[i,j]$ is available in $\mathcal{C}$. Otherwise, if $j - i > k + \lfloor k/\beta \rfloor$, by Lemma 8(1), $C_k[i,j] = C_k[j - (k + \lfloor k/\beta \rfloor), j]$, which is available in $\mathcal{C}$. $\square$

4.2 Computing D_k's and E_k's

Similar to C_k, many D_k's and E_k's entries are redundant. We actually require to compute the following two sets of values: $\mathcal{D} = \{D_k[i,j] \mid 1 \leq k \leq n \text{ and } j - i \leq (k + \lfloor k/\beta \rfloor)\}$ and $\mathcal{E} = \{E_k[i,j] \mid 1 \leq k \leq n \text{ and } j - i \leq (k + \lfloor k/\beta \rfloor)\}$. The other D_k and E_k values can be derived using Lemma 8(2) and 8(3). The lemma below indicates that $\mathcal{D}$ and $\mathcal{E}$ can be found in $O(mn^2)$ time.

Lemma 11. *We can compute the values in $\mathcal{D}$ and $\mathcal{E}$ in $O(mn^2)$ time and $O(mn)$ space.*

Proof. By Lemmas 9 and 10, after an $O(mn^2)$ time preprocessing, each $C_k[i,j]$ can be retrieved in $O(1)$ time. Then given C_k, all entries in $\mathcal{D}$ and $\mathcal{E}$ can be computed in $O(mn^2)$ time and $O(mn)$ space, using dynamic programming based on Lemmas 6 and 7. $\square$

Given $\mathcal{D}$ and $\mathcal{E}$, the following lemmas imply that $D_k[i,j]$ and $E_k[i,j]$ can be retrieved efficiently after a preprocessing.

Lemma 12. *Given $\mathcal{D}$, we can do an $O(mn)$ time preprocessing so that, for all $1 \leq k \leq n$ and $1 \leq i \leq j \leq m$, $D_k[i,j]$ can be retrieved in $O(1)$ time.*

Proof. For $j - i \leq k + \lfloor k/\beta \rfloor$, the values $D_k[i,j]$'s are available in $\mathcal{D}$ and they can be retrieved in $O(1)$ time.

For $j - i > k + \lfloor k/\beta \rfloor$, by Lemma 8(2), $D_k[i,j] = \max_{i \leq i' \leq j - (k + \lfloor k/\beta \rfloor)} D_k[i', i' + (k + \lfloor k/\beta \rfloor)]$. This is a range maximum query over an array of m numbers. By Lemma 7 of [2], after an $O(m)$ time preprocessing, we obtain an auxiliary data-structure so that $D_k[i,j]$ can be retrieved in $O(1)$ time. We need to do such preprocessing for $k = 1, 2, \ldots, n$. Thus, the total preprocessing time is $O(mn)$.

Lemma 13. *Given $\mathcal{E}$, we can do an $O(mn)$ time preprocessing so that, for all $1 \le k \le n$ and $1 \le i \le j \le m$, $E_k[i,j]$ can be retrieved in $O(1)$ time.*

Proof. Similar to Lemma 12.

We conclude this section with the following theorem.

Theorem 3. *The All-Substrings Suffix/Semi-global/Local Alignment Problems can be solved in $O(mn^2)$ time and $O(mn)$ space.*

Proof. The time complexity follows directly from Lemmas 9, 12, and 13. The space complexity follows from the fact that only $O(mn)$ space is needed if we require only the matrices C_n, D_n and E_n.

5   Conclusion

Given two strings $A[1..n]$ and $B[1..m]$, with $n \ll m$, this paper consider the problems of computing the global, local, semi-global, and suffix alignments between A and all m^2 substrings of B. This paper proposes an $O(mn^2)$-time algorithm to solve these problems. Since $n \ll m$, this improves the previous best algorithms which take $O(m^2 n)$ time. The improvement stems from a compact representation of all alignment scores.

One future work is to include the affine or the convex gap penalty into the alignment score scheme. Such enhancement is useful and important to biological applications.

References

1. A. V. Aho and T. G. Peterson. A Minimum Distance Error-Correcting Parser for Context-Free Languages. *SIAM Journal on Computing*, 1(4):305–312, 1972.
2. M. A. Bender and M. Farach-Colton. The LCA Problem Revisited. In *Latin American Symposium on Theoretical Informatics*, pages 88–94, 2000.
3. M. W. Du and S. C. Chang. A Model and a Fast Algorithm for Multiple Errors Spelling Correction. *Acta Informatica*, 29(3):281–302, 1992.
4. D. Gusfield. *Algorithms on Strings, Trees and Sequences: Computer Science and Computational Biology*. Press Syndicate of the University of Cambridge, 1997.
5. J. B. Kruskal. An Overview of Sequences Comparison. In D. Sankoff and J. B. Kruskal, editors, *Time Warps, String Edits and Macromolecules: the Theory and Practice of Sequence Comparison*, pages 1–44. Addison-Wesley, 1983.
6. S. Y. Lu and K. S. Fu. Error-Correcting Tree Automata for Syntactic Pattern Recognition. *IEEE Transactions on Computers*, C-27:1040–1053, 1978.
7. S. B. Needleman and C. D. Wunsch. A General Method Applicable to the Search for Similiarities in the Amino Acid Sequences of Two Proteins. *Journal of Molecular Biology*, 48:443–453, 1970.
8. J. Seitubal and J. Meidanis. *Introduction to Computational Biology*. PWS Publishing Company, 1997.
9. T. F. Smith and M. S. Waterman. Comparison of Biosequences. *Advances in Applied Mathematics*, 2:482–489, 1981.

The Specker-Blatter Theorem Revisited

E. Fischer* and J.A. Makowsky**

Faculty of Computer Science
Technion - Israel Institute of Technology
Haifa, Israel

Abstract. In this paper we study the generating function of classes of graphs and hypergraphs modulo a fixed natural number m. For a class of labeled graphs C we denote by $f_C(n)$ the number of structures of size n. For C definable in Monadic Second Order Logic $MSOL$ with unary and binary relation symbols only, E. Specker and C. Blatter showed in 1981 that for every $m \in \mathbb{N}$, $f_C(n)$ satisfies a linear recurrence relation

$$f_C(n) = \sum_{j=1}^{d_m} a_j^{(m)} f_C(n-j),$$

over $\mathbb{Z}_m$, and hence is ultimately periodic for each m.

In this paper we show how the Specker-Blatter Theorem depends on the choice of constants and relations allowed in the definition of C. Among the main results we have the following:

- For n-ary relations of degree at most d, where each element a is related to at most d other elements by any of the relations, a linear recurrence relation holds, irrespective of the arity of the relations involved.
- In all the results $MSOL$ can be replaced by $CMSOL$, Monadic Second Order Logic with (modular) Counting. This covers many new cases, for which such a recurrence relation was not known before.

1 Introduction and Main Results

Counting objects of a specified kind belongs to the oldest activities in mathematics. In particular, counting the number of (labeled or unlabeled) graphs satisfying a given property is a classic undertaking in combinatorial theory. The first deep results for counting unlabeled graphs are due to J.H. Redfield (1927) and to G. Polya (1937), but were only popularized after 1960. F. Harary, E.M. Palmer and R.C. Read unified these early results, as witnessed in the still enjoyable [HP73].

It is unfortunate that a remarkable theorem due to E. Specker on counting labeled graphs (and more generally, labeled binary relational structures), first

---
* Partially supported by the VPR fund – Dent Charitable Trust – non-military research fund of the Technion-Israeli Institute of Technology.
** Partially supported by a Grant of the Fund for Promotion of Research of the Technion-Israeli Institute of Technology.

T. Warnow and B. Zhu (Eds.): COCOON 2003, LNCS 2697, pp. 90–101, 2003.
© Springer-Verlag Berlin Heidelberg 2003

announced by C. Blatter and E. Specker in 1981, cf. [BS81,BS84,Spe88], has not found the attention it deserves, both for the beauty of the result and the ingenuity in its proof.

E. Specker and C. Blatter look at the function $f_\mathcal{C}(n)$ which counts the number of labeled relational structures of size n with k relations $R_1, \ldots, R_k$, which belong to a class $\mathcal{C}$. We shall call this function the *density function for* $\mathcal{C}$. It is required that $\mathcal{C}$ be definable in Monadic Second Order Logic and that the relations are all unary or binary relations. The theorem says that under these hypotheses the function $f_\mathcal{C}(n)$ satisfies a linear recurrence relation modulo m for every $m \in \mathbb{Z}$. Special cases of this theorem have been studied extensively, cf. [HP73,Ges84, Wil90] and the references therein. However, the possibility of using a formal logical classification as a means to collect many special cases seems to have mostly escaped notice in this case. In the present paper, we shall discuss both the Specker-Blatter theorem, and its variations and limits of generalizabilty. In the long survey version of the paper, [FM] we shall also give numerous examples, mostly taken from [HP73,Ges84,Wil90], which in turn provide combinatorial corollaries to the Specker-Blatter Theorem. Proving directly the linear recurrence relations over every modulus m for all the given examples would have been a nearly impossible undertaking. We should also note that counting structures up to isomorphism is a very different task, cf. [HP73]. From Proposition 8 below one can easily deduce that the Specker-Blatter Theorem does not hold in this setting.

1.1 Counting Labeled Structures

Let $\bar{R} = \{R_1, \ldots, R_\ell\}$ be a set of relation symbols where each R_i is of arity $\rho(i)$. Let $\mathcal{C}$ be a class of relational $\bar{R}$-structures. For an $\bar{R}$-structure $\mathfrak{A}$ with universe A we denote the interpretation of R_i by $R_i(A)$. We denote by $f_\mathcal{C}(n)$ the number of structures in $\mathcal{C}$ over the labeled set $A_n = \{1, \ldots, n\}$, i.e.,

$$f_\mathcal{C}(n) = |\ \{(R_1(A_n), \ldots, R_\ell(A_n)) : \langle A_n, R_1(A_n), \ldots, R_\ell(A_n)\rangle \in \mathcal{C}\}\ |\ .$$

The notion of $\bar{R}$-isomorphism is the expected one: Two structures $\mathfrak{A}, \mathfrak{B}$ are isomorphic, if there is a bijection between their respective universes which preserves relations in both directions.

> **Proviso:** When we speak of a class of structures $\mathcal{C}$, we always assume that $\mathcal{C}$ is closed under $\bar{R}$-isomorphisms. However, we count two isomorphic but differently labeled structures as two different members of $\mathcal{C}$.

1.2 Logical Formalisms

First Order Logic $FOL(\bar{R})$, Monadic Second Order Logic $MSOL(\bar{R})$, and Counting Monadic Second Order Logic $CMSOL(\bar{R})$ are defined as usual, cf. [EF95]. A class of $\bar{R}$-structures $\mathcal{C}$ is is called $FOL(\bar{R})$-*definable* if there exists an $FOL(\bar{R})$ formula ϕ with no free (non-quantified) variables such that for every $\mathfrak{A}$ we have

$\mathfrak{A} \in \mathcal{C}$ if and only if $\mathfrak{A} \models \phi$. Definability for $MSOL(\bar{R})$ and $CMSOL(\bar{R})$ is defined analoguously.

We shall also look at two variations[1] of $CMSOL(\bar{R})$, and analogously for FOL and $MSOL$. The first variation is denoted by $CMSOL_{lab}(\bar{R})$, where the set of relation symbols is extended by an infinite set of constant symbols $c_i, i \in \mathbb{N}$. In a labeled structure over $\{1, \ldots, n\}$ the constant $c_i, i \leq n$ is interpreted as i. If $\phi \in MSOL_{lab}(\bar{R})$ and c_k is the constant occurring in ϕ with largest index, then the universe of a model of ϕ has to contain the set $\{1, \ldots, k\}$.

The second variation is denoted by $CMSOL_{ord}(\bar{R})$, where the set of relation symbols is augmented by a binary relation symbol $R_<$ which is interpreted on $\{1, \ldots, n\}$ as the natural order $1 < 2 < \cdots < n$.

Examples 1. *Let $\bar{R}$ consist of one binary relation symbol R.*

(i) $\mathcal{C} = ORD$, *the class of all linear orders, satisfies $f_{ORD}(n) = n!$. ORD is $FOL(R)$-definable.*

(ii)   *In FOL_{lab} we can look at the above property and additionally require by a formula ϕ_k that the elements $1, \ldots, k \in [n]$ indeed occupy the first k positions of the order defined by R, preserving their natural order. It is easily seen that $f_{ORD \wedge \phi_k}(n) = (n-k)!$. In FOL_{ord} we can express even more stringent compatibilities of the order with the natural order of $\{1, \ldots, n\}$.*

(iii)   *For $\mathcal{C} = GRAPHS$, the class of simple graphs (without loops or multiple edges), $f_{GRAPHS}(n) = 2^{\binom{n}{2}}$. GRAPHS is $FOL(R)$-definable.*

(iv)   *The class REG_r of simple regular graphs where every vertex has degree r is FOL-definable (for any fixed r). Counting the number of labeled regular graphs is treated completely in [HP73, Chapter 7]. For cubic graphs, the function is explicitly given in [HP73, page 175] as $f_{\mathcal{R}_3}(2n+1) = 0$ and*

$$f_{\mathcal{R}_3}(2n) = \frac{(2n)!}{6^n} \sum_{j,k} \frac{(-1)^j (6k-2j)! 6^j}{(3k-j)!(2k-j)!(n-k)!} 48^k \sum_i \frac{(-1)^i j!}{(j-2i)! i!}$$

which is ultimately 0 for every modulus m.

(v)   *The class $CONN$ of all connected graphs is not $FOL(R)$-definable, but it is $MSOL(R)$-definable using a universal quantifier over set variables. For $CONN$ [HP73, page 7] gives the following recurrence:*

$$f_{CONN}(n) = 2^{\binom{n}{2}} - \frac{1}{n} \sum_{k=1}^{n-1} k \binom{n}{k} 2^{\binom{n-k}{2}} f_{CONN}(k).$$

(vi)   *Counting labeled connected graphs is treated in [HP73, Chapter 1] and in [Wil90, Chapter 3]. But our Theorem 5 will give directly, that for every m this function is ultimately 0 modulo m.*

---

[1]   In [Cou90] another version, $MSOL_2$ is considered, where one allows also quantification over sets of edges. The Specker-Blatter Theorem does not hold in this case, as the class $CBIPEQ$ of complete bipartite graphs $K_{n,n}$ with both parts of equal size is definable in $MSOL_2$ and $f_{CBIPEQ}(2n) = \frac{1}{2}\binom{2n}{n}$.

(vii) Let $\mathcal{C} = BIPEQ$ be the class of simple bipartite graphs with m elements on each side (hence $n = 2m$). $BIPEQ$ is not $CMSOL(R)$-definable. However, the class BIP of bipartite graphs with unspecified number of vertices on each side is $MSOL$-definable. Again this is treated in [HP73, Chapter 1].

(viii) Let $\mathcal{C} = EVENDEG$ be the class of simple graphs where each vertex has an even degree. $EVENDEG$ is not $MSOL$-definable, but it is $CMSOL$-definable. $f_{EVENDEG}(n) = 2^{\binom{n-1}{2}}$, cf. [HP73, page 11].

Let $\mathcal{C} = EULER$ be the class of simple connected graphs in $EVENDEG$. $EULER$ is not $MSOL$-definable, but it is $CMSOL$-definable. In [HP73, page 7] a recurrence formula for the number of labeled eulerian graphs is given.

(ix) Let $\mathcal{C} = EQCLIQUE$ be the class of simple graphs which consist of two disjoint cliques of the same size. Then we have $f_{EQCLIQUE}(2n) = \frac{1}{2}\binom{2n}{n}$ and $f_{EQCLIQUE}(2n+1) = 0$. $EQCLIQUE$ is not even $CMSOL(R)$-definable, but it is definable in Second Order Logic SOL, when we allow quantification also over binary relations.

We can modify $\mathcal{C} = EQCLIQUE$ by adding another binary relation symbol R_1 and expressing in $FOL(R_1)$ that R_1 is a bijection between the two cliques. We denote the resulting class of structures by $\mathcal{C} = EQCLIQUE_1$. $f_{EQCLIQUE_1}(2n) = n!\frac{1}{2}\binom{2n}{n}$ and $f_{EQCLIQUE_1}(2n+1) = 0$.

A further modification is $\mathcal{C} = EQCLIQUE_2$, which is $FOL_{ord}(R, R_1)$-definable. We require additionally that the bijection R_1 is such that the first elements (in the order $R_<$) of the cliques are matched, and if $(v_1, v_2) \in R_1$ then the $R_<$- successors $(suc(v_1), suc(v_2)) \in R_1$. This makes the matching unique (if it exists), and we have $f_{EQCLIQUE}(n) = f_{EQCLIQUE_2}(n)$.

Similarly, we can look at $EQ_mCLIQUE$, $EQ_mCLIQUE_1$ and $EQ_mCLIQUE_2$ respectively, where we require m equal size cliques instead of two. Here we also have $f_{EQ_mCLIQUE}(n) = f_{EQ_mCLIQUE_2}(n)$.

The non-definability statements are all relatively easy, using Ehrenfeucht-Fraïssé Games, cf. [EF95].

1.3 The Specker-Blatter Theorem

The following remarkable theorem was announced in [BS81], and proven in [BS84,Spe88]:

Theorem 1 (Specker and Blatter, 1981). Let $\mathcal{C}$ be definable in Monadic Second Order Logic with unary and binary relation symbols only. For every $m \in \mathbb{N}$, there are $d_m, a_j^{(m)} \in \mathbb{N}$ such that the function $f_\mathcal{C}$ satisfies the linear recurrence relation $f_\mathcal{C}(n) \equiv \sum_{j=1}^{d_m} a_j^{(m)} f_\mathcal{C}(n-j)$ (mod m), and hence is ultimately periodic modulo m.

The case of ternary relation symbols, and more generally of arity $k \geq 3$, was left open in [BS84,Spe88] and appears in the list of open problems in Finite Model

Theory, [Mak00, Problem 3.5]. Counterexamples for quaternary relations were first found by E. Fischer, cf. [Fis03].

Theorem 2 (Fischer, 2002). *For every prime p there exists a class of structures $\mathcal{C}_p$ which is definable in first order logic by a formula ϕ_{Im_p}, with one binary relation symbol E and one quaternary relation symbol R, such that $f_{\mathcal{C}_p}$ is not ultimately periodic modulo p.*

From this theorem the existence of such classes are easily deduced also for every non-prime number m (just take p to be a prime divisor of m). The proof of the theorem is based on the class $EQ_pCLIQUE$ from Example (1) above.

1.4   Improvements and Variations

The purpose of this paper is to explore variations and extensions of the Specker-Blatter Theorem.

First, we note that in the case of unary relations symbols, $MSOL_{ord}$ and $CMSOL_{ord}$ have the same expressive power, [Cou90], and $MSOL_{ord}$-sentences define exactly the regular languages. Schützenberger's Theorem characterizes regular languages in terms of the properties of the power series of their generating function. The property in question is $\mathbb{N}$-rationality, which implies rationality. For details the reader should consult [BR84] and for constructive versions [BDFR01]. Hence, the Specker-Blatter Theorem has an important precursor in formal language theory, reformulated for our purposes as:

Theorem 3 (Schützenberger).
For any $\mathcal{C}$ definable in Counting Monadic Second Order Logic with an order, $CMSOL_{ord}(\bar{R})$, where $\bar{R}$ contains only unary relations, the function $f_{\mathcal{C}}$ satisfies a linear recurrence relation $f_{\mathcal{C}}(n) = \sum_{j=1}^{d} a_j f_{\mathcal{C}}(n-j)$ over the integers $\mathbb{Z}$, and in particular satisfies the same relation for every modulus m.

Next we extend the Specker-Blatter Theorem to allow $CMSOL$, rather then $MSOL$.

Theorem 4. *For any $\mathcal{C}$ definable in Counting Monadic Second Order Logic ($CMSOL$) with unary and binary relation symbols only, the function $f_{\mathcal{C}}$ satisfies a linear recurrence relation $f_{\mathcal{C}}(n) \equiv \sum_{j=1}^{d_m} a_j^{(m)} f_{\mathcal{C}}(n-j) \pmod{m}$, for every $m \in \mathbb{N}$.*

The proof is sketched in Section 3.2.

Theorem 4 covers cases not covered by the Specker-Blatter Theorem (Theorem (1). Although $EVENDEG$ is not $MSOL$-definable, it is $CMSOL$-definable, and its function satisfies $f_{EVENDEG}(n+1) = f_{GRAPHS}(n)$. However, it seems not very obvious that the function f_{EULER} satisfies modular recurrence relations.

Finally, we study the case of relations of bounded degree.

Definition 1.

(i)   *Given a structure $\mathfrak{A} = \langle A, R_1^A, \ldots, R_k^A \rangle$, $u \in A$ is called a neighbor of $v \in A$ if there exists a relation R_i^A and some $\bar{a} \in R_i^A$ containing both u and v.*

(ii) We define the Gaifman graph $Gaif(\mathfrak{A})$ of a structure $\mathfrak{A}$ as the graph with the vertex set A and the neighbor relation defined above.

(iii) The degree of a vertex $v \in A$ in $\mathfrak{A}$ is the number of its neighbors. The degree of $\mathfrak{A}$ is defined as the maximum over the degrees of its vertices. It is the degree of its Gaifman graph $Gaif(\mathfrak{A})$.

(iv) A structure $\mathfrak{A}$ is connected if its Gaifman graph $Gaif(\mathfrak{A})$ is connected.

Theorem 5. *For any C definable in Counting Monadic Second Order Logic $CMSOL$, with all relations in all members of C being of bounded degree d, the function f_C satisfies a linear recurrence relation $f_C(n) \equiv \sum_{j=1}^{d_m} a_j^{(m)} f_C(n-j)$ (mod m), for every $m \in \mathbb{N}$. Furthermore, if all the models in C are connected, then $f_C = 0$ (mod m) for $m \in \mathbb{N}$ large enough.*

The proof is given in Section 4.

2 Variations and Counterexamples

2.1 Why Modular Recurrence?

Theorem 1 provides linear recurrence relations modulo m for every $m \in \mathbb{N}$. Theorem 3 provides a uniform linear recurrence relation over $\mathbb{Z}$.

For the following FOL-definable C, with one binary relation symbol, $f_C(n)$ does not satisfy a linear recurrence over $\mathbb{Z}$: the class of all binary relations over any finite set, for which $f_C(n) = 2^{n^2}$, and the class of all linear orders over any finite set, for which $f_C(n) = n!$.

This follows from the well known fact, cf. [LN83], that every function $f : \mathbb{Z} \rightarrow \mathbb{Z}$, which satisfies a linear recurrence relation $f(n+1) = \sum_{i=0}^{k} a_i f(n-i)$ over $\mathbb{Z}$, grows at most exponentially, i.e. there is a constant $c \in \mathbb{Z}$ such that $f(n) \leq 2^{cn}$.

2.2 Trivial Recurrence Relations

We say that a function $f(n)$ satisfies a *trivial modular recurrence* if there are functions $g(n), h(n)$ with $g(n)$ tending to infinity such that $f(n) = g(n)! \cdot h(n)$. We call this a trivial recurrence, because it is equivalent to the statement that for every $m \in \mathbb{N}$ and large enough n, $f(n) \equiv 0$ (mod m). The most obvious example is the number of labeled linear orderings, given by $f_{ord}(n) = n!$ and $g(n) = f(n)$. Clearly, also $f_{EQCLIQUE_1}(n)$ and $f_{REG_3}(n)$ satisfy trivial modular recurrences. For the class of all graphs the recurrences are non-trivial. More generally, for a set of relation symbols $\bar{R}$ with k_j many j-ary relation symbols, the set of all labeled structures on n elements is given by $f_{\bar{R}}(n) = 2^{\sum_j k_j n^j}$ which is only divisible by 2. It follows immediately that

Observation 6. *If C is a class of $\bar{R}$-structures, and $\bar{C}$ its complement, then at least one of $f_C(n)$ or $f_{\bar{C}}(n)$ does not satisfy the trivial modular recurrence relations.*

2.3 Existential Second Order Logic Is too Strong

The example $EQ_pCLIQUE$, cf. Example 1 (1) is definable in Second Order Logic with existential quantification over one binary relation, but it is not $CMSOL$ definable.

Let p be a prime number $b_p(n) = f_{EQ_pCLIQUE}(n) = f_{EQ_pCLIQUE_2}(n)$ the number of graphs with $[n]$ as a set of vertices which are disjoint unions of exactly p same-size cliques, that is, $b_p(n) = f_{EQC_p}(n)$, As an example for $p = 2$, note that $b_2(2k+1) = 0$ and $b_2(2k) = \frac{1}{2}\binom{k}{2}$ for every k.

Proposition 7. *For every n which is not a power of p, we have $b_p(n) \equiv 0$ (mod p), and for every n which is a power of p we have $b_p(n) \equiv 1$ (mod p). In particular, $b_p(n)$ is not ultimately periodic modulo p.*

The proof is given in [Fis03].

Therefore, $f_{EQ_pCLIQUE}$ is not periodic modulo p, and hence does not satisfy a linear recurrence relation modulo p.

2.4 Using the Labels

Labeled structures have additional structure which can not be exploited in defining classes of models in $CMSOL(\bar{R})$. The additional structure consists of the labels. We can import them into our language as additional constants (with fixed interpretation) as in $CMSOL_{lab}(\bar{R})$ or, assuming the labels are linearly ordered, as a linear order with a fixed interpretation, as in $CMSOL_{ord}(\bar{R})$. Theorem 3 states that, when we restrict $\bar{R}$ to unary predicates, adding the linear order still gives us even a uniform recurrence relation. There are $\phi \in FOL_{ord}(R)$ with binary relation symbols only, such that even the non-uniform linear recurrences over $\mathbb{Z}_p$ do not hold. Here we use $EQ_pCLIQUE_2$ from Example 1, with Proposition 7.

Proposition 8. $EQ_pCLIQUE_2$ *is FOL_{ord}-definable, using the order. However $f_{EQ_pCLIQUE}$ is not ultimately periodic modulo p. Therefore $f_{EQ_pCLIQUE_2}$ does not satisfy a linear recurrence relation modulo p.*

In fact, it is not too hard to formulate in FOL_{ord} a property with one binary relation symbol that has the same density function as $EQ_pCLIQUE$.

On the other hand, using the labels as constants does not change the situation, Theorem 4 also holds for $CMSOL_{lab}$. This is proven using standard reduction techniques, and the proof is omitted.

Proposition 9. *For $\phi \in CMSOL_{lab}(\bar{R})$ (resp. $MSOL_{lab}(\bar{R})$, $FOL_{lab}(\bar{R})$), where the arities of the relation symbols in $\bar{R}$ are bounded by r and there are k labels used in ϕ, there exists $\psi \in MSOL(\bar{S})$ (resp. $MSOL(\bar{S})$, $FOL(\bar{S})$) for suitable $\bar{S}$ with the arities of $\bar{S}$ bounded by r such that $f_\phi(n) = f_\psi(n-k)$*

We finally note that in the presence of a fixed order, the modular counting quantifiers are definable in $MSOL_{ord}$. They are, however, not definable in FOL_{ord}. This was already observed in [Cou90].

3   *DU*-Index and Specker Index

Specker's proof of Theorem 1 is based on the analysis of an equivalence relation induced by a class of structures $\mathcal{C}$. It is reminiscent of the Myhill-Nerode congruence relation for words, cf. [HU80], but generalized to graph grammars, and to general structures. Note however, that the Myhill-Nerode congruence is, strictly speaking, not a special case of the Specker equivalence. What one gets is the syntactic congruence relation for formal languages.

3.1   Substitution of Structures

A pointed $\bar{R}$-structure is a pair $(\mathfrak{A}, a)$, with $\mathfrak{A}$ an $\bar{R}$-structure and a an element of the universe A of $\mathfrak{A}$. In $(\mathfrak{A}, a)$, we speak of the structure $\mathfrak{A}$ and the *context* a. The terminology is borrowed from the terminology used in dealing with tree automata, cf. [GS97].

Definition 2. *Given two pointed structures $(\mathfrak{A}, a)$ and $(\mathfrak{B}, b)$ we form a new pointed structure $(\mathfrak{C}, c) = Subst((\mathfrak{A}, a), (\mathfrak{B}, b))$ defined as follows:*

- *The universe of $\mathfrak{C}$ is $A \cup B - \{a\}$.*
- *The context c is given by b, i.e., $c = b$.*
- *For $R \in \bar{R}$ of arity r, R^C is defined by $R^C = (R^A \cap (A - \{a\})^r) \cup R^B \cup I$ where for every relation in R^A which contains a, I contains all possibilities for replacing these occurrences of a with a member of B.*

We similarly define $Subst((\mathfrak{A}, a), \mathfrak{B})$ for a structure $\mathfrak{B}$ that is not pointed, in which case the resulting structure $\mathfrak{C}$ is also not pointed. The disjoint union of two structures $\mathfrak{A}$ and $\mathfrak{B}$ is denoted by $\mathfrak{A} \sqcup \mathfrak{B}$.

Definition 3. *Let $\mathcal{C}$ be a class of, possibly pointed, $\bar{R}$-structures. We define two equivalence relations between $\bar{R}$-structures:*

- *We say that $\mathfrak{A}_1$ and $\mathfrak{A}_2$ are $Su(\mathcal{C})$-equivalent, denoted $\mathfrak{A}_1 \sim_{Su(\mathcal{C})} \mathfrak{A}_2$, if for every pointed structure $(\mathfrak{S}, s)$ we have that $Subst((\mathfrak{S}, s), \mathfrak{A}_1) \in \mathcal{C}$ if and only if $Subst((\mathfrak{S}, s), \mathfrak{A}_2) \in \mathcal{C}$.*
- *Similarly, We say that $\mathfrak{A}_1$ and $\mathfrak{A}_2$ are $DU(\mathcal{C})$-equivalent, denoted $\mathfrak{A}_1 \sim_{DU(\mathcal{C})} \mathfrak{A}_2$, if for every structure $\mathfrak{B}$ we have that $\mathfrak{A}_1 \sqcup \mathfrak{B} \in \mathcal{C}$ iff $\mathfrak{A}_2 \sqcup \mathfrak{B} \in \mathcal{C}$.*
- *The* Specker index *(resp.* DU*-index) of $\mathcal{C}$ is the number of equivalence classes of $\sim_{Su(\mathcal{C})}$ (resp. of $\sim_{DU(\mathcal{C})}$).*

Specker's proof in [Spe88] of Theorem 1 has a purely combinatorial part:

Lemma 10 (Specker's Lemma). *Let $\mathcal{C}$ be a class of $\bar{R}$-structures of finite Specker index with all the relation symbols in $\bar{R}$ at most binary. Then $f_C(n)$ satisfies modular linear recurrence relations for every $m \in \mathbb{N}$.*

In Section 4 we shall prove an analogue of Specker's lemma (Theorem 14) for $\mathcal{C}$ of finite D-index with structures of bounded degree, which generalizes a similar statement du to I. Gessel, [Ges84].

3.2   Classes of Finite Specker Index

Clearly, if C has finite Specker index, then it has finite DU-index. Furthermore, every class of connected graphs has DU-index 2. The class $EQ_2CLIQUE$ has an infinite Specker index.

Let the class $CONN - EQ_2CLIQUE$ be the class of all graphs obtained from members of $EQ_2CLIQUE$ by connecting any two vertices from different cliques. We note that $CONN - EQ_2CLIQUE$ contains structures of arbitrary large degree. The class $CONN - EQ_2CLIQUE$ has DU-index 2, but infinite Specker index. It is an easy exercise to show the same for the class of graphs which contain a hamiltonian cycle. None of these classes with an infinite Specker index are $CMSOL$-definable. This is no accident. Specker noted that all $MSOL$-definable classes of $\bar{R}$-structures (with all relations at most binary) have a finite Specker index. We shall see that this can be extended to $CMSOL$.

Theorem 11. *If C is a class of $\bar{R}$-structures (with no restrictions on the arity) which is $CMSOL$-definable, then C has a finite Specker index.*

The proof is given in [FM]. It uses a form of the Feferman-Vaught Theorem for $CMSOL$ due to Courcelle, [Cou90], see also [Mak01]. Specker[2] noted that there is a continuum of classes (of graphs, of $\bar{R}$-structures) of finite Specker index which are not $CMSOL$-definable.

Without logic, the underlying principle for establishing a finite Specker index of a class C is the following:

Definition 4. *Let C be a class of graphs and $\mathcal{F}$ be a binary operation on $\bar{R}$-structures which is isomorphism invariant. We say that $\mathfrak{A}_0$ and $\mathfrak{A}_1$ are $\mathcal{F}(C)$-equivalent if for every $\mathfrak{B}$, $\mathcal{F}(\mathfrak{A}_0, \mathfrak{B}) \in C$ iff $\mathcal{F}(\mathfrak{A}_1, \mathfrak{B}) \in C$.*
C has a finite $\mathcal{F}$-index if the number of $\mathcal{F}(C)$-equivalence classes is finite.

Proposition 12. *A class of $\bar{R}$-structures C has a finite $\mathcal{F}$-index iff there are $\alpha \in \mathbb{N}$ and classes of $\bar{R}$-structures $\mathcal{K}_j^i$ ($0 \le j \le \alpha, 0 \le i \le 1$) such that $\mathcal{F}(\mathfrak{A}_0, \mathfrak{A}_1) \in C$ iff there exists j such that $\mathfrak{A}_0 \in \mathcal{K}_j^0$ and $\mathfrak{A}_1 \in \mathcal{K}_j^1$.*

Proof. If C is of finite $\mathcal{F}$-index α then we can choose for $\mathcal{K}_j^0$ the equivalence classes and for each $j \le \alpha$

$$\mathcal{K}_j^1 = \{\mathfrak{A} \in Str(\bar{R}) : \mathcal{F}(\mathfrak{A}', \mathfrak{A}) \in C \text{ for } \mathfrak{A}' \in \mathcal{K}_j^0\}$$

Conversely, if the $\mathcal{K}_j^0$ are all disjoint, the pairs $(\mathfrak{A}, \mathfrak{A}')$ with $\mathfrak{A} \in \mathcal{K}_j^0, \mathfrak{A}' \in \mathcal{K}_j^0$ are all in the same equivalence class. But without loss of generality, but possibly increasing α, we can assume that the the the $\mathcal{K}_j^0$ are all disjoint.   □

Corollary 13.

(i) *If C_0, C_1 are classes of finite $\mathcal{F}$-index, then so are all their boolean combinations.*

---

[2] Personal communication

(ii) If C is a class of $\bar{R}$-structures such that $\mathcal{F}(\mathfrak{A}, \mathfrak{B}) \in C$ iff both $\mathfrak{A}, \mathfrak{B} \in C$ then the $\mathcal{F}(C)$-index of C is at most 2.

Proof. For (i) take the coarsest common refinement of the $\mathcal{F}(C_0)$-equivalence and the $\mathcal{F}(C_1)$-equivalence relations.
(ii) is left to the reader. □

4   Structures of Bounded Degree

For an MSOL class C, denote by $f_C^{(d)}(n)$ the number of structures over $[n]$ that are in C and whose degree is at most d. In this section we prove Theorem 5 in the following form:

Theorem 14. *If C is a class of $\bar{R}$-structures which has a finite DU-index, then $f_C^{(d)}(n)$ is ultimately periodic modulo m, hence, $f_C^{(d)}(n)$ satisfies for every $m \in \mathbb{N}$ a linear recurrence relation modulo m.*
Furthermore, if all structures of C are connected, then this modular linear recurrence is trivial.

Lemma 15. *If $\mathfrak{A} \sim_{DU(C)} \mathfrak{B}$, then for every $\mathfrak{C}$ we have $\mathfrak{C} \sqcup \mathfrak{A} \sim_{DU(C)} \mathfrak{C} \sqcup \mathfrak{B}$.*

Proof. Easy, using the associativity of the disjoint union. □

To prove Theorem 14 we define orbits for permutation groups.

Definition 5. *Given a permutation group G that acts on A (and in the natural manner acts on models over the universe A), the orbit in G of a model $\mathfrak{A}$ with the universe A is the set $\mathrm{Orb}_G(\mathfrak{A}) = \{\sigma(\mathfrak{A}) : \sigma \in G\}$.*

For $A' \subset A$ we denote by $S_{A'}$ the group of all permutations for which $\sigma(u) = u$ for every $u \notin A'$. The following lemma is useful for showing linear congruences modulo m.

Lemma 16. *Given $\mathfrak{A}$, if a vertex $v \in A - A'$ has exactly d neighbors in A', then $|\mathrm{Orb}_{S_{A'}}(\mathfrak{A})|$ is divisible by $\binom{|A'|}{d}$.*

Proof. Let N be the set of all neighbors of v which are in A', and let $G \subset S_{A'}$ be the subgroup $\{\sigma_1\sigma_2 : \sigma_1 \in S_N \wedge \sigma_2 \in S_{A'-N}\}$; in other words, G is the subgroup of the permutations in $S_{A'}$ that in addition send all members of N to members of N. It is not hard to see that $|\mathrm{Orb}_{S_{A'}}(\mathfrak{A})| = \binom{|A'|}{|N|}|\mathrm{Orb}_G(\mathfrak{A})|$. □

The following simple observation is used to enable us to require in advance that all structure in C have a degree bounded by d.

Observation 17. *We denote by C_d the class of all members of C that in addition have bounded degree d. If C has a finite DU-index then so does C_d.* □

In the following we fix m and d. Instead of $\mathcal{C}$ we look at $\mathcal{C}_d$, which by Observation 17 also has a finite DU-index. We now note that there is only one equivalence class containing any structures whose maximum degree is larger than d, which is the class $\mathcal{N}_{\mathcal{C}}^{(d)} = \{\mathfrak{A} : \forall_{\mathfrak{B}} (\mathfrak{B} \sqcup \mathfrak{A}) \not\models \mathcal{C}_d)\}$ In order to show that $f_{\mathcal{C}}^{(d)}(n)$ is ultimately periodic modulo m, we show a linear recurrence relation modulo m on the vector function $(f_{\mathcal{E}}(n))_{\mathcal{E}}$ where $\mathcal{E}$ ranges over all other equivalence classes with respect to $\mathcal{C}_d$.

Let $C = md!$. We note that for every $t \in \mathbb{N}$ and $0 < d' \leq d$, m divides $\binom{tC}{d'}$. This with Lemma 16 allows us to prove the following.

Lemma 18. *Let $\mathcal{D} \neq \mathcal{N}_\phi$ be an equivalence class for ϕ, that includes the requirement of the maximum degree not being larger than d. Then*

$$f_{\mathcal{D}}(n) \equiv \sum_{\mathcal{E}} a_{\mathcal{D}, \mathcal{E}, m, (n \bmod C)} f_{\mathcal{E}}(C \lfloor \frac{n-1}{C} \rfloor) \quad (\bmod\ m),$$

for some fixed appropriate $a_{\mathcal{D}, \mathcal{E}, m, (n \bmod C)}$.

Proof. Let $t = \lfloor \frac{n-1}{C} \rfloor$. We look at the set of structures in $\mathcal{D}$ with the universe $[n]$, and look at their orbits with respect to $S_{[tC]}$. If a model $\mathfrak{A}$ has a vertex $v \in [n] - [tC]$ with neighbors in $[tC]$, let us denote the number of its neighbors by d'. Clearly $0 < d' \leq d$, and by Lemma 16 the size of $\mathrm{Orb}_{S_{[tC]}}(\mathfrak{A})$ is divisible by $\binom{tC}{d'}$, and therefore it is divisible by m. Therefore, $f_{\mathcal{D}}(n)$ is equivalent modulo m to the number of structures in $\mathcal{D}$ with the universe $[n]$ that in addition have no vertices in $[n] - [tC]$ with neighbors in $[tC]$.

We now note that any such structure can be uniquely written as $\mathfrak{B} \sqcup \mathfrak{C}$ where $\mathfrak{B}$ is any structure with the universe $[n - tC]$, and $\mathfrak{C}$ is any structure over the universe $[tC]$. We also note using Lemma 15 that the question as to whether $\mathfrak{A}$ is in $\mathcal{D}$ depends only on the equivalence class of $\mathfrak{C}$ and on $\mathfrak{B}$ (whose universe size is bounded by the constant C). By summing over all possible $\mathfrak{B}$ we get the required linear recurrence relation (cases where $\mathfrak{C} \in \mathcal{N}_{\mathcal{C}}^{(d)}$ do not enter this sum because that would necessarily imply $\mathfrak{A} \in \mathcal{N}_{\mathcal{C}}^{(d)} \neq \mathcal{D}$). $\qquad\square$

Proof (of Theorem 14:). We use Lemma 18: Since there is only a finite number of possible values modulo m to the finite dimensional vector $(f_{\mathcal{E}}(n))_{\mathcal{E}}$, the linear recurrence relation in Lemma 18 implies ultimate periodicity for n's which are multiples of C. From this the ultimate periodicity for other values of n follows, since the value of $(f_{\mathcal{E}}(n))_{\mathcal{E}}$ for an n which is not a multiple of C is linearly related modulo m to the value at the nearest multiple of C.

Finally, if all structures are connected we use Lemma 16. Given $\mathfrak{A}$, connectedness implies that there exists a vertex $v \in A'$ that has neighbors in $A - A'$. Denoting the number of such neighbors by d_v, we note that $|\mathrm{Orb}_{S'_A}(\mathfrak{A})|$ is divisible by $\binom{|A'|}{d_v}$, and since $1 \leq d_v \leq d$ (using $|A'| = tC$) it is also divisible by m. This makes the total number of models divisible by m (remember that the set of all models with $A = [n]$ is a disjoint union of such orbits), so $f_{\mathcal{C}}^{(d)}(n)$ ultimately vanishes modulo m. $\qquad\square$

Acknowledgment. We are grateful to E. Specker, for his encouragement and interest in our work, and for his various suggestions and clarifications.

References

[BDFR01] E. Barcucci, A. Del Lungo, A Forsini, and S Rinaldi. A technology for reverse-engineering a combinatorial problem from a rational generating function. *Advances in Applied Mathematics*, 26:129–153, 2001.

[BR84] J. Berstel and C. Reutenauer. *Rational Series and their languages*, volume 12 of *EATCS Monographs on Theoretical Computer Science*. Springer, 1984.

[BS81] C. Blatter and E. Specker. Le nombre de structures finies d'une théorie à charactère fin. *Sciences Mathématiques, Fonds Nationale de la recherche Scientifique, Bruxelles*, pages 41–44, 1981.

[BS84] C. Blatter and E. Specker. Recurrence relations for the number of labeled structures on a finite set. In E. Börger, G. Hasenjaeger, and D. Rödding, editors, *In Logic and Machines: Decision Problems and Complexity*, volume 171 of *Lecture Notes in Computer Science*, pages 43–61. Springer, 1984.

[Cou90] B. Courcelle. The monadic second–order theory of graphs I: Recognizable sets of finite graphs. *Information and Computation*, 85:12–75, 1990.

[EF95] H.D. Ebbinghaus and J. Flum. *Finite Model Theory*. Perspectives in Mathematical Logic. Springer, 1995.

[Fis03] E. Fischer. The Specker-Blatter theorem does not hold for quaternary relations. *Journal of Combinatorial Theory, Series A*, 2003. in press.

[FM] E. Fischer and J.A. Makowsky. The Specker-Blatter theorem revisited. in preparation.

[Ges84] I. Gessel. Combinatorial proofs of congruences. In D.M. Jackson and S.A. Vanstone, editors, *Enumeration and design*, pages 157–197. Academic Press, 1984.

[GS97] F. Gécseg and M. Steinby. Tree languages. In G. Rozenberg and A. Salomaa, editors, *Handbook of formal languages, Vol. 3 : Beyond words*, pages 1–68. Springer Verlag, Berlin, 1997.

[HP73] F. Harary and E. Palmer. *Graphical Enumeration*. Academic Press, 1973.

[HU80] J. E. Hopcroft and J. D. Ullman. *Introduction to Automata Theory, Languages and Computation*. Addison-Wesley Series in Computer Science. Addison-Wesley, 1980.

[LN83] R. Lidl and H. Niederreiter. *Finite Fields*, volume 20 of *Encyclopedia of Mathematics and its Applications*. Cambridge University Press, 1983.

[Mak00] J.A. Makowsky. Specker's problem. In E. Grädel and C. Hirsch, editors, *Problems in Finite Model Theory*. THE FMT Homepage, 2000. Last version: June 2000, http://www-mgi.informatik.rwth-aachen.de/FMT/problems.ps.

[Mak01] J.A. Makowsky. Algorithmic uses of the Feferman-Vaught theorem. Lecture delivered at the Tarski Centenary Conference, Warsaw, May 2001, paper submitted to APAL in January 2003, special issue of the conference, 2001.

[Spe88] E. Specker. Application of logic and combinatorics to enumeration problems. In E. Börger, editor, *Trends in Theoretical Computer Science*, pages 141–169. Computer Science Press, 1988. Reprinted in: Ernst Specker, Selecta, Birkhäuser 1990, pp. 324–350.

[Wil90] H.S. Wilf. *generatingfunctionology*. Academic Press, 1990.

On the Divergence Bounded Computable Real Numbers

Xizhong Zheng

Theoretische Informatik
Brandenburgische Technische Universität Cottbus
D-03044 Cottbus, Germany
`zheng@informatik.tu-cottbus.de`

Abstract. For any function $h : \mathbb{N} \to \mathbb{N}$, we call a real number x *h-bounded computable* (*h*-bc for short) if there is a computable sequence (x_s) of rational numbers which converges to x such that, for any $n \in \mathbb{N}$, there are at most $h(n)$ pairs of non-overlapped indices (i, j) with $|x_i - x_j| \geq 2^{-n}$. In this paper we investigate *h*-bc real numbers for various functions h. We will show a simple sufficient condition for class of functions such that the corresponding *h*-bc real numbers form a field. Then we prove a hierarchy theorem for *h*-bc real numbers. Besides we compare the semi-computability and weak computability with the *h*-bounded computability for special functions h.

1 Introduction

Classically, in order to discuss the effectiveness of a real number x, we consider a computable sequence (x_s) of rational numbers which converges to x. In the optimal situation, the computable sequence (x_s) converges to x *effectively* in the sense that $|x - x_s| \leq 2^{-s}$ for all $s \in \mathbb{N}$. In this case, the real number x can be effectively approximated with an effective error estimation. According to Alan Turing [13], such kind of real numbers are called *computable*. We denote by **EC** the class of all computable real numbers. As shown by Raphael M. Robinson [8], x is computable iff its Dedekind cut $L_x := \{r \in \mathbb{Q} : r < x\}$ is a computable set and iff its binary expansion[1] $x_A := \sum_{i \in A} 2^{-i}$ is computable (i.e., A is a computable set). Of course, not every real number is computable, because there are only countably many computable sequences of rational numbers and hence there are only countably many computable real numbers while the set of real numbers is uncountable. But as shown by Ernst Specker [12], there are also non-computable real numbers which are computably approximable. Here a real number is called *computably approximable* if there is a computable sequence of rational numbers which converges to it. The class of all computably approximable real numbers is denoted by **CA**. Actually, Specker gives an example of computable increasing

---

[1] In this case we consider only the real numbers from the unit interval $[0; 1]$. For other real numbers y, there are an $n \in \mathbb{N}$ and an $x \in [0; 1]$ such that $y = x \pm n$. x and y have obviously the same effectiveness in any reasonable sense.

T. Warnow and B. Zhu (Eds.): COCOON 2003, LNCS 2697, pp. 102–111, 2003.
© Springer-Verlag Berlin Heidelberg 2003

sequence (x_s) defined by $x_s := x_{A_s}$, where (A_s) is an effective enumeration of a non-computable but computably enumerable set $A \subseteq \mathbb{N}$. The limit of an increasing computable sequence of rational numbers is called *left computable* (or *computably enumerable, c.e.,* for short, see [2,4]) and **LC** denotes the class of all left computable real numbers. Thus, we have **EC** $\subsetneq$ **LC**. Similarly, the limit of a decreasing computable sequence of rational numbers is called *right computable*. Left and right computable real numbers are called *semi-computable*. The classes of right and semi-computable real numbers are denoted by **RC** and **SC**, respectively. The arithmetical closure of **LC** is denoted by **WC**, the class of *weakly computable* real numbers. It is shown by Ambos-Spies, Weihrauch and Zheng [1], that x is weakly computable iff there is a computable sequence (x_s) of rational numbers which converges to x and $\sum_{s \in \mathbb{N}} |x_x - x_{s+1}| \leq c$ for some constant c.

Non-computable real numbers can be classified further by, say, Turing reduction by means of binary expansion (see for example Dunlop and Pour-El [5] and Zheng [14]). Namely, $x_A \leq_T x_B$ iff A is Turing reducible to B (denoted by $A \leq_T B$) for any $A, B \subseteq \mathbb{N}$. In recursion theory, the Turing degree $\deg(A)$ of a set A is defined as the class of all subsets of $\mathbb{N}$ which are Turing equivalent to A. For real number x_A, we can define its Turing degree simply by $\deg_T(x_A) := \deg_T(A)$. However, the classification of real numbers by Turing degrees is very coarse and is not related to the analytical property of real numbers very well. For example, Zheng [14] has shown that there are real numbers x, y of c.e. Turing degrees such that their difference $x - y$ does not have even an ω-c.e. Turing degree. Here a Turing degree is ω-c.e. if it contains an ω-c.e. set which is the limit of a computable sequence (A_s) of finite sets such that $|\{s : n \in (A_s \setminus A_{s+1}) \cup (A_{s+1} \setminus A_s)\}| \leq f(n)$ for all n and some computable function f.

A much finer classification of non-computable real numbers is introduced by so-called "Solovay reduction" [11] which can be applied to the class **LC**. Here, for any c.e. real numbers x, y, we say that x is *Solovay reducible* to y (denoted by $x \leq_S y$) if there are a constant c and a partial computable function $f :\subseteq \mathbb{Q} \to \mathbb{Q}$ such that $(\forall r \in \mathbb{Q})(r < y \implies c \cdot (y - r) > x - f(r))$. Very interestingly, the Solovay reduction gives a natural description of the c.e. random real numbers. Namely, a real number x is c.e. random iff it is Solovay complete, i.e., $y \leq_S x$ for any c.e. real number y (see [2] for the details about this result).

Essentially, Solovay reduction compares the convergence speed of the (increasing) approximations to different c.e. real numbers. Based on the approximation speed, Calude and Hertling [3] discuss the c-monotonic computability of real numbers which is extended further to the *h-monotonic computability* of real numbers by Rettinger, Zheng, Gengler and von Braunmühl [7]. For any function $h : \mathbb{N} \to \mathbb{Q}$, a real number x is called *h-monotonic computable* (*h-mc*, for short) if there is a computable sequence (x_s) of rational numbers which converges to x h-monotonically in the sense that, $h(n)|x - x_n| \geq |x - x_m|$ for any $n < m$. Obviously, if $h(n) \leq c < 1$, then h-mc reals are computable. For the constant function $h \equiv c \geq 1$, a dense hierarchy theorem is shown in [6]. Unfortunately, the classes of real numbers defined by h-monotonic computability do not have good analytic property too. For example, even the class of ω-monotonic computable real numbers does not closed under the addition and subtraction, here a real number is ω-mc if it is h-mc for a computable function h.

The convergence speed of an approximation (x_s) to x can also be described by counting jumps of certain length. In [15], a real number is called h-*Cauchy computable* (h-cec, for short) if there is a computable sequence (x_s) of rational numbers which converges to x such that, for any $n \in \mathbb{N}$, there are at most $h(n)$ pairs of indices (i, j) with $n \leq i < j$ and $2^{-n} \leq |x_i - x_j| < 2^{-n+1}$. Denote by h-c$\mathbf{EC}$ the class of all h-cec real numbers. Then, we have obviously that $\mathbf{EC} = 0$-c$\mathbf{EC}$. Furthermore, a hierarchy theorem of [15] shows that g-c$\mathbf{EC} \nsubseteq f$-c$\mathbf{EC}$ for any computable functions f, g such that $(\exists^{\infty} n)(f(n) < g(n))$. Intuitively, if $f(n) < g(n)$ for all $n \in \mathbb{N}$, then an f-cec real number is easier to be approximated than a g-cec number. Thus, h-Cauchy computability introduces a series of classes of non-computable real numbers which have different levels of (non)computability.

In this paper, we explore another approach to describe the approximation speed. For any sequence (x_s) which converges to x, if the number of non-overlapped index pairs (i, j) such that $|x_i - x_j| \geq 2^{-n}$ is bounded by $h(n)$, then we say that (x_s) converges to x h-*bounded effectively*. A real number x is h-*bounded computable* (h-bc, for short) if there is a computable sequence of rational numbers which converges to x h-bounded effectively. Comparing with the h-effective convergence, h-bounded effective convergence consider all jumps which are larger than 2^{-n} instead of only jumps between 2^{-n} and 2^{-n+1}. This tolerance introduces much better analytic properties of h-bounded computable real numbers. For example, a quite simple property about the class C of functions guarantees that the class of all C-bc real numbers is a field, where a real number is C-bc if it is h-bc for some $h \in C$. Obviously, the hierarchy theorem like the case of h-cec real numbers does not hold any more. For example, for any constant function $h \equiv c$, only rational numbers are h-bc. Nevertheless, we can show another natural version of hierarchy theorem that, there is a g-bc real number which is not f-bc, if for any constant c, there are infinitely many $n \in \mathbb{N}$ such that $f(n) + c < g(n)$. Also the weak computability of [1] can be well located in the hierarchy of h-bounded computable real numbers.

2   Divergence Bounded Computability

In this section, we give the precise definition of h-bounded computability of real numbers at first. Then we discuss the basic properties of this notion. Especially, we show a simple condition on the function class C such that corresponding h-bounded real number class is closed under the arithmetical operations.

Definition 2.1. Let $h : \mathbb{N} \to \mathbb{N}$ be a total function, x a real number and C a class of total functions.

1. A sequence (x_s) converges to x h-*bounded effectively* if, for any $n \in \mathbb{N}$, there are at most $h(n)$ non-overlapped pairs (i, j) of indices with $|x_i - x_j| \geq 2^{-n}$.
2. x is h-*bounded computable* (h-bc, for short) if there is a computable sequence (x_s) of rational numbers which converges to x h-bounded effectively.

3. x is *C-bounded computable* (*C-bc*, for short) if it is h-bc for some $h \in C$.

The classes of all h-bc and C-bc real numbers are denoted by h-**BC** and C-**BC** respectively. Especially, if C is the class of all computable total functions, then C-**BC** is denoted also by ω-**BC**. Reasonably, we consider only the h-bounded computability for the non-decreasing functions $h : \mathbb{N} \to \mathbb{N}$. The next lemma is straightforward from the definition.

Lemma 2.2. *Let x be a real number and $f, g : \mathbb{N} \to \mathbb{N}$ total functions.*

1. *x is rational iff x is f-bc and $\liminf_{n \to \infty} f(n) < \infty$;*
2. *If x is computable, then x is id-bc for the identity function $id(n) := n$.*
3. *If $f(n) \leq g(n)$ for almost all $n \in \mathbb{N}$, then f-**BC** $\subseteq g$-**BC**.*
4. *$(f + c)$-**BC** $= f$-**BC** for any constant $c \in \mathbb{N}$.*

Theorem 2.3. *Let C be a class of functions $f : \mathbb{N} \to \mathbb{N}$. If, for any $f, g \in C$ and constant $c \in \mathbb{N}$, the function h defined by $h(n) := f(n + c) + g(n + c)$ is bounded above by some function of C, then the class C-**BC** is a closed field.*

Proof. Let $f, g \in C$. If (x_s) and (y_s) are computable sequences of rational numbers which converge to x and y f- and g-bounded effectively, respectively, then by triangle inequations the computable sequences $(x_s + y_s)$ and $(x_s - y_s)$ converge to $x + y$ and $x - y$ h-bounded effectively, respectively, for the function h defined by $h_1(n) := f(n + 1) + g(n + 1)$.

Let $N \in \mathbb{N}$ such that $|x_n|, |y_n| \leq 2^N$ and $h_2(n) := f(N+n+1)+g(N+n+1)$ for any $n \in \mathbb{N}$. If $|x_i - x_j| \leq 2^{-n}$ and $|y_i - y_j| \leq 2^{-n}$, then we have

$$|x_i y_i - x_j y_j| \leq |x_i||y_i - y_j| + |y_j||x_i - x_j| \leq 2^N \cdot 2^{-n+1} = 2^{-(n-N-1)}.$$

This means that $(x_s y_s)$ converges to xy h_2-bounded effectively.

Now suppose that $y \neq 0$ and w.l.o.g. that $y_s \neq 0$ for all s. Let N be a natural number such that $|x_s|, |y_s| \leq 2^N$ and $|y_s| \geq 2^{-N}$ for all $s \in \mathbb{N}$. If $|x_i - x_j| \leq 2^{-n}$ and $|y_i - y_j| \leq 2^{-n}$, then we have

$$\left| \frac{x_i}{y_i} - \frac{x_j}{y_j} \right| = \left| \frac{x_i y_j - x_j y_i}{y_i y_j} \right| \leq \frac{|x_i||y_i - y_j| + |y_j||x_i - x_j|}{|y_i y_j|}$$
$$\leq 2^{3N} \cdot 2^{-n+1} = 2^{-(n-3N-1)}.$$

That is, the sequence (x_s/y_s) converges to (x/y) h_3-bounded effectively for the function $h_3 : \mathbb{N} \to \mathbb{N}$ defined by $h_3(n) := f(3N + n + 1) + g(3N + n + 1)$. Since the functions h_1, h_2, h_3 are bounded by some functions of C, the class C-**BC** is closed under arithmetical operations $+, -, \times$ and $\div$.

Corollary 2.4. *The classes C-**BC** are fields for any classes C of functions defined in the following:*

1. *$Lin := \{f : f(n) = c \cdot n + d \text{ for some } c, d \in \mathbb{N}\}$;*
2. *$Log^{(k)} := \{f : f(n) = c \log^{(k)}(n) + d \text{ for some } c, d \in \mathbb{N}\}$;*
3. *$Poly := \{f : f(n) = c \cdot n^d \text{ for some } c, d \in \mathbb{N}\}$;*
4. *$Exp_1 := \{f : f(n) = c \cdot 2^n \text{ for some } c \in \mathbb{N}\}$.*

3   Hierarchy Theorem

In this section we will prove a hierarchy theorem for the h-bounded computable real numbers. By definition, the inclusion $f\text{-}\mathbf{BC} \subseteq g\text{-}\mathbf{BC}$ holds obviously, if $f(n) \leq g(n)$ for almost all n. On the other hand, as shown in Lemma 2.2, it does not suffice to separate the classes $f\text{-}\mathbf{BC}$ from $g\text{-}\mathbf{BC}$ if the functions f and g have at most a constant distance. The next hierarchy theorem shows that more than a constant distance suffices for the separation in fact.

Theorem 3.1. Let $f, g : \mathbb{N} \to \mathbb{N}$ be two computable functions which satisfy the condition that $(\forall c \in \mathbb{N})(\exists^{\infty} m \in \mathbb{N})(c + f(m) < g(m))$, then there exists a g-bc real number which is not f-bc. Therefore, $g\text{-}\mathbf{BC} \not\subseteq f\text{-}\mathbf{BC}$.

Proof. We will construct a computable sequence (x_s) of rational numbers which converges g-bounded effectively to some real number x such that x satisfies for all $e \in \mathbb{N}$ the following requirements

$$R_e : \quad (\varphi_e(s)) \text{ converges } f\text{-bounded effectively to } y_e \Longrightarrow y_e \neq x,$$

where (φ_e) is an effective enumeration of partial computable functions $\varphi_e :\subseteq \mathbb{N} \to \mathbb{Q}$. To satisfy a single requirement R_e, we choose an interval I and an m such that $f(m) < g(m)$. Choose further two subintervals $I_e, J_e \subset I$ of the distant 2^{-m}. Then we can find a real number x either from I_e or J_e to avoid the limit y_e of the sequence $(\varphi_e(s))$. To satisfy all the requirements simultaneously, we use a finite injury priority construction as follows.

Stage $s = 0$: Let $m_0 := \min\{m : m \geq 3 \,\&\, f(m) < g(m)\}$, $I_0 := [2^{-m_0}; 2 \cdot 2^{-m_0}]$, $J_0 := [3 \cdot 2^{-m_0}; 4 \cdot 2^{-m_0}]$, $x_0 := 3 \cdot 2^{-(m_0+1)}$ and $t_{e,0} := -1$ for all $e \in \mathbb{N}$.

Stage $s + 1$: Given $t_{e,s}$, x_s and the rational intervals $I_0, I_1, \cdots, I_{k_s}$ and $J_0, \cdots J_{k_s}$ for some $k_s \geq 0$ such that $I_e, J_e \subsetneqq I_{e-1}$, $l(I_e) = l(J_e) = 2^{-m_e}$ and the distance between the intervals I_e and J_e is also 2^{-m_e}, for any $0 \leq e \leq k_s$. We say that a requirement R_e *requires attention* if $e \leq k_s$ and there is a natural number $t > t_{e,s}$ such that $\varphi_{e,s}(t) \in I_{e,s}$ and $\max G_{e,s}(m_e, t) \leq f(m_e)$, for the finite set $G_{e,s}(n, t) := \{m : (\exists v_0 < \cdots < v_m \leq t)(\forall i < m)(|\varphi_{e,s}(v_i) - \varphi_{e,s}(v_{i+1})| \geq 2^{-n})\}$.

Let R_e be the requirement of the minimal index which requires attention and t the corresponding natural number. Then we exchange the intervals I_e and J_e, that is, define $I_{e,s+1} := J_{e,s}$ and $J_{e,s+1} := I_{e,s}$. All intervals I_i and J_i for $i > e$ are set to be undefined. Besides, define $x_{s+1} := \mathrm{mid}(I_{e,s+1})$, $t_{e,s+1} := t$ and $k_{s+1} := e$.

Otherwise, if no requirement requires attention at this stage, then let $e := k_s$ and n_s the maximal $m_{i,t}$ which are defined so far for some $i \in \mathbb{N}$ and $t \leq s$. Denote by $j(s)$ the maximal number of the non-overlapped index pairs (i, j) such that $i < j \leq s$ and $|x_i - x_j| \geq 2^{-n_s}$. Then define

$$m_{e+1} := (\mu m)(m \geq n_s + 3 \,\&\, j(s) + f(m) < g(m)). \tag{1}$$

Choose four rational numbers a_i ($i < 4$) by $a_0 := x_s - 2^{-(m_{e+1}+1)}$ and $a_i := x_s + i \cdot 2^{-m_{e+1}}$ for $i := 1, 2, 3$. Then define $I_{e+1,s+1} := [a_0; a_1]$, $J_{e+1,s+1} := [a_2; a_3]$ and $x_{s+1} := x_s$.

We can show that, for any $e \in \mathbb{N}$, the requirement R_e requires and receives attention only finitely many times and the sequence (x_s) converges g-bounded effectively to some x which satisfies all requirements R_e. Therefore, x is g-bounded computable but not f-bounded computable.

Corollary 3.2. *If* $f, g : \mathbb{N} \to \mathbb{N}$ *are computable functions such that* $f \in o(g)$, *then* $f\text{-}\mathbf{BC} \subsetneqq g\text{-}\mathbf{BC}$.

4   Semi-Computability and Weakly Computability

This section discusses the relationship between h-bounded computability and other known computability of real numbers. Our first result shows that, the classical computability of real numbers cannot be described directly by h-bounded computability for any monotone function h.

Theorem 4.1. *Let* $h : \mathbb{N} \to \mathbb{N}$ *be an unbounded nondecreasing computable function. Then any computable real number* x *is also* h-bc *but there is an* h-bc *real number which is not computable. That is,* $\mathbf{EC} \subsetneqq h\text{-}\mathbf{BC}$.

Proof. Suppose that the computable function h is nondecreasing and unbounded. Then we can define a strictly increasing computable function $g : \mathbb{N} \to \mathbb{N}$ inductively by

$$\begin{cases} g(0) := 0 \\ g(n+1) := (\mu t)\,(t > g(n) \ \& \ h(t) > h(g(n))). \end{cases} \tag{2}$$

This implies that, for any natural numbers n, m, if $g(n) \leq m < g(n+1)$, then $n \leq h(g(n)) = h(m) < h(g(n+1))$.

If x is a computable real number, then there is a computable sequence (x_s) of rational numbers which converges to x such that $|x_t - x_s| < 2^{-(s+1)}$ for all $t \geq s$. Suppose without loss of generality that $|x_0 - x| < 1$. Define a computable sequence (y_s) by $y_s := x_{g(s)}$ for any $s \in \mathbb{N}$.

For any natural number n, we can choose an $i_0 \in \mathbb{N}$ such that $g(i_0) \leq n < g(i_0 + 1)$. Then we have $i_0 \leq hg(i_0) = h(n)$ by the definition (2). If (i, j) is a pair of indices such that $i < j$ and $|y_i - y_j| = |x_{g(i)} - x_{g(j)}| \geq 2^{-n}$, then, by the assumption on (x_s), this implies that $g(i) < n$ and hence $i < i_0$. This means that there are at most i_0 non-overlapped pairs of indices (i, j) such that $|y_i - y_j| \geq 2^{-n}$. Therefore, the sequence (y_s) converges to x h-bounded effectively and hence x is a h-bc real number.

To show the inequality, we can construct a computable sequence (x_s) of rational numbers which converges h-bounded effectively to a non-computable real number x, i.e., x satisfies, for all $e \in \mathbb{N}$, the following requirements

$$R_e : \qquad (\forall s)(\forall t \geq s)(|\varphi_e(s) - \varphi_e(t)| \leq 2^{-s}) \Longrightarrow x \neq \lim_{s \to \infty} \varphi_e(s)$$

where (φ_e) is an effective enumeration of partial computable functions $\varphi_e :\subseteq \mathbb{N} \to \mathbb{Q}$. This construction can be easily implemented by finite injury priority technique. Actually, this result can also be followed directly from a more general result that $h\text{-}\mathbf{BC} \not\subseteq \mathbf{SC}$ of Theorem 4.3.

To prove h-**BC** $\not\subseteq$ **SC**, we apply a criterion of non-semi-computability. Let $A \oplus B := \{2n : n \in A\} \cup \{2n + 1 : n \in B\}$ be the join of sets A and B.

Theorem 4.2 (Ambos-Spies, Weihrauch and Zheng [1]). *For any Turing incomparable c.e. sets $A, B \subseteq \mathbb{N}$, the real number $x_{A \oplus \overline{B}}$ is not semi-computable.*

Let $h : \mathbb{N} \to \mathbb{N}$ be a function. A set $A \subseteq \mathbb{N}$ is called h-*sparse* if, for any $n \in \mathbb{N}$, A contains at most $h(n)$ elements which are less than n, namely, $|A \upharpoonright n| \leq h(n)$. Applying a finite injury priority construction similar to the original proof of the classical Friedberg-Muchnik Theorem (cf. Soare [10], page 118) we can show that, if $h : \mathbb{N} \to \mathbb{N}$ is an unbounded and nondecreasing computable function, then there are Turing incomparable h-sparse c.e. sets $A, B \subseteq \mathbb{N}$, i.e., $A \not\leq_T B$ & $B \not\leq_T A$. Using this observation we can show that h-**BC** $\not\subseteq$ **SC** for any unbounded and nondecreasing computable h.

Theorem 4.3. *Let $h : \mathbb{N} \to \mathbb{N}$ be an unbounded nondecreasing computable function. Then there exists an h-bc real number which is not semi-computable.*

Proof. For any unbounded nondecreasing computable function h, let $A, B \subseteq \mathbb{N}$ be Turing incomparable h-sparse c.e. sets. Then $x_{A \oplus \overline{B}}$ is not semi-computable. If (A_s) and (B_s) are effective enumerations of sets A and B, respectively and $x_s := x_{A_s \oplus \overline{B}_s}$, then (x_s) is a computable sequence of rational numbers which converges to $x_{A \oplus \overline{B}}$. If $i < j$ are two indices such that $|x_i - x_j| \geq 2^{-n}$, then there is some $m \leq n$ such that either $m/2$ enters A or $(m-1)/2$ enters B between stages i and j. Because both A and B are h-sparse, there are at most $h(n)$ such non-overlapped index pairs (i, j). Therefore, $x_{A \oplus \overline{B}}$ is h-bounded computable. ∎

The Theorem 4.3 shows that the class **SC** does not contain all h-bc real numbers if h is unbounded no matter how slowly the function h increases. However, as observed by Soare [9], the set A must be 2^n-c.e. if x_A is a semi-computable real number. Here a set $A \subseteq \mathbb{N}$ is called h-c.e. for some function h means that there is a computable sequence (A_s) of finite sets such that $\lim_{s \to \mathbb{N}} A_s = A$ and, for any $n \in \mathbb{N}$, there are at most $h(n)$ stages s with $n \in A_{s+1} \setminus A_s$ or $n \in A_s \setminus A_{s+1}$. This implies immediately that **SC** $\subseteq 2^n$-**BC**.

On the other hand, the next result shows that **SC** is not contained completely in any class f-**BC** any more, if f is a computable function such that $f \in o(2^n)$.

Theorem 4.4. *Let $o_e(2^n)$ be the class of all computable functions $h : \mathbb{N} \to \mathbb{N}$ such that $h \in o(2^n)$. Then there exists a left computable real number x which is not $o_e(2^n)$-bounded computable. Thus, **SC** $\not\subseteq o_e(2^n)$-**BC**.*

Proof. We will construct an increasing computable sequence (x_s) of rational numbers which converges to some real number x and x satisfies, for all natural numbers $e = \langle i, j \rangle$, the following requirements

$$R_e : \left. \begin{array}{l} \varphi_i \text{ and } \psi_j \text{ are total functions and } \psi_j \in o(2^n) \\ (\varphi_i(s)) \text{ converges } \psi_j\text{-bounded effectively} \end{array} \right\} \Longrightarrow x \neq \lim_{s \to \infty} \varphi_i(s),$$

where (φ_e) and (ψ_e) are effective enumerations of partial computable functions $\varphi_e :\subseteq \mathbb{N} \to \mathbb{Q}$ and $\psi_e :\subseteq \mathbb{N} \to \mathbb{N}$, respectively.

To satisfy a single requirement R_e ($e = \langle i, j \rangle$), we choose a rational interval I_{e-1} of the length $2^{-m_{e-1}}$ for some natural number m_{e-1}. Then look for a witness interval $I_e \subseteq I_{e-1}$ such that each element of I_e satisfies R_e.

At the beginning, the interval I_{e-1} is divided into four subintervals J_e^t for $t < 4$ and let $I_e := J_e^1$ as the (default) candidate of witness interval of R_e. If $\psi_j \in o(2^n)$, then there exists a natural number $m_e > m_{e-1} + 2$ such that $2(\psi_j(m_e) + 2) \cdot 2^{-m_e} \leq 2^{-(m_{e-1}+2)}$. In this case, we divide the interval J_e^3 (which is of length $2^{-(m_{e-1}+2)}$) into subintervals I_e^t of the length 2^{-m_e} for $t < 2^{m_e - (m_{e-1}+2)}$ and let $I_e := I_e^1$ as the new candidate of witness interval of R_e. If the sequence $(\varphi_i(s))$ does not enter the interval I_e^1 at all, then it is a correct witness interval. Otherwise, suppose that $\varphi_i(s_0) \in I_e^1$ for some $s_0 \in \mathbb{N}$. Then we change the witness interval to be I_e^3. If $\varphi_i(s_1) \in I_e^3$ for some $s_1 > s_0$, then let $I_e := I_e^5$, and so on. This can happen at most $\psi_j(m_e)$ times if the sequence $(\varphi_i(s))$ converges ψ_j-bounded effectively.

To satisfy all the requirements R_e simultaneously, we apply the finite injury priority construction which is precisely described as follows.

Stage $s = 0$: Let $m_0 := 2$, $J_0^k := [k/4; (k+1)/4]$ for $k < 4$, $I_0 := J_0^1$ and $x_0 := 1/4$. Set the requirement R_0 into the state of "default" and all other requirements R_e for $e > 0$ into the state of "waiting".

Stage $s + 1$: Given $e_s \in \mathbb{N}$ such that, for all $e \leq e_s$, the natural number m_e, the rational intervals I_e and J_e^k for $k < 4$ (if R_e is in the state "default") or I_e^t for some t's (if R_e is in the state "waiting" of "satisfied") are defined.

A requirement R_e for $e = \langle i, j \rangle$ requires attention if $e \leq e_s$ and one of the following situations appear.

(R1) R_e is in the state of "default" and there is an $m \in \mathbb{N}$ such that

$$m > m_{e,s} + 2 \ \& \ (\psi_{j,s}(m) + 2) \cdot 2^{-m+1} \leq 2^{-m_{e,s}}. \tag{3}$$

(R2) R_e is in the state of "ready" and there is a $t \in \mathbb{N}$ such that $\varphi_{i,s}(t) \in I_e$.

If no requirement requires attention, then we define $e_{s+1} := e_s + 1$ and $m_{e_{s+1}} := m_{e_s} + 2$. Then divide the interval I_{e_s} into four subintervals $J_{e_{s+1}}^k$ for $k < 4$ and let $I_{e_{s+1}} := J_{e_{s+1}}^1$. Finally, set $R_{e_{s+1}}$ into the state of "default".

Otherwise, let R_e ($e = \langle i, j \rangle$) be the requirement of the highest priority (i.e., of the least index e) which requires attention and consider the following cases.

Case 1. The requirement R_e is in the state of "default" at the stage s. Define $m_{e,s+1}$ as the minimal natural number m which satisfies the condition (3). Then we divide the interval J_e^3 into subintervals I_e^t of length $2^{-m_{e,s+1}}$ for $t < 2^{m_{e,s+1} - m_{e,s}}$. Let $I_{e,s+1} := I_e^1$ be the new witness interval of R_e. The requirement R_e is set into the state of "ready" and all requirements $R_{e'}$ for $e' > e$ are set back into the state of "waiting".

Case 2. The requirement R_e is in the state of "ready". If $I_{e,s} = I_{e,s}^t$ for some $t \in \mathbb{N}$ and $I_{e,s}^{t+1}$ is also defined, then let $e_{s+1} := e$ and $I_{e,s+1} := I_{e,s}^{t+1}$ and set all requirements $R_{e'}$ for $e' > e$ into the state of "waiting". Otherwise, if $I_{e,s} = I_{e,s}^t$ and $I_{e,s}^{t+1}$ is not defined any more, then set simply the requirement R_e into the state of "satisfied" and go directly to the next stage.

At the end of stage $s + 1$, we define x_{s+1} as the left endpoint of the rational intervals $I_{e_{s+1}}$. Then the limit x of the non-decreasing computable sequence (x_s) satisfies all the requirements R_e hence is not $o_e(2^n)$-bounded computable.

For the class $o(2^n)$ the situation is different as shown in the next results.

Lemma 4.5. *If x is a semi-computable real number, then there is a function $h \in o(2^n)$ such that x is h-bc. Thus, $\mathbf{SC} \subseteq o(2^n)$-$\mathbf{BC}$.*

Proof. We consider only the left computable x. For right computable real number the proof is similar. Let (x_s) be a strict increasing computable sequence of rational numbers which converges to x. Define a function $g : \mathbb{N} \to \mathbb{N}$ by

$$g(n) := \left| \{ s \in \mathbb{N} : 2^{-n} \leq (x_{s+1} - x_s) < 2^{-n+1} \} \right|.$$

Then we have $\sum_{n \in \mathbb{N}} g(n) \cdot 2^{-n} \leq \sum_{s \in \mathbb{N}} |x_s - x_{s+1}| = x_0 - x$. This implies that $g \in o(2^n)$. Especially, there is an $N_0 \in \mathbb{N}$ such that $g(n) \leq 2^n$ for all $n \geq N_0$. Let $h(n) := \sum_{i=0}^{n} g(i + N_0)$. Given any constant $c > 0$, there is an $N_1 \in \mathbb{N}$ such that $g(n) \leq c/4 \cdot 2^n$ for all $n \geq N_1$. Thus, for any enough large n such that $2^n \geq 2^{N_1+2}/c$, we have $h(n) := \sum_{i=0}^{n} g(i + N_0) \leq \sum_{i=0}^{N_1} 2^i + \sum_{i=N_1}^{n} c/4 \cdot 2^i \leq 2^{N_1+1} + c/4 \cdot 2^{n+1} = 2^n \left(2^{(N_1+1)-n} + c/2 \right) \leq c \cdot 2^n$. Thus, $h \in o(2^n)$ and the sequence (x_s) converges h-bounded effectively. Hence x is an h-bc real number. ∎

By Theorem 2.3, the class $o(2^n)$-$\mathbf{BC}$ is a field which contains all semi-computable real numbers. Therefore, we have

Corollary 4.6. *If x is a weakly computable real number, then there is a function $h \in o(2^n)$ such that x is h-bounded computable. Namely, $\mathbf{WC} \subseteq o(2^n)$-$\mathbf{BC}$.*

Our next result shows that the inclusion $\mathbf{WC} \subseteq o(2^n)$-$\mathbf{BC}$ is proper.

Theorem 4.7. *There is an $o(2^n)$-bc real number which is not weakly computable. Therefore, $\mathbf{WC} \subsetneq o(2^n)$-$\mathbf{BC}$*

Proof. We construct a computable sequence (x_s) of rational numbers and a (non-computable) function $h : \mathbb{N} \to \mathbb{N}$ such that (x_s) converges h-bounded effectively to x which satisfies all the following requirements

$$R_e : \left. \begin{array}{l} \varphi_e \text{ is a total function, and} \\ \sum_{s \in \mathbb{N}} |\varphi_e(s) - \varphi_e(s+1)| \leq 1 \end{array} \right\} \implies \lim_{s \to \infty} \varphi_e(s) \neq x$$

where (φ_e) is an effective enumeration of all partial computable functions $\varphi_e :\subseteq \mathbb{N} \to \mathbb{Q}$. To satisfy a single requirement R_e, we choose two rational intervals I_e and J_e of distance 2^{-m_e} for some m_e. Then we choose the middle point of I_e as x whenever the sequence $(\varphi_e(s))$ does not enter the I_e. Otherwise, we choose the middle of J_e. This can be changed later again if the sequence $(\varphi_e(s))$ enters the interval J_e, and so on. Because of the condition $\sum_{s \in \mathbb{N}} |\varphi_e(s) - \varphi_e(s+1)| \leq 1$, we need at most 2^{m_e} changes. By a finite injury priority construction, this works for all requirements simultaneously. However, the real number x constructed in this way is only a 2^n-bounded computable real number. To guarantee the $o(2^n)$-bounded computability of x, we need several m_e's instead of just one. That is, we choose at first a natural number $m_e > e$, two rational intervals I_e and J_e and implement the above strategy, but at most $2^{m_e - e}$ times. Then we look for a new $m'_e > m_e$ and apply the same procedure up to $2^{m'_e - e}$ times, and so on. This means that, in worst case, we need 2^e different m_e's to satisfy a single requirement R_e. This technique works also in a finite injury priority construction. We omit the detail here.

References

[1] K. Ambos-Spies, K. Weihrauch, and X. Zheng. Weakly computable real numbers. *Journal of Complexity*, 16(4):676–690, 2000.

[2] C. S. Calude. A characterization of c.e. random reals. *Theoretical Computer Science*, 271:3–14, 2002.

[3] C. S. Calude and P. Hertling. Computable approximations of reals: An information-theoretic analysis. *Fundamenta Informaticae*, 33(2):105–120, 1998.

[4] R. G. Downey. Some computability-theoretic aspects of real and randomness. Preprint, September 2001.

[5] A. J. Dunlop and M. B. Pour-El. The degree of unsolvability of a real number. In J. Blanck, V. Brattka, and P. Hertling, editors, *Computability and Complexity in Analysis*, volume 2064 of *LNCS*, pages 16–29, Berlin, 2001. Springer. CCA 2000, Swansea, UK, September 2000.

[6] R. Rettinger and X. Zheng. Hierarchy of monotonically computable real numbers. In *Proceedings of MFCS'01, Mariánské Lázně, Czech Republic, August 27-31, 2001*, volume 2136 of *LNCS*, pages 633–644. Springer, 2001.

[7] R. Rettinger, X. Zheng, R. Gengler, and B. von Braunmühl. Monotonically computable real numbers. *Math. Log. Quart.*, 48(3):459–479, 2002.

[8] R. M. Robinson. Review of "Peter, R., Rekursive Funktionen". *The Journal of Symbolic Logic*, 16:280–282, 1951.

[9] R. Soare. Cohesive sets and recursively enumerable Dedekind cuts. *Pacific J. Math.*, 31:215–231, 1969.

[10] R. I. Soare. *Recursively enumerable sets and degrees. A study of computable functions and computably generated sets*. Perspectives in Mathematical Logic. Springer-Verlag, Berlin, 1987.

[11] R. M. Solovay. Draft of a paper (or a series of papers) on chaitin's work .... manuscript, IBM Thomas J. Watson Research Center, Yorktown Heights, NY, p. 215, 1975.

[12] E. Specker. Nicht konstruktiv beweisbare Sätze der Analysis. *The Journal of Symbolic Logic*, 14(3):145–158, 1949.

[13] A. M. Turing. On computable numbers, with an application to the "Entscheidungsproblem". *Proceedings of the London Mathematical Society*, 42(2):230–265, 1936.

[14] X. Zheng. On the Turing degrees of weakly computable real numbers. *Journal of Logic and Computation*, 13, 2003. (to appear).

[15] X. Zheng, R. Rettinger, and R. Gengler. Weak computability and representation of real numbers. Computer Science Reports 02/03, BTU Cottbus, 2003.

Sparse Parity-Check Matrices over Finite Fields
(Extended Abstract)

Hanno Lefmann

Fakultät für Informatik, TU Chemnitz,
D-09107 Chemnitz, Germany
lefmann@informatik.tu-chemnitz.de

Abstract. For fixed positive integers k, q, r with q a prime power and large m, we investigate matrices with m rows and a maximum number $N_q(m, k, r)$ of columns, such that each column contains at most r nonzero entries from the finite field $GF(q)$ and each k columns are linearly independent over $GF(q)$. For even k we prove the lower bounds $N_q(m, k, r) = \Omega(m^{kr/(2(k-1))})$, and $N_q(m, k, r) = \Omega(m^{(k-1)r/(2(k-2))})$ for odd $k \geq 3$. For $k = 2^i$ and $\gcd(k-1, r) = k-1$ we obtain $N_q(m, k, r) = \Theta(m^{kr/(2(k-1))})$, while for any even $k \geq 4$ and $\gcd(k-1, r) = 1$ we have $N_q(m, k, r) = \Omega(m^{kr/(2(k-1))} \cdot (\log m)^{1/(k-1)})$. For char $(GF(q)) > 2$ we prove that $N_q(m, 4, r) = \Theta(m^{\lceil 4r/3 \rceil/2})$, while for $q = 2^l$ we only have $N_q(m, 4, r) = O(m^{\lceil 4r/3 \rceil/2})$. We can find matrices, fulfilling these lower bounds, in polynomial time. Our results extend and complement earlier results from [4,14], where the case $q = 2$ was considered.

1 Introduction

For a prime power q, let $GF(q)$ be the finite field with q elements. We consider matrices over $GF(q)$ with k-wise independent columns, i.e. each k columns are linearly independent over $GF(q)$, and each column contains at most r nonzero entries from $GF(q) \setminus \{0\}$. Such matrices are called (k, r)-*matrices*. For a given number m of rows, let $N_q(m, k, r)$ denote the maximum number of columns a (k, r)-matrix can have. Matrices with k-wise independent columns are just parity-check matrices for linear codes with minimum distance at least $k + 1$, hence we investigate sizes of sparse parity-check matrices over $GF(q)$.

Here, k, r, q are always fixed positive integers and m is large. By monotonicity, we have $N_q(m, k + 1, r) \leq N_q(m, k, r)$ for $k = 2, 3, \ldots$. For $q = 2$ it was shown in [14] that $N_2(m, 2k + 1, r) \geq 1/2 \cdot N_2(m, 2k, r)$, hence asymptotically it suffices here to consider even independences. The values of $N_2(m, k, 2)$ are asymptotically equal to the maximum number of edges in a graph on m vertices without any cycles of length at most k. Although graphs without short cycles are widely used, not that much is known on the exact growth rate of $N_2(m, k, 2)$: known is $N_2(m, 4, 2) = \Theta(m^{3/2})$, see [8,9], as well as $N_2(m, 6, 2) = \Theta(m^{4/3})$ and $N_2(m, 10, 2) = \Theta(m^{6/5})$, see [3,18]. Improving results by Margulis [17] and Lubotzky, Phillips and Sarnak [15], Lazebnik, Ustimenko and Woldar [13] showed that $N_2(m, 2k, 2) = \Omega(m^{1+2/(3k-3+\varepsilon)})$ with $\varepsilon = 0$ for k odd, and $\varepsilon = 1$ for k

T. Warnow and B. Zhu (Eds.): COCOON 2003, LNCS 2697, pp. 112–121, 2003.
© Springer-Verlag Berlin Heidelberg 2003

even. Concerning upper bounds we have $N_2(m, 2k, 2) = O(m^{1+1/k})$ by the work of Bondy and Simonovits [7]. For $q = 2$ and any fixed $r \geq 1$, $k \geq 4$ even, the following bounds were given by Pudlák, Savický and this author [14]: $N_2(m, k, r) = \Omega(m^{kr/(2(k-1))})$, and, for $k = 2^i$ it is $N_2(m, k, r) = O(m^{\lceil k \cdot r/(k-1) \rceil/2})$. Thus for $k = 2^i$ and $\gcd(k - 1, r) = k - 1$ lower and upper bounds match. However, for $\gcd(k - 1, r) = 1$ the lower bound was improved by Bertram-Kretzberg, Hofmeister and this author [4] to $N_2(m, k, r) = \Omega(m^{kr/(2(k-1))} \cdot (\ln m)^{1/(k-1)})$.

Here we generalize and extend some of these earlier results to arbitrary finite fields $GF(q)$: we obtain the lower bounds $N_q(m, k, r) = \Omega(m^{kr/(2(k-1))})$ for even k, and $N_q(m, k, r) = \Omega(m^{(k-1)r/(2(k-2))})$ for odd $k \geq 3$. For $k = 2^i$ we show that $N_q(m, k, r) = \Theta(m^{kr/(2(k-1))})$ for $\gcd(k-1, r) = k-1$, while for every even $k \geq 4$ with $\gcd(k - 1, r) = 1$ we have $N_q(m, k, r) = \Omega(m^{kr/(2(k-1))} \cdot (\log m)^{1/(k-1)})$. Also, for $k = 4$ and char $(GF(q)) > 2$ we prove that $N_q(m, 4, r) = \Theta(m^{\lceil 4r/3 \rceil/2})$, while for $q = 2^l$ we can only show that $N_q(m, 4, r) = O(m^{\lceil 4r/3 \rceil/2})$. Matrices, which fulfill these lower bounds, can be found in polynomial time. In Section 4 we discuss some applications.

2 Upper Bounds

Since we only care about independencies, *here* in every matrix M all columns are pairwise distinct, in each column the first nonzero entry is 1 and M does not contain an all zeros column, hence $N_q(m, 2, r) = \sum_{i=1}^r \binom{m}{i} \cdot (q-1)^{i-1} = \Theta(m^r)$.

Lemma 1. *Let M be an $m \times n$-matrix over $GF(q)$ with at most $r \geq 1$ nonzero entries in each column. Then the matrix M contains an $m \times n'$-submatrix M' with the following properties:*

(i) $n' \geq r!/(r^r \cdot q^r) \cdot n$, and

(ii) *there is a partition $\{1, \ldots, m\} = R_1 \cup \ldots \cup R_r$ of the set of row-indices of M' and a sequence $(f_1, f_2, \ldots, f_r) \in (GF(q))^r$ such that for each column in M' for $j = 1, \ldots, r$ the following holds: if $f_j = 0$ then there are only zeros within the rows of R_j, and if $f_j \neq 0$ then there is exactly one entry f_j within the rows of R_j and the other entries in R_j are equal to zero, and*

(iii) *the columns of M' are 3-wise independent.*

Proof. Partition at random the set $\{1, \ldots, m\}$ of row-indices of the $m \times n$-matrix M into r parts $R_1, \ldots, R_r$ all of nearly equal size $\lfloor m/r \rfloor$ or $\lceil m/r \rceil$. It is easily seen that the expected number of columns of M with at most one nonzero entry in each row-set R_j, $j = 1, \ldots, r$, is at least $r!/r^r \cdot n$. Take such a subset of columns of M with corresponding partition $\{1, \ldots, m\} = R_1 \cup \ldots \cup R_r$ and call the resulting matrix M^*. For each column in M^* record for $j = 1, \ldots, r$ the occurring nonzero entry f_j within row-set R_j, and set $f_j = 0$, if all entries from R_j are zero. As there are less than q^r such sequences, there are $n' \geq r!/(r^r \cdot q^r) \cdot n$ columns with the same pattern $(f_1, \ldots, f_r)$. These columns form a submatrix M', which fulfills (i) and (ii). Property (iii) easily follows from property (ii). □

Lemma 1 can be made constructive in polynomial time by applying derandomization techniques. From Lemma 1 for fixed $r \geq 1$ and q a prime power we obtain $N_q(m, 3, r) = \Theta(m^r)$ and $N_q(m, 5, r) = \Theta(N_q(m, 4, r))$.

Theorem 1. *Let $k \geq 4$ be even, $r \geq 1$, and q a prime power. Then, for a constant $c > 0$ and $s = 0, \ldots, r - 1$, and for integers $m \geq 1$,*

$$N_q(m, k, r) \leq c \cdot \sqrt{m^s \cdot N_q(m, k/2, 2r - 2s)} + c \cdot m^s. \tag{1}$$

Proof. The proof is similar, but different, to that in [14], where $q = 2$ was considered. Let M be a (k, r)-matrix of dimension $m \times n$, $n = N_q(m, k, r)$, with entries from $GF(q)$. By Lemma 1, the matrix M contains an $m \times n'$-submatrix M', $n' \geq c^* \cdot n$ with $c^* = r!/(r^r \cdot q^r)$, which satisfies (ii), (iii) there.

Put the columns of M' into two matrices M_1 and M_2 of dimensions $m \times n_1$ and $m \times n_2$, respectively, with $n' = n_1 + n_2$. In M_1 put those columns of M' which have with another column from M' at least s nonzero entries at the same positions. In M_2 put the remaining columns, thus $n_2 \leq \sum_{i=1}^{s} \binom{m}{i}$.

Set $[m] := \{1, \ldots, m\}$, let $|c|$ denote the number of nonzero entries in any column c, and for each subset $S \in [[m]]^s$ of row-indices, let $n(S)$ be the number of columns c in M_1 with $c_s \neq 0$ for each position $s \in S$. Using $|c| \geq s$ for each column c in M_1, and then the Cauchy-Schwartz inequality we infer

$$n_1 \leq L := \sum_{S \in [[m]]^s} n(S) = \sum_{c \in M_1} \binom{|c|}{s},$$

$$\sum_{S \in [[m]]^s} \binom{n(S)}{2} \geq \frac{1}{2} \cdot \frac{L \cdot (L - \binom{m}{s})}{\binom{m}{s}}. \tag{2}$$

Form a matrix M_1^* by taking all differences $c_i - c_j$, $i < j$, of those columns of M_1, which share at least at s positions the same nonzero entries. Each column in M_1^* contains at most $2r - 2s$ nonzero entries and these columns are $k/2$-wise independent, hence M_1^* has at most $N_q(m, k/2, 2r - 2s)$ columns. The sum $\sum_{S \in [[m]]^s} \binom{n(S)}{2}$ counts every pair of distinct columns at most $\binom{r-1}{s}$ times, thus

$$\sum_{S \in [[m]]^s} \binom{n(S)}{2} \leq \binom{r-1}{s} \cdot N_q(m, k/2, 2r - 2s)$$

$$n_1 \leq L \leq \binom{m}{s} + \sqrt{2 \cdot \binom{m}{s} \cdot \binom{r-1}{s} \cdot N_q(m, k/2, 2r - 2s)} \qquad \text{(by (2))}.$$

With $n_1 + n_2 \geq c^* \cdot n$ and $n_2 \leq \sum_{i=1}^{s} \binom{m}{i}$, the upper bound (1) follows. □

Corollary 1. *Let $k, r \geq 1$ and q a prime power. For positive integers m,*

$$N_q(m, k, 2) = O\left(m^{1 + 2/2^{\lfloor \log_2 k \rfloor}}\right) \tag{3}$$

$$N_q(m, k, r) = O\left(m^{\lceil kr/(k-1) \rceil/2}\right) \qquad \text{if } k = 2^j. \tag{4}$$

Proof. Inequality (3) follows by induction on $\lfloor \log_2 k \rfloor$ from (1) for $s := 1$, and (4) follows by induction on j and (1) for $s := \lfloor \lceil kr/(k-1) \rceil /2 \rfloor$, compare [14].   □

Lemma 2. *Let $k \geq 2$ and q a prime power. For positive integers m,*
$$N_q(m, k, 2) \geq (1 - o(1)) \cdot N_2(m, k, 2) \ .$$

Proof. This follows by considering the incidence matrix of a graph without cycles of length at most k as a matrix over $GF(q)$. Details are in the full version.   □

With Lemma 2 we obtain the following (constructive) lower bounds from graphs, see [3,13,18], for every $k \geq 1$ and q a prime power: $N_q(m, 4, 2) = \Theta(m^{3/2})$, $N_q(m, 5, 2) = \Theta(m^{3/2})$, $N_q(m, 6, 2) = \Omega(m^{4/3})$, $N_q(m, 10, 2) = \Omega(m^{6/5})$, and $N_q(m, 2k, 2) = \Omega(m^{1+2/(3k-3+\varepsilon)})$ with $\varepsilon \in \{0,1\}$ and $\varepsilon = 1$ iff k is even.

The next lemma shows that asymptotically it suffices to consider $N_q(m, k, r)$ for q a prime. We omit the proof, which uses linear algebra.

Lemma 3. *Let $k \geq 2$, and $l, r \geq 1$, and p a prime. For positive integers m,*
$$N_{p^l}(m, k, r) = \Theta(N_p(m, k, r)) \ .$$

From the results for graphs [3,7,18] and by Lemma 3 we obtain the following:

Corollary 2. *Let $q = 2^l$ and $k \geq 1$. For positive integers m,*
$$N_q(m, 6, 2) = \Theta(m^{4/3}) \quad and \quad N_q(m, 10, 2) = \Theta(m^{6/5})$$
$$N_q(m, 2k, 2) = O(m^{1+1/k}) \ .$$

Next we consider $(4, r)$-matrices over $GF(q)$.

Lemma 4. *Let $GF(q)$ be a finite field with char $(GF(q)) > 2$. Let M' be an $m \times n$-matrix over $GF(q)$ with exactly r nonzero entries in each column, such that Lemma 1 (ii), (iii) are satisfied. Let $F'_1, \ldots, F'_n$ be the sets of positions of the nonzero entries in the n columns of M'. If for no four sets both $F'_g \cup F'_h = F'_i \cup F'_j$ and $F'_g \cap F'_h = F'_i \cap F'_j$ are fulfilled, then the columns of M' are 4-wise independent.*

Proof. Assume that some columns $a_1, \ldots, a_4$ of M', with corresponding sets $F'_1, \ldots, F'_4$, are linearly dependent over $GF(q)$, i.e. $\sum_{i=1}^4 \lambda_i \cdot a_i = 0$ for some $\lambda_1, \ldots, \lambda_4 \in GF(q) \setminus \{0\}$. Let $F_i := F'_i \setminus (F'_1 \cap \ldots \cap F'_4)$ for $i = 1, \ldots, 4$.

Lemma 5. *For $1 \leq i < j < k \leq 4$ it is $F_i \cap F_j \cap F_k = \emptyset$.*

Proof. This follows with Lemma 1 (ii). Details are in the full version.   □

Assume w.l.o.g. that $F_1 \cap F_2 \neq \emptyset$. By Lemma 5 we have $F_3 \not\subseteq F_1 \cup F_2$, hence $F := F_3 \setminus (F_1 \cup F_2) \neq \emptyset$. The dependence of $a_1, \ldots, a_4$ implies $F_4 \setminus (F_1 \cup F_2) = F$, thus $\lambda_3 = -\lambda_4$. If $F_3 \cap (F_1 \setminus F_2) \neq \emptyset$ and $F_4 \cap (F_1 \setminus F_2) \neq \emptyset$, then $\lambda_3 = -\lambda_1$ and $\lambda_4 = -\lambda_1$, which shows $\lambda_3 = \lambda_4 = 0$ for char $(GF(q)) > 2$, a contradiction. Hence, $F_3 \cap (F_1 \setminus F_2) = \emptyset$ or $F_4 \cap (F_1 \setminus F_2) = \emptyset$. Then w.l.o.g. $F_3 = F \cup (F_2 \setminus F_1)$ and $F_4 = F \cup (F_1 \setminus F_2)$, hence $F_1 \cup F_3 = F_2 \cup F_4$ and $F_1 \cap F_3 = F_2 \cap F_4$, which contradicts the assumption.   □

Frankl and Füredi [11] constructed a family $\mathcal{F} \subseteq [[m]]^r$ containing no four sets $F_1, \ldots, F_4 \in \mathcal{F}$ with $F_1 \cup F_2 = F_3 \cup F_4$ and $F_1 \cap F_2 = F_3 \cap F_4$, where $|\mathcal{F}| = \Theta(m^{\lceil 4r/3 \rceil / 2})$, as follows. Let $r = 3t + 1$ (For other values of $(r \bmod 3)$ the construction is similar.), and let K be any field with $m/2 \leq |K| \leq m$. For a subset $X = \{x_1, \ldots, x_g\} \subseteq K$ let $s_i(X) := \sum_{I \in [[g]]^i} \prod_{j \in I} x_j$. For $h \geq 1$ define an $h \times h$-matrix $M_h(X)$ with entries $m_{i,j} = s_{2i-j}(X)$ (where $s_l(X) = 0$ for $l < 0$ or $l > |X|$). Using an averaging argument, for suitable $c_2, c_4, \ldots, c_{2t} \in K$ the family $\mathcal{F} \subseteq [K]^r$ is defined as follows: $X = \{x_1, \ldots, x_r\} \in \mathcal{F}$ if $s_{2i}(X) = c_{2i}$ for $i = 1, \ldots, t$ and $\det(M_h(S)) \neq 0$ for every subset $S \subseteq X$ and $h = 1, \ldots, |S| - 1$. This yields a polynomial time (semi-) construction.

Theorem 2. *Let $r \geq 1$ and q a prime power. For positive integers m,*

$$N_q(m, 4, r) = \begin{cases} \Theta\left(m^{\lceil 4r/3 \rceil / 2}\right) & \text{if char } (GF(q)) > 2 \\ O\left(m^{\lceil 4r/3 \rceil / 2}\right) & \text{if char } (GF(q)) = 2. \end{cases}$$

For $q = 2$ this was shown in [14]. For $q = 2^l$ and $r \equiv 0 \bmod 3$ lower and upper bound match by Theorem 5, compare also Theorem 3.

Proof. For the upper bounds see (4). For the lower bound, let $\mathcal{F} = \{F_1, \ldots, F_n\} \subseteq [[m]]^r$ with $n = \Theta(m^{\lceil 4r/3 \rceil / 2})$, such that for no four sets $F_1, \ldots, F_4 \in \mathcal{F}$ it is $F_1 \cup F_2 = F_3 \cup F_4$ and $F_1 \cap F_2 = F_3 \cap F_4$. Such a family exists by the results in [11]. Define an $m \times n$-matrix M with columns $c_1, \ldots, c_n \in \{0, 1\}^m$: in column c_i put a 1 at position s iff $s \in F_i$, $i = 1, \ldots, n$. By Lemma 1 we obtain an $m \times n'$-submatrix M' of M such that (i) – (iii) there are satisfied. By Lemma 4, the columns of M' are 4-wise independent and the lower bound follows. $\square$

3   Lower Bounds

For proving lower bounds on $N_q(m, k, r)$, we will look for a large independent set in a suitable hypergraph. A *hypergraph* $\mathcal{G} = (V, \mathcal{E})$ has vertex set V and edge set $\mathcal{E}$ where $E \subseteq V$ for every edge $E \in \mathcal{E}$. A hypergraph $\mathcal{G} = (V, \mathcal{E})$ is *l-uniform*, if the edge set $\mathcal{E}$ contains only l-element edges, i.e. $\mathcal{E} \subseteq [V]^l$. An *independent set* in $\mathcal{G} = (V, \mathcal{E})$ is a subset $I \subseteq V$ which contains no edges from $\mathcal{E}$.

Theorem 3. *Let $k \geq 4$, $r \geq 2$, and q a prime power. For positive integers m,*

$$N_q(m, k, r) = \Omega\left(m^{\frac{kr}{2(k-1)}} \cdot (\log m)^{\frac{1}{k-1}}\right) \quad \text{for } k \text{ even, } \gcd(k-1, r) = 1 \quad (5)$$

$$N_q(m, k, r) = \Omega\left(m^{\frac{(k-1)r}{2(k-2)}} \cdot (\log m)^{\frac{1}{k-2}}\right) \quad \text{for } k \text{ odd, } \gcd(k-2, r) = 1. \quad (6)$$

For $q = 2$ this was shown in [4], thus (5), (6) hold for $q = 2^l$ by Lemma 3.

Proof. Partition the set $\{1, \ldots, m\}$ of row-indices into r subsets $R_1, \ldots, R_r$ of sizes $\lfloor m/r \rfloor$ or $\lceil m/r \rceil$, and fix a sequence $(f_1 = 1, f_2, \ldots, f_r) \in (GF(q) \setminus \{0\})^r$ of nonzero elements. Let $C_q(m, r)$ be the set of all column vectors of length m which contain within each row-set R_j exactly one nonzero entry $f_j \in GF(q) \setminus \{0\}$, $j = 1, \ldots, r$. Hence $|C_q(m, r)| = c \cdot m^r$ for some constant $c > 0$.

Form a hypergraph $\mathcal{G} = (V, \mathcal{E}_3 \cup \ldots \cup \mathcal{E}_k)$ with vertex set $V = C_q(m, r)$, where an i-element subset $\{a_1, \ldots, a_i\}$ of V, $i = 3, \ldots, k$, is an edge in this hypergraph $\mathcal{G}$, that is $\{a_1, \ldots, a_i\} \in \mathcal{E}_i$, if and only if $a_1, \ldots, a_i$ are linearly dependent over $GF(q)$ but any $(i - 1)$ columns of $a_1, \ldots, a_i$ are linearly independent.

We will prove a lower bound on the size of a maximum independent set in $\mathcal{G}$, which yields a set of k-wise independent columns. First we will bound from above the numbers $|\mathcal{E}_i|$, $i = 3, \ldots, k$, of i-element edges in $\mathcal{G}$. For a subset $E \in [C_q(m, r)]^i$ of i column vectors, the $m \times i$-matrix $M(E)$ formed by these columns has exactly $i \cdot r$ nonzero entries. If $E \in \mathcal{E}_i$, then in $M(E)$ each row-set R_j contains at most $\lfloor i/2 \rfloor$ non-zero rows, hence $M(E)$ contains at most $\lfloor i/2 \rfloor \cdot r$ nonzero rows. Thus, for some constants $c_i > 0$, $i = 3, \ldots, k$, we have

$$|\mathcal{E}_i| \le \binom{\lceil m/r \rceil}{\lfloor i/2 \rfloor}^r \cdot \binom{i \cdot \lfloor i/2 \rfloor \cdot r}{ir} \le c_i \cdot m^{\lfloor i/2 \rfloor \cdot r}. \tag{7}$$

Next we consider only edges in $\mathcal{E}_l$, where $l := k$ for k even and $l := k - 1$ for k odd, hence $l \ge 4$ is *always even*. For a subset $J \in [C_q(m, r)]^j$, $j = 2, \ldots, l - 1$, let $p(J)$ be the number of nonzero rows in the matrix $M(J)$, and let $p_1(J)$ be the number of rows in $M(J)$ with exactly one nonzero entry. Let $b(J)$ be the number of subsets $S \in [C_q(m, r)]^{l-j}$ where $J \cup S \in \mathcal{E}_l$. For $J \cup S \in \mathcal{E}_l$ in each row in $M(J)$ with exactly one nonzero entry there must be at least one such nonzero entry in the same row in $M(S)$. Let $M(J)$ have the nonzero rows $1, \ldots, p(J)$, say. Each nonzero row $s > p(J)$ in $M(S)$ has at least two nonzero entries, as the columns in $J \cup S$ are linearly dependent, hence there are at most $\lfloor ((l - j)r - p_1(J))/2 \rfloor$ such rows in $M(S)$, i.e. at most $\binom{m - p(J)}{\lfloor ((l-j)r - p_1(J))/2 \rfloor}$ choices for these, thus for a constant $c_p > 0$:

$$b(J) \le c_p \cdot m^{\lfloor \frac{(l-j)r - p_1(J)}{2} \rfloor}. \tag{8}$$

A *2-cycle* in an l-uniform hypergraph $\mathcal{G} = (V, \mathcal{E})$ is a pair $\{E, E'\}$ of distinct edges $E, E' \in \mathcal{E}$ with $|E \cap E'| \ge 2$. We will apply a result of Ajtai, Komlós, Pintz, Spencer and Szemerédi [1], originally an existence result, in the sequel extended and turned into a deterministic polynomial time algorithm in [10,5].

Theorem 4. *Let $l \ge 3$ be a fixed integer. Let $\mathcal{G} = (V, \mathcal{E})$ be an l-uniform hypergraph on $|V| = N$ vertices with average degree $t^{l-1} := l \cdot |\mathcal{E}|/|V|$. If the hypergraph $\mathcal{G} = (V, \mathcal{E})$ contains no 2-cycles, then one can find for any fixed $\delta > 0$ in $\mathcal{G}$ in time $O(N \cdot t^{l-1} + N^3/t^{3-\delta})$ an independent set of size $\Omega(N/t \cdot (\log t)^{1/(l-1)})$. The assertion also holds, if t^{l-1} is only an upper bound on the average degree.*

For $j = 2, \ldots, l - 1$ and $u = 0, \ldots, jr$, let $s_{2,j}(u; \mathcal{G}_l)$ be the number of 2-cycles $\{E, E'\}$ in $\mathcal{G}_l = (V, \mathcal{E}_l)$ with $|E \cap E'| = j$ and $p_1(E \cap E') = u$. The numbers $p_{j,u}(V)$ of subsets $J \in [C_q(m, r)]^j$ with $p_1(J) = u$ satisfy for constants $c_{j,u}^*, c_{j,u} > 0$:

$$p_{j,u}(V) \le c_{j,u}^* \cdot \binom{m}{u} \cdot \binom{m - u}{\lfloor (jr - u)/2 \rfloor} \le c_{j,u} \cdot m^{u + \lfloor \frac{jr - u}{2} \rfloor}, \tag{9}$$

since the matrix $M(J)$ has u rows with exactly one nonzero entry and the remaining $jr - u$ nonzero entries are contained in rows with at least two nonzero entries, thus by (8) and (9) for some constant $C_1 > 0$:

$$s_{2,j}(u;\mathcal{G}_l) \leq \sum_{J\in[C_q(m,r)]^j; \, p_1(J)=u} \binom{b(J)}{2} \leq C_1 \cdot m^{2\cdot\lfloor\frac{(l-j)r-u}{2}\rfloor+u+\lfloor\frac{jr-u}{2}\rfloor}. \quad (10)$$

By (7) the *average degree* t^{l-1} of the l-uniform hypergraph $\mathcal{G}_l = (V,\mathcal{E}_l)$ satisfies $t^{l-1} = (l\cdot|\mathcal{E}_l|)/|V| \leq (l\cdot c_l \cdot m^{\lfloor l/2\rfloor\cdot r})/(c\cdot m^r)$, hence $t \leq t_0 := C_2\cdot m^{(\lfloor l/2\rfloor\cdot r-r)/(l-1)}$ for a constant $C_2 > 0$. To apply Theorem 4, we choose a random subset $V^* \subseteq V$ by picking vertices at random from V, independently of each other with probability $p := t_0^{-1}\cdot m^\varepsilon$ for some constant $\varepsilon > 0$, heading for an l-uniform hypergraph without any 2-cycles. We estimate the expected values $E(\cdot)$ of some parameters of the induced random hypergraph $\mathcal{G}^* = (V^*, \mathcal{E}_3^* \cup \ldots \cup \mathcal{E}_k^*)$ with $\mathcal{E}_i^* := \mathcal{E}_i \cap [V^*]^i$, $i = 3,\ldots,k$, using (7) for some constants $c^*, c_i^* > 0$ as follows:

$$E(|V^*|) = p\cdot|C_q(m,r)| \geq c^*\cdot m^{r-\frac{\lfloor l/2\rfloor\cdot r-r}{l-1}+\varepsilon} \quad (11)$$

$$E(|\mathcal{E}_i^*|) = p^i\cdot|\mathcal{E}_i| \leq p^i\cdot c_i\cdot m^{\lfloor\frac{i}{2}\rfloor\cdot r} \leq c_i^*\cdot m^{\lfloor\frac{i}{2}\rfloor\cdot r-\frac{\lfloor l/2\rfloor\cdot r-r}{l-1}\cdot i+i\cdot\varepsilon}. \quad (12)$$

With (9), (10) we infer for $j = 2,\ldots,l-1$ and $u = 0,\ldots,j\cdot r$, where $\mathcal{G}_l^* = (V^*, \mathcal{E}_l^*)$, for some constants $C_0^*, C_1^* > 0$:

$$E(p_{j,u}(V^*)) = p^j\cdot p_{j,u}(V) \leq C_0^*\cdot m^{u+\lfloor\frac{jr-u}{2}\rfloor-\frac{\lfloor l/2\rfloor\cdot r-r}{l-1}\cdot j+j\cdot\varepsilon} \quad (13)$$

$$E(s_{2,j}(u;\mathcal{G}_l^*)) = p^{2l-j}\cdot s_{2,j}(u;\mathcal{G}_l) \leq$$
$$\leq C_1^*\cdot m^{2\cdot\lfloor\frac{(l-j)r-u}{2}\rfloor+u+\lfloor\frac{jr-u}{2}\rfloor-\frac{\lfloor l/2\rfloor\cdot r-r}{l-1}\cdot(2l-j)+(2l-j)\cdot\varepsilon}. \quad (14)$$

By (11)–(14), Markov's and Chebychev's inequality, there exists a subhypergraph $\mathcal{G}^* = (V^*, \mathcal{E}_3^* \cup \ldots \cup \mathcal{E}_k^*)$ of $\mathcal{G}$ which satisfies for $i = 2,\ldots,k$, and $j = 2,\ldots,l-1$, and $u = 0,\ldots,jr$, (using simply the same notation for the constant factors)

$$|V^*| \geq c^*\cdot m^{r-\frac{\lfloor l/2\rfloor\cdot r-r}{l-1}+\varepsilon} \quad (15)$$

$$|\mathcal{E}_i^*| \leq c_i^*\cdot m^{\lfloor\frac{i}{2}\rfloor\cdot r-\frac{\lfloor l/2\rfloor\cdot r-r}{l-1}\cdot i+i\cdot\varepsilon} \quad (16)$$

$$s_{2,j}(u;\mathcal{G}_l^*) \leq C_1^*\cdot m^{2\cdot\lfloor\frac{(l-j)r-u}{2}\rfloor+u+\lfloor\frac{jr-u}{2}\rfloor-\frac{\lfloor l/2\rfloor\cdot r-r}{l-1}\cdot(2l-j)+(2l-j)\cdot\varepsilon} \quad (17)$$

$$p_{j,u}(V^*) \leq C_0^*\cdot m^{u+\lfloor\frac{jr-u}{2}\rfloor-\frac{\lfloor l/2\rfloor\cdot r-r}{l-1}\cdot j+j\cdot\varepsilon}. \quad (18)$$

Lemma 6. *For $k \geq 4$ and $\varepsilon = r/(2k^2)$ it holds:*

$$|\mathcal{E}_i^*| = o(|V^*|) \text{ for every } i \neq l. \quad (19)$$

Proof. By (15) and (16), since l is even, we have $|E_i^*| = o(|V^*|)$, $i \neq l$, if

$$r - \lfloor i/2\rfloor\cdot r + (i-1)(l-2)r/(2(l-1)) - (i-1)\cdot\varepsilon > 0. \quad (20)$$

Inequality (20) holds, if $(l-i)r/(2(l-1)) - (i-1)\cdot\varepsilon > 0$, which is fulfilled for $i = 2,\ldots,l-1$ and $0 < \varepsilon < r/(2l^2)$. For $i > l$, which is only possible for $i = k$ odd and $l = k-1$, inequality (20) is equivalent to $(k-3)r/(2(k-2)) - (k-1)\cdot\varepsilon > 0$, which with $k \geq 4$ holds for $0 < \varepsilon \leq r/(2k^2)$, thus (19) is satisfied. $\square$

We have the following consequence:

Theorem 5. *Let $k \geq 4$, $r \geq 1$, and q a prime power. For positive integers m,*

$$N_q(m, k, r) = \begin{cases} \Omega\left(m^{\frac{kr}{2(k-1)}}\right) & \text{if } k \text{ is even,} \\ \Omega\left(m^{\frac{(k-1)r}{2(k-2)}}\right) & \text{if } k \text{ is odd.} \end{cases}$$

Thus, for $k = 2^i$ and $\gcd(k-1, r) = k-1$ lower and upper bound (4) match, i.e. $N_q(m, k, r) = \Theta(m^{kr/(2(k-1))})$. Matrices satisfying the bounds in Theorem 5 can be found in polynomial time by using the method of conditional probabilities.

Proof. For every $i \neq l$ and $\varepsilon = r/(2k^2)$ we have $|\mathcal{E}_i^*| = o(|V^*|)$ by (19), and we remove one vertex from each edge $E \in \mathcal{E}_i^*$, and we obtain a subset $V^{**} \subseteq V^*$ with $|V^{**}| \geq (c^*/2) \cdot m^{lr/(2(l-1))+\varepsilon}$, such that the subhypergraph $\mathcal{G}^{**} = (V^{**}, [V^{**}]^l \cap \mathcal{E}_l^*)$ of $\mathcal{G}^*$ is l-uniform with $|[V^{**}]^l \cap \mathcal{E}_l^*| \leq c_l^* \cdot m^{lr/(2(l-1))+l \cdot \varepsilon}$. Again we choose a random subset $V^{***} \subseteq V^{**}$ by picking vertices from V^{**} uniformly at random, independently of each other with probability $p := c_h \cdot m^{-\varepsilon}$, $c_h := (c^*/(4c_l^*))^{1/(l-1)}$. We estimate $E(|V^{***}|) - E(|[V^{***}]^l \cap \mathcal{E}_l^*|)$, hence there exists a subset $V^{***} \subseteq V^{**}$ such that

$$|V^{***}| - |[V^{***}]^l \cap \mathcal{E}_l^*| \geq (c_h \cdot c^*/4) \cdot m^{lr/(2(l-1))}.$$

We delete one vertex from every edge in $[V^{***}]^l \cap \mathcal{E}_l^*$ and we obtain an independent set I in $\mathcal{G}$ with $|I| = \Omega(m^{lr/(2(l-1))})$, and the lower bounds follow by inserting $l := k$ for k even, and $l := k-1$ for k odd.

We remark, that we could have derived the bounds in Theorem 5 already from (7), by picking vertices right away from the set V at random, independently from each other, each with probability $p := c_g \cdot t_0^{-1}$ for some constant $c_g > 0$. □

Lemma 7. *For $\varepsilon = 1/(4k^2)$, $j = 2, \ldots, l-1$, and $\gcd(l-1, r) = 1$ it is*

$$min_{u=0,\ldots,jr} \{p_{j,u}(V^*), s_{2,j}(u; \mathcal{G}_l^*)\} = o(|V^*|). \tag{21}$$

Proof. By (15) and (17), we have $s_{2,j}(u; \mathcal{G}_l^*) = o(|V^*|)$, $j = 2, \ldots, l-1$, if

$$0 > (l-1) \cdot r - jr/2 - u/2 - (l-2)(2l-j-1)r/(2(l-1)) + (2l-j-1) \cdot \varepsilon$$

which holds for $u > (l-j)r/(l-1) + 2 \cdot (2l-j-1) \cdot \varepsilon$.

For l even by (15) and (18) we have $p_{j,u}(V^*) = o(|V^*|)$, $j = 2, \ldots, l-1$, if

$$u/2 + jr/2 - r - (l-2)(j-1)r/(2(l-1)) + (j-1) \cdot \varepsilon < 0$$

which holds for $u < (l-j)r/(l-1) - 2 \cdot (j-1) \cdot \varepsilon$.

For $j = 2, \ldots, l-1$ and $\gcd(l-1, r) = 1$, the quotients $(l-j)r/(l-1)$ are never integers, in particular, they are at least $1/(l-1)$ apart from the nearest integer. For $\varepsilon := 1/(4k^2)$, the statement 'for $u > (l-j)r/(l-1) + 2 \cdot (2l-j-1) \cdot \varepsilon$ or $u < (l-j)r/(l-1) - 2 \cdot (j-1) \cdot \varepsilon$' reads as 'for $u = 0, \ldots, jr$', and (21) follows. □

Let $\varepsilon := 1/(4k^2)$ and $u_0(j) := \lfloor (l-j)r/(l-1) \rfloor$. For $i \neq l$ we delete one vertex from each edge $E \in \mathcal{E}_i^*$ and, by Lemma 6, we obtain a subset $V^{**} \subseteq V^*$ with $|V^{**}| = (1 - o(1)) \cdot |V^*|$. The resulting induced subhypergraph of $\mathcal{G}^{**}$ on V^{**} is l-uniform. For $j = 2, \ldots, l-1$ and $u > u_0(j)$ we delete one vertex from each 2-cycle $\{E, E'\} \in [\mathcal{E}_l^* \cap [V^{**}]^l]^2$ with $|E \cap E'| = j$ and $p_1(E \cap E') = u$. For $u \leq u_0(j)$ we remove one vertex from each subset $J \in [V^{**}]^j$ with $p_1(J) = u$. By Lemma 7 we obtain a subset $V^{***} \subseteq V^{**}$, which does not contain any 2-cycles and satisfies $|V^{***}| = (1 - o(1)) \cdot |V^*|$. We apply Theorem 4 to our hypergraph $\mathcal{G}^{***} = (V^{***}, [V^{***}]^l \cap \mathcal{E}_l^*)$ with average degree $t_*^{l-1} \leq c_0 \cdot (1 + o(1)) \cdot (p \cdot t_0)^{l-1}$ for some constant $c_0 > 0$, and we obtain an independent set $I \subseteq V^{***}$ of size

$$|I| = \Omega\left(|V^{***}|/(p \cdot t_0) \cdot (\log(p \cdot t_0))^{\frac{1}{l-1}}\right) = \Omega\left(m^{\frac{lr}{2(l-1)}} \cdot (\log m)^{\frac{1}{l-1}}\right),$$

which yields the desired lower bounds (5), (6) by inserting the appropriate value of l, i.e. $l := k$ for k even, and $l := k - 1$ for k odd.

With the method of conditional probabilities, the running time is essentially bounded by the number $s_{2,2}(u_0(2) + 1; \mathcal{G}_l) = O(m^{(l-3)r+r/(l-1)})$ of 2-cycles. Compared with $N^3/t_0^{3-3\delta}$ in Theorem 4, with $t_0 = \Theta(m^\varepsilon)$ (else, for $t_0 = o(m^\varepsilon)$, we can improve (5), (6)), we get the time bound $O(m^{(l-3)r+r/(l-1)})$ for $l \geq 6$.

Remark: All calculations are valid, if we pick the columns uniformly at random according to a $(2l - 2)$-wise independent distribution. For simulating this, there exist sample spaces of sizes $O(m^{r(4l-4)})$, see [12], hence with this we also obtain polynomial running time. This finishes the proof of Theorem 3.   □

4   Concluding Remarks

Some of the following possible applications have been stated already in [14] for the case $q = 2$ and can be extended to arbitrary prime powers q. For example, we can extend the length of a linear code, but we reduce its minimum distance:

Proposition 1. *Let A be an $l \times m$ parity-check matrix over $GF(q)$ of a linear code of length m with minimum distance at least $kr + 1$, and let B be a (k, r)-matrix of dimension $m \times n$ over $GF(q)$. Then the matrix-product $A \times B$ is a parity-check matrix of a code of length n with minimum distance at least $k + 1$.*

Proposition 2. *Let A be an $l \times m$-matrix over $GF(q)$ with kr-wise independent columns, and let B be a (k, r)-matrix over $GF(q)$ with dimension $m \times n$. Then the columns in the matrix $A \times B$ are k-wise independent.*

We can use sparse matrices to construct small probability spaces, see [2,6]:

Proposition 3. *Let M be a (k, r)-matrix over $GF(q)$ of dimension $m \times n$.*

(i) *If $X = (X_1, \ldots, X_m)$ is a kr-wise ε-biased random vector over $GF(q)$, then the vector $Y = (Y_1, \ldots, Y_n) = X \times M$ is k-wise ε-biased.*

(ii) *If $S \subseteq (GF(q))^m$ is a kr-wise ε-biased sample space, then the sample space $T = \{s \times M \mid s^T \in S\} \subseteq (GF(q))^n$ is k-wise ε-biased, hence also $(2 \cdot \varepsilon \cdot (1 - q^{-k})/q, k)$-independent.*

Remarks: We can construct large (k, r)-matrices in polynomial time, but our time bound increases (in the exponent) with k, r. For possibly practical applications (quite often algorithms run fast on such matrices) and further studies explicit constructions of (k, r)-matrices are desirable. The constructions in [14] for $q = 2$ can be adapted to arbitrary finite fields, however they yield weaker lower bounds on $N_q(m, k, r)$ than we proved. Apart from $(4, 2)$-matrices over $GF(q)$, the only known asymptotically optimal construction is for $(4, r)$-matrices. Also one might like to investigate possible connections to low-density codes [16].

References

1. M. Ajtai, J. Kómlos, J. Pintz, J. Spencer and E. Szemerédi, Extremal uncrowded hypergraphs, J. Comb. Theory A 32, 1982, 321–335.
2. N. Alon, O. Goldreich, J. Håstad and R. Peralta, Simple constructions of almost k-wise independent random variables, Rand. Struct. & Algorithms 3, 1992, 289–304, and 4, 1993, 119–120.
3. C. T. Benson, Minimal regular graphs of girth eight and twelve, Canad. J. Mathematics 18, 1966, 1091–1094.
4. C. Bertram-Kretzberg, T. Hofmeister and H. Lefmann, Sparse 0-1-matrices and forbidden hypergraphs, Comb., Prob. and Computing 8, 1999, 417–427.
5. C. Bertram-Kretzberg and H. Lefmann, The algorithmic aspects of uncrowded hypergraphs, SIAM J. Computing 29, 1999, 201–230.
6. C. Bertram-Kretzberg and H. Lefmann, MOD_p-tests, almost independence and small probability spaces, Rand. Struct. & Algorithms 16, 2000, 293–313.
7. A. Bondy and M. Simonovits, Cycles of even length in graphs, J. Comb. Theory Ser. B 16, 1974, 97–105.
8. W. G. Brown, On graphs that do not contain a Thomsen graph, Canad. Math. Bulletin 9, 1966, 281–289.
9. P. Erdös, A. Rényi and V. T. Sós, On a problem of graph theory, Stud. Sci. Math. Hungarica 1, 1966, 213–235.
10. A. Fundia, Derandomizing Chebychev's inequality to find independent sets in uncrowded hypergraphs, Rand. Struct. & Algorithms 8, 1996, 131–147.
11. P. Frankl and Z. Füredi, Union-free families of sets and equations over fields, J. Numb. Theory 23, 1986, 210–218.
12. H. Karloff and Y. Mansour, On construction of k-wise independent random variables, Proc. 26th STOC, 1994, 564–573.
13. F. Lazebnik, V. A. Ustimenko and A. J. Woldar, A new series of dense graphs of high girth, Bull. (New Series) of the AMS 32, 1995, 73–79.
14. H. Lefmann, P. Pudlák and P. Savický, On sparse parity-check matrices, Designs, Codes and Cryptography 12, 1997, 107–130.
15. A. Lubotzky, R. Phillips and P. Sarnak, Ramanujan graphs, Combinatorica 8, 1988, 261–277.
16. M. Luby, M. Mitzenmacher, A. Shokrollahi and D. Spielman, Analysis of low-density codes and improved designs using irregular graphs, Proc. 30th STOC, 1998, 249–258.
17. G. A. Margulis, Explicit group theoretical construction of combinatorial schemes and their application to the design of expanders and concentrators, J. Probl. Inform. Transm. 24, 1988, 39–46.
18. R. Wenger, Extremal graphs with no C_4's, C_6's or C_{10}'s, J. Comb. Theory Ser. B 52, 1991, 113–116.

On the Full and Bottleneck Full Steiner Tree Problems[*]

Yen Hung Chen[1], Chin Lung Lu[2], and Chuan Yi Tang[1]

[1] Department of Computer Science, National Tsing Hua University,
Hsinchu 300, Taiwan, R.O.C.
{dr884336,cytang}@cs.nthu.edu.tw
[2] Department of Biological Science and Technology, National Chiao Tung University,
Hsinchu 300, Taiwan, R.O.C.
cllu@mail.nctu.edu.tw

Abstract. Given a graph $G = (V, E)$ with a length function on edges and a subset R of V, the full Steiner tree is defined to be a Steiner tree in G with all the vertices of R as its leaves. Then the full Steiner tree problem is to find a full Steiner tree in G with minimum length, and the bottleneck full Steiner tree problem is to find a full Steiner tree T in G such that the length of the largest edge in T is minimized. In this paper, we present a new approximation algorithm with performance ratio 2ρ for the full Steiner tree problem, where ρ is the best-known performance ratio for the Steiner tree problem. Moreover, we give an exact algorithm of $O(|E| \log |E|)$ time to solve the bottleneck full Steiner tree problem.

1 Introduction

Given an arbitrary graph $G = (V, E)$, a subset $R \subseteq V$ of vertices, and a length (or weight) function $d : E \to R^+$ on the edges, a *Steiner tree* is an acyclic subgraph of G spanning all vertices in R. The vertices of R are usually referred to as *terminals* and the vertices of $V \setminus R$ as *Steiner* (or *optional*) vertices. The length of a Steiner tree is defined to be the sum of the lengths of all its edges. The *Steiner tree problem* (STP for short) is to find a Steiner tree of minimum length in G [4,7,12]. This problem has been shown to be NP-complete [9] and many approximation algorithms have been proposed [1,2,11,14,18,19,20,21,22]. However, the *bottleneck Steiner tree problem*, which is to find a Steiner tree T in G such that the length of the largest edge in T is minimized, can be solved in polynomial time [5,8]. It has been shown that these two problems have many important applications in VLSI design, network communication, computational biology and so on [3,4,7,10,12,13,15].

Motivated by the reconstruction of evolutionary tree in biology, Lu, Tang and Lee studied a variant of the Steiner tree problem, called as the full Steiner tree problem (FSTP for short) [17]. Independently, motivated by VLSI global

---

[*] This work was partly supported by the National Science Council of the Republic of China under grants NSC91-2321-B-007-002 and NSC91-2213-E-321-001.

T. Warnow and B. Zhu (Eds.): COCOON 2003, LNCS 2697, pp. 122–129, 2003.
© Springer-Verlag Berlin Heidelberg 2003

routing and telecommunications, Lin and Xue defined the terminal Steiner tree problem, which is equal to the FSTP [16]. A Steiner tree is *full* if all terminals are the leaves of the Steiner tree [2,12,17]. The *full Steiner tree problem* is to find a full Steiner tree for R in G with minimum length. The problem is shown to be NP-complete and MAX SNP-hard [16], even when the lengths of edges are restricted to be either 1 or 2 [17]. However, Lu, Tang and Lee [17] gave a $\frac{8}{5}$-approximation algorithm for the FSTP with the restriction that the lengths of edges are either 1 or 2, and Lin and Xue [16] presented a $(\rho+2)$-approximation algorithm for the FSTP if the length function is *metric* (i.e., the lengths of edges satisfy the triangle inequality), where ρ is the best-known performance ratio for the STP whose performance ratio is $1 + \frac{\ln 3}{2} \approx 1.55$ [19]. The *bottleneck full Steiner tree problem* (BFSTP for short) is to find a full Steiner tree T for R in G such that the length of the largest edge in T is minimized. In this paper, we present a new approximation algorithm with performance ratio 2ρ for the FSTP, which improves the previous result $\rho+2$ of Lin and Xue [16]. Moreover, we give an $O(|E| \log |E|)$ time algorithm to optimally solve the BFSTP.

The rest of this paper is organized as follows. In Section 2, we describe a 2ρ-approximation algorithm for the FSTP. In section 3, we give an $O(|E| \log |E|)$ time algorithm for optimally solving the BFSTP. Finally, we make a conclusion in Section 4.

2 A 2ρ-Approximation Algorithm for the Full Steiner Tree Problem

FSTP (Full Steiner Tree Problem)
Instance: A complete graph $G = (V, E)$ with $d : E \to R^+$, and a proper subset $R \subset V$ of terminals, where the length function d is metric.
Question: Find a full Steiner tree for R in G with minimum length.

In this section, we will give a 2ρ approximation algorithm for solving the above FSTP, whose length function is metric, in polynomial time. By definition, any full Steiner tree T for R in $G = (V, E)$ contains no edge in $E_R = \{(u,v)|u,v \in R, u \neq v\}$. Hence, throughout the rest of this paper, we assume that G contains no edge in E_R (i.e., $E \cap E_R = \phi$). Let $\mathcal{A}_{STP}$ be the best-known approximation algorithm for the STP, whose performance ratio is $1 + \frac{\ln 3}{2} \approx 1.55$ [19]. Then we apply algorithm $\mathcal{A}_{STP}$ to G and obtain a Steiner tree $S_{APX} = (V', E')$ for R in G. Note that if all vertices of R are leaves in S_{APX}, then S_{APX} is also a full Steiner tree of G. If not, then we use Algorithm 1 to transform it into a full Steiner tree. For convenience, we let $N(r)$ be the set of the neighbors of $r \in R$ in S_{APX} (i.e., $N(r) = \{v|(r,v) \in E'\}$ and its members are all Steiner vertices) and $N_1(r)$ be the nearest neighbor of r in S_{APX} (i.e., $d(r, N_1(r)) = \min \{d(r,v)|v \in N(r)\}$).

Algorithm 1: Method of transforming S_{APX} into a full Steiner tree
  For each $r \in R$ with $|N(r)| \geq 2$ in S_{APX} **do**

1. Let $star(r)$ be the subtree of S_{APX} induced by $\{(r,v)|v \in N(r)\}$. Then remove all the edges in $star(r) \setminus \{(r, N_1(r))\}$ from S_{APX}.
2. Let $G[N(r)]$ be the subgraph of G induced by $N(r)$. Then find a minimum spanning tree of $G[N(r)]$, denoted by $MST(N(r))$, and add all the edges of $MST(N(r))$ into S_{APX}.

end for

The purpose of each iteration in Algorithm 1 is to transform one non-leaf terminal in S_{APX} into a leaf by deleting all the edges, except $(r, N_1(r))$, incident with r and adding all the edges in $MST(N(r))$. After running Algorithm 1, S_{APX} becomes a full Steiner tree. Since there are at most $|R|$ non-leaf terminals in S_{APX}, there are at most $|R|$ iterations in Algorithm 1. For step 1, its total cost is $O(|E|)$ time since for any two non-leaf terminals r and r' in S_{APX}, we have $\{(r,v)|v \in N(r)\} \cap \{(r',v)|v \in N(r')\} = \phi$. In step 2, we need to compute a minimum spanning tree of $G[N(r)]$ in each iteration, which can be done by Prim's Algorithm in $O(|N(r)|^2)$ time [6]. Hence, its total cost is $O(|V|^3)$ time. As a result, the time-complexity of Algorithm 1 is $O(|V|^3)$.

Now, for clarification, we describe our approximation algorithm for the FSTP as follows.

Algorithm APX-FSTP
Input: A complete graph $G = (V, E)$ and a set $R \subset V$ of terminals, where we assume that G contains no edge in E_R and the length function is metric.
Output: A full Steiner tree T_{APX} for R in G.
1. /* **Find a Steiner tree in G** */
   Use the currently best-known approximation algorithm $\mathcal{A}_{STP}$ for the STP to find a Steiner tree S_{APX} in G.
2. /* **Transform S_{APX} into a full Steiner tree T_{APX}** */
   If S_{APX} is not a full Steiner tree **then**
      Use Algorithm 1 to transform S_{APX} into a full Steiner tree and let T_{APX} be such a full Steiner tree.
   else Let $T_{APX} = S_{APX}$.

Theorem 1. *Algorithm APX-FSTP is a 2ρ-approximation algorithm for the FSTP.*

Proof. Note that the time-complexity of Algorithm APX-FSTP is dominated by the cost of the step 1 for running the currently best-known approximation algorithm for the STP [19]. For convenience, we use $len(H)$ to denote the length of any subgraph H of G (i.e., $len(H)$ equals to the sum of the lengths of all the edges of H). Let T_{OPT} and S_{OPT} be the optimal full Steiner tree and Steiner tree for R in G, respectively. Note that we use the currently best-known approximation algorithm $\mathcal{A}_{STP}$ for the STP to find a Steiner tree S_{APX} for R in G. Hence, we have $len(S_{APX}) \leq \rho \cdot len(S_{OPT})$, where ρ is the performance ratio of $\mathcal{A}_{STP}$. Since T_{OPT} is also a Steiner tree for R in G, we have $len(S_{OPT}) \leq len(T_{OPT})$ and hence $len(S_{APX}) \leq \rho \cdot len(T_{OPT})$. Recall that in each iteration of Algorithm 1, we transform a non-leaf terminal r in S_{APX} into a leaf by first removing all the edges, except $(r, N_1(r))$, and then adding all the edges in $MST(N(r))$. Let $N_2(r)$ denote the second nearest neighbor of r in $N(r)$ and let $P = (v_1 \equiv N_1(r), v_2, \ldots, v_k \equiv N_2(r))$ be any arbitary path which exactly visits each vertex in $N(r)$ and both $N_1(r)$ and $N_2(r)$ are its end-vertices, where $k = |N(r)|$. By triangle inequality, we have the following inequalities.

$$d(v_1, v_2) \le d(r, v_1) + d(r, v_2)$$
$$d(v_2, v_3) \le d(r, v_2) + d(r, v_3)$$
$$\vdots$$
$$d(v_{k-1}, v_k) \le d(r, v_{k-1}) + d(r, v_k)$$

By above inequalities, we have $d(v_1, v_2) + d(v_2, v_3) + \ldots + d(v_{k-1}, v_k) \le 2 \cdot len(star(r)) - d(r, v_1) - d(r, v_k)$ and hence $len(P) \le 2 \cdot len(star(r)) - d(r, N_1(r)) - d(r, N_2(r))$. Since $MST(N(r))$ is a minimum spanning tree of $G[N(r)]$, we have $len(MST(N(r))) \le len(P)$. In other words, we have $len(MST(N(r))) \le 2 \cdot len(star(r)) - d(r, N_1(r)) - d(r, N_2(r))$. By construction of T_{APX}, we have

$$len(T_{APX}) = len(S_{APX}) + \sum_{r \in R}(len(MST(N(r))) - len(star(r)) + d(r, N_1(r)))$$

$$\le len(S_{APX}) + \sum_{r \in R}(len(star(r)) - d(r, N_2(r)))$$

$$\le len(S_{APX}) + \sum_{r \in R} len(star(r)).$$

Note that for any two terminals $r, r' \in R$, $star(r)$ and $star(r')$ are edge-disjoint in S_{APX}. Hence, we have $\sum_{r \in R} len(star(r)) \le len(S_{APX})$. As a result, we have $len(T_{APX}) \le 2 \cdot len(S_{APX}) \le 2\rho \cdot len(T_{OPT})$, which implies that the performance ratio of Algorithm APX-FSTP is 2ρ. □

3 An Exact Algorithm for the Bottleneck Full Stenier Tree Problem

BFSTP (Bottleneck Full Steiner Tree Problem)
Instance: A complete graph $G = (V, E)$ with $d : E \to R^+$, and a proper subset $R \subset V$ of terminals.
Question: Find a full Steiner tree for R in G such that the length of the largest edge is minimized.

In the section, we will present an exact algorithm for solving the above BFSTP in $O(|E| \log |E|)$ time. Recall that we assume that G contains no edge in E_R. Then we call an edge in E as *Steiner edge* if both its end-vertices are Steiner vertices; otherwise, as *terminal edge*. Without loss of generality, we use $\{e_1, e_2, \ldots, e_\beta\}$ to denote the set of all Steiner edges in E and assume that their lengths are different (i.e., $d(e_1) < d(e_2) < \ldots < d(e_\beta)$). For any arbitrary tree T, we let $E_b(T)$ denote the bottleneck edge with the largest length in T and let $len_b(T)$ denote the length of $E_b(T)$.

In the following, we describe our algorithm to find an optimal bottleneck full Steiner tree for R in G. In the beginning, we construct a star T_u for each Steiner vertex $u \in V \setminus R$ by letting u as the center of T_u and all terminal vertices of R as the leaves of T_u. Then we let $\mathcal{C}$ denote the collection of all such stars (i.e., $\mathcal{C} = \{T_u | u \in V \setminus R\}$). Clearly, each tree in $\mathcal{C}$ is a feasible solution for the BFSTP, but may not be the optimal one. For convenience, we let $T^\star(\mathcal{C})$ denote the tree in $\mathcal{C}$ with the minimum bottleneck edge (i.e., $len_b(T^\star(\mathcal{C})) = \min\{len_b(T)|T \in \mathcal{C}\}$). It is not hard to see that if $len_b(T^\star(\mathcal{C})) \leq d(e_1)$, then $T^\star(\mathcal{C})$ is the optimal solution for the BFSTP, since we have $len_b(T^\star(\mathcal{C})) \leq len_b(T)$ for any full Steiner tree T in G containing at least one Steiner edge. On the other hand, if $len_b(T^\star(\mathcal{C})) > d(e_1)$, then we perform the following Algorithm 2 to merge some trees in $\mathcal{C}$ into larger trees containing more Steiner vertices, and then we can show later that the full Steiner tree in the resulting $\mathcal{C}$ with the smallest bottleneck edge is an optimal bottleneck full Steiner tree in G. For each terminal vertex $r \in R$, we use $e_T(r, T)$ to denote the terminal edge in a full Steiner tree T incident with r.

Algorithm 2: Repeatedly merge the trees in $\mathcal{C}$

1. Let $e_i = e_1$.
2. Repeat the following steps until the condition (1) $len_b(T^\star(\mathcal{C})) \leq d(e_i)$ or (2) $|\mathcal{C}| = 1$ holds.
    2.1. Let $v_1(e_i)$ and $v_2(e_i)$ be the end-vertices of e_i. Find the trees in $\mathcal{C}$, say T_{i_1} and T_{i_2} respectively, containing $v_1(e_i)$ and $v_2(e_i)$.
    2.2. **If** $T_{i_1} = T_{i_2}$ **then** let $T_i = T_{i_1}$ **else**
        2.2.1. Connect T_{i_1} with T_{i_2} by e_i and let T_i be the resulting tree.
        2.2.2. /* **Make sure each terminal vertex $r \in R$ is adjacent to the nearest Steiner vertex v in T_i (i.e., $d(r, v) = \min\{(r, v')|v'$ is a Steiner vertex in $T_i\}$). */**
           For each $r \in R$, **if** $d(e_T(r, T_{i_1})) \leq d(e_T(r, T_{i_2}))$ **then** delete $e_T(r, T_{i_2})$ in T_i **else** delete $e_T(r, T_{i_1})$ in T_i.
    2.3. Let $i = i + 1$.

It is not hard to see that in each iteration of Algorithm 2, we have $len_b(T_i) \leq len_b(T_{i_1})$ and $len_b(T_i) \leq len_b(T_{i_2})$, which implies that the merged tree T_i may have a better bottleneck edge with smaller length. After performing Algorithm 2, we use $T^\star(\mathcal{C})$ in the resulting $\mathcal{C}$ as the output of our algorithm for the BFSTP. For clarification, we describe our algorithm for the BFSTP as follows.

Algorithm Exact-BFSTP

Input: A complete graph $G = (V, E)$ and a set $R \subset V$ of terminals, where we assume that G contains no edge in E_R.
Output: An optimal bottleneck full Steiner tree T_{OPT} for R in G.
1. For each Steiner vertex $u \in V \setminus R$, construct a star T_u by letting u as its center and all terminal vertices of R as its leaves, and let $\mathcal{C} = \{T_u | u \in V \setminus R\}$.
2. **If** $len_b(T^\star(\mathcal{C})) > d(e_1)$, **then** perform the following steps.

2.1. Sort all Steiner edges of G into an increasing order, say $e_1, e_2, \ldots, e_\beta$, with $d(e_1) < d(e_2) < \ldots < d(e_\beta)$.

2.2. Perform Algorithm 2.

2.3. Repeatedly remove all Steiner vertices of $T^\star(\mathcal{C})$ whose degrees are 1.

3. Let $T_{OPT} = T^\star(\mathcal{C})$.

Theorem 2. *Algorithm Exact-BFSTP is an $O(|E| \log |E|)$ time algorithm for solving the BFSTP.*

Proof. We first analyze the time-complexity of Algorithm Exact-BFSTP as follows. It is not hard to see that its time-complexity is dominated by the steps 2. Clearly, step 2.1 can be done in $O(|E| \log |E|)$ time by heap sorting algorithm [6]. For step 2.2 of executing Algorithm 2, it can be implemented by the disjoint-set operations (i.e., MAKE-SET, UNION and FIND-SET operations), since for each full Steiner tree of $\mathcal{C}$, we can represent it by using the set of its Steiner vertices and clearly, any two such sets are disjoint. Initially, we make $|V \setminus R|$ singleton sets with each consisting of a Steiner vertex u and representing an initial T_u in $\mathcal{C}$. After that, in each iteration of Algorithm 2, we can use two FIND-SET operations to identify T_{i_1} and T_{i_2}, and then use a UNION operation to merge them into T_i. Totally, we have $|V \setminus R|$ MAKE-SET operations, at most $2 \cdot |E|$ FIND-SET operations, and at most $|V \setminus R| - 1$ UNION operations. Note that the cost of m disjoint-set operations on n elements takes $O(m \cdot \alpha(n))$, where $\alpha(n)$ is an inverse Ackermann's function of n which is almost a constant for all practical purposes (i.e., $\alpha(n) \leq 4$ for all $n \leq 10^{80}$) [6]. In other words, above operations can be implemented in $O(|V \setminus R| + 2 \cdot |E| + |V \setminus R| - 1) = O(|E|)$ time. Clearly, the total cost of step 2.2.2 of Algorithm 2 is at most $O(|V \setminus R| \times |R|) = O(|V|^2)$ time. Hence, the total time-complexity of Algorithm 2 is $O(|E|)$ time. As mentioned above, the time-complexity is $O(|E| \log |E|)$.

Next, we prove the correctness of Algorithm Exact-BFSTP by showing that for any full Steiner tree T for R in G, we have $len_b(T_{OPT}) \leq len_b(T)$, where T_{OPT} is the output of our algorithm. Let $V_s(T)$ be the set of the Steiner vertices in T. Suppose that $|V_s(T)| = 1$. Then T must be a full Steiner tree in the initial $\mathcal{C}$ of Algorithm Exact-BFSTP. Since after performing Algorithm 2, $len_b(T^\star(\mathcal{C}))$ may be improved and hence, $len_b(T_{opt}) \leq len_b(T)$. Suppose that $|V_s(T)| \geq 2$. Let e_γ be the edge with the largest length in Steiner edges of T. Assume that Algorithm Exact-BFSTP outputs its result T_{OPT} after its Algorithm 2 considers Steiner edge e_k, where $1 \leq k \leq \beta$. Then we consider the following two cases. Case 1: $k \leq \gamma$. In this case, we have either $len_b(T_{OPT}) \leq d(e_k)$ or the resulting $\mathcal{C}$ has size 1. For the former case, we have $len_b(T_{OPT}) \leq len_b(T)$ since $d(e_k) \leq d(e_\gamma) \leq len_b(T)$. For the latter case, it is clear that e_{k-1} is the largest Steiner edge of T_{OPT}, whose length is less than $d(e_\gamma)$. Moreover, we have $V_s(T) \subseteq V_s(T_{OPT})$ since $V_s(T_{OPT}) = V \setminus R$, which implies that for each terminal $r \in R$, its corresponding terminal edge in T_{OPT} has length smaller than or equal to that in T. The reason is that our algorithm always connects r to the nearest Steiner vertex in T_{OPT}, which contains all Steiner vertices. As discussed above, we have $len_b(T_{OPT}) \leq len_b(T)$. Case 2: $k > \gamma$. For each Steiner edge of

T being considered by our algorithm, it is not hard to see that there is a tree in C containing its two end-vertices. Since the subgraph of T induced by $V_s(T)$ is connected, T_γ in C contains all vertices in $V_s(T)$ after e_γ is considered (i.e., $V_s(T) \subseteq V_s(T_\gamma)$). Then for each terminal $r \in R$, its corresponding terminal edge in T_γ has length smaller than or equal to that in T. In addition, the largest Steiner edge of T_γ has length less than or equal to $d(e_\gamma)$. Hence, we have $len_b(T_\gamma) \leq len_b(T)$. Since $len_b(T_{OPT}) \leq len_b(T_\gamma)$, we have $len_b(T_{OPT}) \leq len_b(T)$.   □

4   Conclusion

In this paper, we first presented a new approximation algorithm with performance ratio 2ρ for the full Steiner tree problem under the metric space, and then we gave an exact algorithm of $O(|E| \log |E|)$ time for solving the bottleneck full Steiner tree problem. It would be interesting to find a better (approximation) algorithm for the (bottleneck) full Steiner tree problem. Note that if the length function is not metric, then whether there exists an approximation algorithm with constant ratio for the full Steiner tree problem is still open.

References

1. Berman, P., Ramaiyer, V.: Improved approximations for the Steiner tree problem. Journal of Algorithms **17** (1994) 381–408.
2. Borchers, A., Du, D.Z.: The k-Steiner ratio in graphs. SIAM Journal on Computing **26** (1997) 857–869.
3. Caldwell, A., Kahng, A., Mantik, S., Markov, I., Zelikovsky, A.: On wirelength estimations for row-based placement. In: Proceedings of the 1998 International Symposium on Physical Design (ISPD 1998) 4–11.
4. Cheng, X., Du, D.Z.: Steiner Tree in Industry. Kluwer Academic Publishers, Dordrecht, Netherlands (2001).
5. Chiang, C., Sarrafzadeh, M., Wong, C.K.: Global router based on Steiner min-max trees. IEEE Transaction on Computer-Aided Design **9** (1990) 1318–1325.
6. Cormen, T.H., Leiserson, C.E., Rivest, R.L., Stein, C.: Introduction to Algorithm. 2nd edition MIT Press, Cambridge (2001).
7. Du, D.Z., Smith, J.M., Rubinstein, J.H.: Advances in Steiner Tree. Kluwer Academic Publishers, Dordrecht, Netherlands (2000).
8. Duin, C.W., Volgenant, A.: The partial sum criterion for Steiner trees in graphs and shortest paths. European Journal of Operations Research **97** (1997) 172–182.
9. Garey, M.R., Graham, R.L., Johnson, D.S.: The complexity of computing Steiner minimal trees. SIAM Journal of Applied Mathematics **32** (1997) 835–859.
10. Graur, D., Li, W.H.: Fundamentals of Molecular Evolution. 2nd edition Sinauer Publishers, Sunderland, Massachusetts (2000).
11. Hougardy, S., Prommel, H.J.: A 1.598 approximation algorithm for the Steiner problem in graphs. In: Proceedings of the 10th Annual ACM-SIAM Symposium on Discrete Algorithms (SODA 1999) 448–453.
12. Hwang, F.K., Richards, D.S., Winter, P.: The Steiner Tree Problem. Annuals of Discrete Mathematics 53, Elsevier Science Publishers, Amsterdam (1992).

13. Kahng, A.B., Robins, G.: On Optimal Interconnections for VLSI. Kluwer Publishers (1995).
14. Karpinski, M., Zelikovsky, A.: New approximation algorithms for the Steiner tree problems. Journal of Combinatorial Optimization **1** (1997) 47–65.
15. Kim, J., Warnow, T.: Tutorial on Phylogenetic Tree Estimation. Manuscript, Department of Ecology and Evolutionary Biology, Yale University (1999).
16. Lin, G.H., Xue, G.L.: On the terminal Steiner tree problem. Information Processing Letters **84** (2002) 103–107.
17. Lu, C.L., Tang, C.Y., Lee, R.C.T.: The full Steiner tree problem. Theoretical Computer Science (to appear).
18. Prommel, H.J., Steger, A.: A New Approximation Algorithm for the Steiner Tree Problem with Performance Ratio 5/3. Journal of Algorithms **36** (2000) 89–101.
19. Robins, G., Zelikovsky A.: Improved Steiner tree approximation in graphs. In: Proceedings of the 11th Annual ACM-SIAM Symposium on Discrete Algorithms (SODA 2000) 770–779.
20. Zelikovsky, A.: An 11/6-approximation algorithm for the network Steiner problem. Algorithmica **9** (1993) 463–470.
21. Zelikovsky, A.: A faster approximation algorithm for the Steiner tree problem in graphs. Information Processing Letters **46** (1993) 79–83.
22. Zelikovsky, A.: Better approximation bounds for the network and Euclidean Steiner tree problems. Technical report CS-96-06, University of Virginia (1996).

The Structure and Number of Global Roundings of a Graph

Tetsuo Asano[1], Naoki Katoh[2], Hisao Tamaki[3], and Takeshi Tokuyama[4]

[1] School of Information Science,
Japan Advanced Institute of Science and Technology,
Tatsunokuchi, Japan
t-asano@jaist.ac.jp
[2] Graduate School of Engineering, Kyoto University,
Kyoto, Japan.
naoki@archi.kyoto-u.ac.jp
[3] Meiji University, Kawasaki, Japan,
tamaki@cs.meiji.ac.jp
[4] GSIS, Tohoku University, Sendai, Japan.
tokuyama@dais.is.tohoku.ac.jp

Abstract. Given a connected weighted graph $G = (V, E)$, we consider a hypergraph $H_G = (V, \mathcal{P}_G)$ corresponding to the set of all shortest paths in G. For a given real assignment $\mathbf{a}$ on V satisfying $0 \leq \mathbf{a}(v) \leq 1$, a global rounding α with respect to H_G is a binary assignment satisfying that $|\sum_{v \in F} \mathbf{a}(v) - \alpha(v)| < 1$ for every $F \in \mathcal{P}_G$. We conjecture that there are at most $|V| + 1$ global roundings for H_G, and also the set of global roundings is an affine independent set. We give several positive evidences for the conjecture.

1 Introduction

Given a real number a, an integer k is a *rounding* of a if the difference between a and k is strictly less than 1, or equivalently, if k is the floor $\lfloor a \rfloor$ or the ceiling $\lceil a \rceil$ of a. We extend this usual notion of rounding into that of *global rounding* on hypergraphs as follows.

Let $H = (V, \mathcal{F})$, where $\mathcal{F} \subset 2^V$, be a hypergraph on a set V of n nodes. Given a real valued function $\mathbf{a}$ on V, we say that an integer valued function α on V is a *global rounding* of $\mathbf{a}$ with respect to H, if $w_F(\alpha)$ is a rounding of $w_F(\mathbf{a})$ for each $F \in \mathcal{F}$, where $w_F(f)$ denotes $\sum_{v \in F} f(v)$. We assume in this paper that the hypergraph contains all the singleton sets as hyperedges; thus, $\alpha(v)$ is a rounding of $\alpha(v)$ for each v, and we can restrict our attention to the case where the ranges of $\mathbf{a}$ and α are $[0, 1]$ and $\{0, 1\}$ respectively.

This notion of global roundings on hypergraphs is closely related to that of *discrepancy* of hypergraphs[6,10,11,4]. Given $\mathbf{a}$ and $\mathbf{b} \in [0, 1]^V$, define the discrepancy $D_H(\mathbf{a}, \mathbf{b})$ between them on H by

$$D_H(\mathbf{a}, \mathbf{b}) = \max_{F \in \mathcal{F}} |w_F(\mathbf{a}) - w_F(\mathbf{b})|.$$

T. Warnow and B. Zhu (Eds.): COCOON 2003, LNCS 2697, pp. 130–138, 2003.
© Springer-Verlag Berlin Heidelberg 2003

The supremum $\sup_{\mathbf{a} \in [0,1]^V} \min_{\alpha \in \{0,1\}^V} D_H(\mathbf{a}, \alpha)$ is called the linear (or inhomogeneous) discrepancy of H, and it is a quality measure of approximability of a real vector with an integral vector to satisfy a constraint given by a linear system corresponding to H.

Thus, the set of global roundings of $\mathbf{a}$ is the set on integral points in the open unit ball around $\mathbf{a}$ by using the discrepancy D_H as the distance function. It is known that the open ball always contains an integral point for any "input" $\mathbf{a}$ if and only if the hypergraph is unimodular (see [4,7]). The fact is utilized in digital halftoning applications [1,2]. It is NP-hard to decide whether the ball is empty (i.e. containing no integral point) or not even for some very simple hypergraphs [3].

In this paper, we are interested in the maximum number $\nu(H)$ of intergral points in an open unit ball under the discrepancy distance.

This direction of research is initiated by Sadakane et al.[13] where the authors discovered a somewhat surprising fact that $\nu(I_n) \leq n + 1$ where I_n is a hypergraph on $V = \{1, 2, .., n\}$ with edge set $\{[i, j]; 1 \leq i \leq j \leq n\}$ consisting of all subintervals of V. We can also see that $\nu(H) \geq n + 1$ for any any hypergraph H: if we let $\mathbf{a}(v) = \epsilon$ for every v, where $\epsilon < 1/n$, then any binary assignment on V that assigns 1 to at most one vertex is a global rounding of H, and hence $\nu(H) \geq n + 1$.

Given this discovery, it is natural to ask for which class of hypergraphs this property $\nu(H) = n + 1$ holds. The understanding of such classes may well be related to algorithmic questions mentioned above. In fact, Sadakane et al. give an efficient algorithm to enumerate all the global roundings of a given input on I_n.

In this paper, we show that $\nu(H) = n + 1$ holds for a considerably wider class of hypergraphs. Given a connected G in which edges are possibly weighted by a positive value, we define a *shortest-path hypergraph* H_G generated by G as follows: a set F of vertices of G is an edge of H_G, if and only if F is the set of vertices of some shortest path in G with respect to the given edge weights. In this notation, $I_n = H_{P_n}$ for the path P_n on n vertices. Note that we permit more than one shortest path between a pair of nodes if they have the same length.

Theorem 1. $\nu(H_G) = n + 1$ *holds for the shortest-path hypergraph H_G, if G is a tree, a cycle, a tree of cycles, or an unweighted mesh.*

Based on the positive evidence above and some failed attempts in creating counterexamples, we conjecture that the result holds for general connected graphs.

Conjecture 1. $\nu(H_G) = n + 1$ for any connected graph G with n nodes.

2   Preliminaries

We start with the following easy observations:

Lemma 1. *For hypergraphs $H = (V, \mathcal{F})$ and $H' = (V, \mathcal{F}')$ such that $\mathcal{F} \subset \mathcal{F}'$, $\nu(H) \geq \nu(H')$.*

Definition 1. *A set A of binary functions on V is called H-compatible if, for each pair α and β in A, $|w_F(\alpha) - w_F(\beta)| \leq 1$ holds for every hyperedge F of H.*

Lemma 2. *For a given input real vector $\mathbf{a}$, the set of global roundings with respect to H is H-compatible.*

Proof. Suppose that α and β are two different global roundings of an input $\mathbf{a}$ with respect to a hypergraph H. We have $|w_F(\alpha) - w_F(\beta)| \leq |w_F(\mathbf{a}) - w_F(\alpha)| + |w_F(\mathbf{a}) - w_F(\beta)| < 2$. Since the value must be integral, we have the lemma.

Let $\mu(H)$ be the maximum cardinality of an H-compatible set. Because of the above lemma, $\mu(H) \geq \nu(H)$. Indeed, instead of an open unit ball, $\mu(H)$ gives the largest cardinality of a unit-diameter set with respect to D_H.

For a vector $\mathbf{q}$ in the n-dimensional real space $\mathcal{R}^n$, $\tilde{\mathbf{q}}$ is the vector in $\mathcal{R}^{n+1}$ obtained by appending 1 as the last coordinate value: i.e., $\tilde{\mathbf{q}} = (q_1, q_2, \ldots, q_n, 1)$ if $\mathbf{q} = (q_1, q_2, \ldots, q_n)$. Vectors $\mathbf{q}_1, \mathbf{q}_2, \ldots \mathbf{q}_s$ are called *affine independent* in $\mathcal{R}^n$ if $\tilde{\mathbf{q}}_1, \tilde{\mathbf{q}}_2, \ldots, \tilde{\mathbf{q}}_s$ are linearly independent in $\mathcal{R}^{n+1}$. If every H-compatible set is an affine independent set regarded as a set of vectors in the n-dimensional space, we call H satisfies the *affine independence property*.

The definition of H-compatible set does not include the input vector $\mathbf{a}$, and facilitates the combinatorial analysis. Thus, instead of Conjecture 1, we consider the following (possibly stronger) variants:

Conjecture 2. $\mu(H_G) = n + 1$ for any connected graph G with n nodes.

Conjecture 3. For any connected graph G, H_G satisfies the affine independence property.

It is clear that Conjecture 3 implies Conjecture 2, and that Conjecture 2 implies Conjecture 1.

3   Properties for General Graphs

For a binary assignment α on V and a subset X of V, $\alpha|_X$ denotes the restriction of α on X. Let $V = X \cup Y$ be a partition of V into nonintersecting subsets X and Y of vertices. For binary assignments α on X and β on Y, $\alpha \oplus \beta$ is a binary assignment on V obtained by concatenating α and β: That is, $\alpha \oplus \beta(v) = \alpha(v)$ if $v \in X$, otherwise it is $\beta(v)$.

The following lemma is a key lemma for our theory:

Lemma 3. *Let $G = (V, E)$ be a connected graph, and let $V = X \cup Y$ be a partition of V. Let α_1 and α_2 be different assignments on X and let β_1 and β_2 be different assignments on Y. Then, the set $\mathcal{F} = \{ \alpha_1 \oplus \beta_1, \alpha_1 \oplus \beta_2, \alpha_2 \oplus \beta_1, \alpha_2 \oplus \beta_2 \}$ cannot be H_G-compatible.*

Proof. Consider $x \in X$ satisfying $\alpha_1(x) \neq \alpha_2(x)$ and $y \in Y$ satisfying $\beta_1(y) \neq \beta_2(y)$. We choose such x and y with the minimum shortest path length. Thus, on each internal node of a shortest path $\mathbf{p}$ from x to y, all four assignments in $\mathcal{F}$ take the same value. Without loss of generality, we assume $\alpha_1(x) = \beta_1(y) = 0$ and $\alpha_2(x) = \beta_2(y) = 1$. Then, $w_{\mathbf{p}}(\alpha_2 \oplus \beta_2) = w_{\mathbf{p}}(\alpha_1 \oplus \beta_1) + 2$, and hence violate the compatibility.

Corollary 1. *Let A be a set of H_G-compatible set, and let $A|_X$ and $A|_Y$ be the set obtained by restricting assignments of A to X and Y, respectively, for a partition of V. If $A|_X$ and $A|_Y$ has μ_X and μ_Y elements, respectively, and $\mu_X \geq \mu_Y$, $|A| \leq \min\{\mu_Y(\mu_Y - 1)/2 + \mu_X, \mu_Y \sqrt{\mu_X} + \mu_X\}$.*

Proof. If we construct a bipartite graph with node sets corresponding to $A|_X$ and $A|_Y$, in which two nodes $\alpha \in A|_X$ and $\beta \in A|_Y$ are connected by an edge if $\alpha \oplus \beta \in A$, Lemma3 implies the K_{22}-free property of the graph. Thus, the corollary follows from a famous result in extremal graph theory ([5], Lemma 9).

Although the recursion $f(n) \leq f(1)(f(1) - 1)/2 + f(n - 1)$ gives a linear upper bound of $f(n)$, this does not imply that $|A| = O(n)$, since the restriction $A|_X$ is not always an $H_{G'}$-compatible set where G' is the induced subgraph by X of G.

The affine independence of an H-compatible set $A = \{\alpha_1, \alpha_2, \ldots, \alpha_m\}$ means that any linear relation $\sum_{1 \leq i \leq m} c_i \alpha_i = 0$ satisfying that $\sum_{1 \leq i \leq m} c_i = 0$ implies that $c_i = 0$ for $1 \leq i \leq m$. We can prove its special case as follows: [1]

Proposition 1. *If α, β, α', and β' are mutually distinct elements of an H_G-compatible set for some graph G, then it cannot happen that $\alpha - \beta = \alpha' - \beta'$.*

Proof. Let X be the subset of V consisting of u satisfying $\alpha(u) = \beta(u)$, and let Y be its complement in V. Let $\alpha = \xi \oplus \eta$, where ξ and η are the parts of α on X and Y, respectively. Thus, $\beta = \xi \oplus \bar{\eta}$, where $\bar{\eta}$ is obtained by flipping all the entries of η.

Let $\alpha' = \xi' \oplus \eta'$. Then, since $\alpha - \beta = \alpha' - \beta'$, $\beta' = \xi' \oplus \bar{\eta}'$. Moreover, $\eta - \bar{\eta}$ and $\eta' - \bar{\eta}'$ are vectors whose entries are 1 and -1, and hence $\eta - \bar{\eta} = \eta' - \bar{\eta}'$ implies that $\eta = \eta'$.

Thus, all of $\xi \oplus \eta$, $\xi \oplus \bar{\eta}$, $\xi' \oplus \eta$, and $\xi' \oplus \bar{\eta}$ are in A; this contradicts with Lemma 3 if they are different to each other.

4   Graphs for which the Conjectures Hold

4.1   Graphs with Path-Preserving Ordering

Given a connected graph $G = (V, E)$, consider an ordering $v_1, v_2, \ldots, v_n$ of nodes of V. Let $V_i = \{v_1, v_2, \ldots, v_i\}$, and let G_i be the induced subgraph of G by V_i. The ordering is path-preserving if G_i is connected for each i, and a shortest

---

[1] This fact was suggested by Günter Rote.

path in G_i is always a shortest path in G. It is clear that a tree with arbitrary edge lengths and a complete graph with a uniform edge length have path-preserving orderings. More generally, a k-tree with a uniform edge length has a path-preserving ordering by its definition. A d-dimensional mesh, where each edge has unit length, is also a typical example.

Theorem 2. *If G has a path-preserving ordering, H_G satisfies the affine independence property.*

Proof. We prove the statement by induction on $|V|$. If $n = 1$, the statement is trivial, since $(0, 1)$ and $(1, 1)$ are linearly independent. If G has a path-preserving ordering, it gives a path-preserving ordering for G_{n-1} that has $n-1$ nodes. Thus, from the induction hypothesis, we assume that any $H_{G_{n-1}}$-compatible set of binary assignments is an affine independent set. Let π be the restriction map from $\{0, 1\}^V$ to $\{0, 1\}^{V_{n-1}}$ corresponding to restriction of a binary assignment on V to the one on V_{n-1}. Let A be an H_G-compatible set, and let $\pi(A) = \{\pi(\alpha) : \alpha \in A\}$ be the set obtained by restricting A to V_{n-1} and removing the multiplicities. The set $\pi(A)$ must be an $H_{G_{n-1}}$-compatible set: otherwise, there must be a shortest path in G_{n-1} violating the compatibility condition for A, which cannot happen since the path is also a shortest path in G.

For each $\beta \in \pi(A)$, let $\beta \oplus 0$ and $\beta \oplus 1$ be its extension in $\{0, 1\}^V$ obtained by assigning 0 and 1 to v_n, respectively. Naturally, $\pi^{-1}(\beta)$ is a subset of $\{\beta \oplus 0, \beta \oplus 1\}$. For any two different assignments β and γ in $\pi(A)$, it cannot happen that all of $\beta \oplus 0$, $\beta \oplus 1$, $\gamma \oplus 0$, and $\gamma \oplus 1$ are in A. Indeed, this is a special case of Lemma 3 for $X = V_{n-1}$ and $Y = \{v_n\}$. Thus, there is at most one rounding in $\pi(A)$ satisfying that its inverse image by π contains two elements.

Let the elements of A listed as $\alpha_1, \ldots, \alpha_k$ where $\alpha_1 = \beta \oplus 0$ and $\alpha_2 = \beta \oplus 1$ for some $\beta \in \pi(A)$. Suppose a linear relation $\sum_{1 \le i \le k} c_i \alpha_i = 0$ holds with $\sum_{1 \le i \le k} = 0$. By the induction hypothesis that $\pi(A)$ is affine independent, we have $c_1 + c_2 = 0$ and $c_i = 0$ for $3 \le i \le k$. Because of the last components of the vectors, it follows that $c_1 = c_2 = 0$ as well.

Corollary 2. *For a connected graph G, if we consider the hypergraph H associated with the set of all paths in G (irrespective of their lengths), H satisfies the affine independence property.*

Proof. Consider a spanning tree T of G. Then the hypergraph associated with the set of all paths in G has the same node set as H, and its hyperedge set is a subset of that of H. Hence, it suffices to prove the statement for T, which has a path-preserving ordering. Every path in T is a shortest path in T; hence, the set is H_T-compatible, and consequently, affine independent.

A graph G is *series connection* of two graphs G_1 and G_2 if $G = G_1 \cup G_2$ and $G_1 \cap G_2 = \{v\}$ (implying that they share no edge), where v is called the *separator*. We have the following (proof is omited in this version):

Theorem 3. *Suppose that a graph G is a series connection of two connected graphs G_1 and G_2. Then, $\mu(G) \le \mu(G_1) + \mu(G_2) - 2$. If both of H_{G_1} and H_{G_2} satisfy the affine independence property, so does H_G.*

4.2   The Case of a Cycle

Let C_n be a cycle on n vertices $V = \{1, 2, \ldots, n\}$ with edge set $\{e_1, \ldots, e_n\}$ where $e_i = (i, i+1)$, $1 \le i \le n$. The arithmetics on vertices are cyclic, i.e., $n+1 = 1$. We sometimes refer to the edge e_n as e_0 as well. For $i, j \in V$, let $P(i, j)$ denote the path from i to j containing the nodes $v_i, v_{i+1}, \ldots v_j$ in this cyclic order. Note that $P(j, i)$ is different from $P(i, j)$ if $i \ne j$. $P(i, i)$ is naturally interpreted as an empty path consisting of a single vertex and no edge. Let $\mathcal{P} = \mathcal{P}_{C_n}$ be a set of shortest paths on C_n. Note that for any given edge lengths, $\mathcal{P}$ satisfies the following conditions: (1) $P(i, i) \in \mathcal{P}$ for every $i \in V$, (2) if $P \in \mathcal{P}$ then every subpath of P is in $\mathcal{P}$, and (3) for every pair i, j of distinct vertices of C_n, at least one of $P(i, j)$ and $P(j, i)$ is in $\mathcal{P}$.

Theorem 4. $\mu(H_{C_n}) = n + 1$.

As a corollary of Theorem 4 and Theorem 3, we have the following:

Corollary 3. $\mu(H_G) = n + 1$ if G is a tree of cycles with n vertices.

We often write $\mathcal{P}$ for H_{C_n} identifying the hypergraph and the set of hyperedges in this section for abbreviation. We devote the rest of this section for proving Theorem 4. We omit proofs of some lemmas because of space limitation.

For $n \le 2$ the theorem is trivial to verify, so we will assume $n \ge 3$ in the sequel. For an assignment α, we define $w(\alpha) = w_V(\alpha) = \sum_{v \in C_n} \alpha(v)$ to be the weight of α over all vertices in C_n.

Lemma 4. Let α and β be $\mathcal{P}$-compatible assignments on C_n. Then, $w(\alpha)$ and $w(\beta)$ differ by at most 1.

Lemma 5. Suppose $w(\alpha) = w(\beta)$ for assignment α and β. Then, if α and β are $\mathcal{P}$-compatible they are compatible on every path of C_n.

From the above observations and Corollary 2, it is clear that $\mu(H_{C_n}) \le 2(n + 1)$. We reduce it to $n + 1$ by using a pair of equivalence relations on the set of edges of C_n each of which is generated from the $\mathcal{P}$-compatible set of assignments with a uniform weight.

The following notion of edge opposition is one of our main tools. Let e_i and e_j be two edges of C_n. We say e_i opposes e_j (and vice versa) if paths $P(i+1, j)$ and $P(j+1, i)$ are both in $\mathcal{P}$. Note that when $P(i+1, i) \in \mathcal{P}$, e_i opposes itself in this definition. However, in this case, the length of e_i is so large that it does not appear in any shortest path, and we can cut the cycle into a path at e_i to reduce the problem into the sequence rounding problem. Thus, we assume this does not happen. It is routine to verify the following lemma

Lemma 6. For every edge e_i of C_n, there is at least one edge e_j that opposes e_i. Moreover, if edges e_i and e_j oppose each other, then, either e_{i+1} opposes e_j or e_{j+1} opposes e_i.

Define the *opposition graph*, denoted by $opp(\mathcal{P})$, to be the graph on $E(C_n)$ in which $\{e_i, e_j\}$ is an edge if and only if e_i and e_j oppose each other. By Lemma 6, we obtain the following:

Lemma 7. *The opposition graph opp($\mathcal{P}$) is connected.*

We next prove a lemma regarding two equivalence relations on the vertex set of a graph. Let G be a graph. We say that a pair (R_1, R_2) of equivalence relations on $V(G)$ *honors* G, if for every edge $\{u, v\}$ of G, u and v are equivalent either in R_1 or in R_2. For an equivalence relation R, denote by $ec(R)$ the number of equivalence classes of R.

Lemma 8. *Let G be a connected graph on n vertices and suppose a pair (R_1, R_2) of equivalence relations on $V(G)$ honors G. Then $ec(R_1) + ec(R_2) \leq n + 1$.*

Proof. Fix an arbitrary spanning tree T of G. We assume (R_1, R_2) honors G and hence it honors T. We grow tree S from a singleton tree towards T, and consider $f(S)$ that is the sum of the number of equivalence classes for R_1 and R_2 among the nodes of S. Initially, we have one node, and hence $f(S) = 2$. If we add an edge and a vertex, $f(S)$ increases by at most one, since the vertex is equivalent to an existing class for at least one of the equivalence relations. Hence, $f(T) \leq n + 1$.

The following equivalence relation on the edge set of C_n plays a central role in our proof.

Let A be a set of assignments of uniform weight. We say that two edges e_i, e_j of C_n are *A-equivalent* and write $e_i \sim_A e_j$ if and only if either $i = j$ or $w_{P(i+1,j)}(\alpha)$ is the same for every $\alpha \in A$. This relation is symmetric since the assignments in A have the same weight on the entire cycle and $P(i + 1, j)$ and $P(j + 1, i)$ are complement to each other in terms of their vertex sets. It is indeed straightforward to check the transitivity to confirm that A-equivalence is an equivalence relation for any assignment set A of uniform weight.

Lemma 9. *Let A and B be sets of assignments on C_n such that $A \cup B$ is a $\mathcal{P}$-compatible set and, for some fixed integer w, $w(a) = w$ for every $a \in A$ and $w(b) = w + 1$ for every $b \in B$. Then, for any pair of edges e_i and e_j opposing each other with respect to $\mathcal{P}$, either $e_i \sim_A e_j$ or $e_i \sim_B e_j$; in other words, the pair $(\sim_A, \sim_B)$ honors the opposition graph opp($\mathcal{P}$).*

Consider a set A of of $\mathcal{P}$-compatible assignments in which all assingments have the same weight. From Lemma 5, the set A is an I_n-compatible assignment, where I_n is the hypergraph on V associated with all the intervals on the graph obtained by cutting C_n at the edge $e_0 = (v_n, v_1)$. Let $V_i = \{1, 2, \ldots, i\} \subset V$, and let $A(V_i)$ be the set of assignments on V_i obtained by restricting A to V_i.

Lemma 10. $|A(V_i)| \leq |A(V_{i-1})| + 1$.

Proof. $V_i = V_{i-1} \cup \{v_i\}$. Applying Lemma 3, there is at most one assignment α in $A(V_{i-1})$ such that both of $\alpha \oplus 0$ and $\alpha \oplus 1$ are in $A(V_i)$. Thus, we obtain the lemma.

We call the index i a *branching index* of A if $|A(V_i)| = |A(V_{i-1})| + 1$ holds. Note that for a branching index, there must be an assignment α in $A(V_{i-1})$ such that both of $\alpha \oplus 0$ and $\alpha \oplus 1$ are in $A(V_i)$.

Lemma 11. *Let A be a set of pairwise $\mathcal{P}$-compatible assignments in which all assingments have the same weight. Then, i is a branching index in A only if the edge $e_i = (i, i+1)$ is A-equivalent to none of $e_0, e_1, e_2, \ldots, e_{i-1}$.*

Proof. Suppose level i is a branching index. Then, we have $\alpha \in A(V_{i-1})$ such that $\alpha \oplus 0$ and $\alpha \oplus 1$ are in $A(V_i)$. If e_i is A-equivalent to e_j for $j < i$, the assignments $\alpha \oplus 0$ and $\alpha \oplus 1$ must have the same total weight on $V_i \setminus V_j = \{j+1, \ldots, i\}$. This is impossible, since two assignments are the same on $V_i \setminus V_j$ except on i.

Consider the number $ec(\sim_A)|_{V_i}$ of equivalence classes in V_i. Lemma 11 implies that $|A(V_i)| - |A(V_{i-1})| \leq ec(\sim_A)|_{V_i} - ec(\sim_A)|_{V_{i-1}}$. Thus, we have the following corollary:

Corollary 4. *Let A be a set of assignments that are pairwise compatible and have the same weight. Then $|A| \leq ec(\sim_A)$.*

We are now ready to prove Theorem 4. Let $\mathcal{P}$ be an arbitrary shortest path system on C_n and let A be an arbitrary set of pairwise $\mathcal{P}$-compatible assignments on C_n. By Lemma 4, the assignments of A have at most two weights. If there is only one weight, then $|A| \leq ec(\sim_A)$ by Corollary 4 and hence $|A| \leq n + 1$. Suppose A consists of two subsets A_1 and A_2, with the assignments in A_1 having weight w and those in A_2 having weight $w + 1$. By Lemma 9, the pair of equivalence relations $(\sim_{A_1}, \sim_{A_2})$ honors the opposition graph $opp(\mathcal{P})$. Since $opp(\mathcal{P})$ is connected (Lemma 7), we have $ec(\sim_{A_1}) + ec(\sim_{A_2}) \leq n+1$ by Lemma 8. We are done, since $|A_i| \leq ec(\sim_{A_i})$ for $i = 1, 2$ by Corollary 4.

5   Concluding Remarks

We have proven the conjectures only for special graphs. It will be nice if the conjectures are proven for wider classes of graphs such as outer-planar graphs and series parallel graphs [2]. Also, the affine independence property for the cycle graph has not been proven in this paper.

For a general graph, we do not even know whether $\nu(H_G)$ is polynomially bounded by the number of vertices. It is plausible that the number of roundings can become large if the entries have some middle values (around 0.5). However, for a special input **a** consisting of entries with a same value $0.5 + \epsilon$, we can show that the number of global roundings of **a** is bounded by $n + 1$ if G is bipartite; otherwise by $m + 1$, where m is the number of edges in G [9].

Acknowledgement. The authors thank Jesper Jansson, Günter Rote, and Akiyoshi Shioura for fruitful discussion.

---

[2] One of the authors recently proved it for outer-planar graphs

References

1. T. Asano, N. Katoh, K. Obokata, and T. Tokuyama, Matrix Rounding under the L_p-Discrepancy Measure and Its Application to Digital Halftoning, *Proc. 13th ACM-SIAM SODA* (2002) pp. 896–904.
2. T. Asano, T. Matsui, and T. Tokuyama, Optimal Roundings of Sequences and Matrices, *Nordic Journal of Computing* **7** (2000) pp.241–256. (Preliminary version in SWAT00).
3. T. Asano and T. Tokuyama, How to Color a Checkerboard with a Given Distribution – Matrix Rounding Achieving Low 2×2 Discrepancy, *Proc. 12th ISAAC, LNCS 2223*(2001) pp. 636–648.
4. J. Beck and V. T. Sós, *Discrepancy Theory*, in *Handbook of Combinatorics Volume II* (ed. T. Graham, M. Grötshel, and L. Lovász) 1995, Elsevier.
5. B. Bollobás. *Modern Graph Theory*, GTM 184, Springer-Verlag, 1998.
6. B. Chazelle, *The Discrepancy Method: Randomness and Complexity*, Princeton University, 2000.
7. B. Doerr, Lattice Approximation and Linear Discrepancy of Totally Unimodular Matrices, *Proc. 12th ACM-SIAM SODA* (2001) pp.119–125.
8. A. Hoffman and G. Kruskal, Integral Boundary Points of Convex Polyhedra, In *Linear Inequalities and Related Systems* (ed. W. Kuhn and A. Tucker) (1956) pp. 223–246.
9. J. Jansson and T. Tokuyama, Semi-Balanced Coloring of Graphs– 2-Colorings Based on a Relaxed Discrepancy Condition, Submitted.
10. J. Matoušek, *Geometric Discrepancy*, Algorithms and Combinatorics 18, Springer Verlag 1999.
11. H. Niederreiter, *Random Number Generations and Quasi Monte Carlo Methods*, CBMS-NSF Regional Conference Series in Applied Math., SIAM, 1992.
12. J. Pach and P. Agarwal, *Combinatorial Geometry*, John-Wiley & Sons, 1995.
13. K. Sadakane, N. Takki-Chebihi, and T. Tokuyama, Combinatorics and Algorithms on Low-Discrepancy Roundings of a Real Sequence, *Proc. 28th ICALP, LNCS 2076* (2001) pp. 166–177.

On Even Triangulations of 2-Connected Embedded Graphs*

Huaming Zhang and Xin He

Department of Computer Science and Engineering,
SUNY at Buffalo, Amherst, NY, 14260, USA

Abstract. Recently, Hoffmann and Kriegel proved an important combinatorial theorem [4]: Every 2-connected bipartite plane graph G has a triangulation in which all vertices have even degree (it's called an *even triangulation*). Combined with a classical Whitney's Theorem, this result implies that every such a graph has a 3-colorable plane triangulation. Using this theorem, Hoffmann and Kriegel significantly improved the upper bounds of several art gallery and prison guard problems. In [7], Zhang and He presented a linear time algorithm which relies on the complicated algebraic proof in [4]. This proof cannot be extended to similar graphs embedded on high genus surfaces. It's not known whether Hoffmann and Kriegel's Theorem is true for such graphs.
In this paper, we describe a totally independent and much simpler proof of the above theorem, using only graph-theoretic arguments. Our new proof can be easily extend to show the existence of even triangulations for similar graphs on high genus surfaces. Hence we show that Hoffmann and Kriegel's theorem remains valid for such graphs. Our new proof leads to a very simple linear time algorithm for finding even triangulations.

1 Introduction

Let $G = (V, E)$ be a 2-connected bipartite plane graph. A *triangulation* of G is a plane graph obtained from G by adding new edges into the faces of G so that all of its faces are triangles. A triangulation G' of G is called *even* if all vertices of G' have even degree. Recently, Hoffmann and Kriegel proved an important combinatorial theorem [3,4]:

Theorem 1. *Every 2-connected bipartite plane graph has an even triangulation.*

Combined with the following classical Whitney's Theorem:

Theorem 2. *A plane triangulation is 3-colorable iff all of its vertices have even degree.*

Theorem 1 implies that every 2-connected bipartite plane graph has a 3-colorable plane triangulation.

Theorem 1 was proved in [4] by showing that a linear equation system derived from the input graph G has a solution. An even triangulation of G is found by

---

* Research supported in part by NSF Grant CCR-9912418.

T. Warnow and B. Zhu (Eds.): COCOON 2003, LNCS 2697, pp. 139–148, 2003.
© Springer-Verlag Berlin Heidelberg 2003

solving this linear equation system. In [7], a linear time algorithm for solving this problem was presented. This algorithm relies on the complicated algebraic techniques in [3,4]. It cannot be extended to graphs on high genus surfaces.

In this paper, we present a new proof of Theorem 1. It is totally different and much simpler than the proof in [4]. Our new proof is based on newly revealed properties of G and its dual graph G^*, which leads to a *very simple* $O(n)$ time algorithm for constructing an even triangulation of G.

Because a crucial property that is true for plane graphs does not hold for graphs on high genus surfaces, the proof of Theorem 1 in [4] fails for such graphs. It's not known whether Theorem 1 is valid for such graphs. We show that our new proof of Theorem 1 and the algorithm for constructing even triangulations of 2-connected bipartite plane graphs also apply to similar graphs on high genus surfaces. On the other hand, because the above-mentioned crucial difference between the plane graphs and graphs on high genus surfaces, we can only prove a lower bound on the number of distinct even triangulations of such graphs. The problem of determining the exact value of this number remains an open problem.

The rest of the paper is organized as follows. In Section 2, we introduce the definitions and preliminary results in [3,4]. In Section 3, we provide new proofs of Theorem 1 and another key theorem in [3,4]. In Section 4, we investigate the problem for graphs on high genus surfaces. Details of this paper are in technical report 2002-13 at CSE department of SUNY at Buffalo.

2   Preliminaries

In this section, we give definitions and preliminary results. All definitions are standard and can be found in [1].

Let $G = (V, E)$ be a plane graph with $n = |V|$ vertices and $m = |E|$ edges. The *degree* of a vertex $v \in V$, denoted by $deg_G(v)$ (or simply by $deg(v)$.), is the number of edges incident to v. A *diagonal* of G is an edge which does not belong to E and connects two vertices of a facial cycle. G is *triangulated* if it has no diagonals. (Namely all of its facial cycles, including the exterior face, are triangles). A *triangulation* of G is obtained from G by adding a set of diagonals such that the resulting graph is plane and triangulated. An *even triangulation* is a triangulation in which all vertices have even degree. In this case, we also call the set of diagonals added into G an even triangulation of G.

Without loss of generality, let G be a 2-connected bipartite plane graph all of whose facial cycles have length 4 (Add diagonals if necessary). Such a graph will be called a 2-connected maximal bipartite plane graph (2MBP graph for short). By Euler's formula, a 2MBP graph with n vertices always has $n - 2$ faces and $m = 2n - 4$ edges. We denote the faces of G by $Q(G) = \{q_1, q_2, \ldots, q_{n-2}\}$. When G is clearly understood, we simply use Q to denote $Q(G)$.

Since G is bipartite, we can fix a 2-coloring of G with colors 0 and 1. Denote the color of a vertex v by $c(v)$. For any face $q_i \in Q$, the set of the four vertices on the boundary of q_i is denoted by V_{q_i} and we set $Q_v = \{q_i \in Q | v \in V_{q_i}\}$. Since every facial cycle of G is a 4-cycle, every face $q_i \in Q$ has two diagonals:

the diagonal joining the two 0-colored (1-colored, respectively) vertices in V_{q_i} is called the 0-diagonal (1-diagonal, respectively). Thus a triangulation of G is nothing but choosing for each face q_i either the 0-diagonal or the 1-diagonal and adding it into q_i. To *flip a diagonal of* q_i means to choose the other diagonal of q_i. We associate each face $q_i \in Q$ with a $\{0,1\}$-valued variable x_i, and each triangulation T of G with a vector $\boldsymbol{x} = (x_1, x_2, \ldots, x_{n-2}) \in GF(2)^{n-2}$, where: T contains the 0-diagonal of the face $q_i \Longleftrightarrow x_i = 1$

This mapping defines a one-to-one correspondence between the set of triangulations of G and $GF(2)^{n-2}$. Hoffmann and Kriegel [3,4] proved that a vector $\boldsymbol{x} = (x_i)_{1 \le i \le n-2} \in GF(2)^{n-2}$ represents an even triangulation of G iff $\boldsymbol{x}$ is a solution of the following linear equation system over $GF(2)$:

$$\sum_{q_i \in Q_v} x_i = deg(v) + |Q_v| c(v) \pmod 2 \quad (\forall v \in V) \tag{1}$$

In [4], Hoffmann and Kriegel showed that Equation (1) always has a solution, and hence proved Theorem 1.

To obtain all solutions of Equation (1) (i.e. all even triangulations of G), [4] introduced the concept of *straight walk*. For a 2MBP graph G, its dual graph G^* is 4-regular and connected. Consider a walk S in G^*. Since G^* is 4-regular, at every vertex of S, we have four possible choices to continue the walk: go back, turn left, go straight, or turn right. A closed walk of G^* consisting of only straight steps at every vertex is called a *straight walk* or an *S-walk*. The edge set of G^* can be uniquely partitioned into S-walks. We use $\mathcal{S}(G^*) = \{S_1, \ldots, S_k\}$ to denote this partition, where each S_i $(1 \le i \le k)$ is an S-walk of G^*. Each vertex of G^* (i.e. each face of G) occurs either twice on one S-walk or on two different S-walks. If a face f occurs on one S-walk twice, it is called a *1-walk face*. If f occurs on two different S-walks, it is called a *2-walk face*.

Figure 1 shows a 2MBP graph and its dual graph G^*. The edges of G are represented by solid lines. The edges of G^* are represented by dotted lines. The vertices of G^* (i.e. the faces of G) are represented by small circles. $\mathcal{S}(G^*)$ contains 3 S-walks S_1, S_2 and S_3. The face q_1 is a 1-walk face since it occurs on S_3 twice. The face q_2 is a 2-walk face since it occurs on both S_2 and S_3. (The S-walks in this figure are directed. Its meaning will be discussed in the next section).

The following theorem was proved in [3,4] by showing another linear equation system derived from G has a solution:

Theorem 3. *Let G be a 2MBP graph and $\mathcal{S}(G^*) = \{S_1, \ldots, S_k\}$ be the S-walks of G^*. The following statements hold:*

1. *If T is an even triangulation of G and the diagonals of T are flipped along an S-walk S_i, we obtain another even triangulation of G. (If a face f occurs on S twice, its diagonal is flipped twice and hence remains unchanged.)*
2. *If T_1 and T_2 are two even triangulations of G, then there is a collection of S-walks such that by flipping the diagonals of T_1 along these S-walks we obtain T_2.*

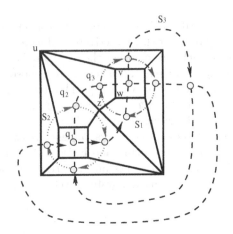

Fig. 1. A 2MBP graph G, the dual graph $G^{\square}$ and one of its S-orientation $\mathcal{O}$.

3   Even Triangulations of 2MBP Graphs

In this section, we provide new proofs of Theorems 1 and 3. First, we introduce a few definitions. An *orientation* of an undirected graph is an assignment of directions to its edges.

Definition 1. Let G be a 2MBP graph.

1. A *G-orientation* of G is an orientation of G such that every facial cycle of G is decomposed into two directed paths, one in clockwise direction and another in counterclockwise direction, each of length 2.
2. Let $\mathcal{G}_1, \mathcal{G}_2$ be two G-orientations of G. If every edge of G has reverse direction in $\mathcal{G}_1$ and $\mathcal{G}_2$, we say $\mathcal{G}_2$ is the *reverse* of $\mathcal{G}_1$. In this case, $\mathcal{G}_1$ and $\mathcal{G}_2$ are called a *G-orientation pair* of G.
3. Fix a G-orientation $\mathcal{G}$ of G. For each face q of G, the starting and the ending vertices of the two directed paths on the boundary of q are called the *primary vertices* of q. The diagonal of q connecting the two primary vertices is called its *primary diagonal* (with respect to $\mathcal{G}$). The other diagonal of q is called its *secondary diagonal* (with respect to $\mathcal{G}$).

Note: for a G-orientation pair $\mathcal{G}_1, \mathcal{G}_2$ and any face q of G, the primary diagonal of q with respect to $\mathcal{G}_1$ is the same as its primary diagonal with respect to $\mathcal{G}_2$. If $\mathcal{G}_1, \mathcal{G}_2$ do not form a G-orientation pair, then for some faces q of G, the primary diagonal of q with respect to $\mathcal{G}_1$ is different from that with respect to $\mathcal{G}_2$.

Definition 2. Let G be a 2MBP graph and G^* be its dual graph with $\mathcal{S}(G^*) = \{S_1, \ldots, S_k\}$.

1. An *S-orientation* of G^* is an orientation of G^* such that every S-walk S_i $(1 \leq i \leq k)$ is a directed closed walk.

2. Let $\mathcal{O}_1, \mathcal{O}_2$ be two S-orientations of G^*. If every S_i ($1 \leq i \leq k$) has reverse direction in $\mathcal{O}_1$ and $\mathcal{O}_2$, we say $\mathcal{O}_2$ is the *reverse* of $\mathcal{O}_1$. In this case, $\mathcal{O}_1$ and $\mathcal{O}_2$ are called an *S-orientation pair* of G^*.

3. Fix an S-orientation $\mathcal{O}$ of G^*. For each face q of G, if an S-walk S_i of G^* steps out of (into, respectively) q through an edge e on the boundary of q, then e is called an *out-edge* (*in-edge*, respectively) of q (with respect to $\mathcal{O}$).

We can assign two different directions to each S-walk. So, if G^* has k S-walks, it has 2^k distinct S-orientations and 2^{k-1} distinct S-orientation pairs.

Consider an arbitrary S-orientation $\mathcal{O}$ of G^*. It is easy to check that every face q of G has two out-edges and two in-edges with respect to $\mathcal{O}$, and the two in-edges of q are always incident to a common vertex on the boundary of q. Thus there are always two non-adjacent vertices on the boundary of q that are incident to both in-edges and out-edges of q. For example, consider the face q_3 in Fig. 1. An S-orientation $\mathcal{O}$ of G^* is shown in the figure. With respect to $\mathcal{O}$, the edge (u, v) is an out-edge of q_3 and the edge (w, z) is an in-edge of q_3. The vertices u and w are incident to both in-edges and out-edges of q_3.

Next we show that there exists a natural one-to-one mapping between the set of S-orientations of G^* and the set of G-orientations of G which preserves the pair relation.

Definition 3. Let G be a 2MBP graph and $\mathcal{O}$ be an S-orientation of G^*. We define an orientation of G from $\mathcal{O}$ as follows. Let e be any edge of G and e^* be its dual edge in G^*. Let q_1 and q_2 be the two faces of G with e on their boundaries. Suppose that e^* is directed from q_2 to q_1 in $\mathcal{O}$. When traveling e^* from q_2 to q_1, we direct e from right to left. (In other words, e is directed counterclockwise on the boundary of q_2 and clockwise on the boundary of q_1). This orientation of G will be called the *orientation induced* from $\mathcal{O}$ and denoted by $\pi(\mathcal{O})$.

An example of the induced orientation is shown in Fig. 2 (a). The S-walks S_i, S_j, S_m pass through the faces q_1, q_2 in the shown directions. The induced directions of the edges on the boundaries of q_1 and q_2 are also shown.

We omit the proof of the following lemma:

Lemma 1. *Let G be a 2MBP graph. If $\mathcal{O}$ is an S-orientation of G^*, then $\pi(\mathcal{O})$ is a G-orientation of G. If $\mathcal{O}$ and $\mathcal{O}'$ form an S-orientation pair, then $\pi(\mathcal{O})$ and $\pi(\mathcal{O}')$ form a G-orientation pair of G.*

Let $\mathcal{O}$ be an S-orientation of G^* and $\mathcal{G} = \pi(\mathcal{O})$. For any face q of G, the *primary diagonal* d of q with respect to $\mathcal{G}$ is the diagonal connecting the two vertices on the boundary of q that are incident to both the in-edges and the out-edges of q with respect to $\mathcal{O}$. We also call d the primary diagonal of q with respect to $\mathcal{O}$. For example, consider the face q_1 in Fig. 2 (a). The vertices v_1 and v_3 are incident to both in-edges and out-edges of q_1. Thus the primary diagonal of q_1 with respect to $\mathcal{O}$ is (v_1, v_3).

Definition 4. Let G be a 2MBP graph and $\mathcal{G}$ be a G-orientation of G. We define an orientation of G^* as follows. Consider any edge e^* in G^*. Let e be the edge in G corresponding to e^*. When traveling e along its direction in $\mathcal{G}$, we

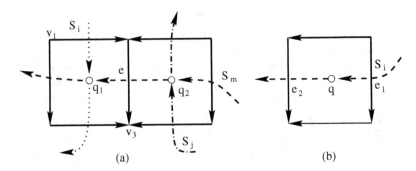

Fig. 2. (a) An S-orientation $\mathcal{O}$ induces a G-orientation $\mathcal{G}$; (b) A G-orientation $\mathcal{G}$ derives an S-orientation $\mathcal{O}$.

direct e^* from the face q_2 on the left to the face q_1 on the right. This orientation of G^* will be called the orientation *derived from* $\mathcal{G}$, and denoted by $\delta(\mathcal{G})$.

Let $\mathcal{S}(G^*) = \{S_1, S_2, \cdots, S_k\}$ be S-walks of G^*. Let $\mathcal{G}$ be a G-orientation of G. Let e_1 and e_2 be the two edges on the boundary of a face q that are walked through by S_i. Note that e_1 and e_2 are opposite on the boundary of q. Let e_1^* and e_2^* be the dual edges corresponding to e_1 and e_2, respectively. Because $\mathcal{G}$ is a G-orientation, e_1 and e_2 must have different direction. Thus e_1^* and e_2^* on S_i are assigned consistent directions in $\delta(\mathcal{G})$. (see Fig. 2 (b)). So, we have:

Lemma 2. *Let G be a 2MBP graph. If $\mathcal{G}$ is a G-orientation G, then $\delta(\mathcal{G})$ is an S-orientation of G^*. If $\mathcal{G}, \mathcal{G}'$ form a G-orientation pair, then $\delta(\mathcal{G})$ and $\delta(\mathcal{G}')$ form an S-orientation pair.*

By Definitions 3 and 4, it's straightforward to verify that π and δ are inverse mappings to each other. Hence we have:

Theorem 4. *For a 2MBP G, the mapping π (and δ) is a one-to-one correspondence between the set of G-orientations of G and the set of S-orientations of G^*, which preserves the pair relation.*

The following theorem describes how to obtain an even triangulation from a G-orientation of G.

Theorem 5. *Let G be a 2MBP graph.*

1. *Let $\mathcal{G}$ be a G-orientation of G. Then adding the primary diagonals into each face of G results an even triangulation of G, which is called the even triangulation determined by $\mathcal{G}$, and denoted by $\mathcal{T}(\mathcal{G})$.*
2. *Let $\mathcal{G}$ and $\mathcal{G}'$ be two G-orientations of G. If $\mathcal{G}, \mathcal{G}'$ form a G-orientation pair of G, they determine the same even triangulation of G, i.e. $\mathcal{T}(\mathcal{G}) = \mathcal{T}(\mathcal{G}')$. If $\mathcal{G}, \mathcal{G}'$ do not form a G-orientation pair of G, they determine different even triangulations of G, i.e. $\mathcal{T}(\mathcal{G}) \neq \mathcal{T}(\mathcal{G}')$.*

Proof. Statement 1: Let G' be the graph resulting from adding the primary diagonals (with respect to $\mathcal{G}$) into the faces of G. Consider any vertex v of G.

Let $e_1, e_2, \cdots, e_t$ be the edges in G incident to v in clockwise direction. Thus $deg_G(v) = t$. Let q_i $(1 \leq i \leq t)$ be the face incident to v and with e_i and e_{i+1} on its boundary (where $e_{t+1} = e_1$).

We call e_j a *go-in* (*go-out*, respectively) edge of v if e_j is directed into (out of, respectively) v in $\mathcal{G}$. A face q_i is called a *gap face* of v if one of the two edges e_i and e_{i+1} is a go-in edge of v and another is a go-out edge of v. q_i is called a *good face* of v if e_i, e_{i+1} are both go-in edges of v or both go-out edges of v. Denote the number of gap faces of v by $gap(v)$. Since a gap face is always between a block of consecutive go-out edges and a block of consecutive go-in edges around v, it is easy to see that $gap(v)$ is always an even number.

The primary diagonal (with respect to $\mathcal{G}$) of each face q is the diagonal connecting the starting and the ending vertices of the two directed paths on the boundary of q. Thus v is incident to the primary diagonal of the face q_i iff q_i is a good face of v. Hence $deg_{G'}(v) = deg_G(v) + t - gap(v) = 2t - gap(v)$ is always even. Therefore $G' = \mathcal{T}(\mathcal{G})$ is an even triangulation of G.

Figure 3 (a) shows an example of this construction. The solid lines are edges in G. The dotted lines are primary diagonals which are added into G'. The black dots indicate the gap faces. We have $deg_{G'}(v) = 2 \times 6 - 4 = 8$.

Statement 2: Suppose $\mathcal{G}$ and $\mathcal{G}'$ form a G-orientation pair of G. For any face q of G, the primary diagonal of q with respect to $\mathcal{G}$ is the same as that with respect to $\mathcal{G}'$. So $\mathcal{T}(\mathcal{G})$ is the same as $\mathcal{T}(\mathcal{G}')$.

Suppose $\mathcal{G}, \mathcal{G}'$ do not form a G-orientation pair. Then they have different sets of primary diagonals. So $\mathcal{T}(\mathcal{G})$ is different from $\mathcal{T}(\mathcal{G}')$. □

If $\mathcal{G}$ is induced from an S-orientation $\mathcal{O}$ of G^*, we also call $\mathcal{T}(\mathcal{G})$ the even triangulation *determined by* $\mathcal{O}$, and denote it by $\mathcal{T}(\mathcal{O})$.

Figure 4 shows an even triangulation constructed for the graph G in Fig. 1. Based on the discussion above, we have the following:

New Proof of Theorem 1: Let $\mathcal{S}(G^*) = \{S_1, S_2, \cdots, S_k\}$ be the S-walks of G^*. Arbitrarily assign a direction to each S_i. This gives an S-orientation $\mathcal{O}$ of G^*. By Lemma 1, we get an G-orientation $\mathcal{G} = \pi(\mathcal{O})$. By Theorem 5, we get an even triangulation $\mathcal{T}(\mathcal{G})$ of G. □

Theorem 5 states that every G-orientation pair determines an even triangulation of G. Next we show that every even triangulation of G is determined by a G-orientation pair of G.

Theorem 6. *Let G be a 2MBP graph and $\mathcal{S}(G^*) = \{S_1, S_2, \cdots, S_k\}$ be the S-walks of G^*.*

1. *Every even triangulation G' of G is determined by a G-orientation pair.*
2. *G has exactly 2^{k-1} distinct even triangulations.*

Proof. Statement 1: Let G' be any even triangulation of G. We want to show there exists a G-orientation $\mathcal{G}$ of G such that $G' = \mathcal{T}(\mathcal{G})$.

Let G'^* be the dual graph of G'. For each face q of G, the diagonal of q from G' splits q into two faces, which will be called the *subfaces* of G. So q contains two subfaces.

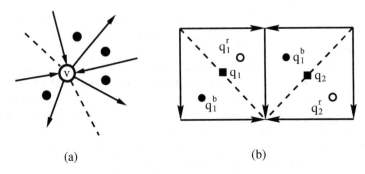

Fig. 3. (a) Adding primary diagonals with respect to a G-orientation $\mathcal{G}$ results even degree at vertex v; (b) An even triangulation of G induces a G-orientation $\mathcal{G}$ on faces q_1 and q_2.

Since each vertex has even degree in G', each facial cycle of G'^* is of even length. Thus G'^* is a bipartite plane graph. So we can color its vertices by 2 colors. In other words, we can color the subfaces of G by red and black so that no two red (black, respectively) subfaces are adjacent. Note that each face q of G contains one red and one black subface.

Consider any edge e of G. Let q_1 and q_2 be the two faces of G with e on their common boundary. Let q_1^r and q_1^b be the red and the black subfaces contained in q_1, respectively. Let q_2^r and q_2^b be the red and the black subfaces contained in q_2, respectively. Note that exactly one red subface and one black subface from the four subfaces $q_1^r, q_1^b, q_2^r, q_2^b$ have e on their boundaries. We direct e clockwise on its neighboring red subface. (Hence, e is directed counterclockwise on its black neighboring subface). This way, each edge e of G is consistently assigned a direction. Let $\mathcal{G}$ denote this orientation of G.

Consider any face q of G. Let q^r and q^b be the red and black subfaces contained in q, respectively. Two boundary edges of q are the boundary edges of q^r. Hence they are directed clockwise and form a directed path of length 2 in $\mathcal{G}$. The other two boundary edges of q are the boundary edges of q^b. Hence they are directed counterclockwise and form another directed path of length 2 in $\mathcal{G}$. Thus $\mathcal{G}$ is indeed a G-orientation of G.

Figure 3 (b) shows an example of this construction. Two faces q_1 and q_2 are shown in the figure by black squares. Each of them contains two subfaces. The red subfaces are indicated by empty circles. The black subfaces are indicated by solid circles. The edges of G are directed as described above.

For each face q of G, the diagonal d of q from the even triangulation G' is the diagonal which separates the red and the black subfaces of q. So it is the diagonal of q connecting the starting and the ending vertices of the two directed paths on the boundary of q. Hence d is the primary diagonal of q with respect to $\mathcal{G}$. Therefore, the even triangulation $\mathcal{T}(\mathcal{G})$ determined by $\mathcal{G}$ is exactly G'.

Statement 2: It follows directly from the statement 1 and Theorem 5.   □

In order to provide a new proof of Theorem 3, we need the following technical lemma, which can be easily proved:

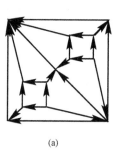

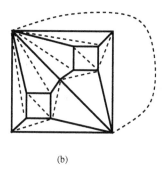

(a)   (b)

Fig. 4. (a) A G-orientation $\mathcal{G}$ of G corresponding to the S-orientation $\mathcal{O}$ in Fig. 1; (b) An even triangulation of G determined by $\mathcal{G}$.

Lemma 3. *Let G be a 2MBP and $\mathcal{S}(G^*) = \{S_1, \cdots, S_k\}$ be the S-walks of G^*. Let $\mathcal{O}_1$ be an S-orientation of G^*. Let $\mathcal{O}_2$ be the S-orientation of G^* obtained from $\mathcal{O}_1$ by reversing the direction of an S-walk S_i.*

Then $\mathcal{T}(\mathcal{O}_2)$ can be obtained from $\mathcal{T}(\mathcal{O}_1)$ by flipping the diagonals of all faces on S_i. (If q occurs on S_i twice, its diagonal is flipped twice and hence remains unchanged).

We are now ready to give the following:

New Proof of Theorem 3:

Statement 1: Let T be an even triangulation of a 2MBP G. Then T is determined by an S-orientation $\mathcal{O}$ of G^* by Theorem 6, i.e. $T = \mathcal{T}(\mathcal{O})$. Let $\mathcal{O}'$ be the S-orientation obtained from $\mathcal{O}$ by reversing the direction of an S-walk S_i. Let T' be the even triangulation $\mathcal{T}(\mathcal{O}')$ of G determined by $\mathcal{O}'$. By Lemma 3, T' is obtained from T be flipping the diagonals of the faces on S_i.

Statement 2: Let T_1 and T_2 be any two even triangulations of G. By Theorem 6, $T_1 = \mathcal{T}(\mathcal{O}_1)$ is determined by an S-orientation $\mathcal{O}_1$ and $T_2 = \mathcal{T}(\mathcal{O}_2)$ is determined by another S-orientation $\mathcal{O}_2$. Let $S_{i_1}, S_{i_2}, \cdots, S_{i_t}$ $(t \leq k)$ be those S-walks whose directions in $\mathcal{O}_1$ and $\mathcal{O}_2$ are reversed. By repeated applications of Lemma 3, if we start with T_1 and flip the diagonals of the faces on the S-walks S_{i_j} $(1 \leq j \leq t)$ one by one, we get T_2.   □

Based on the discussion above, we obtain the following:

Theorem 7. *Given a 2-connected bipartite plane graph G, an even triangulation of G can be constructed in $O(n)$ time.*

4   Even Triangulations of 2-Connected Graphs on High Genus Surfaces

Let $\mathcal{F}_g$ be a closed, connected, orientable surface without boundary, where g $(g \geq 0)$ is the *genus* of the surface. A graph G is said to be *embedded* on $\mathcal{F}_g$ if G is drawn on $\mathcal{F}_g$ such that no two edges cross and each of its faces is an open disc on $\mathcal{F}_g$. Let G be embeded on $\mathcal{F}_g$. If all of the facial cycles of G are of length 4, we call such a graph a *2-connected maximal even face graph on $\mathcal{F}_g$* (or 2MEFg graph for short).

There is a crucial difference between graphs on the plane and graphs on $\mathcal{F}_g$: If every face of a plane graph G is an even cycle, then G is bipartite. In contrast, even if every face of a graph G embedded on $\mathcal{F}_g$ ($g > 0$) is an even cycle, G is not necessarily bipartite. This means that a 2MEFg graph is not necessarily bipartite, so Equation (1) does not make sense any more. However, our proof of Theorem 5 can be adopted without much modification, we have the following:

Theorem 8. *Let G be a 2MEFg graph on $\mathcal{F}_g$.*

1. *Let $\mathcal{G}$ be a G-orientation of G. Then adding the primary diagonals to each face of G results an even triangulation of G, which is called the even triangulation determined by $\mathcal{G}$, and denoted by $\mathcal{T}(\mathcal{G})$.*
2. *Let $\mathcal{G}$ and $\mathcal{G}'$ be two G-orientations of G. If $\mathcal{G}, \mathcal{G}'$ form a G-orientation pair of G, they determine the same even triangulation of G, i.e. $\mathcal{T}(\mathcal{G}) = \mathcal{T}(\mathcal{G}')$. If $\mathcal{G}, \mathcal{G}'$ do not form a G-orientation pair of G, they determine different even triangulations of G, i.e. $\mathcal{T}(\mathcal{G}) \neq \mathcal{T}(\mathcal{G}')$.*

Similarly, the statement 1 of Theorem 3 is still true. However, the statement 2 of Theorem 3 and Theorem 6 are *false* for 2MEFg graphs due to the difference mentioned above. We have the following weaker result:

Theorem 9. *Let G be a 2MEFg graph on $\mathcal{F}_g$ and $\mathcal{S}(G^*) = \{S_1, \ldots, S_k\}$ be all S-walks of G^*. Then G has at least $\max(2^{k-1}, 2^{2g-2})$ distinct even triangulations.*

References

1. J. A. Bondy and U. S.R.Murty, *Graph theory with applications*, North Holland, New York, 1979.
2. T. H. Cormen, C. E. Leiserson and R. L. Rivest, *An introduction to algorithms*, McGraw-Hill, New York, 1990
3. F. Hoffmann and K. Kriegel, A graph-coloring result and its consequences for polygon-guarding problems, Technical Report TR-B-93-08, Inst. f. Informatik, Freie Universität, Berlin, 1993.
4. F. Hoffmann and K. Kriegel, A graph-coloring result and its consequences for polygon-guarding problems, *SIAM J. Discrete Math*, Vol 9(2): 210–224, 1996.
5. N. Jacobson, *Lectures in Abstract Algebras*, Springer-Verlag, New York, 1975.
6. R. J. Lipton, J. Rose and R. E. Tarjan, Generalized Nested Dissection, *SIAM J. Numer. Anal.*, 16: 346–358, 1979.
7. Huaming Zhang and Xin He, *A simple linear time algorithm for finding even triangulations of 2-connected bipartite plane graphs*, in Proceedings of ESA'02, LNCS 2461, pp. 902–913.

Petri Nets with Simple Circuits

Hsu-Chun Yen and Lien-Po Yu

Dept. of Electrical Eng., National Taiwan University,
Taipei, Taiwan, R.O.C.
yen@cc.ee.ntu.edu.tw

Abstract. We study the complexity of the reachability problem for a
new subclass of Petri nets called *simple-circuit Petri nets*, which properly
contains several well known subclasses such as *conflict-free, BPP, normal* Petri nets and more. A new *decomposition* approach is applied to
developing an integer linear programming formulation for characterizing
the reachability sets of such Petri nets. Consequently, the reachability
problem is shown to be NP-complete. The model checking problem for
some temporal logics is also investigated for simple-circuit Petri nets.

1 Introduction

Petri nets (*PNs*, for short) have been a popular model for reasoning about the
behaviors of concurrent systems [13]. The *reachability* problem is among the
most important problems in the study of PNs. Reachability analysis is key to
the solutions of such PN problems as *liveness, fairness, controllability, model
checking* and more. In addition, identifying a tight complexity bound for the
reachability problem remains a great challenge in the community of theoretical
computer science. Although known to be decidable, the existing algorithm for
the problem remains not even primitive recursive [11] (see also [9]), while the
problem is also known to be exponential space hard [10].

Integer linear programming(ILP) has long been a tool for the reachability
analysis of PNs. It is well known that in a PN $\mathcal{P}$ with initial marking μ_0, a
marking μ is reachable from μ_0 *only if* there exists a column vector x such that
the *state equation* $\mu_0 + A \cdot x = \mu$ holds, where A is the *addition matrix* of $\mathcal{P}$.
Although the converse does not necessarily hold, there are restricted classes of
PNs for which the state equation is sufficient and necessary to capture reachability. Most notable is the class of *circuit-free* PNs as well as the class of PNs
without *token-free* circuits in every reachable marking [15]. Other subclasses for
which reachability has been thoroughly studied and solved include *conflict-free,
normal* [7,15], *sinkless* [7,15], *BPP-net* [1,18], *trap-circuit* [8,17], and *extended
trap-circuit* PN [17], etc. For each of them, deciding reachability can be equated
with solving an ILP problem. A question arises: Can we enlarge the PN class
while retaining the nice property of reachability being characterizable by ILP?
Affirmative answer to this question is one of the contributions of this paper.

Circuits in BPP-nets are referred to as ⊕-circuits with every transition in
the net having exactly one input place, and the firing of a transition removing

T. Warnow and B. Zhu (Eds.): COCOON 2003, LNCS 2697, pp. 149–158, 2003.
© Springer-Verlag Berlin Heidelberg 2003

exactly one token from it's sole input place. In normal PNs, no transition is capable of decreasing the token count of a minimal circuit, and such circuits are called ⊙-circuits. Our new PN class, called *simple-circuit Petri nets* (*sc-PNs*, for short), consists of those in which each minimal circuit is either a ⊙-circuit, or a ⊕-circuit which is not properly included in any non-⊕-circuit. By relaxing the constraints on circuits, our sc-PNs properly contain that of conflict-free, normal, trap-circuit, extended trap-circuit, and BPP-nets as Figure 1 indicates.

To analyze sc-PNs, the technique of the so-called *decomposition approach* is used. Given a computation σ of a PN, the basic idea is to rearrange σ into some canonical form $\sigma_1 \sigma_2 \cdots \sigma_n$ with each of them being of some 'simpler form'. By a 'simpler form' we mean the sub-PN induced by each of the segments has its reachability set characterizable by certain well-understood and easily solvable formulations, such as ILP. For cases, we can also place a bound on the number of segments in the above canonical computation. Demonstrating the applicability of the decomposition approach to sc-PNs is another contribution of our work. It is worthy of noting that our analysis yields an ILP formulation for the reachability problem in which the initial and final markings are regarded as *parameters*, as opposed to being constants as in many of the traditional reachability analysis of PNs. The complexity of model checking with respect to a number of temporal logics is also investigated.

2   Preliminaries

A *Petri net* is a 3-tuple (P, T, φ), where P is a finite set of *places*, T is a finite set of *transitions*, and φ is a *flow function* $\varphi : (P \times T) \cup (T \times P) \to \{0, 1\}$. A *marking* is a mapping $\mu : P \to N$. Pictorially, Petri net is a directed, bipartite graph consisting of two kinds of nodes: *places* (represented by circles within which each small black dot denotes a *token*) and *transitions* (represented by bars or boxes). See Figure 1 for an example.

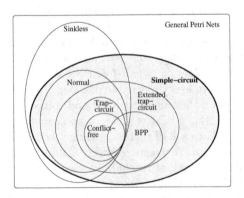

Fig. 1. Containment relationships among various Petri net classes.

A transition $t \in T$ is *enabled* at a marking μ iff $\forall p \in P, \varphi(p,t) \le \mu(p)$. If a transition t is enabled, it may *fire* by removing a token from each input place and putting a token in each output place. We then write $\mu \xmapsto{t} \mu'$, where $\mu'(p) = \mu(p) - \varphi(p,t) + \varphi(t,p) \; \forall p \in P$. A sequence of transitions $\sigma = t_1 ... t_n$ is a *firing sequence* from μ_0 iff $\mu_0 \xmapsto{t_1} \mu_1 \xmapsto{t_2} \cdots \xmapsto{t_n} \mu_n$ for some markings $\mu_1, ..., \mu_n$. (We also write '$\mu_0 \xmapsto{\sigma} \mu_n$'.) A *marked* PN is a pair $((P,T,\varphi), \mu_0)$, where (P,T,φ) is a PN, and μ_0 is called the *initial marking*. Throughout the rest of this paper, the word 'marked' will be omitted if it is clear from the context. By establishing an ordering on the elements of P and T (i.e., $P = \{p_1, ..., p_k\}$ and $T = \{r_1, ..., r_m\}$), we can view a marking μ as a k-dimensional column vector with its i-th component being $\mu(p_i)$, and $\#_\sigma$ as an m-dimensional vector with its jth entry denoting the number of occurrences of transition r_j in σ. The *reachability set* of $\mathcal{P}$ with respect to μ_0 is the set $R(\mathcal{P}, \mu_0) = \{\mu \mid \exists \sigma \in T^*, \mu_0 \xmapsto{\sigma} \mu\}$. The *reachability problem* is that of, given a marked PN $\mathcal{P}$ (with initial marking μ_0) and a marking μ, deciding whether $\mu \in R(\mathcal{P}, \mu_0)$.

For ease of expression, the following notations will be used. Let σ, σ' be transition sequences, p a place, and t a transition. $\Delta(\sigma) = [T] \cdot \#_\sigma$ defines the *displacement* of σ. For $p \in P$, $\Delta(\sigma)(p)$ denotes the component of $\Delta(\sigma)$ corresponding to place p. $Tr(\sigma) = \{t \mid t \in T, \#_\sigma(t) > 0\}$ denotes the set of transitions used in σ. $\sigma \div \sigma'$ represents the sequence resulting from removing each transition of σ' from the leftmost occurrence of such a transition in σ. For instance, if $\sigma = t_1 t_2 t_3 t_4 t_5$ and $\sigma' = t_1 t_3 t_4$, then $\sigma \div \sigma' = t_2 t_5$. Intuitively, $\sigma \div \sigma'$ $p^\bullet = \{t \mid \varphi(p,t) \ge 1, t \in T\}$ is the set of output transitions of p; $t^\bullet = \{p \mid \varphi(t,p) \ge 1, p \in P\}$ is the set of output places of t; $^\bullet p = \{t \mid \varphi(t,p) \ge 1, t \in T\}$ is the set of input transitions of p; $^\bullet t = \{p \mid \varphi(p,t) \ge 1, p \in P\}$ is the set of input places of t. Given $\mu_0 \xmapsto{\sigma} \mu$, a sequence σ' is said to be a *rearrangement* of σ if $\#_\sigma = \#_{\sigma'}$ and $\mu_0 \xmapsto{\sigma'} \mu$. A *circuit* c of a PN is a sequence $p_1 t_1 p_2 t_2 \cdots p_n t_n p_1$ ($p_i \in P$, $t_i \in T$, $p_i \in {}^\bullet t_i$, $t_i \in {}^\bullet p_{i+1}$), such that $p_i \ne p_j, \forall i \ne j$. We write $P_c = \{p_1, p_2, \cdots, p_n\}$ (resp., $T_c = \{t_1, t_2, \cdots, t_n\}$) to denote the set of places (resp., transitions) in c, and $tr(c)$ to represent the sequence $t_1 t_2 \cdots t_n$. We define the token count of circuit c in marking μ to be $\mu(c) = \sum_{p \in P_c} \mu(p)$. A circuit c is said to be *token-free* in μ iff $\mu(c) = 0$. Given two circuits c and c', c is said to be *included* (resp., *properly included*) in c' iff $P_c \subseteq P_{c'}$ (resp., $P_c \subset P_{c'}$). We say c is *minimal* iff it does not properly include any other circuit. Circuit c is said to be a

- $\oplus$-circuit iff $\forall i, 1 \le i \le n, {}^\bullet t_i = \{p_i\}$
- $\odot$-circuit iff $\forall t \in T, (\sum_{p \in P_c} (\varphi(t,p) - \varphi(p,t))) \ge 0$

A set of circuits $\mathcal{C} = \{c_1, c_2, ..., c_n\}$ is said to be *connected* iff $\forall i, j, 1 \le i, j \le n$, there exist $1 \le h_1, h_2, ..., h_r \le n$, for some r, such that $h_1 = i$, $h_r = j$, and $\forall l, 1 \le l < r, P_{c_{h_l}} \cap P_{c_{h_{l+1}}} \ne \emptyset$. In words, every pair of neighboring circuits in $c_{h_1}, c_{h_2}, ..., c_{h_r}$ share at least one place. σ is said to *cover* circuit c if $\#_{tr(c)} \le \#_\sigma$.

A PN $\mathcal{P} = (P, T, \varphi)$ is a *sc-PN* if $\forall$ minimal circuit c in $\mathcal{P}$, (1) c is a $\odot$-circuit, or (2) c is a $\oplus$-circuit and if c' properly includes c, c' must be a $\oplus$-circuit as well. To the best of our knowledge, the class of sc-PNs defined in this paper is new.

Notice that being *simple-circuit* is a 'structural property' which is independent of the initial marking.

The interested reader is referred to [13] for more about PNs.

3   Decomposition Approach for Reachability

Decomposition is one of the few useful techniques for analyzing various subclasses of PNs, as demonstrated in [4,5,7,14,17,18] and more recently in [3]. Given a computation $\mu_0 \overset{\sigma}{\longmapsto} \mu$ of a PN $\mathcal{P}=(P,T,\varphi)$, the idea is to rearrange σ into $\sigma_1\sigma_2\cdots\sigma_n$ such that $\mu_0 \overset{\sigma_1}{\longmapsto} \mu_1 \overset{\sigma_2}{\longmapsto} \mu_2 \cdots \mu_{n-1} \overset{\sigma_n}{\longmapsto} \mu_n = \mu$ and each of $\sigma_1, ..., \sigma_n$ is of some 'simpler form.' What a 'simpler form' means is that if we define $\mathcal{P}_i = (P, T_i, \varphi|_{T_i})$ (where $T_i = Tr(\sigma_i)(\subseteq T)$, and $\varphi|_{T_i}$ is the restriction of φ to $(P \times T_i) \cup (T_i \times P)$) to be a sub-PN induced by segment σ_i ($1 \le i \le n$), then reachability in $\mathcal{P}_i$ is easily solvable. Two notable examples are the classes of *normal* PNs and *BPP nets*, for which the decomposition approach gives rise to an ILP formulation for reachability. Since sc-PNs admit both circuit types found in *normal* and *BPP* PNs, a detailed description of how the decomposition is performed for these two sub-classes is in order. For simplicity, we write $ILP(\mathcal{P}, \mu_0, \mu)$ to denote an instance of ILP for checking whether μ is reachable from μ_0 in $\mathcal{P}$.

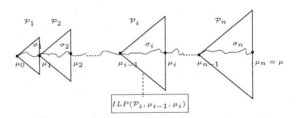

Fig. 2. Decomposition of a normal Petri net. (Note the number of distinct transitions that T_i contains is reflected by the size of respective triangle associated with $\mathcal{P}_i$)

3.1   Decomposition Approach for Normal PNs

The idea behind the decomposition analysis of normal PNs is illustrated in Figure 2. The rearrangement $\sigma_1\sigma_2\cdots\sigma_n$ of σ is such that if $\sigma = t_1\sigma_1't_2\sigma_2'\cdots t_n\sigma_n'$ where $t_1, t_2, ..., t_n$ mark the first occurrences of the respective transitions in σ (i.e., $t_i, 1 < i \le n$, is not in the prefix $t_1\sigma_1'\cdots t_{i-1}\sigma_{i-1}'$) then σ_i is a permutation of $t_i\sigma_i'$. Furthermore, by letting (1) $T_0 = \varnothing$; (2) $\forall 1 \le i \le n$, $T_i = T_{i-1} \cup \{t_i\}$, and φ_i is the restriction of φ to $(P \times T_i) \cup (T_i \times P)$, reachability in $\mathcal{P}_i$ can be captured by an instance $ILP(\mathcal{P}_i, \mu_{i-1}, \mu_i)$. This, in conjunction with the fact that $n \le |T|$ allows the reachability problem to be solved by ILP. See [7] for more details.

3.2   Decomposition Approach for BPP-Nets

The idea of rearranging an arbitrary computation in a BPP-net into a canonical one is explained using Figure 3.

(1) Referring to Figure 3(a), suppose the letters a, b, d, e, f depict those transitions that form a $\oplus$-circuit c in the PN with $\mu(c) > 0$.

(2) We use c as a 'seed' to grow the largest collection of connected circuits that are covered by σ (In Figure 3(b), circuit c together with c' (consisting of transitions x, y, and z) forms such a collection.)

(3) We then follow a 'shortest' circuit-free transition sequence of the remaining computation (actually, a rearrangement of $\sigma \doteq abdefxyz$ in Figure 3(b)) until reaching a marking (see marking μ' in Figure 3(c)) in which a non-token-free circuit (i.e., c'' in Figure 3(c)) is covered by the subsequent computation.

(4) Using the above non-token-free circuit as a new seed and repeating the above procedures until the remaining computation becomes null, we are able to rearrange an arbitrary computation of a BPP-net into a canonical one.

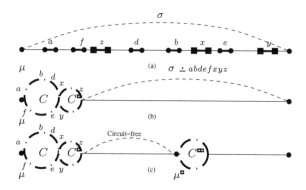

Fig. 3. Decomposition of a BPP-net.

It was shown in [18] that the above decomposition allows us to formulate reachability as ILP. A similar strategy has been applied to the so-called *extended trap-circuit PNs* which subsume BPP-nets [17].

4   Characterizing sc-PNs Computations Using ILP

(1) Stemming from the idea of Section 3.1, the decomposition is constructed stage-by-stage (Figure 4) as a sequence of sub-PNs $\mathcal{P}_1, \cdots, \mathcal{P}_n$ where $\mathcal{P}_i = ((P, T_i, \varphi|_{T_i}), \mu_{i-1}))$, $T_0 = \emptyset$, $T_i = T_{i-1} \cup \{t_i\}$, for some $t_i \notin T_{i-1}$ enabled at μ_{i-1}, and $\sigma_i \in T_i^*$. ($t_1, ..., t_n$ are chosen the same way) Unlike normal PNs where reachability in $\mathcal{P}_i$ can be completely captured by the state equation, a more involved procedure to further decompose σ_i is needed.

(2) For stage i, we carry out the following steps:

(2.1) Apply a strategy similar to that of Section 3.2 to rearrange σ_i such that once a $\oplus$-circuit c covered by σ_i is enabled , then use c as a 'seed' to grow the largest collection C of connected circuits covered by σ_i (see Figure 3 for a similar demonstration). (Guaranteed by Lemma 2).

(2.2) If the set of transitions in the remaining computation, together with its associated places, forms a normal PN, it's done; otherwise, similar to Section 3.2, we follow a sequence along which all the $\oplus$-circuits covered by the computation remain token-free (α_1^i in Figure 4) until reaching a marking where a $\oplus$-circuit is not token-free. Due to the nature of the decomposition (elaborated in Lemma 3), the $(P, Tr(\alpha_j^i), \varphi|_{Tr(\alpha_j^i)})$ is guaranteed to be a normal sub-PN.

(2.3) Taking the newly found $\oplus$-circuit as a new 'seed,' the above procedures repeat anew until no more $\oplus$-circuit is covered by the remaining computation of σ_i (see α_h^i in Figure 4).

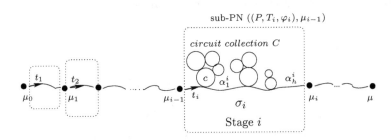

Fig. 4. Rearranging an sc-PN computation into a canonical form.

Lemma 1. *(from Lemma 1 in [18]) Let $\mathcal{C} = \{c_1, c_2, ..., c_n\}$ be a set of connected $\oplus$-circuits in an sc-PN $\mathcal{P}$ and μ be a marking with $\mu(c_i) > 0$, for some i. For arbitrary integers $a_1, a_2, ..., a_n > 0$, there exists a sequence σ such that $\mu \stackrel{\sigma}{\longmapsto}$ and*

$$\#_\sigma = \sum_{j=1}^{n} a_j(\#_{c_j}).$$ *(In words, from μ there exists a firable sequence σ utilizing circuit c_j exactly a_j times, for every j.)*

Lemma 2. *Consider a computation $\mu_0 \stackrel{\tau}{\longmapsto} \mu_1 \stackrel{\sigma}{\longmapsto} \mu_2$ in an sc-PN $\mathcal{P} = (P, T, \varphi)$. Let $\mathcal{C} = \{c_1, c_2, ..., c_z\}$ be a set of connected $\oplus$-circuits and $a_1, a_2, ..., a_z$ be positive integers such that*

(a) *$(\exists i, 1 \leq i \leq z)\ (\mu_1(c_i) > 0)$ (i.e., some circuit c_i is not token-free in μ_1.),*

(b) *$\sigma \doteq (c_1^{a_1} \cdots c_z^{a_z})$ does not cover any $\oplus$-circuit that shares some place with $\oplus$-circuits in $\mathcal{C}$ (i.e., $\mathcal{C}$ is a largest collection of connected circuits.),*

(c) *$\sum_{j=1}^{z} a_j(\#_{c_j}) \leq \#_\sigma$ (i.e., all the circuits $c_1, c_2, ..., c_z$ are covered by σ.), and*

(d) *$\forall t \in Tr(\sigma)$, t is enabled in some marking along $\mu_0 \stackrel{\tau}{\longmapsto} \mu_1$,*

then there exist δ_1 and δ_2 such that (1) $\#_{\delta_1} = \sum_{j=1}^{z} a_j(\#_{c_j})$, (2) $\#_{\delta_2} = \#_{\sigma \doteq \delta_1}$, and (3) $\mu_1 \stackrel{\delta_1}{\longmapsto} \mu_3 \stackrel{\delta_2}{\longmapsto} \mu_2$, for some μ_3. (In words, σ can be rearranged into $\delta_1 \delta_2$

such that δ_1 consists of the largest collection of connected $\oplus$-circuits with at least one of them marked in μ_1.)

Proof. First notice that $\mu_1 \xrightarrow{\delta_1} \mu_3$ is guaranteed by Lemma 1; it suffices to prove that $\mu_3 \xrightarrow{\delta_2} \mu_2$, for some δ_2 which is a rearrangement of $\sigma \doteq \delta_1$. Suppose, to the contrary, that none of the permutations of $\sigma \doteq \delta_1$ is firable in μ_3. We let α be the longest sequence such that $\#_\alpha < \#_{\sigma \doteq \delta_1}$ and $\mu_3 \xrightarrow{\alpha} \mu_4$, for some μ_4. (By 'longest' we mean that for all α' with $\#_{\alpha'} < \#_{\sigma \doteq \delta_1}$ and $\mu_3 \xrightarrow{\alpha'}$, it must be the case that $|\alpha'| \leq |\alpha|$.) Let $\beta = (\sigma \doteq \delta_1) \doteq \alpha$. We let X be $\{p | p \in {}^\bullet t, \mu_4(p) = 0, t \in Tr(\beta)\}$. i.e., X consists of those input places of transitions in $Tr(\beta)$ that are token-free in μ_4. We now make the following observations:

1. $\forall p \in X, \exists t' \in Tr(\beta)$, such that $p \in t'^\bullet$. (This is because $\mu_4(p) + \Delta(\beta)(p) = \mu_2(p) \geq 0$ and $\mu_4(p) = 0$.)
2. There must be some place r in X such that either (i) $\mu_1(r) > 0$, or (ii) ($\exists$ $t_1 \in Tr(\delta_1\alpha)$) ($r \in t_1^\bullet$). And for each such r, $\exists t_2 \in Tr(\delta_1\alpha)$ such that $r \in {}^\bullet t_2$. (Assume, to the contrary, that neither (i) nor (ii) holds. In σ, let f be the first transition depositing a token into some place in X. Since $f \notin Tr(\delta_1\alpha)$, one of f's input places, say g, must be in X. In this case, place g could never have possessed a token along σ to the marking at which f is fired – a contradiction. The existence of a t_2 results from $\mu_4(r) = 0$.)

Let R be the set of all places r satisfying Observation 2(i) or (ii) above. We are to show that at least one place in R must be along a circuit consisting of some places in X and some transitions in $Tr(\beta)$. Suppose, to the contrary, that none of R is on a circuit; then there must be an $s \in R$ such that s cannot be reached from the remaining places in R through places in X and transitions in $Tr(\beta)$. For s, let t_3 be a transition guaranteed by Observation 1 above. Due to the selection of s, t_3 could never have been fired in σ since ${}^\bullet t_3 \cap X$ would never possess a token (because none of the input places of t_3 is in R, and due to the definition of s, none of R is capable of supplying a token to ${}^\bullet t_3$ directly or indirectly) – a contradiction. Intuitively, one can think of R as places through which tokens are 'pumped' into the sub-PN consisting of places in X and transitions in $Tr(\beta)$.

Let $r \in R$ be a place on a circuit, say c, and t_2 (guaranteed by Observation 2) be a transition in $\delta_1\alpha$ removing a token from r. (Note c is token-free in μ_4.) Due to Assumption (d) of the lemma, c is not a $\odot$-circuit; otherwise, c would not have become token-free in μ_4. Now c is a $\oplus$-circuit whether it is a minimal circuit or not. (It clearly holds if c is minimal; otherwise, due to Condition (2) in the definition of sc-PNs, c again must be a $\oplus$-circuit.) If t_2 is in δ_1 (which comprises only circuits from $\mathcal{C}$), then c must have shared some place with one of the circuits in $\mathcal{C}$ – violating Assumption (b) of the lemma. If t_2 is in α, then r is marked during the course of the computation α, which implies that c should have been added to α – violating the assumption about α being longest.

$\square$

By repeatedly applying the above cut-and-paste strategy, we can construct the decomposition within a stage (e.g., stage i in Figure 4) as the following lemma indicates. One of the keys in this lemma lies in that the sub-computation linking two neighboring $\oplus$-circuit collections (see α_1^i in Figure 4, for instance) constitutes a normal sub-PN. Due to space limitations, the proof details are omitted.

Lemma 3. *Let* $\mathcal{P} = ((P, T, \varphi), \mu_0)$ *be an sc-PN and* $\mu_0 \overset{\delta}{\longmapsto} \mu_1$ *for some* $\delta \in T^*$. *Suppose* $\mathcal{P}' = ((P, T', \varphi|_{T'}), \mu_1)$ *is a sub-PN such that each* $t \in T' (\subseteq T)$ *is enabled at some point along* $\mu_0 \overset{\delta}{\longmapsto} \mu_1$. *Then,* μ *is reachable from* μ_1 *in* $\mathcal{P}'$ *iff there exists a sequence* $\sigma = \pi_1 \alpha_1 \pi_2 \alpha_2 \cdots \pi_h \alpha_h$ ($\alpha_i, \pi_i \in T^*$*, and* $1 \leq h \leq |T|$*) which witnesses* $\mu_1 \overset{\sigma}{\longmapsto} \mu$ *and satisfies the following conditions:*

1. $\forall i, 1 \leq i \leq h$,
    a) $\exists$ *a set* $\mathcal{C}_i = \{c_1^i, ..., c_{r_i}^i\}$ ($r_i \leq m$) *of connected* $\oplus$-*circuits such that*
    $$\Delta(\pi_i) = \sum_{j=1}^{r_i} a_j^i \Delta(c_j^i) \text{ for some integers } a_1^i, a_2^i, ..., a_{r_i}^i > 0,$$
    b) *the remaining sequence* $\alpha_i \cdots \pi_h \alpha_h$ *does not cover any circuit which shares some place with circuits in* $\mathcal{C}_i$, *and*
    c) $\sum_{i=1}^{h} |\mathcal{C}_i| \leq |T|$, *i.e., the total number of distinct circuits considered above is bounded by the number of transitions of the PN.*
2. $\forall i, 1 \leq i \leq h, \alpha_i \in T^+$, $(P, Tr(\alpha_i), \varphi|_{Tr(\alpha_i)})$ *forms a normal sub-PN.*

Theorem 1. *Each computation* $\mu_0 \overset{\sigma}{\longmapsto} \mu$ *of an sc-PN* $\mathcal{P} = (P, T, \varphi)$ *can be rearranged into a canonical one* $\mu_0 \overset{\sigma_1}{\longmapsto} \mu_1 \overset{\sigma_2}{\longmapsto} \mu_2 \cdots \overset{\sigma_n}{\longmapsto} \mu_n(= \mu)$*, for some* n*,* $1 \leq n \leq m$*, such that* $\forall i, 1 < i \leq n$

1. $Tr(\sigma_i) - Tr(\sigma_{i-1}) = \{t_i\}$*, for some* $t_i \in T$*,*
2. $\sigma_i = \pi_1^i \alpha_1^i \pi_2^i \alpha_2^i \cdots \pi_{h_i}^i \alpha_{h_i}^i$*, where* π_j^i *and* $\alpha_j^i (1 \leq j \leq h_i)$ *satisfy those conditions stated in Lemma 3.*

Proof. Omitted.

$\square$

Lemma 4. *(From Lemma 4.3 in [7]) Given a normal PN* $\mathcal{P} = (P, T, \varphi)$ *and a marking* μ_0*, we can construct, in nondeterministic polynomial time, a system of linear inequalities* $ILP(\mathcal{P}, \mu_0, \mu)$ *(of size bounded by a polynomial in the size of* $\mathcal{P}$*) such that* μ *is reachable from* μ_0 *iff* $ILP(\mathcal{P}, \mu_0, \mu)$ *has an integer solution.*

Theorem 2. *Given an sc-PN* $\mathcal{P} = (P, T, \varphi)$ *and a marking* μ_0*, we can construct, in nondeterministic polynomial time, a system of linear inequalities* $ILP(\mathcal{P}, \mu_0, \mu)$ *(of size bounded by a polynomial in the size of* $\mathcal{P}$*) such that* μ *is reachable from* μ_0 *iff* $ILP(\mathcal{P}, \mu_0, \mu)$ *has an integer solution.*

Proof. (Sketch) By Theorem 1, $\mu \in R(\mathcal{P}, \mu_0)$ iff $\exists$ a computation $\mu_0 \overset{\sigma_1}{\longmapsto} \mu_1 \overset{\sigma_2}{\longmapsto} \mu_2 \cdots \overset{\sigma_n}{\longmapsto} \mu_n(= \mu)$ meeting Conditions (1) and (2) of the theorem. In addition, those sub-computations π_j^i and α_j^i ($1 \leq j \leq h_i$) within σ_i

$(\sigma_i=\pi_1^i\alpha_1^i\pi_2^i\alpha_2^i\cdots\pi_{h_i}^i\alpha_{h_i}^i)$ satisfy those conditions stated in Lemma 3. To set up the $ILP(\mathcal{P},\mu_0,\mu)$, we begin by guessing the sequence $t_1,\cdots,t_n$ to capture $t_i = head(\sigma_i)$. The associated inequalities are: $\mu_{i-1}(p) \geq \varphi(p, t_i), \forall p \in P, \forall i, 1 < i \leq n$. The desired system of linear inequalities associated with $\mu_{i-1} \xrightarrow{\sigma_i} \mu_i$ is set up as follows. (σ_i uses transitions taken from $\{t_1, ..., t_i\}$.)

1. For $1 \leq j \leq h_i$, guess the set of connected $\oplus$-circuits $\mathcal{C}_{i,j}$ ($=\{c_1^{i,j}, ..., c_{r_{i,j}}^{i,j}\}$) and verify the conditions as stated in (a), (b) and (c) of Conditions 1 of Lemma 3; for $1 \leq j \leq h_i - 1$, guess the the sequence α_j^i and check the sub-PN $\mathcal{P}_{i,j} = (P, Tr(\alpha_j^i), \varphi|_{Tr(\alpha_j^i)})$ forms a normal sub-PN. It is not hard to see that checking each of the above can be done in polynomial time.

2. As $\mu_{i-1} \xrightarrow{\sigma_i} \mu_i$ and $\sigma_i=\pi_1^i\alpha_1^i\pi_2^i\alpha_2^i\cdots\pi_{h_i}^i\alpha_{h_i}^i$, there shall be some markings, say $\mu_{i,j}$ and $\mu'_{i,j}$ ($\geq \mathbf{0}$), $\forall j, 1 \leq j \leq h_i$, such that $\mu_{i-1} = \mu_{i,1} \xrightarrow{\pi_1^i} \mu'_{i,1} \xrightarrow{\alpha_1^i}$ $\mu_{i,2} \xrightarrow{\pi_2^i} \mu'_{i,2} \xrightarrow{\alpha_2^i} \cdots \mu_{i,h_i} \xrightarrow{\pi_{h_i}^i} \mu'_{i,h_i} \xrightarrow{\alpha_{h_i}^i} \mu_i$. Now, we are able to set up the following linear inequalities to capture the above PN computation:

$$\begin{cases} ILP(\mathcal{C}_{i,j}, \mu_{i,j}, \mu'_{i,j}), \forall j, 1 \leq j \leq h_i & -----(due\ to\ Lemma\ 1) \\ ILP(\mathcal{P}_{i,j}, \mu'_{i,j}, \mu_{i,j+1}), \forall j, 1 \leq j \leq h_i & -----(due\ to\ Lemma\ 4) \end{cases}$$

Due to space limitations, the remaining details are omitted.

$\square$

It is also known that the reachability problem for either normal or BPP-nets is NP-hard, and hence the following holds.

Theorem 3. *The reachability problem for sc-PNs is NP-complete.*

5   Model Checking

EF is the fragment of *unified system of branching time* allowing only EF operators (and their duals), but not EG operators. For *labeled PNs* $\mathcal{P}=(P,T,\varphi,l)$, l is a *labeling function* $l : T \to \Sigma$ (a set of *labels*), each formula ϕ in logic *EF* is of the form: $\phi ::= \mathbf{true} \mid \neg\phi \mid \phi_1 \wedge \phi_2 \mid E(a)\phi \mid EF\phi$, where $a \in \Sigma$. A marking μ satisfying a formula ϕ, denoted by $\mu \models \phi$, is defined inductively as follows.

$\mu \models true$ always holds
$\mu \models \neg\phi$ iff $\neg(\mu \models \phi)$
$\mu \models \phi_1 \wedge \phi_2$ iff $\mu \models \phi_1$ and $\mu \models \phi_2$
$\mu \models E(a)\phi$ iff $\exists \mu'$ such that $\mu \xrightarrow{t} \mu'$ and $\mu' \models \phi$, for some t with $l(t) = a$
$\mu \models EF\phi$ iff $\exists$ a path $\mu_1 \longmapsto \mu_2 \longmapsto \mu_3 \cdots$ s.t. $\mu = \mu_1$ and $\exists i \geq 1$ $\mu_i \models \phi$

A labeled PN $\mathcal{P}=((P,T,\varphi,l),\mu_0)$ is said to satisfy a formula ϕ if $\mu_0 \models \phi$. In [2], *EF* has been augmented with *Presburger formulas* and is called *EF+Pres* for which model checking is decidable for BPP-nets ([2]). For *EF*, model checking for BPP-nets is PSPACE-complete ([12]). In what follows, we supplement the results of [2,12] by showing that for sc-PNs, model checking a fragment of *EF+Pres* can be equated with solving a systems of linear inequalities, thus yielding an NP algorithm. Let $\tilde{EF} + Pres$ be a fragment of $EF+Pres$ with the $\neg$ operator being applied only to formulas without $E(a)$ and EF operators.

Theorem 4. *For $\tilde{EF} + Pres$, the model checking problem for sc-PNs is NP-complete.*

Proof. Omitted.

$\square$

Similar strategies can be used to derive complexity of model checking for a linear-time temporal logic defined in [6], and *path formulas* defined in [16].

References

1. Esparza, J. Petri nets, commutative context-free grammars and basic parallel processes, *Fundamenta Informaticae* **30**, 24–41, 1997.
2. Esparza, J. Decidability of model checking for infinite-state concurrent systems, *Acta Inform.* **34**, 85–107, 1997.
3. Fribourg, L. Petri nets, flat languages and linear arithmetic. *9th Int. Workshop. on Functional and Logic Programming*, pp. 344–365, 2000.
4. Fribourg, L. and Olsén, H. Proving safety properties of infinite state systems by compilation into Presburger arithmetic, *LNCS* **1243**, 213–227, 1997.
5. Fribourg, L. and Olsén, H. A decompositional approach for computing least fixed-points of datalog programs with z-counters, *Constraints, An International Journal* **2**, 305–335, 1997.
6. Howell, R. and Rosier, L. On questions of fairness and temporal logic for conflict-free Petri nets, In G. Rozenberg, editor, Advances in Petri Nets, *LNCS* **340**, 200–226, Springer-Verlag, Berlin, 1988.
7. Howell, R., Rosier, L. and Yen, H. Normal and sinkless Petri nets, *J. of Computer and System Sciences* **46**, 1–26, 1993.
8. Ichikawa, A. and Hiraishi, K. Analysis and control of discrete event systems represented by Petri nets, *LNCIS* **103** ,115–134, 1987.
9. Kosaraju, R. Decidability of reachability in vector addition systems, *Proc. the 14th Annual ACM Symposium on Theory of Computing*, 267–280, 1982.
10. Lipton, R. *The reachability problem requires exponential space*, Technical Report 62, Yale University, Dept. of CS., Jan. 1976.
11. Mayr, E. An algorithm for the general Petri net reachability problem, *SIAM J. Comput.* **13**, 441–460, 1984.
12. Mayr, R. Weak bisimulation and model checking for basic parallel processes, *Proc. FSTTCS'96, LNCS* **1180**, 88–99, 1996.
13. Murata, T. Petri nets: properties, analysis and applications, *Proc. Of the IEEE* **77(4)**, 541–580, 1989.
14. Olsén, H. Automatic verification of Petri nets in a CLP framework, Ph.D. Thesis, Dept. of Computer and Information Science, IDA, Linköping Univ., 1997.
15. Yamasaki, H. Normal Petri nets, *Theoretical Comput. Science* **31**, 307–315, 1984.
16. Yen, H. A unified approach for deciding the existence of certain Petri net paths, *Inform. and Comput.*, **96(1)**, 119–137, 1992.
17. Yen, H. On the regularity of Petri net languages, *Inform. and Comput.*, **124(2)**, 168–181, 1996.
18. Yen, H. On reachability equivalence for BPP-nets, *Theoretical Computer Science*, **179**, 301–317, 1997.

Automatic Verification of Multi-queue Discrete Timed Automata

Pierluigi San Pietro[1]* and Zhe Dang[2]

[1] Dipartimento di Elettronica e Informazione
Politecnico di Milano, Italia
pierluigi.sanpietro@polimi.it
[2] School of Electrical Engineering and Computer Science
Washington State University
Pullman, WA 99164, USA
zdang@eecs.wsu.edu

Abstract. We propose a new infinite-state model, called the Multi-queue Discrete Timed Automaton *MQDTA*, which extends Timed Automata with queues, but only has integer-valued clocks. Due to careful restrictions on queue usage, the binary reachability (the set of all pairs of configurations (α, β) of an *MQDTA* such that α can reach β through zero or more transitions) is effectively semilinear. We then prove the decidability of a class of Presburger formulae defined over the binary reachability, allowing the automatic verification of many interesting properties of a *MQDTA*. The *MQDTA* model can be used to specify and verify various systems with unbounded queues, such as a real-time scheduler.

Keywords: Timed Automata, infinite-state model-checking, real-time systems.

1 Introduction

Real-time systems are widely regarded as a natural application area of formal methods, since the presence of the time variable makes them more difficult to specify, design and test. The limited expressiveness of finite automata has recently sparkled much research into the automated verification of infinite state systems. Most research in the field has concentrated on finding good abstractions or approximations that map infinite state systems into finite ones (e.g., parametrized model checking [19] and generalized model checking [15]). A complementary approach to abstraction is the definition and study of infinite-state models for which "interesting" properties are still decidable. Most of the works have concentrated on very few models, such as Petri Nets (*PN*), Pushdown Automata (*PA*) and Timed Automata (*TA*), and have studied the decidability and complexity of model-checking various temporal and modal logics. A *TA* [4] is basically a finite-state automaton with a certain number of unbounded clocks that can be tested and reset. Since their introduction and the definition of appropriate model checking algorithms [17], *TA* have become a useful model to investigate the verification of real-time systems and have been extensively studied. The expressive power of *TA* has many limitations in modeling, since many real-time systems are simply not finite-state, even when time is ignored.

* Supported in part by MIUR grants FIRB RBAU01MCAC and COFIN 2001015271.

T. Warnow and B. Zhu (Eds.): COCOON 2003, LNCS 2697, pp. 159–171, 2003.
© Springer-Verlag Berlin Heidelberg 2003

Other infinite-state models for which forms of automatic verification are possible are based on *PN* (e.g., [16]), on various versions of counter machines (e.g., [10]), on *PA* (e.g., [5]), or on process calculi (e.g., [21]), but, at least in their basic versions, they do not consider timing requirements and are thus not amenable for modeling real-time systems. Among the infinite-state models that consider time, there are many timed extensions of Petri Nets but their binary reachability is typically undecidable if the net is unbounded (i.e., it is not finite state). A recent notable example of model checking a timed version of Petri Nets is [2], where it is shown that coverability properties are decidable, using well-quasi orderings techniques. A more general result holds for an extension of *TA*, Timed Networks [1], for which safety properties have been shown to be decidable. However, Timed Networks consist of an arbitrary set of identical timed automata, which is a very special case, although potentially useful in modeling infinite-state timed systems. Recently, Timed Pushdown Automata (*TPA*) [13,12] have been proposed, extending pushdown processes with unbounded discrete clocks. Considering that both the region techniques [4] and the flattening techniques [11] for *TA* can not be used for *TPA*, a totally different technique is proposed to show that safety and binary reachability analysis are still decidable [13,12]. *Queues* are a good model of many interesting systems, such as schedulers, for which automatic verification has rarely been attempted. Queues are usually regarded as hopeless for verification, since it is well known that a finite-state automaton equipped with one unbounded queue can simulate a Turing machine. However, there are restricted models with queues for which reachability is decidable (e.g., [9]). Here, we consider the Generalized Context-free Grammars (*GCG*) of [8], which use both queues and stacks with suitable constraints to generate only semilinear languages, and which are well suited to modeling of scheduling policies. However, automatic verification of *GCG* has never been investigated, and *GCG* do not consider time.

In this paper, we study how to couple a timed automaton with a multi-queue automaton (inspired by the *GCG* model) so that the resulting machine can be effectively used for modeling, while retaining the decidability of a class of Presburger formulae over the binary reachability set, with control-state variables, clock value variables and count variables. Hence, such machines are amenable for modeling and automatic verification of many infinite-state real-time systems, such as real-time process schedulers. The paper is structured as follows. Section 2 defines the *MQDTA*, introduces its untimed version (called Multi-queue-stack machine, *MQSM*) and proves the effective semilinearity of the model, by using a *GCG*. Section 3 proves the main result of the paper, i.e., the effective semilinearity of the binary reachability for *MQDTA*, by showing that clocks may be eliminated and an *MQDTA* may be translated into an equivalent *MQSM*. Section 4 proves the decidability of a class of Presburger formulae over the binary reachability, showing their applicability to an example.

2   Multi-queue Discrete Timed Automata

In this section we introduce the *MQDTA* model, which extends Discrete Timed Automata *DTA* by allowing a number of queues. The presentation is self-contained abd does not require previous knowledge of *DTA*. A *clock constraint* is a Boolean combination of *atomic clock constraints* in the following form: $x \# c, x - y \# c$ where c is an integer, x, y are integer-valued clocks and $\#$ denotes $\leq, \geq, <, >,$ or $=$. Let $\mathcal{L}_X$ be the set of all clock constraints on clocks X. Let $\mathbf{Z}$ be the set of integers and $\mathbf{Z}^+$ be the set of nonnegative integers.

Definition 1 (*MQDTA*). *A Multi-queue Discrete Timed Automaton (MQDTA) with* $n \geq 0$ *FIFO queues is a tuple* $\langle S, X, \Gamma, s_f, \mathbf{R}, E, Q_1, \cdots, Q_n \rangle$ *where: Γ is the queue alphabet; $Q_1, \cdots, Q_n$ are queues; S is a finite set of (control) states; X is a finite set of clocks with values in $\mathbf{Z}^+$; $s_f \in S$ is the final state; $\mathbf{R} \subseteq \Gamma \times S$ is the restart set; E is a finite set of edges, such that each edge $e \in E$ is in the form of $\langle s, \lambda, (\eta_1, \cdots, \eta_n), l, s' \rangle$ where $s, s' \in S$ with $s \neq s_f$ (the final state s_f does not have a successor); $\lambda \subseteq X$ is the set of clock resets; $l \in \mathcal{L}_X$ is the enabling condition.*

The queue operation is characterized by a tuple $(\eta_1, \cdots, \eta_n)$ with $\eta_1, \cdots, \eta_n \in \Gamma^*$, to denote that each η_i is put at the end of the queue Q_i, $1 \leq i \leq n$.

Let $\mathcal{A}$ be an *MQDTA* with n queues. Intuitively, the queues are totally ordered from Q_1 to Q_n and for a pair $(\gamma, s) \in \mathbf{R}$, s will be the next start state of $\mathcal{A}$ if the head of the first nonempty queue is γ. Notice that, for $n = 0$, the *MQDTA* reduces to a *DTA*.

The semantics is defined as follows. A *configuration* α of $\mathcal{A}$ is a tuple $\langle s, \pi_1, \cdots, \pi_n, c_1, \cdots, c_k \rangle$ where $s \in S, c_1, \cdots, c_k \in \mathbf{Z}^+$ are the state and the clock values respectively. $\pi_1, \cdots, \pi_n \in \Gamma^*$ are the contents of each queue, with the leftmost character being the head and rightmost character being the tail. We use α_{Q_i} to denote each π_i in α, with $\alpha_{\mathbf{q}}, \alpha_{x_1}, \cdots, \alpha_{x_k}$ to denote $s, c_1, \ldots, c_k$ respectively.

Let $\alpha \overset{\langle s, \lambda, (\eta_1, \cdots, \eta_n), l, s' \rangle}{\longrightarrow} \alpha'$ denote a one-step transition along an edge $\langle s, \lambda, (\eta_1, \cdots, \eta_n), l, s' \rangle$ in $\mathcal{A}$ satisfying the following conditions:

- The state s is set to a new location s', i.e., $\alpha_{\mathbf{q}} = s, \alpha'_{\mathbf{q}} = s'$.
- Each clock changes according to λ. If there are no clock resets on the edge, i.e., $\lambda = \emptyset$, then clocks progress by one time unit, i.e., for each $x \in X, \alpha'_x = \alpha_x + 1$. If $\lambda \neq \emptyset$, then for each $x \in \lambda, \alpha'_x = 0$ while for each $x \notin \lambda, \alpha'_x = \alpha_x$.
- The enabling condition is satisfied, i.e., $l(\alpha)$ is true.
- The content of each queue is updated: $\alpha'_{Q_i} = \alpha_{Q_i} \eta_i$ for each $1 \leq i \leq n$.

Besides the above defined one-step transition, an *MQDTA* $\mathcal{A}$ can fire a restart transition when it is in the final state s_f. Let $\alpha \overset{restart}{\longrightarrow} \alpha'$ denote a *restart transition* in $\mathcal{A}$ satisfying the following four conditions:

1) $\alpha_{\mathbf{q}} = s_f$, i.e., this restart transition only fires at the final state.

2) Some queue in α is not empty. Let $\gamma \in \Gamma$ be the head of the first (in the order from Q_1 to Q_n) nonempty queue. The next state should be indicated in the restart set $\mathbf{R}$. That is, $(\gamma, \alpha'_{\mathbf{q}}) \in \mathbf{R}$.

3) Let γ be the head of the j-th queue where $1 \leq j \leq n$ and α_{Q_j} is not empty, and for all $1 \leq i < j, \alpha_{Q_i}$ is empty. Assume $\alpha_{Q_j} = \gamma \pi$ for some $\pi \in \Gamma^*$. Then, $\alpha'_{Q_j} = \pi$, and for all $1 \leq i \leq n$ with $i \neq j, \alpha'_{Q_i} = \alpha_{Q_i}$. That is, the head γ must be removed from the queue, while the other queues are not modified.

4) Clocks are reset, i.e., $\alpha'_x = 0$ for all $x \in X$.

From now on, $\mathcal{A}$ is a *MQDTA* specified as above. We simply write $\alpha \to_{\mathcal{A}} \alpha'$ if α can reach α' by either a one-step transition or a restart transition. The *binary reachability*

$\leadsto^{\mathcal{A}}$ is the reflexive and transitive closure of $\to_{\mathcal{A}}$ A configuration $\alpha = \langle s, \pi_1, \cdots, \pi_n, c_1, \cdots, c_k \rangle$ can be encoded as a string $[\alpha]$ by concatenating the symbol representation of s, the strings $\pi_1, \cdots, \pi_n$, and the (unary) string representation of $c_1, \cdots, c_k$ with a delimiter "$". The binary reachability $\leadsto^{\mathcal{A}}$ can be considered as the language: $\{[\alpha]\$[\beta] : \alpha \leadsto^{\mathcal{A}} \beta\}$.

An Example. Consider a LAN printer, which may accept two types of jobs: Large (L) and Small (S). When a job is being printed, no other job can interrupt it. However, if a job takes too long to be completed, then the printer preempts it and puts it into a special, lower priority queue, called the batch queue. The timeout for the L jobs is 200 seconds, while for the S jobs is only 100. The jobs in the batch queue (called batch jobs, B) can be printed without time limits, but they are overridden (i.e., put at the end of the batch queue) whenever L or S job arrives. The arrival of new jobs is not completely random: if the printer is busy printing, then the interval between the arrival of new jobs is at least 50 seconds. The specification of the example is formalized with an *MQDTA* with two clocks (called *timeout* and *last* respectively) and two queues. The set of states is: $\{start, printL, printS, printB\}$. The alphabet of the queues is: $\{L, S, B\}$. The graph of the transition function is shown in Fig. 1. Multiple transitions from one state to another state are denoted by multiple labels instead of by multiple arcs. The labels used on the transitions have the following syntax: [clock condition] / [queue update] [clock assignment]. The notation for clock conditions and assignments is obvious. A queue update such as $(L)1(B)2$ means that L and B are written on queue 1 and queue 2, respectively. The automaton starts the execution in the *start* state. When either an S or an L job is put into queue 1, the automaton enters s_f: it reads the queue content and executes a *restart* transition (denoted by the dashed arrows). A *restart* transition goes from s_f to the next state depending on the queue content: if the front of the queue is S it enters state *printS*, if it is L it enters state *printL*, if it is B it enters state *printB*. When a *restart* transition is executed, all the clocks are reset (i.e., *timeout* := 0 and *last* := 0 are executed).

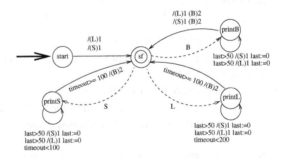

Fig. 1. The state transition graph of the *DTMQ* $\mathcal{A}$ of the example.

The proof of the main result of this paper, i.e., $\leadsto^{\mathcal{A}}$ is a semilinear language, is based on the following untimed version of *MQDTA*, which allows both queues and stacks. A *Multi-queue-Stack Machine (MQSM)* M with n (FIFO) queues, m (LIFO) stacks and a one-way input tape is a tuple: $\langle S, \Sigma, \Gamma, \Theta, s_0, s_f, \delta, Q_1, \cdots, Q_n, C_1, \cdots, C_m \rangle$, where S is a finite set of states with the *initial state* $s_0 \in S$ and the *final state* $s_f \in S$, Σ is the input alphabet, Γ and Θ are two disjoint alphabets for the queues $Q_1, \cdots, Q_n$ and the stacks $C_1, \cdots, C_m$ respectively. The queues and the stacks are arranged so that $Q_1, \cdots, Q_n$ are followed by $C_1, \cdots, C_m$. δ is a finite set of *transitions*. We distinguish three kinds of transitions:

A *push-transition* has the form $\langle s, \sigma, (\eta_1, \cdots, \eta_n, \xi_1, \cdots, \xi_m), s' \rangle$. That is, from state $s \in S$, M moves its input head to the right and reads an input symbol $\sigma \in \Sigma$ (if $\sigma = \epsilon$, however, M the input head does not move, i.e., M executes an ϵ-move), puts $\eta_1, \cdots, \eta_n \in \Gamma^*$ at the end of queues $Q_1, \cdots, Q_n$, and pushes $\xi_1, \cdots, \xi_m \in \Theta^*$ into the stacks $C_1, \cdots, C_m$. Thus, a push-transition is an element of $S \times (\Sigma \cup \{\epsilon\}) \times (\Gamma^*)^n \times (\Theta^*)^m \times S$. A *pop-stack-transition* has the form $\langle s_f, \sigma, \theta, s' \rangle$. That is, from the final state $s_f \in S$, on the input symbol $\sigma \in \Sigma \cup \{\epsilon\}$, M pops the top of the first nonempty stack and transits to state $s' \in S$. Thus, a pop-stack-transition is an element of $\{s_f\} \times (\Sigma \cup \{\epsilon\}) \times \Theta \times S$. A *pop-queue-transition* has the form of $\langle s_f, \sigma, \gamma, s' \rangle$. That is, from the final state s_f, on the input symbol $\sigma \in \Sigma \cup \{\epsilon\}$, M pops the top of the first nonempty queue and transits to state $s' \in S$. Thus, a pop-queue-transition is an element of $\{s_f\} \times (\Sigma \cup \{\epsilon\}) \times \Gamma \times S$. Therefore, δ is a finite subset of $(S \times (\Sigma \cup \{\epsilon\}) \times (\Gamma^*)^n \times (\Theta^*)^m \times S) \cup (\{s_f\} \times (\Sigma \cup \{\epsilon\}) \times \Theta \times S) \cup (\{s_f\} \times (\Sigma \cup \{\epsilon\}) \times \Gamma \times S)$.

A *configuration* of M is a tuple $\langle s, w; \gamma_1, \cdots, \gamma_n; c_1, \cdots, c_m \rangle$, where $s \in S$ is the state, $w \in \Sigma^*$ is the input word, $\gamma_1, \cdots, \gamma_n \in \Gamma^*$ are the contents of the queues (with the leftmost character being the head and the rightmost character being the tail), $c_1, \cdots, c_m \in \Theta^*$ are the contents of the stacks (with the leftmost character being the top and rightmost character being the bottom). The one-step transition $\Rightarrow_M$ of M is a binary relation over configurations. That is,

$$\langle s, w; \gamma_1, \cdots, \gamma_n; c_1, \cdots, c_m \rangle \Rightarrow_M \langle s', w'; \gamma_1', \cdots, \gamma_n'; c_1', \cdots, c_m' \rangle$$

iff one of the conditions is satisfied:

- The transition is a push-transition $\langle s, \sigma, (\eta_1, \cdots, \eta_n, \xi_1, \cdots, \xi_m), s' \rangle \in \delta$. Then: $w = \sigma w'$, for each $1 \leq i \leq n$, $\gamma_i' = \gamma_i \eta_i$, and for each $1 \leq j \leq m$, $c_j' = \xi_j c_j$.
- The transition is a pop-stack-transition $\langle s_f, \sigma, \theta, s' \rangle \in \delta$. Then: $s = s_f$, $w = \sigma w'$, for each $1 \leq i \leq n$, $\gamma_i' = \gamma_i = \epsilon$. There exists $1 \leq j \leq m$ such that $c_j = \theta c_j'$ and for all $1 \leq k \leq j - 1$, $c_k' = c_k = \epsilon$, and for all $j + 1 \leq k \leq m$, $c_k' = c_k$.
- The transition is a pop-queue-transition $\langle s_f, \sigma, \gamma, s' \rangle \in \delta$. Then: $s = s_f$, $w = \sigma w'$, there is a $1 \leq i \leq n$ such that $\gamma_i = \gamma \gamma_i'$, for all $1 \leq k \leq i - 1$, $\gamma_k = \gamma_k' = \epsilon$, for all $i + 1 \leq k \leq n$, $\gamma_k = \gamma_k'$, and for all $1 \leq k \leq n$, $c_k' = c_k$.

The *transition relation* $\Rightarrow_M^*$ is the transitive closure of the binary relation $\Rightarrow_M$ over configurations. A string $w \in \Sigma^*$ is *accepted* by M if $\langle s_0, w; \gamma_0, \epsilon, \cdots, \epsilon; \epsilon, \cdots, \epsilon \rangle \Rightarrow_M^* \langle s_f, \epsilon; \epsilon, \cdots, \epsilon; \epsilon, \cdots, \epsilon \rangle$.

Theorem 1. *Languages accepted by Multi-queue-stack machines are semilinear.*

Proof. The proof of this result is based on the fact that the language accepted by *MQSM* may be generated by a Generalized Context-free Grammar. In fact, an *MQSM* is actually a *GCG* in disguise, and it is just a variant of the accepting device of *GCG*, the multi-queue-pushdown automaton [7]. *GCG* only generate suitable permutations of context-free languages, and hence their languages are semilinear and the semilinear sets are effectively constructible.   □

3   Main Results

To prove the main result of the paper, we need a few more definitions. Let $\Sigma = \{a_1, \ldots, a_n\}$ be an alphabet, for some $n \geq 1$. A *Parikh transform* P translates each $w \in \Sigma^*$ into $a_1^{\#a_1(w)} \cdots a_k^{\#a_k(w)}$. Let $\$ \notin \Sigma$ be a symbol. For every $n \geq 1$, a language L is *segmented* if every word of L has the form $w_1 \$ w_2 \cdots \$ w_n$, where each $w_i \in \Sigma^*$. P is abused on every word $w = w_1 \$ \cdots \$ w_n$ of a segmented language L, with $P(\mathbf{w}) = P(w_1) \$ \cdots \$ P(w_n)$, and $P(L) = \{P(\mathbf{w}) : \mathbf{w} \in L\}$. A segmented language L is *locally commutative* if $\mathbf{w} \in L$ iff $P(\mathbf{w}) \in P(L)$.

Lemma 1. *For all languages L_1 and L_2, with L_2 segmented and locally commutative, the following statements hold: 1. $P(L_1)$ is a semilinear language iff L_1 is. 2. If L_1 and L_2 are semilinear languages then so is $L_1 \cap L_2$.*

Proof. 1. Let p be the Parikh mapping (see [18] for this traditional definition). Simply notice that $p(P(L_1)) = p(L_1)$. 2. Suppose that each word L_2 has $n \geq 1$ occurrences of $\$$, and let C be the language of all the words in $(\Sigma \cup \{\$\})^*$ with exactly n occurrences of $\$$. Let L_3 be the segmented language $L_1 \cap C$, which is semilinear because L_1 is semilinear and C is obviously semilinear and commutative. Since $L_2 \subseteq C$, then $L_1 \cap L_2 = L_1 \cap (C \cap L_2) = L_3 \cap L_2$. Hence, $P(L_1 \cap L_2) = P(L_3 \cap L_2) = P(L_3) \cap P(L_2)$, since L_2 is locally commutative and L_3 is segmented. Then, from the proof of part (1) above, $p(L_1 \cap L_2) = p(P(L_1 \cap L_2)) = p(P(L_3) \cap P(L_2))$. Since elements in $P(L_3)$ and $P(L_2)$ are made of tuples, and from part (1), $P(L_3)$ and $P(L_2)$ are semilinear languages, also $P(L_3) \cap P(L_2)$ is a semilinear language [14]. The result follows from part (1).   □

Lemma 2. *Let $L \subseteq \{[\alpha]\$[\beta] : \alpha, \beta$ are configurations of $\mathcal{A}\}$ be a semilinear language. Then, given a clock constraint $l \in \mathcal{L}_X$, $L' := \{[\alpha]\$[\beta] \in L : l(\alpha_{x_1}, \cdots, \alpha_{x_k})\}$ is also a semilinear language.*

The proof of Lemma 2 is immediate from Lemma 1.

Let $\mathcal{A} = $ be an *MQDTA* with clocks $x_1, \cdots, x_k$ and queues $Q_1, \cdots, Q_n$. We now show a technique to eliminate the *tests* (which are the enabling conditions, i.e., Boolean combinations of $x_i \# c$, $x_i - x_j \# c$ with c an integer constant) in an *MQDTA* $\mathcal{A}$. Let m be one plus the maximal absolute value of all the integer constants that appear in the tests in $\mathcal{A}$. Denote the finite set $[m] =_{def} \{-m, \cdots, 0, \cdots, m\}$. *Entries* a_{ij} and b_i for $1 \leq i, j \leq k$ are finite state variables with values in $[m]$. Intuitively, a_{ij} and b_i are used to record $x_i - x_j$ and x_i respectively. However, during the computation of $\mathcal{A}$, when $x_i - x_j$ (or x_i) goes beyond m or below $-m$, a_{ij} (or b_i) stays the same as m or $-m$. On executing an edge e, the new entry values a'_{ij} and b'_i can be expressed by only using the old values a_{ij} and b_i through *entry updating instructions*. For instance, suppose the set of clock resets is $\lambda = \{x_3\}$ for e. The entry updating instructions on this edge are: for all $1 \leq i,j \leq k$ with $i,j \neq 3, a'_{33} := 0; a'_{3j} = -b_j; a'_{i3} := b_i; a'_{ij} := a_{ij}; b'_i := b_i; b'_3 := 0$. The detailed construction of the entry updating instructions on any edge e is omitted, since it is a variant of the one presented in [13]. The addition of appropriate entry updating instructions on each edge guarantees that: after $\mathcal{A}$ executes any transition, (1). $x_i - x_j \# c$ iff $a_{ij} \# c$, (2). $x_i \# c$ iff $b_i \# c$, for all $1 \leq i,j \leq k$ and for each integer $-m < c < m$. The proof of this statement (omitted here) is similar to one in [13], though here we deal with different machine models and different clock behaviors. Thus, by adding entry updating instructions, each $x_i - x_j \# c$ (or $x_i \# c$) can be replaced by $a_{ij} \# c$ (or $b_i \# c$). The resulting automaton is denoted by $\mathcal{A}'$.

The *replaced tests* and the entry updating instructions in $\mathcal{A}'$ can be further eliminated by expanding the states S. The result automaton is called $\mathcal{A}^2$ with states $S_2 \subseteq S \times [m]^{(k^2+k)}$. In short, each expanded state in S_2 indicates the *original* state $s \in S$ and values (totally $k^2 + k$ many) of entries a_{ij} and b_i. Each edge e (connecting a pair of states s, s' in S) in $\mathcal{A}'$ is thus split into a finite number of edges in $\mathcal{A}^2$. Each split edge in $\mathcal{A}^2$ connects two expanded states $\bar{s}, \bar{s}'$ in S_2 with the their original states being s and s' respectively, and the values of the entries in $\bar{s}$ satisfying the test on e (thus the replaced tests are eliminated in $\mathcal{A}^2$), and the values of the entries in $\bar{s}$ and $\bar{s}'$ being consistent with the entry updating instructions on e (thus the entry updating instructions are eliminated in $\mathcal{A}^2$). Recall that, for an *MQDTA*, there is only one final state s_f. So, in $\mathcal{A}^2$, all the expanded states with their original state being s_f are merged into one state s_f – doing this will not change the behavior of $\mathcal{A}^2$ since the final state is used to initiate a restart transition and its enabling condition cannot depend on clock values. The restart set $\mathbf{R}_2$ of $\mathcal{A}^2$ is modified as the set of all pairs $(\gamma, \bar{s})$ such that $(\gamma, s) \in \mathbf{R}$ in $\mathcal{A}'$ and $\bar{s}$ has the original state s and all the entry values in $\bar{s}$ are 0 (since after a restart transition from s_f to s, all the clocks $x_1, \cdots, x_k$ become zero.). Now, $\mathcal{A}^2$ is an *MQDTA* with states S_2, restart set $\mathbf{R}_2$, clocks $x_1, \cdots, x_k$, and queues $Q_1, \cdots, Q_n$. Each enabling condition in $\mathcal{A}^2$ is simply *true*.

Based upon the above simulations, we can establish the following theorem, noticing that $\mathcal{A}'$ simulates $\mathcal{A}$ as we indicated before.

Theorem 2. *For all configurations α and β of $\mathcal{A}$, $\alpha \leadsto^{\mathcal{A}} \beta$ iff there are configurations α^2 and β^2 of $\mathcal{A}^2$ such that $\alpha^2 \leadsto^{\mathcal{A}^2} \beta^2$, and the following conditions hold:*

- *The original states of the extended states $\alpha_{\mathbf{q}}^2$ and $\beta_{\mathbf{q}}^2$ are $\alpha_{\mathbf{q}}$ and $\beta_{\mathbf{q}}$ respectively. The entry values a_{ij} and b_i in the extended state $\alpha_{\mathbf{q}}^2$ are the initial entry values constructed from α as: if $|\alpha_{x_i} - \alpha_{x_j}| \leq m$, $a_{ij} := \alpha_{x_i} - \alpha_{x_j}$; if $\alpha_{x_i} - \alpha_{x_j} > m$, $a_{ij} := m$; if $\alpha_{x_i} - \alpha_{x_j} < -m$, $a_{ij} := -m$; if $\alpha_{x_i} \leq m$, $b_i := \alpha_{x_i}$; if $\alpha_{x_i} > m$, $b_i := m$; for each $1 \leq i,j \leq k$.*
- *Clock values and queue contents are the same in α and α^2, and in β and β^2.*

Theorem 3. *If $\leadsto^{\mathcal{A}^2}$ is semilinear, then so is $\leadsto^{\mathcal{A}}$.*

Proof. From Theorem 2, the entry values in the extended states $\alpha_{\mathbf{q}}^2$ and $\beta_{\mathbf{q}}^2$ can be dropped by applying a homomorphism. However, $\leadsto^{\mathcal{A}}$ is the homomorphic image not only of

$\leadsto^{\mathcal{A}^2}$, but of a proper subset of $\leadsto^{\mathcal{A}^2}$. In fact, as stated in Theorem 2, the entry values in the extended state $\alpha_{\mathbf{q}}^2$ are the initial entry values constructed from α. This condition can be expressed as a clock constraint, since the entry values a_{ij} and b_i are bounded. The result is immediate by applying Lemma 2 on $\mathcal{A}^2$, and applying the homomorphism to L' as in the lemma.   □

Theorem 4. *The language $\leadsto^{\mathcal{A}^2}$ is semilinear.*

Proof. The language $\leadsto^{\mathcal{A}^2}$ is $\{[\alpha]\$[\beta] : \alpha \leadsto^{\mathcal{A}^2} \beta\}$. An *MQSM* M to simulate $\mathcal{A}^2$ has an input alphabet including all the following symbols:

- symbols $\dot{s}$ and $\ddot{s}$ for each s in the state set of $\mathcal{A}^2$. $\dot{s}$ is used to encode the state $\alpha_{\mathbf{q}}$ of α, and $\ddot{s}$ is used to encode the state $\beta_{\mathbf{q}}$ of β.
- symbols $\dot{u}_i$ and $\ddot{u}_i$, $1 \le i \le k$. $\dot{u}_i$ is used to encode the unary string representation of the clock value α_{x_i}, and $\ddot{u}_i$ is for β_{x_i}.
- symbols $\dot{\gamma}$ and $\ddot{\gamma}$ for each $\gamma \in \Gamma$. Letters $\dot{\gamma}$ are used to encode queue words $\alpha_{Q_1}, \cdots, \alpha_{Q_n}$ of α. Letters $\ddot{\gamma}$ are for those of β.
- $3k + 2n + 3$ delimiters $\$, \dot{\&}, \ddot{\&}, \ddot{\&}_i, \dot{\#}_i, \dot{?}_j, \ddot{\#}_i, \ddot{?}_j$, for $1 \le i \le k$ and $1 \le j \le n$.
- *padding symbols* $\dot{@}$ and $\ddot{\%}_i$ for $1 \le i \le k$.

The format of the input to M is:

$$\alpha_{\mathbf{q}} \dot{\&} \dot{\alpha}_{Q_1} \dot{?}_1 \cdots \dot{\alpha}_{Q_n} \dot{?}_n \dot{u}_1^{\alpha_{x_1}} \dot{\#}_1 \cdots \dot{u}_k^{\alpha_{x_k}} \dot{\#}_k \dot{@}^t \$ \beta_{\mathbf{q}} \ddot{\&} \ddot{\beta}_{Q_1} \ddot{?}_1 \cdots \ddot{\beta}_{Q_n} \ddot{?}_n \ddot{u}_1^{\beta_{x_1}} \ddot{\&}_1 \ddot{\%}_1^{t_1} \ddot{\#}_1 \cdots \ddot{u}_k^{\beta_{x_k}} \ddot{\&}_k \ddot{\%}_k^{t_k} \ddot{\#}_k$$

The part before $\$$ is the encoding for α, and the part after $\$$ is for the encoding for β. The first part has four segments, from left to right:

- $\dot{\alpha}_{\mathbf{q}}$ is a symbol encoding the state $\alpha_{\mathbf{q}}$, followed by a delimiter $\dot{\&}$.
- $\dot{\alpha}_{Q_1} \dot{?}_1 \cdots \dot{\alpha}_{Q_n} \dot{?}_n$ is the concatenation of the queue words α_{Q_i}, using the delimiters $\dot{?}_1, \cdots, \dot{?}_n$. Note that, instead of using α_{Q_i} for a queue word, we use $\dot{\alpha}_{Q_i}$ by replacing each $\gamma \in \Gamma$ with $\dot{\gamma}$.
- $\dot{u}_1^{\alpha_{x_1}} \dot{\#}_1 \cdots \dot{u}_k^{\alpha_{x_k}} \dot{\#}_k$ is the unary string representations $\dot{u}_i^{\alpha_{x_i}}$ of the clock value α_{x_i} using the symbol $\dot{u}_i$, concatenated by delimiters $\dot{\#}_1, \cdots, \dot{\#}_k$.
- a *padding word* $\dot{@}^t$ is a unary string over character $\dot{@}$. The number t is used to indicate the number of transitions in $\mathcal{A}^2$ that lead from α to β.

The second part has three segments from left to right, the first two being defined similarly, while the third one $\ddot{u}_1^{\beta_{x_1}} \ddot{\&}_1 \ddot{\%}_1^{t_1} \ddot{\#}_1 \cdots \ddot{u}_k^{\beta_{x_k}} \ddot{\&}_k \ddot{\%}_k^{t_k} \ddot{\#}_k$ is the unary string representation $\ddot{u}_i^{\beta_{x_i}}$ of the clock value β_{x_i} using the symbol $\ddot{u}_i$, concatenated by delimiters $\ddot{\#}_1, \cdots, \ddot{\#}_k$. But we do not simply use $\ddot{u}_i^{\beta_{x_i}}$: instead, there is a padding $\ddot{\%}_i^{t_i}$ (a unary word of length t_i over the character $\ddot{\%}_i$) after each $\ddot{u}_i^{\beta_{x_i}}$, separated by a delimiter $\ddot{\&}_i$. These *clock padding words* will be made clear later.

Besides queues $Q_1, \cdots, Q_n$, M has stacks $C_1, \cdots, C_k$. Each stack C_i is used to store the clock value of x_i of $\mathcal{A}^2$. At start, M first pushes a new symbol Z_i twice onto each stack C_i – these symbols are used as indicate the bottom of each stack. M then reads the input tape up to the padding word $\dot{@}^t$. During the process, M stores the queue contents and clock values into $Q_1, \cdots, Q_n$ and $C_1, \cdots, C_k$ respectively. Then, M starts to simulate $\mathcal{A}^2$ from the state α_q read from the input tape. Each move of $\mathcal{A}^2$ causes M to read a symbol @ and to simulate the queue operations using its own queues. The clock changes in $\mathcal{A}^2$ are simulated by using the stacks of M. Suppose that currently $\mathcal{A}^2$ executes an edge with a set λ of clock resets. If $\lambda = \emptyset$, then, after firing the transition, all clocks progress by one time unit. M simulates this by pushing the symbol $\ddot{u}_i$ onto the stack C_i for each $1 \leq i \leq k$. Otherwise, if $\lambda \neq \emptyset$, then, after the firing of the transition, the clocks in λ are reset and the others are left unchanged. M simulates this by pushing a special symbol Z_i onto the stack C_i for each $x_i \in \lambda$, and by pushing nothing (ϵ) on the other stacks. During the simulation of $\mathcal{A}^2$, M never pops the stacks, and in fact it need not, since all the enabling conditions in $\mathcal{A}^2$ are simply *true*. After having read all the padding word @^{t}, when M reads the delimiter $\$$, it must make sure that the current state of $\mathcal{A}^2$ corresponds to the symbol $\ddot{\beta}_q$ on the input tape. M also pushes a new symbol Y_i onto each queue Q_i, in order to use them later to decide whether a queue is empty. M then moves to the final state s_f, which is also the final state of $\mathcal{A}^2$. There, M starts checking that the rest of the input tape is consistent with its current queue and stack contents. Such check requires M to pop repeatedly from its queues and stacks; these operations require, from the definition of an *MQSM*, that pop-queue-transitions and pop-stack-transitions occur in a final state and that the next state after a pop operation only depends on the current input character and the queue or stack symbol just read. But we use different sets of alphabets in the encoding of the rest of the input tape. Therefore, a pop-queue-transition executed now cannot be confused with a normal pop-queue-transition in M's simulating $\mathcal{A}^2$ when reading the padding word @^{t}.

M proceeds by emptying each i-th queue, from Q_1 to Q_n, while checking the correspondence between the current top symbol of a queue and the symbol on the input tape. M can also check when the queue Q_i becomes empty by checking that the current input character is the delimiter $\ddot{?}_i$ and that the current top of the queue is Y_i (the symbol M pushed before). After all the queues are successfully compared and emptied, M starts to compare the clock values $\ddot{u}_i^{\beta_{x_i}}$ on the input tape with the stack C_i, from C_1 to C_k. For each C_i, M reads the input $\ddot{u}_i^{\beta_{x_i}}$ and pops a symbol from C_i. Once the bottom symbol Z_i becomes the current top symbol, the current input character must be the delimiter $\ddot{\&}_i$. After this, M empties C_i by reading through the clock padding word $\ddot{\%}_i^{t_i}$, but it makes sure that the delimiter $\ddot{\#}_i$, right after the clock padding word, is correspondent to the last symbol Z_i on the stack (remember that initially we pushed two Z_i's onto the stack. Thus, t_i is a guess of how many symbols there are between the first Z_i and the last Z_i in the stack. What if such a guess is wrong? In that case, since we use different $\ddot{u}_i$ for each i to represent both the stack word and the clock values on the input tape, M always knows, assuming the guess is wrong, whether a stack symbol $\ddot{u}_i$ hits an unexpected symbol like $\ddot{u}_j$ – either the guess of t_i is too small or it is too large. In this case, and in all the other cases where comparisons fail, M moves into a deadlock state – a special state where no further transition is possible.

M accepts the input iff all comparisons are successful and the input head is at the end of the tape. Notice that M has no ϵ-moves. Denote with $L(M)$ the language accepted by M. Thus, $L(M)$ is a semilinear language from Theorem 1. Notice that M does not check whether the input is in a correct format. Let L' be the regular language composed of all the strings in the correct format-it is a segmented language with $3k + 2n + 3$ segments. It is easy to check that L' is also a semilinear, locally commutative language. Thus, $L'' = L(M) \cap L'$, i.e., the set of all input strings accepted by M and in the correct format, is also a semilinear language from Lemma 1. L'' is different from the language $\leadsto^{\mathcal{A}^2}$, but not too much. From the previous construction, $\alpha \leadsto^{\mathcal{A}^2} \beta$ iff there are $t, t_1, \cdots, t_k$, such that the input word given as in the beginning of this proof can be accepted by M. Thus, define a homomorphism h such that, $h(@) = h(\%_i) = \epsilon, h(\$) = h(\#_i) = h(\&) = h(\ddot{\&}) = h(\ddot{\&}_i) = h(?_j) = h(\#_i) = h(\ddot{?}_j) = \$, h(\dot{s}) = h(\ddot{s}) = s, h(\dot{\gamma}) = h(\ddot{\gamma}) = \gamma, h(\dot{u}_i) = h(\ddot{u}_i) = 1$, for all $1 \leq i \leq k$ and $1 \leq j \leq n$, for all s being a state of $\mathcal{A}^2$, for all $\gamma \in \Gamma$. Obviously, $h(L'') = \leadsto^{\mathcal{A}^2}$. Thus, $\leadsto^{\mathcal{A}^2}$ is a semilinear language (since the homomorphic image of a semilinear language is still semilinear).   □

The following main theorem can be shown by combining Theorems 3 and 4.

Theorem 5. $\leadsto^{\mathcal{A}}$ *is a semilinear language for any MQDTA $\mathcal{A}$.*

An *MQDTA* $\mathcal{A}$ has no input tape, i.e., there is no event label on edges. However, if each edge is labeled, we can extend the states of $\mathcal{A}$ by combining a state with a label. In this case, a configuration contains only the current event label instead of the whole input word consumed. This may make applications more convenient to be dealt with, though all results still hold.

4   Verification Results

In this section, we formulate properties that can be verified for an *MQDTA*. Given an *MQDTA* $\mathcal{A}$, let $\alpha, \beta \cdots$ denote variables ranging over configurations, and let $\alpha_{\mathbf{q}}$ (state variables), α_{x_i} (clock value variables) and α_{Q_j} (queue content variables) denote, respectively, the state, the clock x_i's value and the content of the queue Q_j of α, $1 \leq i \leq k, 1 \leq j \leq n$. We use a count variable $\#_\gamma(\alpha_{Q_j})$ to denote the number of occurrences of a character $\gamma \in \Gamma$ in the content of the queue Q_j in α, $1 \leq j \leq n$. An *MQDTA*-term t is defined as follows: $t \; ::= \; n \mid \alpha_{x_i} \mid \#_\gamma(\alpha_{Q_j}) \mid t - t \mid t + t$, where n is an integer, $\gamma \in \Gamma$, $1 \leq i \leq k, 1 \leq j \leq n$. An *MQDTA*-formula f is defined as follows: $f \; ::= \; t > 0 \mid t \bmod n = 0 \mid \neg f \mid f \vee f \mid a_{\mathbf{q}} = q$, where $n \neq 0$ is an integer and q is a state of $\mathcal{A}$. Thus, f is a quantifier-free Presburger formula over control state variables, clock value variables and count variables. For $m \geq 1$, let F be a formula in the following format: $\bigvee_{1 \leq i \leq m} (f_i \wedge \alpha^i \leadsto^{\mathcal{A}} \beta^i)$, where each f_i is a *MQDTA*-formula and all α^i and β^i are configuration variables. Let $\exists F$ be a closed formula such that each free variable in F is existentially quantified. Then, the property $\exists F$ can be verified.

Theorem 6. *The truth value of $\exists F$ with respect to an MQDTA $\mathcal{A}$ is decidable for any MQDTA-formula F.*

Proof. Let $L(E)$ be the language of the string encodings of the tuples of all the configurations that satisfy a *MQDTA*-formula E. Thus, $L(F) = \bigcup_i \left(L(\alpha^i \leadsto^{\mathcal{A}} \beta^i) \cap L(f_i) \right)$. We will show that $L(F)$ is a semilinear language. Since all the proofs are constructive, the semilinear set of $L(F)$ can be effectively constructed from F and $\mathcal{A}$. Thus, testing whether $\exists F = false$, which is equivalent to testing the emptiness of $L(F)$, is decidable [14]. Since semilinearity is closed under union, without loss of generality we show that $L_1 \cap L_2$ is a semilinear language, where $L_1 = L(\alpha \leadsto^{\mathcal{A}} \beta)$ and $L_2 = L(f)$. From Theorem 5, L_1 is a semilinear language. Notice that L_2 is locally commutative – the reason is that the contents of the queues are used only by count variables. $P(L_2)$, the result of applying the Parikh transform P, is a semilinear language. Thus, from Lemma 1, part (1), L_2 is a semilinear language. Hence, $L_1 \cap L_2$ is a semilinear language from Lemma 1, part (2). □

For instance, the following property: "for all configurations α and β with $\alpha \leadsto^{\mathcal{A}} \beta$, clock x_2 in β is the sum of clocks x_1 and x_2 in α, and symbol γ_1 appears in the first queue Q_1 in β is twice as many as symbol γ_2 does in the second queue Q_2 in α." can be expressed as, $\forall\alpha\forall\beta(\alpha \leadsto^{\mathcal{A}} \beta \rightarrow (\beta_{x_2} = \alpha_{x_1} + \alpha_{x_2} \wedge \#_{\gamma_1}(\beta_{Q_1}) = 2\#_{\gamma_2}(\alpha_{Q_2})))$. The negation of this property is equivalent to $\exists F$ for some $MQDTA$-formula F. Thus, it can be verified.

Verification of an Example. Consider the LAN printer example of Section 2. A property verifiable with our model is that the first queue is actually bounded: it can never contain more than 4 elements. This can be formalized as follows: for every α, β, such that $\alpha \leadsto^{\mathcal{A}} \beta : \alpha_{start} \wedge \#_S(\alpha_{Q1}) = \#_L(\alpha_{Q1}) = \#_B(\alpha_{Q2}) = 0 \wedge \#_S(\beta_{Q1}) + \#_L(\beta_{Q2}) \leq 4$. Notice that the binary queue of $\mathcal{A}$ need not be bounded, but the boundedness is a decidable property of $\mathcal{A}$. If this is the case, the implementation of the system might rely on a small buffer of size 4 to implement the queue. More sophisticated properties can also be verified, and the system itself could be made more complex.

5 Conclusions

We introduced a new version of Timed Automata augmented with queues ($MQDTA$), and we proved that its binary reachability is effectively semilinear, allowing the automatic verification of a class of Presburger formulae over control state variables, clock value variables and queue content.

An $MQDTA$ is more powerful than the other timed (finite-state or pushdown) models, and it can be used for modeling various systems where $FIFO$ policies are used. Such models are based on discrete–rather than dense–time. This choice is perfectly adequate for *synchronous real-time systems*, where there is always an underlying discrete model of time, but it is also suitable for modeling various asynchronous systems where discrete time is a good approximation of a dense one. Since this is the first paper introducing and investigating the model, we did not develop explicitly a verification algorithm. Using an automata-theoretic approach, we reduced the problem of checking reachability properties to checking certain Presburger formulae over integer values. Hence, the the the complexity of the verification has a very high upper bound (nondeterministic double exponential). This is not as hopeless as it may seem. For instance, the Omega-library of [20] could be used to implement a verification algorithm, since the library is usually reasonably efficient for formulae without alternating quantifiers (as in our case). Also, a very high upper bound is typical of the automata-theoretic approach, but often the upper bound may be reduced by using a (more complex) process algebra approach.

The *MQDTA* has some limitations in expressivity: for instance, it cannot check how long a symbol has been stored in a queue before being consumed. Moreover, when a queue is read, all the clocks are reset. However, the model could be powerful enough to describe and verify useful, real-life infinite-state systems (such as the simple job scheduler with timeouts and a priority queue of Section 2) that, at the best of our knowledge, cannot be modeled and automatically verified by any other formalism. The model only considers queues, but in general stacks could be used instead or together with queues, since the *GCG* model (on which *MQDTA* are based) allows both kinds of rewriting policies. This can be useful, since for instance a stack can model recursive procedure calls, and the queues may model a process scheduler.

Our results imply that it is decidable to verify whether the paths of the reachability satisfy constraints expressed in a fragment of Presburger arithmetic. This can be easily achieved by recording, in one additional queue of the automaton, the history of the moves. For instance, it is possible to verify that the total time a symbol has been waiting in a queue does not exceed a given threshold, even though, as remarked above, this control cannot be done by the *MQDTA* itself.

References

1. P. Abdulla and B. Jonsson. Model checking of systems with many identical timed processes. *Theoretical Computer Science*, 290(1):241–264, 2002.
2. P. Abdulla and A. Nylén. Timed petri nets and bqos. In *ICATPN'2001, 22nd Int. Conf. on application and theory of Petri nets*, 2001.
3. R. Alur, C. Courcoubetis, and D. Dill. Model-checking in dense real-time. *Information and Computation*, 104(1):2–34, May 1993.
4. R. Alur and D. L. Dill. A theory of timed automata. *Theoretical Computer Science*, 126(2):183–235, April 1994.
5. M. Benedikt P. Godefroid and T. Reps. Model checking of unrestricted hierarchical state machines. In *ICALP 2001*, of *LNCS* 2076, pp. 652–666. Springer, 2001.
6. L. Breveglieri, A. Cherubini, C. Citrini, and S. Crespi Reghizzi. Multiple pushdown languages and grammars. *Int. Journal of Found. of Computer Science*, 7:253–291, 1996.
7. L. Breveglieri, A. Cherubini, and S. Crespi Reghizzi. Real-time scheduling by queue automata. In *FTRTFT'92*, vol/ 571 of *LNCS*, pages 131–148. Springer, 1992.
8. L. Breveglieri, A. Cherubini, and S. Crespi Reghizzi. Modelling operating systems schedulers with multi-stack-queue grammars. In *Fundamentals of Computation Theory*, volume 1684 of *LNCS*, pages 161–172. Springer, 1999.
9. G. Cece and A. Finkel. Programs with quasi-stable channels are effectively recognizable. In *CAV'97*, volume 1254 of *LNCS*, pages 304–315. Springer, 1997.
10. H. Comon and Y. Jurski. Multiple counters automata, safety analysis and Presburger arithmetic. In *CAV'98*, volume 1427 of *LNCS*, pages 268–279. Springer, 1998.
11. H. Comon and Y. Jurski. Timed automata and the theory of real numbers. In *CONCUR'99*, volume 1664 of *LNCS*, pages 242–257. Springer, 1999.
12. Zhe Dang. Binary reachability analysis of pushdown timed automata with dense clocks. In *CAV'01*, volume 2102 of *LNCS*, pages 506–517. Springer, 2001.
13. Zhe Dang, O. H. Ibarra, T. Bultan, R. A. Kemmerer, and J. Su. Binary reachability analysis of discrete pushdown timed automata. In *CAV'00*, *LNCS* 1855, pages 69–84. Springer, 2000.
14. S. Ginsburg and E. Spanier. Semigroups, presburger formulas, and languages. *Pacific J. of Mathematics*, 16:285–296, 1966.

15. P. Godefroid and R. Jagadeesan. Automatic abstraction using generalized model checking. In *CAV'02*, volume 2404 of *LNCS*, pages 137–150. Springer, 2002.
16. B. Grahlmann. The state of pep. In *AMAST'98*, *LNCS* 1548, pages 522–526. Springer, 1998.
17. T. A. Henzinger, X. Nicollin, J. Sifakis, and S. Yovine. Symbolic model checking for real-time systems. *Information and Computation*, 111(2):193–244, June 1994.
18. R. Parikh. On context-free languages. *Journal of the ACM*, 13:570–581, 1966.
19. A. Pnueli and E. Shahar. Livenss and acceleraiton in parameterized verification. In *CAV'00*, volume 1855 of *LNCS*. Springer, 2000.
20. W. Pugh. The omega test: a fast and practical integer programming algorithm for dependence analysis. *Communications of the ACM*, 35(8):102–114, 1992.
21. B. Steffen and O. Burkart. Model checking the full modal mu-calculus for infinite sequential processes. In *ICALP'97*, volume 1256 of *LNCS*, pages 419–429.

List Total Colorings of Series-Parallel Graphs

Xiao Zhou, Yuki Matsuo, and Takao Nishizeki

Graduate School of Information Sciences, Tohoku University
Aoba-yama 05, Sendai 980-8579, JAPAN
{zhou,matsuo,nishi}@nishizeki.ecei.tohoku.ac.jp

Abstract. A total coloring of a graph G is a coloring of all elements of G, i.e. vertices and edges, in such a way that no two adjacent or incident elements receive the same color. Let $L(x)$ be a set of colors assigned to each element x of G. Then a list total coloring of G is a total coloring such that each element x receives a color contained in $L(x)$. The list total coloring problem asks whether G has a list total coloring. In this paper, we first show that the list total coloring problem is NP-complete even for series-parallel graphs. We then give a sufficient condition for a series-parallel graph to have a list total coloring, that is, we prove a theorem that any series-parallel graph G has a list total coloring if $|L(v)| \geq \min\{5, \Delta + 1\}$ for each vertex v and $|L(e)| \geq \max\{5, d(v) + 1, d(w) + 1\}$ for each edge $e = vw$, where Δ is the maximum degree of G and $d(v)$ and $d(w)$ are the degrees of the ends v and w of e, respectively. The theorem implies that any series-parallel graph G has a total coloring with $\Delta + 1$ colors if $\Delta \geq 4$. We finally present a linear-time algorithm to find a list total coloring of a given series-parallel graph G if G satisfies the sufficient condition.

1 Introduction

In this paper a graph means a finite "simple" graph without multiple edges and selfloops. We denote by $G = (V, E)$ a graph with a vertex set V and an edge set E. We often denote V by $V(G)$, E by $E(G)$, and $V(G) \cup E(G)$ by $VE(G)$. An edge joining vertices v and w is denoted by vw. For each vertex $v \in V$, we denote by $d(v, G)$ or simply $d(v)$ the *degree* of v, that is, the number of edges incident to v. We denote by $\Delta(G)$ or simply Δ the *maximum degree* of G. A graph is *series-parallel* if it contains no subgraph isomorphic to a subdivision of a complete graph K_4 of four vertices [7,10]. Thus a series-parallel graph is a "partial 2-tree," and its "tree-width" is bounded by 2. A series-parallel graph represents a network obtained by repeating "series connection" and "parallel connection." The graph in Fig. 1 is series-parallel.

A *total coloring* of a graph G is to color all vertices and edges in G so that any two adjacent vertices and any two adjacent edges receive different colors, and any vertex receives a color different from the colors of all edges incident to it [3,13]. Figure 1(b) depicts a total coloring of a graph, where a color is attached to each element. The minimum number of colors necessary for a total coloring of G is called the *total chromatic number* of G, and is denoted by $\chi_t(G)$. Clearly

T. Warnow and B. Zhu (Eds.): COCOON 2003, LNCS 2697, pp. 172–181, 2003.
© Springer-Verlag Berlin Heidelberg 2003

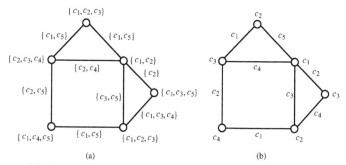

Fig. 1. (a) A series-parallel graph with lists, and (b) a list total coloring.

$\chi_t(G) \geq \Delta(G) + 1$. It is conjectured that $\chi_t(G) \leq \Delta(G) + 2$ for any simple graph G. However, this "total coloring conjecture" has not been verified [8,13].

In this paper we deal with a generalized type of a total coloring called a "list total coloring," which has some applications to a scheduling problem such as timetabling, routing in optical networks, and frequency assignment in cellular networks [9,13]. Suppose that a set $L(x)$ of colors, called a *list of x*, is assigned to each element $x \in VE(G)$, as illustrated in Fig. 1(a). Then a total coloring φ of G is called a *list total coloring of G for L* if $\varphi(x) \in L(x)$ for each element $x \in VE(G)$, where $\varphi(x)$ is the color assigned to x by φ. The list total coloring φ is simply called an *L-total coloring*. The coloring in Fig. 1(b) is indeed an L-total coloring of the graph in Fig. 1(a). An ordinary total coloring is an L-total coloring for which all lists $L(x)$ are same. Thus an L-total coloring is a generalization of a total coloring. The *list total coloring problem* asks whether a graph G has an L-total coloring for given G and L. The problem is NP-complete in general, because the ordinary total coloring problem is NP-complete [12]. The list vertex-coloring problem and the list edge-coloring problem are similarly defined. The list vertex-coloring problem can be solved in polynomial time for partial k-trees and hence for series-parallel graphs [6].

In this paper we first show that both the list edge-coloring problem and the list total coloring problem are NP-complete even for series-parallel graphs. Thus it is unlikely that there is a polynomial-time algorithm to solve the list total coloring problem even for series-parallel graphs. We then obtain a sufficient condition for a series-parallel graph G to have an L-total coloring. That is, we prove a theorem that a series-parallel graph G has an L-total coloring if $|L(v)| \geq \min\{5, \Delta + 1\}$ for each vertex v and $|L(e)| \geq \max\{5, d(v) + 1, d(w) + 1\}$ for each edge $e = vw$. The theorem implies that the total chromatic number $\chi_t(G)$ of a series-parallel graph G satisfies

$$\chi_t(G) \begin{cases} \leq \Delta(G) + 2 & \text{if } \Delta(G) \leq 3; \\ = \Delta(G) + 1 & \text{if } \Delta(G) \geq 4. \end{cases}$$

We finally present a linear-time algorithm to find an L-total coloring of a given series-parallel graph if L satisfies the sufficient condition above.

The rest of the paper is organized as follows. Section 2 presents a proof of the NP-completeness. Section 3 presents a proof of our sufficient condition. Section

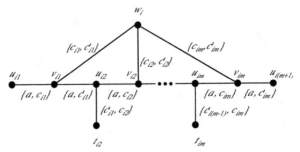

Fig. 2. Gadget G_{x_i}.

4 presents an algorithm. Finally Section 5 concludes with discussion and an open problem.

2   NP-Completeness

The "edge-disjoint paths problem" is one of a few problems which are NP-complete even for series-parallel graphs [11]. We first show that the list edge-coloring problem is another example of such problems although the list vertex-coloring problem can solved in polynomial time for series-parallel graphs [6].

Theorem 1. *The list edge-coloring problem is NP-complete for series-parallel graphs.*

Proof. [1] It suffices to show that the SAT problem can be transformed in polynomial time to the list edge-coloring problem for a series-parallel graph G. Let f be a formula in conjunctive normal form with n variables $x_1, x_1, \cdots, x_n$ and m clauses $C_1, C_2, \cdots, C_m$. For each variable x_i, $1 \leq i \leq n$, we construct a gadget G_{x_i} as illustrated in Fig. 2. G_{x_i} consists of a path $u_{i1}, v_{i1}, u_{i2}, v_{i2}, \cdots, u_{im}, v_{im}$, $u_{i(m+1)}$ of length $2m$, a vertex w_i of degree m, and $m-1$ vertices t_{ij}, $2 \leq j \leq m$, of degree 1. The lists of edges in G_{x_i} are assigned as follows:

- for each edge $w_i v_{ij}$, $1 \leq j \leq m$, $L(w_i v_{ij}) = \{c_{ij}, c'_{ij}\}$;
- for each edge $u_{ij} v_{ij}$, $1 \leq j \leq m$, $L(u_{ij} v_{ij}) = \{a, c_{ij}\}$;
- for each edge $v_{ij} u_{i(j+1)}$, $1 \leq j \leq m$, $L(v_{ij} u_{i(j+1)}) = \{a, c'_{ij}\}$; and
- for each edge $u_{ij} t_{ij}$, $2 \leq j \leq m$, $L(w_i t_{ij}) = \{c'_{i(j-1)}, c_{ij}\}$.

Clearly, if the gadget G_{x_i} has a list edge-coloring φ, then either $\varphi(w_i v_{ij}) = c_{ij}$ for all j, $1 \leq j \leq m$, or $\varphi(w_i v_{ij}) = c'_{ij}$ for all j, $1 \leq j \leq m$. The color c_{ij} for edge $w_i v_{ij}$ corresponds to $x_i = $ False, and the color c'_{ij} for edge $w_i v_{ij}$ corresponds to $x_i = $ True. Now a graph G is constructed as follows (see Fig. 3):

- identify all the vertices w_i for the n gadgets G_{x_i}, $1 \leq i \leq n$, as a single vertex w;
- add m new vertices v_{C_j}, $1 \leq j \leq m$, and m edges $w v_{C_j}$; and

---

[1] We thank Dániel Mark for discussion on the proof.

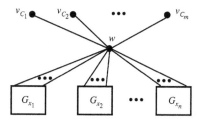

Fig. 3. Series-parallel graph G.

- let $L(v_{C_j}w)$, $1 \leq i \leq m$, be a list $\subseteq \{c_{1j}, c'_{1j}, c_{2j}, c'_{2j}, \cdots, c_{nj}, c'_{nj}\}$ such that $c_{ij} \in L(v_{C_j}w)$ iff the positive literal of x_i is in clause C_j and $c'_{ij} \in L(v_{C_j}w)$ iff the negative literal of x_i is in C_j.

It is clear that G is a series-parallel graph and both the number of vertices in G and the total number of colors in lists are bounded by a polynomial in n and m, and that f is satisfiable iff G has a list edge-coloring. □

A graph is *outer planar* if it contains no subgraph isomorphic to a subdivision of a complete bipartite graph $K_{2,3}$. We then immediately have the following corollary from Theorem 1.

Corollary 1.

(a) *The list total coloring problem is NP-complete for series-parallel graphs.*
(b) *Both the list edge-coloring problem and the list total coloring problem are NP-complete for 2-connected outer planar graphs.*

Proof. (a) The list edge-coloring problem for a graph can be easily transformed in polynomial time to the list total coloring problem for the same graph, in which the list of each vertex is a set of a single new color.

(b) One can add new edges to the graph G in the proof of Theorem 1 so that the resulting graph is a 2-connected outer planar graph; the list of each new edge is a set of a single new color. □

3   Sufficient Condition

Although the list edge-coloring problem is NP-complete for series-parallel graphs, several sufficient conditions for a series-parallel graph to have a list edge-coloring are known [4,7,14]. In this section we present a sufficient condition for a series-parallel graph to have a list total coloring, that is, we prove the following theorem.

Theorem 2. *Let G be a series-parallel graph, and let L be a list of G such that*

$$|L(u)| \geq \min\{5, \Delta(G) + 1\} \tag{1}$$

for each vertex $u \in V(G)$ and

$$|L(e)| \geq \max\{5, d(v) + 1, d(w) + 1\} \tag{2}$$

for each edge $e = vw \in E(G)$. Then G has an L-total coloring.

An odd cycle C_n, $n \geq 3$, has no L-total coloring for some list L such that $|L(u)| = \Delta(C_n) = 2$ for each vertex $u \in V(C_n)$, and hence one cannot decrease the second term $\Delta(G) + 1$ of the right side of Eq. (1). A series-parallel graph G has no L-total coloring if G has a vertex v such that $|L(v)| = d(v) = 5$, and each neighbor $w \in N(v)$ of v satisfies $d(w) \leq d(v)$ and $L(vw) = L(v)$. Thus one cannot decrease the second term $d(v) + 1$ of the right side of Eq. (2). Similarly one cannot decrease the third term $d(w) + 1$.

Considering the case where each list consists of the same $\Delta + 1$ or $\Delta + 2$ colors, one can easily observe that Theorem 2 implies the following corollary.

Corollary 2. *The total chromatic number $\chi_t(G)$ of a series-parallel graph G satisfies*

$$\chi_t(G) \begin{cases} \leq \Delta(G) + 2 & \text{if } \Delta(G) \leq 3; \\ = \Delta(G) + 1 & \text{if } \Delta(G) \geq 4. \end{cases}$$

Any "s-degenerate" graph G, $s \geq 1$, satisfies $\chi_t(G) = \Delta + 1$ if either $\Delta(G) \geq 2s^2$ [1] or $\Delta(G) \geq 4s + 3$ [5]. Since a series-parallel graph G is 2-degenerate, $\chi_t(G) = \Delta(G) + 1$ if $\Delta(G) \geq 8$. The corollary above improves this result.

Before presenting a proof of Theorem 2, we present notations and lemmas which will be used in the proof.

The *size* of a graph $G = (V, E)$ is $|V| + |E|$. We say that a graph G' is smaller than a graph G if the size of G' is smaller than that of G. We denote by $G - e$ the graph obtained from G by deleting an edge e, and by $G - v$ the graph obtained from a graph G by deleting a vertex v and all edges incident to v. We denote by $N(v)$ the set of neighbors of v in G. Since G is a simple graph, $d(v) = |N(v)|$ for any vertex $v \in V$.

Let L be a list of a graph G. Let G' be a subgraph of G, and let L' be a list of G' such that $L'(x) = L(x)$ for each element $x \in VE(G')$. Suppose that we have already obtained an L'-total coloring φ' of G', and that we are going to extend φ' to an L-total coloring φ of G without altering the colors in G'. Let U be the set of all uncolored vertices, that is, $U = V(G) - V(G')$. For a vertex $v \in V(G)$, we denote by $C(v, \varphi')$ the set of all colors that φ' have assigned to v and the edges incident to v in G', that is,

$$C(v, \varphi') = \{\varphi'(v)\} \cup \{\varphi'(vx) \mid vx \in E(G')\}.$$

Then

$$|C(v, \varphi')| = \begin{cases} d(v, G') + 1 & \text{if } v \in V(G'); \\ 0 & \text{if } v \in U. \end{cases} \tag{3}$$

Let H be a subgraph of G induced by the set $E(G) - E(G')$ of all uncolored edges. For each uncolored edge $vw \in E(H)$, let

$$L_{av}(vw, \varphi') = L(vw) - (C(v, \varphi') \cup C(w, \varphi')). \tag{4}$$

Then $L_{av}(vw, \varphi')$ is the set of all colors in $L(vw)$ that are *available* for vw when φ' is extended to φ, and we have

$$|L_{av}(vw, \varphi')| \geq |L(vw)| - |C(v, \varphi')| - |C(w, \varphi')|. \tag{5}$$

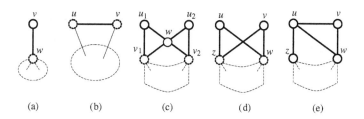

(a)   (b)   (c)   (d)   (e)

Fig. 4. Five substructures in series-parallel graphs.

For each uncolored vertex $v \in U \subseteq V(H)$, let

$$L_{av}(v, \varphi') = L(v) - \{\varphi'(u) \mid u \in N(v) \cap V(G')\}. \qquad (6)$$

Then $L_{av}(v, \varphi')$ is the set of all colors that are *available* for v when φ' is extended to φ. An L_{av}-*coloring* φ_H of H is defined to be a coloring of all elements in $U \cup E(H)$ such that

(a) $\varphi_H(x) \in L_{av}(x, \varphi')$ for each element $x \in U \cup E(H)$;
(b) $\varphi_H(v) \neq \varphi_H(w)$ if $v, w \in U$ and $vw \in E(H)$;
(c) $\varphi_H(e) \neq \varphi_H(e')$ if edges e and e' of H share a common end; and
(d) $\varphi_H(v) \neq \varphi_H(e)$ if $v \in U$ and $e = vw \in E(H)$.

The vertices in $V(H) - U$ are not colored by φ_H. One can easily observe the following lemma.

Lemma 1. *If H has an L_{av}-coloring φ_H, then φ' and φ_H can be extended to an L-total coloring φ of G as follows:*

$$\varphi(x) = \begin{cases} \varphi'(x) & \text{if } x \in VE(G'); \\ \varphi_H(x) & \text{if } x \in U \cup E(H). \end{cases}$$

The following lemma proved by Juvan *et al.* [7] will be used in the proof of Theorem 2.

Lemma 2. *Every non-empty series-parallel graph G satisfies one of the following five conditions* (a)–(e) *(see Fig. 4):*

(a) *there is a vertex v of degree zero or one;*
(b) *there are two distinct vertices u and v which have degree two and are adjacent to each other;*
(c) *there are five distinct vertices u_1, u_2, v_1, v_2 and w such that $N(u_i) = \{v_i, w\}$, $i = 1, 2$, and $N(w) = \{u_1, u_2, v_1, v_2\}$;*
(d) *there are two distinct vertices u and v of degree two such that $N(u) = N(v)$; and*
(e) *there are four distinct vertices u, v, w and z such that $N(u) = \{v, w, z\}$ and $N(v) = \{u, w\}$.*

In the remainder of this paper, we denote the substructures described in (a), (b), $\cdots$, (e) of Lemma 2 simply by *substructures* (a), (b), $\cdots$, (e), respectively.
One can easily prove the following Lemma 3 on L_{av}-colorings of the graphs in Fig. 5.

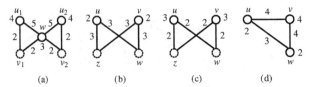

Fig. 5. Graphs H and the lower bounds on $|L_{av}(x)|$.

Lemma 3. *Let H be any of the four graphs in Fig. 5, where all vertices in U are drawn by solid circles and all vertices in $V(H) - U$ by dotted circles. If $|L_{av}(x)|$ is no smaller than the number attached to x in Fig. 5 for each element $x \in U \cup E(H)$, then H has an L_{av}-coloring.*

We are now ready to give a proof of Theorem 2.

(Proof of Theorem 2)

Suppose for a contradiction that there exists a series-parallel graph G which has no L-total coloring for a list L satisfying Eqs. (1) and (2). Assume that G is the smallest one among all such series-parallel graphs. One can easily observe that G is connected and $\Delta(G) \geq 3$. By Lemma 2 G has one of the five substructures (a)–(e) in Fig. 4, and hence there are the following three cases to consider.

Case a: G has a substructure (a).

In this case, there are two distinct vertices v and w in G such that $d(v) = 1$ and $vw \in E(G)$, as illustrated in Fig. 4(a). Let $G' = G - v$, and let L' be a list of G' such that $L'(x) = L(x)$ for each element $x \in VE(G')$. Then G' is a series-parallel graph smaller than G, and L' satisfies Eqs. (1) and (2). Therefore G' has an L'-total coloring φ' by the inductive assumption. Let H be a subgraph of G induced by the uncolored edge vw. The uncolored vertex in H is v, and hence $U = \{v\}$. In Fig. 4(a) H is drawn by thick lines, the uncolored vertex v by a thick solid circle and the colored vertex u by a thick dotted circle. By Eq. (2) $|L(vw)| \geq d(w, G)+1$, and by Eq. (3) $|C(v, \varphi')| = 0$ and $|C(w, \varphi')| = d(w, G')+1 = d(w, G)$. Therefore, by Eq. (5), we have $|L_{av}(vw, \varphi')| \geq |L(vw)| - |C(w, \varphi')| \geq 1$, and hence there is a color $c_1 \in L_{av}(vw, \varphi')$. Since $\Delta(G) \geq 3$, by Eq. (1) we have $|L(v)| \geq \min\{5, \Delta(G) + 1\} \geq 4$, and hence by Eq. (6) $|L_{av}(v, \varphi')| \geq |L(v)| - \{\varphi'(w)\}| \geq 3$. Therefore there is a color $c_2 \in L_{av}(v, \varphi') - \{c_1\}$. Thus H has an L_{av}-coloring φ_H such that $\varphi_H(vw) = c_1$ and $\varphi_H(v) = c_2$. Hence by Lemma 1 φ' and φ_H can be extended to an L-total coloring φ of G:

$$\varphi(x) = \begin{cases} c_1 & \text{if } x = vw; \\ c_2 & \text{if } x = v; \\ \varphi'(x) & \text{otherwise.} \end{cases}$$

This is a contradiction to the assumption that G has no L-total coloring.

Case b: G has a substructure (b).

In this case there are two adjacent vertices u and v of degree 2 in G, as illustrated in Fig. 4(b). Let G' be a graph obtained from G by deleting edge uv, and let $L'(x) = L(x)$ for each element $x \in VE(G')$. Then G' has an L'-total coloring φ' similarly as in Case a. In this case $U = \emptyset$, and H is a subgraph of

G induced by the uncolored edge uv. H is drawn by thick solid and dotted lines in Fig. 4(b). By Eq. (2) we have $|L(uv)| \geq 5$, and hence by Eqs. (3) and (5) we have

$$\begin{aligned}|L_{\mathrm{av}}(uv, \varphi')| &\geq |L(uv)| - |C(u, \varphi')| - |C(v, \varphi')| \\ &= |L(uv)| - (d(u, G') + 1) - (d(v, G') + 1) \\ &= |L(uv)| - d(u, G) - d(v, G) \\ &\geq 1. \end{aligned}$$

Thus there is a color $c \in L_{\mathrm{av}}(uv, \varphi')$, and hence H has an L_{av}-coloring φ_H: $\varphi_H(uv) = c$. By Lemma 1 φ' and φ_H can be extended to an L-total coloring of G, a contradiction.

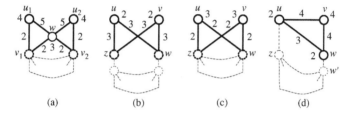

(a)   (b)   (c)   (d)

Fig. 6. Subgraphs H and the lower bounds on $|L_{\mathrm{av}}(x, \varphi^\circ)|$.

Case c: G has one of substructures (c), (d) and (e).

The proof is omitted in this extended abstract due to the page limitation.   □

4   Linear Algorithm

Our proof of Theorem 2 is constructive, and hence immediately yields the following recursive algorithm to find an L-total coloring of a given series-parallel graph G if L satisfies Eqs. (1) and (2). One may assume without loss of generality that both Eqs. (1) and (2) hold in equality; otherwise, delete an appropriate number of colors from each list.

Algorithm Color(G, L)

Step 1. Find one of the substructures (a)–(e) contained in G;

Step 2. Construct appropriate subgraphs G' and H according to the found substructure;

Step 3. Find recursively an L'-total coloring φ' of G';

Step 4. Find an L_{av}-coloring φ_H of H; and

Step 5. Extend φ' and φ_H to an L-total coloring of G.

In the remainder of this section we show the algorithm takes linear time. Since $G = (V, E)$ is a series-parallel simple graph, $|E| \leq 2n - 2$. Let $l(G) =$

$\sum_{x \in VE(G)} |L(x)|$ be the *total list size*. Then the input size is $N = n + |E| + l(G)$, where $n = |V|$. Since Eq. (2) holds,

$$\sum_{u \in V} d^2(u) \le 2 \sum_{e \in E} |L(e)| \le 2l(G) = O(N). \tag{7}$$

We show that the algorithm takes time $O(N)$, as follows.

We represent a graph G by adjacency lists. One may assume that the set $\bigcup_{x \in VE(G)} L(x)$ of colors is totally ordered. For each element $x \in VE(G)$, we store the colors $L(x)$ in a linked list in increasing order. This can be done in time $O(N)$ by the radix sorting [2]. We next analyze the computation time of Steps 1–5.

Clearly each of Steps 1–5 is executed at most $n + |E| \le 3n - 2 = O(n)$ times.

Each vertex in the substructures (a)–(e) is within distance 2 from a vertex of degree at most 2 through a vertex of degree at most 3. Therefore, all the executions of Step 1 can be done total in time $O(n)$ by a standard bookkeeping method to maintain the degrees of all vertices together with all paths of length two with an intermediate vertex of degree two.

Since the size of H is $O(1)$, one execution of Step 2 can be done in time $O(1)$. Since Step 2 is executed $O(n)$ times, the total execution time of Step 2 is $O(n)$.

In Step 3 the algorithm recursively calls **Algorithm Color** for a smaller graph G' to find an L'-total coloring φ' of G'. Clearly the total execution time of Step 3 other than the time required for recursive calls is $O(n)$.

For each each vertex u in G', we store the colors in set $C(u, \varphi')$ in increasing order in a linked list. In Step 4 the algorithm finds a set $L_{av}(e, \varphi')$ of colors available for each edge $e = vw \in E(H)$. This can be done in time $O(|L(e)|)$ by simultaneously traversing three linked lists $C(v, \varphi')$, $C(w, \varphi')$ and $L(e)$, because $|C(v, \varphi')| = d(v, G') \le d(v, G) \le |L(e)|$ and $|C(w, \varphi')| = d(w, G') \le d(w, G) \le |L(e)|$ by Eq. (2). Hence one can find $L_{av}(e, \varphi')$ for all edges e in H in time $O(\sum_{e \in E(H)} |L(e)|) = O(l(H))$. Similarly, one can find $L_{av}(u, \varphi')$ for all vertices u in H in time $O(\sum_{u \in V(H)} d(u, G))$, because $|C(u, \varphi')| \le d(u, G)$, and Eq. (1) holds in equality and hence $|L(u)| \le 5$. For each vertex $u \in V(H)$ there is an edge $e \in E(H)$ incident to u, and hence $\sum_{u \in V(H)} d(u, G) = O(\sum_{e \in E(H)} |L(e)|) = O(l(H))$. Clearly an L_{av}-coloring φ_H of H can be found in time $O(l(H))$. All subgraphs H's are edge-disjoint, but are not always vertex-disjoint with each other. However, the same vertex u is contained in at most $d(u, G)$ of all subgraphs H's. Therefore, by Eq. (7), the total execution time of Step 4 is bounded by $\sum_H l(H) \le l(G) + \sum_{u \in V} d^2(u, G) = O(N)$.

In Step 5 the algorithm extends φ' and φ_H to an L-total coloring φ of G. For each vertex u of G' to which an edge of H is incident, the ordered color lists $C(u, \varphi')$ and $C(u, \varphi_H)$ must be merged to a single ordered color list $C(u, \varphi)$. This can be done total in time $O(l(H))$. One can update $L_{av}(u, \varphi')$ to $L_{av}(u, \varphi)$ in time $O(1)$ for any vertex u in H, because $|L(u)| \le 5$. Thus one execution of Step 5 can be done in time $O(l(H))$, and hence the total execution time of Step 5 is $O(N)$.

Thus the algorithm takes time $O(N)$.

5 Conclusions

In this paper we first show that both the list edge-coloring problem and the list total coloring problem are NP-complete for series-parallel graphs. We then obtain a sufficient condition for a series-parallel graph G to have an L-total coloring. We finally give a linear algorithm to find an L-total coloring of G if L satisfies the condition. The algorithm finds an ordinary total coloring of G using the minimum number $\Delta + 1$ of colors in time $O(n\Delta)$ if $\Delta \geq 4$. It is remaining as an open problem to improve the computation time $O(n\Delta)$.

References

1. O. V. Borodin, A. V. Kostochka and D. R. Woodall, List edge and list total colorings of multigraphs, *J. Combinatorial Theory, Series B*, 71, pp. 184–204, 1997.
2. T. H. Cormen, C. E. Leiserson and R. L. Rivest, *Introduction to Algorithms*, MIT Press, Cambridge, MA, 1990.
3. R. Diestel, *Graph Theory*, Springer-verlag, New York, 1997.
4. T. Fujino, X. Zhou and T. Nishizeki, List edge-colorings of series-parallel graphs, *IEICE Trans. on Fundamentals*, E86-A, 5, pp. 191–203, 2003.
5. S. Isobe, X. Zhou and T. Nishizeki, Total colorings of degenerated graphs, *Proc. of ICALP 2001, Lect. Notes in Computer Science, Springer*, 2076, pp. 506–517, 2001.
6. K. Jansen and P. Scheffler, Generalized coloring for tree-like graphs, *Discrete Applied Math.*, 75, pp. 135–155, 1997.
7. M. Juvan, B. Mohar and R. Thomas, List edge-coloring of series-parallel graphs, *The Electronic Journal of Combinatorics*, 6, pp. 1–6, 1999.
8. T. R. Jensen and B. Toft, *Graph Coloring Problems*, John Wiley & Sons, New York, 1995.
9. M. Kubale, Introduction to Computational Complexity and Algorithmic Graph Coloring, *Gdańskie Towarzystwo Naukowe*, Gdańsk, Poland, 1998.
10. T. Nishizeki and N. Chiba, *Planar Graphs: Theory and Algorithms*, North-Holland, Amsterdm, 1988.
11. T. Nishizeki, J. Vygen and X. Zhou, The edge-disjoint paths problem is NP-complete for series-parallel graphs, *Discrete Applied Math.*, 115, pp. 177–186, 2001.
12. A. Sánchez-Arroyo, Determining the total colouring number is NP-hard, *Discrete Math.*, 78, pp. 315–319, 1989.
13. H. P. Yap, *Total Colourings of Graphs*, Lect. Notes in Math., 1623, Springer-verlag, Berlin, 1996.
14. J. L. Wu, List edge-coloring of series-parallel graphs, *Shandong Daxue Xuebao Kexue Ban*, 35, 2, pp. 144–149, 2000 (in Chinese).

Finding Hidden Independent Sets in Interval Graphs

Therese Biedl[1], Broňa Brejová[1], Erik D. Demaine[2], Angèle M. Hamel[3], Alejandro López-Ortiz[1], and Tomáš Vinař[1]

[1] School of Computer Science, University of Waterloo,
Waterloo, ON N2L 3G1, Canada,
{biedl,bbrejova,alopez-ortiz,tvinar}@uwaterloo.ca
[2] MIT Laboratory for Computer Science, 200 Technology Square,
Cambridge, MA 02139, USA,
edemaine@mit.edu
[3] Department of Physics and Computing, Wilfrid Laurier University,
Waterloo, ON, N2L 3C5, Canada,
ahamel@wlu.ca

Abstract. Consider a game in a given set of intervals (and their implied interval graph G) in which the adversary chooses an independent set X in G. The goal is to discover this hidden independent set X by making the fewest queries of the form "Is point p covered by an interval in X?" Our interest in this problem stems from two applications: experimental gene discovery and the game of Battleship (in a 1-dimensional setting). We provide adaptive algorithms for both the verification scenario (given an independent set, is it X?) and the discovery scenario (find X without any information). Under some assumptions, these algorithms use an asymptotically optimal number of queries in every instance.

1 Introduction

An *interval graph* is an intersection graph of intervals on the real line, i.e. vertices are represented by intervals and there is an edge between two vertices if and only if their corresponding intervals intersect. An *independent set* in G is a set of vertices such that no two vertices share an edge.

In this paper we study how to determine, given a set of intervals (with their implied interval graph G), an unknown (hidden) independent set X in G chosen by an adversary. We determine X by playing an interactive game against an adversary using queries of the following type: "Is a point p on the real line covered by an interval in X?" The adversary always answers the query truthfully. The goal is to use the smallest possible number of queries to determine set X. This problem is motivated by two applications: recovering gene structure with experimental techniques and the game of Battleship. We explain the connections to our problem after stating it precisely.

While there is a wide literature regarding games in graphs (e.g., [3,8,12]), our problem appears to be new in this area. Several games involving finding a hidden

T. Warnow and B. Zhu (Eds.): COCOON 2003, LNCS 2697, pp. 182–191, 2003.
© Springer-Verlag Berlin Heidelberg 2003

object using queries have also been studied in the bioinformatics literature. Xu et al. [20] discuss the problem of locating hidden exon boundaries in cDNA. This leads to a game in which the hidden object is a subset $A \subseteq \{1, \ldots, n\}$ and the queries are of the type "Given an interval I, does it contain an element of A?". Beigel et al. [2] discuss the problem of finding a hidden perfect matching in a complete graph applied to the problem of closing gaps in DNA sequencing data. McConnell and Spinrad [13] consider the tangentially related problem of reconstructing an interval graph given probes about the neighbors of only a partial set of vertices.

Terminology. An interval graph may have a number of different representations by intervals. In what follows, when we say "interval graph," we presume that one representation has been fixed. Without loss of generality, we may assume that in this representation all intervals are closed, have length at least one, and their end points are integers between 1 and $2n$, where n is the number of intervals. We assume that the input graph has no two identical intervals, but intervals are allowed to have the same start point or the same end point. We denote the interval of the ith vertex by $I_i = [s_i, f_i]$, where $s_i < f_i$ are integers. An edge (i, j) thus exists if $I_i \cap I_j \neq \emptyset$.

The complement $\overline{G}$ of an interval graph G has a special structure. Assume that (i, j) is not an edge in G, i.e., $I_i \cap I_j = \emptyset$. Then either $f_i < s_j$ or $f_j < s_i$, and thus we can orient the edge in $\overline{G}$ as $i \to j$ or $j \to i$. Thus, $\overline{G}$ has a natural orientation of the edges, and this orientation is well-known to be acyclic and transitive. For this and other results about interval graphs, see e.g. [11].

We refer to the initially unknown independent set in G chosen by an adversary, and refer to this set as the *hidden independent set*. If V' is an independent set in G, then it is a clique in the complement graph $\overline{G}$. If G is an interval graph, then any clique in $\overline{G}$ has a unique topological order consistent with orientation of its edges. We can thus consider V' as a (directed) path π in $\overline{G}$, and will speak of a *hidden (directed) path* instead of a hidden independent set. We will generally omit the word "directed" as we will not be talking about any other kind of path.

We determine the hidden independent set through *probes* and *queries*. A *probe* is an interval $(a, a + 1)$ where a is integer. A *query* is the use of a probe to determine information about the hidden independent set. Specifically, a query is a statement of the form: "Is there some vertex in the hidden independent set whose interval intersects the probe?" A query can be answered either "yes" or "no".

Our Results. We study two versions of the problem. First, the verification problem consists of verifying, via probe queries, that a purported independent set Y is the hidden independent set. Second, the discovery problem consists of identifying the set X.

For the verification problem, we give a protocol to determine whether $X = Y$ using the exact optimal number of queries for that specific instance. For the discovery problem, we give a linear-time algorithm for discovering X. Different graphs may require different number of queries to discover the hidden independent set. If at most a constant number of intervals start at a common point, then our protocol is within a constant factor of the optimal number of queries for that specific graph. That is, our algorithm is instance-optimal in the sense of [10] and optimally adaptive in the sense of [6]. If this assumption is not satisfied, then the number of queries may be larger than the information-theoretic lower bound; however, we also prove stronger lower bounds for some of these cases.

Applications to Gene Finding. Recent advances in molecular biology have resulted in genomic sequences of several organisms. These sequences need to be annotated, i.e., biological meaning needs to be assigned to particular regions of the sequence. An important step in the annotation process is the identification of genes, which are the portions of the genome producing the organism's proteins. A gene is a set of non-overlapping exons, each exon being an interval of the DNA sequence.

There are a number of computational tools for gene prediction (e.g., [4,16]); however, experimental studies (e.g., [14,5]) show that the best of them predicts, on average, only about 50% of the entire genes correctly. It is therefore important to have alternative methods that can produce or verify such predictions by using experimental data.

While genes cannot be reliably predicted by purely computational means, we can use these methods to provide us with a set of candidate exons. Algorithms for gene prediction have to balance sensitivity (i.e., how many real exons they discover) with specificity (i.e., how many false exons they predict), and usually it is possible to increase sensitivity at the expense of a decrease in specificity. By using a highly sensitive method, we may generate a candidate set that contains many false exons but has only a very small probability of excluding a real exon.

To apply our algorithms, we may view the set of candidate exons as the set of intervals defining an interval graph. The gene we want to discover then corresponds to a hidden independent set in this interval graph. Queries in our algorithms correspond to the question: "Is a given short region of DNA sequence contained in a real exon?" In order to use our method for finding genes, we need to answer this question by appropriate biological experiments.

Thousands of such queries can be answered simultaneously by an expression array experiment [18]. Shoemaker et al. [19] have used expression arrays to verify gene predictions in annotation of human chromosome 22 [7]. They probed DNA sequence at short regular intervals (every 10 nucleotides). Using our algorithm for the independent set verification (Section 2), we can design a smaller set of queries which can verify the gene prediction, thus reducing the cost.

Queries similar to ours can be also implemented using polymerase chain reaction (PCR) technology [17]. The PCR can answer the query of the following form: "Given two short regions of the DNA sequence, do both of them occur in the same gene (possibly in two different exons)?" An answer to our query can be obtained by PCR provided that we already know at least one short region of DNA which occurs in our gene. Our algorithm for the independent set discovery (Section 3) then yields an experimental protocol for finding genes. However, many aspects of the real experimental domain further restrict the set of possible queries and would need to be addressed to apply this technique in practice (see e.g., [5]). This application of PCR technology was inspired by open problem

12.94 in [15]. PCR queries were also used in similar way to determine the exon boundaries in cDNA clones [20].

Applications to 1-Dimensional Battleship. The game of Battleship (also known as Convoy and Sinking-Ships) is a well-known two-person game. Both players have an $n \times n$ grid and a fixed set of ships, where each ship is a $1 \times k$ rectangle for some $k \leq n$. Each player arranges the ships on his/her grid in such a way that no two ships intersect. Then players take turns shooting at each other's ships by calling the coordinates of a grid position. The player that first sinks all ships (by hitting all grid positions that contain a ship) wins. There are many variants of Battleship (see e.g., [1]) involving other ship shapes or higher dimensions. Battleship becomes an interval graph game in the 1-dimensional version. Here the ships are intervals with integral end points, and, as before, no two intersecting ship positions may be taken. The allowed operations are now exactly our queries: given an open unit interval $(a, a + 1)$, does one ship overlap this interval?

2 Independent Set Verification

Let Y be the candidate set to be verified. There are two types of probes: the ones for which the probe intersects some interval in Y (we call this a *positive probe*) and the ones for which it does not (we call this a *negative probe*). For a probe the *expected answer* is the answer that is consistent with $X = Y$. Thus, a positive probe has expected answer "yes," while a negative probe has expected answer "no."

Consider an algorithm to solve the verification problem. If for some query it does not get the expected answer, then $X \neq Y$ and the algorithm can terminate. Otherwise the algorithm must continue until enough queries are made to determine that $X = Y$. Thus the worst case for any optimal verification algorithm is when $X = Y$ (i.e., all answers are as expected).

This implies that we can rephrase the verification problem as follows: given a graph G and an independent set Y, produce a set of queries U such that Y is the only independent set in G consistent with the expected answers to all queries in U. We say that a set of queries U *verifies* that $X = Y$ if every independent set $Z \neq Y$ is inconsistent with the expected answer of at least one query in U; we say this query *eliminates* Z.

Finding a Minimum Set of Positive Probes. We first study a special case in which only queries with positive probes are allowed. This case is then used as a subroutine for the general case. Note that for some inputs it is impossible to verify $Y = X$ using only positive probes.

Let $G[a, b]$ denote the subgraph of G induced by intervals completely contained in the region $[a, b]$ and for any independent set Z, let $Z[a, b]$ denote the subset of Z of intervals completely contained in $[a, b]$. The minimum set of positive probes for a graph G will be computed using a directed acyclic graph H. Graph H contains one vertex for every positive probe. Let a_{min} be the smallest start point and a_{max} be the largest end point of an interval in G. Two additional

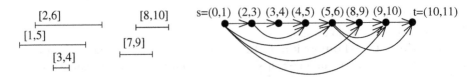

Fig. 1. An interval graph and its corresponding graph H for $Y = \{[2, 6], [8, 10]\}$. For example, edge $(5, 6) \to (9, 10)$ exists because the independent set $\{[7, 9]\} \in G[6, 9]$ intersects all positive probes between 6 and 9. Edge $(0, 1) \to (2, 3)$ exists because there are no positive probes in $G[1, 2]$ and thus empty independent set satisfies the definition.

vertices s and t are added to H, where s corresponds to probe $(a_{min} - 1, a_{min})$ and t corresponds to probe $(a_{max}, a_{max} + 1)$. Note that these two probes are negative for G.

Intuitively, H contains a directed edge from one probe to another if no positive probe between them can distinguish Y from some other independent set. More precisely, for any $a < b$, graph H contains an edge $e_{a,b}$ from $(a, a + 1)$ to $(b, b + 1)$ if and only if there is an independent set $Z_{a,b}$ in $G[a + 1, b]$ that intersects all positive probes $(c, c + 1)$ with $a < c < b$ and that is different from Y. See Figure 1 for an example of graph H. Graph H has $O(n)$ vertices and $O(n^2)$ edges, where n is the number of intervals. Using dynamic programming, it can be constructed in $O(n^2)$ time. The following lemma shows the connection between graph H and the optimal set of positive queries.

Lemma 1. *A set of positive probes U verifies that $X = Y$ if and only if vertices s and t become disconnected in graph H after removal of all vertices in U.*

Proof. (Sketch.) The crucial observation is that a path from s to t in graph H through a set of positive probes U exists if and only if there is an independent set other than Y which is consistent with all positive probes except the ones in U. For a given path π such independent set can be constructed as a union of sets $Z_{a,b}$ over all edges $e_{a,b} \in \pi$. On the other hand, for a given independent set Z there is always a path from s to t in graph H using exactly all positive probes inconsistent with Z. □

Thus the minimal set of positive probes to verify $X = Y$ corresponds to the smallest set of vertices in H that disconnect s and t. This vertex-connectivity problem can be solved in $O(n^{8/3})$ time using network flows [9]. Since we want to use this as a subroutine in the general case, we expand the result to any subgraph $G[a, b]$ of G. On such a subgraph we need to verify that $X[a, b] = Y[a, b]$.

Lemma 2. *Let $A_+[a, b]$ be the smallest number of positive probes needed to verify that $X[a, b] = Y[a, b]$ in $G[a, b]$, or $A_+[a, b] = \infty$ if this is not possible. Then $A_+[a, b]$ can be computed in $O(n^{8/3})$ time.*

Finding a Minimum Set of Probes in the General Case. The general case, in which both positive and negative probes are allowed, is solved by a dynamic programming algorithm that has the result of Lemma 2 as a base case.

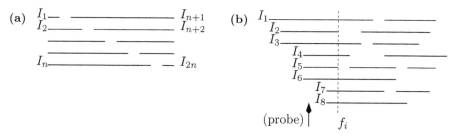

Fig. 2. (a) **The staircase** consists of $2n$ intervals, with interval $I_i = [0, 2i - 1]$ for $i = 1, \ldots, n$ and $I_i = [2(i - n), 2n + 1]$ for $i = n + 1, \ldots, 2n$. The staircase contains $p = n(n + 1)/2 + 2n + 1$ independent sets, therefore Theorem 2 gives a lower bound of $2 \log_2 n + O(1)$ queries. However, a better lower bound can be achieved. In the worst case, each query eliminates at most one pair $\{I_i, I_{n+i}\}$ and thus we need at least $n - 1$ queries. (b) **Paths eliminated:** $p_1 + \ldots + p_6$ ("no") or $p_7 + p_8 + p_{rest}$ ("yes").

Lemma 3. *Let $A[a]$ be the smallest number of queries needed to verify that $X[1, a] = Y[1, a]$ in the interval graph $G[1, a]$. Then*

$$
A[a] = \min \begin{cases} A_+[1, a], \\ \min_b A[b] + A_+[b + 1, a] + 1, \text{ where } (b, b+1) \text{ is a negative probe} \\ \qquad\qquad\qquad\qquad\qquad\qquad \text{intersecting } [1, a] \end{cases}
$$

Note, that we have at most n^2 subproblems $A_+[a, b]$ to solve, and hence obtain the overall result.

Theorem 1. *Given an n-vertex interval graph G and an independent set Y in G, we can find in $O(n^{14/3})$ time the minimum set of queries that verifies whether Y is the hidden independent set chosen by an adversary.*

3   Independent Set Discovery

In this section we give an interactive protocol to find an independent set X. In this case the next query depends on the outcome of the previous query. A simple information-theoretic argument yields the following lower bound.

Theorem 2. *Assume that G is a graph that contains p independent sets. Regardless of the types of yes/no queries allowed, we need at least $\lceil \log_2 p \rceil$ queries to find a hidden independent set X in the worst case.*

This lower bound is not always tight, even for an interval graph (see the so-called *staircase* example in Figure 2a). However, we give an algorithm which matches asymptotically the lower bound of $\lceil \log_2 p \rceil$ queries under the assumption that at most a constant number of intervals start at the same point.
Overview of the Algorithm. The algorithm to detect the hidden path in the complement graph $\overline{G}$ is recursive. The crucial idea is that with a constant number of queries we eliminate at least a constant fraction of the remaining

paths. This would be straightforward if the set of paths always had a central element allowing us to eliminate readily the desired constant fraction of paths. However there are configurations with no such element. Surprisingly, they appear infrequently. Therefore, after $O(\log p)$ queries, we know the correct path.

For ease of notation, assume that the intervals $I_1, \ldots, I_n$ are sorted by increasing start point, breaking ties arbitrarily. Let I_i be the interval with the leftmost right end point f_i . The intervals that intersect I_i are called *clique intervals* (they are the intervals $I_1, \ldots, I_k$ with k such that $s_k \le f_i$ and $s_{k+1} > f_i$). Note that as the name suggests, they form a clique in G, and at most one of them is in any path. Our algorithm operates under two different scenarios. Let a *legal path* be a path that could be the solution even under the following added restrictions. In the *unrestricted scenario*, any path is a legal path; this is the scenario at the beginning of the algorithm. In the *restricted scenario*, only a path that intersects (f_{i-1}, f_i) is legal (this information is obtained through previous queries). Any legal path thus uses a clique interval that starts strictly before f_i.

Effects of Queries. The algorithm always queries at $(a, a+1)$ for some $a \le f_i$. Only clique intervals can intersect the probe. If the answer to the query is "no", then we eliminate all clique intervals that intersect $(a, a + 1)$. If the original scenario was unrestricted, then all remaining paths are consistent with this query and we can solve the problem recursively. If the original scenario was restricted, we already know that one of the clique intervals $I_1, \ldots, I_k$ is in the hidden path X. Eliminating some clique intervals may increase the value of f_i and therefore add some more intervals to the clique intervals. None of these new clique intervals can be in X, and thus they can also be eliminated. Then we solve the restricted scenario recursively on the new graph. Assume now that the answer to the query is "yes". Since X contains at most one clique interval, all clique intervals not intersecting $(a, a + 1)$ can be eliminated. One of the remaining clique intervals will be part of the solution, so the next scenario will be restricted. We also can eliminate all intervals that become clique intervals due to an increase in f_i.

If in the new situation we are now in the restricted scenario with only one clique interval I_1, then I_1 belongs to X. Therefore, I_1 can be eliminated from the graph and we solve the unrestricted scenario on the resulting graph recursively. Afterwards we add I_1 to get the hidden path X.

Some Definitions and Observations. Consider a specific point in time when we want to find the next query. Let P_{legal} be the set of all legal paths. Since every legal path contains at most one clique interval, we can partition P_{legal} as $P_{legal} = P_1 \cup \cdots \cup P_k \cup P_{rest}$, where P_j is the set of legal paths that use clique interval I_j, and P_{rest} denotes the legal paths that do not use a clique interval. (P_{rest} is empty in the restricted scenario.) Define $p_\beta = |P_\beta|$ for all subscripts β. One can show the following properties of these sets of paths:

Lemma 4. a) *In the unrestricted scenario, $p_i = p_{rest}$.*
b) $p_{rest} \le \frac{1}{2} p_{legal}$.
c) *If I_{j_1} and I_{j_2} are clique intervals with $f_{j_1} \le f_{j_2}$ then $p_{j_1} \ge p_{j_2}$.*

The following lemma summarizes the effects of a query (see Figure 2b for an illustration):

Lemma 5. *If we query at $(s_j, s_j + 1)$ for some j with $s_j < f_i$, then we can eliminate either $p_1 + \cdots + p_{j'}$ paths or $p_{j'+1} + \cdots + p_k + p_{rest}$ paths, where $j' \geq j$ is the largest index with $s_{j'} = s_j$.*

Choosing Queries. In light of Lemma 5 we will try to find a j such that both sets of possibly eliminated paths contain a constant fraction of the paths. To find such a j, define $1 \leq \ell \leq k$ to be the index such that

$$p_1 + \cdots + p_{\ell-1} < \tfrac{1}{2}p_{legal} \quad \text{and} \quad p_1 + \cdots + p_{\ell-1} + p_\ell \geq \tfrac{1}{2}p_{legal}; \qquad (1)$$

this is well-defined by Lemma 4b. Define ℓ^- and ℓ^+ to be the smallest/largest index such that $s_{\ell^-} = s_\ell = s_{\ell^+}$, thus $\ell^- \leq \ell \leq \ell^+$. We distinguish three cases:

C1: $p_1 + \cdots + p_{\ell^- - 1} \geq \tfrac{1}{4}p_{legal}$ and $p_{\ell^-} + \cdots + p_k + p_{rest} \geq \tfrac{1}{4}p_{legal}$: The algorithm queries at the beginning of $I_{\ell^- - 1}$, i.e., at $(s_{\ell^- - 1}, s_{\ell^- - 1} + 1)$. By Lemma 5 this eliminates at least $\tfrac{1}{4}p_{legal}$ paths.

C2: $p_1 + \cdots + p_{\ell^+} \geq \tfrac{1}{4}p_{legal}$ and $p_{\ell^+ + 1} + \cdots + p_k + p_{rest} \geq \tfrac{1}{4}p_{legal}$: In this case, query at $(s_{\ell^+}, s_{\ell^+} + 1)$. By Lemma 5 this eliminates at least $\tfrac{1}{4}p_{legal}$ paths.

C3: All remaining cases: We query with probe $(f_i, f_i + 1)$. Note that this query is not covered by Lemma 5, and we will analyze it separately.

In case (C1) and (C2) we eliminate at least a constant fraction of the legal paths, and hence the number of such queries is at most $O(\log p)$. The analysis is more intricate in case (C3). Using Lemma 4, one can show the following in case of a positive answer.

Lemma 6. *Let θ denote the maximum number of intervals that have a common start point (i.e., $\ell^+ - \ell^- + 1 \leq \theta$). Then in case (C3), $p_{\ell^-} + \cdots + p_{\ell^+} > \tfrac{1}{2}p_{legal}$, and a positive answer to the query eliminates at least $p_i \geq \tfrac{1}{2\theta}p_{legal}$ paths.*

If the query in (C3) yields a negative answer, then possibly less than a constant fraction of paths is eliminated, but we account for this query in a different way.

Lemma 7. *During all recursive calls, we have at most $\log_2 p$ times a negative answer in case (C3), where p is the number of paths in the original graph.*

Proof. (Sketch.) Let s be the number of such queries. In case (C3) at least one interval intersects $(f_i, f_i + 1)$ (otherwise Lemma 6 implies that we are in a restricted scenario with only one clique interval, which is a contradiction). The negative answer eliminates all intervals intersecting $(f_i, f_i + 1)$ and therefore we will not return to (C3) until the value of f_i has changed. Thus for each negative answer in case (C3), we have a different value of f_i. Let $f_{i_1} < \cdots < f_{i_s}$ be these values, and let I_{i_j} be a clique interval that ends at f_{i_j} and was not eliminated when we queried at $(f_{i_j}, f_{i_j} + 1)$. Then $\{I_{i_1}, \ldots, I_{i_s}\}$ is an independent set and so is every subset of this set. Therefore $p \geq 2^s$ as desired. $\qquad \square$

The above analysis of cases C1, C2, and C3 yields the following result.

Lemma 8. *Assume we are given a set of n intervals that define p paths, and at most θ intervals start at the same point. Then any hidden path X can be found with at most $\log_2 p + \max\{\log_{2\theta/(2\theta-1)} p, \log_{4/3} p\}$ queries.*

Note that as long as θ is a constant, we use $O(\log_2 p)$ queries, which is asymptotically optimal. Our algorithm can easily be implemented in polynomial time and indeed, with the right data structure its running time is $O(n + m)$, where m is the number of edges in the complement of the interval graph. Details are omitted due to space limitations.

Theorem 3. *Given an n-vertex interval graph G with m edges in its complement, we can find the hidden independent set in G using q queries, where q is asymptotically optimal if at most a constant number of intervals start in any one point. The overall computation time and space is $O(n + m)$.*

4   Conclusions and Future Work

In this paper we studied a problem motivated by applications in bioinformatics and game playing: given an interval graph, how can we find an independent set chosen by an adversary with as few queries as possible? We gave polynomial-time algorithms both for verifying whether some independent set is the one chosen by the adversary, and for discovering what set the adversary has chosen. The algorithm for verification gives the optimal number of queries for all instances. The algorithm for independent set discovery gives a number of queries that is optimal to within constant factor, provided that no more than a constant number of intervals start at the same point. This algorithm is optimal in the adaptive sense as well as in the worst case sense. We also proved a stronger lower bound than the one implied by a simple information theory argument.

The main open question is whether our adaptive algorithm can extend to instances in which many intervals may start at a common point, and still achieve a number of queries that is within a constant factor of optimal. Another open problem relates to the motivation of this work from gene finding using PCR techniques. Here we need to consider that obtaining probing material is often done via an external provider, and the turnaround time between each request might dominate the total time. We might thus consider performing several probes in parallel rounds. What is the minimum number of queries required if the entire computation must be done in a given number of rounds? In the application to gene finding, we might also be able to eliminate certain edges of $\overline{G}$ using biological background information. Can we adapt our algorithm to take advantage of this, i.e., use an optimal number of queries subject to knowing this information? (Note that G is now no longer necessarily an interval graph.)

Acknowledgments. We thank Dan Brown and the participants at the Bioinformatics problem sessions at the University of Waterloo for many useful comments on this problem. All authors were partially supported by NSERC.

References

1. Battleships variations. Mountain Vista Software. Web page.
   See http://www.mountainvistasoft.com/variations.htm
2. R. Beigel, N. Alon, M. S. Apydin, and L. Fortnow. An optimal procedure for gap closing in whole genome shotgun sequencing. In *5th Annual International Conference on Computational Molecular Biology (RECOMB)*, pages 22–30, 2001.
3. H. L. Bodlaender and D. Kratsch. The complexity of coloring games on perfect graphs. *Theoretical Computer Science*, 106(2):309–326, 1992.
4. C. Burge and S. Karlin. Prediction of complete gene structures in human genomic DNA. *Journal of Molecular Biology*, 268(1):78–94, 1997.
5. M. Das, C. B. Burge, E. Park, J. Colinas, and J. Pelletier. Assessment of the total number of human transcription units. *Genomics*, 77(1-2):71–78, 2001.
6. E. D. Demaine, A. López-Ortiz, and J. I. Munro. Adaptive set intersections, unions, and differences. In *11th Annual ACM-SIAM Symposium on Discrete Algorithms (SODA)*, pages 743–752, 2000.
7. I. Dunham et al. The DNA sequence of human chromosome 22. *Nature*, 402(6761):489–495, 1999.
8. P. Erdős and J. L. Selfridge. On a combinatorial game. *Journal of Combinatorial Theory – Series A*, 14:298–301, 1973.
9. S. Even and R. E. Tarjan. Network flow and testing graph connectivity. *SIAM Journal on Computing*, 4:507–518, 1975.
10. R. Fagin, A. Lotem, and M. Naor. Optimal aggregation algorithms for middleware. In *20th ACM Symposium on Principle of Database Systems (PODS)*, pages 102–113, 2001.
11. M. C. Golumbic. *Algorithmic graph theory and perfect graphs*. Academic Press, New York, 1980.
12. L. M. Kirousis and C. H. Papadimitriou. Searching and pebbling. *Theoretical Computer Science*, 47(2):205–218, 1986.
13. R. M. McConnell and J. P. Spinrad. Construction of probe interval models. In *13th Annual ACM-SIAM Symposium on Discrete Algorithms (SODA)*, pages 866–875, 2002.
14. N. Pavy, S. Rombauts, P. Dehais, C. Mathe, D. V. Ramana, P. Leroy, and P. Rouze. Evaluation of gene prediction software using a genomic data set: application to Arabidopsis thaliana sequences. *Bioinformatics*, 15(11):887–889, 1999.
15. P. A. Pevzner. *Computational molecular biology: an algorithmic approach*. MIT Press, 2000.
16. A. A. Salamov and V. V. Solovyev. Ab initio gene finding in Drosophila genomic DNA. *Genome Research*, 10(4):516–522, 2000.
17. S. J. Scharf, G. T. Horn, and H. A. Erlich. Direct cloning and sequence analysis of enzymatically amplified genomic sequences. *Science*, 233(4768):1076–1078, 1986.
18. M. Schena, D. Shalon, R. W. Davis, and P. O. Brown. Quantitative monitoring of gene expression patterns with a complementary DNA microarray. *Science*, 270(5235):467–470, 1995.
19. D. D. Shoemaker, E. E. Schadt, et al. Experimental annotation of the human genome using microarray technology. *Nature*, 409(6822):922–927, 2001.
20. G. Xu, S. H. Sze, C. P. Liu, P. A. Pevzner, and N. Arnheim. Gene hunting without sequencing genomic clones: finding exon boundaries in cDNAs. *Genomics*, 47(2):171–179, 1998.

Matroid Representation of Clique Complexes

Kenji Kashiwabara[1], Yoshio Okamoto[2*], and Takeaki Uno[3]

[1] Department of Systems Science, Graduate School of Arts and Sciences,
The University of Tokyo, 3–8–1, Komaba, Meguro, Tokyo, 153–8902, Japan.
kashiwa@graco.c.u-tokyo.ac.jp
[2] Institute of Theoretical Computer Science, Department of Computer Science,
ETH Zurich, ETH Zentrum, CH-8092, Zurich, Switzerland.
okamotoy@inf.ethz.ch
[3] National Institute of Informatics, 2–1–2, Hitotsubashi, Chiyoda,
Tokyo, 101-8430, Japan.
uno@nii.jp

Abstract. In this paper, we approach the quality of a greedy algorithm for the maximum weighted clique problem from the viewpoint of matroid theory. More precisely, we consider the clique complex of a graph (the collection of all cliques of the graph) and investigate the minimum number k such that the clique complex of a given graph can be represented as the intersection of k matroids. This number k can be regarded as a measure of "how complex a graph is with respect to the maximum weighted clique problem" since a greedy algorithm is a k-approximation algorithm for this problem. We characterize graphs whose clique complexes can be represented as the intersection of k matroids for any $k > 0$. Moreover, we determine the necessary and sufficient number of matroids for the representation of all graphs with n vertices. This number turns out to be $n - 1$. Other related investigations are also given.

1 Introduction

A lot of combinatorial optimization problems can be seen as optimization problems on the corresponding independence systems. For example, for the minimum cost spanning tree problem the corresponding independence system is the collection of all forests of a given graph; for the maximum weighted matching problem the corresponding independence system is the collection of all matchings of a given graph; for the maximum weighted clique problem the corresponding independence system is the collection of all cliques of a given graph, which is called the clique complex of the graph. More examples are provided by Korte–Vygen [9].

It is known that any independence system can be represented as the intersection of some matroids. Jenkyns [7] and Korte–Hausmann [8] showed that a greedy algorithm is a k-approximation algorithm for the maximum weighted base

---

* Supported by the Joint Berlin/Zürich Graduate Program "Combinatorics, Geometry, and Computation" (CGC), financed by ETH Zurich and the German Science Foundation (DFG).

T. Warnow and B. Zhu (Eds.): COCOON 2003, LNCS 2697, pp. 192–201, 2003.
© Springer-Verlag Berlin Heidelberg 2003

problem of an independence system which can be represented as the intersection of k matroids. (Their result can be seen as a generalization of the validity of the greedy algorithm for matroids, shown by Rado [13] and Edmonds [3].) So the minimum number of matroids which we need to represent an independence system as their intersection is one of the measures of "how complex an independence system is with respect to the corresponding optimization problem."

In this paper, we investigate how many matroids we need to represent the clique complex of a graph as their intersection, while Fekete–Firla–Spille [5] investigated the same problem for matching complexes. We will show that the clique complex of a given graph G is the intersection of k matroids if and only if there exists a family of k stable-set partitions of G such that every edge of $\overline{G}$ is contained in a stable set of some stable-set partition in the family. This theorem implies that the problem to determine the clique complex of a given graph have a representation by k matroids or not belongs to NP (for any fixed k). This is not a trivial fact since in general the size of an independence system will be exponential when we treat it computationally.

The organization of this paper is as follows. In Sect. 2, we will introduce some terminology on independence systems. The proof of the main theorem will be given in Sect. 3. In Sect. 4, we will consider an extremal problem related to our theorem. In Sect. 5, we will investigate the case of two matroids more thoroughly. This case is significantly important since the maximum weighted base problem can be solved exactly in polynomial time for the intersection of two matroids [6]. (Namely, in this case, the maximum weighted clique problem can be solved in polynomial time for any non-negative weight vector by Frank's algorithm [6].) From the observation in that section, we can find the algorithm by Protti–Szwarcfiter [12] checks that a given clique complex has a representation by two matroids or not in polynomial time. We will conclude with Sect. 6.

2   Preliminaries

We will assume the basic concepts in graph theory. If you find something unfamiliar, see a textbook of graph theory (Diestel's book [2] or so). Here we will fix our notations. In this paper, all graphs are finite and simple unless stated otherwise. For a graph $G = (V, E)$ we denote the subgraph induced by $V' \subseteq V$ by $G[V']$. The complement of G is denoted by $\overline{G}$. The vertex set and the edge set of a graph $G = (V, E)$ are denoted by $V(G)$ and $E(G)$, respectively. A complete graph and a cycle with n vertices are denoted by K_n and C_n, respectively. The maximum degree, the chromatic number and the edge-chromatic number (or the chromatic index) of a graph G are denoted by $\Delta(G)$, $\chi(G)$ and $\chi'(G)$, respectively. A *clique* of a graph $G = (V, E)$ is a subset $C \subseteq V$ such that the induced subgraph $G[C]$ is complete. A *stable set* of a graph $G = (V, E)$ is a subset $S \subseteq V$ such that the induced subgraph $G[S]$ has no edge.

Now we introduce some notions of independence systems and matroids. For details of them, see Oxley's book [11]. Given a non-empty finite set V, an *independence system* on V is a non-empty family $\mathcal{I}$ of subsets of V satisfying: $X \in \mathcal{I}$

implies $Y \in \mathcal{I}$ for all $Y \subseteq X \subseteq V$. The set V is called the *ground set* of this independence system. In the literature, an independence system is also called an *abstract simplicial complex*. A *matroid* is an independence system $\mathcal{I}$ additionally satisfying the following *augmentation axiom*: for $X, Y \in \mathcal{I}$ with $|X| > |Y|$ there exists $z \in X \setminus Y$ such that $Y \cup \{z\} \in \mathcal{I}$. For an independence system $\mathcal{I}$, a set $X \in \mathcal{I}$ is called *independent* and a set $X \notin \mathcal{I}$ is called *dependent*. A *base* and a *circuit* of an independence system is a maximal independent set and a minimal dependent set, respectively. We denote the family of bases of an independence system $\mathcal{I}$ and the family of circuits of $\mathcal{I}$ by $\mathcal{B}(\mathcal{I})$ and $\mathcal{C}(\mathcal{I})$, respectively. Note that we can reconstruct an independence system $\mathcal{I}$ from $\mathcal{B}(\mathcal{I})$ or $\mathcal{C}(\mathcal{I})$ as $\mathcal{I} = \{X \subseteq V : X \subseteq B \text{ for some } B \in \mathcal{B}(\mathcal{I})\}$ and $\mathcal{I} = \{X \subseteq V : C \not\subseteq X \text{ for all } C \in \mathcal{C}(\mathcal{I})\}$. In particular, $\mathcal{B}(\mathcal{I}_1) = \mathcal{B}(\mathcal{I}_2)$ if and only if $\mathcal{I}_1 = \mathcal{I}_2$; similarly $\mathcal{C}(\mathcal{I}_1) = \mathcal{C}(\mathcal{I}_2)$ if and only if $\mathcal{I}_1 = \mathcal{I}_2$. We can see that the bases of a matroid have the same size from the augmentation axiom, but it is not the case for a general independence system.

Let $\mathcal{I}$ be a matroid on V. An element $x \in V$ is called a *loop* if $\{x\}$ is a circuit of $\mathcal{I}$. We say that $x, y \in V$ are *parallel* if $\{x, y\}$ is a circuit of the matroid $\mathcal{I}$. The next is well known.

Lemma 2.1 (see [11]). *For a matroid without a loop, the relation that "x is parallel to y" is an equivalence relation.*

Let $\mathcal{I}_1, \mathcal{I}_2$ be independence systems on the same ground set V. The *intersection* of $\mathcal{I}_1$ and $\mathcal{I}_2$ is just $\mathcal{I}_1 \cap \mathcal{I}_2$. The intersection of more independence systems is defined in a similar way. Note that the intersection of independence systems is also an independence system. Also note that the family of circuits of $\mathcal{I}_1 \cap \mathcal{I}_2$ is the family of the minimal sets of $\mathcal{C}(\mathcal{I}_1) \cup \mathcal{C}(\mathcal{I}_2)$, i.e., $\mathcal{C}(\mathcal{I}_1 \cap \mathcal{I}_2) = \min(\mathcal{C}(\mathcal{I}_1) \cup \mathcal{C}(\mathcal{I}_2))$. The following well-known observation is crucial in this paper.

Lemma 2.2 (see [4,5,9]). *Every independence system can be represented as the intersection of finitely many matroids on the same ground set.*

Proof. Denote the circuits of an independence system $\mathcal{I}$ by $C^{(1)}, \ldots, C^{(m)}$, and consider the matroid $\mathcal{I}_i$ with a unique circuit $\mathcal{C}(\mathcal{I}_i) = \{C^{(i)}\}$ for each $i \in \{1, \ldots, m\}$. Then, the family of the circuits of $\bigcap_{i=1}^{m} \mathcal{I}_i$ is nothing but $\{C^{(1)}, \ldots, C^{(m)}\}$. Therefore, we have $\mathcal{I} = \bigcap_{i=1}^{m} \mathcal{I}_i$. □

Due to Lemma 2.2, we are interested in representation of an independence system as the intersection of matroids. From the construction in the proof of Lemma 2.2, we can see that the number of matroids which we need to represent an independence system $\mathcal{I}$ by the intersection is at most $|\mathcal{C}(\mathcal{I})|$. However, we might do better. In this paper, we investigate such a number for a clique complex.

3   Clique Complexes and the Main Theorem

A graph gives rise to various independence systems. Among them, we will investigate clique complexes.

The *clique complex* of a graph $G = (V, E)$ is the collection of all cliques of G. We denote the clique complex of G by $\mathfrak{C}(G)$. Note that the empty set is a clique and $\{v\}$ is also a clique for each $v \in V$. So we can see that the clique complex is actually an independence system on V. We also say that an independence system is a clique complex if it is isomorphic to the clique complex of some graph. Notice that a clique complex is also called a *flag complex* in the literature.

Here we give some subclasses of the clique complexes. (We omit necessary definitions.) (1) The family of the stable sets of a graph G is nothing but the clique complex of $\overline{G}$. (2) The family of the matchings of a graph G is the clique complex of the complement of the line graph of G, which is called the *matching complex* of G. (3) The family of the chains of a poset P is the clique complex of the comparability graph of P, which is called the *order complex* of P. (4) The family of the antichains of a poset P is the clique complex of the complement of the comparability graph of P.

The next lemma may be a folklore.

Lemma 3.1. *Let $\mathcal{I}$ be an independence system on a finite set V. Then, $\mathcal{I}$ is a clique complex if and only if the size of every circuit of $\mathcal{I}$ is two. In particular, the circuits of the clique complex of G are the edges of $\overline{G}$.*

Proof. Let $\mathcal{I}$ be the clique complex of $G = (V, E)$. Since a single vertex $v \in V$ forms a clique, the size of each circuit is greater than one. Each dependent set of size two is an edge of the complement. Observe that they are minimal dependent sets since the size of each dependent set is greater than one. Suppose that there exists a circuit C of size more than two. Then each two elements in C form an edge of G. Hence C is a clique. This is a contradiction.

Conversely, assume that the size of every circuit of $\mathcal{I}$ is two. Then construct a graph $G' = (V, E')$ with $E' = \{\{u, v\} \in \binom{V}{2} : \{u, v\} \notin \mathcal{C}(\mathcal{I})\}$. Consider the clique complex $\mathfrak{C}(G')$. By the opposite direction which we have just shown, we can see that $\mathcal{C}(\mathfrak{C}(G')) = \mathcal{C}(\mathcal{I})$. Therefore $\mathcal{I}$ is the clique complex of G'. □

Now we start studying the number of matroids which we need for the representation of a clique complex as their intersection. First we characterize the case in which we need only one matroid. (namely the case in which a clique complex is a matroid). To do this, we define a partition matroid. A *partition matroid* is a matroid $\mathcal{I}(\mathcal{P})$ associated with a partition $\mathcal{P} = \{P_1, P_2, \ldots, P_r\}$ of V defined as $\mathcal{I}(\mathcal{P}) = \{I \subseteq V : |I \cap P_i| \leq 1 \text{ for all } i \in \{1, \ldots, r\}\}$. Observe that $\mathcal{I}(\mathcal{P})$ is a clique complex. Indeed if we construct a graph $G_\mathcal{P} = (V, E)$ from $\mathcal{P}$ as $u, v \in V$ are adjacent in $G_\mathcal{P}$ if and only if u, v are elements of distinct partition classes in $\mathcal{P}$, then we can see that $\mathcal{I}(\mathcal{P}) = \mathfrak{C}(G_\mathcal{P})$. Note that $\mathcal{C}(\mathcal{I}(\mathcal{P})) = \{\{u, v\} \in \binom{V}{2} : \{u, v\} \subseteq P_i \text{ for some } i \in \{1, \ldots, r\}\}$. Also note that $G_\mathcal{P}$ is a complete r-partite graph with the partition $\mathcal{P}$. In the next lemma, the equivalence of (1) and (3) is also noticed by Okamoto [10].

Lemma 3.2. *Let G be a graph. Then the following are equivalent. (1) The clique complex of G is a matroid. (2) The clique complex of G is a partition matroid. (3) G is complete r-partite for some r.*

Proof. "(2) $\Rightarrow$ (1)" is clear, and "(3) $\Rightarrow$ (2)" is immediate from the discussion above. So we only have to show "(1) $\Rightarrow$ (3)." Assume that the clique complex $\mathfrak{C}(G)$ is a matroid. By Lemma 3.1, every circuit of $\mathfrak{C}(G)$ is of size two, which corresponds to an edge of $\overline{G}$. So the elements of each circuit are parallel. Lemma 2.1 says that the parallel elements induce an equivalence relation on $V(G)$, which yields a partition $\mathcal{P} = \{P_1, \dots, P_r\}$ of $V(G)$. Thus, we can see that G is a complete r-partite graph with the vertex partition $\mathcal{P}$. $\square$

For the case of more matroids, we use a stable-set partition. A *stable-set partition* of a graph $G = (V, E)$ is a partition $\mathcal{P} = \{P_1, \dots, P_r\}$ of V such that each P_i is a stable set of G. The following theorem is the main result of this paper. It tells us how many matroids we need to represent a given clique complex.

Theorem 3.3. *Let $G = (V, E)$ be a graph. Then, the clique complex $\mathfrak{C}(G)$ can be represented as the intersection of k matroids if and only if there exist k stable-set partitions $\mathcal{P}^{(1)}, \dots, \mathcal{P}^{(k)}$ such that $\{u, v\} \in \binom{V}{2}$ is an edge of $\overline{G}$ if and only if $\{u, v\} \subseteq S$ for some $S \in \bigcup_{i=1}^{k} \mathcal{P}^{(i)}$ (in particular, $\mathfrak{C}(G) = \bigcap_{i=1}^{k} \mathcal{I}(\mathcal{P}^{(i)})$).*

To show the theorem, we use the following lemmas.

Lemma 3.4. *Let $G = (V, E)$ be a graph. If the clique complex $\mathfrak{C}(G)$ can be represented as the intersection of k matroids, then there exist k stable-set partitions $\mathcal{P}^{(1)}, \dots, \mathcal{P}^{(k)}$ such that $\mathfrak{C}(G) = \bigcap_{i=1}^{k} \mathcal{I}(\mathcal{P}^{(i)})$.*

Proof. Assume that $\mathfrak{C}(G)$ is represented as the intersection of k matroids $\mathcal{I}_1$, $\dots$, $\mathcal{I}_k$. Choose $\mathcal{I}_i$ arbitrarily ($i \in \{1, \dots, k\}$). Then the parallel elements of $\mathcal{I}_i$ induce an equivalence relation on V. Let $\mathcal{P}^{(i)}$ be the partition of V arising from this equivalence relation. Then the two-element circuits of $\mathcal{I}_i$ are the circuits of $\mathcal{I}(\mathcal{P}^{(i)})$. Moreover, there is no loop in $\mathcal{I}_i$ (otherwise $\bigcap \mathcal{I}_i$ cannot be a clique complex). Therefore, we have that $\min(\bigcup_{i=1}^{k} \mathcal{C}(\mathcal{I}_i)) = \min(\bigcup_{i=1}^{k} \mathcal{C}(\mathcal{I}(\mathcal{P}^{(i)})))$, which means that $\mathfrak{C}(G) = \bigcap_{i=1}^{k} \mathcal{I}_i = \bigcap_{i=1}^{k} \mathcal{I}(\mathcal{P}^{(i)})$. $\square$

Lemma 3.5. *Let $G = (V, E)$ be a graph and $\mathcal{P}$ be a partition of V. Then $\mathfrak{C}(G) \subseteq \mathcal{I}(\mathcal{P})$ if and only if $\mathcal{P}$ is a stable-set partition of G.*

Proof. Assume that $\mathcal{P}$ is a stable-set partition of G. Take $I \in \mathfrak{C}(G)$ arbitrarily. Then we have $|I \cap P| \leq 1$ for each $P \in \mathcal{P}$ by the definitions of cliques and stable sets. Hence $I \in \mathcal{I}(\mathcal{P})$, namely $\mathfrak{C}(G) \subseteq \mathcal{I}(\mathcal{P})$. Conversely, assume that $\mathfrak{C}(G) \subseteq \mathcal{I}(\mathcal{P})$. Take $P \in \mathcal{P}$ and a clique C of G arbitrarily. From our assumption, we have $C \in \mathcal{I}(\mathcal{P})$. Therefore, it holds that $|C \cap P| \leq 1$. This means that P is a stable set of G, namely $\mathcal{P}$ is a stable-set partition of G. $\square$

Now it is time to prove Theorem 3.3.

Proof (of Theorem 3.3). Assume that a given clique complex $\mathfrak{C}(G)$ is represented as the intersection of k matroids $\mathcal{I}_1, \dots, \mathcal{I}_k$. From Lemma 3.4, $\mathfrak{C}(G)$ can be represented as the intersection of k matroids associated with stable-set partitions $\mathcal{P}^{(1)}, \dots, \mathcal{P}^{(k)}$ of G. We will show that these partitions $\mathcal{P}^{(1)}, \dots, \mathcal{P}^{(k)}$ satisfy the condition in the statement of the theorem. By Lemma 3.1, $\{u, v\}$ is an edge of $\overline{G}$ if and only if $\{u, v\}$ is a circuit of the clique complex $\mathfrak{C}(G)$, namely $\{u, v\} \in C(\mathfrak{C}(G))$ $= \min(\bigcup_{i=1}^{k} C(\mathcal{I}_i)) = \min(\bigcup_{i=1}^{k} C(\mathcal{I}(\mathcal{P}^{(i)}))) = \bigcup_{i=1}^{k} C(\mathcal{I}(\mathcal{P}^{(i)}))$. So this means that there exists at least one $i \in \{1, \dots, k\}$ such that $\{u, v\} \in C(\mathcal{I}(\mathcal{P}^{(i)}))$. Hence, $\{u, v\} \subseteq S$ for some $S \in \mathcal{P}^{(i)}$ if and only if $\{u, v\}$ is an edge of $\overline{G}$.

Conversely, assume that we are given a family of stable-set partitions $\mathcal{P}^{(1)}$, ..., $\mathcal{P}^{(k)}$ of V satisfying the condition in the statement of the theorem. We will show that $\mathfrak{C}(G) = \bigcap_{i=1}^{k} \mathcal{I}(\mathcal{P}^{(i)})$. By Lemma 3.5, we can see that $\mathfrak{C}(G) \subseteq \mathcal{I}(\mathcal{P}^{(i)})$ for all $i \in \{1, \dots, k\}$. This shows that $\mathfrak{C}(G) \subseteq \bigcap_{i=1}^{k} \mathcal{I}(\mathcal{P}^{(i)})$. In order to show that $\mathfrak{C}(G) \supseteq \bigcap_{i=1}^{k} \mathcal{I}(\mathcal{P}^{(i)})$, we only have to show that $C(\mathfrak{C}(G)) \subseteq \bigcup_{i=1}^{k} C(\mathcal{I}(\mathcal{P}^{(I)}))$. Take $C \in C(\mathfrak{C}(G))$ arbitrarily. By Lemma 3.1 we have $|C| = 2$. Set $C = \{u, v\} \in E(\overline{G})$. From our assumption, it follows that $\{u, v\} \subseteq S$ for some $S \in \bigcup_{i=1}^{k} \mathcal{P}^{(i)}$. This means that $\{u, v\} \in \bigcup_{i=1}^{k} C(\mathcal{I}(\mathcal{P}^{(i)}))$. $\qquad\square$

4   An Extremal Problem for Clique Complexes

Let $\mu(G)$ be the minimum number of matroids which we need for the representation of the clique complex of G as their intersection, and $\mu(n)$ be the maximum of $\mu(G)$ over all graphs G with n vertices. Namely, $\mu(G) = \min\{k : \mathfrak{C}(G) = \bigcap_{i=1}^{k} \mathcal{I}_i$ where $\mathcal{I}_1, \dots, \mathcal{I}_k$ are matroids$\}$, and $\mu(n) = \max\{\mu(G) : G$ has n vertices$\}$. In this section, we will determine $\mu(n)$. From Lemmas 2.2 and 3.1 we can immediately obtain $\mu(n) \leq \binom{n}{2}$. However, the following theorem tells us this is far from the truth.

Theorem 4.1. *For every $n \geq 2$, it holds that $\mu(n) = n - 1$.*

First we will prove that $\mu(n) \geq n - 1$. Consider the graph $K_1 \cup K_{n-1}$.

$K_1 \cup K_5$

Lemma 4.2. *For $n \geq 2$, we have $\mu(K_1 \cup K_{n-1}) = n - 1$, particularly $\mu(n) \geq n - 1$.*

Proof. $\overline{K_1 \cup K_{n-1}}$ has $n - 1$ edges. From Lemma 3.1, the number of the circuits of $\mathfrak{C}(K_1 \cup K_{n-1})$ is $n-1$. By the argument below the proof of Lemma 2.2, we have $\mu(K_1 \cup K_{n-1}) \leq n-1$. Now, suppose that $\mu(K_1 \cup K_{n-1}) \leq n-2$. By Theorem 3.3 and the pigeon hole principle, there exists a stable-set partition $\mathcal{P}$ of $K_1 \cup K_{n-1}$ such that some class P in $\mathcal{P}$ contains at least two edges of $\overline{K_1 \cup K_{n-1}}$. However, this is impossible since P is stable. Hence, we have $\mu(K_1 \cup K_{n-1}) = n - 1$. $\quad\square$

Next we will prove that $\mu(n) \leq n - 1$. To do that, we will look at the relationship of $\mu(G)$ with the edge-chromatic number.

Lemma 4.3. *We have $\mu(G) \leq \chi'(\overline{G})$ for any graph G with n vertices. Particularly, if n is even we have $\mu(G) \leq n - 1$ and if n is odd we have $\mu(G) \leq n$. Moreover, if $\mu(G) = n$ then n is odd and the maximum degree of $\overline{G}$ is $n-1$ (i.e., G has an isolated vertex).*

Proof. Consider a minimum edge-coloring of $\overline{G}$, and we will construct $\chi'(\overline{G})$ stable-set partitions of a graph G with n vertices from this edge-coloring. We have the color classes $C^{(1)}, \ldots, C^{(k)}$ of the edges where $k = \chi'(\overline{G})$. Let us take a color class $C^{(i)} = \{e_1^{(i)}, \ldots, e_{l_i}^{(i)}\}$ ($i \in \{1, \ldots, k\}$) and construct a stable-set partition $\mathcal{P}^{(i)}$ of G from $C^{(i)}$ as follows: S is a member of $\mathcal{P}^{(i)}$ if and only if either (1) S is a two-element set belonging to $C^{(i)}$ (i.e., $S = e_j^{(i)}$ for some $j \in \{1, \ldots, l_i\}$) or (2) S is a one-element set $\{v\}$ which is not used in $C^{(i)}$ (i.e., $v \notin e_j^{(i)}$ for all $j \in \{1, \ldots, l_i\}$). Notice that $\mathcal{P}^{(i)}$ is actually a stable-set partition. Then we collect all the stable-set partitions $\mathcal{P}^{(1)}, \ldots, \mathcal{P}^{(k)}$ constructed by the procedure above. Moreover, we can check that these stable-set partitions satisfy the condition in Theorem 3.3. Hence, we have $\mu(G) \leq k = \chi'(\overline{G})$.

Here, notice that $\chi'(\overline{G}) \leq \chi'(K_n)$. So if n is even, then $\chi'(K_n)$ is $n-1$, which concludes $\mu(G) \leq n-1$. If n is odd, then $\chi'(K_n)$ is n, which concludes $\mu(G) \leq n$.

Assume that $\mu(G) = n$. From the discussion above, n should be odd. Remark that Vizing's theorem says for a graph H with maximum degree $\Delta(H)$ we have $\chi'(H) = \Delta(H)$ or $\Delta(H)+1$. If $\Delta(\overline{G}) \leq n-1$, then we have that $\mu(G) \leq \chi'(\overline{G}) \leq \Delta(\overline{G}) + 1 \leq n$. So $\mu(G) = n$ holds only if $\Delta(\overline{G}) + 1 = n$. $\square$

Now, we will show that if a graph G with an odd number of vertices has an isolated vertex then $\mu(G) \leq n - 1$. This completes the proof of Theorem 4.1.

Lemma 4.4. *Let n be odd and G be a graph with n vertices which has an isolated vertex. Then $\mu(G) \leq n - 1$.*

Proof. Let v^* be an isolated vertex of G. Consider the subgraph of G induced by $V(G) \setminus \{v^*\}$. Call this induced subgraph G'. Since G' has $n-1$ vertices, which is even, we have $\mu(G') \leq n-2$ from Lemma 4.3.

Now we will construct $n-1$ stable-set partitions of G which satisfy the condition in Theorem 3.3 from $n-2$ stable-set partitions of G' which also satisfy the condition in Theorem 3.3. Denote the vertices of G' by $1, \ldots, n-1$, and the stable-set partitions of G' by $\mathcal{P}'^{(1)}, \ldots, \mathcal{P}'^{(n-2)}$. Then construct stable-set partitions $\mathcal{P}^{(1)}, \ldots, \mathcal{P}^{(n-2)}, \mathcal{P}^{(n-1)}$ of G as follows. For $i = 1, \ldots, n-2$, $S \in \mathcal{P}^{(i)}$ if and only if either (1) $S \in \mathcal{P}'^{(i)}$ and $i \notin S$ or (2) $v^* \in S$, $S \setminus \{v^*\} \in \mathcal{P}'^{(i)}$ and $i \in S$. Also $S \in \mathcal{P}^{(n-1)}$ if and only if either (1) $S = \{v^*, n-1\}$ or (2) $S = \{i\}$ ($i = 1, \ldots, n-2$). We can observe that the stable-set partitions $\mathcal{P}^{(1)}, \ldots, \mathcal{P}^{(n-1)}$ satisfy the condition in Theorem 3.3 since v^* is an isolated vertex of G. $\square$

5   Characterization for Two Matroids

In this section, we will look more closely at a clique complex which can be represented as the intersection of two matroids. Note that Fekete–Firla–Spille [5] gave a characterization of the graphs whose matching complexes can be represented as the intersections of two matroids. So the theorem in this section is a generalization of their result.

To do this, we introduce another concept. The *stable-set graph* of a graph $G = (V, E)$ is a graph whose vertices are the maximal stable sets of G and two vertices of which are adjacent if the corresponding two maximal stable sets share a vertex in G. We denote the stable-set graph of a graph G by $\mathcal{S}(G)$.

Lemma 5.1. *Let G be a graph. Then the clique complex $\mathfrak{C}(G)$ can be represented as the intersection of k matroids if the stable-set graph $\mathcal{S}(G)$ is k-colorable.*

Proof. Assume that we are given a k-coloring c of $\mathcal{S}(G)$. Then gather the maximal stable sets of G which have the same color with respect to c, that is, put $C_i = \{S \in V(\mathcal{S}(G)) : c(S) = i\}$ for all $i = 1, \dots, k$. We can see that the members of C_i are disjoint maximal stable sets of G for each i.

Now we construct a graph G_i from C_i as follows. The vertex set of G_i is the same as that of G, and two vertices of G_i are adjacent if and only if either (1) one belongs to a maximal stable set in C_i and the other belongs to another maximal stable set in C_i, or (2) one belongs to a maximal stable set in C_i and the other belongs to no maximal stable set in C_i. Remark that G_i is complete r-partite, where r is equal to $|C_i|$ plus the number of the vertices which do not belong to any maximal stable set in C_i. Then consider $\mathfrak{C}(G_i)$, the clique complex of G_i. By Lemma 3.2, we can see that $\mathfrak{C}(G_i)$ is actually a matroid. Since an edge of G is also an edge of G_i, we have that $\mathfrak{C}(G) \subseteq \mathfrak{C}(G_i)$.

Here we consider the intersection $\mathcal{I} = \bigcap_{i=1}^{k} \mathfrak{C}(G_i)$. Since $\mathfrak{C}(G) \subseteq \mathfrak{C}(G_i)$ for any i, we have $\mathfrak{C}(G) \subseteq \mathcal{I}$. Since each circuit of $\mathfrak{C}(G)$ is also a circuit of $\mathfrak{C}(G_i)$ for some i (recall Lemma 3.1), we also have $C(\mathfrak{C}(G)) \subseteq C(\mathcal{I})$, which implies $\mathfrak{C}(G) \supseteq \mathcal{I}$. Thus we have $\mathfrak{C}(G) = \mathcal{I}$. $\qquad\qquad\square$

Note that the converse of Lemma 5.1 does not hold even if $k = 3$. A counterexample is the graph $G = (V, E)$ defined as $V = \{1, 2, 3, 4, 5, 6\}$ and $E = \{\{1, 2\}, \{3, 4\}, \{5, 6\}\}$. Here $\mathfrak{C}(G)$ is represented as the intersection of three matroids $C(\mathcal{I}_1) = \{\{1, 3, 5\}, \{2, 4, 6\}\}$, $C(\mathcal{I}_2) = \{\{1, 3, 6\}, \{2, 4, 5\}\}$ and $C(\mathcal{I}_3) = \{\{1, 4, 5\}, \{2, 3, 6\}\}$ while $\mathcal{S}(G)$ is not 3-colorable but 4-colorable. However, the converse holds if $k = 2$.

Theorem 5.2. *Let G be a graph. The clique complex $\mathfrak{C}(G)$ can be represented as the intersection of two matroids if and only if the stable-set graph $\mathcal{S}(G)$ is 2-colorable (i.e., bipartite).*

Proof. The if-part is straightforward from Lemma 5.1. We will show the only-if-part. Assume that $\mathfrak{C}(G)$ is represented as the intersection of two matroids. Thanks to Theorem 3.3, we assume that these two matroids are associated with stable-set partitions $\mathcal{P}^{(1)}, \mathcal{P}^{(2)}$ of G satisfying the condition in Theorem 3.3.

Let S be a maximal stable set of G. Now we will see that $S \in \mathcal{P}^{(1)} \cup \mathcal{P}^{(2)}$. From the maximality of S, we only have to show that $S \subseteq P$ for some $P \in \mathcal{P}^{(1)} \cup \mathcal{P}^{(2)}$. (Then, the maximality of S will tell us that $S = P$.) This claim clearly holds if $|S| = 1$. If $|S| = 2$, the claim holds from the condition in Theorem 3.3.

Assume that $|S| \geq 3$. Consider the following independence system $\mathcal{I} = \{I \subseteq S : I \subseteq P$ for some $P \in \mathcal{P}^{(1)} \cup \mathcal{P}^{(2)}\}$. Take a base B of $\mathcal{I}$ arbitrarily. Since $B \subseteq S$, B is a dependent set of $\mathfrak{C}(G)$. So B contains a circuit of $\mathfrak{C}(G)$. By Lemma 3.1, we have $|B| \geq 2$. Suppose that $S \setminus B \neq \emptyset$ for a contradiction. Pick up $u \in S \setminus B$. Assume that $B \subseteq P$ for some $P \in \mathcal{P}^{(1)}$, without loss of generality. Then $\{u, v\}$ is a circuit of $\mathfrak{C}(G)$ for any $v \in B$ since S is a stable set of G. Moreover, u and v belong to different sets of $\mathcal{P}^{(1)}$ (otherwise, it would violate the maximality of B). From the condition in Theorem 3.3, there should exist some $P' \in \mathcal{P}^{(2)}$ such that $\{u, v\} \subseteq P'$ for all $v \in B$. By the transitivity of the equivalence relation induced by $\mathcal{P}^{(2)}$, we have $\{u\} \cup B \subseteq P'$. This contradicts the maximality of B. Therefore, we have $S = B$, which means that $S \in \mathcal{P}^{(1)} \cup \mathcal{P}^{(2)}$.

Now we color the vertices of $\mathcal{S}(G)$, i.e., the maximal stable sets of G. If a maximal stable set S belongs to $\mathcal{P}^{(1)}$, then S is colored by 1. Similarly if S belongs to $\mathcal{P}^{(2)}$, then S is colored by 2. (If S belongs to both, then S is colored by either 1 or 2 arbitrarily.) This coloring certainly provides a proper 2-coloring of $\mathcal{S}(G)$ since $\mathcal{P}^{(1)}$ and $\mathcal{P}^{(2)}$ are stable-set partitions of G. $\square$

Some researchers already noticed that the bipartiteness of $\mathcal{S}(G)$ is characterized by other properties. We gather them in the following proposition.

Proposition 5.3. *Let G be a graph. Then the following are equivalent. (1) The stable-set graph $\mathcal{S}(G)$ is bipartite. (2) G is the complement of the line graph of a bipartite multigraph. (3) G has no induced subgraph isomorphic to $K_1 \cup K_3$, $K_1 \cup K_2 \cup K_2$, $K_1 \cup P_3$ or $\overline{C_{2k+3}}$ $(k = 1, 2, \dots)$.*

$$K_1 \cup K_3 \qquad K_1 \cup K_2 \cup K_2 \qquad K_1 \cup P_3$$

Proof. "(1) $\Leftrightarrow$ (2)" is immediate from a result by Cai–Corneil–Proskurowski [1]. Also "(1) $\Leftrightarrow$ (3)" is immediate from a result by Protti–Szwarcfiter [12]. $\square$

Remark that we can decide whether the stable-set graph of a graph is bipartite or not in polynomial time using the algorithm described by Protti–Szwarcfiter [12].

6   Concluding Remarks

In this paper, motivated by the quality of a natural greedy algorithm for the maximum weighted clique problem, we characterized the number k such that the clique complex of a graph can be represented as the intersection of k matroids (Theorem 3.3). This implies that the problem to determine the clique complex of a given graph has a representation by k matroids or not belongs to NP. Also, in Sect. 5 we observed that the corresponding problem for two matroids can be solved in polynomial time. However, the problem for three or more matroids is not known to be solved in polynomial time. We leave the further issue on computational complexity of this problem as an open problem.

Moreover, we showed that $n - 1$ matroids are necessary and sufficient for the representation of the clique complexes of all graphs with n vertices (Theorem 4.1). This implies that the approximation ratio of the greedy algorithm for the maximum weighted clique problems is at most $n - 1$.

Furthermore, we can show the following theorem. In this theorem, we see a graph itself as an independence system: namely a subset of the vertex set is independent if and only if it is (1) the empty set, (2) a one-element set, or (3) a two-element set which forms an edge of the graph. A proof is omitted due to the page limitation.

Theorem 6.1. *A graph G can be represented as the intersection of k matroids if and only if the clique complex $\mathfrak{C}(G)$ can be represented as the intersection of k matroids.*

In this paper, we approached the quality of a greedy algorithm for the maximum weighted clique problem from the viewpoint of matroid theory. This approach must be useful for other combinatorial optimization problems.

Acknowledgements. The authors thank Emo Welzl for suggesting the problem discussed in Sect. 4 and anonymous referees for useful comments.

References

1. L. Cai, D. Corneil and A. Proskurowski: A generalization of line graphs: (X, Y)-intersection graphs. Journal of Graph Theory **21** (1996) 267–287.
2. R. Diestel: Graph Theory (2nd Edition). Springer Verlag, New York, 2000.
3. J. Edmonds: Matroids and the greedy algorithm. Mathematical Programming **1** (1971) 127–136.
4. U. Faigle: Matroids in combinatorial optimization. In: Combinatorial Geometries (N. White, ed.), Cambridge University Press, Cambridge, 1987, pp. 161–210.
5. S.P. Fekete, R.T. Firla and B. Spille: Characterizing matchings as the intersection of matroids. Preprint, December 2002, arXiv:math.CO/0212235.
6. A. Frank: A weighted matroid intersection algorithm. Journal of Algorithms **2** (1981) 328–336.
7. T.A. Jenkyns: The efficacy of the "greedy" algorithm. Proceedings of the 7th Southeastern Conference on Combinatorics, Graph Theory, and Computing, Utilitas Mathematica, Winnipeg, 1976, pp. 341–350.
8. B. Korte and D. Hausmann: An analysis of the greedy algorithm for independence systems. In: Algorithmic Aspects of Combinatorics; Annals of Discrete Mathematic **2** (B. Alspach et al., eds.), North-Holland, Amsterdam, 1978, pp. 65–74.
9. B. Korte and J. Vygen: Combinatorial Optimization (2nd Edition). Springer Verlag, Berlin Heidelberg, 2002.
10. Y. Okamoto: Submodularity of some classes of the combinatorial optimization games. Mathematical Methods of Operations Research **58** (2003), to appear.
11. J. Oxley: Matroid Theory. Oxford University Press, New York, 1992.
12. F. Protti and J.L. Szwarcfiter: Clique-inverse graphs of bipartite graphs. Journal of Combinatorial Mathematics and Combinatorial Computing **40** (2002) 193–203.
13. R. Rado: Note on independence functions. Proceedings of the London Mathematical Society **7** (1957) 300–320.

On Proving Circuit Lower Bounds against the Polynomial-Time Hierarchy: Positive and Negative Results[*]

Jin-Yi Cai and Osamu Watanabe

[1] Computer Sci. Dept., Univ. of Wisconsin, Madison, WI 53706, USA
jyc@cs.wisc.edu
[2] Dept. of Math. and Comp. Sci., Tokyo Inst. of Technology
watanabe@is.titech.ac.jp

Abstract. We consider the problem of proving circuit lower bounds against the polynomial-time hierarchy. We give both positive and negative results. For the positive side, for any fixed integer $k > 0$, we give an explicit Σ_2^p language, acceptable by a Σ_2^p-machine with running time $O(n^{k^2+k})$, that requires circuit size $> n^k$. For the negative side, we propose a new stringent notion of relativization, and prove under this stringent relativization that every language in the polynomial-time hierarchy has polynomial circuit size. (For technical details, see also [CW03].)

1 Introduction

Proving circuit lower bounds is a most basic problem in complexity theory. The class P has polynomial size circuits. It is also widely believed that NP does not share this property, i.e., that some specific set requires super polynomial circuit size. While this remains the most concrete approach to the NP vs. P problem, we cannot even prove this for any fixed $k > 1$ that any set $L \in$ NP requires circuit size $> n^k$.

If we relax the restriction from NP to the second level of the polynomial-time hierarchy Σ_2^p, Kannan [Kan82] did prove that for any fixed polynomial n^k, there is some set L in Σ_2^p which requires circuit size $> n^k$. Kannan in fact proved the existence theorem for some set in $\Sigma_2^p \cap \Pi_2^p$. This result has been improved by Köbler and Watanabe [KW98] who showed, based on the technique developed in [BCGKT], that such a set exists in ZPP^{NP}. The work in [Cai01] implies that a yet lower class S_2^p contains such a set. (See [BFT98,MVW99] for related topics.)

However, Kannan's proof for Σ_2^p, and all the subsequent improvements mentioned above, are not "constructive" in the sense that it does not identify a single Σ_2^p machine whose language requires circuit size $> n^k$. At the top level, all these

[*] The first author is supported in part by NSF grants CCR-0208013 and CCR-0196197 and U.S.-Japan Collaborative Research NSF SBE-INT 9726724. The second author is supported in part by by the JSPS/NSF Collaborative Research 1999 and by the Ministry for Education, Grant-in-Aid for Scientific Research (C), 2001.

proofs are of the following type: *Either* SAT does not have n^k size circuit, in which case we are done, *or* SAT has n^k size circuit, then we can define some other set, which by the existence of the hypothetical circuit for SAT can be shown in Σ_2^p, and it requires circuit size $> n^k$. Constructively, Kannan gave a set in $\Sigma_4^p \cap \Pi_4^p$. In [MVW99] a set in Δ_3^p was constructively given. In this paper we improve this to Σ_2^p.

Theorem 1. *For any integer $k > 0$, we can construct a Σ_2^p-machine with $O(n^{k^2} \log^{k+1} n)$ running time that accepts a set with no n^k size circuits.*

A similar constructive proof is still open for the stronger statements on such sets in $\Sigma_2^p \cap \Pi_2^p$ (resp., $\mathrm{ZPP}^{\mathrm{NP}}$, and S_2^p).

Our main result in this paper deals with the difficulty in proving super polynomial circuit size lower bound for any set in the polynomial-time hierarchy, PH. While it is possible to prove lower bound above any fixed polynomial, at least for some sets in Σ_2^p, the real challenge is to prove super polynomial circuit size lower bound for a single language. Not only have we not been able to do this for any set in NP, but also no super polynomial lower bound is known for any set in PH. In this paper we propose a new notion of relativization, which is more stringent than the usual notion of relativization. Under this stringent relativization we prove that every language in the polynomial-time hierarchy has polynomial circuit size. Thus, in a strong sense, one cannot prove relativizable super polynomial lower bound for any set in the polynomial-time hierarchy.

We note that a relativization where EXP (thus, PH) has polynomial size circuits is known [He84]. Relativization results can be generally classified as either separation or collapsing/containment results. We deal with relativized collapsing results here since we are interested in demonstrating the difficulty of proving unconditional circuit lower bound for PH. By surveying existing relativized collapsing results, we found the following asymmetry is often present. In almost all of these relativized collapsing results, the proof is achieved by allowing stronger access of oracles to the simulating computation than the simulated computation. For example, in the usual proof of $\mathrm{P}^A = \mathrm{NP}^A$ or $\mathrm{P}^A = \mathrm{PSPACE}^A$, we encode QBF in the oracle. In terms of the simulation by the $\mathrm{P}^{\mathrm{QBF}}$ machine $\mathcal{M}$ simulating an $\mathrm{NP}^{\mathrm{QBF}}$ or $\mathrm{PSPACE}^{\mathrm{QBF}}$ computation $\mathcal{M}'$ on an input x, $\mathcal{M}$ will access an oracle location polynomially longer than where the corresponding access $\mathcal{M}'$ makes. That is, P^A machines are given more powerful oracle access. The simulated machine is denied access to certain segments of the oracle where the simulating machine can access. In the result of Heller [He84] for $\mathrm{EXP}^A \subseteq \mathrm{P}^A/\mathrm{poly}$, an EXP machine is supposed to have running time 2^{n^k} for some constant k, and thus "most" of the segment of A at length n^{k+1} is untouched for computation at length n. Then one can simply code this EXP^A computation at a suitable location at length n^{k+1}, where the poly-size circuit can access.

We would like to disallow this type of "hiding" of computation afforded by unequal oracle access.

In this paper, we propose the following more "stringent" oracle computation model. Essentially, we require that, for any n, the simulating and simulated machine or circuit should have the same access to oracle.

We can formalize this as follows. We generalize the definition of $\mathcal{C}$/poly by Karp and Lipton [KL80] to allow the underlying machines to have random access to an advice string. Fix any "length function" ℓ, then a function $s : n \mapsto \{0,1\}^{\ell(n)}$ is called an *advice function of size* $\ell(n)$. Given any advice function s, we say a language L is in the class $\mathcal{C}/s$ *via random access to advice* if there is some machine $\mathcal{M}$ representing the class $\mathcal{C}$, such that $x \in L$ iff $\mathcal{M}(x; s(|x|))$ accepts, where we denote the computation $\mathcal{M}$ on x with random access to $s(|x|)$ by $\mathcal{M}(x; s(|x|))$. (The notion of *random access* is the usual one: A machine $\mathcal{M}$ can write down an index to a bit of $s(|x|)$ on a special tape and then it gets that bit in unit time.) We denote this language as $L(\mathcal{M}; s)$. Roughly speaking the string $s(n)$ represents that segment of the oracle accessible for both sides at length n.

This notion is extended to circuits as follows. We consider a circuit for input length n consists of standard AND, OR, and NOT gates and oracle query gates. An oracle query gate takes m input bits $z = b_1 b_2 \ldots b_m$ as an index to $s(n)$ and gets output $s(n)[z]$.

When we compare two classes, we insist the respective machines (or circuits) have access to the same advice string s. We say a relativized containment is "stringent" (w.r.t. s), if for every machine $\mathcal{M}_1$ representing $\mathcal{C}_1$, there is a machine $\mathcal{M}_2$ (or circuit family) representing $\mathcal{C}_2$, such that $L(\mathcal{M}_1; s) = L(\mathcal{M}_2; s)$. Note that s is fixed for all machines (or circuits) from the classes $\mathcal{C}_1$ and $\mathcal{C}_2$.

Theorem 2 (Main Theorem). *There is a stringent relativization such that PH is contained in P/poly. More specifically, there is an advice function s of length 2^{cn}, for some constant $c > 2$, such that for any integer $d > 0$ and any real $k \geq 1$, if $\mathcal{M}$ is an oracle Σ_d^p-machine with running time $O(n^k)$, then there is a sequence of Boolean circuits $\{C_n\}_{n \geq 0}$, such that $L(C_n; s(n)) = L(\mathcal{M}; s)^{=n}$. For all sufficiently large n, the size of C_n is bounded by n^{cdk}, for some universal constant $c > 0$.*

Our proof technique for the main theorem is based on the decision tree version of the Switching Lemma for constant depth circuits and Nisan-Wigderson pseudorandom generator.

2   Proof of Theorem 1

Kannan [Kan82] proved that for any fixed polynomial n^k, there is some set L in $\Sigma_2^p \cap \Pi_2^p$ with circuit size $> n^k$. However, in terms of explicit construction, he only gave a set in $\Sigma_4^p \cap \Pi_4^p$. An improvement to Δ_3^p was stated in [MVW99].

In this section we give a constructive proof of Kannan's theorem for Σ_2^p. That is, we give a description of a Σ_2^p machine $\mathcal{M}$ accepting a set with the n^k circuit size lower bound.

For any $n \geq 0$, a binary sequence χ of length $l \leq 2^n$ is called a *partial characteristic sequence*, which will specify the membership of lexicographically the first l strings of $\{0,1\}^n$. We denote this subset of $\{0,1\}^n$ by $L(\chi)$. We say that χ is consistent with a circuit C with n input gates, iff $\forall i, 1 \leq i \leq l$, $C(x_i)$ outputs the ith bit of χ, where x_i is the ith string of $\{0,1\}^n$.

We can encode every circuit C of size $\le s$ as a string u of length $\text{len}(s)$, where $\text{len}(s)$ is defined as $\text{len}(s) = c_{\text{circ}}\lfloor s \log s \rfloor$ with some constant c_{circ}. We may consider every u with $|u| = \text{len}(s)$ encodes some circuit of size $\le s$; if a string u is not a proper code or the encoded circuit has size $> s$, we assume that this u encodes the constant 0 circuit. The following is immediate by counting.

Claim 1 *For any $s > 1$, there exists a partial characteristic sequence of length $l = \text{len}(s) + 1$ that is not consistent with any circuit of size $\le s$.*

Our goal is to define a set L that has no n^k size circuit but that is recognized by some explicitly defined Σ_2^p machine $\mathcal{M}$. Our construction follows essentially the same outline as the one given in [MVW99], which in turn uses ideas given in Kannan's original proof. The further improvement is mainly an even more efficient use of alternation.

For a given n, denote by $\ell = \text{len}(n^k) + 1$, and we try to construct a partial characteristic sequence χ of length ℓ that is consistent with *no* circuit of size $\le n^k$. We will introduce an auxiliary set PreCIRC that is in NP. With this PreCIRC, some Σ_2^p machine can *uniquely* determine the desired characteristic sequence χ_{non} (on its accepting path). We would like to define our set L (partially) consistent with this sequence χ_{non}. But Σ_2^p computation using some auxiliary NP set cannot be implemented, in general, by any Σ_2^p machine. Suppose here that PreCIRC has n^k size circuits; then some Σ_2^p machine can guess such circuits, verify them, and use them for computing χ_{non} and recognizing strings according to χ_{non}. What if there are no such circuits for PreCIRC? We will define L so that one part of L is consistent with PreCIRC (while the other part is consistent with χ_{non} if PreCIRC is computable by some n^k size circuits). If PreCIRC has no n^k size circuit, then the part of L that is consistent with PreCIRC can guarantee the desired hardness of L.

Now we describe our construction in detail. Denote by $\ell = \text{len}(n^k) + 1$. By "$v \succ u$" we mean that u is a prefix of v. To compute the "hard" characteristic sequence χ_{non}, we want to determine, for a given pair of a partial characteristic sequence χ and a string u, whether u can be extended to some description v of a circuit that is consistent with χ. The set PreCIRC is defined for this task. More precisely, for any $n > 0$, and for any strings χ of length ℓ and u of length $\le \text{len}(n^k)$, we define PreCIRC as follows.

$$1^n 0 \chi u 0 1^{\text{len}(n^k)-|u|} \in \text{PreCIRC}$$
$$\Leftrightarrow (\exists v \succ u)\,[\,|v| = \text{len}(n^k) \;\&\; \text{the circuit encoded by } v \text{ is consistent with } \chi\,].$$

Strings of any other form are not contained in PreCIRC. For simplifying our notation, we will simply write (χ, u) for $1^n 0 \chi u 0 1^{\text{len}(n^k)-|u|}$. Since n determines ℓ, and the length of χ is ℓ, χ and u are uniquely determined from $0^n 1 \chi u 1 0^{\text{len}(n^k)-|u|}$. The length of (χ, u) is $\tilde{n} = n + 2\ell + 1$. Note that $\tilde{n}$ is $O(n^k \log n)$.

Finally define our machine $\mathcal{M}$. Informally we want $\mathcal{M}$ to accept an input x if and only if either $x \in 1\{0,1\}^{n-1}$ and $x \in \text{PreCIRC}$, or $x \in 0\{0,1\}^{n-1}$ and $x \in L$, where $L^{=n}$ is a set with no n^k size circuits, for all sufficiently large n, if $\text{PreCIRC}^{=n}$ has n^k size circuits for all sufficiently large n. More formally, for any given input x of length n, if x starts with 1, then $\mathcal{M}$ accepts it iff $x \in$ PreCIRC. Suppose otherwise; that is, x starts with 0. Then first $\mathcal{M}$ existentially guesses a partial characteristic sequence χ_{non} of length ℓ and a circuit of size $\widetilde{n}^k$, more precisely, a string v_{pre} of length len$(\widetilde{n}^k)$ encoding a circuit for PreCIRC$^{=\widetilde{n}}$ of size $\leq \widetilde{n}^k$. After that, $\mathcal{M}$ enters the universal stage, where it checks the following items: (Here we identify a circuit with a string coding it.)

(1) $\forall \chi$, $|\chi| = \ell$, and $\forall u$, $|u| \leq$ len(n^k), check that v_{pre} is "locally consistent" on (χ, u) as follows:

$$v_{\mathrm{pre}}(\chi, u) = 1 \ \& \ |u| = \text{len}(n^k) \implies \text{circuit } u \text{ is consistent with } \chi, \text{ and}$$
$$v_{\mathrm{pre}}(\chi, u) = 1 \ \& \ |u| < \text{len}(n^k) \implies v_{\mathrm{pre}}(\chi, u0) = 1 \ \vee \ v_{\mathrm{pre}}(\chi, u1) = 1.$$

(2) $\forall u$, $|u| =$ len(n^k), compute the χ_u of length ℓ defined by circuit u, and verify that v_{pre} works for χ_u and all prefix u' of u, i.e., $v_{\mathrm{pre}}(\chi_u, u') = 1$.

(3) The guessed χ_{non} is lexicographically the first string of length ℓ such that no circuit of size s is consistent with it, *according to* v_{pre}. That is, check $v_{\mathrm{pre}}(\chi_{\mathrm{non}}, \epsilon) = 0$, where ϵ is the empty string, and $\forall \chi$ if $|\chi| = \ell$ and χ is lexicographically smaller than χ_{non} then $v_{\mathrm{pre}}(\chi, \epsilon) = 1$ holds.

Finally on each universal branch, if $\mathcal{M}$ passes the particular test of this branch, then $\mathcal{M}$ accepts the input $x \in 0\{0,1\}^{n-1}$ iff χ_{non} has bit 1 for the string x.

From our discussion, it is not hard to show that this machine $\mathcal{M}$ accepts a set with the n^k circuit size lower bound.

3   Proof of Theorem 2

Consider any Σ_d^{p} oracle alternating Turing machine $\mathcal{M}$ with time bound n^k. Let $m = n^k$ and $I \geq (2 + \delta)n$, for constant $\delta > 0$, and denote $N = 2^n$ and $M = 2^m$. We want to define $s(n)$ of length 2^I such that a polynomial size circuit $C_{\mathcal{M}}$ simulates $\mathcal{M}$ at length n with random access to $s(n)$. For notational convenience we will prove only for $I = m$, and we will assume $k > 2$ and $d \geq 7$.

It can be easily adapted from [FSS81] that the computation of a Σ_d^{p} machine $\mathcal{M}(x, s(|x|))$ when the input x is of length n, gives rise to a bounded depth Boolean circuit C_x of the following type: The inputs are Boolean variables representing the bits $s(n)[z]$. With a slight abuse of notation we denote them by z and $\overline{z}$ as well. The Boolean circuit C_x starts with an OR gate at the top, and alternate with AND's and OR's with depth $d + 1$, where the bottom level gates have bounded fan-in at most m, and all other AND and OR gates are unbounded fan-in, except by the overall circuit size, which is bounded by $m2^m$. Without loss of generality we may assume the circuit is tree like, except for the input level.

Our first idea is to use random restrictions to "kill" the circuit. For any Boolean function f over variables $x_1, \ldots, x_n$, a *random restriction* ρ (for some specified parameter p) is a random function that assigns each x_i either 0, 1, or $*$, with probability $\Pr[\rho(x_i) = *] = p$ and $\Pr[\rho(x_i) = 0] = \Pr[\rho(x_i) = 1] = (1-p)/2$, for each i independently. Assigning $*$ means to leave it as a variable. Let $f \mid_\rho$ denote a function obtained by this random restriction. It is known that after a random restriction ρ (for a suitably chosen parameter p), a constant depth circuit of suitable size is sufficiently weakened so as to have small min-terms and max-terms. Results of this type are generally known as Switching Lemmas, and the strongest form known is due to Håstad [Hås86a]. However it turns out that we need a different form, namely a decision tree type Switching Lemma (see, e.g., [Cai86]), which is sated as follows.

Lemma 1. *For any depth $d + 1$ Boolean circuit C on M inputs $z_1, z_2, \ldots, z_M$, of size at most s and bottom fan-in at most t, we have*

$$\Pr_\rho[\mathrm{DC}(C \mid_\rho) \geq t] \ \leq \ s 2^{-t},$$

where the random restriction ρ is defined for $p = 1/(10t)^d$, and DC *stands for decision tree complexity.*

To reduce all circuits C_x, $x \in \{0,1\}^n$, to small depth decision trees, we apply a random restriction with $p_0 = 1/(20m)^d$ to these circuits. Then,

Claim 2 $\Pr_\rho[\bigvee_{x \in \{0,1\}^n}[\mathrm{DC}(C_x \mid_\rho) \geq 2m]] \ \leq \ 2^n \cdot (mM)2^{-2m} \ = \ m/2^{m-n}.$

Assume some random restriction ρ that reduces all circuits C_x, $x \in \{0,1\}^n$, to depth $\leq 2m$ decision trees. Then by assigning 0 every time to the "next variable" asked by each decision tree, we can fix the values of all decision trees, which are in fact the results of $\mathcal{M}$ on all length n inputs. The all-0 answers maintain consistency (this is one place why we need decision tree type Switching Lemma, rather than min/max term bounds.) Since random restrictions assign $*$ to $p_0 M$ variables *on average*, we may assume that the restriction ρ assigns $*$ to at least $p_0 M/2$ variables. Hence, after the random restriction and the decision tree assignments, we still have $p_0 M/2 - 2mN \gg N = 2^n$ unassigned variables, which we want to use to encode the (already fixed) values of C_x, i.e., the result of $\mathcal{M}$ on input x, for every x in $\{0,1\}^n$.

The problem with this idea is that after we have coded the values of all the $N = 2^n$ circuits, there does not seem to be any easy way to recover this information. We have N computations to code, and it is infeasible for the final polynomial-size oracle circuit to "remember" more than a polynomial number of bits as the address of the coding region.

To overcome this difficulty, our next idea is to use not true random restrictions, but pseudorandom restrictions via the Nisan-Wigderson generator (see, e.g., [NW94]). The NW generator generates pseudorandom bits provably indistinguishable from true random bits by polynomial size constant depth circuits. While our circuits are not of polynomial size, this can be scaled up easily. Our idea is then to use the output of some NW generator to perform the "random" restriction, and to argue that all $N = 2^n$ circuits are "killed" with high probability, just as before with true random restrictions. The basic argument is that no constant depth circuits of an appropriate size can tell the difference under either a true random assignment or a pseudorandom assignment coming from the NW generator. However, for our purpose in this paper, we wish to say that a certain behavior of these N constant depth circuits — namely they are likely to possess small depth decision trees after a "random" restriction with 0, 1 and $*$'s

— is preserved when "pseudorandom restrictions" are substituted for "random restrictions". In order to use the NW generator, it is vitally important that the property to be preserved be expressible itself by a constant depth circuit of an appropriate size. The idea then is to design such a circuit D that "expresses" this. However, the property of having small depth decision trees is not easy to express in constant depth. So the next idea is to use a *consequence* of this property, namely all $C_x|\rho$ are determined, by setting a subset of $*$'s of cardinality $\leq 2mN$ all to 0. This property in turn is sufficient to the construction of $s(n)$.

More specifically, we design the following circuit D:

(a) Let $L = \lceil \log_2 \frac{1}{p_0} \rceil \approx dk \log_2(20n)$. D takes Boolean inputs $(a_{z,1}, \ldots, a_{z,L}, b_z)$, for $z \in \{0,1\}^m$; that is, D has $(L+1)M$ input variables.

(b) For each $z \in \{0,1\}^m$, a tuple $(a_{z,1}, \ldots, a_{z,L}, b_z)$ is used to simulate a random restriction ρ. That is, we define $\rho(z) = *$ if all $a_{z,1}, \ldots, a_{z,L}$ are 0; otherwise, define $\rho(z) = b_z$.

(c) For given input bits (which define ρ as explained above), $D = 1$ iff $\exists \sigma$ assigning $\leq 2mN$ many $*$'s to 0, such that $\forall x \in \{0,1\}^n$, $C_x|\rho|\sigma \equiv 0$ or 1.

We can construct such a circuit with reasonable size and depth.

Claim 3 *There exists a circuit D satisfying the above (a) $\sim$ (c) with size(D) $< 2^{3m^2}$ and depth$(D) \leq 2d + 6 \leq 3d - 1$. Furthermore, from our construction and Claim 2, we have $\Pr[D = 1] \geq 1 - m/2^{m-n}$, where the probability is over uniform input bits for D. (Below we will use $\hat{d}$ to denote $3d - 1$.)*

Now we apply a NW generator to this circuit D. First we recall some basic notions on NW generators from [NW94].

Let U, M, m and q be positive integers. Let $[U]$ be some set of cardinality U, e.g., $\{1, 2, \ldots, U\}$. A collection of subsets $\mathcal{S} = \{S_1, \ldots, S_M\}$ of some domain $[U]$ is called a (m, q)-*design* if it satisfies the following conditions: (i) $\forall i$, $1 \leq i \leq M$ $[\ |S_i| = q\ ]$, and (ii) $\forall i$, $\forall j$, $1 \leq i \neq j \leq M$ $[\ |S_i \cap S_j| \leq m\ ]$. Based on a given (m, q)-design $\mathcal{S} = \{S_1, \ldots, S_M\}$ with domain $[U]$, we define the following function $g_{\mathcal{S}} : \{0,1\}^U \to \{0,1\}^M$, which we call a (parity based) *NW generator*.

$$g_{\mathcal{S}}(x_1 \cdots x_U) = y_1 \cdots y_M,$$
$$\text{where each } y_i, \ 1 \leq i \leq M, \text{ is defined}$$
$$\text{by } y_i = x_{s_1} \oplus \cdots \oplus x_{s_q} \quad (\text{where } S_i = \{s_1, \ldots, s_q\} \subseteq [U]).$$

For the pseudorandomness of this generator, we have the following lemma [NW94].

Lemma 2. *For any positive integers U, M, m, q, s and e, and positive real ϵ, let $g_{\mathcal{S}}$ be the NW generator defined as above. Suppose for any depth $e + 1$ circuit C on q input bits and of size at most $s + c_{\mathrm{nw}} 2^m M$ (where c_{nw} is some constant), the q bit parity function $u_1 \oplus \cdots \oplus u_q$ has the following bias:*

$$\left| \Pr_{(u_1, \ldots, u_q) \in \{0,1\}^q}[C(u_1, \ldots, u_q) = u_1 \oplus \cdots \oplus u_q] - \frac{1}{2} \right| \leq \frac{\epsilon}{M}.$$

Then g_S has the following pseudorandomness against any depth e circuit E on M input bits and of size at most s.

$$\left| \mathrm{Pr}_{\boldsymbol{y} \in \{0,1\}^M}[E(\boldsymbol{y}) = 1] - \mathrm{Pr}_{\boldsymbol{x} \in \{0,1\}^U}[E(g_S(\boldsymbol{x})) = 1] \right| \; \leq \; \epsilon.$$

To apply the NW generator to our depth $\widehat{d}$ circuit D constructed above, we set the parameters and define an (m, q)-design as follows. We take a finite field $\mathbf{F}$, and set $q = |\mathbf{F}|$ and $U = q^2$. More specifically, the finite field we use is $\mathbf{F} = \mathbf{Z}_2[X]/(X^{2 \cdot 3^u} + X^{3^u} + 1)$ [vL91], where each element $\alpha \in \mathbf{F}$ takes $K = 2 \cdot 3^u$ bits, and $q = |\mathbf{F}| = 2^K$. We choose u so that $q \geq (3m^2 + 1)^{\widehat{d}+2}$. Then $q^{1/(\widehat{d}+2)} \geq \log_2(2^{3m^2} + c_{nw} 2^m M)$, where c_{nw} is the constant in the above lemma. Clearly $q \leq n^{ckd}$ will do, for some universal constant c, for example $c = 7$. Then $K = O(dk \log n)$. Thus, this field has polynomial size and each element is represented by $O(\log n)$ bits. All arithmetic operations in this field $\mathbf{F}$ are easy.

We consider precisely $M = 2^m$ polynomials $f_z(\xi) \in \mathbf{F}[\xi]$, each of degree at most m, where each f_z is indexed by its coefficients, concatenated as a bit sequence of length exactly m. The precise manner in which this is done is not very important, but for definiteness, we can take the following. We take polynomials of degree $\delta = \lfloor m/K \rfloor$ with exactly $\delta + 1$ coefficients,

$$f_z(\xi) \; = \; c_\delta \xi^\delta + \ldots + c_1 \xi + c_0,$$

where all c_j varies over $\mathbf{F}$, except c_δ is restricted to exactly $2^{m-K \cdot \delta}$ many values. Note that $0 \leq m - K \cdot \delta < K$. The concatenation $z = \langle c_\delta \cdots c_0 \rangle$ has exactly m bits. Each f_z defines a subset of $\mathbf{F} \times \mathbf{F}$ of cardinality q, $\{(\alpha, f_z(\alpha)) \mid \alpha \in \mathbf{F}\}$, which we denote by S_z. A (m, q)-design that we use is defined as $\mathcal{S} = \{S_1, \ldots, S_M\}$, indexed by $z \in \{0,1\}^m$, which is identified with $\{1, \ldots, M\}$. Note that $\mathbf{F} \times \mathbf{F}$ is a domain $[U]$ with $U = q^2$. It is easy to see that this (m, q)-design satisfies the conditions (i) and (ii).

Then by a standard argument we can prove the following

Claim 4 *Our NW generator g_S has the following pseudorandomness against any circuit E of size at most 2^{3m^2} and depth $\widehat{d}$:*

$$\left| \mathrm{Pr}_{\boldsymbol{y} \in \{0,1\}^M}[E(\boldsymbol{y}) = 1] - \mathrm{Pr}_{\boldsymbol{x} \in \{0,1\}^U}[E(g_S(\boldsymbol{x})) = 1] \right| \; \leq \; 2^{m-3m^2}.$$

Note that our NW generator g_S generates a pseudo random sequence of length $M = 2^m$ from a seed of length $U = q^2$. On the other hand, recall that the circuit D takes $(L+1)M$ Boolean inputs, i.e., $(a_{z,1}, \ldots, a_{z,L}, b_z)$, for $z \in \{0,1\}^m$, where $M = 2^m$ and $L = \lceil \log_2 \frac{1}{p_0} \rceil$. Hence, to provide these input values by our NW generator, we run the generator for $L + 1$ times, using $(L+1)q^2$ bits, a sequence of independently and uniformly distributed bits $\{u_{\alpha,\beta}^{(0)}, u_{\alpha,\beta}^{(1)}, \ldots, u_{\alpha,\beta}^{(L)}\}$, for each $\alpha, \beta \in \mathbf{F}$. That is, for each $j = 1, \ldots, L$, we use q^2 bits $\{u_{\alpha,\beta}^{(j)} \mid \alpha, \beta \in \mathbf{F}\}$ to generate the M Boolean values of $a_{z,j}$, for $z \in \{0,1\}^m$. Similarly, the set $\{u_{\alpha,\beta}^{(0)} \mid \alpha, \beta \in \mathbf{F}\}$ of q^2 bits is used to generate the M Boolean values of b_z, for $z \in \{0,1\}^m$. More specifically, for each $z \in \{0,1\}^m$ and $j = 1, \ldots, L$, we define

$a_{z,j}$ and b_z by $a_{z,j} = \bigoplus_{\alpha \in \mathbf{F}} u^{(j)}_{\alpha, f_z(\alpha)}$, and $b_z = \bigoplus_{\alpha \in \mathbf{F}} u^{(0)}_{\alpha, f_z(\alpha)}$. Then from Claim 2 and Claim 4, we have

Claim 5 *Let $g_S^{(i)}$ denote the pseudorandom output sequence of g_S on random seed bits $\{u^{(i)}_{\alpha,\beta} \mid \alpha, \beta \in \mathbf{F}\}$, for $0 \leq i \leq L$. Then $\Pr[D(g_S^{(1)}, \ldots, g_S^{(L)}, g_S^{(0)}) = 1] \geq 1 - o(1)$, where the probability is over independently and uniformly distributed bits $\{u^{(0)}_{\alpha,\beta}, u^{(1)}_{\alpha,\beta}, \ldots, u^{(L)}_{\alpha,\beta}\}$, for $\alpha, \beta \in \mathbf{F}$.*

This claim states that with high probability, a pseudorandom sequence satisfies D, meaning that the random restriction induced from the pseudorandom sequence reduces *all* C_x to simple functions (e.g., small decision trees); more precisely, after the restriction, all their values can be determined by assigning at most $2mN$ additional variables to 0. Next we will argue that, for such a pseudorandom restriction, one can find some space to encode the determined value of each C_x.

Consider a restriction induced by a pseudorandom sequence satisfying D. Apply this restriction to all variables z of circuits C_x, and fix further the value of some set Y of variables to 0 in order to determine the value of circuits C_x for all $x \in \{0, 1\}^n$. We may assume that the size of Y is at most $2m2^n$, which is guaranteed by the fact that $D = 1$ with our pseudorandom sequence. Then there exists y_0 of length $(1 + \epsilon)n < m/2$ such that a segment $T_{y_0} = \{z \in \{0, 1\}^m \mid y_0 \text{ is a prefix of } z\}$ has no intersection with Y; that is, all variables in T_{y_0} are free from any variables used to fix the value of circuits C_x. This is simply because $2m2^n \ll 2^{(1+\epsilon)n}$. Our plan is to code the results of C_x by Boolean variables z of the form $z = y_0 x w$, for some "padding" w, where the pseudorandom ρ had assigned a $*$. These z are assigned $*$ by the pseudorandom restriction, and unlike true random restrictions, we can actually compute their location in polynomial time from the polynomially many seed bits.

Claim 6 *With high probability a sequence of random source bits $u_{\alpha,\beta}$ satisfies the following: for all $c_\delta, \ldots, c_1 \in \mathbf{F}$, there exists $c_0 \in \mathbf{F}$ such that z^* is assigned $*$ by the pseudorandom restriction induced from the source bits, where z^* is a string in $\{0, 1\}^m$ that represents $\langle c_\delta \cdots c_1 c_0 \rangle$. Furthermore, this c_0 can be computed in polynomial time from $u_{\alpha,\beta}$ and $c_\delta, \ldots, c_1$.*

We now summarize our construction of the advice string $s(n)$. Choose any setting of the random bits $\omega = u_{\alpha,\beta}$, such that it generates $(L+1)M$ pseudorandom bits Ω satisfying both $D = 1$ and the condition of Claim 6. Let ρ_Ω be the restriction induced by this pseudorandom sequence. We set $s(n)[z] = 0$ or 1 if $\rho_\Omega(z) = 0$ or 1, respectively. Next, choose a set $Y \subseteq \{0, 1\}^m$, where $\rho_\Omega(z) = *$ and $|Y| \leq 2mN$, such that setting all these $s(n)[z] = 0$ determines the value of circuits C_x for all $x \in \{0, 1\}^n$. This set Y is guaranteed by $D = 1$. Then we set these $s(n)[z] = 0$. Then, fix one y_0 such that $T_{y_0} \cap Y = \emptyset$. This y_0 exists by the pigeonhole principle. Then for any $x \in \{0, 1\}^n$, set $s(n)[z] = 0$ (resp. 1), for any z of the form $y_0 x w$ for some w, and $\rho_\Omega(z) = *$, if and only if the (already determined) value of C_x is 0 (resp. 1). Set all remaining $s(n)[z] = 0$.

Finally we explain how to design a polynomial-size circuit $C_\mathcal{M}$ simulating $\mathcal{M}(x; s(|x|))$. We may assume that the information on the seed ω (of length $(L+1)q^2 = n^{O(kd)}$) and y_0 are hardwired into the circuit and they can be used in the computation. For a given input x, the circuit computes a location z^* represented by $c_\delta, \ldots, c_0$ corresponding to $y_0 x w$ satisfying the condition of Claim 6. When z^* is obtained, the circuit queries the bit $s(|x|)[z^*]$ and accepts the input if and only if $s(|x|)[z^*] = 1$. Therefore, the whole computation can be implemented by some circuit of size n^{ckd} for some constant $c > 0$.

References

[BCGKT] N. Bshouty, R. Cleve, R. Gavaldà, S. Kannan, and C. Tamon, Oracles and queries that are sufficient for exact learning, *J. Comput. and System Sci.* 52(3), 421–433, 1996.

[BFT98] H. Buhrman, L. Fortnow, and T. Thierauf, Nonrelativizing separations, in *Proc. the 13th IEEE Conference on Computational Complexity* (CCC'98), IEEE, 8–12, 1998.

[Cai86] J-Y. Cai, With probability one, a random oracle separates PSPACE from the polynomial-time hierarchy, in *Proc. 18th ACM Symposium on Theory of Computing* (STOC'86), ACM, 21–29, 1986. See also the journal version appeared in *J. Comput. and System Sci.* 38(1): 68–85 (1989).

[Cai01] J-Y. Cai, $S_2^P \subseteq ZPP^{NP}$, in *Proc. 42th IEEE Symposium on Foundations of Computer Science* (FOCS'01), IEEE, 620–628, 2001.

[CW03] J-Y. Cai and O. Watanabe, Research Report C-167, Dept. of Math. Comput. Sci., Tokyo Inst. of Tech., 2003. Available from www.is.titech.ac.jp/research/research-report/C/index.html.

[FSS81] M. Furst, J. Saxe, and M. Sipser, Parity, circuits, and the polynomial time hierarchy, in *Proc. 22nd IEEE Symposium on Foundations of Computer Science* (FOCS'81), IEEE, 260–270, 1981.

[Hås86a] J. Håstad, Almost optimal lower bounds for small depth circuits, in *Proc. 18th ACM Symposium on Theory of Computing* (STOC'86), ACM, 6–20, 1986.

[He84] H. Heller, On relativized polynomial and exponential computations, *SIAM J. Comput.* 13(4), 717–725, 1984.

[Kan82] R. Kannan, Circuit-size lower bounds and non-reducibility to sparse sets, *Information and Control*, 55, 40–56, 1982.

[KL80] R.M. Karp and R.J. Lipton, Some connections between nonuniform and uniform complexity classes, in *Proc. 12th ACM Symposium on Theory of Computing* (STOC'80), ACM, 302–309, 1980.

[KW98] J. Köbler and O. Watanabe, New collapse consequences of NP having small circuits, *SIAM J. Comput.*, 28, 311–324, 1998.

[MVW99] P.B. Miltersen, N.V. Vinodchandran, and O. Watanabe, Super-Polynomial versus half-exponential circuit size in the exponential hierarchy, in *Proc. 5th Annual International Conference on Computing and Combinatorics* (COCOON'99), Lecture Notes in Computer Science 1627, 210–220, 1999.

[NW94] N. Nisan and A. Wigderson, Hardness vs randomness, *J. Comput. Syst. Sci.* 49, 149–167, 1994.

[vL91] J. van Lint, *Introduction to Coding Theory*, Springer-Verlag, 1991.

The Complexity of Boolean Matrix Root Computation

Martin Kutz[*]

Freie Universität Berlin, Germany
kutz@math.fu-berlin.de

Abstract. We show that finding roots of Boolean matrices is an $\mathcal{NP}$-hard problem. This answers a twenty year old question from semigroup theory. Interpreting Boolean matrices as directed graphs, we further reveal a connection between Boolean matrix roots and graph isomorphism, which leads to a proof that for a certain subclass of Boolean matrices related to subdivision digraphs, root finding is of the same complexity as the graph-isomorphism problem.

1 Introduction

Multiplication of Boolean zero-one matrices is defined as ordinary matrix multiplication with $+$ and $\cdot$ replaced by the Boolean operations $\vee$ and $\wedge$. So the matrix product $C = AB$ is given by

$$c_{ij} = \bigvee_{h=1}^{n} a_{ih} \wedge b_{hj},$$

and as with matrices over fields, the kth power A^k of a Boolean $n \times n$ matrix A is simply the k-fold product of A with itself.

Besides its theoretical relevance for semigroup theory, Boolean matrix algebra serves as a fundamental tool in algorithmic graph theory. Efficient algorithms for transitive-closure or shortest-path computations rely on the interpretation of directed graphs as Boolean matrices [16,1,3].

In this work, we investigate the computational complexity of finding roots of a given Boolean matrix. A kth *root* of a square Boolean matrix B is some other matrix A whose kth power A^k equals B. Twenty years ago, in the open problems section of his book [9], Kim asked if given a matrix B, such a root A can be computed in polynomial time or whether this problem is perhaps $\mathcal{NP}$-complete. (Actually, he inquired for the case $k = 2$ only.) We give an answer to that question.

Theorem 1. *Deciding whether a square Boolean matrix has a kth root is $\mathcal{NP}$-complete for each single parameter $k \geq 2$.*

---

[*] Member of the European graduate school "Combinatorics, Geometry, and Computation" supported by the Deutsche Forschungsgemeinschaft, grant GRK 588/2.

T. Warnow and B. Zhu (Eds.): COCOON 2003, LNCS 2697, pp. 212–221, 2003.
© Springer-Verlag Berlin Heidelberg 2003

With the "right" computational problem for the reduction, the proof of this result turns out surprisingly simple. This is quite remarkable since it thus relates Boolean matrix roots to a well-known $\mathcal{NP}$-complete problem, which yields insight in the local structure of Boolean matrices.

In the second, technically more challenging part of our work, we reveal further properties of matrix roots which show a close relation to graph isomorphism. This eventually leads to a proof that for a certain subclass of Boolean matrices kth root computation is graph-isomorphism complete. Before we can state this result precisely, we have to switch from matrices to the graph theoretic point of view. Actually, throughout this whole exposition we shall interpret Boolean matrices as adjacency matrices of directed graphs.

Boolean Matrices and Graph Theory. Any Boolean $n \times n$ matrix $A = (a_{ij})$ can be interpreted as a directed graph D on the vertex set $\{1, \ldots, n\}$ with an arc from j to i iff $a_{ij} = 1$. So D may have loops but no multiple arcs. The kth power of D, $k \in \mathbb{N}$, is the directed graph D^k defined on the same vertex set and with an arc from a to b if and only if there is a directed walk of length exactly k from a to b in D (possibly visiting some vertices several times); compare the figure. It is easy to see that the adjacency matrix of D^k is in fact the kth power of the adjacency matrix of D (see, for example, [18]).[1]

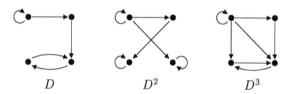

$$D \qquad D^2 \qquad D^3$$

So taking the graph theoretic point of view, we investigate the kth-root problem for digraphs: *given a directed graph D, does there exist another digraph R (on the same vertex set) such that $R^k = D$.* Our answer to the guiding question then reads as follows.

Theorem 1 (digraph version). *Deciding whether a digraph has a kth root is $\mathcal{NP}$-complete for each single parameter $k \geq 2$.*

Our second main result, which relates roots to isomorphisms, is based on subdivisions, defined as follows.

Definition 1. *The* complete subdivision *of a digraph D is the digraph obtained from D by replacing each arc $a \to b$ of D by a new vertex x_{ab} and the two arcs $a \to x_{ab} \to b$. We call a digraph a* subdivision *digraph if it is (isomorphic to) the complete subdivision of some digraph.*

---

[1] Alternatively, one might view a Boolean matrix as a binary relation. Then the kth matrix power is simply the k-fold composition of this relation.

Subdivisions are a fundamental notion in graph theory. But opposed to their common usage in relation with topological minors, we employ them here to equip our graphs with a certain stiffness that makes root finding computationally simpler. In fact, under an additional minor degree condition on which we shall comment later, we can show that finding roots of such graphs is of the same complexity as the graph-isomorphism problem.

Theorem 2. *Deciding whether a subdivision digraph with positive minimal indegree and outdegree has a kth root, is graph-isomorphism complete for each parameter $k \geq 2$.*

Graph Isomorphism. The graph-isomorphism problem asks: are two given (di)-graphs[2] isomorphic or not? It is neither known to have a polynomial-time solution nor is it known to be $\mathcal{NP}$-complete. On the contrary, it is a prime candidate for a problem strictly between $\mathcal{P}$ and $\mathcal{NP}$-completeness (cf. [10] and [12]). A computational problem of the same complexity as the graph-isomorphism problem is called *graph-isomorphism complete*, or simply *isomorphism complete* because isomorphism problems for several algebraic or combinatorial structures fall into this class. For example, isomorphism of semigroups and finite automata [2], finitely represented algebras, or convex polytopes [8]. Other problems ask for properties of the automorphism group of a graph, for example, computing the size of the automorphism group or its orbits [14].[3] Finally, several restrictions of the graph-isomorphism problem are known to remain isomorphism complete, as for example isomorphism of regular graphs [2].

As the above list indicates, actually all problems known to be isomorphism complete are more or less obviously isomorphism problems of various combinatorial structures. Hence, the relation between digraph roots and graphs isomorphism established through Theorem 2 may come quite as surprise.

Theorem 2 rests on a structural result which states that any kth root of a subdivision digraph D essentially establishes a one-to-one correspondence between k isomorphic subgraphs of D (Theorem 3). Due to space constraints, we shall only sketch the proofs of Theorems 1, 2, and 3, stating some of the central lemmas that pave the way.

2   Related Work—Related Questions

Over the field of complex numbers or the reals, matrix roots are a well-studied and still up-to-date topic of linear algebra [11,7,17]. But results from that field of research do generally not apply to Boolean matrices. While it is known, for example, that every regular matrix over the complex numbers has a kth root for

---

[2] One usually considers undirected graphs but it is well-known and easily seen that with respect to their computational complexity the undirected and directed version of the problem are equivalent.

[3] The latter two problems are known to be isomorphism complete only in the weaker sense of Turing reduction, as opposed to the concept of many-one reduction.

any $k \geq 2$ [17], this is not true for Boolean matrices, as the invertible matrix $\left(\begin{smallmatrix} 0 & 1 \\ 1 & 0 \end{smallmatrix}\right)$ shows. Further, complex or real matrices are amenable to numerical methods like Newton iteration [6], whereas such techniques clearly do not apply to Boolean matrices. When it comes to roots, Boolean matrices don't seem to have much in common with matrices over $\mathbb{C}$ since the former behave much more rigidly than the latter.

The situation is, however, slightly different if we ask for powers of a matrix instead of roots. There are theoretical results on Boolean matrix powers [4] and in practice we can of course compute the kth power of a Boolean matrix A by treating it as a matrix over the reals. We calculate A^k over $\mathbb{R}$ and afterwards replace each positive entry with 1. This simple reformulation allows us, for example, to apply fast matrix multiplication methods such as Strassen's to path problems in graphs [16,1]. But this simulation through matrices over the reals clearly only works because there cannot happen cancellation between positive and negative entries. For root finding, such simulation over $\mathbb{R}$ or $\mathbb{C}$ would lead into major problems.

Alternative Notions of Graph Powers. A problem similar to the one at hand has been discussed by Motwani and Sudan. In [15] they showed that computing square roots of undirected graphs is $\mathcal{NP}$-hard. But their notion of graph powers differs from ours in two important points.

They consider undirected graphs only, which corresponds to having bidirectional edges in our setting. This not only restricts the set of possible inputs but also—and this is the decisive difference—the solutions. For example, the four-vertex graph consisting of two disjoint bidirectional arcs has the directed 4-cycle as a square root, but no undirected graph can be a root.

Further, Motwani and Sudan define squaring to maintain existing edges, which in our setting would corresponds to attaching loops to all vertices. This monotonicity ensures that much information of the underlying graph can be read off from its square and the hardness proof of [15] makes essential use of this property. In contrast to this, squaring a digraph under the rules derived from Boolean matrix multiplication can almost completely destroy the neighborhood information and may even decompose the digraph. Actually, most of our arguments depend crucially on such "vanishing edges." So apparently, the squares in [15] and our notion of powers are essentially different concepts.

3 $\mathcal{NP}$-Completeness

We now sketch the proof of Theorem 1, presenting the two main ingredients for our $\mathcal{NP}$-completeness result: a suitable $\mathcal{NP}$-complete problem and a many-one reduction from that problem to digraph roots.

Surprisingly, the reduction is very straightforward, it goes without any sophisticated gadgets. This simplicity indicates that the two problems are actually very closely related. While skipping the details of the correctness proof here, we elaborate a bit on the reduction to visualize and endorse this claim. The appropriate problem for our reduction is the *set-basis problem:*

Let $\mathfrak{C}$ be a collection of subsets of some finite set S. A *set basis* of $\mathfrak{C}$ is another collection $\mathfrak{B}$ of subsets of S such that each $C \in \mathfrak{C}$ can be written as a union of sets from $\mathfrak{B}$. Given a finite set S together with such a collection $\mathfrak{C}$ of subsets of S and an integer $r \le |S|$, the *set-basis problem* asks whether there exists a set basis $\mathfrak{B}$ for $\mathfrak{C}$ consisting of at most r sets. This problem is known to be $\mathcal{NP}$-complete [19].

The Reduction. The key idea for the reduction stems from the following general observations about digraph square roots.

Consider some set X of vertices of a digraph D and let Z denote all outneighbors of vertices in X. Let us assume for simplicity that X and Z are disjoint, so in particular, there are no loops or cycles. Then in a square root of the digraph D, any of the arcs from X to Z must be realized as walks of length two and since there are no loops, these walks are actually paths. Hence, in the root there must exist a set Y of "intermediate vertices" through which all these paths can pass. If now—for whatever reason—there is only a small number of such intermediate vertices available, $|Y| \le r$, say, with r a little smaller than $|Z|$, these paths have to interact in order to ship all their information from X to Z.

We claim that the the square-root problem for these sets X, Y, and Z is nothing but a set-basis instance. This is easily seen by interpreting Z as the ground set S and X as a collection of subsets of S, defined trough containment relations given by the original D-arcs. The vertices in Y represent the set basis where the root arcs from Y to Z define the subsets and the arcs from X to Y tell us how to represent sets in X as unions of Y-sets.

In order to turn a set-basis instance into a digraph, we simply draw the containment graph of the set system $\mathfrak{C}$ on S and provide the right number of intermediate vertices. In the general case of kth roots that would be $k - 1$ times r vertices, which we leave almost isolated except for some framework arcs to ensure that any root uses them as intended.

Interpretation. We emphasize that the given set-basis instance is completely maintained by our reduction. Its containment relations are encoded one-to-one by arcs of the digraph. Moreover, the preceding discussion shows that an instance of the digraph-root problem can be seen as a large collection of interacting set-basis problems. One might well argue that finding digraph roots is actually a generalized set-basis problem.

As a corroboration for this point of view we mention that the set-basis problem already appeared before in connection with Boolean matrix algebra. Markowsky [13] used it in a very economic proof for the $\mathcal{NP}$-completeness of Schein-rank computation.[4]

---

[4] Analogous to the matrix rank over fields, the *Schein rank* of a Boolean matrix A is the minimal integer ρ such that A can be represented as a Boolean sum $A = \bigvee_{i=1}^{\rho} c_i r_i$, where the c_i are column and the r_i row vectors with zero-one entries [9, Sec. 1.4].

4 Roots and Isomorphism

In this second part, we establish the isomorphism-completeness result of Theorem 2. Our considerations are guided by the following fundamental connection between digraph roots and digraph isomorphism.

Proposition 1. *Let* $D = D_1 \dot{\cup} D_2 \dot{\cup} \cdots \dot{\cup} D_k$ *be the disjoint union of* k *isomorphic digraphs* $D_1, \ldots, D_k$. *Then* D *has a* kth *root.*

Because the proof is short, instructive, and of importance for the general understanding of Theorem 2, we briefly sketch the ideas. We construct the sought-after root R on the vertices of D from the isomorphisms $\varphi_i : D_1 \to D_i$, $1 \leq i \leq k$ (φ_1 being simply the identity). For each vertex a of D_1, we let R contain the path $\varphi_1(a) \to \varphi_2(a) \to \cdots \to \varphi_k(a)$ and additionally the arcs $\varphi_k(a) \to b$ for all D-outneighbors b of a. The following figure shows a local picture of this construction. (The continuous lines form the root, the dashed lines the given D.) One easily verifies that in fact, $R^k = D$.

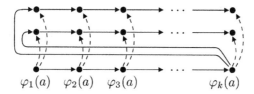

$$\varphi_1(a) \quad \varphi_2(a) \quad \varphi_3(a) \qquad\qquad \varphi_k(a)$$

Obviously it was essential to switch from matrices to digraphs. While it might be possible to carry out our $\mathcal{NP}$-completeness proof in terms of matrices, the statement and proof of Proposition 1 clearly belong to the realm of graph theory.

Identifying Subdivision Vertices

The crucial step towards our isomorphism-completeness result is to show that subdivision digraphs almost satisfy a converse of Proposition 1. That is, any root of such a digraph carries an isomorphism structure of its components. However, we have to take care of some degenerate cases that do not fit into this picture.

Usually in a subdivision digraph, one can easily distinguish the original vertices, commonly called the *branching vertices*, from the *subdivision vertices*. In fact, a subdivision digraph is obviously bipartite and as soon as every weakly connected component contains at least one vertex whose indegree or outdegree differs from 1, the two classes can be uniquely identified.

A problem arises with subdivision digraphs that contain isolated cycles (of even length). In such components, all vertices look like subdivision vertices and this absence of clearly identifiable branching vertices leads to untypical behavior with respect to root finding. Fortunately, isolated cycles are simple objects and we can completely describe their powers.

Lemma 1. *The kth power of a directed cycle of length r is the disjoint union of* $\gcd(r, k)$ *cycles of length* $r/\gcd(r, k)$.

As a consequence, isolated cycles clearly cannot have the isomorphism property we are looking for. But this is no problem. It turns out that a vertex that lies on an isolated cycle of a subdivision digraph D must also lie on an isolated cycle in any root of D. Thus, with respect to roots, cycle vertices do not interact with the other vertices of a subdivision digraph and so we may in the following restrict our attention to subdivision digraphs without cycles. Then each vertex can really be uniquely identified as subdivision or branching vertex.

From Roots to Isomorphisms

With all isolated cycles removed, subdivision digraphs now bear the desired isomorphism structure, under the unfortunately indispensable additional condition that each vertex has at least one inneighbor and one outneighbor.

Theorem 3. *A subdivision digraph without isolated cycles and with positive minimal indegree and outdegree has a kth root if and only if it is the disjoint union of k isomorphic digraphs.*

The proof of Theorem 3 is lengthy and rather technical and we have to omit the details due to space constraints, but the key ideas are easily explained: Recall the long root paths in the proof of Proposition 1. Each inner vertex had indegree and outdegree exactly 1 in R. Theorem 3 rests on the fact that conversely, any root of a subdivision digraph satisfying the preconditions has exactly this structure, i.e., it consists mainly of paths of length $k - 1$ on which no other paths enter or leave. From those paths one can read off the desired isomorphisms. Let us substantiate these ideas by stating some of the central lemmas.

Long Paths. Our aim is to assign each vertex of R to a path of exactly k vertices, the beginning and end of which shall be uniquely determined.

Lemma 2. *Let R be a kth root of a subdivision digraph D without isolated cycles. Then any subdivision vertex of D lies on an R-path $a_1 \rightarrow a_2 \rightarrow \cdots \rightarrow a_k$ of length $k - 1$ where each a_i, $1 \leq i \leq k$, is a subdivision vertex of D. Moreover, such a path is maximal in the sense that the inneighbors of a_1 and the outneighbors of a_k are branching vertices of D.*

There exists an analog of Lemma 2 for branching vertices, which looks almost the same, with the exception that we have to forbid isolated vertices. So here the degree condition of Theorem 3 enters the first time, in a weakened form.

An attempt to prove Lemma 2 directly, faces a principal problem: *subdivision vertex* and *branching vertex* are global notions. A branching vertex with indegree and outdegree 1 is locally indistinguishable from a subdivision vertex. We resolve this ambiguity by ignoring the global picture for a moment, calling a vertex *thin* if it looks like subdivision vertex, i.e., if it has indegree and outdegree 1 in D. For such vertices, we can prove a preliminary version of Lemma 2.

Lemma 3. *Let R be a kth root of a subdivision digraph D and let $a_0 \to a_1 \to \cdots \to a_l$ be an R-walk of length $l \leq k$ between two D-thin vertices a_0 and a_l. Then all a_i, $1 \leq i < l$, are also thin (with respect to D).*

The arguments employed in the proof of Lemma 3 are typical for most of our intermediate results on the way to Theorem 3. Thinness in D tells us that all R-walks starting from a_0 and a_l have to meet again after exactly k steps. We use these confluent walks to "sandwich" R-walks that start from one of the intermediate vertices a_i, showing that those walks also meet again after a certain number of steps. Thinness of a_i in D finally follows from the simple but important observation that in a subdivision digraph, two different vertices cannot have common inneighbors and common outneighbors at the same time.

Combining Lemma 3 with its analog for non-thin vertices eventually leads to a proof of Lemma 2 and its counterpart for branching vertices.

Unique Arcs. In order to use the paths from the preceding paragraph for isomorphism construction, we have to make sure that they indeed establish one-to-one correspondences. Therefore we show that those paths do not interfere, i.e., they must only touch at their end vertices. As above, we resort to the technical notion of thinness.

Lemma 4. *Let R be a kth root of a subdivision digraph D and let a, b be two thin vertices of D with $a \to b$ in R. Then there are no further R-arcs leaving a or entering b.*

Again, we have an analog of this lemma for pairs of non-thin vertices but once more we have to be careful about the existence of neighbors, which was trivially guaranteed for thin vertices. In the next lemma, the additional degree condition of Theorem 3 is indispensable.

Lemma 5. *Let R be a kth root of a subdivision digraph D. Let a, b be two non-thin vertices of D with $a \to b$ in R such that a has an outneighbor and b has an inneighbor in D. Then there are no further R-arcs leaving a or entering b.*

The two preceding lemmas naturally lift to statements about subdivision and branching vertices, the only problem being to show that of two adjacent branching vertices either both are thin or neither is, which is not too difficult.

Concluding the Proofs. The statement of Theorem 3 is now obtained by "inverting" the proof of Proposition 1. The paths provided by Lemma 2 establish correspondences between k disjoint subgraphs of the digraph D and with the help of Lemmas 4 and 5 this can be done in a unique and consistent way.

Theorem 2 then comes as a direct consequence. Isolated cycles are computationally easy to deal with, as we already argued. We should specify that the reduction from root finding to isomorphism finding is actually a Turing reduction, which means that we can find roots in polynomial time on a Turing machine that may call an isomorphism oracle several times at unit cost. Note that we cannot turn it into a stronger many-one reduction (Karp reduction) by simply checking whether k copies of one component of the given digraph D are isomorphic to D itself because the k isomorphic subgraphs of D need not be connected. The other reduction, from isomorphisms to roots, however, can be done in a many-one fashion as Proposition 1 shows.

Dropping the Degree Condition

Let us indicate what can happen in a subdivision digraph that contains vertices without in– or outneighbors. The following figure shows such a digraph D together with a square root R.

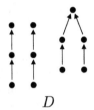
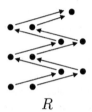

$$D \qquad\qquad R$$

The two topmost root arcs can touch since the precondition of Lemma 5 does not hold. Observe that instead of being the disjoint union of two isomorphic subgraphs, the left component can be decomposed into two parts, A and B (the former consisting of the two paths on the left, the latter containing the remaining five vertices) such that there exists a surjective *homomorphism* from A onto B (i.e., an arc-preserving map). This homomorphism corresponds exactly to those arcs of R that go from A to B.

This example is only meant to indicate the phenomena that might show up. The general situation is more difficult to analyze and it is not clear whether the digraph root problem remains isomorphism complete under these relaxed conditions since the general homomorphism problem for graphs is $\mathcal{NP}$-complete [5].

5   Outlook

While the original problem, the open complexity status of Boolean matrix root computation is now settled, the discovered relation to graph isomorphism raises new questions. First of all, it would be desirable to get rid of the degree condition of Theorem 2. Though this restriction turned out indispensable for underlying structural statement of Theorem 3, it is not clear whether it might be possible to eliminate it from the complexity result since that would lead to special homomorphism problems which take us closer to the world of $\mathcal{NP}$-hardness.

More generally, we may ask for relaxations of the concept of subdivisions that still serve the task of "deactivating" computationally hard aspects of root finding, thus keeping the problem isomorphism complete. If one traces the details of our proofs, notions like bounded tree width appear promising and might lead to weaker conditions for isomorphism completeness. Eventually, the problem of Boolean matrix root computation could turn out to be a suitable object for analyzing the "boundary" between isomorphism completeness and $\mathcal{NP}$-hardness.

References

1. A. V. Aho, M. R. Garey, and J. D. Ullman. The transitive reduction of a directed graph. *SIAM Journal on Computing*, 1(2):131–137, 1972.
2. Kellogg S. Booth. Isomorphism testing for graphs, semigroups, and finite automata are polynomially equivalent problems. *SIAM Journal on Computing*, 7(3):273–279, 1978.
3. Thomas H. Cormen, Charles E. Leiserson, and Ronald L. Rivest. *Introduction to Algorithms*, chapter 26. MIT Press, 1990.
4. Bart de Schutter and Bart de Moor. On the sequence of consecutive powers of a matrix in a Boolean algebra. *SIAM Journal on Matrix Analysis and Applications*, 21(1):328–354, 1999.
5. Pavol Hell and Jaroslav Nešetřil. On the complexity of H-coloring. *Journal of Combinatorial Theory, Series B*, 48:92–110, 1990.
6. Nicholas J. Higham. Newton's method for the matrix square root. *Mathematics of Computation*, 46(174):537–549, 1986.
7. C. R. Johnson, K. Okubo, and R. Reams. Uniqueness of matrix square roots and an application. *Linear Algebra and Applications*, 323:52–60, 2001.
8. Volker Kaibel and Alexander Schwartz. On the complexity of isomorphism problems related to polytopes. To appear in *Graphs and Combinatorics*.
9. Ki Hang Kim. *Boolean matrix theory and applications*. Marcel Dekker, Inc., 1982.
10. Johannes Köbler, Uwe Schöning, and Jacobo Torán. *The graph isomorphism problem*. Birkhäuser, 1993.
11. Ya Yan Lu. A Padé approximation method for square roots of symmetric positive definite matrices. *SIAM Journal on Matrix Analysis and Applications*, 19(3):833–845, 1998.
12. Anna Lubiw. Some NP-complete problems similar to graph isomorphism. *SIAM Journal on Computing*, 1981.
13. George Markowsky. Ordering D-classes and computing Shein rank is hard. *Semigroup Forum*, 44:373–375, 1992.
14. Rudolph Mathon. A note on the graph isomorphism counting problem. *Information Processing Letters*, 8(3):131–132, 1979.
15. Rajeev Motwani and Madhu Sudan. Computing roots of graphs is hard. *Discrete Applied Mathematics*, 54(1):81–88, 1994.
16. Ian Munro. Efficient determination of the transitive closure of a directed graph. *Information Processing Letters*, 1:56–58, 1971.
17. Panayiotis J. Psarrakos. On the mth roots of a complex matrix. *The Electronic Journal of Linear Algebra*, 9:32–41, 2002.
18. Kenneth A. Ross and Charles R. B. Wright. *Discrete mathematics*, chapter 7.5. Prentice-Hall, second edition, 1988.
19. Larry L. Stockmeyer. The minimal set basis problem is NP-complete. IBM Research Report No. RC-5431, IBM Thomas J. Watson Research Center, 1975.

A Fast Bit-Parallel Algorithm for Matching Extended Regular Expressions*

Hiroaki Yamamoto[1] and Takashi Miyazaki[2]

[1] Department of Information Engineering, Shinshu University,
4-17-1 Wakasato, Nagano-shi, 380-8553 Japan.
yamamoto@cs.shinshu-u.ac.jp
[2] Nagano National College of Technology,
716 Tokuma, Nagano-shi, 381-8550 Japan.
miya@ee.nagano-nct.ac.jp

Abstract. This paper addresses the extended regular expression matching problem: given an extended regular expression (a regular expression with intersection and complement) r of length m and a text string $x = a_1 \cdots a_n$ of length n, find all occurrences of substrings which match r. We will present a new bit-parallel pattern matching algorithm which runs in $O((mn^2 + ex(r)n^3)/W)$ time and $O((mn + ex(r)n^2)/W)$ space, where $ex(r)$ is the number of extended operators (intersection and complement) occurring in r, and W is word-length of a computer. In addition, we actually implement the proposed algorithm and evaluate the performance.

1 Introduction

Regular expression matching algorithms have been extensively studied and implemented in many application softwares [1,3,4,5,8,9,10,11]. Extended regular expressions (EREs) are REs with intersection and complement operators, and seem to offer a more flexible searching system. We here consider the ERE matching problem as follows: Given an ERE r of length m and a text string $x = a_1 \cdots a_n$ of length n, find all occurrences of substrings which match r. Namely, we are searching for the set $F = \{(i,j) \mid a_i \cdots a_j \in L(r)\}$, where $L(r)$ denotes the language denoted by r. For example, given $r = ((a \vee b)^* aa(a \vee b)^*) \wedge (\neg((a \vee b)^* bb(a \vee b)^*))$ and $x = aabbaabb$, the desired set F becomes $\{(1,2),(1,3),(4,6),(4,7),(5,6),(5,7)\}$ because r denotes the set of strings over $\{a,b\}$ such that aa appears but not bb.

It is widely known that such a pattern matching problem can be solved by using a recognition algorithm. The standard recognition algorithm for REs runs in $O(mn)$ time and $O(m)$ space, based on nondeterministic finite automata (NFAs for short) [1,2,7]. Myers [8] has improved this algorithm so that it runs in $O(mn/\log n)$ time and space. Thus, for REs, efficient algorithms based on NFAs have been presented, and have been used for the RE matching problem.

As for EREs, Hopcroft and Ullman [7] first gave a recognition algorithm based on the inductive definition of EREs, which runs in $O(mn^3)$ time and

---

* This research has been supported by the REFEC

$O(mn^2)$ space. Knight and Myers [5] and Hirst [6] gave other recognition algorithms, which run faster than Hopcroft's one for some kinds of EREs. Recently, Yamamoto [12] has given a new recognition algorithm for EREs which runs in $O(mn^2+kn^3)$ time and $O(mn+kn^2)$ space in the worst case, where k is a number less than the number of extended operators (that is, intersection and complement) in r. His algorithm is a natural extension of the NFA-based algorithm for REs. It is obvious that we can easily solve the ERE matching problem by applying such recognition algorithms to all the suffixes $a_i \cdots a_n$ of $x = a_1 \cdots a_n$, although this causes an n-fold slowdown comparing to recognition algorithms.

In this paper, we will extend Yamamoto's algorithm to the ERE matching problem and design a faster algorithm by introducing bit-parallelism. Many pattern matching algorithms employ bit parallelism to improve the running time because this technique can achieve a speed-up depending on word-length. Our algorithm is as follows. As in [12], we first partition the parse tree of a given ERE into modules, translate each module into augmented NFAs (A-NFAs), and then simulate the obtained A-NFAs on all the suffixes of x using a data structure called *a directed computation graph* (DCG). This time, Yamamoto's algorithm must compute a transition from a state many times at a time. We improve the algorithm by introducing bit parallelism so that such a transition is computed in parallel. Furthermore, since DCGs can share the common parts, we can simulate A-NFAs using DCGs with almost same size as a DCG generated by simulating A-NFAs on just one string x. Finally, our algorithm can achieve an almost W-fold speed-up of Yamamoto's recognition algorithm by using W-bit parallel operations in the best case.

The main result is as follows: Let r be an ERE of length m over an alphabet Σ, and let x be a text string of length n in Σ^*. Then we can design an ERE matching algorithm which runs in $O((mn^2 + ex(r)n^3)/W)$ time and $O((mn + ex(r)n^2)/W)$ space, where $ex(r)$ is the number of extended operators in r.

In addition, we actually implement the proposed algorithm and evaluate the performance.

2   Extended Regular Expressions

Let Σ be an alphabet. The extended regular expressions (EREs) over Σ are recursively defined by five operators, union ($\vee$), concatenation ($\cdot$), closure (star $*$), intersection ($\wedge$) and complement ($\neg$). Thus EREs are REs with intersection and complement. By $ex(r)$, we denote the number of extended operators (that is, $\wedge$ and $\neg$) occurring in an ERE r. Furthermore, we introduce the parse tree T_r of r, which is the tree such that each leaf is labeled with a symbol in Σ and each of the other nodes is labeled with an operator. See [12] for the detail.

3   Modules and Augmented NFAs

As in [12], we define modules and augmented NFAs for EREs. Let r be an ERE over an alphabet Σ and let T_r be the parse tree of r. Then, we partition T_r by nodes labeled with intersection $\wedge$ and complement $\neg$ into subtrees such that (1) the root of each subtree is either a child of a node labeled with $\wedge$ or $\neg$ in T_r or the root of T_r, (2) each subtree does not contain any interior nodes labeled by $\wedge$ or $\neg$, (3) each leaf is labeled by $\emptyset$, ϵ, $a \in \Sigma$, $\wedge$ or $\neg$. If it is labeled by $\wedge$ ($\neg$, respectively), then it is called *a universal leaf* (*a negating leaf*, respectively). These leaves are also called *a modular leaf*. We call such a subtree *a module*.

Let R and R' be modules in the parse tree T_r. If a modular leaf u of R becomes the parent of the root of R' in T_r, then R is called *a parent of* R', and conversely R' is called *a child of* R or *a child of* R *at* u. Thus there are two children, called *a universal pair*, at each universal leaf, while one child, called *a negating module*, at each negating leaf. If the root of a module R is the root of T_r, then R is called *a root module*. If a module R does not have any children, then R is called *a leaf module*. It is clear that such a parent-child relationship induces *a modular tree* $\mathcal{T}_r = (\mathcal{R}, \mathcal{E})$ such that (1) $\mathcal{R}$ is a set of modules, (2) $(R, R') \in \mathcal{E}$ if and only if R is the parent of R'. The depth of each module and the height of a modular tree $\mathcal{T}_r$ is defined in the standard way.

Now, for each module R, we relabel every modular leaf u of R with a new symbol σ_u called *a modular symbol*. By this relabeling, R can be viewed as an RE over $\Sigma \cup \{\sigma_u \mid u$ is a modular leaf of $R\ \}$. Then, by the standard linear translation from REs into NFAs (for example, see [7]), we can construct an NFA M_R for a module R. Let us call this M_R *an augmented NFA* (A-NFA for short). It is clear that we can define a parent-child relationship among A-NFAs according to the corresponding modules. We call a transition back to a previous state generated by a star operator *a back transition*. By removing such back transitions, we number each state of M_R from the initial state in the topological order. By the topological order, we mean that for any states q and p, $q \leq p$ if and only if there is a path from q to p. This order is used to efficiently simulate A-NFAs. In addition, we call a state q in M_R *a universal state* if there is $\delta(q, \sigma_u) = q'$ for a universal leaf u, and call a state q in M_R *a negating state* if there is $\delta(q, \sigma_u) = q'$ for a negating leaf u. The other states are called *an existential state*. Furthermore, if a module R' is a child of R at u, then A-NFA $M_{R'}$ is said to be *associated with* σ_u (or *associated with* q). It is clear that the following theorem holds.

Theorem 1. *Let r be an ERE of length m, let $R_0, \ldots, R_l$ be modules produced by partitioning T_r, and let m_j be the length of the subexpression of r corresponding to the module R_j. Then we can construct an A-NFA M_j for each module R_j such that the number of states of M_j is at most $2m_j$.*

4   Directed Computation Graphs

We will give the definition of *a directed computation graph* (*DCG* for short), and then give an implementation by bit-vector arrays.

4.1   Definition of a DCG

Let r be an ERE over an alphabet Σ and let $x = a_1 \cdots a_n$ be an input string in Σ^*. We first partition T_r into modules $R_0, \ldots, R_l$ and construct A-NFAs

$M_0, \ldots, M_l$ for each module as described in Theorem 1. Here R_0 is the root module. After that, to determine if $x \in L(r)$, we simulate the set $\{M_0, \ldots, M_l\}$ of A-NFAs on x. This time, for each A-NFA M_j $(0 \le j \le l)$, we introduce variables, called *an existential-element set* (*an EE-set* for short), which take a subset of states of M_j. To simulate each module M_j $(0 \le j \le l)$, we use at most $n+1$ EE-sets U_j^i $(0 \le i \le n)$. An EE-set U_j^i is used to simulate M_j on $a_i \cdots a_n$ using a simple state-transition simulation. Namely, U_j^i always maintains states reachable from the initial state q_j of M_j after M_j has read $a_i \cdots a_{i'}$ for any $i \le i' \le n$. To simulate the set $\{M_0, \ldots, M_l\}$, we will construct a DCG $\mathcal{G} = (\mathcal{U}, \mathcal{E})$ such that (1) $\mathcal{U}$ is the set of *nodes*, which consists of EE-sets, and $\mathcal{E}$ is the set of *edges*, which consists of pairs (U, U') of nodes, (2) a node U_0^i is called *the i-th source node*, which has no incoming edges, (3) Nodes with no outgoing edges are called *a sink node*, (4) let $U_{j_1}^i$, $U_{j_2}^{i'}$ and $U_{j_3}^{i'}$ be nodes of $\mathcal{U}$ for A-NFAs M_{j_1}, M_{j_2} and M_{j_3}, respectively. Then there exist directed edges $(U_{j_1}^i, U_{j_2}^{i'})$ and $(U_{j_1}^i, U_{j_3}^{i'})$ in $\mathcal{E}$ if and only if R_{j_2} and R_{j_3} are two children of R_{j_1} at a universal leaf u and M_{j_1} reaches the universal state corresponding to u while processing $a_{i'}$, (5) let $U_{j_1}^i$ and $U_{j_2}^{i'}$ be nodes of $\mathcal{U}$ for A-NFAs M_{j_1} and M_{j_2}, respectively. Then there exists a directed edge $(U_{j_1}^i, U_{j_2}^{i'})$ in $\mathcal{E}$ if and only if R_{j_2} is the child of R_{j_1} at a negating leaf u and M_{j_1} reaches the negating state corresponding to u while processing $a_{i'}$.

4.2   Implementation of DCGs by Bit-Vector Arrays

For each A-NFA M_j, $n+1$ EE-sets $U_j^0, \ldots, U_j^n$ are used to simulate M_j. Hence these EE-sets contain only states of M_j. If states in these EE-sets can be processed in parallel, then we have a possibility to achieve a speed-up and a space-saving. We will achieve a speed-up by taking advantage of W-bit parallel boolean operations. For this purpose, we introduce bit-vector arrays $V_j[q, Z]$ for each A-NFA M_j and bit-vector arrays $P_{(j_0, j_1)}[t, Z]$ for each pair (M_{j_0}, M_{j_1}) of A-NFAs such that M_{j_1} is a child of M_{j_0}.

For each A-NFA M_j, $V_j[q, Z]$ is defined as follows, where q is a state of M_j and $0 \le Z \le \lfloor n/W \rfloor$, where $\lfloor n/W \rfloor$ denotes the maximum integer less than or equal to n/W. An array $V_j[q, Z] = v$, where v is a W-bit vector, if and only if for any $i \in BIT(v)$, $q \in U_j^{ZW+i}$. We assume that bits of v are numbered from 0, and we define $BIT(v) = \{i \mid$ the ith bit of v is 1$\}$. Note that the total size of arrays $V_0, \ldots, V_n$ is $O(mn/W)$ because by Theorem 1, the total number of states over all A-NFAs translated from r is $O(m)$.

Next, for each pair (M_{j_0}, M_{j_1}) of A-NFAs such that M_{j_1} is a child of M_{j_0}, $P_{(j_0, j_1)}[t, Z]$ is defined as follows, where $0 \le t \le n$ and $0 \le Z \le \lfloor n/W \rfloor$. An array $P_{(j_0, j_1)}[t, Z] = v$, where v is a W-bit vector, if and only if for any $i \in BIT(v)$, $U_{j_1}^t$ is a child of $U_{j_0}^{ZW+i}$.

These arrays express the parent-child relationship of a DCG. Note that if a pair (R_{j_1}, R_{j_2}) is a universal pair, then it suffices that either $P_{(j_0, j_1)}[t, Z]$ or $P_{(j_0, j_2)}[t, Z]$ is defined. Furthermore note that the total size of arrays $P_{(j_0, j_1)}$

is $O(ex(r)n^2/W)$ because the total number of pairs (R_{j_1}, R_{j_0}) is at most the number of extended operators, that is, $ex(r)$.

5   Bit-Parallel ERE Matching Algorithm

In an algorithm, the operations |, & and ~ denote boolean operations bit-wise OR, bit-wise AND, and bit-wise NOT, respectively. Furthermore, we use two operations *BitSet* and *BitCheck*. $BitSet(v, i)$ sets the ith bit of v to 1. $BitCheck(v, i)$ checks if the ith bit of v is equal to 1, and if equal, it returns 1; otherwise returns 0. Since these operations can simply be implemented so that they run in $O(1)$ time, their details are here omitted. The main algorithm is the algorithm *MATCHING* given in Fig. 1, which consists of two main parts. In the first part (Step 1 and Step 2), we translate a given ERE into A-NFAs. In the second part (Step 3), we simulate A-NFAs by using a DCG and bit-parallelism. In what follows, let r be an ERE of length m and let $x = a_1 \cdots a_n$ be an input string of length n.

5.1   Translation an ERE into A-NFAs

The translation from ERE r into A-NFAs can be done by standard parsing algorithms, such as $LL(1)$ and $LR(1)$. Hence the detail of the translation is here omitted. In fact, we have implemented the translation by using an $SLR(1)$ parsing algorithm. Therefore, for any ERE r of length m, we can translate r into A-NFAs in $O(m)$ time and space.

5.2   Matching Algorithm Using A-NFAs

In order to find all substrings of x which match r, for each position i of x, we generate a DCG to simulate the computation of A-NFAs on the suffix $a_i \cdots a_n$. Hence we have n DCGs during the computation. Since there are many common parts among these DCGs, we will simulate A-NFAs in parallel by sharing them. The basic idea of the simulation is similar to that of Yamamoto [12], although we need several new ideas due to compute n DCGs in parallel by using bit-parallel operations.

Let M_0 be an A-NFA for the root module. Then *MATCHING* checks if DCGs includes the final state of M_0 each time it processes input symbols. If, after processing the ith symbol, the DCG simulating $a_{i_1} \cdots a_n$ includes the final state, then *MATCHING* outputs a pair (i_1, i) of positions because substring $a_{i_1} \cdots a_i$ matches r.

The procedure for the simulation is *FindMatch* given in Fig. 2. The procedure *FindMatch* simulates A-NFAs on all the suffixes $a_i \cdots a_n$ in parallel. The simulation for each suffix is invoked by *BitSet* in Step 1-1 and Step 2-2. The simulation by *FindMatch* is an extension of the standard state-transition simulation of NFAs, and consists of two procedures, the procedure *EClosure* which computes ϵ-moves and the procedure *GoTo* which computes the transition by a

---

Algorithm MATCHING(r, x)
Input: an ERE r, a text string $x = a_1 \cdots a_n$.
Output: all pairs (i, j) such that $a_i \cdots a_j \in L(r)$.

Step 1. Partition T_r into modules $R_0, \ldots, R_l$.
Step 2. Translate each module R_j $(0 \leq j \leq l)$ to an A-NFA M_j.
Step 3. Let M_0 be an A-NFA for the root module R_0 and let q_0 be the initial state and the
   final state of M_0, respectively. Then, for $i = 0$ to n, do the following:
   1. do $FindMatch(M_0, x, q_0, q_f, i)$,
   2. /* The following code outputs $(ZW + i' + 1, i)$ such that $i' \in BIT(V_0[q_f, Z])$. This means
      $a_{ZW + i' + 1} \cdots a_i \in L(r)$. Note that $i < ZW + i' + 1$ means that the empty string matches
      r. */
      for all $Z \in [0, \lfloor i/W \rfloor]$ such that $V_0[q_f, Z] \neq 0$do
         if $Z < \lfloor i/W \rfloor$ then $MAX = W - 1$; else set $MAX = i \bmod W$;
            for $i' = 0$ to MAX do
               if $BitCheck(V_0[q_f, Z], i') = 1$, then output $(ZW + i' + 1, i)$.

Fig. 1. The algorithm $MATCHING$

Procedure FindMatch(M, x, q_0, q_f, i)
M: an A-NFA M; x: a text string;
q_0: the initial state q_0 of M; q_f: the final state of M;
i: a position of x;

Step 1. If $i = 0$, then do the following:
   1. $BitSet(V_0[q_0, 0], 0)$. /* Start up the simulation on $a_1 \cdots a_n$. */
   2. $\mathcal{R}' = \{R_0\}$ and then $EClosure(\mathcal{R}', 0)$.
Step 2. If $i \geq 1$, then do the following:
   1. $GoTo(\mathcal{R}, i)$.
   2. $BitSet(V_0[q_0, \lfloor i/W \rfloor], i - \lfloor i/W \rfloor W)$. /* Start up the simulation on $a_{i+1} \cdots a_n$. */
   3. $EClosure(\mathcal{R}, i)$.

Fig. 2. The procedure $FindMatch$

Procedure EClosure$(\mathcal{R}', i)$
$\mathcal{R}'$: a subset of $\mathcal{R}$;
i: a position of x;

Step 1. For $h = h_{max}$ to h_{min}, do the following:
   1. $EpsilonMove(\mathcal{R}^h, i)$.
   2. $ModCheck(\mathcal{R}^h, i)$.

Fig. 3. The procedure $EClosure$

---

symbol a. The different point from the standard state-transition simulation is in
the processing of intersection and complement in $EClosure$.

Now we explain the process for intersection and complement. For the simplicity, let us consider an expression $r_1 \wedge r_2$, where r_1 and r_2 are REs. Then
we have three A-NFAs, that is, M_{j_0} for $\wedge$, M_{j_1} for r_1 and M_{j_2} for r_2, and M_{j_1}
and M_{j_2} are children of M_{j_0}. In this case, M_{j_0} has two states q and q' with the
transition from q to q' by a modular symbol σ. We invoke M_{j_1} and M_{j_2} when
M_{j_0} reaches state q, and then move M_{j_0} from q to q' when M_{j_1} and M_{j_2} both
reach the final states at same time. This guarantees that M_{j_0} can move from
q to q' by a string if and only if the string is in $L(r_1 \wedge r_2)$. Thus we need to
perform a transition by a modular symbol when the children reach their final

---

Procedure EpsilonMove($\mathcal{R}'$, i)
$\mathcal{R}'$: a subset of $\mathcal{R}$;
i: an input position;

1. repeat the following twice:
2.   for all $R_j \in \mathcal{R}'$ do
3.   for all $q \in M_j$ in the topological order do
      Case 1: if q is an existential state, then
         for all $q' \in \delta(q, \epsilon)$ do
            for all $Z \in [0, \lfloor i/W \rfloor]$ such that $V_j[q, Z] \neq 0$ do $V_j[q', Z] := V_j[q, Z] \mid V_j[q', Z]$,
      Case 2: if q is a universal state, then do the following: Here, let M_{j_1} and M_{j_2} be A-NFAs which
      are associated with q, and let q_{j_1} and q_{j_2} be initial states of these A-NFAs, respectively,
        a) if $BitCheck(V_j[q_{j_1}, \lfloor i/W \rfloor], i - \lfloor i/W \rfloor W) = 1$, then
           for all $Z \in [0, \lfloor i/W \rfloor]$ such that $V_j[q, Z] \neq 0$ do
            $P_{(j,j_1)}[i, Z] := P_{(j,j_1)}[i, Z] \mid V_j[q, Z]$,
        b) if $BitCheck(V_j[q_{j_1}, \lfloor i/W \rfloor], i - \lfloor i/W \rfloor W) = 0$, then
          i. $BitSet(V_j[q_{j_1}, \lfloor i/W \rfloor], i - \lfloor i/W \rfloor W)$ and $BitSet(V_{j_2}[q_{j_2}, \lfloor i/W \rfloor], i - \lfloor i/W \rfloor W)$,
          ii. for all $Z \in [0, \lfloor i/W \rfloor]$ such that $V_j[q, Z] \neq 0$ do $P_{(j,j_1)}[i, Z] := V_j[q, Z]$,
          iii. do $EClosure(\{R_{j_1}, R_{j_2}\}, i)$,
      Case 3: if q is a negating state, then do the following: Here, let M_{j_1} be the A-NFA which is
      associated with q, and let q_{j_1} be the initial state of M_{j_1}.
        a) if $BitCheck(V_{j_1}[q_{j_1}, \lfloor i/W \rfloor], i - \lfloor i/W \rfloor W) = 1$, then
           for all $Z \in [0, \lfloor i/W \rfloor]$ such that $V_j[q, Z] \neq 0$ do
            $P_{(j,j_1)}[i, Z] := P_{(j,j_1)}[i, Z] \mid V_j[q, Z]$,
        b) if $BitCheck(V_{j_1}[q_{j_1}, \lfloor i/W \rfloor], i - \lfloor i/W \rfloor W) = 0$, then
          i. $BitSet(V_{j_1}[q_{j_1}, \lfloor i/W \rfloor], i - \lfloor i/W \rfloor W)$,
          ii. for all $Z \in [0, \lfloor i/W \rfloor]$ such that $V_j[q, Z] \neq 0$ do $P_{(j,j_1)}[i, Z] := V_j[q, Z]$,
          iii. do $EClosure(\{R_{j_1}\}, i)$.

Fig. 4. The procedure *EpsilonMove*

Procedure GoTo($\mathcal{R}'$, i)
$\mathcal{R}'$: a subset of $\mathcal{R}$;
i: an input position;

1. for all $R_j \in \mathcal{R}'$, do the following:
   a) for all states q of M_j in the reverse topological order do
      i. if there is a state q' such that $\delta(q, a_i) = \{q'\}$, then
         for all $Z \in [0, \lfloor i/W \rfloor]$ such that $V_j[q, Z] \neq 0$ do $V_j[q', Z] := V_j[q, Z]$,
      ii. otherwise
         for all $Z \in [0, \lfloor i/W \rfloor]$ such that $V_j[q, Z] \neq 0$ do $V_j[q', Z] := 0$.

Fig. 5. The procedure *GoTo*

states at same time. Similarly, for complement, we need to perform a transition
by a modular symbol when the child does not reach the final state. The procedure *EClosure* contains such a processing as well as computes transitions by
ϵ. Namely, *EClosure* computes transitions by ϵ by the procedure *EpsilonMove*
and checks the conditions for intersection and complement by the procedures
ModCheck. The detail of *EClosure* is given in Fig. 3.

We must pay attention to the order in which A-NFAs are simulated. As seen
above, transitions of an A-NFA depend on transitions of children A-NFAs of the
A-NFA. Therefore, to simulate efficiently, *EClosure* proceeds in order from leaves
to the root in the modular tree. Now let h_{max} and h_{min} be the maximum depth
and the minimum depth over $\mathcal{R} = \{R_0, \ldots, R_l\}$, respectively. We can partition
$\mathcal{R}$ into some subsets $\mathcal{R}^{h_{min}}, \cdots, \mathcal{R}^{h_{max}}$ by the depth of each module R such that

---

Procedure ModCheck($\mathcal{R}', i$)
$\mathcal{R}'$: a subset of $\mathcal{R}$;
i: an input position;

1. for all universal pairs (R_{j_1}, R_{j_2}) in $\mathcal{R}'$ do
   for all $Z \in [0, \lfloor i/W \rfloor]$ such that $V_{j_1}[p_1, Z] \neq 0$ do
   a) $v := V_{j_1}[p_1, Z]$ & $V_{j_2}[p_2, Z]$, where M_{j_1} and M_{j_2} are associated with a modular symbol σ, and p_1 and p_2 are the final states of them, respectively,
   b) if $Z < \lfloor i/W \rfloor$ then $MAX = W - 1$; else set $MAX = i \bmod W$;
       for $i' = 0$ to MAX do
          if $BitCheck(v, i') = 1$, then
             for all $Z' \in [0, \lfloor i/W \rfloor]$ such that $P_{(j_0, j_1)}[i' + ZW, Z'] \neq 0$ do
                $V_{j_0}[q, Z'] := V_{j_0}[q, Z'] \mid P_{(j_0, j_1)}[i' + ZW, Z']$, where q is a state such that $\delta(q', \sigma) = q$
       for a universal state q'.
2. for all negating modules R_{j_1} in $\mathcal{R}'$ do
   for $Z = 0$ to $\lfloor i/W \rfloor$ do /* Note that, in this case, we cannot limit the range to Z such that $V_{j_1}[p_1, Z] \neq 0$. */
   a) $v := \tilde{\ }V_{j_1}[p_1, Z]$, where p_1 is the final state of M_{j_1},
   b) if $Z < \lfloor i/W \rfloor$ then $MAX = W - 1$; else $MAX = i \bmod W$;
       for $i' = 0$ to MAX do
          if $BitCheck(v, i') = 1$, then
             for all $Z' \in [0, \lfloor i/W \rfloor]$ such that $P_{(j_0, j_1)}[i' + ZW, Z'] \neq 0$ do
                $V_{j_0}[q, Z'] := V_{j_0}[q, Z'] \mid P_{(j_0, j_1)}[i' + ZW, Z']$, where q is a state such that $\delta(q', \sigma) = q$
       for a negating state q'.

Fig. 6. The procedure *ModCheck*

$\mathcal{R}^h$ consists of modules with the depth h. Then *EClosure* simulates ϵ-moves in the order from $\mathcal{R}^{h_{max}}$ to $\mathcal{R}^{h_{min}}$.

The procedure *EpsilonMove*, given in Fig. 4, performs transitions by ϵ. The job of *EpsilonMove* is classified into three cases according to kinds of states. Let M_j, q and i be an A-NFA, a state and an input position currently processed, respectively. If q is existential, then *EpsilonMove* simply computes the next states reachable by ϵ. If q is universal, then *EpsilonMove* performs as follows. Let M_{j_1} and M_{j_2} be children associated with q. *EpsilonMove* generates $U_{j_1}^i$ and $U_{j_2}^i$, which include only the initial states of M_{j_1} and M_{j_2}, respectively, to simulate M_{j_1} and M_{j_2} on $a_{i+1} \cdots a_n$. In addition, for $U_j^{i'}$ $(0 \leq i' \leq i)$ which includes q, it generates two directed edges from $U_j^{i'}$ to $U_{j_1}^i$ and from $U_j^{i'}$ to $U_{j_2}^i$. These jobs are performed on bit-vector arrays in parallel. In case of a negating state, the similar process is performed. In order to compute a state included in several EE-sets $U_j^{i_1}, \ldots, U_j^{i_l}$ in parallel, *EpsilonMove* computes the states of M_j in the topological order. This time, if there is a back transition in M_j, then the computation in the topological order may not be able to compute all reachable states. However, as in [11], we can overcome this difficulty by computing the states of M_j twice in the topological order.

As mentioned before, the procedure *ModCheck*, given in Fig. 6, checks if transitions from universal and negating states are possible, and if possible, then it performs the transitions. Finally we have the following theorem.

Theorem 2. *Given an ERE r of length m and a text string $x = a_1 \cdots a_n$ of length n, the algorithm MATCHING can find the set $F = \{(i, j) \mid a_i \cdots a_j \in L(r)\}$ in $O((mn^2 + ex(r)n^3)/W)$ time and $O((mn + ex(r)n^2)/W)$ space.*

Table 1. Computation time (in seconds) for strings over $\{0, 1, a, b\}$

ERE	length of string	translation time	initial setting time	matching time	total time
ERE1	200	0.0004	0.013	0.056	0.0694
	2000	0.0004	0.391	3.452	3.8434
	8000	0.0004	5.096	55.591	60.6874
ERE2	200	0.0005	0.013	0.012	0.0255
	2000	0.0005	0.383	0.152	0.5355
	8000	0.0005	5.175	1.078	6.2535
ERE3	200	0.0009	0.023	0.086	0.1099
	2000	0.0009	0.625	3.899	4.5249
	8000	0.0009	8.208	55.887	64.0959
ERE4	200	0.0009	0.014	0.075	0.0899
	2000	0.0009	0.383	1.246	1.6299
	8000	0.0009	5.097	4.96	10.0579

Table 2. Computation time (in seconds) for a string from a log file

ERE	length of string	translation time	initial setting time	matching time	total time
ERE1	8000	0.0004	5.069	54.429	59.4984
ERE2	8000	0.0005	5.093	2.908	8.0015
ERE3	8000	0.0009	8.129	54.443	62.5729
ERE4	8000	0.0009	5.096	1.292	6.3889

6   Experimental Results

We have implemented the proposed algorithm using C++ and evaluated the performance. Our machine is a LINUX PC with 32-bit Pentium-4 of 1.7GHz and 256MB main memory. Hence we have fixed the parameter $W = 32$. We measure the running time in seconds. Table 1 and Table 2 show experimental results for a string over $\{0, 1, a, b\}$ which we have appropriately chosen and a string from a log file of our machine, respectively. Note that the string from a log file contains many kinds of characters. In tables, *translation time* shows the time to translate an ERE into A-NFAs, *initial setting time* shows the time to initialize bit-vector arrays, and *matching time* shows the time to find all substrings which match a given ERE. We use the following EREs.

- ERE1 : $((0^* \wedge (0 \vee 1^*))^*0) \wedge (0^* \wedge (\neg((00)^*)))$
  This denotes strings over $\{0\}$ such that the number of 0's is odd.
- ERE2 : $0(((0^* \wedge (0 \vee 1^*))^*0) \wedge (0^* \wedge (\neg((00)^*))))$
  This denotes strings over $\{0\}$ such that the number of 0's is even.
- ERE3 : $(((a \vee b)^*aa(a \vee b)^*) \wedge (\neg((a \vee b)^*bb(a \vee b)^*)))1(((0^* \wedge (0 \vee 1^*))^*0) \wedge (0^* \wedge (\neg((00)^*))))$
  This is a concatenation of the ERE discussed in Introduction, a 1, and ERE1.

- ERE4 : $(((a \vee b)^*aa(a \vee b)^*) \wedge (((a \vee b)^*bb(a \vee b)^*)))1(((0^* \wedge (0 \vee 1^*))^*0) \wedge (0^* \wedge (((00)^*))))$
  This is an ERE obtained by deleting complement $\neg$ from ERE3.

We can see the following observations from experiments

- The running time of our algorithm does not theoretically depend on the size of an alphabet. In fact, the matching time for the same ERE in Table 1 and Table 2 is almost same.
- Although the difference between ERE1 and ERE2 is small (only the leading 0 of ERE2), the matching time for ERE2 is much faster than that for ERE1. This is because the size of a DCG generated by ERE2 becomes smaller. On the other hand, sizes of DCGs for ERE1 and ERE3 become larger. Thus *matching time* heavily depends on the form of a given ERE.
- Processing complement seems to require more time than intersection. This is because there is a possibility that a negating module takes more time in *ModCheck*. In fact, by removing complement from ERE3, the matching time is much improved.
- The translation time of each ERE can be almost ignored, compared with the matching time. Therefor, we can expect a further speed-up by translating an A-NFA with a small size into a deterministic finite automaton, as in [8, 10].

References

1. A.V. Aho, Algorithms for finding patterns in strings, In J.V. Leeuwen, ed. Handbook of theoretical computer science, Elsevier Science Pub., 1990.
2. A. Apostolico, Z. Galil ed., Pattern Matching Algorithms, Oxford University Press, 1997.
3. R. Baeza-Yates and G. Gonnet, Fast Text Searching for Regular Expressions or Automaton Searching on Tries, J. of the ACM, 43,6, 915–936, 1996.
4. R. Baeza-Yates and B. Ribeiro-Neto, Modern Information Retrieval, Addison Wesley, 1999.
5. J.R. Knight and E.W. Myers, Super-Pattern matching, Algorithmica, 13, 1–2, 211–243, 1995.
6. S.C. Hirst, A New Algorithm Solving Membership of Extended Regular Expressions, Tech. Report, The University of Sydney, 1989.
7. J.E. Hopcroft and J.D. Ullman, Introduction to automata theory language and computation, Addison Wesley, Reading Mass, 1979.
8. G. Myers, A Four Russians Algorithm for Regular Expression Pattern Matching, J. of the ACM, 39,4, 430–448, 1992.
9. E. Myers and W. Miller, Approximate Matching of Regular Expressions, Bull. of Mathematical Biology, 51, 1, 5–37, 1989.
10. G. Navarro and M. Raffinot, Compact DFA Representation for Fast Regular Expression Search, Proc. WAE2001, LNCS 2141, 1–12, 2001.
11. S. Wu, U. Manber and E. Myers, A Sub-Quadratic Algorithm for Approximate Regular Expression Matching, J. of Algorithm, 19, 346–360, 1995.
12. H. Yamamoto, A New Recognition Algorithm for Extended Regular Expressions, Proc. ISAAC2001, LNCS 2223, 267–277, 2001.

Group Mutual Exclusion Algorithms Based on Ticket Orders

Masataka Takamura and Yoshihide Igarashi

Department of Computer Science, Gunma University, Kiryu, Japan 376-8515
{takamura,igarashi}@comp.cs.gunma-u.ac.jp

Abstract. Group mutual exclusion is an interesting generalization of the mutual exclusion problem. This problem was introduced by Joung, and some algorithms for the problem have been proposed by incorporating mutual exclusion algorithms. Group mutual exclusion occurs naturally in a situation where a resource can be shared by processes of the same group, but not by processes of different groups. It is also called the congenial talking philosophers problem. In this paper we propose three algorithms based on ticket orders for the group mutual exclusion problem in the asynchronous shared memory model. The first algorithm is a simple modification of the Bakery algorithm. It satisfies group mutual exclusion, but does not satisfy lockout freedom. The second and the third algorithms are further modifications from the first one in order to satisfy lockout freedom and to improve the concurrency performance. They use single-writer shared variables, together with two multi-writer shared variables that are never concurrently written. The third algorithm has another desirable property, called smooth admission. By this property, during the period that the resource is occupied by the leader, a process wishing to join the same group as the leader's group can be granted use of the resource in constant time.

1   Introduction

Mutual exclusion is a problem of managing access to a single indivisible resource that can only support one user at a time. An early algorithm for the mutual exclusion problem was proposed by Dijkstra [4]. Since then it has been widely studied [1,2,3,7,11,13,15,16,17]. The k-exclusion problem is a natural generalization of the mutual exclusion problem. In k-exclusion, some number of users, specified by parameter k, are allowed to use the resource concurrently [3,5,14].

Group mutual exclusion is another natural generalization of the mutual exclusion problem. This problem was introduced by Joung in [8], and some algorithms for the problem have been proposed [6,8,9,10]. Group mutual exclusion is required in a situation where a resource can be shared by processes of the same group, but not by processes of different groups. A combination of k-exclusion and group mutual exclusion was also studied [18,19]. The algorithm by Joung in [8] uses multi-writer/multi-reader shared variables in the asynchronous shared memory model. Joung also gave solutions for the group mutual exclusion problem in the message passing model [9].

T. Warnow and B. Zhu (Eds.): COCOON 2003, LNCS 2697, pp. 232–241, 2003.
© Springer-Verlag Berlin Heidelberg 2003

As described in [8], group mutual exclusion can be described as the *congenial talking philosophers* problem. We assume that there are n philosophers. They spend their time thinking alone. When a philosopher is tired of thinking, he/she attempts to attend a forum and to talk at the forum. We assume that there is only one meeting room. A philosopher wishing to attend a forum can do so if the meeting room is empty, or if some philosophers interested in the same forum as the philosopher in question are already in the meeting room. The *congenial talking philosophers* problem is to design an algorithm such that a philosopher wishing to attend a forum will eventually succeed in doing so. Philosophers interested in the same forum as the current forum in the meeting room should be encouraged to attend it. This type of performance is measured as concurrency of attending a forum. It is undesirable that the maximum waiting time for a philosopher wishing to enter a forum is too long. If we request that the maximum waiting time for a philosopher should be as short as possible, then it is usually difficult to achieve a high degree of concurrency.

In this paper, we propose three algorithms based on ticket orders for the group mutual exclusion problem in the asynchronous shared memory model. The first algorithm is a simple modification of the Bakery algorithm, but its concurrency performance is poor. Furthermore, it does not satisfy lockout freedom although it satisfies progress for the trying region. The second and third algorithms are further modifications of the first algorithm in order to satisfy lockout freedom and to improve the concurrency performance. The third algorithm has another desirable feature, called smooth admission, i.e., while the resource is occupied by the leader of a group, a process wishing to join the same group can smoothly enter it in constant time.

2   Preliminaries

The computational model used in this paper is the asynchronous shared memory model. It is a collection of processes and shared variables. Processes take steps at arbitrary speeds, and there is no global clock. Interactions between a process and its corresponding philosopher are by input actions from the philosopher to the process and by output actions from the process to the philosopher. Each process is considered to be a state machine with arrows entering and leaving the process, representing its input and output actions. All communication among the processes is via shared memory. Lamport [12] defined three categories, *safe*, *regular*, and *atomic*, for shared variables according to possible assumptions about what can happen in the concurrent case of read operations and write operations. A shared variable is said to be regular if every read operation returns either the last value written to the shared variable before the start of the read operation or a value written by one of the overlapping write operations. A shared variable is said to be atomic if it is regular and has the additional property that read operations and write operations behave as if they occur in some total order. In this paper, we assume that all shared variables are atomic. The first algorithm in this paper uses only single-writer/multi-reader shared variables. The second and the third algorithms use single-writer/multi-reader shared variables, together with two multi-writer/multi-reader shared variables that are never concurrently written.

A philosopher with access to a forum is modeled as being in the talking region. When a philosopher is not involved in any forum, he/she is said to be in the thinking region. In order to gain admittance to the talking region, his/her corresponding process executes a trying protocol. The duration from the start of execution of the trying protocol to the entrance to the talking region is called the trying region. After the end of talking by a philosopher at a forum in the meeting room, his/her corresponding process executes an exit protocol. The duration of execution of the exit protocol is called the exit region. These regions are followed in cyclic order, from the thinking region to the trying region, to the talking region, to the exit region, and then back again to the thinking region. The congenial talking philosophers problem is to devise protocols for the philosophers to efficiently and fairly attend a forum when they wish to talk under the conditions that there is only one meeting room and that only a single forum can be held in the meeting room at the same time.

We assume n philosophers, $P_1, ..., P_n$, who spend their time either thinking alone or talking in a forum. We also assume that there are m different fora. Each philosopher P_i ($1 \leq i \leq n$) corresponds to process i. The inputs to process i from philosopher P_i are $try_i(f)$ which means a request by P_i for access to forum $f \in \{1, ..., m\}$ to talk there, and $exit_i$ which means an announcement of the end of talking by P_i. The outputs from process i to philosopher P_i are $talk_i$ which means granting attendance at the meeting room to P_i, and $think_i$ which means that P_i can continue with his/her thinking alone. These are external actions of the shared memory system. We assume that a philosopher in a forum spends an unpredictable but finite amount of time in the forum. The system to solve the congenial talking philosophers problem should satisfy the following conditions:

(1) **group mutual exclusion:** If some philosopher is in a forum, then no other philosopher can be in a different forum at the same time .

(2) **lockout freedom:** Any philosopher wishing to attend a forum eventually does so if any philosopher in any forum always leaves the forum after he/she spends a finite amount time in the forum.

(3) **progress for the exit region:** If a philosopher is in the exit region, then at some later point he/she enters the thinking region.

Waiting time and occupancy are important criteria to evaluate solutions to the congenial talking philosophers problem. Waiting time is the amount of time from when a philosopher wishes to attend a forum until he/she attends the forum. Concurrency is important to increase system performance concerning the resource. It is desirable for a solution to the congenial talking philosophers problem to satisfy the following property called *concurrent occupancy* [6,9,10].

(4) **concurrent occupancy:** If some philosopher P requests to attend a forum and no philosopher is currently attending or requesting a different forum, then P can smoothly attend the forum without waiting for other philosophers to leave the forum.

3 A Simple Modification of the Bakery Algorithm

The following procedure (n, m)-$SGBakery$ is a simple modification of the Bakery algorithm, where n is the number of philosophers, m is the number of different fora, and N is the set of natural numbers. If $n = m$ and each philosopher i is interested in just forum i ($1 \leq i \leq n$), then the procedure is the same as the Bakery algorithm for the mutual exclusion problem. Relation $(a, b) \leq (a', b')$ in the procedure means $a < a'$, or $a = a'$ and $b \leq b'$.

procedure (n,m)-$SGBakery$
shared variables
    for every $i \in \{1, ..., n\}$:
        $transit(i) \in \{0, 1\}$, initially 0, writable by process i and readable
            by all processes $j \neq i$;
        $ticket(i) \in N$, initially 0, writable by process i and readable by
            all processes $j \neq i$;
        $forum(i) \in \{0, 1, 2, ..., m\}$, initially 0, writable by process i and
            readable by all processes $j \neq i$;

process i
    input actions {inputs to process i from philosopher P_i}:
      $try_i(f)$ for every $1 \leq f \leq m$, $exit_i$;
    output actions {outputs from process i to philosopher P_i}:
      $talking_i$, $thinking_i$;

    ** thinking region **

01: $try_i(f)$:
02: $transit(i) := 1$;
03: $ticket(i) := 1 + max_{j \neq i} ticket(j)$;
04: $forum(i) := f$;
05: $transit(i) := 0$;
06: **for each** $j \neq i$ **do**
07:    **waitfor** $transit(j) = 0$;
08:    **waitfor** $([ticket(j) = 0]$
        or $[(ticket(i), f) \leq (ticket(j), forum(j))]$
        or $[ticket(i) \geq ticket(j)$ **and** $forum(j) = f])$;
09: $talking_i$;

    ** talking region **

10: $exit_i$:
11: $ticket(i) := 0$;
12: $forum(i) := 0$;
13: $thinking_i$;

Theorem 1. *Procedure (n, m)-$SGBakery$ guarantees group mutual exclusion, progress for the trying region, and progress for the exit region.*

Proof. For process i ($1 \leq i \leq n$), just after receiving an input signal $try_i(f)$ from P_i, it enters the entrance part called doorway (line 02 to line 05), where shared variable $ticket(i)$ is set to be larger by 1 than the maximum ticket number observed by process i at line 03 and shared variable $forum(i)$ is set to be f at line 04. From the condition in the **waitfor** statement at line 08, if process i is allowed to move to the talking region, any other process who wishes to attend a different forum must have a larger pair of (ticket, forum) than the pair held by process i. Hence, such a process wishing to attend a different forum cannot move to the talking region unless process i leaves the talking region and resets the contents of $ticket(i)$ and $forum(i)$ in the exit region. Therefore, group mutual exclusion is guaranteed. Since we assume that each philosopher spends only a finite amount of time in a forum and any ticket obtained in the doorway is always larger than tickets issued before, some philosopher wishing to attend a forum will eventually be able to do so. Hence, progress for the trying region is guaranteed. For each i $(1 \leq i \leq n)$, the exit region consists of two reset statements, $ticket(i) := 0$ and $forum(i) := 0$, and a sending output signal to P_i, $thinking_i$. Hence, progress for the exit region is also guaranteed. ☐

If many philosophers wish simultaneously to attend the same forum, high concurrency can be achieved. However, concurrency of (n, m)-$SGBakery$ is, in general, poor. Furthermore, it does not satisfy lockout freedom although it satisfies progress for the trying region and for the exit region. This fact can be easily shown by a scenario such that lockout freedom is not satisfied.

4   A Highly Concurrent Algorithm

A capturing technique is used to improve the concurrency performance of the algorithms for the group mutual exclusion problem in [8,9,10]. By this technique, a philosopher attending a forum sends a message to all philosophers interested in the same forum, asking them to join the forum. The algorithm proposed in this section, (n, m)-$HCGME$, uses a door in the doorway. A chair is chosen in each current forum. When the chair wishes to leave the forum, he/she closes the door to prevent other philosophers from entering the same forum. While the door is open, philosophers wishing to enter the same forum as the current forum held in the meeting room are allowed to enter it. In this way, concurrency can be improved. All shared variables, except for $door$ and $chair$, are single-writer/multi-reader shared variables. Both $door$ and $chair$ are multi-writer/multi-reader shared variables, but no concurrent writing operations take place for either of these two shared variables.

> **procedure** (n,m)-$HCGME$
> **shared variables**
>   for every $i \in \{1, ..., n\}$:
> $transit(i) \in \{0, 1\}$, initially 0, writable by process i and readable
>   by all processes $j \neq i$;
> $checkdw(i) \in \{0, 1\}$, initially 0, writable by process i and readable
>   by all processes $j \neq i$;
> $ticket(i) \in N$, initially 0, writable by process i and readable by
>   all processes $j \neq i$;

$forum(i) \in \{0, 1, ..., m\}$, initially 0, writable by process i and
   readable by all processes $j \neq i$;
$forum(0)$, always 0, readable by all processes;
$door \in \{open, close\}$, initially $open$, writable and readable by
   all processes (but never concurrently written);
$chair \in \{0, 1, ..., n\}$, initially 0, writable and readable by all processes
   (but never concurrently written);

process i
   input/output actions: the same as the input/output actions of
                (n, m)-$SGBakery$;

** thinking region **

01: $try_i(f)$:
02: $transit(i) := 1$;
03: **if** $door = close$ **then begin**
04:     $checkdw(i) := 1$;
05:     **waitfor** $door = open$;
06:     $checkdw(i) := 0$ **end**;
07: $ticket(i) := 1 + max_{j \neq i} ticket(j)$;
08: $forum(i) := f$;
09: $transit(i) := 0$;
10: **for** each $j \neq i$ **do begin**
11:     **waitfor** $transit(j) = 0$ **or** $checkdw(j) = 1$
           or $f = forum(chair)$;
12:     **waitfor** $ticket(j) = 0$
           or $(ticket(i), f, i) < (ticket(j), forum(j), j)$
           or $f = forum(chair)$ **end**;
13: **if for** each $j \neq i$, $ticket(j) = 0$
           or $(ticket(i), f, i) < (ticket(j), forum(j), j)$
           then $chair := i$;
14: $talking_i$;

** talking region **

15: $exit_i$:
16: **if** $chair = i$ **then begin**
17:     $door := close$;
18:     **for** each $j \neq i$ **do begin**
19:        **waitfor** $transit(j) = 0$ **or** $checkdw(j) = 1$;
20:        **waitfor** $f \neq forum(j)$ **or** $ticket(j) = 0$ **end**;
21:     $chair := 0$;
22:     $door := open$;
23:     **for** each $j \neq i$ **do**
24:        **waitfor** $checkdw(j) = 0$ **end**;
25: $forum(i) := 0$;
26: $ticket(i) := 0$;
27: $thinking_i$;

In the procedure above, the order among triples is also lexicographical. The process with the smallest triple comprising non-zero ticket number, forum, and process identifier is chosen as the chair of the forum held in the meeting room. If P_i in a forum is not the chair, P_i can smoothly leave the talking region by resetting $ticket(i)$ and $forum(i)$. However, when the chair wishes to leave the forum, he/she must close the door in the doorway and wait for all philosophers in the same forum to leave. After the chair observes that all philosophers in the forum have left, he/she resigns the chair and opens the door. Then after confirming that all philosophers waiting at line 05 have noticed that the door has opened, he/she who resigned the chair leaves the forum.

Theorem 2. *In any execution by (n,m)-HCGME, shared variable chair is never concurrently written, and once chair is set to be i ($1 \le i \le n$), the contents of chair remains i until it is reset to be 0 by process i at line 21 of the program. In any execution by (n,m)-HCGME, shared variable door is also never concurrently written.*

Proof. For each process i ($1 \le i \le n$) in the trying region, there are two ways of setting its ticket number. One way is that it observes *door* to be *open* at line 03 and then sets $ticket(i)$ at line 07. The other way is that it observes *door* to be *close* at line 03 and waits until *door* is opened at line 05. Once it observes *door* to be *open* at line 05, it sets $ticket(i)$ at line 07.

Process i at line 12 is waiting until the triple of its ticket, forum and identifier becomes the smallest among the triples of processes in the trying region or in the talking region. During the jth loop ($1 \le j \le n$) of the **waitfor** statement at line 12 of the execution by process i, if process i observes that the second condition (i.e., $checkdw(j) = 1$) is satisfied then $ticket(j)$ is 0 or greater than $ticket(i)$ at line 12 of the execution by process i, because the execution at line 07 by process j is after the contents of $ticket(i)$ was set by process i. The mechanism for choosing smallest triple among the triples of the processes including process i itself such that process i observes the first condition (i.e., $transit(j) = 0$) of the **waitfor** statement at line 12 is the same as the mechanism for choosing the smallest one by the Bakery algorithm for the mutual exclusion problem. Hence, *chair* is never concurrently written, and once *chair* is set to be the identifier of a process, it remains until it is reset by the process at line 21. Since *door* can be written by a chair process, it is also never concurrently written.   □

Theorem 3. (n,m)-HCGME *guarantees group mutual exclusion, lockout freedom, and progress for the exit region.*

Proof. The chair philosopher is uniquely determined by *chair*. While a philosopher is the chair, only philosophers wishing to attend the same forum as the chair's forum are allowed to enter the forum. When the chair philosopher wishes to leave the forum, he/she closes *door* in his/her exit region to prevent a newcomer to the trying region from getting a ticket. After observing that all other philosophers in the forum have left, the chair philosopher resigns the chair and opens *door*. Hence, group mutual exclusion is guaranteed. The time from when a philosopher enters his/her trying region until he/she enters the talking region is bounded by $2tc + O(nt)l$, where $t = min\{n, m\}$, c is an upper bound on the time that any philosopher spends in the talking region, and l is an upper bound on the time between two successive atomic steps by a process. Hence, lockout freedom is guaranteed. Progress for the exit region is obvious.   □

5   A Technique for Smooth Admission

Here, we show a technique to reduce the waiting time for processes wishing to attend the forum of the chair philosopher. A modified algorithm, (n, m)-$SAHCGME$, using this technique is given below. In an execution by (n, m)-$SAHCGME$, when the chair philosopher is chosen, the forum number is stored in *shared* variable *door*. This shared variable is never concurrently written. At the beginning of the trying region each philosopher checks whether his/her forum of interest is the same as the forum indicated in *door*. If a philosopher notices that he/she wishes to attend the same forum as the forum shown in *door*, the philosopher sets his/her ticket number as well as forum number to be the same as the ticket number and forum number of the chair philosopher, and then he/she is granted attendance to the forum. In this way, such philosophers can attend the forum in $O(1)$ atomic steps after the entrance to the trying region.

procedure (n,m)-$SAHCGME$
shared variables: the same as the shared variables of (n, m)-$HCGME$
                   except for *door*, the domain of *door* is
                   $\{close, 0, 1, ..., m\}$, and it is initially 0;
process i
   input/output actions: the same as the input/output actions of
                   (n, m)-$SGBakery$;

   ** thinking region **

```
01:   tryᵢ(f):
02:   transit(i) := 1;
03:   if f ≠ door then begin
04:       if door = close then begin
05:          checkdw(i) := 1;
06:          waitfor door ≠ close;
07:          checkdw(i) := 0 end;
08:       ticket(i) := 1 + maxⱼ≠ᵢticket(j);
09:       forum(i) := f;
10:       transit(i) := 0;
11:       for each j ≠ i do begin
12:          waitfor transit(j) = 0 or checkdw(j) = 1
                 or f = forum(chair);
13:          waitfor ticket(j) = 0
                 or (ticket(i), f, i) < (ticket(j), forum(j), j)
                 or f = forum(chair) end;
14:       if for each j ≠ i, ticket(j) = 0
                 or (ticket(i), f, i) < (ticket(j), forum(j), j)
```

```
15:        then begin
16:          chair := i;
17:          door := f end
18:      end else then begin
19:      ticket(i) := ticket(chair);
20:      forum(i) := f;
21:      transit(i) := 0 end;
22:  talking_i;
```

** talking region **

```
23:  exit_i:
24:  if chair = i then begin
25:      door := close;
26:      for each j ≠ i do begin
27:          waitfor transit(j) = 0 or checkdw(j) = 1;
28:          waitfor f ≠ forum(j) or ticket(j) = 0 end;
29:      chair := 0;
30:      door := 0;
31:      for each j ≠ i do
32:          waitfor checkdw(j) = 0 end;
33:  forum(i) := 0;
34:  ticket(i) := 0 ;
35:  thinking_i;
```

Except for the method of smooth admission for philosophers wishing to attend the current forum, in an execution by (n, m)-$SAHCGME$, the mechanism for satisfying group mutual exclusion and lockout freedom is essentially the same as the mechanism in an execution by (n, m)-$HCGME$. We therefore omit the proof of the correctness of the algorithm here.

Theorem 4. (n, m)-$SAHCGME$ *guarantees group mutual exclusion, lockout freedom, and progress for the exit region.*

6 Concluding Remarks

We have proposed two lockout-free algorithms, (n, m)-$HCGME$ and (n, m)-$SAHCGME$, for the group mutual exclusion problem on the asynchronous shared memory model. The algorithms given in [8,18,19] use multi-writer/multi-reader shared variables that may be concurrently written. The algorithms given in [6,10] use n^2 and n multi-writer/multi-reader shared variables, respectively that are never concurrently written. On the other hand, (n, m)-$HCGME$ and (n, m)-$SAHCGME$ use single-writer/multi-reader shared variables with only two multi-writer/multi-reader shared variables that are never concurrently written. The concurrency performances of these algorithms are superior to the algorithms in [6,10]. The ticket domain of each of our algorithms is unbounded, as in the Bakery algorithm. At present we do not know whether we can simply modify our algorithms by using similar techniques given in [1,7,17] so that the the ticket domain is bounded. This problem is worthy of further investigation.

References

1. U. Abraham, "Bakery algorithms", Technical Report, Dept. of Mathematics, Ben Gurion University, Beer-Sheva, Israel, 2001.
2. J. H. Anderson, "Lamport on mutual exclusion: 27 years of planting seeds", *Proceedings of the 27th Annual ACM Symposium on Principles of Distributed Computing*, Newport, Rhode Island, pp.3–12, 2001.
3. H. Attiya and J. Welch, "Distributed Computing: Fundamentals, Simulations and Advanced Topics", McGraw-Hill, New York, 1998.
4. E. W. Dijkstra, "Solution of a problem in concurrent programming control", *Communications of the ACM*, vol.8, p.569, 1965.
5. M. J. Fischer, N. A. Lynch, J. E. Burns, and A. Borodi, "Resource allocation with immunity to limited process failure", *20th Annual Symposium on Foundations of Computer Science*, San Juan, Puerto Rico: 234–254, 1979.
6. V. Hadzilacos, "A note on group mutual exclusion", *Proceedings of 12th Annual ACM Symposium on Principles of Distributed Computing*, Newport, Rhode Island, pp.100–106, 2001.
7. P. Jayanti, K. Tan, G. Friedland, and A. Katz, "Bounded Lamport's bakery algorithm", *Proceedings of SOFSEM'2001, Lecture Notes in Computer Science*, vol.2234, Springer-Verlag, Berlin, pp.261–270, 2001.
8. Yuh-Jzer Joung, "Asynchronous group mutual exclusion", *Distributed Computing*, vol.13, pp.189–206, 2000.
9. Yuh-Jzer Joung, "The congenial talking philosophers problem in computer networks", *Distributed Computing*, vol.15, pp.155–175, 2002.
10. P. Keane and M. Moir, "A simple local-spin group mutual exclusion algorithm", *IEEE Transactions on Parallel and Distributed Systems*, vol.12, 2001.
11. L. Lamport, "A new solution of Dijkstra's concurrent programming problem", *Communications of the ACM*, vol.17, pp.453–455, 1974.
12. L. Lamport, "The mutual exclusion problem. Part II : Statement and solutions", *J. of the ACM*, vol. 33, pp.327–348, 1986.
13. N. A. Lynch, "Distributed Algorithms", Morgan Kaufmann, San Francisco, California, 1996.
14. M. Omori, K. Obokata, K. Motegi and Y. Igarashi, "Analysis of some lockout avoidance algorithms for the k-exclusion problem", *Interdisciplinary Information Sciences*, vol.8, pp.187–192, 2002.
15. G. L. Peterson, "Myths about the mutual exclusion problem", *Information Processing Letters*, vol.12, pp.115–116, 1981.
16. G. L. Peterson and M. J. Fischer, "Economical solutions for the critical section problem in a distributed system", *Proceedings of the 9th Annual ACM Symposium on Theory of Computing*, Boulder, Colorado, pp.91–97, 1977.
17. M. Takamura and Y. Igarashi, "Simple mutual exclusion algorithms based on bounded tickets on the asynchronous shared memory model", *IEICE Transactions on Information and Systems*, vol.E86-D, pp.246–254, 2003.
18. K. Vidyasankar, "A highly concurrent group mutual l-exclusion algorithm", *Proceedings of the 12th Annual ACM Symposium on Principles of Distributed Computing*, Monterey, California, p.130, 2002.
19. K. Vidyasankar, "A simple group mutual l-exclusion algorithm", *Information Processing Letters*, vol.85, pp.79–85, 2003.

Distributed Algorithm for Better Approximation of the Maximum Matching*

A. Czygrinow[1] and M. Hańćkowiak[2]

[1] Department of Mathematics and Statistics
Arizona State University
Tempe, AZ 85287-1804, USA
andrzej@math.la.asu.edu

[2] Faculty of Mathematics and Computer Science
Adam Mickiewicz University, Poznań, Poland
mhanckow@main.amu.edu.pl

Abstract. Let G be a graph on n vertices that does not have odd cycles of lengths $3, \ldots, 2k - 1$. We present an efficient distributed algorithm that finds in $O(\log^D n)$ steps $(D = D(k))$ matching M, such that $|M| \geq (1 - \alpha)|M^{\blacksquare}|$, where $M^{\blacksquare}$ is a maximum matching in G, $\alpha = \frac{1}{k+1}$.

1 Introduction

We study the problem of finding an approximation to a maximum matching by a distributed algorithm. We consider the distributed network model introduced by Linial in [Li92]. In this model a distributed network is represented by an undirected graph. Vertices of the graph correspond to processors and edges to communication links between processors. The network is synchronized and in a single step each vertex can send messages to its neighbors, can receive the messages from its neighbors, and can perform some local computations. We make no explicit assumptions about the amount of local computations and the lengths of messages. The general problem in this setting is the following. Let G be a distributed network described above. Can vertices of G compute a global function of G in a relatively short time? In particular, the global function that we consider is a "large" matching, which we want to find in the poly-logarithmic (in $|V(G)|$) number of steps. Recent result of Hańćkowiak, Karoński, and Panconesi [HKP99] provides a breakthrough in this area. In [HKP99], authors presented an efficient distributed algorithm that finds a maximal matching. Note that, when M is a maximal matching, and M^* a maximum matching then $|M| \geq \frac{1}{2}|M^*|$. Using the techniques from [HKP99], Czygrinow, Hańćkowiak, and Szymańska [CHS02] designed an efficient distributed algorithm that finds a matching M such that $|M| \geq \frac{2}{3}|M^*|$. In this paper, we propose a distributed algorithm, that finds a better approximation of the maximum matching in case the graph does not have "short" odd cycles.

* Research supported by KBN grant no. 7 T11C 032 20

T. Warnow and B. Zhu (Eds.): COCOON 2003, LNCS 2697, pp. 242–251, 2003.
© Springer-Verlag Berlin Heidelberg 2003

Theorem 1. *Let k be an integer larger than one and let G be a graph on n vertices that does not have odd cycles of lengths $3, \ldots, 2k - 1$. Then there is a constant $D = D(k)$ and distributed algorithm, that finds in $\log^D n$ steps matching M such that*

$$|M| \geq \frac{k}{k+1}|M^*|,$$

where M^* is a maximum matching in G.

If M is a matching in graph G, then a path of length $2l-1$ is called M-augmenting if it has length $2l - 1$, edges alternate between M and $E(G) \setminus M$, and both endpoints are not M-saturated (see the next section for necessary definitions). Proof of Theorem 1 is based on the following fact (see [FGHP93]).

Theorem 2. *Let M^* be a maximum matching in graph G and let M be a matching such that there are no M-augmenting paths of lengths $1, 3, 5, \ldots, 2k-1$. Then*

$$|M| \geq \frac{k}{k+1}|M^*|.$$

In particular, the assumption that there are no M-augmenting paths of length one means that M is maximal. The algorithm from Theorem 1 proceeds as follows. Fist, we find a maximal matching M using the algorithm from [HKP99]. Then we iterate with $l = 2, \ldots, k$ and "improve" matching M by finding a maximal set of M-augmenting paths of lengths $2l - 1$. Once the maximal set of paths is found, the edges in paths that belong to M and those that are not in M are exchanged. It is easy to see that after this exchange "new" matching does not have augmenting paths of lengths at most $2l - 1$. Consequently its size is at least $\frac{l}{l+1}|M^*|$. Finding the maximal set of M-augmenting paths is however not trivial and in fact it occupies the central part of the paper. The set is found by considering an auxiliary, virtual, graph constructed from G and matching M. It turns out that using, so-called spanners, a maximal set of M-augmenting paths can be found in this virtual graph (procedure MAXIMAL-PATHS-IN-LAYERED-GRAPH). Once found, it is translated to a maximal set of M-augmenting paths in the original graph G. To be able to claim that paths in the virtual graph correspond to paths (not cycles) in the original graph G the assumption about the cycle-structure of G is used. Finally, note that although there is a rich literature on distributed algorithms for matching problems most of the papers are focused on a different computational model than the one considered in this paper.

The rest of the paper is structured as follows. In the next section, we introduce necessary notation and auxiliary facts. Section 3 contains procedure that finds a maximal set of augmenting paths in the virtual graph. In Section 4, we describe the translation method and the main algorithm.

2 Definitions and Notation

In this section, we introduce necessary definitions and notation. Let M be a matching in graph G. We say that a vertex v is M-saturated if v is an endpoint of some edge from M. An edge $e = \{u, v\}$ is M-saturated if either u or v is M-saturated.

Let P be a path in G and let M_1 and M_2 be two maximal matchings in P that partition the edges of P, i.e. $E(P) = M_1 \cup M_2$. We say that P is M-alternating if either $M_1 \subset M$ or $M_2 \subset M$. Path P of length $2k - 1$, $k \geq 1$, augments M if P is M-alternating and both ends of P are not M-saturated. These M-alternating paths that augment M will play a crucial role in our approach.

Definition 1. *Let M be a matching and let k be a positive integer. A path is called an $(M, 2k - 1)$-path if it augments M and has length $2k - 1$.*

Paths are called *disjoint* if they are pairwise vertex-disjoint. Next we define the notion of a substantial matching and a substantial set of paths.

Definition 2. *A matching M is γ-substantial in graph $G = (V, E)$ if the number of M-saturated edges in G is at least $\gamma|E|$.*

Using an algorithm from [HKP99] we can find a γ-substantial matching for any constant γ.

Lemma 1. *Let n be the number of vertices of a graph. For any $\gamma > 0$ there is a distributed algorithm that finds in $O(\log^3 n)$ steps a γ-substantial matching.*

The notion of a substantial matching extends in a natural way to a substantial set of paths.

Definition 3. *For matching M and positive integer k let $\mathcal{M}$ be the set of all $(M, 2k - 1)$-paths in graph G. A set $\mathcal{P}$ of $(M, 2k - 1)$-paths is γ-path-substantial if the number of $(M, 2k - 1)$-paths that have at least one common vertex with some path from $\mathcal{P}$ is at least $\gamma|\mathcal{M}|$.*

Next, we define the **modification** of G with respect to a set of $(M, 2k-1)$-paths $\mathcal{P}$. The modification is composed of 2 steps:
A) remove all edges of G that are incident to some path from $\mathcal{P}$,
B) remove all edges that do not belong to some augmenting paths of length $2k - 1$; ("new" edges of this sort can appear after step A).
Our algorithm uses the so called spanners in bipartite graphs. A bipartite graph $H = (A, B, E)$ is called a *D-block* if for every vertex $a \in A$,

$$\frac{D}{2} < deg_H(a) \leq D.$$

A key ingredient used in rendering a "substantial" set of disjoint paths will be a spanner.

Definition 4. *An (α, β)-spanner of a D-block $H = (A, B, E)$ (from A to B) is a subgraph $S = (A', B, E')$ of H such that the following conditions are satisfied.*

1. *$|A'| \geq \alpha|A|$.*
2. *For every vertex $a \in A'$, $deg_S(a) = 1$.*
3. *For every vertex $b \in B$, $deg_S(b) < \frac{\beta}{D} deg_H(b) + 1$.*

In other words, a spanner is a collection of stars such that degrees of centers of stars are bounded. Note that spanners played an important role in designing an algorithm for finding a maximal matching in [HKP99]. In particular the following fact is proved in [HKP99].

Lemma 2. *Let $H = (A, B, E)$ be a D-block and let $n = |A| + |B|$. There is a distributed algorithm that finds in $O(\log^3 n)$ steps a $(\frac{1}{2}, 16)$-spanner.*

One of the main parts of the algorithm will be a procedure that finds a set of disjoint paths in an auxiliary graph which has a layered structure. The number of layers in our auxiliary graphs will always be even and will always be a constant independent of n.

Definition 5. *Graph G is called a layered graph with $2L$-layers if the vertex set of G can be partitioned into sets $X_1, \ldots, X_{2L}$ so that*

- $\forall i = 1..2L : X_i$ *is an independent set.*
- $\forall i = 1..2L - 1 :$ *vertices from X_i are connected only with vertices from X_{i+1}.*
- $\forall i = 1..L - 1 : |X_{2i}| = |X_{2i+1}|$ *and the edges between X_{2i} and X_{2i+1} form a perfect matching.*

We will often write $G = (X_1, X_2, \ldots, X_{2L})$ to denote a $2L$-layered graph with layers $X_1, \ldots, X_{2L}$. Let G be as above, for $i = 1, \ldots, 2L - 1$, we denote by $G_i = G[X_i, X_{i+1}]$ the bipartite graph induced by X_i, X_{i+1}. In particular, for even i, G_i is a perfect matching between X_i and X_{i+1}. In addition, let G_{left} be the graph induced by $(X_1, \ldots, X_{2L-2})$ and, to make the notation uniform, let $G_{right} = G_{2L-1}$.

Definition 6. *Let $G = (X_1, X_2, \ldots, X_{2L})$ be a layered graph and let $M = \bigcup_{i=1}^{L-1} E(G_{2i})$ be the matching in G. Then G is called a (D_1, D_2)-block if*

- $\forall v \in X_{2L-1} : d_{G_{2L-1}}(v) \in [\frac{D_2}{2}, D_2]$.
- $\forall v \in X_{2L-2} :$ *the number of $(M, 2L - 3)$-paths that have one endpoint in X_1 another in v is larger than $\frac{D_1}{2}$ and is at most D_1.*

3   Maximal Set of Paths in a Layered Graph

In this section, we present an algorithm that finds a maximal set of augmenting paths in a layered graph. Layered graph will appear in a natural way as a virtual (auxiliary) graph in the course of the algorithm. The algorithm for finding a maximal set of augmenting paths consists of two main procedures: SUBSTANTIAL-PATHS-IN-BLOCK and MAXIMAL-PATHS-IN-LAYERED-GRAPH. The first one finds a set of disjoint augmenting paths in a (D_1, D_2)-block. The procedure is recursive and in the recursive step it invokes MAXIMAL-PATHS-IN-LAYERED-GRAPH in a layered graph with a smaller number of layers. Procedure MAXIMAL-PATHS-IN-LAYERED-GRAPH iterates over all (D_1, D_2)-blocks for $D_1 = D_1(i) = \frac{n^{L-1}}{2^i}$, $D_2 = D_2(j) = \frac{n}{2^j}$. In each block it invokes SUBSTANTIAL-PATHS-IN-BLOCK $O(\log n)$ times to find a maximal set of paths in the block.

Since every augmenting path in the layered graph is a path of some block (and the blocks are disjoint) the union of the sets of paths will be maximal in the layered graph G. Let $G = (X_1, \ldots, X_{2L})$ be a (D_1, D_2)-block. Recall that G_i denotes the bipartite graph $G[X_i, X_{i+1}]$ and that G_i is in fact a perfect matching between X_i and X_{i+1} when i is even. To find a substantial set of disjoint paths in G, SUBSTANTIAL-PATHS-IN-BLOCK finds a $(\frac{1}{2}, 16)$-spanner in $G_{right} = G_{2L-1}$ with centers of stars in X_{2L}. Each star $(z, y_1, \ldots y_k)$, with $z \in X_{2L}$ and $y_j \in X_{2L-1}$ is extended by considering the vertices $x_j \in X_{2L-2}$ that are matched with y_j by edges of matching G_{2L-2}. In the next step of the procedure new layered graph G'_{left} is constructed (in fact as explained later, G'_{left} is a multi-graph) with $2L-2$ layers $(X_1, X_2, \ldots, X_{2L-3}, X')$ where X' contains so-called "super-vertices" that correspond to stars of the spanner. Then procedure calls MAXIMAL-PATHS-IN-LAYERED-GRAPH with G'_{left} as an input graph, and the paths obtained in G'_{left} are extended using the stars of the spanner. The resulting set of paths is substantial in graph G. More precisely we can describe the above procedure as follows.

PROCEDURE SUBSTANTIAL-PATHS-IN-BLOCK
Input: a (D_1, D_2)-block $G = (X_1, \ldots X_{2L})$ with $2L$ layers, $L > 1$.
Output: a substantial set of disjoint paths connecting X_1 and X_{2L} in block G.

1. Find a $(\frac{1}{2}, 16)$-spanner S in G_{right}.
2. Construct an auxiliary $(2L - 2)$-layered multi-graph $G'_{left} = (X_1, \ldots, X_{2L-3}, X')$ as follows.
   - For every star in the spanner S let $x(1), \ldots, x(l)$ be the vertices in X_{2L-1} that have degree one in S and let $x'(1), \ldots, x'(l)$ be the vertices in X_{2L-2} that are matched with $x(1), \ldots, x(l)$ by the matching M between X_{2L-2} and X_{2L-1}. Create a *super-vertex* $s = s(x'(1), \ldots, x'(l)) = \{x'(1), \ldots, x'(l)\}$. The vertex set X' contains all super-vertices.
   - For every vertex $x \in X_{2L-3}$ and every $x'(j) \in s(x'(1), \ldots, x'(l))$ put an edge between x and the super-vertex $s = s(x'(1), \ldots, x'(l))$ in G'_{left} if x and $x'(j)$ are connected in G_{2L-3}.
   - Every edge of G that is present in G_i for some $i \in \{1, \ldots, 2L-4\}$ is also present in G'_{left}
3. Consider a maximal simple subgraph of G'_{left} (i.e. discard parallel edges).
   - If $L > 2$ then invoke MAXIMAL-PATHS-IN-LAYERED-GRAPH to find a maximal set of $(M, 2L - 3)$-paths in the subgraph of G'_{left}.
   - If $L = 2$ then find a maximal matching (maximal set of $(M, 1)$-paths) in the subgraph using the procedure from [HKP99].
4. Extend paths found in (3) to $(M, 2L - 1)$-paths in the block G using the matching edges of G_{2L-2} and spanner S.

There is a small technical point that must be addressed at this moment. In the third step of the algorithm, parallel edges are discarded and then a maximal set of paths is found. Note that parallel edges only add the repetitions of $(M, 2L-3)$-paths. Consequently the set of paths found in step 3 is also maximal in G'_{left}.

Next we shall show that the set of paths obtained from SUBSTANTIAL-PATHS-IN-BLOCK is substantial in G. Let $\mathcal{P}$ be the set of paths found by the procedure, $\mathcal{P}_{left}$ be the set of paths in G_{left} obtained from $\mathcal{P}$ by restricting to G_{left}, and $\mathcal{P}_{right}$ set of paths (of length one) in G_{right} obtained from $\mathcal{P}$ by restricting to G_{right}.

Lemma 3. *For every* $0 < \kappa < \frac{1}{4}$ *either* $\mathcal{P}_{left}$ *is* $(1-4\kappa)/4$*-path-substantial in* G_{left} *or* $\mathcal{P}_{right}$ *is* $\kappa/16$*-substantial in* G_{right}.

Proof of Lemma 3 is rather long and it mimics the proof of a similar fact from [CHS02]. It is an easy consequence of Lemma 3 that the set of paths found by SUBSTANTIAL-PATHS-IN-BLOCK is substantial in G.

Lemma 4. *If* $\mathcal{P}_{left}$ *is* γ*-path-substantial in* G_{left} *or* $\mathcal{P}_{right}$ *is* γ*-substantial in* G_{right} *then* $\mathcal{P}$ *is* $\gamma/2$*-path-substantial in* G.

We omit a routine proof.

PROCEDURE MAXIMAL-PATHS-IN-LAYERED-GRAPH
Input: a layered graph G with $2L$-layers $(X_1, \ldots X_{2L})$.
Output: a maximal set of disjoint augmenting paths connecting X_1 and X_{2L} in G.

1. for $i := 1, \ldots, (L-1)\lceil \log n \rceil$ do:
   for $j := 1, \ldots, \lceil \log n \rceil$ do:
   a) $D_1 := \frac{n^{L-1}}{2^i}$; $D_2 := \frac{n}{2^j}$;
   b) iterate $O(\log n)$-times:
      – invoke SUBSTANTIAL-PATHS-IN-BLOCK(D_1, D_2) in order to find a substantial set of paths in the block and modify the graph G with respect to that set.

Lemma 5. *Procedure* MAXIMAL-PATHS-IN-LAYERED-GRAPH *finds a maximal set of disjoint paths in* G *in* $\log^{O(L)} n$ *steps*.

Proof. First note that due to the modification done in each step of the procedure the resulting paths are disjoint. We shall show that the set of paths is also maximal. First, fix a (D_1, D_2)-block. By Lemma 4, SUBSTANTIAL-PATHS-IN-BLOCK(D_1, D_2) finds, for some fixed constant $\alpha > 0$, an α-path-substantial set of disjoint paths in a (D_1, D_2)-block. Thus in the modification step, at least α fraction of all paths in the block will be deleted from it. However the total number of the paths in whole G is at most $n^{O(L)}$ and so after $O(\log n)$ iterations all the paths will be deleted from the block. In other words, set of paths obtained after iterations in (b) will be maximal in the block. Note that there are $O(\log^2 n)$ disjoint blocks , defined by parameters D_1, D_2 in (a), and each path of G belongs to exactly one of them. Thus after iterations in step 1 the set found by the procedure will be maximal in every block and consequently maximal in G. Finally let us estimate the running time of the procedure. First note that the order of the virtual graph is $O(n)$ and each step in the virtual graph can be simulated in graph G in

$O(1)$ steps. For $i > 1$, let R_i be the running time of the procedure invoked in the layered graph with $2i$ blocks and let $R_1 = O(\log^4 n)$ be the time needed to find the maximal matching using the algorithm from [HKP99]. To estimate R_i notice that there are $O(\log^2 n)$ iterations over all possible blocks and in each block $O(\log n)$ iterations are needed. In each iteration in a block SUBSTANTIAL-PATHS-IN-BLOCK(D_1, D_2) is invoked. This procedure finds a spanner in $O(\log^3 n)$ steps (Lemma 2) and then invokes MAXIMAL-PATHS-IN-LAYERED-GRAPH in a layered graph with $2i-2$ layers (or finds a maximal matching in the bipartite graph when $i = 2$). Thus, $R_i = O(\log^3 n(\log^3 n + R_{i-1}))$. Consequently, $R_L = \log^{O(L)} n$.

4   Graphs without Odd Cycles of Length Less than or Equal to c

In this section, we shall present our main procedure for computing a matching M from Theorem 1. As explained in the introduction our approach is based on finding M-augmenting paths. Finding a substantial set of disjoint M-augmenting paths in the input graph G seems to be difficult. However, as we saw in the last section, if a graph has a layered structure, augmenting paths can be found by MAXIMAL-PATHS-IN-LAYERED-GRAPH. The main problem that we must address is how to reduce graph G to a layered graph (Procedure REDUCE) and how to translate disjoint paths in the layered graph to the original graph G preserving the property that the set of paths is substantial (Procedure TRANSLATE). Next procedure constructs a virtual, layered graph from G.

PROCEDURE REDUCE
Input: a graph G, a matching M in G, a number L.
Output: a layered graph $Lay(G) = (X_1, \ldots, X_{2L})$.

1. Lets denote by N vertices of G not saturated by M.
   $X_1 := N$; $X_i := \bigcup_{e \in M} e$, for $i = 2, \ldots, 2L - 1$; $X_{2L} := N$.
2. For $k = 1, \ldots, 2L - 1$ define $G_k = Lay(G)[X_k, X_{k+1}]$ as follows.
   For $i = 1, \ldots, L$: $e \in G_{2i-1}$ if and only if there exist an augmenting path in G that has an edge e on the $2i - 1$ position (counting from whatever end).
   For $i = 1, \ldots, L - 1$: $e \in G_{2i}$ if and only if $e \in M$.

Notice, that layers $X_2, \ldots X_{2L-1}$ in $Lay(G)$ contain all vertices that are contained in edges of M. Thus every edge $\{a_1, a_2\} \in M$ appears twice in G_{2i} (see Figure 1). One important property of $Lay(G)$ is that every $(M, 2L-1)$-path in G is present (has a corresponding $(M, 2L-1)$-path) in $Lay(G)$. Using assumptions about G, we can show that the opposite direction is also true. Namely that every $(M, 2L - 1)$-path in $Lay(G)$ corresponds to a $(M, 2L - 1)$-path in G.

Lemma 6. *Let G be a graph without odd cycles of lengths $3, 5, \ldots, 2L-1$ and let M be such that there are no augmenting paths in G of lengths $1, 3, 5, \ldots, 2L - 3$. Then every augmenting path connecting X_1 with X_{2L} in the layered graph corresponds to an augmenting path in graph G of length $2L - 1$.*

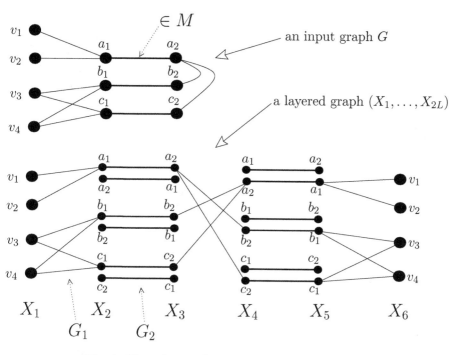

Fig. 1. The reduction (augmenting paths of length 5).

We omit the proof. Note that assumptions in Lemma 6 will be satisfied during the execution of our algorithm. First, G does not have odd cycles of lengths at most $2L - 1$ by the assumption. Second, G will not have shorter augmenting paths as we will iterate with $k = 3, 5, \ldots, 2L - 1$ and in the kth iteration, we shall eliminate all (M, k)-augmenting paths. Finally, the initial matching M will be maximal and so no $(M, 1)$-augmenting paths will be present. Let $\mathcal{P}$ be a set of disjoint $(M, 2L - 1)$-paths in $Lay(G)$. By Lemma 6 $\mathcal{P}$ is also a set of $(M, 2L - 1)$-paths in G. However the paths do not need to be disjoint in G. Define the **path graph** $G_{\mathcal{P}}$ by $V(G_{\mathcal{P}}) = \mathcal{P}$ and put an edge between two vertices of $G_{\mathcal{P}}$ when the corresponding paths intersect in G. In addition, we define the identifiers of vertices of $V(G_{\mathcal{P}})$. Let $Id(v)$ denote the identifier of vertex v in G. If $p = v_1, v_2, \ldots, v_{2L}$ is a path in G which is a vertex in $V(G_{\mathcal{P}})$ and if $Id(v_1) < Id(v_{2L})$ then the identifier of p in $G_{\mathcal{P}}$ is the vector $[Id(v_1), Id(v_2), \ldots, Id(v_{2L})]$. Although paths from $\mathcal{P}$ don't need to be disjoint in G the maximum degree of $G_{\mathcal{P}}$ will be a constant that depends only on L.

Lemma 7. *Let $\mathcal{P}$ be a set of paths obtained by* MAXIMAL-PATHS-IN-LAYERED-GRAPH. *The maximum degree of $G_{\mathcal{P}}$ is a constant independent of n.*

Before we describe the procedure that translates disjoint paths in $Lay(G)$ to disjoint paths in G we need to introduce the notion of a heavy Maximal Independent Set (MIS). Let $H = (W, F)$ be a graph with weights on vertices, i.e.

$w \in W$ has a weight $weight(w)$. The weight of a set of vertices A is then defined as $weight(A) := \sum_{w \in A} weight(w)$. Similarly the weight of a subgraph Q is defined as $weight(Q) := weight(V(Q))$. The α-**heavy MIS** in H is a maximal independent set A of vertices such that $weight(A) \geq \alpha \, weight(H)$. We note here, that for a constant d a $\frac{1}{2(d+1)^2}$-heavy MIS can be found by a distributed algorithm in a graph G with $\Delta(G) = d$, in time $O(\log^2 n)$. Details of this procedure will appear in the journal version of the paper.

Now we can describe procedure TRANSLATE:

PROCEDURE TRANSLATE
Input: a set $\mathcal{P}$ of disjoint, augmenting paths connecting sets X_1 and X_{2L} in the layered graph; $\mathcal{P}$ is maximal in the layered graph.
Output: a set $\mathcal{P}'$ of disjoint, augmenting paths of lengths $2L - 1$ in the input graph G; $\mathcal{P}'$ is path-substantial in G.

1. For each vertex v of the path graph $G_\mathcal{P}$ assign the value $weight(v)$ equal to the number of augmenting paths of length $2L - 1$ in G that are incident to v (or in other words "touch" v).
2. Compute a heavy MIS, X in $G_\mathcal{P}$.
3. The set $\mathcal{P}'$ contains paths that correspond to vertices of X.

Theorem 3. *Let G be as in Lemma 6 and let $Lay(G)$ be the layered graph with $2L$ layers obtained from G by procedure* REDUCE. *Let $\mathcal{P}$ be the set of paths in $Lay(G)$ obtained by* MAXIMAL-PATHS-IN-LAYERED-GRAPH. *Then the set $\mathcal{P}'$ obtained from $\mathcal{P}$ by* TRANSLATE *is a β-path-substantial set of disjoint paths in G for certain constant β. In addition,* TRANSLATE *runs $O(L \log^2 n)$ steps.*

We omit the proof. Next, we describe the procedure MAXIMAL-PATHS that finds a maximal, set of $(M, shortest)$-paths in G for an odd, positive integer, $shortest$. Note that G and M are as in Lemma 6 with $shortest = 2L - 1$. In the final procedure MAIN we iterate with $shortest = 3, 5, \ldots, c$ to obtain a maximal set of (M, c)-paths in G.

PROCEDURE MAXIMAL-PATHS
Input: an input graph G; a matching M; a number $shortest$.
Output: a maximal, set of $(M, shortest)$-paths in G.

- Let $\mathcal{P} := \emptyset$.
- Repeat $O(\log n)$ times:
  1. Call procedure REDUCE to build a layered graph with $2L$-layers $(X_1, \ldots, X_{2L})$, where $L := (shortest + 1)/2$.
  2. Call MAXIMAL-PATHS-IN-LAYERED-GRAPH on layered graph to find maximal set of disjoint, augmenting paths connecting sets X_1 and X_{2L}.
  3. Call TRANSLATE to get a path-substantial set $\mathcal{P}'$ of disjoint, augmenting paths in G.
  4. Modify the current graph with respect to $\mathcal{P}'$. Let $\mathcal{P} := \mathcal{P} \cup \mathcal{P}'$
- The set $\mathcal{P}$ is the result of that procedure.

Theorem 4. *Let G be a graph without odd cycles of lengths $3, 5, \ldots,$ shortest and let M be such that there are no M-augmenting paths in G of lengths at most shortest $- 2$. Then procedure* MAXIMAL-PATHS *finds a maximal set of disjoint $(M, shortest)$-paths in G. In addition,* MAXIMAL-PATHS *runs in $\log^{O(shortest)} n$ steps.*

PROCEDURE MAIN

Input: an odd number $c \geq 3$; graph G without odd cycles of length less than or equal to c.
Output: a matching in G such that there is no augmenting path of length less than or equal to c.

1. Compute a maximal matching in G using the procedure from [HKP99].
2. For *shortest* $:= 3$ to c step 2 do:
   a) Call MAXIMAL-PATHS to find a maximal set of disjoint, augmenting paths of length *shortest*.
   b) Augment all augmenting paths.

Note that during the execution of the algorithm we try to "improve" the matching computed in step 1: some of the edges of the matching will be deleted, and some will be added to it as we augment paths in 2(b). We can summarize the discussion in the following theorem.

Theorem 5. *Let c be an integer larger than two and let G be a graph on n vertices that does not have odd cycles of lengths less than or equal to c. Then procedure* MAIN *finds a matching M in G such that there are no M-augmenting paths in G of lengths less than or equal to c. In addition,* MAIN *runs in $\log^{O(c)} n$ steps.*

Proof of Theorem 1. By Theorem 5, PROCEDURE MAIN finds a matching M so that there are no M-augmenting paths of lengths $1, 3, 5, \ldots, c$, where $c = 2k - 1$. Thus, by Theorem 2, $|M| \geq \frac{k}{k+1} |M^*|$.

References

[CHS02]   A. Czygrinow, M. Hańćkowiak, E. Szymańska, *Distributed algorithm for approximating the maximum matching*, submitted, 2002.

[FGHP93] T. Fischer, A. V. Goldberg, D. J. Haglin, S. Plotkin, *Approximating matchings in parallel*, Information Processing Letters, 1993, 46, pp. 115–118.

[HKP99]   M. Hańćkowiak, M. Karoński, A. Panconesi, *A faster distributed algorithm for computing maximal matching deterministically*, Proceedings of PODC 99, the Eighteen Annual ACM SIGACT-SIGOPS Symposium on Principles of Distributed Computing, pp. 219–228.

[Li92]   N. Linial, *Locality in distributed graph algorithms*, SIAM Journal on Computing, 1992, 21(1), pp. 193–201.

Efficient Mappings for Parity-Declustered Data Layouts

Eric J. Schwabe* and Ian M. Sutherland

DePaul University
eschwabe@cs.depaul.edu, sans_peur2001@yahoo.com

Abstract. The joint demands of high performance and fault tolerance in a large array of disks can be satisfied by a parity-declustered data layout. Such a data layout is generated by partitioning the data on the disks into stripes and choosing a part of each stripe to hold redundant information. Thus the data layout can be represented as a table of stripes. The data mapping problem is the problem of translating a data address into a disk identifier and an offset on that disk. Recent work has yielded mappings that compute disks and offsets directly from data addresses without the need to store tables. In this paper, we show that parity-declustered data layouts based on commutative rings yield mappings with improved computational efficiency and wider applicability.

1   Introduction

1.1   Data Layouts for Disk Arrays

Disk arrays provide increased I/O throughput for large data sets by distributing the data over a collection of smaller disks (instead of a single larger disk) and allowing parallel access [7]. Since each disk in the array may fail independently with some probability per unit time, the probability that some disk in a large array will fail in unit time is greatly increased. Thus, the ability to reconstruct the contents of a failed disk is important to the feasibility of large disk arrays.

One technique to achieve fault tolerance in an array of v disks is called RAID5 (thus named by Patterson, Gibson, and Katz [7]). This technique is illustrated in Figure 1. Each disk is divided into *units*, and in each row, one of the units holds the bitwise exclusive "or" (i.e., parity) of the remaining $v - 1$ units. This allows the disk array to recover from a single disk failure, as the contents of each unit on the failed disk can be reconstructed by taking the bitwise exclusive "or" of the $v - 1$ surviving units from that row. Thus, by dedicating $1/v$ of the total space in the array to redundant information, the array can recover from any single disk failure by reading the entire contents of each of the surviving disks.

In general, we can achieve fault tolerance by constructing a *data layout* — an arrangement of data and redundant information that allows the array to reconstruct the contents of one or more failed disks. A data layout is created by

---

* Supported in part by NSF Grant CCR-9996375.

T. Warnow and B. Zhu (Eds.): COCOON 2003, LNCS 2697, pp. 252–261, 2003.
© Springer-Verlag Berlin Heidelberg 2003

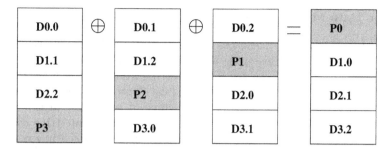

Fig. 1. RAID5 on a four-disk array. In each row **i**, the parity unit **Pi** is the bitwise exclusive "or" of the three data units **Di.0**, **Di.1**, and **Di.2**.

partitioning the units in the array into a collection of non-overlapping *stripes*. (In the RAID5 example, the stripes used are precisely the rows.) The number of units in each stripe is called the *stripe size*. Some subset of the units in each stripe will hold users' data; however, one or more units per stripe will instead hold redundant information computed from the data stored in the other units of the stripe. (In the RAID5 example, the stripe size is v, and one unit per stripe stores the parity of the remaining units.) This redundant information stored for each stripe enables the array to recover from disk failures.

If an array must remain available during the reconstruction of lost data, or must be taken off-line for as little time as possible for failure recovery, we may wish to reduce the time spent on failure recovery at the cost of dedicating more space to redundant information. This tradeoff of additional redundant space for reduced recovery time can be achieved using a technique called *parity declustering*, in which the stripe size k is smaller than the array size v. Parity-declustered data layouts have been considered by, among others, Holland and Gibson [4], Muntz and Lui [6], Schwabe and Sutherland [8], Stockmeyer [9], and Alvarez, Burkhard, and Cristian [1].

Many constructions of parity-declustered layouts use *balanced incomplete block designs* (BIBDs). A BIBD is a collection of b subsets (called *tuples*) of k elements, each drawn from a set of v elements, that satisfies the following two properties (see, e.g., Hanani [3]): First, each element appears the same number of times (called r) among the b tuples. Second, each pair of elements appears the same number of times (called λ) among the b tuples. (In fact, as long as $k \geq 2$, the second property implies the first, since $r = \lambda \cdot \frac{v-1}{k-1}$.)

In order to construct a data layout from a BIBD, we consider the v elements to be the disks in the array. Each tuple in the BIBD corresponds to a stripe containing one unit from each of the disks that appear in that tuple. Therefore each stripe will contain units from exactly k disks, and each disk will contain exactly r units. (We call r the *size* of the layout.) For each pair of disks, there are exactly λ stripes that contain a unit from both disks. If each stripe contains $k - 1$ units of data and one of redundant information, then when one disk fails, exactly λ units from each of the remaining disks (i.e., a $\frac{k-1}{v-1}$ fraction of their contents) will have to be read to reconstruct the lost data.

1.2   The Data Mapping Problem

A disk array appears to its clients as a single logical disk with a linear address space. Data addresses in this space are mapped to disks and offsets on those disks. One way to do this is to use a table derived from a BIBD. The tuples of the BIBD make up the rows of the table, and each entry in a row is an element of that tuple. In this table, each row will represent one stripe in the layout, and each entry in a row will represent a disk from which that stripe contains a unit. Addresses from the linear address space can be assigned in row-major order to the entries of the table, ignoring the last entry in each row, which is a redundancy unit. (Thus $k - 1$ data units are assigned to each row in the table.) This associates a disk identifier with each address. The offset for an address is the number of times the corresponding disk identifier appears in rows above the row where that address appears. This mapping is due to Holland and Gibson [4], and is illustrated in Figure 2 for the complete block design with $v = 5$ and $k = 4$. To compute a disk and offset from a given address, we first map it to a row and column in the table, setting $row = address / (k - 1)$ and $column = address \bmod k - 1$. Next, we use the contents of the table to determine the disk and offset where that address is located.

The disk number can be obtained with a single lookup in the table of stripes, since it is the value stored in the computed row and column. The offset is a bit more difficult to compute, as it depends on the number of occurrences of the discovered disk number that appear in rows above the computed row. Offsets can be precomputed while the table is being constructed (requiring additional work proportional to the size of the table), and if this is done then the offset can also be determined with a single table lookup. However, the resulting table could be quite large. For instance, in the case of a data layout derived from a *complete* block design, which consists of all subsets of size k of the set of v disks, the table will have $\binom{v}{k}$ rows and k columns.

1.3   Our Results

This paper considers ways to reduce this space requirement by using data layouts that do not require the storage of tables of stripes. Alvarez, Burkhard, and Cristian [1] proposed the DATUM layouts for this purpose, but did not consider the computational complexity of their data mappings nor the usability of the layouts for large arrays. We review these layouts in Section 2.

We present an alternative in Section 3: ring-based data layouts. Both DATUM layouts and ring-based layouts take advantage of their mathematical structure to compute disks and offsets directly. We analyze the computational complexity of the data mappings of ring-based layouts as well as those of DATUM, and show in Section 4 that ring-based layouts have smaller time complexity than DATUM layouts ($O(k \log v)$ versus $O(kv)$). Ring-based layouts are also applicable to a wider range of array configurations than DATUM layouts.

A:	1 (0) $^{(0)}$	2 (0) $^{(1)}$	3 (0) $^{(2)}$	0 (0)
B:	0 (1) $^{(3)}$	2 (1) $^{(4)}$	4 (0) $^{(5)}$	1 (1)
C:	0 (2) $^{(6)}$	1 (2) $^{(7)}$	3 (1) $^{(8)}$	4 (1)
D:	0 (3) $^{(9)}$	3 (2) $^{(10)}$	4 (2) $^{(11)}$	2 (2)
E:	1 (3)$^{(12)}$	2 (3) $^{(13)}$	4 (3) $^{(14)}$	3 (3)

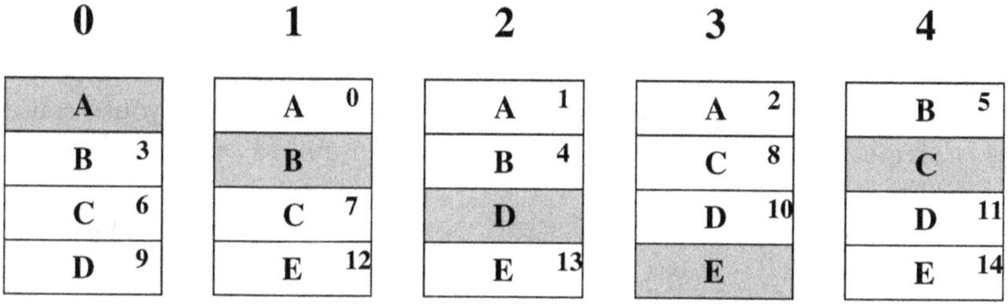

Fig. 2. A table of stripes derived from the complete block design with $v = 5$ and $k = 4$, and the parity-declustered data layout derived from it. Each table entry shows "**disk number** (offset) $^{(\text{address})}$" for one unit. Shaded units store redundant information, and so have no data addresses assigned to them.

2   DATUM Layouts

Alvarez, Burkhard, and Cristian [1] developed the first parity-declustered data layouts, called DATUM layouts, for which mappings of data addresses to disks and offsets are not computed using table lookup. Instead, disks and offsets are computed directly from addresses. In the following, we describe their construction and analyze the complexity of their data mappings.

2.1   Layout Construction and Data Mapping Complexity

DATUM layouts are based on complete block designs, with a particular ordering of their tuples. The set of tuples in the complete block design is the set of all subsets of k of the v disks (we assume that the v disks are labeled $\{0, 1, \ldots, v - 1\}$). Within each tuple, disks appear in increasing order. The ordering of the tuples is as follows: $(X_1, \ldots, X_k)$ precedes $(Y_1, \ldots, Y_k)$ if and only if for some $j \leq k$, $X_j < Y_j$ and for all i satisfying $j < i \leq k$, $X_i = Y_i$. The number of tuples that precede a given tuple in this ordering is called the *rank* of that tuple.

   Given this order, Alvarez et al. defined two functions: $\texttt{loc}(X_1, \ldots, X_k)$, which computes the rank of an input tuple, and its inverse, $\texttt{invloc}(rank)$, which generates the k elements of the tuple with a particular rank. If we are given a row *rank* and column *col* in the table, we can compute the disk number stored at that location by taking X_{col}, the col^{th} element in the tuple returned by $\texttt{invloc}(rank)$. Once the tuple $(X_1, X_2, \ldots, X_k)$ in row *rank* has been found, the offset of the unit from that stripe on disk X_{col} can also be computed.

Alvarez et al. established the correctness of their functions and formulas, but did not analyze their computational complexity. In fact, the entire process of computing the disk and offset in a DATUM layout takes $O(kv)$ steps.

2.2   Usability of Layouts for Large Disk Arrays

DATUM layouts eliminate the need to store a table of size polynomial in v in exchange for enough space to store the tuple $(X_1, X_2, \ldots, X_k)$ and $O(kv)$ time to compute disks and offsets. These layouts can be constructed for all possible values of v and k, but they may be too large to use. DATUM layouts contain $b = \binom{v}{k}$ stripes, so each disk in the array must contain, $bk/v = \binom{v-1}{k-1}$ units in order for the layout to be used.

Consider an array of 10GB disks with a unit size of 4KB. The number of units on each disk is 2,621,440, so any layout with size greater than this amount cannot be used. We are only looking for a rough guideline for usability, so we will consider any layout with size at most 10 million to be usable.

Even for moderately sized arrays that are commercially available (e.g., $v = 64$ disks), DATUM layouts are too large to be usable for more than 80% of the values of k. For arrays of 128 and 256 disks, the percentage of k values that are ruled out rises to more than 93% and 96% respectively. This is admittedly a very rough guideline for usability, but the same pattern of rapidly decreasing usability applies even for much more generous usability guidelines. Even if the disks in an array are sufficiently large to use a particular layout, using a smaller layout may still improve performance by leading to better local load balancing across the array and a smaller amount of wasted space when the disk size is not an integral multiple of the layout size.

3   Ring-Based Layouts

In this section, we extend the ring-based data layouts of Schwabe and Sutherland [8] to eliminate the need to store tables to describe their stripes. We use the algebraic structure of a ring-based block design to develop functions to map data addresses to the corresponding disks and offsets. These ring-based data layouts have two advantages over DATUM layouts:

1. They are smaller, and therefore applicable to a wider range of arrays – they contain only $v(v-1)$ stripes rather than $\binom{v}{k}$;
2. The functions to compute disks and offsets from data addresses are more efficiently computable – they have worst-case running time $O(k \log v \log \log v)$ rather than $O(kv)$.

In the following, we review the ring-based data layout construction of Schwabe and Sutherland [8], and present algorithms to compute disks and offsets without explicitly storing tables of stripes.

3.1   Layout Construction

Ring-based data layouts are derived from a class of block designs called *ring-based block designs*. The elements of a ring-based block design are taken from a *commutative ring with a unit* (hereafter referred to as simply a "ring"). A ring is an algebraic object consisting of a set of elements, an addition operation (associative, commutative, and having an identity element 0 and additive inverses), and a multiplication operation (associative, commutative, and having an identity element $1 \neq 0$) that distributes over addition. The *order* of a ring R is the number of elements in R.

A set of elements $\{g_0, \ldots, g_{k-1}\}$ of a ring R are called *generators* of a ring-based block design if, whenever $i \neq j$, $g_i - g_j$ has a multiplicative inverse. The tuples of a ring-based block design are indexed by pairs (y, x), where x is an arbitrary ring element and y is an arbitrary non-zero ring element. Given a ring R of order v and a set of generators as above, the tuple indexed by (y, x) is the set $T_{(y,x)} = \{y(g_i - g_0) + x \mid i = 0, \ldots, k - 1\}$. The ring-based block design for R and a set of k generators is $\{T_{(y,x)} \mid x \in R, y \in R - \{0\}\}$. This set of tuples is a BIBD with $v(v - 1)$ tuples [8].

If $v = \prod_{i=1}^{m} p_i^{n_i}$, where $p_1, p_2, \ldots, p_m$ are distinct primes, there exists a ring R of order v and a set of k generators in R if and only if $k \leq \min\{p_i^{n_i} \mid i = 1, \ldots, m\}$ [8]. Schwabe and Sutherland showed that R can be taken to be the cross product of finite fields $\mathrm{GF}(p_1^{n_1}) \times \mathrm{GF}(p_2^{n_2}) \times \ldots \times \mathrm{GF}(p_m^{n_m})$, with operations defined component-wise. The ring will contain k generators for every k that satisfies the above condition. From this point forward, R will denote such a ring.

A ring-based data layout is obtained from a ring-based block design by ordering the tuples of the block design from 0 to $b - 1$. To do this, we will first define a bijection f from the ring R to the set $\{0, 1, \ldots, v - 1\}$ that will identify each ring element with a unique integer. This will allow us to associate the index (y, x) of a tuple with the pair of integers $(f(y), f(x))$, so we can regard the tuples as being indexed by integers (j, i) where $i \in \{0, 1, \ldots, v - 1\}$ and $j \in \{1, 2, \ldots, v - 1\}$. To avoid confusion, when we are using such a pair of integers as a tuple index, we will write the pair as $\langle j, i \rangle$ rather than (j, i). We then order the tuples by their indices $\langle 1, 0 \rangle, \langle 1, 1 \rangle, \ldots, \langle 1, v - 1 \rangle, \langle 2, 0 \rangle, \langle 2, 1 \rangle, \ldots, \langle 2, v - 1 \rangle, \ldots, \langle v - 1, 0 \rangle, \langle v - 1, 1 \rangle, \ldots, \langle v - 1, v - 1 \rangle$.

The bijection f will use the following representation of the ring elements. The elements of the field $\mathrm{GF}(p^n)$ can be represented as polynomials of degree at most $n - 1$ in a variable x with coefficients being integers mod p. Thus, a ring element is represented as an m–tuple of polynomials $(P_1(\mathsf{x}), \ldots, P_m(\mathsf{x}))$, where $P_i(\mathsf{x})$ has degree at most n_i and coefficients that are integers mod p_i. The bijection f is defined as follows: Evaluate each polynomial P_i at p_i, to obtain an m-tuple $(P_1(p_1), \ldots, P_m(p_m))$ of non-negative integers. The value of $f(P_1(\mathsf{x}), \ldots, P_m(\mathsf{x}))$ will be the rank of $(P_1(p_1), \ldots, P_m(p_m))$ in the lexicographic order. (Clearly, this yields a bijective mapping f between the ring elements and the integers from 0 to $v - 1$). This rank can be computed by the following algorithm f:

```
f(P₁(x), ..., Pₘ(x))
    total = 0
    for i = 1 to m
        total = total * pᵢⁿⁱ
        total = total + Pᵢ(pᵢ)
    return total
```

Each iteration of the "for" loop takes $O(n_i)$ steps (for the polynomial evaluation), so the total time for the m loop iterations is $\sum_{i=1}^{m} O(n_i) = O(\log v)$. Therefore the time to compute f is $O(\log v)$.

To compute the inverse of f, we must take an integer x and determine the m–tuple of polynomials $(P_1(x), \ldots, P_m(x))$ for which $(P_1(p_1), \ldots, P_m(p_m))$ will have rank x in the lexicographic order. The outer loop of the following algorithm invf computes the values $P_m(p_m), P_{m-1}(p_{m-1}), \ldots, P_1(p_1)$, and the inner loop computes the coefficients of each P_i from $P_i(p_i)$:

```
invf(x)
    for i = m to 1
        xᵢ = x mod pᵢⁿⁱ
        x = x div pᵢⁿⁱ
        for j = 0 to nᵢ − 2
            a(i, j) = xᵢ mod pᵢ^{j+1}
            xᵢ = xᵢ div pᵢ^{j+1}
        a(i, nᵢ − 1) = xᵢ
    return a
```

The ith iteration of the outer loop takes constant time, plus the time required for the inner loop, which is $\Theta(n_i)$. This yields a total of $\sum_{i=1}^{m} \Theta(n_i) = O(\log v)$ steps to compute the inverse of f. Therefore we can convert ring elements into integers in $\{0, 1, \ldots, v − 1\}$ and vice versa in $O(\log v)$ steps.

The ordering of the tuples in the ring-based block design by their indices $\langle 1, 0 \rangle, \langle 1, 1 \rangle, \ldots, \langle 1, v − 1 \rangle, \langle 2, 0 \rangle, \langle 2, 1 \rangle, \ldots, \langle 2, v − 1 \rangle, \ldots, \langle v − 1, 0 \rangle, \langle v − 1, 1 \rangle, \ldots, \langle v − 1, v − 1 \rangle$ defines the ring-based data layout. A table of these tuples would consist of $v(v − 1)$ rows and k columns. We now describe how to use this order to compute disks and offsets without using table lookups.

3.2   Computational Complexity of Data Mappings

In order to compute a disk and offset for a particular data address, we must:

1. Convert the address to a rank (row) and position (column) in the table;
2. Compute the numerical values of $f(y)$ and $f(x)$ corresponding to that rank;
3. Compute the ring elements y and x that index the tuple of that rank;

4. Compute the ring element in the desired position of that tuple;
5. Convert that ring element (which represents the disk identifier) to its numerical label;
6. Compute the offset of the desired address on that disk.

Steps 1 and 2 can be done in constant time with simple arithmetic operations. The conversion to ring elements in Step 3 and back to numerical values in Step 5 both require a constant number of applications of the function f and its inverse, which take a total of $O(\log v)$ steps.

Step 4 must compute the element in the given position of the tuple indexed by ring elements y and x. This element is given by $y(g_j - g_0) + x$, where j is the given position. Computing this element from y, x, and the two generators g_j and g_0 requires one subtraction, one multiplication, and one addition of ring elements.

Addition in the field is polynomial addition, with coefficients added mod p_i. Clearly, adding or subtracting two field elements will take $O(n_i)$ steps. Multiplication in the field is polynomial multiplication, where the product is taken modulo some fixed irreducible polynomial of degree n_i (which must be stored), and all coefficients are computed mod p_i. Multiplying two field elements will therefore take $O(n_i \log n_i)$ steps for the initial multiplication (using, e.g., a Discrete Fourier Transform); evaluating the resulting product modulo an irreducible polynomial adds only $O(n_i \log n_i)$ more steps (see, e.g., von zur Gathen and Gerhard [10]).

Therefore, addition and subtraction of ring elements take $O(\sum_{i=1}^{m} n_i) = O(\log v)$ steps, and multiplication of ring elements takes $O(\sum_{i=1}^{m} n_i \log n_i) = O(\log v \log \log v)$ steps. Thus, Step 4 takes a total of $O(\log v \log \log v)$ steps.

The offset computed in Step 6 is given by the number of occurrences of the disk in tuples with rank lower than the rank computed in Step 1. First we note that given the ordering of the tuples, each set $S_y = \{T_{(y,x)} \mid x \in R\}$ of tuples contains exactly k occurrences of each disk (once in each possible position in a tuple), so that the number of occurrences of disk d in tuples of rank lower than r is $k \lfloor r/v \rfloor$ (which is $k \cdot f(y)$), plus the number of occurrences of d in tuples of the form $T_{(y,x')}$, where $f(x') < f(x)$.

To compute this last term, we note that there are at most $k - 1$ possible positions in which d can appear: $i \in \{0, \dots, j - 1, j + 1, \dots, k - 1\}$. In each case, we must have $d = y(g_i - g_0) + x'$, or solving for x', $x' = d - y(g_i - g_0)$. If $f(x') < f(x)$, then $T_{(y,x')}$ contains disk d and has rank lower than that of $T_{(y,x)}$. So we compute x' for each of the $k-1$ positions other than j, and compare $f(x')$ to $f(x)$, keeping track of the number of positions for which the result is smaller than $f(x)$. This amount is added to $k \cdot f(y)$ to obtain the offset. This takes a total of $O(k \log v \log \log v)$ steps. Therefore, computing the disk and offset for a particular data address takes a total of $O(k \log v \log \log v)$ steps.

Since a polynomial of degree n_i can be stored in $O(n_i)$ space, the space required to store the m polynomials that make up a ring element is $O(\sum_{i=1}^{m} n_i) = O(\log v)$. The computation of the disk and offset requires $O(\log v)$ space for the various ring elements involved. In addition, $O(k \log v)$ space is needed to store the

k generators, and $O(\log v)$ space is needed to store the m irreducible polynomials, n_i's, and p_i's. This yields a total of $O(k \log v)$ space. The space required can be reduced to $O(\log v)$ with no increase in the running time if we use a particular canonical set of generators that can be constructed as needed rather than stored.

3.3   Reducing the Time Complexity

We can improve the $O(k \log v \log \log v)$ running time that we obtained using a polynomial representation of ring elements by using a different representation of ring elements and storing at most an additional $v - 1$ integers. As before, each ring element is an m–tuple of field elements, with addition and multiplication defined component–wise. However, the individual field elements are represented differently.

By definition, the non-zero elements of a field form a group under multiplication. In a finite field $\mathrm{GF}(p^n)$, this group is actually a *cyclic* group of order $p^n - 1$. That is, the elements of the group are exactly $\{1, \alpha, \alpha^2, \ldots, \alpha^{p^n-2}\}$ for some non-zero field element α. (See Koblitz [5] for further discussion of these concepts.) We can define a bijection $i \mapsto \bar{i}$ from $\{0, \ldots, p^n - 1\}$ onto $\mathrm{GF}(p^n)$ as follows: $\bar{i} = 0$ if $i = 0$, and α^{i-1} otherwise.

This bijection allows us to represent field elements as integers from 0 to $p^n - 1$; we will call this representation the *exponent* representation. Using the exponent representation of field elements, addition, subtraction and multiplication in the field can be performed in constant time. Addition and subtraction also require that a list of $p^n - 1$ precomputed integers be stored. Using this representation, the time and space requirements are $O(k \log v)$ and $O(v)$, respectively.

4   Comparisons between DATUM and Ring-Based Layouts

DATUM layouts require $O(kv)$ time to compute disk numbers and offsets with space requirements of $\Theta(k)$. The implementation of ring-based layouts using the polynomial representation of ring elements requires less time – only $O(k \log v \log \log v)$ – to compute disk numbers and offsets, and the space requirements are $O(\log v)$. If $k = \Omega(\log v)$, then the space requirements for ring-based layouts are no greater than those of DATUM layouts. The implementation of ring-based layouts using the exponent representation of ring elements requires even less time, $O(k \log v)$, but more space, $O(v)$. Here, we must have $k = \Omega(v)$ for the space requirements not to exceed those of DATUM layouts.

Recall that a DATUM layout has size $\binom{v-1}{k-1}$ for an array of v disks and stripe size k. This potentially large size rules out the use of these layouts for many values of v and k that include commercially available array configurations. As arrays grow larger and/or their constituent disks smaller, DATUM layouts will work for even fewer values of v and k.

On the other hand, when they exist for a particular v and k, ring-based layouts have size $k(v - 1)$. Thus, for any v smaller than $\sqrt{10,000,000}$ (roughly

3,000), all existing ring-based layouts are usable (using the rough definition of usability discussed earlier). For larger v, the layouts are usable as long as k is at most $10,000,000/(v-1)$. Thus, the usability of ring-based layouts will not be limited by disk sizes until array sizes grow by more than an order of magnitude. It is also more efficient to compute disks and offsets from data addresses in ring-based layouts than in DATUM layouts.

References

1. Alvarez, G.A., Burkhard, W.A., Cristian, F.: Tolerating Multiple Failures in RAID Architectures with Optimal Storage and Uniform Declustering. Proceedings of the 24th ACM/IEEE International Symposium on Computer Architecture (1997) 62–71
2. Alvarez, G.A., Burkhard, W.A., Stockmeyer, L.J., Cristian, F.: Declustered Disk Array Architectures with Optimal and Near-Optimal Parallelism. Proceedings of the 25th ACM/IEEE International Symposium on Computer Architecture (1998) 109–120
3. Hanani, H.: Balanced Incomplete Block Designs and Related Designs. Discrete Mathematics 11 (1975) 255–369
4. Holland, M., Gibson, G.A.: Parity Declustering for Continuous Operation in Redundant Disk Arrays. Proceedings of the 5th International Conference on Architectural Support for Programming Languages and Operating Systems (1992) 23–35
5. Koblitz, N.: A Course in Number Theory and Cryptography. Springer-Verlag (1987)
6. Muntz, R.R., Lui, J.C.S.: Performance Analysis of Disk Arrays Under Failure. Proceedings of the 16th Conference on Very Large Data Bases (1990) 162–173
7. Patterson, D.A., Gibson, G.A., Katz, R.H.: A Case for Redundant Arrays of Inexpensive Disks (RAID). Proceedings of the Conference on Management of Data (1988) 109–116
8. Schwabe, E.J., Sutherland, I.M.: Improved Parity-Declustered Data Layouts for Disk Arrays. Journal of Computer and System Sciences 53 (No. 3) (1996) 328–343
9. Stockmeyer, L.: Parallelism in Parity-Declustered Layouts for Disk Arrays. Technical Report RJ9915, IBM Almaden Research Center (1994)
10. von zur Gathen, J., Gerhard, J.: Modern Computer Algebra. Cambridge University Press (1999)

Approximate Rank Aggregation
(Preliminary Version)

Xiaotie Deng[1], Qizhi Fang[2], and Shanfeng Zhu[1]

[1] Department of Computer Science, City University of Hong Kong,
Hong Kong, P. R. China
{deng,zhusf}@cs.cityu.edu.hk
[2] Department of Mathematics, Ocean University of Qingdao,
Qingdao 266071, Shandong, P. R. China
fangqizhi@public.qd.sd.cn

Abstract. In this paper, we consider algorithmic issues of the rank aggregation problem for information retrieval on the Web. We introduce a weighted version of the metric of the normalized Kendall-τ distance, originally proposed for the problem by Dwork, et al.,[7] and show that it satisfies the extended Condorcet criterion. Our main technical contribution is a polynomial time approximation scheme, in addition to a practical heuristic algorithm with ratio 2 for the NP-hard problem.

Keywords: Rank aggregation, Kendall-τ distance, coherence, weighted ECC.

1 Introduction

How to compute a "consensus" ranking of the alternatives based on preferences of individual voters is called the "rank aggregation problem". It is widely discussed in the literatures of social choice theory and finds its application to election, and most recently, meta-search over the Web. Dwork, et al., [7] pioneered the study of the rank aggregation problem in the context of Web searching with an eye toward reducing spam in meta-search. Comparing with traditional voting problem, the rank aggregation problem on the web has some distinct features. Firstly, the number of voters is much less than the number of alternatives. Secondly, each voter ranks a different set of alternatives, determined by the different coverage of web search engines.

Dwork, et al., made use of Kendall-τ distance as a criterion: Given a collection of partial rankings $\tau_1, \cdots, \tau_k$ of alternative web pages, they want to find a complete ranking π which minimize the average of the Kendall-τ distance between π and τ_i $(i = 1, \cdots, k)$. The Kendall-τ distance between two ranking lists is the total number of pairs of alternatives that are assigned to different relative orders in the two ranking lists. They showed that this problem, the Kemeny aggregation problem, is NP-hard for fixed even $k \geq 4$ and developed an effective procedure "local Kemenization" to obtain a local Kemeny optimal ranking which

T. Warnow and B. Zhu (Eds.): COCOON 2003, LNCS 2697, pp. 262–271, 2003.
© Springer-Verlag Berlin Heidelberg 2003

satisfies the extended Condorcet criterion. A 2-approximation algorithm was obtained for full list rank aggregation but no proven approximation algorithm was known for partial list rank aggregation [7].

In fact, the metric of Kendall-τ distance may not be the best for partial rankings. If two partial rankings overlap over a small number of alternatives (and thus their Kendall-τ distance is small), one may not have full confidence to conclude that the two rankings differ a little. We propose to consider both Kendall-τ distance and size of overlap of the partial ranking lists for an alternative measure for partial ranking aggregation. For a given collection of partial rankings $\tau_1, \cdots, \tau_k$ with different ranking lengths, we are interested in finding a final ranking π such that the sum of $|N_{\tau_i} \cap N_\pi|(1 - \frac{D(\tau_i, \pi)}{C^2_{|N_{\tau_i} \cap N_\pi|}})$ is maximized, where N_{τ_i} is the set of alternatives in τ_i, N_π is the set of alternatives in π, $D(\tau_i, \pi)$ is the Kendall-τ distance between π and τ_i ($i = 1, 2, \cdots, k$). We note that this problem is equivalent to Kemeny aggregation problem [7] in a weighted version. Here, the weight of each partial ranking is determined by its overlap with the final ranking.

In this paper, we focus on the new aggregation method (we call it the Coherence aggregation problem) and related issues, as well as their complexity and algorithmic problems. In section 2, we introduce the formal definitions. We generalize the extended Condorcet criterion (ECC) to the weighted case, and show that the Coherence optimal ranking for partial ranking aggregation satisfies the weighted ECC. In section 3, we discuss the NP-hardness of the Coherence aggregation problem and present a heuristic algorithm with performance ratio 2, and with a proof that the heuristic solution satisfies the weighted ECC. We note that although the Kemeny aggregation problem and the Coherence aggregation problem are equivalent in the weighted case, there is no any approximation algorithm with constant ratio for Kemeny aggregation for partial rankings. In Section 4, we derive a PTAS for the Coherence aggregation problem. Our approach is motivated by techniques developed by Arora, et. al, [1,2]. Our solution further extends and exploits their general methodology and provides new insight into design and analysis of polynomial time approximation scheme. In Section 5, we conclude with remarks.

2 Definitions

Given a set of alternatives $N = \{1, 2, \cdots, n\}$, a ranking π with respect to N is a permutation of some elements of N which represents a voter's or a judge's preference on these alternatives. If π orders all the elements in N, it is called a complete ranking; otherwise, a partial ranking. For a ranking π, let N_π denote the set of elements presented in π, $|\pi| = |N_\pi|$ denote the number of elements in π, or the length of π. For each $i \in N_\pi$, $\pi(i)$ denote the position of the element i in π, and for any two elements $i, j \in N_\pi$, $\pi(i) < \pi(j)$ implies that i is ranked higher than j by π.

The rank aggregation problem is to combine a number of different rank orderings on a set of alternatives, in order to obtain a 'better' ranking. The notion of 'better' depends on what objective we strive to optimize. Among numerous ranking criteria, the methods based on *Kendall-τ distance* are accepted and studied extensively [10,3,4,5,6]. The Kendall-τ distance between two rankings π and σ is defined as

$$D(\pi, \sigma) = |\{(i, j) : \pi(i) < \pi(j), \text{but } \sigma(i) > \sigma(j), \forall i, j \in N_\pi \cap N_\sigma\}|.$$

Given a collection of partial rankings $\tau_1, \tau_2, \cdots, \tau_k$, the Kemeny optimal aggregation is a complete ranking π with respect to the union of the elements of $\tau_1, \tau_2, \cdots, \tau_k$ which minimizes the total Kendall-τ distance $D(\pi; \tau_1, \cdots, \tau_k) = \sum_{i=1}^{k} D(\pi, \tau_i)$.

In the definition of Kendall-τ distance, if it is not the case that the elements i and j appear in both rankings, the pair (i, j) contributes nothing to their Kendall-τ distance. Kendall-τ distance ignores the effect of the size of "overlap" in the measure of the discrepancy of two rankings. Based on this point, we will define another measurement, called *coherence*, to further characterize the relationship of two rankings. This measurement will be used as a criterion in our rank aggregation model.

Definition 2.1 *For two partial rankings τ and σ with $|N_\tau \cap N_\sigma| \geq 2$, the coherence of τ and σ is defined as*

$$\Phi(\tau, \sigma) = |N_\tau \cap N_\sigma| \left(1 - \frac{D(\tau, \sigma)}{C_{|N_\tau \cap N_\sigma|}^2}\right).$$

When $|N_\tau \cap N_\sigma| \leq 1$, we define the coherence $\Phi(\tau, \sigma) = 0$.

Definition 2.2 *For a collection of partial rankings $\tau_1, \tau_2, \cdots, \tau_K$ and a complete ranking π with respect to $N = N_{\tau_1} \cup \cdots \cup N_{\tau_K}$, $|\tau_i| = n_i \geq 2$ $(i = 1, 2, \cdots, K)$, we denote the total coherence by*

$$\Phi(\pi; \tau_1, \tau_2, \cdots, \tau_K) = \sum_{i=1}^{K} \Phi(\pi, \tau_i) = \sum_{i=1}^{K} n_i \left(1 - \frac{D(\pi, \tau_i)}{C_{n_i}^2}\right).$$

The Coherence optimal aggregation is a complete ranking σ of the elements in N which maximizes the total coherence $\Phi(\pi; \tau_1, \tau_2, \cdots, \tau_K)$ over all complete rankings π. The problem of finding the Coherence optimal aggregations is called Coherence aggregation problem.

From the definition of coherence, the contribution of partial ranking τ_i $(i = 1, 2, \cdots, K)$ to the total coherence $\Phi(\pi; \tau_1, \tau_2, \cdots, \tau_K)$ is

$$n_i \left(1 - \frac{D(\tau_i, \pi)}{C_{n_i}^2}\right) = \frac{2}{n_i - 1} \left[C_{n_i}^2 - D(\tau_i, \pi)\right].$$

Let

$$\omega_i = \frac{2}{n_i - 1}, \quad i = 1, 2, \cdots, K.$$

If ω_i is considered as the weight of the corresponding ranking, the Coherence aggregation problem is equivalent to the Kemeny aggregation problem in the weighted version, where the weight of each partial ranking is determined by its overlap with the final ranking. When the lengths of all partial rankings are equal, the Coherence aggregation problem is equivalent to the Kemeny aggregation problem proposed by Dwork et al., [7]. Kemeny optimal rankings are of particular interest because they satisfy the extended Condorcet criterion (ECC): if there is a partition $(P, \bar{P})$ of the elements in N such that for any $i \in P$ and $j \in \bar{P}$, the majority prefers i to j, then i must be ranked higher than j. Recently, Dwork, et al.,[7] studied the Kemeny optimal aggregation problem in the context of the Web and showed that ECC has excellent "spam-fighting" properties in the context of meta-search. When the weights are imposed upon the rankings, we can generalize the ECC to the following weighted version.

Weighted Extended Condorcet Criterion (Weighted ECC):

Given partial rankings $\tau_1, \cdots, \tau_K$ and the corresponding weights $\alpha_1, \cdots, \alpha_K$. Let π be a complete ranking of their aggregation. For any partition $(P, \bar{P})$ of the elements of N, and for all $i \in P$ and $j \in \bar{P}$, if we have $\sum_{l:\tau_l(i)<\tau_l(j)} \alpha_l > \sum_{l:\tau_l(i)>\tau_l(j)} \alpha_l$, then in the aggregation π, i is ranked higher than j. We call π satisfying the weighted extended Condorcet criterion (weighted ECC).

Proposition 2.1 *Let π be a coherence optimal aggregation for partial rankings $\tau_1, \cdots, \tau_K$. Then π satisfies the weighted extended Condorcet criterion with respect to $\tau_1, \cdots, \tau_K$ and their weights $\omega_1, \cdots, \omega_K$.*

Proof. Suppose that there is a partition $(P, \bar{P})$ of N such that for all $i \in P$ and $j \in \bar{P}$ we have that $\sum_{l:\tau_l(i)<\tau_l(j)} \omega_l > \sum_{l:\tau_l(i)>\tau_l(j)} \omega_l$, but there exist two elements $i^* \in P$ and $j^* \in \bar{P}$ such that $\pi(j^*) < \pi(i^*)$. Let (i^*, j^*) be an adjacent such pair in π. Let π' be the ranking obtained by transposing the positions of i^* and j^*. Then we have that

$$\Phi(\pi'; \tau_1, \cdots, \tau_K) - \Phi(\pi; \tau_1, \cdots, \tau_K) = \sum_{l:\tau_l(i^*)<\tau_l(j^*)} \omega_l - \sum_{l:\tau_l(j^*)<\tau_l(i^*)} \omega_l > 0,$$

which contradicts to the optimality of π.   □

3   Complexity and Heuristic Algorithm

For partial rankings of length 2, finding Coherence optimal aggregation is exactly the same problem as finding an acyclic subgraph with maximum weight in a weighted digraph, and hence is NP-hard [9]. Dwork, et al.,[7] discussed the hardness in the setting of interest in meta-search: many alternatives and very few voters. They showed that computing a Kemeny optimal ranking is still NP-hard for any fixed even $K \geq 4$. Their result derives directly the NP-hardness of the Coherence aggregation problem for all integer $K \geq 4$, since odd number of partial rankings can be obtained from even number of complete rankings by splitting one complete ranking into two partial rankings.

Theorem 3.1 *The Coherence aggregation problem for a given collection of K partial rankings, for integer $K \geq 4$, is NP-hard.*

The difficulty of computing the Coherence optimal ranking arises from its NP-hardness. Since for any aggregation π and its reversal π^r with respect to

$\tau_1, \cdots, \tau_K$, the sum of their total coherence is a constant $\Phi(\pi) + \Phi(\pi^r) = \sum_{l=1}^{K} n_l$, a simple 2-approximation algorithm can be obtained by comparing the values of π and π^r. In this section, we investigate heuristic procedures that construct a better aggregation while taking into account the data of the given instance of the problem. The algorithm consists of two parts: Initial Aggregation and Adjustment. When the given collection of rankings is clear from the context, we will denote $\Phi(\pi; \tau_1, \cdots, \tau_K)$ by $\Phi(\pi)$.

Given a collection of partial rankings $\tau_1, \cdots, \tau_K$, with $|\tau_l| = n_l \geq 2$ ($l = 1, 2, \cdots, K$) and $N = N_{\tau_1} \cup \cdots \cup N_{\tau_K} = \{1, 2, \cdots, n\}$. The weight of each partial ranking is defined as $\omega_l = \dfrac{2}{n_l - 1}$, $l = 1, 2, \cdots, K$. For each ordered pair (i, j) ($i, j \in N$), we define the preference value r_{ij} as the sum of weights of the partial rankings which rank i higher than j, that is,

$$r_{ij} = \sum_{l:\tau_l(i) < \tau_l(j)} \omega_l.$$

The Coherence aggregation problem is to find a ranking π of N which maximizes the total Coherence $\Phi(\pi) = \sum_{i,j:\pi(i) < \pi(j)} r_{ij}$.

For each element $i \in N$, denote

$$P(i) = \sum_{j \neq i} r_{ji} \quad \text{and} \quad Q(i) = \sum_{j \neq i} r_{ij}.$$

We note that $P(i)$ and $Q(i)$ are the corresponding contributions to the total Coherence by assigning element i in the lowest position and the highest position of the ranking, respectively. The main idea of Initial Ranking procedure is, in every iteration, to arrange an element to the lowest or highest position, according to their contributions $P(i)$ and $Q(i)$. In Adjustment procedure, if there are two adjacent ordered elements i_k and i_{k+1} such that $r_{i_k i_{k+1}} < r_{i_{k+1} i_k}$ in the ranking obtained already, we transpose the positions of i_k and i_{k+1} to get a better ranking.

Initial Ranking Procedure

1. Set $S = N$, $u = 1$ and $v = n$.
2. Compute $\gamma = \max_{i \in S} \{|P(i) - Q(i)|\}$, and denote i^* the element with the largest γ. If $P(i^*) \leq Q(i^*)$, set $\pi(i^*) = u$, $u \leftarrow u + 1$; if $P(i^*) > Q(i^*)$, set $\pi(i^*) = v$, $v \leftarrow v - 1$. For each element $j \in S \setminus \{i^*\}$, let

$$P(j) \leftarrow P(j) - r_{i^* j} \quad \text{and} \quad Q(j) \leftarrow Q(j) - r_{j i^*}.$$

And set $S \leftarrow S \setminus \{i^*\}$.
3. If $v > u$, go to Step 2; else, stop and output the ranking π.

Adjustment Procedure. Given a ranking $\pi = i_1, i_2, \cdots, i_n$.

1. Set $\pi^* = j_1$, $j_1 = i_1$ and $l = 1$.
2. Compute $k^* = \begin{cases} 0 & \forall 1 \leq k \leq l, \ r_{j_k i_{l+1}} \leq r_{i_{l+1} j_k} \\ \max\{k : 1 \leq k \leq l, \ r_{j_k i_{l+1}} > r_{i_{l+1} j_k}\} & \text{otherwise} \end{cases}$

   For $i \leq k^*$, set $j_i \leftarrow j_i$ (when $k^* = 0$, there is no element being arranged);

For $i = k^* + 1$, set $j_i \leftarrow i_{l+1}$;
For $k^* + 1 < i \leq l + 1$, set $j_i \leftarrow j_{i-1}$.
Set $\pi^* \leftarrow j_1, \cdots, j_{l+1}$, and $l \leftarrow l + 1$.
3. If $l < n$, go to Step 2; else, stop and output the ranking π^*.

The coherence preserved by the Initial Ranking procedure is at least one half of the total value $\sum_l n_l$, since this property holds in every iteration with respect to the coherence incurred by the element i^*. We remark that there may be some other rules for choosing the element i^* in the Initial Ranking procedure, such as, according to the value (1) $\gamma = \max_{i \in S}\{P(i)\} = P(i^*)$; or (2) $\gamma = \max_{i \in S}\{Q(i)\} = Q(i^*)$ for choosing and ranking the corresponding element. From the definition of weighted ECC and Proposition 2.1, we have

Proposition 3.2 *Let π^* be a ranking obtained from Adjustment procedure with respect to $\tau_1, \cdots, \tau_K$ and their weights $\omega_1, \cdots, \omega_K$. Then π^* satisfies the weighted ECC.*

4   Polynomial Time Approximation Schemes

Arora, et al., [2] presented a unified framework for developing into polynomial time approximation schemes (PTASs) for "dense" instances of many NP-hard optimization problems, such as, maximum cut, graph bisection and maximum 3-satisfiability. Their unified framework begins with the idea of exhaustive sampling: picking a small random set of elements, guessing where they go on the optimum solution, and then using their placement to determine the placement of other elements. Arora, et al., [1] applied this technique to assignment problems by shrinking the space of possible placements of the random sample. They designed PTASs for some 'smooth' dense subcases of many well known NP-hard arrangement problems, including minimum linear arrangement, d-dimensional arrangement, betweenness, maximum acyclic subgraph, etc. In this section, we show that the same techniques in [1] can also derive a PTAS for the Coherence aggregation problem, though the coefficients do not satisfy the 'smoothness' condition.

In this section, we consider the Coherence aggregation problem for K partial rankings $\tau_1, \cdots, \tau_K$, where K is an integer indifference of $n = |N| = |N_{\tau_1} \cup \cdots \cup N_{\tau_K}|$, $|\tau_s| = n_s \geq 3$ $(s = 1, 2, \cdots, K)$. The weight of each partial ranking ω_s and the preference value r_{ij} are defined as in Section 3. Since for any complete ranking π and its reversal π^r, $\Phi(\pi) + \Phi(\pi^r) = \sum_{s=1}^{K} n_s \geq n$, the optimal value of this problem is no less than $n/2$. Therefore to obtain an optimal ranking with at least the value $(1 - \gamma)$ times the optimum, where $\gamma > 0$ is arbitrary, it suffices to find a ranking whose value is within an additional factor of ϵn from the optimal value of the optimal ranking for a suitable $\epsilon > 0$. We present our result in the following theorem.

Theorem 4.1 *Suppose the ranking π^* is the optimal solution of the Coherence aggregation problem. Then for any fixed $\epsilon > 0$, in time $n^{O(1/\epsilon^2)}$ we can find a ranking π of N such that*

$$\Phi(\pi) \geq \Phi(\pi^*) - \epsilon n.$$

Several Chernoff-style tail bounds are important in the analysis of randomized procedure. The following result is needed repeatedly in this paper, which we present as a lemma for completeness.

Lemma 4.2 *Let $X_1, X_2, \cdots, X_n$ be n independent random variables such that $0 \leq X_i \leq 1$. Then for $X = \sum_{i=1}^{n} X_i$, $\mu = E[X]$ and $\lambda \geq 0$,*

$$\Pr[|X - \mu| > \lambda] \leq 2e^{-2\lambda^2/n}.$$

Let ϵ be a given small positive, and $t = c/\epsilon$ for some suitable large constant $c > 0$. Here we assume for simplicity that n is a multiple of t. Construct t sequential groups $I_1, I_2, \cdots, I_t$. A *placement* is a mapping $g : N \to \{1, 2, \cdots, t\}$ from the set N to the set of groups $\{I_1, I_2, \cdots, I_t\}$. It is *proper* if it maps n/t elements of N to each group, that is, for every $1 \leq j \leq t$, $|\{i \in N | g(i) = j\}| = n/t$. The value of a placement g, denoted by $\phi(g)$, is defined as

$$\phi(g) = \sum_{\forall i,j \in N, g(i) < g(j)} r_{ij} = \sum_{s=1}^{K} w_s |\{(i,j) : \tau_s(i) < \tau_s(j) \text{ and } g(i) < g(j)\}|.$$

Lemma 4.3 *If π is a ranking and g is its induced proper placement, then*

$$\phi(g) \leq \Phi(\pi) \leq \phi(g) + \frac{3Kn}{t}.$$

Let π^* be an optimal ranking and let g^* be its induced placement, and let $\epsilon' = (1 - \frac{3K}{c})\epsilon$. Assume that g is a proper placement such that

$$\phi(g) \geq \phi(g^*) - \epsilon'n,$$

and π is an arbitrary ranking such that g is the placement induced by π. By Lemma 4.3, we have that

$$\Phi(\pi) \geq \phi(g) \geq \phi(g^*) - \epsilon'n \geq \Phi(\pi^*) - \frac{3Kn}{t} - \epsilon'n = \Phi(\pi^*) - \epsilon n.$$

Therefore, finding an optimal ranking to our problem can be reduced to the problem of finding a proper placement within an additive factor of $\epsilon'n$ from the optimal placement.

The optimal placement problem can be formulated as the quadratic arrangement problem:

$$\text{Max } \sum_{s=1}^{K} [\sum_{ijkl} c_{ijkl}^s x_{ik} x_{jl}]$$

$$\text{s.t.} \begin{cases} \sum_{j=1}^{n} x_{ik} = n/t & k = 1, 2, \cdots, t \\ \sum_{k=1}^{t} x_{ik} = 1 & i = 1, 2, \cdots, n \\ x_{ik} = 0, 1 & i = 1, \cdots, n; \ k = 1, \cdots, t \end{cases}$$

Here, $c_{ijkl}^s = \begin{cases} w_s & \text{if } \tau_s(i) < \tau_s(j) \text{ and } 0 < k < l \\ 0 & \text{otherwise} \end{cases}$. Let g^* be the optimal placement, and let

$$\hat{e}_{ik}^s = \sum_{jl} c_{ijkl}^s g_{jl}^* = w_s \left| \{j \in N_{\tau_s} : \tau_s(i) < \tau_s(j) \text{ and } g^*(j) > k\} \right|.$$

Then g^* is an integral solution to the linear program

$$\text{Max } \sum_{s=1}^{K} [\sum_{i=1}^{n} \sum_{k=1}^{t} \hat{e}_{ik}^s x_{ik}]$$

$$\text{s.t. } \begin{cases} \sum_{i=1}^{n} x_{ik} = n/t & k = 1, 2, \cdots, t \\ \sum_{k=1}^{t} x_{ik} = 1 & i = 1, 2, \cdots, n \\ \sum_{jl} c_{ijkl}^s x_{jl} = \hat{e}_{ik}^s \ s = 1, \cdots, K; \ i \in N_{\tau_s}; \ k = 1, \cdots, t \\ 0 \le x_{ik} \le 1 & i = 1, \cdots, n; \ k = 1, \cdots, t \end{cases}$$

We will use the method of exhaustively sampling [1,2] to estimate $\hat{e}_{ik}^s$'s. However, since the lengths of K given partial rankings may be quite different from each other, the coefficients of above quadratic arrangement problem do not satisfy the "smooth" condition. To make a more accurate estimate, we make randomly sampling and estimation for each given partial ranking separately.

Randomly picking with replacement a multi-set T_s of $O(\log n_s/\delta^2)$ elements (where δ is a sufficiently small fraction of ϵ' which we will determine later) from the set N_{τ_s} $(s = 1, \cdots, K)$ respectively, we estimate $\hat{e}_{ik}^s$ by the sum $(n_s/|T_s|)\omega_s|\{j \in T_s : \tau_s(i) < \tau_s(j)$ and $g^*(j) > k\}|$. Thus, we chose randomly a multi-set $T = T_1 \cup \cdots \cup T_K$ with size $|T| = O(\log n)$. Since the optimal placement g^* is not known in advance, we enumerate all possible function $h : T \to \{1, 2, \cdots, t\}$ that assign elements in T to groups $I_1, \cdots, I_t$. For each such function, we solve a linear program $\mathcal{M}_h$ described below and round the (fractional) optimal solution to construct a proper placement. Among all these placements, we pick up one with maximum value. When the function h we considered is the same as h^* which is the restriction of an optimal placement g^* to T, the placement g we get from the linear program $\mathcal{M}_h$ will satisfy $\phi(g) \ge \phi(g^*) - \epsilon'n$ with high probability, over the random choice of T.

Let h be a given function $h : T \to \{1, 2, \cdots, t\}$. For simplicity, we will identify h with its restrictions on T_s's $(s = 1, 2, \cdots, K)$ in the rest of this section. For the given partial ranking τ_s, each element $i \in N_{\tau_s}$ and group I_k $(k = 1, 2, \cdots, t)$, we compute an estimate e_{ik}^s of the value (derived from τ_s) of assigning i to I_k in any placement g whose restriction to T_s is h:

$$e_{ik}^s = \frac{n_s}{|T_s|}\omega_s \left| \{j \in T_s : \tau_s(i) < \tau_s(j) \text{ and } h(j) > k\} \right|.$$

Lemma 4.4 *Pick uniformly at random with replacement a multi-set T_s of $O(\log n_s/\delta^2)$ elements from N_{τ_s}. Let g be a placement and h be the restrictions of g on T_s. Then with high probability (over the choice of sample T_s),*

$$\left| e_{ik}^s - \omega_s \left| \{j \in N_{\tau_s} : \tau_s(i) < \tau_s(j) \text{ and } g(j) > k\} \right| \right| \le 3\delta. \tag{4.1}$$

Consider the following linear program $\mathcal{M}_h$:

$$\text{Max } Z(x) = \sum_{s=1}^{K} (\sum_{i \in N_{\tau_s}} \sum_{k=1}^{t} e_{ik}^s x_{ik})$$

$$\mathcal{M}_h : \text{ s.t. } \begin{cases} \sum_{i=1}^{n} x_{ik} = n/t & k = 1, 2, \cdots, t \\ \sum_{k=1}^{t} x_{ik} = 1 & i = 1, 2, \cdots, n \\ \left| \sum_{j:\tau_s(i)<\tau_s(j)} \sum_{l>k} \omega_s x_{jl} - e_{ik}^s \right| \le 3\delta & s = 1, \cdots, K; i \in N_{\tau_s}; \\ 0 \le x_{ik} \le 1 & i = 1, \cdots, n; k = 1, \cdots, t \end{cases}$$

Let x^h be the optimal solution for $\mathcal{M}_h$. We round x^h_{ik} using randomized rounding techniques of Raghavan and Thompson [12] to obtain a placement $\tilde{r}$ and corresponding proper placement r^h as follows: (1) for each element i, independently take $\tilde{r}(i) = k$ with probability x^h_{ik}; (2) construct a proper placement r^h from $\tilde{r}$ by moving elements from groups with more than n/t elements assigned to them to groups with less than n/t elements assigned to them arbitrarily. We will discuss the relation between the optimal value $Z(x^h)$ of $\mathcal{M}_h$ and the value of corresponding placement r^h, $\phi(r^h)$. Let

$$Z_s(x^h) = \sum_{i \in N_{\tau_s}} \sum_{k=1}^{t} e^s_{ik} x^h_{ik}, \qquad Z(x^h) = \sum_{s=1}^{K} Z_s(x^h);$$
$$\phi_s(\tilde{r}) = \omega_s |\{(i,j) : \tau_s(i) < \tau_s(j) \text{ and } \tilde{r}(i) < \tilde{r}(j)\}|, \quad \phi(\tilde{r}) = \sum_{s=1}^{K} \phi_s(\tilde{r}).$$

Lemma 4.5 *Let h be a function that assigns elements of T to groups $I_1, \cdots, I_t$, and r^h be the proper placement constructed from the optimal fractional solution x^h of $\mathcal{M}_h$. Then*

$$\phi(r^h) \geq Z(x^h) - 4K\delta n. \tag{4.2}$$

Lemma 4.6 *Let g^* be the optimal placement, h^* be the restriction of g^* to the sample T and r^* be the proper placement constructed from the optimal solution x^* of $\mathcal{M}_{h^*}$. Then*

$$\phi(r^*) \geq \phi(g^*) - \epsilon' n.$$

In this procedure, we enumerate all possible function $h : T \rightarrow \{1, 2, \cdots, t\}$ and choose a placement with maximum value among all placement r^h constructed. Since r^* is a candidate for our chosen placement r^h, and we choose the placement with maximum value which is no less than the value of r^*, therefore, we obtain the desired result.

The PTAS described above uses randomization in picking the sample set of elements T and in rounding the optimal solution to linear program $\mathcal{M}_h$. For the procedure of rounding the optimal solution x^h of linear program $\mathcal{M}_h$, we can derandomize it in a standard way using the method of conditional probabilities [11]. As discussed in [1] (also in [8]), the procedure of sampling the set of elements T_s can be substituted by an alternative way of picking random walks of length $|T_s|$ on a constant degree expander graph. Since there are only polynomial many random walks of length $|T_s| = O(\log n_s/\delta^2)$ on this expander, the procedure of sampling the total set T can be substituted by picking polynomial many random walks of length $O(\log n/\delta^2)$. Thus, we can derandomize the algorithm by exhaustedly going through all possibilities, i.e., $t^{|T|} = t^{O(\log n/\delta^2)} = n^{O(1/\epsilon^2)}$ placements of the elements in the sample. The running time of our algorithm is $n^{O(1/\epsilon^2)}$.

5   Conclusion and Further Work

Considering the distinct features in the context of meta-search on the web, we have developed a new rank aggregation method based on the criterion of Coherence. We have proposed not only a practical heuristic algorithm with the solution satisfying the weighted extended Condorcet criterion, but also a theoretical polynomial time approximation scheme (PTAS) for the Coherence aggregation problems. Our algorithm extends and exploits the general framework of Arora, et al., [1,2], for design and analysis of polynomial time approximation schemes. Other metrics in social choice theory are also worth of further exploration with the algorithmic approach.

Acknowledgement. This work is supported by a joint research grant (N_CityU 102/01) of Hong Kong RGC and NNSFC of China.

References

1. S. Arora, A. Frieze and H. Kaplan, A new rounding procedure for the assignment problem with applications to dense graph arrangement problems, FOCS96:21–30
2. S. Arora, D. Karger and M. Karpinski, Polynomial-time approximation schemes for dense instances of NP-hard optimization problems, STOC95:284–293
3. J.P. Barthelemy, A. Guenoche and O. Hudry, Median linear orders: Heuristics and a branch and bound algorithm, European Journal of Operational Research 42(1989): 313–325.
4. J.P. Barthelemy and B. Monjardet, The median Procedure in cluster analysis and social choice theory, Mathematical Social Sciences 1(1981): 235–267.
5. J.J. Bartholdi, D.A. Tovey and M.A. Trick, Voting schemes for which it can be difficult to tell who won the election, Social Choice and Welfare, 6(1989): 157–165.
6. I. Charon, A. Guenoche, O. Hudry and F. Woirgard, New results on the computation of median orders, Discrete Mathematics 165/166(1997): 139–153.
7. C. Dwork, R. Kumar, M. Naor and D. Sivakumar, Rank aggregation methods for the web, WWW10 (2001), 613–622.
8. D. Gillman, A Chernoff bound for random walks on expanders, SIAM J. Comput. 27(1998): 1203–1220.
9. R.M. Karp, Reducibility among combinatorial problems, in: R.E. Miller and J.W. Thatcher, eds., Complexity of Computer Computations (Plenue, New York, 1972) 85–103.
10. J.G. Kemeny, Mathematics without numbers, Daedalus 88(1959): 577–591.
11. P. Raghavan, Probabilistic construction of deterministic algorithms: Approximating packing integer programs, Journal of Computer and System Sciences 37(1988): 130–143.
12. P. Raghavan and C. Thompson, Randomized rounding: a technique for provably good algorithms and algorithmic proofs, Combinatorica 7(1987): 365–374.

Perturbation of the Hyper-Linked Environment

Hyun Chul Lee and Allan Borodin

Department of Computer Science
University of Toronto
Toronto, Ontario, M5S3G4
{leehyun,bor}@cs.toronto.edu

Abstract. After the seminal paper of Kleinberg [1] and the introduction of PageRank [2], there has been a surge of research activity in the area of web mining using link analysis algorithms. Subsequent to the first generation of algorithms, a significant amount of improvements and variations appeared. However, the issue of stability has received little attention in spite of its practical and theoretical implications. For instance, the issue of "link spamming" is closely related to stability: is it possible to boost up the rank of a page by adding/removing few nodes to/from it? In this paper, we study the stability aspect of various link analysis algorithms concluding that some algorithms are more robust than others. Also, we show that those unstable algorithms may become stable when they are properly "randomized".

1 Introduction

The use of link analysis algorithms for different web mining purposes became quite popular after the first introduction of algorithms to identify authoritative sources in the web [1,2]. Different attempts [6,7,3,9,10,12,13,14] to improve these algorithms were taken. A simple evaluation of query results using human judgement is normally employed to measure the performance of algorithms. Ng et al. [4] and Borodin et al. [3] take a slightly different path from other papers: Ng et al. study the stability aspect of some link analysis algorithms like PageRank and HITS, providing some insight into ways of designing stable link analysis methods. Borodin et al. introduce some formal definitions of stability and rank stability along with the analysis of some algorithms.

Stability is an important feature to consider in a such highly dynamic environment as World Wide Web. The World Wide Web is continuously evolving, so if a link analysis is to provide a robust notion of authoritativeness of pages, then it is natural to ask for a link analysis algorithm to be stable under small perturbations on the web topology. Intuitively, a small change of the web topology should not affect the overall link structure, and a proper definition of stability should reflect this intuition properly. The stability issue also has some practical implications such as that of "link spamming", i.e. a good link analysis algorithm should be robust to any malicious attempt of web designers to promote the rank of their pages by adding/removing few links to/from them.

T. Warnow and B. Zhu (Eds.): COCOON 2003, LNCS 2697, pp. 272–283, 2003.
© Springer-Verlag Berlin Heidelberg 2003

The current link analysis algorithms can be classified into two categories. The first class of algorithms are *algebraic methods* such as HITS [1],PageRank [2], SALSA [6] and various hybrid algorithms of the first two [3,9,11]. These methods essentially compute principal eigenvectors of particular matrices related to the adjacency matrix of a certain web graph to identify the most relevant pages on that web graph. The second class of algorithms are *probabilistic methods* such as PHITS [10] and Bayesian [3] algorithms. These algorithms, using some probabilistic assumptions and techniques, estimate the rank of pages on a specific topic. *Algebraic methods* are the most popular ones, thus in this paper we only study the *algebraic methods*. More specifically, after introducing our revised definition of stability, we show the following results regarding the stability of algebraic methods: 1) PageRank is stable on the class of all directed graphs. 2) SALSA is stable on the class of authority connected graphs but not stable on the class of all directed graphs. 3) HITS is not stable on the class of authority connected graphs. Finally, we introduce randomized versions of HITS and SALSA showing stability for these algorithms.

2   Overview of Algorithms

We begin by reviewing some algebraic link analysis algorithms, the reader familiar with this material may wish to skip ahead to Section 3.

2.1   HITS

Created by Kleinberg [1], HITS is the first link analysis algorithm used for web mining. In contrast to PageRank, it was never implemented in a commercial search engine until a new search engine Teoma[1] integrated a variation of HITS[2] as part of its ranking system. First, this algorithm constructs a *Root Set* of pages consisting of a short list of webpages returned by the search engine. Later, this *Root Set* is augmented by pages that are pointed to by pages in the *Root Set*, and also by pages that point to pages in the *Root Set* to form a larger set called *Base Set*, which makes HITS a query dependent method. With the *Base set*, HITS forms the adjacency matrix A where $A_{ij} = 1$ if there is a link from i to j and 0 otherwise. Next, it assigns to each page i an authority weight a_i and a hub weight h_i, then the equations $a_i^{(t+1)} = \sum_{j \to i} h_j^{(t)}$ and $h_i^{(t)} = \sum_{i \to k} a_k^{(t)}$ are iterated until $a_i^{(t)}$ and $h_i^{(t)}$ converge to the fixed points a_i^* and h_i^* respectively (with the vectors renormalized to unit length at each iteration). Also, it is easily seen than the fixed points a^* and h^* are principal eigenvectors of $A^t A$ and $A A^t$ respectively. The authority value of a page i is taken to be a_i^*, and the hub value of page i is taken to be h_i^* in a similar manner.

---

[1] http://www.teoma.com
[2] teoma vs. Google, Round Two, Siliconvalley.internet.com, April 2, 2002

2.2 PageRank

The popularity of PageRank is due to the commercial success search engine Google[3] created by Brin and Page [2]. PageRank simulates a random surfer who jumps to a randomly chosen web page with probability ϵ, and follows one of the forward-links on the current page with probability $1 - \epsilon$. This process defines a markov chain on the web pages. The transition probability matrix of this markov chain is given by $(\epsilon U + (1 - \epsilon)A_{row})$ where A_{row} is constructed by renormalizing each row of the adjacency matrix A to sum to 1[4] and U is the transition matrix of uniform transition probabilities. The vector p that represents PageRank scores of pages is then defined to be the stationary distribution of this markov chain. PageRank does not make distinction between hub values and authority values, rather it assigns a single value(PageRank) to each page. In this paper, the PageRank score p_i of page i is taken to be both authority and hub values of the page for the sake of our analysis.

2.3 SALSA

As an alternative algorithm to HITS(an algorithm to avoid "topic-drift"), SALSA is proposed by Lempel and Moran [6]. SALSA performs two random walks on web pages; a random walk by following a backward-link and then a forward-link alternately, and another one by following a forward-link and then a backward-link alternately. The authority weights are defined to be the stationary distribution of the former random walk, and the hub weights are defined to be the stationary distribution of the latter random walk. Thus, SALSA assigns separate hub and authority scores to each page. The transition probability matrices of the markov chains for the authorities and hubs are given by $\tilde{A} = A_{col}^t A_{row}$, $\tilde{H} = A_{row} A_{col}^t$, where A_{col} is constructed by renormalizing each column of the adjacency matrix A to sum to 1, and A_{row} is constructed by renormalizing each row of the adjacency matrix A to sum to 1. One attractive aspect of SALSA is that its stationarity distributions have explicit forms [6].

3   Definitions and Notations

In this section, we introduce some basic definitions and notations used throughout the rest of paper. Given $G = (V, E)$ a directed graph representing a set of pages and their interconnecting links, we define the *co-citation graph* of G as an undirected graph $G_a = (V', E')$ such that $V' = V$ and $E' = \{(p, q)|$ if there exists a node r that links to both p and q }. A directed graph $G = (V, E)$ is called *authority connected* if its co-citation graph is connected. The edge

---

[3] http://www.google.com

[4] It is not clear from the original definition how to deal with the situation where the current page has no forward-link from it. In this paper, we use the simplest approach, i.e. when a page has no forward-link(a row of A has all zero entries), then the corresponding row of A_{row} is constructed to have all entries equal to $1/n$.

distance d_e between two graphs $G_1 = (V, E_1)$ and $G_2 = (V, E_2)$ is defined as $d_e(G_1, G_2) = |(E_1 \cup E_2)\backslash(E_1 \cap E_2)|$. We define a link analysis algorithm $T = (a^T(G), h^T(G))$ as a pair of functions that map a directed graph G of size N to a N-dimensional vector. We call the vector $a^T(G)$ the authority weight vector of algorithm T on graph G and $h^T(G)$ the hub weight vector of algorithm T on graph G. The value of the entry $a_i^T(G)$ of vector $a^T(G)$ denotes the authority weight assigned by the algorithm T to the page i. Similarly, the value of the entry $h_i^T(G)$ of vector $h^T(G)$ denotes the hub weight assigned by the algorithm T to the page i. If the algorithm T does not make distinction between hub and authority values, then we treat the single weight of page as both hub and authority weights. If it is clear in the context, then we simply use a instead of $a^T(G)$ to denote the authority vector of algorithm on graph G, and a_i instead of $a_i^T(G)$ to denote the authority weight assigned by the algorithm T to the page i. A similar convention is used for the hub vector. Given a graph G, we can view a perturbation on graph G, as an operation ∂ on graph G, that adds and/or removes links to produce a new graph $G' = \partial G$. We denote by $\tilde{a}^T(G) = a^T(\partial G)$ the new authority vector of the perturbed graph ∂G, and by $\tilde{a}_i^T$ its respective new authority weight assigned by the algorithm T to page i.

Let BP and FP denote the set of pages whose backward-links are perturbed and the set of pages whose forward-links are perturbed respectively. Let BU denote the set of pages whose backward-links remain unperturbed even after the perturbation, and let FU be the set of pages whose forward-links remain unperturbed even after the perturbation. Let $\tilde{\mathcal{G}}$ be the set of all directed graphs, let $\mathcal{G}^{\mathcal{N}}$ be the class of all directed graphs of size N, let $\mathcal{G}_{AC}$ be the class of all authority connected graphs, and let $\mathcal{G}_{AC}^{\mathcal{N}}$ be the class of authority connected graphs of size N. Therefore, $\mathcal{G}_{AC} \subset \tilde{\mathcal{G}}$, $\mathcal{G}^{\mathcal{N}} \subset \tilde{\mathcal{G}}$ and $\mathcal{G}_{AC}^{\mathcal{N}} \subset \mathcal{G}_{AC}$ hold. It is our particular interest to study the stability issues of link analysis algorithms on the class $\mathcal{G}_{AC}$ because an authority connected graph can be viewed as representation of topical web graphs (a set of pages that pertain to the same topic).

Before introducing our definition of stability, the original definition of stability will be introduced, so that the reader who is not familiar with [3] can understand the motivation driving a new definition.

Previous Definition: An algorithm T is stable if for every fixed K, we have

$$\lim_{N \to \infty} \max_{G_1 \in \mathcal{G}^{\mathcal{N}}, d_e(G, \partial G) \leq K} \min_{\gamma_1, \gamma_2 \geq 1} ||\gamma_1 a^T(G) - \gamma_2 a^T(\partial G)||_1 = 0$$

Based on this definition, Borodin et al. show 1) HITS is not stable on $\mathcal{G}^{\mathcal{N}}$. 2) SALSA is not stable on $\mathcal{G}_{AC}^{\mathcal{N}}$ but it is stable on $\mathcal{G}^{\mathcal{N}}$. We think this definition of stability is not sufficiently robust to reflect the realistic stability of link analysis algorithms, i.e. *the impact of perturbation depends on both the number and the weights of perturbed nodes, but rather, the definition only considers the number of perturbed links.* Thus, motivated by [4] which bounds the magnitude of perturbation for PageRank by a linear function of the aggregated PageRank scores of all perturbed pages, we define our notion of stability.

Definition 1. *Let c_i be the number of backward-links of page i that are perturbed. We say that an algorithm T is stable on $S \subseteq \tilde{\mathcal{G}}$ if we have a fixed constant value k such that for any $G \in S$ and $\partial G \in S$*

$$||a^T(G) - a^T(\partial G)||_1 \leq k \left(\sum_{i \in BP} c_i a_i + \sum_{j \in FP} h_j \right)$$

holds.

The intuitive idea behind this definition is as follows. Each time we add/remove a link there are two pages involved with this action, namely a page(call it j) whose forward-link is perturbed and another page(call it i) whose backward-link is perturbed. Roughly speaking, the "cost" of this addition/removal is $a_i + h_j$. Both a_i and h_j will contribute to the impact of perturbation, but the contribution of h_j would not be as considerable as that of a_i since the authority weight of a page is mainly due to backward-links rather than forward-links. Therefore, in our definition, a_i is more heavily weighted by the number of perturbed backward-links of page i. Although it is also possible to have a definition in terms of the eigengap of particular matrix related to the link analysis algorithm, it presents some difficulties. For some link analysis algorithms like probabilistic ones, there is no natural way of formalizing eigengap since its role in the algorithm is obscure.

4   Results

In this section, we present our results regarding the stability of PageRank, SALSA, and HITS.

4.1   Stability of PageRank

It is proven in [4] that $||\tilde{p} - p||_1 \leq 2/\epsilon \cdot (\sum_{i \in P} p_i)$, where P denotes the set of perturbed nodes. Slightly adapting this result to our definiton of stability, the following proposition is obtained.

Proposition 1. *PageRank is stable on the class of all directed graphs $\tilde{\mathcal{G}}$. Specifically, given a graph $G = (V, E) \in \tilde{\mathcal{G}}$, G is perturbed producing a new graph ∂G. Let p be the original PageRank score, then the new PageRank score $\tilde{p}$ satisfies:*

$$||\tilde{p} - p||_1 \leq \frac{2(1 - \epsilon)}{\epsilon} \cdot \left(\sum_{i \in FP} p_i \right)$$

Note that our proposition only focuses on the set FP rather than the entire set of perturbed nodes.

4.2   Stability/Instability of SALSA

Proposition 2. *Let G, $G' \in \mathcal{G}_{AC}$, s the original SALSA authority vector, and c_i the number of perturbed backward-links of page i, then the new SALSA authority vector $\tilde{s}$ obtained after the perturbation satisfies:*

$$||s - \tilde{s}||_1 \leq 2 \left(\sum_{i \in BP} \frac{c_i}{w} \right)$$

where w denotes the number of links (edges) in G. Moreover, if we only perturb those pages whose $|B(i)| > 0$ (†), then

$$||s - \tilde{s}||_1 \leq 2 \left(\sum_{i \in BP} c_i s_i \right)$$

Note that proposition 2 states that SALSA is stable on the class of authority connected graphs $\mathcal{G}_{AC}$ under the assumption (†).

Proposition 3. *SALSA is not stable on the class of all directed graphs $\tilde{\mathcal{G}}$*

Proof: Consider a graph that consists of complete graphs C_1 and C_2 of size 2 and n respectively. Also, there exists an extra hub h that points to two authority nodes p and q of the component C_1 and C_2 respectively. Now, we perturb the graph removing the link from hub h to authority p. Then, we observe that for the node $s \in C_1 \setminus p$, we have $a_s = 1/(n(n-1)+4)$, $\tilde{a}_s = (2/(n+2))(1/2) = 1/(n+2)$, and for p, we have $a_p = \frac{2}{n(n-1)+4}$, $\tilde{a}_p = (2/(n+2))(1/2) = 1/(n+2)$. Moreover, $h_h = 2/(n(n-1)+4)$. Thus, $|a_p - \tilde{a}_p| + |a_s - \tilde{a}_s| = (2n^2 - 5n + 2)/((n+2)(n(n-1)+4)) > (n-6)/4 \cdot (2/(n(n-1)+4) + 2/(n(n-/1)+4)) = (n-6)/4 \cdot (a_p + h_h)$. Consequently $||a - \tilde{a}||_1 > (n-6)/4 \cdot (a_p + h_h)$ which proves the proposition.

4.3   Instability of HITS

To illustrate the high sensitivity of HITS to the topology of graph, we start with an example.

Example 1. Consider a graph $G = (V, E)$ that consists of n nodes that form a cycle. More precisely, G has links $\{1 \rightarrow 2, 2 \rightarrow 3, \ldots, n-1 \rightarrow n, n \rightarrow 1\}$. Next, G is perturbed by removing $1 \rightarrow 2$ and adding $1 \rightarrow 3$. The weight is evenly distributed among all nodes in G, i.e. for all $i \in V$, we have $a_i = 1/n$, $h_i = 1/n$. On the other hand, we have $\tilde{a}_3 = 1$ and $\tilde{a}_i = 0$ for the rest of nodes. Moreover, we have $\tilde{h}_1 = 1/2$, $\tilde{h}_2 = 1/2$ and $\tilde{h}_i = 0$ for the rest of nodes. Hence, $||a - \tilde{a}||_1 = (n-1)/n + 1 - 1/n > 1 = n/3(a_2 + a_3 + h_1)$.

Note that this example shows that HITS fails to be stable even under small perturbation of a connected graph.[5] Therefore, the following proposition is not surprising.

---

[5] It is not answered in [3] whether HITS is stable or not when the perturbed graph remains connected after the perturbation.

Proposition 4. *HITS is not stable on the class of authority connected graphs* $\mathcal{G}_{AC}$

Proof: Consider the graphs G and ∂G that consist of 2n+1 nodes. Let $\mathcal{A} = \{1, \ldots, n\}$ denote the first n nodes, let $\mathcal{B} = \{n+1, \ldots, 2n\}$ denote the next n nodes, and let s denote the last node. Both graphs contain the links $\{s \rightarrow i | i \in \mathcal{B}\}$ and $\{i \rightarrow j | i \in \mathcal{A}, j \in \mathcal{B}\}$. The perturbed graph ∂G additionally contains links $\{j \rightarrow i | j \in \mathcal{B}, i \in \mathcal{A}\}$ and $\{s \rightarrow i : i \in \mathcal{A}\}$. G and ∂G are authority connected graphs. For all $i \in \mathcal{A}$, we have $a_i = 0$, $h_i = 1/(n+1)$, $\tilde{a}_i = 1/(2n)$. For all $j \in \mathcal{B}$, we have $a_j = 1/n$, $h_j = 0$, $\tilde{a}_j = 1/(2n)$. Finally, we have $a_s = 0$, $h_s = 1/(n+1)$, $\tilde{a}_s = 0$. Therefore, $||a - \tilde{a}|| \geq \sum_{i \in A} ||a_i - \tilde{a}_i|| = 1/2 > n/2(\sum_{i \in A} 2 \cdot a_i + \sum_{j \in B \cup \{s\}} h_j)$ proving instability.

Example 1 and Proposition 4 show some extreme scenarios where HITS fails to be stable. Apparently, addition/removal of even small number of links may alternate substantially the whole weight distribution under HITS, and the experimental study about the stability of HITS appears in Section 6.

5   Improvement of Algorithms

In the previous section, some limitations of SALSA and HITS in terms of stability were shown. In this section, we explore how the randomization of the algorithms can eliminate their instability.

5.1   Randomized HITS

The first version of randomized HITS is introduced in [9] under the name of *two-level reputation rank*. Also, a slightly different version is proposed by Ng et al. [7] This randomization of HITS consists of the following random surfer model: the random surfer picks uniformly a random page with probability ϵ and follows a link with probability $1 - \epsilon$. If he decides to follow a link then he checks if it is odd time step or even time step. If it is odd time step, then he follows uniformly at random a forward-link. If it is even time step, then he follows uniformly at random a backward-link. Note that this process defines a random walk on pages which is similiar in spirit to HITS. The stationary distribution on odd time steps is defined to be the authority weights of pages and the stationary distribution on even time steps is defined to be the hub weights of pages. Formally, the authority weights and hub weights of pages are calculated by updating the equations $a^{(t+1)} = \epsilon \cdot U + (1 - \epsilon) \cdot A_{row}^t$, $h^{(t)}$ and $h^{(t+1)} = \epsilon \cdot U + (1 - \epsilon) \cdot A_{col} a^{(t+1)}$ where each entry of U is $1/n$, A_{row} is the same as the adjacency matrix of the graph A with its rows normalized to sum to 1, A_{col} is the the same as the adjacency matrix of the graph A with its rows normalized to sum to 1.[6] The equations in (1) are iterated until they converge to the fixed points a^* and h^*. The convergence of these iterations is proved in [9]. We refer this version of HITS as *randomized HITS* or simply *RHITS*. Under *RHITS* each node is treated as both authority and hub. Next, we investigate stability aspect of *RHITS*.

Proposition 5. *RHITS is stable on the class of all directed graphs $\tilde{\mathcal{G}}$. Specifically, given a graph $G = (V, E) \in \tilde{\mathcal{G}}$,the graph G is perturbed producing a new graph ∂G, then we have*

$$||\tilde{a} - a||_1 \leq \frac{2(1 - \epsilon)}{\epsilon} \cdot \left(\sum_{j \in FP} h_i + \frac{1}{2 - \epsilon} \sum_{i \in BP} a_i \right)$$

5.2   Randomized SALSA

In a similar manner as that of HITS, it is possible to overcome the limitation of SALSA by randomizing the algorithm. We call this algorithm *Randomized SALSA* or simply *RSALSA*. Let be two random surfers; the first random surfer picks uniformly a random page with probability ϵ, and it follows a backward-link then a foward link with probability $1 - \epsilon$. This random surfer model defines the random walk on the authority nodes. The second random surfer picks uniformly a random page with probability ϵ, and it follows a forward-link then a backward-link with probability $1 - \epsilon$, defining a random walk on the hub nodes. More precisely, the markov chain for the authorities and hubs have the transition probabilities $P_a(i, j) = \frac{\epsilon}{n} + (1 - \epsilon) \sum_{\{k:k \in B(i) \cap B(j)\}} \frac{1}{|B(i)|} \frac{1}{|F(k)|}$ and $P_h(i, j) = \frac{\epsilon}{n} + (1 - \epsilon) \sum_{\{k:k \in F(i) \cap F(j)\}} \frac{1}{|F(i)|} \frac{1}{|B(k)|}$. The convergence of markov chains to unique distributions are guaranteed from the fact that a markov chain that has transition probabilities P(x,y) of the form $P(x, y) = \epsilon \mu(y) + (1 - \epsilon)Q(x, y)$ for some distributions μ and Q is *uniformly ergodic* [16]. Hence, the powers of transition probabilities converge geometrically to the unique distributions. Similar to *RHITS*, *RSALSA* treats each page as both authority and hub.

Proposition 6. *RSALSA is stable on the class of all directed graphs $\tilde{\mathcal{G}}$. Specifically, given a graph $G = (V, E) \in \tilde{\mathcal{G}}$ representing a web subgraph,the graph G is perturbed producing a new graph ∂G, then we have*

$$||\tilde{a} - a||_1 \leq \frac{4(1 - \epsilon)}{\epsilon} \cdot \left(\sum_{i \in BP} a_i \right)$$

---

[6] When a row of A_{row} has all zero entries, then the corresponding row of A_{row} is constructed to have all entries equal to 1/n. Similarly, if a col of A_{col} has all zero entries, then the corresponding col of A_{col} is constructed to have all entries equal to 1/n.

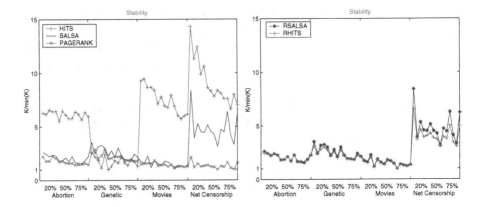

Fig. 1. Sensitivity Analysis

6   Experimental Results

Although the study of stability from the previous sections gives some useful theoretical insight about the robustness of algebraic link analysis algorithms, the notion of stability introduced in this paper is a *worst-case* notion. Hence, the theoretical analysis from previous sections will be complemented with some experimental studies to evaluate the robustness of algorithms in practice. Also, we study the performance of *RHITS* and *RSALSA* relative to some queries. From this study we show that *RHITS* can be, for instance, a way to overcome the limitation of HITS while being robust to perturbations since both *RHITS* and *RSALSA* outperform HITS specially on those queries in which HITS fails because of "Topic Drift" [6]. Stability results will be presented in Section 6.1 while the performance of algorithms relative to various queries will be presented in Section 6.2

6.1   Stability Result

Given the four directed graphs,[7] each one representing the set of pages and their interconnecting links, obtained as results of queries on "Genetic", "Abortion", "Movies" and "Net Censorship" following the guidelines suggested in [1], we perturbed each graph by randomly removing 25%, 50% and 75% of links from each one. We ran five perturbations on each graph to construct fifteen datasets in total for each topic. In order to measure the stability, we compared the magnitude of perturbation $||a - \tilde{a}||_1$ to the weights of perturbed pages. More precisely, we defined our sensitivity measure of link analysis algorithm T as $K = ||a - \tilde{a}||_1/(\sum_{i \in BP} c_i a_i + \sum_{j \in FP} h_j)$ where c_i denotes the number of backward-links that are perturbed. Recall from our definition of stability that

---

[7] These web data are freely available at
   http://www.cs.toronto.edu/~tsap/experiments

when the algorithm is not stable then K would be unbounded. Thus, volatile K values would be a possible indication of instability of the algorithm. In fact, HITS seems to present this kind of behavior as shown later on. We computed K for each dataset respect to all algebraic algorithms considered in this paper.[8] Let min(K) be the smallest K for each algebraic algorithm over all queries and for all trials. We divided each K by min(K) which corresponds to the Y axis in Figure 2. One can observe from Figure 2 the high sensitivity of HITS from its volatile K/min(K) values on different queries. On the other hand, the stability of PageRank is remarkable showing a stable behavior regardless of the dataset. Finally, the sensitivity of SALSA, *RSALSA* and *RHITS* can be seen to be between PageRank and HITS.

Table 1. Top 10 pages on "Abortion" (ϵ=0.1, Base Set Size=2293)

Index	URL	Title	Index	URL	Title
(1165)	DimeClicks.com -...	http://www5.dime clicks.com	(368)	Current Events - Law	http://law.miningco.com
(1193)	HitBox.com - ...	http://rd1.hitbox.com/	(1769)	Priests for Life Index	http://www.priestsforlife.org
(1184-92)	Amazon.com-...	http://www.amazon.com/	(0)	Abortion Clinics OnLine	http://www.gynpages.com
(1948)	Politics1: Hot Politics...	http://www.politics1.com/	(925)	Pregnancy Centers Online	http://www.pregnancy centers.org
(962)	ProlifeInfo	http://www.proli fe.org/ultimate	(1461)	Planned Parenthood Federation	http://www.planned parenthood.org
(1769)	Priests for Life Index	http://www.priests forlife.org	(666)	RoevWade.org	http://www.roevwade.org
(719)	Abortion and reprod. Res.	http://www.naral.org	(2)	The Abortion Rights Activist	http://www.cais.com /agm/main
(925)	Pregnancy Centers Online	http://www.pregnanc ycenters.org	(1984)	The John Birch Society	http://www.jbs.org
(0)	Abortion Clinics OnLine	http://www.gynpages.com	(1983)	American Opinion Book Service	http://www.aobs-store.com
(666)	RoevWade.org	http://www.roevwade.org	(2375)	About	http://home.about.com
(718)	Human Life International	http://www.hli.org	(1985)	TRIMonline	http://www.trimonline.org
(1325)	Feminists For Life of A.	http://www.serve.com /fem4life	(2382)	AllExperts.com	http://www.allexperts.com
(2262)	The Ultimate Pro-Life Resources	http://www.prolifeinfo.org	(46)	Project Rachel,...	http://manaco.simplenet.com/
(717)	National Right to Life Organization	http://www.nrlc.org	(2501)	The March For Life Fund	http://www.marchforlife.org
(1139)	Simple Catholicism	http://www.geocities.com/			

6.2   Performance Evaluation

In this section, we report results of a series of experiments that we conducted to evaluate the ranking quality of algebraic link analysis algorithms considered in this paper. We ran each algorithm on four different queries using the same datasets as those of [3]. For the sake of brevity, we only present the top 10 pages on the query "Abortion" (Table 1). The full set of experimental results can be found at the web page http://www.cs.toronto.edu/~leehyun/experiment.htm. The "Tightly Knit Community(TKC)" [6] effect for HITS is clearly observed with this particular query since its returned pages contain many irrelevant pages from "Amazon.com" in its top 10 pages. All top 10 pages produced by *RSALSA*

---

[8] Notice that when $\epsilon \rightarrow 1$, the stability of PageRank, *RHITS* and *RSALSA* is increased as the algorithms are reduced into simple uniform random jumps. Thus, the value of $\epsilon = 0.1$ was chosen to minimize the influence of ϵ on the stability of algorithms even though $0.1 < \epsilon < 0.2$ is the most widely used value of ϵ for PageRank.

and *RHITS*, in contrast, are relevant to the topic "Abortion". It should be noted that in spite of the identical top 10 pages between SALSA and RSALSA, a more careful study of low ranked pages revealed substantial difference between these algorithms.

	HITS	SALSA	RHITS	RSALSA	PageRank
	1165	717	717	717	1984
	1193	962	962	962	1983
	1184	1769	1769	1769	2375
	1188	719	719	719	1985
	1191	925	0	925	2382
	1189	0	925	0	46
	1187	666	1461	666	2501
	1192	718	718	718	717
	1190	1325	666	1325	1139
	1948	2262	2	2262	368
precision	0.1	1	1	1	0.3

	HITS	SALSA	RHITS	RSALSA	PageRank
HITS	10	0	0	0	0
SALSA	0	10	8	10	0
RHITS	0	8	10	8	0
RSALSA	0	10	8	10	0
PageRank	0	0	0	0	10

7   Conclusions

We studied the stability aspect of different algebraic link analysis algorithms. We gave a new definition of stability motivated by the definition of stability given in [3] and some bounds for $||a - \tilde{a}||_1$ found in [4]. In this paper, we showed that PageRank is stable, HITS is not stable, and SALSA is stable under certain circumstances according to our new definition of stability. Also, we reexamined *Randomized HITS* introduced in [9,7] showing that the algorithm is stable. Also, we proposed *Randomized SALSA* as a way to overcome the instability of SALSA on the class of all directed graphs $\tilde{\mathcal{G}}$. Finally, stability of link analysis algorithms were studied experimentally. Our work leads toward some practical and theoretical open questions to be attacked for future work. Above all, a more detailed studies about the stability aspect of link analysis algorithms on a large set of queries will be required. Also, it would be interesting to study if the notion of rank stability in [3] and our notion of stability are related.

References

1. J. Kleinberg, "Authoritive sources in a hyperlinked environment", *Journal of the ACM*, 46 1999.
2. S. Brin and L. Page, "The natomy of a large-scale hypertextual(Web) search engine", *Proc. 7th International World Wide Web Conference*, 1998.
3. A. Borodin, G.O. Roberts, J.S. Rosenthal, and P. Tsaparas, "Finding authorities and hubs from link structures on the world wide web", *Proc. 10th International WWW conference*, 2001.
4. A.Ng, A.Zheng and M.Jordan, "Link Analysis, Eigenvectors and Stability", *Proc. 7th International Conference on Artificial Intelligence*, 2001.
5. R. Lempel and S. Moran, "Rank-Stabiltiy and Rank-Similarity of Web Link-Based Ranking Algorithms", Technion CS Department technical report, CS-2001-22, 2001.
6. R. Lempel and S. Moran, The stochastic approach for link-structure analysis(SALSA) and the TKC effect. *Proc. 9th International World Wide Web Conference*, May 2000.
7. A.Ng, A.Zheng and M.Jordan, "Stable Algorithms for Link Analysis". *Proc. 24th ACM-SIGIR Conference on research and development in Information Retrieval* 415–429, May 2001.
8. T. Lindvall, Lectures on the coupling method, wiley series in probability and mathematica statistics, 1992.
9. D. Rafiei and A. Mendelzon, "What is this page known for? Computing web page reputations", *Proc. the 9th International World Wide Web Conference*, Amsterdam, Netherlands, 2000.
10. D. Cohn and H. Chang, "Probabilistically Identifying Authoritative Documents", *Proc. 17th International Conference on Machine Learning*, 2000.
11. K. Bharat and M. Henzinger, "Improved algorithms for topic distillation in a hyperlinked environment", *Proc. of 21st International Conference on Research and Development in Information Retrieval (SIGIR 1998)*.
12. S. Chakrabarti, M. Joshi and V. Tawde, "Enhanced topic distillation using text, markup tags, and hyperlinks", *Proc. 24th ACM-SIGIR Conference on Research and Development in Information Retrieval*, 2001.
13. D. Achilioptas, A. Fiat, A. Karlin, and F. McSherry, "Web search through hub synthesis", *Proc. 42nd Foundation of Computer Science*, Las Vegas, Nevada, 2001.
14. Y. Azar, A. Fiat, A. Karlin, F. McSherry, J. Saia, "Spectral analysis of data", *Proc. 33rd Symposium on Theory of Computing*, Hersonissos, Crete, Greece, 2001.
15. T. Haveliwala, "Efficient Computation of PageRank", Technical report, Stanford University Database Group, 1999.
16. S. Meyn and R. Tweedie, "Markov chains and stochastic stability", Springer, 1993.

Fast Construction of Generalized Suffix Trees over a Very Large Alphabet

Zhixiang Chen[1], Richard Fowler[1], Ada Wai-Chee Fu[2], and Chunyue Wang[1]

[1] Department of Computer Science, University of Texas-Pan American,
Edinburg TX 78539 USA.
chen@cs.panam.edu, {fowler,cwang}@panam.edu
[2] Department of Computer Science, Chinese University of Hong Kong,
Shatin, N.T., Hong Kong.
adafu@cse.cuhk.edu.hk

Abstract. The work in this paper is motivated by the real-world problems such as mining frequent traversal path patterns from very large Web logs. Generalized suffix trees over a very large alphabet can be used to solve such problems. However, traditional algorithms such as the Weiner, Ukkonen and McCreight algorithms are not sufficient assurance of practicality because of large magnitudes of the alphabet and the set of strings in those real-world problems. Two new algorithms are designed for fast construction of generalized suffix trees over a very large alphabet, and their performance is analyzed in comparison with the well-known Ukkonen algorithm. It is shown that these two algorithms have better performance, and can deal with large alphabets and large string sets well.

1 Introduction and Problem Formulation

1.1 Introduction

Recently, suffix trees have found many applications in bio-informatics, data mining and knowledge discovery. The first linear-time algorithm for constructing suffix trees was given by Weiner in [17] in 1973. A different but more space efficient algorithm was given by McCreight in [13] in 1976. Almost twenty years later Ukkonen gave a conceptually different linear time algorithm that allows on-line construction of a suffix tree and is much easier to understand. These algorithms build, in their original design, a suffix tree for a single string S over a given alphabet Σ. However, for any set of strings $\{S_1, S_2, \ldots, S_n\}$ over Σ, those algorithms can be easily extended to build a tree to represent all suffixes in the set of strings in linear time. Such a tree that represents all suffixes in strings $S_1, S_2, \ldots, S_n$, is called a *"generalized"* suffix tree.

Typical applications of generalized suffix trees include the identification of frequent (or longest frequent) substrings in a set of strings. One particular example of such applications is the mining of frequent traversal path patterns of Web users from very large Web logs [7,5], because such patterns are frequent

T. Warnow and B. Zhu (Eds.): COCOON 2003, LNCS 2697, pp. 284–293, 2003.
© Springer-Verlag Berlin Heidelberg 2003

(or longest frequent) substrings in the set of maximal forward references of Web users when maximal forward references are understood as strings of URLs. Such discovered patterns (or knowledge) can be used to predict where the Web users are going, i.e., what they are seeking for, so that it helps the construction and maintenance of real-time intelligent Web servers that are able to dynamically tailor their designs to satisfy users' needs [7]. It has significant potential to reduce, through prefetching and caching, Web latencies that have been perceived by users year after year [12]. It can also help the administrative personnel to predict the trends of the users' needs so that they can adjust their products to attract more users (and customers) now and in the future [2]. Other examples include document clustering, where a short summary of a document is viewed as a string of keywords and a generalized suffix tree is built for a set of such strings to group documents into different clusters.

In the mining of frequent traversal path patterns, specific properties exist for the data set. As investigated in [5], maximal forward references of Web logs exhibit properties as shown in Figure 1. In summary, the following properties hold: (1) The size of the set of strings (or maximal forward references) is very large, ranging from megabyte magnitude to gigabyte magnitude. (2) The size of the alphabet (or the number of unique URLs) is very large, ranging from thousands to tens of thousands or more. (3) All strings have a length ≤ 30 (derived from the given parameter setting of Web log sessionization). (4) More than 90% strings have lengths less than or equal to 4, and the average length is about 2.04.

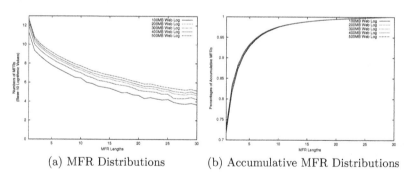

(a) MFR Distributions   (b) Accumulative MFR Distributions

Fig. 1. Properties of Maximal Forward References (MFRs)

1.2   Problem Formulation

Throughout this paper, we use Σ to denote an alphabet, and let $|\Sigma|$ denote the size of Σ. For any string $s \in \Sigma^*$, let $|s|$ denote its size, i.e., the number of all occurrences of letters in s. For any set of strings $\mathcal{S}$, let $|\mathcal{S}| = \sum_{s \in \mathcal{S}} |s|$. When $\mathcal{S}$ is stored as a file, we also refer $|\mathcal{S}|$ as the number of bytes of $\mathcal{S}$. Motivated by real world problems such as mining frequent traversal path patterns, in this paper we study fast construction of generalized suffix trees for a set $\mathcal{S}$ of strings over an alphabet Σ under the following conditions:

Conditions.

1. $|\Sigma|$ is very large, ranging from thousands to tens of thousands or more.

2. $|\mathcal{S}| = \sum_{s \in \mathcal{S}} |s|$, the size of the set of of strings, is very large, ranging from megabyte magnitudes to gigabytes magnitudes or more.

3. For each string $s \in \mathcal{S}$, $|s| \leq \alpha$, where α is a small constant.

We shall point out that depending on concrete applications, more restrictions can be added to the third condition. For example, in the case of mining frequent traversal path patterns, we can further require that 90% of strings in $\mathcal{S}$ have a length ≤ 4 and the average string length in $\mathcal{S}$ is about 2.04.

We now give several formal definitions. Unlike traditional suffix trees, in this paper we additionally require that counting information of substrings are recorded at internal nodes and leaves as well.

Definition 1. *For any string $s \in \Sigma$, we also denote $s = s[1..n]$ where $n = |s|$. For every i, j with $1 \leq i \leq j \leq n$, $s[i]$ is the i-th letter in s, and $s[i..j]$ is the substring from the i-th letter to the j-th letter. Note that $s[i..n]$ is a suffix starting at the i-th letter.*

Definition 2. *A suffix tree $\mathcal{T}$ for a string $s[1..n]$ over a given alphabet Σ is a rooted directed tree with exactly n leaves. Each internal node other than the root has at least two children and each edge is labeled with a nonempty substring of s. No two edges out of a node can have edge labels starting with the same letter. Each internal node or leaf has a counter to indicate the number of times the concatenation of the edge labels on the path from the root to the node or leaf occurs in the string s. The key feature of the suffix tree is that the concatenation of the edge labels on each path from the root to one of the n leaves represents exactly one of n suffixes of s.*

Definition 3. *Given a set of strings $\mathcal{S} = \{s_1, s_1, \ldots, s_m\}$ over an alphabet Σ, a suffix tree $\mathcal{T}$ for first string s_1 can be generalized to represent all suffixes and to record the counting information of substrings in the set of strings. The key feature of such a tree is that the concatenation of the edge labels on each path from the root to one of the leaves represents exactly a distinct suffixes in $\mathcal{S}$, and every suffix in $\mathcal{S}$ is represented by exactly one of such concatenations. Such a tree is called a "generalized" suffix tree. Usually, we assume that $\$ \notin \mathcal{S}$, and $\mathcal{S}$ is represented as $s_1\$s_2\$\ldots\$s_m\$$.*

In Figure 2(a), we illustrate a generalized suffix tree for "*mississippi\$missing\$ sipping\$*".

The goal of this paper is to design algorithms for fast construction of generalized suffix trees under Conditions (1) to (3). A sorting-based algorithm Sb-SfxTree (Sorting-based Suffix Tree) and a hashing-based algorithm HbSfxTree (Hashing-based Suffix Tree) will be devised and their performance will be analyzed comparatively. It is shown that algorithms SbSfxTree and HbSfxTree are substantially faster than Ukkonen's algorithm. Furthermore, these two algorithms have superior space scalable performance and can be easily tuned to parallel or distributed algorithms.

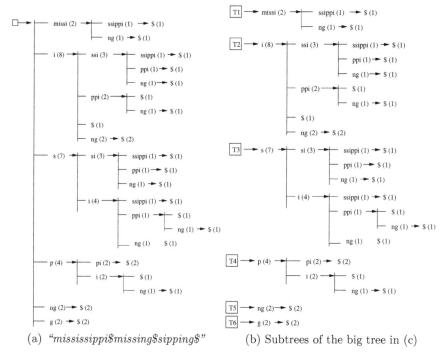

(a) *"mississippi\$missing\$sipping\$"*   (b) Subtrees of the big tree in (c)

Fig. 2. A Generalized Suffix Tree and Its Subtrees

The rest of the paper is organized as follows. In Section 2, we will review practical implementation challenges of suffix tree construction. Some properties of suffix trees are given in Section 3. Algorithms SbSfxTree and HbSfxTree are devised in Section 4. Performance analysis is given in Section 5. Finally, we conclude the paper in Section 6.

2   Practical Implementation Challenges

As well discussed in Gusfield [8] (pages 116 to 119), the Weiner, Ukkonen, and McCreight algorithms [17,13,16] have ignored the size of the alphabet Σ, and have not considered memory paging when trees are large and hence cannot be stored in RAM. When the size of Σ is too large to be ignored, those three algorithms all require $\theta(|\mathcal{S}| \cdot |\Sigma|)$ space, or the linear time bound $O(|\mathcal{S}|)$ should be replaced with $min\{O(|\mathcal{S}| \cdot \log |\mathcal{S}|), O(|\mathcal{S}| \cdot \log |\Sigma|)\}$.

The main design issues in all the three well known algorithms [17,13,16] are how to represent and search the branches out of the nodes of the tree. For example, in the Ukkonen algorithm, in order to achieve linear space complexity, array indexes are used to represent substrings labeling tree edges under the implicit assumption that the whole string (or the set of strings) is kept in RAM and represented as an array; and in order to achieve linear time complexity, suffix links are used to allow quick walks from one part of the tree to another part under the implicit assumption that the entire tree is kept in RAM. Those design techniques are great for theoretical time/space bounds, but are inadequate for paging if the string (or the set of strings) or the entire tree cannot be stored in RAM. Because of those algorithms' dependence on the availability of the entire string (or the set of strings) and the entire tree in RAM, and because of the tree's lack of nice locality properties, those algorithms cannot support parallel or distributed construction of the tree. Therefore, new techniques are needed for implementing generalized suffix trees for very large sets of strings over a very large alphabet.

Gusfield [8] summaries four basic alternative techniques for represent branches in order to balance the constraints of space against the need for speed. The first one is to use an array of size $\theta(|\Sigma|)$ at each non-leaf node to represent branches to children nodes. The second is to use linked list to replace array in the first technique. The third is to replace the linked at each non-leaf node with some balanced tree. Finally, the last is to use hashing at each non-leaf node to facilitate branch search. However, all the above alternative techniques fail to overcome the dependence on the availability of the entire string (or the set of strings) and the entire tree. The first three also increase the burden of space demand when the alphabet is very large. The challenge for the last one is to find a hashing scheme to balance space with speed. These techniques cannot facilitate the parallel or distributed construction of the trees.

3   Some Properties

Let $\mathcal{T}$ be a generalized suffix tree for a set of strings $\mathcal{S}$ over the alphabet Σ. For each child node v of the root of $\mathcal{T}$, we can obtain a subtree for v by simply removing all other children nodes and their descendants as well as their edges. E.g., Six subtrees are shown in Fig.2(b) for the tree in Fig.2(a). For each t of such subtrees, it is obvious that the root of t has exactly one edge leading to its only child node or to its only leaf. Let $t[1]$ denote the first letter in the string on the edge out of the root. We say a string s is contained in a subtree t if s is the concatenation of all edge-labels on a path from the root of t to some leaf of t.

Lemma 1. *Let W be the number of distinct letters that appear in the set of strings $\mathcal{S}$. $\mathcal{T}$ has exactly W many subtrees. Moreover, for any two distinct subtrees t' and t'' of $\mathcal{T}$, $t'[1] \neq t''[1]$.*

Proof Sketch. Directly from Definitions 2 and 3.

We may assume without loss of generality that $\mathcal{S}$ contains every letter in Σ (otherwise, a smaller Σ can be used). Lemma 1 means that $\mathcal{T}$ has exactly $|\Sigma|$ many subtrees, each of which starts with a letter in Σ.

Lemma 2. *For any subtree t of $\mathcal{T}$, and for any suffix s in any string of $\mathcal{S}$, t contains s if and only if $s[1] = t[1]$.*

Proof Sketch. If t contains s, then s is the concatenation of edge labels on a path from the root of t to some leaf of t, thus we have $s[1] = t[1]$. By Definitions

2 and 3, s is the concatenation of edge labels on a path from the root of $\mathcal{T}$ to one of T's leaves. Let t' be the subtree of $\mathcal{T}$ that has the path representing s, then we have $s[1] = t'[1]$. If $s[1] = t[1]$, then $t[1] = t'[1]$, hence it follows from Lemma 1 that $t = t'$, i.e., t contains s.

Corollary 1. *Let t be any subtree of $\mathcal{T}$, and s be any substring in a string of $\mathcal{S}$. Then, s is the concatenation of edge labels on a path from the root of t to some internal node (or leaf) of t, and the frequency of s is recorded at the node (or leaf) counter, if and only if $s[1] = t[1]$.*

Proof Sketch. By Definitions 2 and 3, Lemma 2 and the fact that any substring in a string of $\mathcal{S}$ is a prefix of some suffix in a string of $\mathcal{S}$.

4 New Algorithms

4.1 The Strategy

Lemma 2 and Corollary 1 combined imply a new way of fast construction of generalized suffix trees. The strategy is to organize all suffixes starting with the same letter into a group and build a subtree for each of such groups. A more or less related strategy has been devised in [9], but the strings considered there are over a small alphabet and the method used to build subtrees is of quadratic time complexity.

The task of grouping can be done by means of sorting or hashing. We shall point out that the hashing here is substantially different from other hashing techniques used to improve the performance of suffix tree construction [8]. We use hashing here for the purpose of *"divide-and-conquer"*, while others use hashing to speed up searching the branches out of the nodes. The task of constructing a subtree can be done easily, say, with one phrase execution of the Ukkonen Algorithm. Recall that the Ukkonen algorithm builds a suffix tree for a string $s[1 : n]$ in n phrases with the i-th phrase adding the i-th suffix $s[i : n]$ to the existing (but partially built) suffix tree. Let $ResUkkonen(SuffixTree\ t,\ NewString\ s)$ denote the one phrase execution of the Ukkonen Algorithm to add the only suffix $s[1 : n]$ to t. We additionally require that ResUkkonen records frequencies of substrings at internal nodes and leaves, which can be done easily by tuning the Ukkonen algorithm.

4.2 Algorithm SbSfxTree

The key idea is as follows. Read strings sequentially from an input file, and for every string $s[1 : n]$ output its n suffixes to a temporary file. Sort the temporary file to group all the suffixes starting with the same letter together. Finally, build a subtree for each group of such suffixes.

```
input:
    infile: a set of strings
    tmpfile: a set of suffixes
    outfile: a set of subtrees
Begin
1.      while (infile is not empty)
2.          readString(infile, s[1:n])
3.          for (i = 1; i ≤ n; i + +)
4.              tmpfile.append(s[i : n])
6.      sort(tmpfile); createSuffixTree(t)
7.      while (tmpfile is not empty)
8.          readString(tmpfile, s)
9.          if (t.empty() or t[1] == s[1])
10.             RstUkkonen(sft, s)
11.         else if (t[1]≠ s[1])
12.             Output(t,outfile), ResetTree(t)
13.     Output(t,outfile)
end
```

Fig. 3. Algorithm SbSfxTree

4.3   Algorithm HbSfxMiner

The key idea is to replace sorting with hashing to group all suffixes with the same starting letter together.

```
input:
    infile: a set of strings
    f: a hashing function from letters to integers
    outfile: a set of subtrees
Begin
1.      create subtrees t_1, . . . , t_{|Σ|};
2.      while (infile is not empty)
3.          readString(infile, s[1:n])
4.          for (i = 1; i ≤ n; i + +)
5.              RstUkkonen(t_{f(s[i])}, s[i:n])
6.      for (i = 0; i < |Σ|; i + +)
7.          Output(t_i, outfile)
end
```

Fig. 4. Algorithm HbSfxTree

5   Performance Analysis

Due to space limit, we only give complexity bounds for algorithm SbSfxTree and would like to point out that similar bounds can be given to algorithm HbSfxTree. We will also present experimental results for both algorithms.

Theorem 1. *Let $\mathcal{T}$ be a generalized suffix tree of a set of strings $\mathcal{S}$ over an alphabet Σ satisfying Conditions(1), (2) and (3). Assume the size of each subtree t of $\mathcal{T}$ is $O(|\mathcal{S}|/|\Sigma|)$. Algorithm SbSfxTree builds $\mathcal{T}$ (via building all its subtrees) in time $O(|\mathcal{S}| \cdot \log |\mathcal{S}| + |\mathcal{S}| \cdot \log |\Sigma|)$ and in space $O(|\mathcal{S}|/|\Sigma|)$.*

Proof Sketch. By Conditions (1), (2) and (3), each string $s \in \mathcal{S}$ has at most α suffixes. Hence, the size of the suffix file is at most $\alpha|\mathcal{S}|$, this means that sorting to group suffixes is of $O|\mathcal{S}| \cdot \log |\mathcal{S}|)$ time complexity and of $O(|\mathcal{S}|/|\Sigma|)$ space complexity when a buffer of $O(|\mathcal{S}|/|\Sigma|)$ size is used. It follows from the Ukkonen algorithm that building a subtree requires $O(|\mathcal{S}|/|\Sigma|)$ space and $O((|\mathcal{S}|/|\Sigma|) \cdot \log |\Sigma|)$ time. This means that the total space for building all the subtrees is still $O(|\mathcal{S}|/|\Sigma|)$, but the total time is by Lemma 1 $O((|\Sigma| \cdot |\mathcal{S}|/|\Sigma|) \cdot \log |\Sigma|) = O(|\mathcal{S}| \cdot \log |\Sigma|)$.

In theory, algorithm SbSfxTree has better space complexity, while it has almost the same time complexity bound as the Weiner, Ukkonen, and McCreight algorithms. In Fig.5(a,b,c), we report experimental analysis of SbSfxTree and HbSfxTree in comparison with the Ukkonen algorithm. The computing environment is a Dell PWS 340 with a 1.5 GHz P4 Processor, 512 MB RAM and 18 GB memory. In those experiments, we used alphabets $\Sigma_i, i = 1, 2, 3$, such that $|\Sigma_1| = 10,000$, $|\Sigma_2| = 15,000$ and $|\Sigma_3| = 20,000$. For each Σ_i, we generated three sets of 1 million strings such that sizes of strings follow Poisson distributions with means values of 1, 2 and 3, respectively. It is clear that algorithms SbSfxTree and HbSfxTree have substantially better performance than the Ukkonen algorithm. For the set of strings following Poisson distribution of means value 3, the Ukkonen algorithm ran out of memory. The current version of algorithm HbSfxTree also ran out memory, because all the subtrees were stored in RAM. We shall point out that this can be improved through paging subtrees in the next stage of implementation.

In [5], we has applied algorithms SbSfxTree and HbSfxTree to the mining of frequent traversal path patterns from very large Web logs. Fig.5(d,e,f) shows performance of SbSfxTree and HbSfxTree based mining in comparison with the Ukkonen algorithm based mining. It is clear that SbSfxTree and HbSfxTree are far superior to the Ukkonen algorithm. It is also shown [5] that SbSfxTree and HbSfxTree are far superior to the apriori-like algorithms within the context of mining frequent traversal path patterns.

Remark 1. By Lemma 2, the construction of one subtree has no dependence on any other subtrees. This means that both algorithms SbSfxTree and HbSfxTree can be easily revised to allow parallel or distributed construction of generalized suffix tree.

6   Conclusions

The work in this paper is motivated by the real-world problems such as mining frequent traversal path patterns from very large Web logs. Generalized suffix trees over a very large alphabet can be used to solve such problems. However,

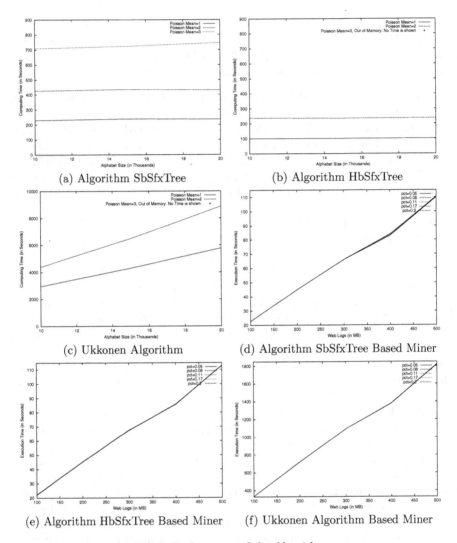

Fig. 5. Performance of the Algorithms

due to large magnitudes of the underlying alphabet and the set of strings, traditional algorithms such as the Weiner, Ukkonen and McCreight algorithms are not sufficient assurance of practicality. We have designed two algorithms for fast construction of generalized suffix trees over very alphabet. We have shown that the two algorithms are efficient in theory and in practice, and applied them to solve the problem of mining frequent traversal path patterns.

Acknowledgment. Thank Prof. Ukkonen for sending his work [16] to us. Thank Yavuz Tor for helping us on experiments in Fig.5(a,b,c). The work of the first two and the last authors is supported in part by the Computing and

Information Technology Center of the University of Texas-Pan American. The work of the third author is supported by the CUHK RGC Research Grant Direct Allocation ID 2050279.

References

1. J. Borges and M. Levene. Data mining of user navigation patterns. *MS99*, 1999.
2. A.G. Buchner and M.D. Mulvenna. Discovering internet marketing intelligence through online analytical web usage mining. *ACM SIGMOD RECORD*, pages 54–61, Dec. 1998.
3. L. Catledge and J. Pitkow. Characterizing browsing behaviors on the world wide web. *Computer Networks and ISDN Systems*, 27, 1995.
4. Z. Chen, R. Fowler, and A. Fu, Linear time algorithms for finding maximal forward references, Proc. of the IEEE Intl. Conf. on Info. Tech.: Coding & computing (ITCC 2003), 2003.
5. Z. Chen, R. Fowler, A. Fu, and C. Wang, Linear and sublinear time algorithms for mining frequent traversal path patterns from very large Web logs, Proceeding of the Seventh International Database Engineering and Applications Symposium, 2003.
6. Z. Chen, A. Fu, and F. Tong, Optimal algorithms for finding user access sessions from very large Web logs, Advances in Knowledge Discovery and Data Mining/PAKDD'02, Lecture Notes in Computer Science 2336, pages 290–296, 2002. (Full version will appear in Journal of World Wide Web: Internet and Information Systems, 2003.)
7. M.S. Chen, J.S. Park, and P.S. Yu. Efficient data mining for path traversal patterns. *IEEE Transactions on Knowledge and Data Engineering*, 10:2:209–221, 1998.
8. D. Gusfield, *Algorithms on Strings, Trees, and Sequences*, Cambridge University Press, 1997.
9. E. Hunt, M.P. Atkinson and R.W. Irving, A database index to large biological sequences, Proceedings of the 27th International Conference on Very Large Data Bases, pages 139–148, 2001.
10. R. Kosala and H. Blockeel, Web mining research: A survey, SIGKDD Explorations, 2(1), pages 1–15, 2000.
11. F. Masseglia, P. Poncelet, and R. Cicchetti, An efficient algorithm for Web usage mining, Networking and Information Systems Journal, 2(5–6), pages 571–603, 1999.
12. J. Pitkow and P. Pirolli, Mining longest repeating subsequences to predict World Wide Web Surfing, Proc. of the Second USENIX Symposium on Internet Technologies & Systems, pages 11–14, 1999.
13. E.M. McCreight, A space-economical suffix tree construction algorithm, Journal of Algorithms, 23(2),pages 262–272, 1976.
14. C. Shababi, A.M. Zarkesh, J. Abidi, and V. Shah. Knowledge discovery from user's web page navigation. *Proceedings of the Seventh IEEE Intl. Workshop on Research Issues in Data Engineering (RIDE)*, pages 20–29, 1997.
15. Z. Su, Q. Yang, Y. Lu, and H. Zhang, WhatNext: A prediction system for Web requests using N-gram sequence models, Proc. of the First International Conference on Web Information Systems Engineering, pages 200–207, 2000.
16. E. Ukkonen, On-line construction of suffix trees, Algorithmica, 14(3), pages 249–260, 1995.
17. P. Weiner, Linear pattern matching algorithms, Proc. of the 14th IEEE Annual Symp. on Switching and Automata Theory, pages 1–11, 1973.

Complexity Theoretic Aspects of Some Cryptographic Functions

Eike Kiltz and Hans Ulrich Simon

Lehrstuhl Mathematik & Informatik,
Ruhr-Universität Bochum, Germany.
{kiltz,simon}@lmi.ruhr-uni-bochum.de

Abstract. In this work, we are interested in non-trivial upper bounds on the spectral norm of binary matrices M from $\{-1, 1\}^{N \times N}$. It is known that the distributed Boolean function represented by M is hard to compute in various restricted models of computation if the spectral norm is bounded from above by $N^{1 \square \varepsilon}$, where $\varepsilon > 0$ denotes a fixed constant. For instance, the size of a two-layer threshold circuit (with polynomially bounded weights for the gates in the hidden layer, but unbounded weights for the output gate) grows exponentially fast with $n := \log N$. We prove sufficient conditions on M that imply small spectral norms (and thus high computational complexity in restricted models). Our general results cover specific cases, where the matrix M represents a bit (the least significant bit or other fixed bits) of a cryptographic decoding function. For instance, the decoding functions of the Pointcheval [9], the El Gamal [6], and the RSA-Paillier [2] cryptosystems can be addressed by our technique. In order to obtain our results, we make a detour on exponential sums and on spectral norms of matrices with complex entries. This method might be considered interesting in its own right.

1 Introduction

Despite the fact that almost all Boolean functions require circuits of exponential size for their computation (which follows from counting arguments), there is amazingly small progress in *proving* lower bounds on the circuit size for *concrete* families of Boolean functions. The best result as yet is still the lower bound of Blum [1] who presented a family of Boolean functions that can only be computed if the circuit contains at least $3n$ gates. Even for some restricted models of computation, research process may get stuck as long as the right tools for proving lower bounds are missing. For instance, there was no progress in proving exponential lower bounds on the size of *monotone* Boolean circuits until Razborov discovered his celebrated lower bound on the size of monotone circuits that decide the Clique problem. Another more recent example concerns a theorem proven by Forster [4]. He was able to show that a distributed Boolean function $f : \{0,1\}^n \times \{0,1\}^n \to \{0,1\}$ has probabilistic communication complexity (in the unbounded error model) at least $n - \log \|M\|_2$, where M is the matrix given by $M_{x,y} = (-1)^{f(x,y)}$ and $\|M\|_2$ denotes the so-called *spectral norm* of M

T. Warnow and B. Zhu (Eds.): COCOON 2003, LNCS 2697, pp. 294–303, 2003.
© Springer-Verlag Berlin Heidelberg 2003

(see Definition 8). Since the spectral norm of the Hadamard matrix H (with ± 1-entries) is $2^{n/2}$, it has communication complexity $n - n/2 = n/2$. This resolved a problem that had been open for roughly twenty years since the invention of the unbounded error model of probabilistic communication complexity by Simon and Paturi [8]. As shown by several authors [5], the new lower bound on the communication complexity could be used to lift existing lower bounds in various other restricted models of computation to a significantly higher level. For instance, one can convert "good" upper bounds on the spectral norm into exponential lower bounds on the size of threshold circuits with two layers, where the gates in the hidden layer have polynomially bounded weights, but the output gate has unbounded weights. Before Forster came up with his result, one could get exponential lower bounds only when the weights of the output gate were polynomially bounded.

Despite this progress, it turned out quite soon that Forster's result is not always easy to apply. It clearly applies directly to orthogonal matrices (whose spectral norm is small and well-known). In general however, the computation of upper bounds on the spectral norm of a family of matrices is a difficult task. As there are not so many examples (besides the orthogonal matrices) so far, we think it is justified to provide some additional tools that allow one to exploit Forster's result in a more powerful manner. In this paper, we show how exponential sums and the analysis of the spectral norm of some auxiliary matrices with entries from $\mathbb{C}$ (the complex numbers) can be used for this purpose. Interestingly, the families of matrices that can be analysed by our methods are natural (and maybe of some interest) in the sense that they represent the bits of cryptographic decoding functions. It comes as no surprise that these bits are hard to compute, but the point is that, without Forster's Theorem and some additional tools that are presented in this paper, there is absolutely no chance to derive lower bounds of this type. We see our paper as a step in the long process of adding new "horse power" to the existing machinery for proving lower bounds.

Related Work. There are various research papers that are closely related to our work. In [3,11,13,14], lower bounds on several complexity measures (but not on the spectral norm) of the Diffie-Hellman function, the related Squaring Exponent function and the discrete logarithm function are given. It is shown, for instance, that any polynomial representation of the Diffie-Hellman functions, even for small subsets of their input, must inevitably have many non-zero coefficients and, thus, high degree. Lower bounds that result from the analysis of the spectral norm typically have a different flavour (and are sometimes stronger). For instance, the aforementioned threshold circuits represent a proper superclass of the class of functions that can be represented by sparse polynomials. The paper with the closest relation to our own is probably [7]. There the spectral norm of matrices that represent the bits of the Diffie-Hellman function is analysed. As in this paper, a good upper bound is proven that leads to lower bounds in various restricted models of computation (as explained above). It should be mentioned, however, that the tools used in [7] and the tools required in this paper are quite different.

2   Functions with Multiplicative Structure

Let $M : \mathbb{Z}_N \times \mathbb{Z}_N$ be the function that satisfies $M(x, y) = xy$ (multiplication in $\mathbb{Z}_N$). In this paper, we want to exploit the fact that some cryptographic functions are close relatives of M. The general concept of a "multiplicative structure" is captured in the following

Definition 1. *Let $\mathcal{X}$ and $\mathcal{Y}$ be two subsets of $\mathbb{Z}_N$. We say H has a multiplicative structure on subdomain $\mathcal{X} \times \mathcal{Y}$ if there exist permutations σ, τ of $\mathbb{Z}_N$ such that $H(x, y) = M_{\sigma,\tau}(x, y) := \sigma(x)\tau(y)$ for all $x \in \mathcal{X}$ and $y \in \mathcal{Y}$.*

We present some examples:

Example 2 (Pointcheval Cryptosystem [9]). Let N be an RSA-modulus, i.e. $N = pq$ for two primes p, q of (roughly) the same bit length. Let $e, d \in \mathbb{Z}_N$ such that $ed = 1 \pmod{\varphi(N)}$. Let $\mathcal{R} = \{r \in \mathbb{Z}_N \mid r + 1 \in \mathbb{Z}_N^*\}$. The following function $F : \mathcal{R} \times \mathbb{Z}_N \to \mathbb{Z}_N \times \mathbb{Z}_N$ such that $F(r, m) = (r^e, (r + 1)^e m)$, where r is chosen at random, encodes a message m. Define $\mathcal{X} := \{r^e \mid r \in \mathcal{R}\}$. As for decoding (given the "secret key" d), we may use any function $H : \mathbb{Z}_N \times \mathbb{Z}_N \to \mathbb{Z}_N$ that satisfies $H(x, y) = y(x^d + 1)^{-e}$ for all $x \in \mathcal{X}$: it is easy to check that this implies that $H(r^e, (r + 1)^e m) = m$ for all $r \in \mathcal{R}$. Outside the subdomain $\mathcal{X} \times \mathbb{Z}_N$ (when H is applied to invalid cyphers), we may however allow H to attain arbitrary values. Since $\tilde{\sigma}(x) = (x^d + 1)^{-e}$ is an injective mapping from $\mathcal{X}$ to $\mathbb{Z}_N$, it can be completed to a permutation σ on $\mathbb{Z}_N$. Then $H(x, y) = \sigma(x)y$ and it follows that any correct decoding function H has a multiplicative structure on subdomain $\mathcal{X} \times \mathbb{Z}_N$. Note that $|\mathcal{X}| = \varphi(N) = (p - 1)(q - 1)$. Thus, the part outside the subdomain has a relatively small size: $N - |\mathcal{X}| = N - \varphi(N) = p + q - 1 \approx N^{1/2}$.

Example 3 (El Gamal Cryptosystem [6]). Let $N = p$ be a prime modulus and let g be a generator of $\mathbb{Z}_N^*$. The "public key" has the form $h = g^s$ for some $s \in \mathbb{Z}_N^*$. A message $m \in \mathbb{Z}_N$ is encrypted with "randomness" $r \in \mathbb{Z}_N^*$ by evaluating the function $F : \mathbb{Z}_N^* \times \mathbb{Z}_N \to \mathbb{Z}_N \times \mathbb{Z}_N$ such that $F(r, m) = (g^r, h^r m)$. As for decoding (given the "secret key" s), we may use any function $H : \mathbb{Z}_N \times \mathbb{Z}_N \to \mathbb{Z}_N$ that satisfies $H(x, y) = yx^{-s}$ for all $x \in \mathbb{Z}_N^*$ and all $y \in \mathbb{Z}_N$: it is easy to check that $H(g^r, h^r m) = m$. Since, for all $x, y \in \mathbb{Z}_N^* \times \mathbb{Z}_N$ and for $\sigma(x) := x^{-s}$, H satisfies $H(x, y) = \sigma(x)y$, we may conclude that H has a a multiplicative structure on subdomain $\mathbb{Z}_N^* \times \mathbb{Z}_N$. Note that $N - |\mathcal{X}| = p - (p - 1) = 1$.

Example 4 (RSA-Paillier Cryptosystem [2]). This is a modification of the original Paillier cryptosystem. Again the function that performs decoding has a multiplicative structure. Details are given in the full paper.

In all these examples $\mathcal{Y} = \mathbb{Z}_N$ and τ is the identity. The generality of Definition 1 will not however cause much trouble in what follows.

In the course of the paper we will prove the following

Theorem 5. *Let $B_k(z)$ denote the k-th least significant bit of (the binary representation) of $z \in \mathbb{Z}_N$. Assume that k is either constant or (as a function in N) satisfies $k(N) = o(\log N)$.[1] Let H be a function with a multiplicative struc-*

---

[1] This rules out the possibility to select bits of high significance.

ture on subdomain $\mathcal{X} \times \mathcal{Y}$ of $\mathbb{Z}_N \times \mathbb{Z}_N$. Assume that $N - |\mathcal{X}| = O(N^\alpha)$ and $N - |\mathcal{Y}| = O(N^\alpha)$ for some fixed constant $0 \leq \alpha < 1$. Then, the spectral norm of the matrix $B_k \circ H$ (defined in section 3) is bounded above by $O(N^{(1+\alpha)/2})$ if $\alpha > 0$ and by $N^{1/2+o(1)}$ for $\alpha = 0$.

Note that this theorem applies to the cryptographic functions from examples 2, 3, and 4, where $N - |\mathcal{Y}| = 0$ and $N - |\mathcal{X}| = 1$ (El-Gamal) or $N - |\mathcal{X}| = \theta(N^{1/2})$ (Pointcheval and RSA-Paillier), respectively.

Theorem 6 ([5], Thrm. 6). *A binary function $A : \mathbb{Z}_N \times \mathbb{Z}_N \to \{-1, 1\}$ having its associated spectral norm upper bounded by $s(N)$ cannot be computed by a threshold circuit with two layers, where the output gate has unbounded weights and the gates in the hidden layer have polynomially bounded weights, unless the number of threshold gates is $N^{1-o(1)}/s(N)$.*

By this lower-bound machinery applied to Theorem 5 we obtain the

Corollary 7. *Given the notation and assumptions from Theorem 5, the binary function $B_k \circ H$ (taking input values $(x, y) \in \mathbb{Z}_N \times \mathbb{Z}_N$) cannot be computed by a threshold circuit with two layers, where the output gate has unbounded weights and the gates in the hidden layer have polynomially bounded weights, unless the number of threshold gates is $N^{\frac{1-\alpha}{2} - o(1)}$.*

In the full paper, we mention also the lower bound on the probabilistic communication complexity (and lower bounds in other restricted models of computation).

3   Some Algebraic Definitions and Facts

From here throughout the paper, a vector u is considered as column vector and u' denotes its transpose. The so-called spectral norm of a matrix with real or complex entries plays a central role in our paper:

Definition 8. *The spectral norm of $M \in \mathbb{R}^{N_1 \times N_2}$ is defined as follows:*

$$\|M\|_2 := \sup_{u:\|u\|_2=1} \|Mu\|_2 .$$

Here, the supremum ranges over vectors with real components. The analogous definition is used for matrices from $\mathbb{C}^{N_1 \times N_2}$. (Clearly, the supremum then ranges over vectors with complex components.)

We briefly mention some well-known facts concerning the spectral norm:

Fact 9 *The spectral norm of matrix $A \in \mathbb{R}^{N \times N}$ satisfies the following equation:*

$$\|A\|_2 = \sup_{u,v:\|u\|_2=\|v\|_2=1} |u'Av| = \sup_{u,v:\|u\|_2=\|v\|_2=1} \left| \sum_{x=0}^{N-1} \sum_{y=0}^{N-1} A_{x,y} u_x v_y \right| . \quad (1)$$

The analogous remark holds for matrices and vectors over the field $\mathbb{C}$ of complex numbers.

As a consequence of the Cauchy-Schwarz inequality, we obtain

$$\sup_{u,v:\|u\|_2=\|v\|_2=1} \left| \sum_{x=0}^{N-1} \sum_{y=0}^{N-1} u_x v_y \right| = N \ . \tag{2}$$

For each $a \in \mathbb{Z}_N$, we define the following function $E_a^N : \mathbb{Z}_N \to \mathbb{C}$:

$$E_a^N(x) = e^{\frac{2\pi i a x}{N}} = \cos \frac{2\pi a x}{N} + i \cdot \sin \frac{2\pi a x}{N} \ . \tag{3}$$

In the sequel, we drop the superscript N when it is obvious from context. Clearly, the following rules are valid:

$$E_a(x \pm y) = E_a(x) \cdot E_{\pm a}(y) \tag{4}$$

$$\sum_{a \in \mathbb{Z}_N} E_{-a}(y) = \sum_{a \in \mathbb{Z}_N} E_a(y) \tag{5}$$

The following identity is well-known:

$$\frac{1}{N} \sum_{a \in \mathbb{Z}_N} E_a(x) = \begin{cases} 0 \text{ if } x \neq 0, \\ 1 \text{ if } x = 0. \end{cases} \tag{6}$$

From (6), we can immediately infer the following identity, which holds for each binary function $B : \mathbb{Z}_N \to \{-1, 1\}$:

$$B(z_0) = \frac{1}{N} \sum_{z \in \mathbb{Z}_N} B(z) \sum_{a \in \mathbb{Z}_N} E_a(z_0 - z) \ . \tag{7}$$

4   Upper Bounds on the Spectral Norm

Let $\mathcal{X}$ and $\mathcal{Y}$ be finite sets. In this section, we identify functions of the form $H : \mathcal{X} \times \mathcal{Y} \to \mathcal{Z}$ with matrices from $\mathcal{Z}^{\mathcal{X} \times \mathcal{Y}}$. Similarly, functions of the form $K : \mathcal{Z} \to \mathcal{K}$ are viewed as (column) vectors from $\mathcal{K}^{\mathcal{Z}}$. Note that the composition $K \circ H : \mathcal{X} \times \mathcal{Y} \to \mathcal{K}$ can then be identified with a matrix from $\mathcal{K}^{\mathcal{X} \times \mathcal{Y}}$. We are mainly interested in the following situation:

- $\mathcal{X} = \mathcal{Y} = \mathcal{Z} = \mathbb{Z}_N$ and $H : \mathbb{Z}_N \times \mathbb{Z}_N \to \mathbb{Z}_N$ is a "cryptographic function".
- For K, we consider functions $B : \mathbb{Z}_N \to \{-1, 1\}$ and the functions $E_a : \mathbb{Z}_N \to \mathbb{C}$ from (3). B represents a (hopefully secure) bit of the cryptographic function H.

Functions E_a are used as a mathematical tool within the analysis of the computational complexity of $B \circ H$. We pursue the strategy to bound the spectral norm of $B \circ H$ in terms of the spectral norm of $E_a \circ H$ (and some other terms). The hope is that "good" upper bounds on the spectral norm of $E_a \circ H$ are known (or easy to compute) and that they can be converted into "good" upper bounds on the spectral norm of $B \circ H$. As indicated in the introduction, we may then conclude that $B \circ H$ is hard to compute on some restricted computational devices.

The remainder of this section is organised as follows. In subsection 4.1, we bound $\|B \circ H\|_2$ in terms of $\|E_a \circ H\|_2$ and some additional terms. The additional terms depend on the binary function B only, but not on H. In this sense, we isolate the effects of B and H on the spectral norm of $B \circ H$ from each other. In subsection 4.2, we are concerned with bounds on $\|E_a \circ H\|_2$. The terms depending on B only are analysed in subsection 4.3. The whole analysis is completed in subsection 4.4.

4.1   The Splitting Lemma

The *bias* of a binary function $B : \mathbb{Z}_N \to \{-1, 1\}$ is given by

$$\mathrm{bias}_B(N) := \left| \sum_{z \in \mathbb{Z}_N} B(z) \right| \in \{0, \ldots, N\} . \tag{8}$$

It measures the degree of "balance" between negative and positive entries in vector B (where value 0 reflects perfect balance). The inner product of B and E_a (both viewed as vectors with complex entries) is denoted as $\langle B, E_a \rangle$.

The main result in this subsection reads as follows:

Lemma 10. $\|B \circ H\|_2 \leq \mathrm{bias}_B(N) + \frac{1}{N} \cdot \left| \sum_{a=1}^{N-1} \langle B, E_a \rangle \|E_{-a} \circ H\|_2 \right|.$

Proof. Choose $u, v \in \mathbb{R}^N$ such that $\|u\|_2 = \|v\|_2 = 1$ and $\|B \circ H\|_2 = |u'(B \circ H)v|$. We obtain

$$|u'(B \circ H)v| = \left| \sum_{x \in \mathbb{Z}_N} \sum_{y \in \mathbb{Z}_N} u_x v_y B(H(x, y)) \right|$$

$$\stackrel{(7)}{=} \frac{1}{N} \left| \sum_{x \in \mathbb{Z}_N} \sum_{y \in \mathbb{Z}_N} u_x v_y \sum_{z \in \mathbb{Z}_N} B(z) \sum_{a \in \mathbb{Z}_N} E_a(H(x, y) - z) \right|$$

$$\stackrel{(4)(5)}{=} \frac{1}{N} \left| \sum_{a \in \mathbb{Z}_N} \sum_{z \in \mathbb{Z}_N} B(z) E_a(z) \sum_{x \in \mathbb{Z}_N} \sum_{y \in \mathbb{Z}_N} u_x v_y E_{-a}(H(x, y)) \right|$$

$$= \frac{1}{N} \left| \sum_{a \in \mathbb{Z}_N} S(a) \right| ,$$

where $S(a)$ is given by

$$S(a) = \sum_{z \in \mathbb{Z}_N} B(z) E_a(z) \sum_{x \in \mathbb{Z}_N} \sum_{y \in \mathbb{Z}_N} u_x v_y E_{-a}(H(x, y))$$

$$= \langle B, E_a \rangle \cdot u'(E_{-a} \circ H)v \leq \langle B, E_a \rangle \cdot \|E_{-a} \circ H\|_2 .$$

The proof is completed by discussing the case $a = 0$ separately. Since $E_0(z) = 1$ for each $z \in \mathbb{Z}_N$, we get

$$\frac{1}{N}|S(0)| = \frac{1}{N}\left|\sum_{z\in\mathbb{Z}_N} B(z) \sum_{x\in\mathbb{Z}_N} \sum_{y\in\mathbb{Z}_N} u_x v_y\right| \overset{(2)}{\leq} \left|\sum_{z\in\mathbb{Z}_N} B(z)\right| = bias_B(N) .$$

The set $\mathbb{Z}_N$ can be cut into $\tau(N)$ slices, where $\tau(N)$ is the number of divisors of N. The d-th slice contains all $a \in \mathbb{Z}_N$ such that $\gcd(a, N) = d$. These are precisely the elements of the form ad for $a \in \mathbb{Z}^*_{N/d}$. This leads to the following

Corollary 11. *Assume that* $\left|\sum_{a\in\mathbb{Z}^*_{N/d}} \langle B, E_{ad}\rangle\right| \leq U_B(N, d)$ *and* $\|E_{ad} \circ H\|_2 \leq U_H(N, d)$ *for all* $a \in \mathbb{Z}^*_{N/d}$. *Define* $EXP_B(N, d) := \sum_{a\in\mathbb{Z}^*_{N/d}} \langle B, E_{ad}\rangle$. *Then,*

$$\|B \circ H\|_2 \leq bias_B(N) + \frac{1}{N} \cdot \left|\sum_{d:d|N,d<N} \sum_{a\in\mathbb{Z}^*_{N/d}} \langle B, E_{ad}\rangle \|E_{-ad} \circ H\|_2\right|$$

$$\leq bias_B(N) + \frac{1}{N} \cdot \sum_{d:d|N,d<N} U_B(N, d) \cdot U_H(N, d) .$$

We refer to the terms $bias_B(N)$ and $EXP_B(N, d)$ as the *balance terms associated with* B and briefly note the well-known fact [10] that

$$\tau(N) = 2^{(1+o(1))\frac{\ln N}{\ln \ln N}} = N^{o(1)} . \tag{9}$$

4.2   The Spectral Norm of the Auxiliary Matrices

In this subsection, we will bound $\|E_a \circ M\|_2$ for the function $M(x, y) = xy$. We will explain later how results dealing with M (multiplication in $\mathbb{Z}_N$) can be extended to arbitrary functions with a multiplicative structure. We make use of the following well-known result (that can be found in [12]):

Lemma 12. *For all* $a \in \mathbb{Z}^*_N$, *the following holds:* $\|E_a \circ M\|_2 \leq N^{1/2}$.

Lemma 12 applies only if $\gcd(a, N) = 1$. However, it is easy to extend the result to arbitrary $a \in \{1, \ldots, N - 1\}$. To this end we present the following lemma that holds for general functions $F : \mathbb{Z}_N \times \mathbb{Z}_N \to \mathbb{Z}_N$.

Lemma 13. *For all* d *such that* $d|N$ *and* $d < N$, *for all* $a \in \mathbb{Z}^*_{N/d}$, *for all permutations* σ, τ *on* $\mathbb{Z}_N$, *and for* $F_{\sigma,\tau}(x, y) = F(\sigma(x), \tau(y))$, *the following holds:* $\|E_{ad} \circ F_{\sigma,\tau}\|_2 = \|E_{ad} \circ F\|_2 \leq d\|E_a^{N/d} \circ F\|_2$.

Before we prove this Lemma we can make the following direct corollary by setting $F = M$ and then using Lemma 12 to obtain

Corollary 14. *For all* d *such that* $d|N$ *and* $d < N$, *for all* $a \in \mathbb{Z}^*_{N/d}$, *the following holds:* $\|E_{ad} \circ M_{\sigma,\tau}\|_2 \leq (dN)^{1/2}$.

Proof (of Lemma 13). The first equality is obvious since the spectral norm is invariant under permutation of rows and columns.

Note that each $x \in \mathbb{Z}_N$ has a unique decomposition $x = x_1 N/d + x_2$ with $x_1 \in \{0, \ldots, d-1\}$ and $x_2 \in \{0, \ldots, N/d - 1\}$. We choose $u, v \in \mathbb{R}^N$ such that $\|u\|_2 = \|v\|_2 = 1$ and $\|E_{ad} \circ F\|_2 = u'(E_{ad} \circ F)v$. For $j = 0, \ldots, d-1$, we define the vectors $\bar{u}_j, \bar{v}_j$ as follows:

$$\bar{u}_j := (u_{jN/d}, \ldots, u_{jN/d + N/d - 1})', \quad \bar{v}_j := (v_{jN/d}, \ldots, v_{jN/d + N/d - 1})'.$$

Define $s_j = \|\bar{u}_j\|_2$ and $t_j = \|\bar{v}_j\|_2$. Note that $\|s\|_2 = \|t\|_2 = 1$ for $s = (s_0, \ldots, s_{d-1})'$ and $t = (t_0, \ldots, t_{d-1})'$. The following calculation completes the proof:

$$
\begin{aligned}
|u'(E_{ad}^N \circ F)v| &= \left| \sum_{x,y \in \mathbb{Z}_N} u_x v_y e^{\frac{2\pi i \, a d F(x,y)}{N}} \right| \\
&= \left| \sum_{x_1, y_1 = 0}^{d-1} \sum_{x_2, y_2 = 0}^{N/d-1} u_{x_1 N/d + x_2} v_{y_1 N/d + y_2} E_a^{N/d}(F(x,y)) \right| \\
&\leq \sum_{j,k=0}^{d-1} |\bar{u}_j'(E_a^{N/d} \circ F)\bar{v}_k| \overset{(1)}{\leq} \sum_{j,k=0}^{d-1} s_j t_k \|E_a^{N/d} \circ F\|_2 \\
&\overset{(2)}{\leq} d \cdot \|E_a^{N/d} \circ F\|_2 .
\end{aligned}
$$

4.3   The Balance Terms

In this section we show upper bounds on the balance terms $\text{bias}_B(N)$ and $EXP_B(N, d)$ for various functions B. Due to space constraints we have to refer the reader to the full version of this work for all proofs of this section.

Function $B : \mathbb{Z}_N \to \{-1, 1\}$ is called *unbiased* if $\text{bias}_B(N) = N^{o(1)}$.

Lemma 15. *Let $B_k(z)$ denote the k-th least significant bit of $z \in \mathbb{Z}_N$. Then, $B_k : \mathbb{Z}_N \to \{-1, 1\}$ is unbiased if $k = k(N) = o(\log N)$.*

Lemma 16. *For every divisor $d < N$ of N,*

$$EXP_{B_k}(N, d) = O\left(N/d \cdot \text{bias}_{B_k}(N) \log(N/d)\right) . \tag{10}$$

In particular, if B_k is unbiased, then $EXP_{B_k}(N, d) = N^{1+o(1)}/d$.

Function $B : \mathbb{Z}_N \to \{-1, 1\}$ is called *semilinear of length k* if there exist parameters $M_i, L_i \in \mathbb{Z}_N$ and $K_i \in \mathbb{Z}_N^*$ such that, for all $z \in \mathbb{Z}_N$, condition

$$B(z) = 1 \Longleftrightarrow z \in \bigcup_{i=1 \ldots k} H_i, \quad H_i := \{K_i z + M_i \bmod N \mid 0 \leq z \leq L_i\},$$

is valid and the sets H_i are pairwise disjoint. We call B *strongly semilinear of length k* if B and $-B$ both are semilinear of length k. The next lemma shows that the property *strongly semilinear* implies a "good" upper bound on $EXP_B(N, d)$. Some concrete strongly semilinear functions are mentioned in the full paper.

Lemma 17. *Let $B : \mathbb{Z}_N \to \{-1,1\}$ be strongly semilinear of length k. Then, for every divisor $d < N$ of N, the following holds:*

$$EXP_B(N,d) = O\left(kN/d \cdot \log(N/d)\right) = kN^{1+o(1)}/d \ . \tag{11}$$

4.4   Putting All Together

We may now apply Corollary 11 with $U_{M_{\sigma,\tau}}(N,d) = (dN)^{1/2}$ (according to Corollary 14) and with the bounds from (10) and (11):

Corollary 18. *For each function $B_k : \mathbb{Z}_N \to \{-1,1\}$ (the k-th least significant bit) and for all permutations σ, τ of $\mathbb{Z}_N$, the following holds:*

$$\|B_k \circ M_{\sigma,\tau}\|_2 = O(bias_{B_k}(N)N^{1/2}\log(N) \sum_{d:d|N, d<N} d^{-1/2}) \overset{(9)}{=} bias_{B_k}(N)N^{1/2+o(1)}$$

In particular, if B_k is unbiased, then $\|B_k \circ M_{\sigma,\tau}\|_2 = N^{1/2+o(1)}$.

Corollary 19. *For each strongly semilinear function $B : \mathbb{Z}_N \to \{-1,1\}$ of length k and for all permutations σ, τ of $\mathbb{Z}_N$, the following holds:*

$$\|B \circ M_{\sigma,\tau}\|_2 = bias_B(N) + kN^{1/2+o(1)} \ .$$

In particular, if B is unbiased and strongly semilinear of length $N^{o(1)}$, then $\|B \circ M_{\sigma,\tau}\|_2 = N^{1/2+o(1)}$.

A function H with a multiplicative structure coincides with $M_{\sigma,\tau}$ on a sub-domain $\mathcal{X} \times \mathcal{Y}$. The "non-multiplicative behaviour" outside the subdomain can however be controlled by showing that whenever two matrices $C, D : \mathbb{Z}_N \times \mathbb{Z}_N \to \{-1,1\}$ coincide on a large enough subdomain $\mathcal{X} \times \mathcal{Y}$, then $|\|C\|_2 - \|D\|_2|$ is small. To be more precise, in the full version of this paper we prove

Lemma 20. *If H has a multiplicative structure on subdomain $\mathcal{X} \times \mathcal{Y}$, i.e., it coincides with $M_{\sigma,\tau}$ on $\mathcal{X} \times \mathcal{Y}$, then*

$$\|B \circ H\|_2 \leq \|B \circ M_{\sigma,\tau}\|_2 + (N \cdot (N - |\mathcal{X}|))^{1/2} +$$
$$(N \cdot (N - |\mathcal{Y}|))^{1/2} + ((N - |\mathcal{X}|) \cdot (N - |\mathcal{Y}|))^{1/2} \ .$$

Now Theorem 5 (even a more general version) follows directly by combining Corollaries 18 and 19 with Lemma 20.

5   Conclusions

From Corollary 19, we know that for each strongly semilinear function $B : \mathbb{Z}_N \to \{-1,1\}$ of length k and for all permutations σ, τ of $\mathbb{Z}_N$,

$$\|B \circ M_{\sigma,\tau}\|_2 = \mathrm{bias}_B(N) + kN^{1/2+o(1)} \ . \tag{12}$$

We will argue that in general there is no hope for improving this upper bound on the spectral norm beyond the value $\mathrm{bias}_B(N)$. To this end, let B be the indicator function on a small interval (say of length εN for a small constant $\varepsilon > 0$). Obviously, B is strongly semilinear of length 1 and has a large bias $(N(1 - 2\varepsilon))$. Now the given upper bound on the spectral norm is dominated by this term. Since the matrix $M_{\sigma,\tau}$ defines permutations in every row and every column, we can conclude that $\mathrm{bias}_B(N) = 1/N \sum_{x,y \in \mathbb{Z}_N} M_{x,y}$. On the other hand, we know that

$$\|B \circ M_{\sigma,\tau}\|_2 \geq \sum_{x,y \in \mathbb{Z}_N} \frac{1}{\sqrt{N}} M_{x,y} \frac{1}{\sqrt{N}} = \frac{1}{N} \sum_{x,y \in \mathbb{Z}_N} M_{x,y} = \mathrm{bias}_B(N) = N(1 - 2\varepsilon).$$

Note that this lower bound on $\|B \circ M_{\sigma,\tau}\|_2$ has the same order of magnitude as the upper bound derived from (12).

References

1. N. Blum. A Boolean function requiring $3n$ network size. *Theoretical Computer Science*, 28(3):337–345, February 1984.
2. D. Catalano, R. Gennaro, N. Howgrave-Graham, and P. Q. Nguyen. Paillier's cryptosystem revisited. In *Proceedings of the 8th ACM Conference on Computer and Communications Security*, pages 206–214, 2001.
3. D. Coppersmith and I. Shparlinski. On polynomial approximation of the Discrete Logarithm and the Diffie-Hellman mapping. *Journal of Cryptology*, 13(3):339–360, March 2000.
4. J. Forster. A linear lower bound on the unbounded error probabilistic communication complexity. In *Proceedings of the Sitheenth Annual Conference on Computational Complexity*, pages 100–106. IEEE Computer Society, 2001.
5. J. Forster, M. Krause, S. V. Lokam, R. Mubarakzjanov, N. Schmitt, and H. U. Simon. Relations between communication complexity, linear arrangements, and computational complexity. In *Proceedings of the Conference on Foundations of Software Technology and Theoretical Computer Science*, pages 171–182, 2001.
6. T. El Gamal. A public key cryptosystem and a signature scheme based on discrete logarithms. *Advances in Cryptology—CRYPTO '84*, pages 10–18, 1984.
7. E. Kiltz. On the representation of boolean predicates of the Diffie-Hellman function. In *Proc. of 20th International Symposium on Theoretical Aspects of Computer Science STACS*, pages 223–233, 2003.
8. R. Paturi and J. Simon. Probabilistic communication complexity. *Journal of Computer and System Sciences*, 33(1):106–123, 1986.
9. D. Pointcheval. New public key cryptosystems based on the dependent — RSA problems. *Lecture Notes in Computer Science*, 1592:239–254, 1999.
10. K. Prachar. *Primzahlverteilung*. Springer-Verlag, Berlin, 1957.
11. I. E. Shparlinski. *Cryptographic Application of Analytic Number Theory*. Birkhäuser Verlag, 2002.
12. I. M. Vinogradov. *Elements of number theory*. Dover Publications., 1954.
13. A. Winterhof. A note on the interpolation of the Diffie-Hellman mapping. In *Bulletin of the Australian Mathematical Society*, volume 64, pages 475–477, 2001.
14. A. Winterhof. Polynomial interpolation of the discrete logarithm. *Designs, Codes and Cryptography*, 25:63–72, 2002.

Quantum Sampling for Balanced Allocations

Kazuo Iwama[1,2], Akinori Kawachi[1,2], and Shigeru Yamashita[1*]

[1] Imai Quantum Computation and Information Project,
ERATO, Japan Sci. and Tech. Corp.
406, Iseya-cho, Kamigyo-ku, Kyoto 602-0873, Japan
yamasita@qci.jst.go.jp
[2] Graduate School of Informatics, Kyoto University
Yoshida Honmachi, Sakyo-ku, Kyoto 606-8501, Japan
{iwama,kawachi}@kuis.kyoto-u.ac.jp

Abstract. It is known that the original Grover Search (GS) can be modified to use a general value for the phase θ of the diffusion transform. Then, if the number of answers is relatively large, this modified GS can find one of the answers with probability one in a single iteration. However, such a quick and error-free GS can only be possible if we can initially adjust the value of θ correctly against the number of answers, and this seems very hard in usual occasions. A natural question now arises: Can we enjoy a merit even if GS is used without such an adjustment? In this paper, we give a positive answer using the balls-and-bins game in which the random sampling of bins is replaced by the quantum sampling, i.e., a single round of modified GS. It is shown that by using the quantum sampling: (i) The maximum load can be improved quadratically for the static model of the game and this improvement is optimal. (ii) That is also improved to $O(1)$ for the continuous model if we have a certain knowledge about the total number of balls in the bins after the system becomes stable.

1 Introduction

Suppose that we are given $a_1, a_2, \cdots, a_N$, such that one a_i has value 1 and all the others have 0. Grover search (GS, [13]), which is apparently one of the most celebrated quantum algorithms, can retrieve this a_i, called an *answer*, in time $O(\sqrt{N})$. The scheme is so general and many applications have been investigated, including quantum counting [8], minimum finding [12], claw finding [9] and so on. One can see, however, that all these applications focus on the case that the number of answers is relatively small, where the goal is how to find one of them quickly.

In this paper, we focus on the opposite case, i.e., the case that there are relatively many answers. If one half of $\{a_1, \cdots, a_N\}$ are answers, for instance,

---

* Current affiliation: Graduate School of Information Science, Nara Institute of Science and Technology.

then one can get an answer easily by repeating a (classical) random sampling, say twice. So, we are no longer interested in the time complexity. Instead, we are interested in the probability that we hit an answer. In the above case, the random sampling hits an answer with probability one half. If we could increase this probability to, for example, $1 - o(1)$ by using quantum mechanisms, then it would be much more desirable especially when we use such a sampling repeatedly. One typical example is the load-balancing problem investigated in this paper.

Suppose that we have N processors. Jobs arrive one after another, each of which should be assigned to one of the N processors. If we do not have enough information such as the loading factor of each machine or do not wish to spend too much time for NP-type scheduling, *randomization* often works well. This has been a popular problem with a long history and widely studied using the so-called balls-and-bins model [16][17][20]: Suppose that M balls are placed into N bins, one by one, uniformly at random. The number of balls stacked in each bin is called the *load* of the bin. Then if $M \approx N$, it is well known that the load of the highest bin is $\Theta(\ln N / \ln \ln N)$ with high probability. It is also known that if we allow a small amount of coordination, i.e., if we select two bins at random and put the ball into the less high bin, then the maximum load drops dramatically, i.e., to $O(\ln \ln N)$ [2][19][23]. Our major objective in this paper is to study what happens if the above random sampling for placing balls is replaced by quantum sampling.

1.1 Modified Grover Search

In this paper, we use the modified Grover Search(GS) introduced in [5][10][18] whose unitary transformation is given by

$$G(\theta, f) = -HS_0 HS_f,$$

where H is the Hadamard transformation and f is a given oracle $\{0, ..., N-1\} \to \{0, 1\}$, and

$$S_0 = I + (e^{i\theta} - 1)|0\rangle\langle 0|, \quad S_f = I + (e^{i\theta} - 1) \sum_{f(i)=1} |i\rangle\langle i|.$$

Here S_0 and S_f are generalization of $I_0 = I - 2|0\rangle\langle 0|$ and $I_f = I - 2\sum_{f(i)=1} |i\rangle\langle i|$, respectively, which are used in the original GS[13]. If we set $\theta = -\pi$, then it is the same as the original GS. Suppose for example that the number, $\#_f$, of answers (i.e., $\#_f = |\{i | f(i) = 1\}|$) is $N/2$. Then if we set $\theta = \pi/2$, we can get an answer with zero error by applying $G(\pi/2, f)$ *only once*. In general, suppose that $M = cN$ for a constant c such that $c \leq 3/4$. Then this single-round, error-free GS is available by setting $\theta = \arccos(1 - \frac{N}{2\#_f})$ [10].

1.2 Load Balancing

There are two different balls-and-bins models. In the *static model*, we put M balls into N bins, one by one, uniformly at random. In the *continuous model*, after we have put M balls as before, we continue putting balls uniformly at random, one at a unit time, and then remove one ball at a unit time. (Thus the total number of the balls in the bins remains unchanged during the game.)

Now suppose that the random sampling, which is used when putting a ball into a bin, is replaced by the quantum sampling based on the (modified) GS. Then the procedure is called *quantum load balancing* (QLB). More precisely, QLB is determined by two parameters $0 \le \theta \le 2\pi$ and an integer T and is denoted by $QLB(\theta, T)$. In each round, QLB selects the bin into which a ball is to be placed as follows;

(i) Initialize a $\log N$-bit quantum register to all 0's and apply the Hadamard transformation H.

(ii) Apply $G(\theta, f_T)$ to the register, where f_T is an oracle defined by

$$
f_T(i) = \begin{cases} 1 & \text{if bin } i \text{ holds } T-1 \text{ balls or less} \\ 0 & \text{otherwise} \end{cases}
$$

(iii) Observe the register to get value j. Put the ball to bin j.

QLB is similarly defined also for the continuous model but when we remove a ball, it is selected uniformly at random.

1.3   Our Contribution

Suppose that we know the number, $m \ge \frac{1}{4}N$, of empty bins out of the all N bins in each round. Then, by setting $T = 1$ and $\theta = \arccos(1 - \frac{N}{2m})$, the quantum search always selects an empty bin, namely, we can achieve the completely balanced allocation or the cost of the highest bin is equal to one. However, this is obviously unfair, since the condition is almost the same as that we know the current round is the i-th one. (If so, we can achieve the same result by putting the ith ball into the ith bin deterministically.) The natural setting, therefore, is to assume that both T and θ have fixed values during the game. Then it is no longer possible to enjoy the above-mentioned ideal selection of the bins during the whole game. Then it is no longer possible to enjoy the above-mentioned error-free selection during the whole game.

To see this, suppose that we would set $T = 1$ as before, i.e., we always wish to select an empty bin. Then, as described later, we can get $\left(\frac{N-m}{N}\right)^3$ as the error probability ($=$ the probability that the quantum sampling selects a nonempty bin) by setting θ appropriately. This appears much better than the classical sampling whose error probability is $\left(\frac{N-m}{N}\right)$. Unfortunately this is not true. The number of nonempty bins can grow to, say, cN for a constant c. Then the above error probability of the quantum sampling becomes c^3, which is essentially the same as c in the sense that both are constant. It is not hard to see that the maximum load is $\Omega(\ln N/\ln \ln N)$, which is the same as the classical case up to a constant factor.

Thus, the primary question is how to select T and θ to minimize the maximum load. In this paper we prove the following results: (i) The maximum load is at most $(2+o(1))\sqrt{\ln N/\ln\ln N}$ with high probability, if we set $T = \sqrt{\ln N/\ln\ln N}$ and $\theta = \pi/3$. Namely, even if we use fixed T and θ, QLB is quadratically better than its classical counterpart. (ii) This selection of T and θ is optimal since we can prove that the maximum load is $\Omega(\sqrt{\ln N/\ln\ln N})$ with high probability for any T and θ. (iii) Apparently QLB is more powerful for the continuous game, since the number of balls does not alter once it becomes stable. Suppose that M^* is the number of balls after the system becomes stable and that we can guess the value of M^* within the error of $\pm O(\sqrt{M^*})$. Then QLB can achieve a maximum load of $O(1)$. Recall that the maximum load does not change essentially between the static and continuous models in classical case.

1.4   Previous Results

As mentioned earlier, many applications of GS have appeared in the literature. Recently, the first nontrivial lower bound of $\Omega(N^{1/5})$ was proven [1] for the collision problem [9], and this result was improved in [21]. See [7][8][12][15] for more applications. We need significantly less iterations if we have some information (e.g., the number of 1's in the vector) of the answer. See [14] for more. It has also been a popular research topic to increase the success probability of GS in several occasions. Other than [10] described before, [18] investigates the case when there are few answers and [4][5] investigates the case when we have no information about the number of answers.

2   Basic Ideas and Useful Lemmas

Intuitively, QLB works as follows: Suppose that we set $T = T_0$ and $\theta = \theta_0$. Then at the beginning (until some fraction of N balls have been placed), the maximum load does not reach T_0 as shown in Fig 1. (The figure illustrates how the load of each bin looks like supposing the bins are sorted with their load.) In this case, $f_{T_0}(i) = 1$ for $1 \leq \forall i \leq N$, and therefore one can see that $G(\theta_0, f_{T_0})$ does not change the amplitude of each quantum state, i.e., $QLB(\theta_0, T_0)$ is exactly the same as the classical load balancing. Now, the load of the highest bin reaches T_0 as shown in Fig. 1 (b). Fig. 1 (d) shows the final distribution. We call the bin whose load is at least T_0 a *high bin* and a *low bin* otherwise. From this point of time, $f_{T_0}(i) = 0$ if bin i is high. Therefore QLB selects a low bin with higher probability as the bin to which a ball is placed. Recall, however, that we set $\theta = \theta_0$, which means that QLB selects a low bin with probability one (or never selects a high bin) only when the number of high bins is equal to X_0 that is determined by θ_0 and N (see Fig. 1 (c)). In other words, QLB selects high bins with some non-zero probability P_e before and after this optimal point of time. We often call this probability P_e *error probability* since selecting high bins obviously increases the maximum height. Our first lemma gives this error probability:

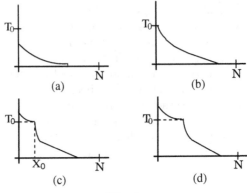

Fig. 1.

Lemma 1   Let k be the number of bins whose height is T or more. Then $QLB(\theta, T)$ selects a high bin with probability

$$P_e(k) = \frac{k}{N}\left(1 - 2\cos\theta - \frac{2k}{N}(1 - \cos\theta)\right)^2.$$

Proof   After applying $G(\theta, f_T)$, the amplitude of each high bin is

$$\frac{1}{\sqrt{N}}\left(-e^{i\theta} - (e^{i\theta} - 1)^2(1 - \frac{k}{N})^2\right),$$

as described in Sec. 1.1. Since we have k high bins, the error probability is

$$\frac{k}{N}\left|\left(-e^{i\theta} - (e^{i\theta} - 1)^2(1 - \frac{k}{N})\right)\right|^2 = \frac{k}{N}\left(1 - 2\cos\theta - \frac{2k}{N}(1 - \cos\theta)\right)^2. \quad \square$$

The next lemma [19] is convenient when we approximate the balls-and-bins game by the independent Poisson process. Suppose that we place M balls into N bins uniformly at random, which is called the *exact case*. We also consider the Poisson case where each bin independently receives a ball with probability M/N in each round (thus its load becomes a Poisson variable with mean M/N). Also, let a *load based event* be an event that depends solely on the loads of the bins.

Lemma 2   Suppose that the probability of a load based event is monotonically increasing with the total number of balls. Then the event which takes place with probability p in the Poisson case takes place with at most $4p$ in the exact case.

The third lemma (see e.g., [19]) is on the maximum load of the classical balls-and-bins process.

Lemma 3   Suppose that M balls are put into N bins uniformly at random. Then the maximum load is $\Omega(\ln N / \ln(N/M))$ with at least probability $1 - 1/N$ if $M < N/\ln N$, and is $\Omega(\ln N / \ln\ln N)$ with probability at least $1 - 1/N$ if $N/\ln N \leq M \leq N$.

3   Static Upper Bounds

As mentioned in Sec. 1.3, we need to use a relatively large value for the threshold T in order to take advantage of the quantum sampling. As for the angle θ, Lemma 1 strongly suggests to use $\theta = \pi/3$, which implies that $1 - 2\cos\theta = 0$.

Theorem 1   Suppose that N balls are put into N bins using QLB $\left(\frac{\pi}{3}, \sqrt{\frac{\ln N}{\ln\ln N}}\right)$. Then the maximum load is at most $(2 + o(1))\sqrt{\ln N/\ln\ln N}$ with high probability.

Proof   Our proof is composed of two stages. In the first stage, we bound from above the number k of the bins whose height is T or more when the game ends. In the second stage, we evaluate the number ν_i of bins whose height is i or more for each $i \geq T$. To do so, we use the method called layered induction [2] with the base condition that $\nu_T \leq k$. It should be noted that we can use the layered induction since the probability that a bin of height T or more receives a ball is very small thanks to the quantum sampling. This condition is not met for the bins whose height is less than T and there is no obvious way of using the layered induction for the first stage. We use notations similar to [2] to represent the state of QLB at time t: Let $h^A(t)$ be the *height* of the t-th ball by an algorithm A, i.e., the number of balls in the bin right after the t-th ball is put into the bin. Also let $\nu_i^A(t)$ be the number of bins that have a load of i or more at time t by an algorithm A. We omit the superscript A when it is clear which algorithm we are considering.

Stage 1: Our basic approach is to approximate the behavior of the game by an independent Poisson process. Consider the moment that the t-th ball is coming. Then each of the low bins receives that ball with probability $p(t) = (1 - (\frac{\nu_T(t)}{N})^3)\frac{1}{N - \nu_T(t)} \leq \frac{3}{N}$ by Lemma 1 for $\theta = \pi/3$. Recall that we are interested in the number of bins whose height is at least T, which is not affected by the behaviour of the bins after their height becomes T. Therefore, we can use the above probability $p(t)$ also for a high bin. Thus, we can use the independent Poisson process with probability $p(t)$. More precisely, we can claim the following lemma using Lemma 2.

Lemma 4   For independent binomially distributed random variables $X_1,\ldots, X_n$ with parameters n and p, let $\kappa(n, p, T)$ be the number of the random variables whose value is larger than T. Then, $Pr[\nu_T^{QLB(\pi/3,T)}(N) \geq k] \leq 4Pr[\kappa(N, \frac{3}{N}, T) \geq k]$ for all T.

Let X be the random variable for $\nu_T(N)$ of $QLB(\pi/3, \sqrt{\ln N/\ln\ln N})$ and X' be the corresponding random variable when we use the above Poisson approximation where a bin receives a ball with probability $\frac{3}{N}$. By Chernoff bounds, for $0 < \delta < 2e - 1$

$$Pr[X' > (1 + \delta)E(X')] < \exp\left(-E(X')\frac{\delta^2}{4}\right).$$

Let p be the probability that a specific bin receives T or more balls in the above Poisson approximation. Since the probability that a specific bin receives a ball at one moment is $\frac{3}{N}$ and since $\sum_{i=T}^{N} \binom{N}{i} q^i(1-q)^{N-i} \leq \binom{N}{T} q^T$ and $\binom{N}{T} \leq (\frac{eN}{T})^T$ [11],

$$p = \sum_{i=T}^{N} \binom{N}{i} \left(\frac{3}{N}\right)^i \left(1 - \frac{3}{N}\right)^{N-i} \leq \binom{N}{T} \left(\frac{3}{N}\right)^T \leq \frac{(3e)^T}{T^T}.$$

Let p' be this final value $(3e)^T/T^T$. Thus,

$$Pr[X' > (1+\delta)Np'] \leq Pr(X' > (1+\delta)Np] \leq \exp\left(-E(X')\frac{\delta^2}{4}\right).$$

Since $p \geq \binom{N}{T} \left(\frac{3}{N}\right)^T \left(1 - \frac{3}{N}\right)^{N-T}$ and we now set $T = \sqrt{\ln N/\ln\ln N}$, one can easily see that there must exist constant $0 < \delta < 2e - 1$ such that

$$\exp\left(-E(X')\frac{\delta^2}{4}\right) = \exp\left(-\frac{\delta^2}{4}Np\right) \leq \frac{1}{4N}.$$

Combining the above inequalities and Lemma 4, we can show

$$Pr[X > (1+\delta)\frac{(3e)^T}{T^T}N] \leq 4Pr[X' > (1+\delta)\frac{(3e)^T}{T^T}N] \leq 1/N.$$

Therefore the number of the bins whose load is at least T is at most $(1 + \delta)$ $(3e)^T N/T^T$ with probability $1 - 1/N$.

Stage 2: Next we define a sequence α_i as follows and show $\nu_i(N) \leq \alpha_i$ for each $i \geq T+1$ with high probability under the condition that $\alpha_T = (1+\delta)\frac{(3e)^T}{T^T}N$:

$$\alpha_i = \left(e\alpha_T^2/N^2\right)^{i-T} \alpha_T \quad (i \geq T)$$

Thus one can obtain the bound by calculating the value i such that α_i less than one. Since the argument from now on is similar to [2], we omit it in this paper.□

4   Static Lower Bounds

Here we show the matching lower bounds.

Theorem 2   Suppose that N balls are put into N bins using $QLB(\theta, T)$. Then the maximum load is $\Omega(\sqrt{\ln N/\ln\ln N})$ with high probability for any $0 \leq \theta \leq 2\pi$ and any $0 \leq T \leq N$.

Proof   If we set $T > \sqrt{\ln N/\ln\ln N}$, then the maximum load clearly reaches the bound as illustrated in Sec. 2. So, without loss of generality we can assume $T \leq \sqrt{\ln N/\ln\ln N}$. Also we can assume $\cos\theta \neq 1$ since by Lemma 1 $\cos\theta \neq 1$ means that $P_e(k) = \frac{k}{N}$ which implies QLB is exactly the same as the classical load balancing (CLB for short). Let us fix T and θ accordingly.

As before our proof consists of two stages. In the first stage, we put $N/18$ balls and calculate the number of high bins. This time, the number of high bins is bounded from below. (It was bounded from above, previously.) Let k_1 be this lower bound. In the second stage, we show that k_1 high bins receive at least, say, l balls during some period of time. Here we need to be careful to select this time period , since we wish to bound the error probability from below. Recall that a ball goes to a high bin by the error of quantum sampling and this error probability can be zero when the setting of θ and T is optimal for the number of high bins. Once we obtain l, then we can conclude that the maximum load is at least $T + \Omega(\frac{\ln k_1}{\ln(k_1/l)})$ by Lemma 3, since the l balls are put into the high k_1 bins uniformly at random.

Stage 1: Suppose that the bins have received the first $N/18$ balls. Since $T \le \sqrt{\frac{\ln N}{\ln \ln N}}$, one can easily calculate that some bins get to height T during this period. Also it is easy to show that the probability that a low bin receives a ball is greater than $1/2N$ during the above period. However, recall that we now wish to obtain a lower bound for the number of high bins and, as before, we are not interested in the behavior of the high bins once they becomes high. Let X the random variable for the number of high bins when we have received $N/18$ balls and let X' be the number of the random independent binomially distributed random variables with parameters $N/18$ and $1/2N$, whose value is larger than T. As before, we can approximate X by X'. More precisely, we use the following lemma.

Lemma 5 For all θ and T, $Pr[\nu_T^{QLB(\theta,T)}(\frac{N}{18}) < k] \le 4Pr[\kappa\left(\frac{N}{18}, \frac{1}{2N}, T\right) < k]$. By Chernoff bounds,

$$Pr[X' < (1-\delta)E(X')] < \exp\left(-\frac{\delta^2}{2}E(X')\right),$$

where $E(X') = \left((\frac{1}{36})^T e^{-\frac{1}{36}} N\right)/T!$ by Poisson distribution. Therefore,

$$Pr[X' < (1-\delta)E(X')] < \exp\left(-\frac{\delta^2}{2}\frac{(\frac{1}{36})^T e^{-\frac{1}{36}}}{T!}N\right).$$

Since $T \le \sqrt{\ln N / \ln \ln N}$, we can select $0 < \delta < 1$ such that

$$\exp\left(-\frac{\delta^2}{2}\frac{\alpha^T e^{-\alpha}}{T!}N\right) \le \frac{1}{4N}.$$

Now by Lemma 5,

$$Pr[X < (1-\delta)\frac{(\frac{1}{36})^T e^{-\frac{1}{36}}}{T!}N] \le 4Pr[X' < (1-\delta)\frac{(\frac{1}{36})^T e^{-\frac{1}{36}}}{T!}N] \le 1/N,$$

namely, at the end of Stage 1, the number of high bins is at least $(1-\delta)$ $\frac{(\frac{1}{36})^T e^{-\frac{1}{36}}}{T!}N$, denoted by k_1 hereafter with probability $1 - 1/N$.

Stage 2: Since k_1 is a lower bound, there must be a point of time, say t, before $N/18$ when the number of high bins becomes k_1. We will calculate the number of bins that these k_1 bins receive during some period of time starting at t. Let k $(\geq k_1)$ be the number of high bins at some moment in this period. Then, by Lemma 1, the probability that a ball goes to one of the k_1 high bins is $P_e(k) = \frac{k_1}{N}\left(1 - 2\cos\theta - \frac{2k}{N}(1 - \cos\theta)\right)^2$. Thus $P_e(k)$ becomes a local minimum $(=0)$ at $k = \frac{1-2\cos\theta}{1-\cos\theta}\frac{N}{2} = k^*$. Here, without loss of generality, we can assume that $|k_1 - k^*| \geq c_1 k^*$ for a constant $c_1 > 0$ (Otherwise, we modify Stage 1 so that we receive, say, $N/5$ balls instead of $N/18$ balls. Then the value of k_1 is shifted and that of k^* unchanged. So, the condition is met.) A simple calculation implies that if $Pe(k^*) = 0$, then for a positive value c,

$$Pe(ck^*) = \frac{k_1}{N}(1 - c)^2(1 - 2\cos\theta)^2 = \frac{k_1}{N}(1 - c)^2\left(\frac{k^*}{N - k^*}\right)^2.$$

Therefore, if $|k_1 - k^*| \geq c_1 k^*$ for a constant $c_1 > 0$, $Pe(dk_1) = c_2(\frac{k_1}{N})^3$ for some constant $d > 0$ and $c_2 > 0$. Now we select the above period so that the bins receive another $c_1 k_1$ balls in it. The number of high bins is obviously less than $(1 + c_1)k_1$ at the end of this period. Accordingly, the number of high bins is dk_1 for some constant d during the above period. The above discussion follows that the probability that a ball goes to one of the k_1 high bins is at least $c_2(\frac{k_1}{N})^3$ for some constant $c_2 > 0$ during the period. Thus, using the similar discussion as before, we can conclude that the k_1 bins receive at least

$$l = (1 - \delta)c_2\left(\frac{k_1}{N}\right)^3 c_1 k_1 = c_3\left(\frac{(\frac{1}{36})^T e^{-\frac{1}{36}}}{T!}\right)^4 N$$

balls during this period for some constant c_3 with probability $1 - 1/N$.

Since those balls are placed into the k_1 bins uniformly at random, we can use Lemma 3 to claim that the maximum load only for this period is at least

$$\frac{\ln k_1}{\ln(k_1/l)} = \Omega\left(\frac{\ln N}{T \ln T}\right)$$

for any θ with high probability. (This is the case when $l < k_1/\ln k_1$. If $l > k_1/\ln k_1$, the maximum load becomes even more.) Thus the overall load is

$$T + \Omega\left(\frac{\ln k_1}{\ln(k_1/l)}\right) \geq T + \Omega\left(\frac{\ln N}{T \ln T}\right),$$

which becomes minimum when $T = O(\sqrt{\ln N/\ln \ln N})$. One can easily see that this minimum value is still $\Omega(\sqrt{\ln N/\ln \ln N})$. $\square$

5   Continuous Model

5.1   Basic Idea

As described in the previous section, the reason why the maximum load for the static model increases is that the value for θ cannot follow the increasing number of balls during the game. In the continuous model, the number of balls does not alter once the system enters a stable state; QLB appears to be much more suited for this model. Let $BB(M, N)$ be the continuous model of N bins which becomes stable when M balls have come. Without otherwise stated, we also assume that M can be written as $M = cN$ for $0 < c \leq \frac{3}{4}$. It is known that if we use CLB, then the maximum load does not differ from the static case, i.e., $\Theta(\ln N / \ln \ln N)$ for simple random sampling and $\Theta(\ln \ln N)$ for the coordination.

Now let us design QLB for $BB(N, M)$. If we know the value of c, say $c = \frac{1}{2}$, then we can set $T = 1$ and also can set θ such that the error probability (= the probability of selecting high bins) of $GS(\theta, f_1)$ becomes zero when the number of high bins is $\frac{N}{2}$. By Lemma 1, $\theta = \arccos(1 - \frac{1}{2(1-c)})$. Then the game proceeds as follows: (i) When cN balls have arrived, the distribution of the load looks like Fig. 2, where one can easily calculate that the maximum load is $\Theta(\ln N / \ln \ln N)$. (ii) After that, however, the number k of empty bins are decreasing because of the positive effect of quantum sampling. This can be shown by a standard analysis using the one-dimensional Markov chain (see later for details). (iii) One can see that the state for $k = 0$ is so-called an absorbing state of the Markov chain (i.e., once the system enters that state, it never leaves). Thus, the system is approaching to the completely balanced state where all the bins have load one.

Remark   The reason for restricting $c \leq \frac{3}{4}$ is that it is no longer possible to achieve the zero-error by a singe iteration of GS if $c > \frac{3}{4}$. However, if we allow multiple iterations, then we can extend the above method for a larger c, for instance, up to $c \approx 0.90$ by two iterations and up to $c \approx 0.95$ by three iterations.

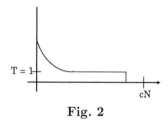

Fig. 2

Thus QLB is really powerful if we know the number M of balls in the stable state correctly. Unfortunately however, this assumption is apparently too strong due to our original motivations. Note that it also loses a lot of interest if we assume that no information is available about M: It is not hard to show using the same argument as in Section 4 that the maximum load becomes similarly large if the estimation of M differs from its real value by ϵN for a constant ϵ. We shall now investigate the case that we have a pretty good estimation about this value M.

5.2   Approximation within Sublinear Bounds

Suppose that we know the value of M within sublinear bounds. Then our QLB works well, i.e., the maximum load is bounded by $O(1)$. The first theorem deals with the case that our estimation is shifted to a smaller value.

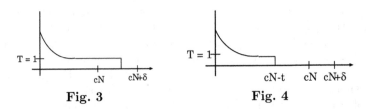

Fig. 3   **Fig. 4**

Theorem 3   Suppose that our QLB uses $GS(\arccos(1 - \frac{1}{2(1-c)}), 1)$, which becomes optimal when $M = cN$, but the real value of M is $cN + \delta$. Then the maximum load is at most k with probability at least $1 - \frac{1}{N}$ if $0 \leq \delta \leq O(N^{1 - \frac{1}{k-1}})$ for any constant $k \geq 2$.

Proof   See Fig. 3. It is not hard to see that the number t of nonempty bins reaches cN after a sufficiently long period of time. After that t can be more than cN but can never be less than cN since balls never go to high bins when $t = cN$. Let $P_e(t)$ be the probability that a ball falls in a nonempty bin. Then, by Lemma 1

$$P_e(t) = \left(\frac{cN - t}{cN - N} \right)^2 \frac{t}{N},$$

for $cN \leq t \leq cN + \delta$ and it becomes maximum when $t = cN + \delta$, whose value is

$$P_e^* = \frac{\delta^2(cN + \delta)}{N^3(1 - c)^2}.$$

Now suppose that the load of some specific bin is k. We investigate how large this value can be using the one-dimensional Markov chain. Let P_k be the probability that the system is in state k (i.e., the load is k) after it becomes stable. Let μ_i (λ_i, resp.) be the probability that the state transition occurs from state i to $i + 1$ (i to $i - 1$, resp.). Then it is well known that P_k can be written as

$$P_k = P_0 \prod_{i=0}^{k-1} \frac{\mu_i}{\lambda_{i+1}}.$$

One can observe that the state transition from i to $i - 1$ occurs if one of the i balls in the bin is selected (at random) to be removed. Thus $\lambda_i = \frac{i}{cN + \delta}$. Similarly, the state transition from i to $i + 1$ occurs if the bin is selected by QLB. Since $k \geq 1$, this occurs with probability $P_e(t)$. Since $P_e(t) \leq P_e^*$, we have $\mu_i \leq P_e^*/(cN + t) \leq P_e^*/cN$.

Recall that the number of empty bins is $N-t$, it follows that $\mu_0 = (1 - P_e(t))/(N - t) \leq 1/(N - cN - \delta)$. We can thus conclude that

$$P_k \leq P_0 \frac{1/(N - cN - \delta)}{1/(cN + \delta)} (\frac{cN + \delta}{cN})^{k-1} (P_e^*)^{k-1} \prod_{i=1}^{k} 1/i$$

$$\leq P_0 \frac{cN + \delta}{(1 - c)N - \delta} \left(\frac{\frac{\delta^2 (cN+\delta)}{N^3 (1-c)^2}(cN + \delta)}{cN} \right)^{k-1},$$

and it is not hard to see $P_k = O(1/N^2)$ if k satisfies $\delta \leq O(N^{1-1/(k-1)})$. Since $\frac{\mu_k}{\lambda_{k+1}} < \frac{1}{2}$, we have

$$\sum_{l=k}^{cN+\delta} P_l < 2P_k < O(1/N^2),$$

which means there is no bin holding k or more balls with probability $1-O(1/N)$.□

We next consider the case that our estimation of the number of balls is shifted to the larger side.

Theorem 4   Suppose that our QLB uses $GS(\arccos(1 - \frac{1}{2\{1-c-(\delta/N)\}}, 1))$; namely it becomes optimal when the number of balls in the stable state is $cN+\delta$. Then if the actual number of balls is cN and if $0 \leq \delta \leq O(N^{1/2})$, then the maximum load is $O(1)$ with probability at least $1 - O(1/N)$.

Proof   See Fig. 4. Let the number of nonempty bins after the system becomes stable be $cN - t$ and $P_e(t)$ be the (error) probability that a ball falls in a nonempty bin. Then by Lemma 1,

$$P_e(t) = \frac{1}{N^2} \left(\frac{\delta + t}{1 - c - \frac{\delta}{N}} \right)^2 (c - \frac{t}{N}).$$

Now we use the Markov chain to analyze how the value t changes. Let P_t be the probability that the system is in state t (i.e., the number of nonempty bins is $cN - t$) and μ_t (λ_t, resp.) be the probability that the state changes from t to $t + 1$ (t to $t - 1$, resp.). One can see that the state changes from t to $t + 1$ if a ball falls in a bin with load two or more and a ball is removed from a bin of load one. Therefore

$$\mu_t = P_e(t) \frac{s}{cN - t} \frac{cN - t}{cN} \leq P_e(t)(1 - \frac{t}{cN}) = \hat{\mu}_t,$$

where s is the number of bins whose load is one. Conversely, the number of nonempty bins decreases if a ball falls in an empty bin and a ball is removed from a bin of load two or more. Thus

$$\lambda_t = (1 - P_e(t))(1 - \frac{cN - t}{cN + \delta}) \geq (1 - P_e(t))\frac{t}{cN} = \hat{\lambda}_t.$$

Exactly as before we have

$$P_t = P_0 \frac{\mu_0 \cdots \mu_{t-1}}{\lambda_1 \cdots \lambda_t} \le \frac{\hat{\mu}_0 \cdots \hat{\mu}_{t-1}}{\hat{\lambda}_1 \cdots \hat{\lambda}_t} = \prod_{i=0}^{t-1} \frac{P_e(i)}{1 - P_e(i+1)} \binom{cN}{t}.$$

Note that $\delta \ll N$ and we can assume that $t \le \delta$ (t cannot be too large as shown later). Consequently, $P_e(t) \le \frac{4\delta^2}{N^2} \frac{c}{(1-c)^2}$, which implies

$$\prod_{i=0}^{t-1} \frac{P_e(i)}{1 - P_e(i+1)} \le \left(\frac{4c}{(1-c)^2} \right)^t \frac{\delta^{2t}}{N^{2t}}.$$

Since $\delta \le \sqrt{N}$, we can conclude

$$P_t \le \left(\frac{2c}{1-c} \frac{\delta}{cN} \right)^{2t} \binom{cN}{t} \le \left(\frac{4c^2 e}{(1-c)^2} \right)^t \left(\frac{\delta^2}{cN} \right)^t \frac{1}{t^t} \le \left(\frac{4ce}{(1-c)^2} \right)^t \frac{1}{t^t}.$$

We next prove that P_t decreases monotonically. To do so, we calculate

$$\frac{\mu_{t-1}}{\lambda_t} \le \frac{\hat{\mu_{t-1}}}{\hat{\lambda}_t} \le \frac{P_e(t-1)(1 - \frac{t-1}{cN})}{(1 - P_e(t)) \frac{t}{cN}} \le \frac{cN}{t \left(\frac{1-c}{\delta+t} \right)^2 N^2 - c} \approx \frac{c}{(1-c)^2 t},$$

since $\delta \le \sqrt{N}$. This value is obviously less than one if $t \ge \frac{c}{(1-c)^2}$, which guarantees that P_t decreases monotonically for almost all range of t. Using this fact and the upper bound of P_t previously calculated, we can claim that

$$\sum_{t=\sqrt{N}}^{cN-1} P_t \le P_{\sqrt{N}} \cdot N \le \frac{1}{N^{\frac{1}{2}\sqrt{N}-2}},$$

which means that the probability that $t \le \sqrt{N}$ is at least $1 - 1/N^{\frac{1}{2}\sqrt{N}-2}$. Namely the number of nonempty bins is at least $cN - O(\sqrt{N})$ with high probability.

Since the number of nonempty bins is bounded below, we can now bound the error probability $P_e(t)$ from above. Namely,

$$P_e(t) \le \frac{4c}{(1-c)^2} \frac{1}{N}.$$

One can see that this error probability is even smaller than the case discussed in the proof of the previous theorem. Therefore, we can use the same argument as before to bound the maximum load. Although details are omitted, the maximum load is at most five with probability $1 - O(1/N)$. $\square$

6   Concluding Remarks

Although we omit details, if we allow to sample two bins at each step, quantum sampling does not help, i.e., the maximum load is $\Theta(\ln \ln N)$ as the same as classical case[2]. The proof is very similar to that of [2]. Since the error probability is already very small in the classical case, its further reduction by quantum sampling does not make any essential difference. The anonymous reviewer informed us that there exists another model, called the "parallel bins-and-balls" [3][19], in which each ball can select $d(\geq 1)$ bins and ask them the situation of collision, all in parallel. The bound is $\Theta\left(\sqrt[r]{\frac{\ln N}{\ln \ln N}}\right)$ if the protocol involves r rounds of communication. This model is also quite powerful and it does not seem that quantum mechanism gives us any significant improvement.

References

1. S. Aaronson, "Quantum Lower Bound for the Collision Problem," Proceedings of the 34th ACM Symposium on Theory of Computing, 635–642, 2002.
2. Y. Azar, A. Z. Broder, A. R. Karlin, and E. Upfal, "Balanced Allocations," SIAM Journal on Computing, Vol. 29, No. 1, 180–200, 1999.
3. M. Adler, S. Chakarabarti, M. Mitzenmacher, and L. Rasmussen, "Parallel Randomized Load Balancing," Proceedings of the 27th ACM Symposium on Theory of Computing, 238–247, 1995.
4. M. Boyer, G. Brassard, P. Høyer, and A. Tapp, "Tight Bounds on Quantum Searching," Fortschritte der Physik, vol. 46(4-5), 493–505, 1998.
5. G. Brassard, P. Høyer, M. Mosca, and A. Tapp, "Quantum amplitude amplification and estimation", Quantum Computation and Quantum Information: A Millennium Volume, AMS Contemporary Mathematics Series, May 2000.
6. P. Berenbrink, A. Czumaj, A. Steger, and B. Vöcking, "Balanced Allocations: The Heavily Loaded Case," Proceedings of the 32nd ACM Symposium on Theory of Computing, 745–754, 2000.
7. G. Brassard, P. Høyer and A. Tapp, "Quantum cryptanalysis of hash and claw-free functions," Proceedings of 3rd Latin American Symposium on Theoretical Informatics (LATIN'98), Vol. 1380 of Lecture Notes in Computer Science, 163–169, 1998.
8. G. Brassard, P. Høyer, and A. Tapp, "Quantum Counting," Proceedings of the 25th International Colloquium on Automata, Languages, and Programming, Vol. 1443 of Lecture Notes in Computer Science, 820–831, 1998.
9. H. Buhrman, C. Dürr, M. Heiligman, P. Høyer, F. Magniez, M. Santha, and R. de Wolf, "Quantum Algorithm for Element Distinctness", Proceedings of the 16th IEEE Conference on Computational Complexity, 131–137, 2001.
10. D. P. Chi and J. Kim, "Quantum Database Searching by a Single Query," Lecture at First NASA International Conference on Quantum Computing and Quantum Communications, 1998.
11. T. H. Cormen, C. E. Leiserson and R. L. Rivest, "Introduction to Algorithms," Cambridge, Mass.: MIT Press, 1990.
12. C. Dürr and P. Høyer "A Quantum Algorithm for Finding the Minimum," LANL preprint, http://xxx.lanl.gov/archive/quant-ph/9607014.
13. L. K. Grover, "A fast quantum mechanical algorithm for database search," Proceedings of the 28th ACM Symposium on Theory of Computing, 212–218, 1996.
14. L. K. Grover, "A framework for fast quantum mechanical algorithms," Proceedings of the 30th ACM Symposium on Theory of Computing, 53–56, 1998.
15. L. K. Grover, "Rapid sampling through quantum computing," Proceedings of the 32nd ACM Symposium on Theory of Computing, 618–626, 2000.

16. N. Johnson, and S. Kotz, "Urn Models and Their Application," John Wily & Sons, New York, 1977.
17. V. Kolchin, B. Sevastyanov, and V. Chistyakov, "Random Allocations," John Wiley & Sons, New York, 1978.
18. G. L. Long, "Grover Algorithm with Zero Theoretical Failure Rate," Physical Review A , Vol. 64 022307, June, 2001.
19. M. Mitzenmacher, "The Power of Two Choices in Randomized Load Balancing," Ph.D. Thesis, 1996.
20. R. Motwani, and P. Raghavan, "Randomized Algorithms," Cambridge University Press, 1995.
21. Y. Shi, "Quantum Lower Bounds for the Collision and the Element Distinctness Problems," Proceedings of the 43rd IEEE Symposium on Foundations of Computer Science, 513–519, 2002.
22. P. W. Shor, "Algorithms for Quantum Computation: Discrete Log and Factoring", SIAM Journal of Computing, Vol. 26, No. 5, 1414–1509, 1997.
23. B. Vöcking, "How Asymmetry Helps Load Balancing," Proceedings of the 40th IEEE Symposium on Foundations of Computer Science, 131–140, 1999.

Fault-Hamiltonicity of Product Graph of Path and Cycle*

Jung-Heum Park[1] and Hee-Chul Kim[2]

[1] The Catholic University of Korea, Puchon, Kyonggi-do 420-743, Korea
j.h.park@catholic.ac.kr
[2] Hankuk University of Foreign Studies, Yongin, Kyonggi-do 449-791, Korea
hckim@hufs.ac.kr

Abstract. We investigate hamiltonian properties of $P_m \times C_n$, $m \geq 2$ and even $n \geq 4$, which is bipartite, in the presence of faulty vertices and/or edges. We show that $P_m \times C_n$ with n even is strongly hamiltonian-laceable if the number of faulty elements is one or less. When the number of faulty elements is two, it has a fault-free cycle of length at least $mn-2$ unless both faulty elements are contained in the same partite vertex set; otherwise, it has a fault-free cycle of length $mn-4$. A sufficient condition is derived for the graph with two faulty edges to have a hamiltonian cycle. By applying fault-hamiltonicity of $P_m \times C_n$ to a two-dimensional torus $C_m \times C_n$, we obtain interesting hamiltonian properties of a faulty $C_m \times C_n$.

1 Introduction

Embedding of linear arrays and rings into a faulty interconnection graph is one of the central issues in parallel processing. The problem is modeled as finding as long fault-free paths and cycles as possible in the graph with some faulty vertices and/or edges. Fault-hamiltonicity of various interconnection graphs were investigated in the literature. Among them, hamiltonian properties of faulty $P_m \times C_n$ and $C_m \times C_n$ were considered in [4,5,6,8]. Here, P_m is a path with m vertices and C_n is a cycle with n vertices. Many interconnection graphs such as tori, hypercubes, recursive circulants[7], and double loop networks have a spanning subgraph isomorphic to $P_m \times C_n$ for some m and n. Hamiltonian properties of $P_m \times C_n$ with faulty elements play an important role in discovering fault-hamiltonicity of such interconnection graphs.

A graph G is called *k-fault hamiltonian* (resp. *k-fault hamiltonian-connected*) if $G - F$ has a hamiltonian cycle (resp. a hamiltonian path joining every pair of vertices) for any set F of faulty elements such that $|F| \leq k$. It was proved in [4, 8] that $P_m \times C_n$, $n \geq 3$ odd, is hamiltonian-connected and 1-fault hamiltonian. Throughout this paper, a hamiltonian path (resp. cycle) in a graph G with faulty elements F means a hamiltonian path (resp. cycle) in $G - F$.

---

* This work was supported by grant No. 98-0102-07-01-3 from the Basic Research Program of the KOSEF.

T. Warnow and B. Zhu (Eds.): COCOON 2003, LNCS 2697, pp. 319–328, 2003.
© Springer-Verlag Berlin Heidelberg 2003

We let G be a bipartite graph with N vertices such that $|B| = |W|$, where B and W are the sets of black and white vertices in G, respectively. We denote by F_v and F_e the sets of faulty vertices and edges in G, respectively. We let $F = F_v \cup F_e$, $f_v^w = |F_v \cap W|$, $f_v^b = |F_v \cap B|$, $f_e = |F_e|$, $f_v = f_v^w + f_v^b$, and $f = f_v + f_e$. When $f_v^b = f_v^w$, a fault-free path of length $N - 2f_v^b - 1$ joining a pair of vertices with different colors is called an L^{opt}-path. For a pair of vertices with the same color, a fault-free path of length $N - 2f_v^b - 2$ between them is called an L^{opt}-path. When $f_v^b < f_v^w$, fault-free paths of length $N - 2f_v^w$ for a pair of black vertices, of length $N - 2f_v^w - 1$ for a pair of vertices with different colors, and of length $N - 2f_v^w - 2$ for a pair of white vertices, are called L^{opt}-paths. Similarly, we can define an L^{opt}-path for a bipartite graph with $f_v^w < f_v^b$. A fault-free cycle of length $N - 2\max\{f_v^b, f_v^w\}$ is called an L^{opt}-cycle. The lengths of an L^{opt}-path and an L^{opt}-cycle are the longest possible. In other words, there are no fault-free path and cycle longer than an L^{opt}-path and an L^{opt}-cycle, respectively.

A bipartite graph with $|B| = |W|$ (resp. $|B| = |W| + 1$) is called hamiltonian-laceable if it has a hamiltonian path joining every pair of vertices with different colors (resp. joining every pair of black vertices). Strong hamiltonian-laceability of a bipartite graph with $|B| = |W|$ was defined in [2]. We extend the notion of strong hamiltonian-laceability to a bipartite graph with faulty elements as follows. For any faulty set F such that $|F| \leq k$, a bipartite graph G which has an L^{opt}-path between every pair of fault-free vertices is called k-fault strongly hamiltonian-laceable.

$P_m \times P_n$, $m, n \geq 4$, is hamiltonian-laceable[3], and $P_m \times P_n$ with $f = f_v \leq 2$ has an L^{opt}-cycle when both m and n are multiples of four[5]. It has been known in [6,8] that $P_m \times C_n$, $n \geq 4$ even, with one or less faulty element is hamiltonian-laceable. We will show in Section 3 that $P_m \times C_n$, $n \geq 4$ even, is 1-fault strongly hamiltonian-laceable, which is an extension of the work in [6,8]. Moreover, we will show that $P_m \times C_n$, $n \geq 4$ even, has an L^{opt}-cycle if $f = 2$ and $f_v \geq 1$. When $f = f_e = 2$, it has a fault-free cycle of length at least $mn - 2$, and has a hamiltonian cycle if $m \geq 3$, $n \geq 6$ even and two faulty edges are not incident to a common vertex of degree three.

It has been known in [4] that a non-bipartite $C_m \times C_n$ is 1-fault hamiltonian-connected and 2-fault hamiltonian, and that a bipartite $C_m \times C_n$ with one or less faulty element is hamiltonian-laceable. $C_m \times C_n$ with $f = f_v \leq 4$ has an L^{opt}-cycle when both m and n are multiples of four[5]. We will show in Section 4, by utilizing hamiltonian properties of faulty $P_m \times C_n$, that a bipartite $C_m \times C_n$ is 1-fault strongly hamiltonian-laceable and has an L^{opt}-cycle when $f \leq 2$.

2   Preliminaries

The vertex set V of $P_m \times C_n$ is $\{v_j^i \mid 1 \leq i \leq m, 1 \leq j \leq n\}$, and the edge set $E = E_r \cup E_c$, where $E_r = \{(v_j^i, v_{j+1}^i) \mid 1 \leq i \leq m, 1 \leq j < n\} \cup \{(v_n^i, v_1^i) \mid 1 \leq i \leq m\}$ and $E_c = \{(v_j^i, v_j^{i+1}) \mid 1 \leq i < m, 1 \leq j \leq n\}$. An edge contained in E_r is called a row edge, and an edge in E_c is called a column edge. We denote by $R(i)$ and $C(j)$ the vertices in row i and column j, respectively. That is, $R(i) = \{v_j^i \mid 1 \leq j \leq n\}$

and $C(j) = \{v_j^i | 1 \leq i \leq m\}$. We let $R(i, i') = \bigcup_{i \leq k \leq i'} R(k)$ if $i \leq i'$; otherwise, $R(i, i') = \emptyset$. Similarly, we let $C(j, j') = \bigcup_{j \leq k \leq j'} C(k)$ if $j \leq j'$; otherwise, $C(j, j') = \emptyset$. v_j^i is a *black* vertex if $i + j$ is even; otherwise, it is a *white* vertex.

In $P_m \times C_n$, every pair of vertices v and w in $R(i) \cup R(m - i + 1)$ for each i, $1 \leq i \leq m$, are *similar*, that is, there is an automorphism ϕ such that $\phi(v) = w$. A pair of edges (v, w) and (v', w') are called *similar* if there is an automorphism ψ such that $\psi(v) = v'$ and $\psi(w) = w'$. Any two row edges in $\{(v, w) |$ either $v, w \in R(i)$ or $v, w \in R(m-i+1)\}$ are similar for each i, $1 \leq i \leq m$, and any two column edges in $\{(v, w) |$ either $v \in R(i), w \in R(i+1)$ or $v \in R(m-i+1), w \in R(m-i)\}$ are also similar for each i, $1 \leq i \leq m$.

We employ lemmas on hamiltonian properties of $P_m \times P_n$ and $P_m \times C_n$. We call a vertex in $P_m \times P_n$ a *corner vertex* if it is of degree two.

Lemma 1. *[1] Let G be a rectangular grid $P_m \times P_n$, $m, n \geq 2$. (a) If mn is even, then G has a hamiltonian path from any corner vertex v to any other vertex with color different from v. (b) If mn is odd, then G has a hamiltonian path from any corner vertex v to any other vertex with the same color as v.*

Lemma 2. *[8] (a) $P_m \times C_n$, $n \geq 3$ odd, is hamiltonian-connected and 1-fault hamiltonian. (b) $P_m \times C_n$, $n \geq 4$ even, is 1-fault hamiltonian-laceable.*

We denote by $H[v, w | X]$ a hamiltonian path in $G\langle X \rangle - F$ joining a pair of vertices v and w, if any, where $G\langle X \rangle$ is the subgraph of G induced by a vertex subset X. A path is represented as a sequence of vertices. If $G\langle X \rangle - F$ is empty or has no hamiltonian path between v and w, $H[v, w | X]$ is an empty sequence.

We let P and Q be two vertex-disjoint paths $(a_1, a_2, \cdots, a_k)$ and $(b_1, b_2, \cdots, b_l)$ on a graph G, respectively, such that (a_i, b_1) and (a_{i+1}, b_l) are edges in G. If we replace (a_i, a_{i+1}) with (a_i, b_1) and (a_{i+1}, b_l), then P and Q are merged into a single path $(a_1, a_2, \cdots, a_i, b_1, b_2, \cdots, b_l, a_{i+1}, \cdots, a_k)$. We call such a replacement a *merge* of P and Q w.r.t. (a_i, b_1) and (a_{i+1}, b_l). If P is a closed path (that is, a cycle), the merge operation results in a single cycle. We denote by $V(P)$ the set of vertices on a path P.

3 $P_m \times C_n$ with Even $n \geq 4$

3.1 $P_m \times C_n$ with One or Less Faulty Element

We will show, in this section, that $P_m \times C_n$, $n \geq 4$ even, is 1-fault strongly hamiltonian-laceable. First of all, we are going to show that $P_m \times C_n$ with a single faulty vertex is strongly hamiltonian-laceable by constructing an L^{opt}-path P joining every pair of fault-free vertices s and t.

Lemma 3. *$P_2 \times C_n$, n even, with a single faulty vertex is strongly hamiltonian-laceable. Furthermore, there is an L^{opt}-path joining every pair of vertices which passes through both an edge in $G\langle R(1) \rangle$ and an edge in $G\langle R(2) \rangle$.*

Proof. W.l.o.g., we assume that the faulty vertex is v_n^1. We let $s = v_i^x$ and $t = v_j^y$, $1 \leq x, y \leq 2$, and assume w.l.o.g. that $i \leq j$.

Case 1 $s, t \in B$ (see Fig. 1 (a)). $P = (H[s, v_1^2|C(1, j-1)], v_n^2, H[v_{n-1}^2, t|C(j, n-1)])$. Note that when $j = n$, $H[v_{n-1}^2, t|C(j, n-1)]$ is an empty sequence. The existence of a nonempty $H[s, v_1^2|C(1, j-1)]$ is due to Lemma 1 (a).

Case 2 $s, t \in W$. First, let us consider the case that $i \neq 1$. If $j = i + 1$ (see Fig. 1 (b)), $P = (s, H[v_{i-1}^x, v_1^2|C(1, i-1)], v_n^2, H[v_{n-1}^2, v_{j+1}^y|C(j+1, n-1)], t)$; otherwise (see Fig. 1 (c)), $P = (H[s, v_i^{3-x}|C(i, j-1)], H[v_{i-1}^{3-x}, v_2^2|C(2, i-1)], v_1^2, v_n^2, H[v_{n-1}^2, v_{j+1}^y|C(j+1, n-1)], t)$. For the case that $i = 1$, $P = (s, H[v_2^2, t|C(2, n-1)])$.

Case 3   Either $s \in B, t \in W$ or $s \in W, t \in B$.

Case 3.1 $i \neq j$. Let us consider the case that $j \neq n$ (see Fig. 1 (d)). Let P' be a hamiltonian path joining s and t in $G\langle C(i, j)\rangle$. P' passes through both edges (v_i^1, v_i^2) and (v_j^1, v_j^2) since P' passes through one of the two vertices v_i^1 and v_i^2 (resp. v_j^1 and v_j^2) of degree 2 in $G\langle C(i, j)\rangle$ as an intermediate vertex. Let $Q' = H[v_{i-1}^1, v_{i-1}^2|C(1, i-1)]$ and $Q'' = H[v_{j+1}^1, v_{j+1}^2|C(j+1, n-1)]$. We let P'' be a resulting path by a merge of P' and Q' w.r.t. (v_i^1, v_{i-1}^1) and (v_i^2, v_{i-1}^2) if Q' is not empty; otherwise, let $P'' = P'$. If Q'' is empty, P'' is an L^{opt}-path; otherwise, by applying a merge of P'' and Q'' w.r.t. (v_j^1, v_{j+1}^1) and (v_j^2, v_{j+1}^2), we can get an L^{opt}-path. Now, we consider the case that $j = n$. If $i \neq n-1$ (see Fig. 1 (e)), $P = (H[s, v_{n-2}^2|C(1, n-2)], v_{n-1}^2, t)$; otherwise (see Fig. 1 (f)), $P = (H[s, v_2^2|C(2, n-1)], v_1^2, t)$.

Case 3.2 $i = j$. We first let i be odd. When $i \neq 1, n-1$ (see Fig. 1 (g)), $P = (v_i^1, H[v_{i-1}^1, v_2^2|C(2, i-1)], v_1^2, v_n^2, H[v_{n-1}^2, v_{i+1}^2|C(i+1, n-1)], v_i^2)$. When $i = 1$, $P = H[v_1^1, v_1^2|C(1, n-1)]$. When $i = n-1$, $P = H[v_{n-1}^1, v_{n-1}^2|C(1, n-1)]$. For the case that i is even (see Fig. 1 (h)), $P = (v_i^1, H[v_{i+1}^1, v_{n-1}^2|C(i+1, n-1)], v_n^2, v_1^2, H[v_2^2, v_{i-1}^2|C(2, i-1)], v_i^2)$.

Unless either (i) $n = 4$ and $s, t \in W$, or (ii) $n = 4$ and $s \in R(1)$ has a color different from $t \in R(1)$, any L^{opt}-path passes through a vertex in $R(1)$ and a vertex in $R(2)$ as intermediate vertices, and thus it passes through both an edge in $G\langle R(1)\rangle$ and an edge in $G\langle R(2)\rangle$. For the cases (i) and (ii), it is easy to see that the L^{opt}-paths constructed here (Case 2 and Case 3.1) always satisfy the condition. $\square$

Lemma 4. $P_m \times C_n$, n even, with a single faulty vertex is strongly hamiltonian-laceable. Furthermore, there is an L^{opt}-path joining every pair of vertices which passes through both an edge in $G\langle R(1)\rangle$ and an edge in $G\langle R(m)\rangle$.

Proof. The proof is by induction on m. We consider the case $m \geq 3$ by Lemma 3. We assume w.l.o.g. that the faulty vertex v_f is white and contained in $G\langle R(1, m-1)\rangle$ due to the similarity of $P_m \times C_n$ discussed in Section 2.

Case 1 $s, t \in R(1, m-1)$. We let P' be an L^{opt}-path joining s and t in $G\langle R(1, m-1)\rangle$ which passes through both an edge in $G\langle R(1)\rangle$ and an edge (x, y) in $G\langle R(m-1)\rangle$. A merge of P' and $G\langle R(m)\rangle - (x', y')$ w.r.t. (x, x') and (y, y') results in an L^{opt}-path P, where x' and y' are the vertices in $R(m)$ adjacent to

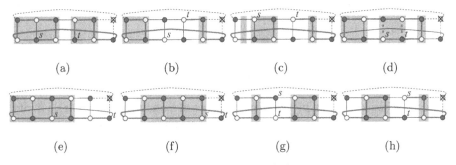

Fig. 1. Illustration of the proof of Lemma 3

x and y, respectively. Obviously, P passes through an edge in $G\langle R(m)\rangle$ as well as an edge in $G\langle R(1)\rangle$.

Case 2 $s, t \in R(m)$. When $s, t \in B$ (see Fig. 2 (a)) or s has a color different from t (see Fig. 2 (b)), we choose s' and t' in $R(m)$ which are not adjacent to v_f such that there are two paths P' joining s and s' and P'' joining t' and t which satisfy $V(P') \cap V(P'') = \emptyset$ and $V(P') \cup V(P'') = R(m)$. When $s, t \in W$ (see Fig. 2 (c)), we choose s' and t' in $R(m)$ which are not adjacent to v_f such that there are two paths P' joining s and s' and P'' joining t' and t which satisfy $V(P') \cap V(P'') = \emptyset$, $V(P') \cup V(P'') \subseteq R(m)$, and $|V(P') \cup V(P'')| = n - 1$. We let s'' and t'' be the vertices in $R(m-1)$ which are adjacent to s' and t', respectively. Observe that $s'', t'' \in B$ if $s, t \in B$; otherwise, s'' has a color different from t''. $P = (P', Q, P'')$ is a desired L^{opt}-path, where Q is an L^{opt}-path in $G\langle R(1, m-1)\rangle$ joining s'' and t'' which satisfies the condition.

Case 3 $s \in R(1, m-1)$ and $t \in R(m)$. When $s, t \in B$ or s has a color different from t (see Fig. 2 (d)), we choose t' in $R(m)$ which is not adjacent to s and v_f such that there is a path P' joining t' and t which satisfies $V(P') = R(m)$. When $s, t \in W$, we choose t' in $R(m)$ such that there is a path P' joining t' and t which satisfies $V(P') \subseteq R(m)$ and $|V(P')| = n - 1$. We let t'' be the vertex in $R(m-1)$ which is adjacent to t'. $P = (Q, P')$ is a desired L^{opt}-path, where Q is an L^{opt}-path in $G\langle R(1, m-1)\rangle$ between s and t'' which satisfies the condition. $\qquad\square$

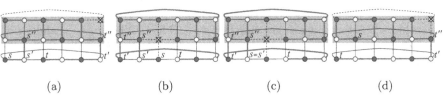

Fig. 2. Illustration of the proof of Lemma 4

Strong hamiltonian-laceability of $P_m \times C_n$, $n \geq 4$ even, with a single faulty edge can be shown by utilizing Lemma 4 as follows.

Lemma 5. $P_m \times C_n$, n even, with a single faulty edge is strongly hamiltonian-laceable.

Proof. By Lemma 2 (b), $P_m \times C_n$ has a hamiltonian path between any two vertices with different colors. It remains to show that there is an L^{opt}-path (of length $mn - 2$) joining every pair of vertices s and t with the same color. Let (x, y) be the faulty edge. We assume w.l.o.g. that x is black and y is white. When s and t are black, we find an L^{opt}-path P between s and t regarding y as a faulty vertex by using Lemma 4. P does not pass through (x, y) as well as y, and the length of P is $mn - 2$. Thus, P is a desired L^{opt}-path. In a similar way, we can construct an L^{opt}-path for a pair of white vertices.   □

We know, by Lemma 4 and Lemma 5, that $P_m \times C_n$, $n \geq 4$ even, with a single faulty element is strongly hamiltonian-laceable, which implies that $P_m \times C_n$ without faulty elements is also strongly hamiltonian-laceable. Thus, we have the following theorem.

Theorem 1. $P_m \times C_n$, $n \geq 4$ even, is 1-fault strongly hamiltonian-laceable.

Corollary 1. $P_m \times C_n$, $n \geq 4$ even, has a hamiltonian cycle passing through any arbitrary edge when $f = f_e \leq 1$.

An m-dimensional hypercube Q_m has a spanning subgraph isomorphic to $P_2 \times C_{2^{m-1}}$. A recursive circulant $G(cd^m, d)$ with degree four or more has a spanning subgraph isomorphic to $P_d \times C_{cd^{m-1}}$. $G(cd^m, d)$ with degree four or more is bipartite if and only if c is even and d is odd[7].

Corollary 2. (a) An m-dimensional hypercube Q_m, $m \geq 3$, is 1-fault strongly hamiltonian-laceable. (b) A bipartite recursive circulant $G(cd^m, d)$ with degree four or more is 1-fault strongly hamiltonian-laceable.

3.2 $P_m \times C_n$ with Two Faulty Elements

A bipartite graph is called 2-vertex-fault L^{opt}-cyclic if it has an L^{opt}-cycle when $f = f_v \leq 2$.

Lemma 6. $P_2 \times C_n$, $n \geq 4$ even, is 2-vertex-fault L^{opt}-cyclic.

Proof. It is sufficient to show that $P_2 \times C_n$ has an L^{opt}-cycle C when $f = f_v = 2$ by Theorem 1. We assume w.l.o.g. that v_1^1 is faulty, and let v_f be the faulty vertex other than v_1^1. Let us consider the case that $v_f \in B$ first. When $v_f = v_i^1$ and $i \neq 1$ (see Fig. 3 (a)), $C = (H[v_1^2, v_{i-1}^2|C(1, i-1)], v_i^2, H[v_{i+1}^2, v_{n-1}^2|C(i+1, n-1)], v_n^2)$. When $v_f = v_1^1$, $C = (v_1^2, H[v_2^2, v_{n-1}^2|C(2, n-1)], v_n^2)$. When $v_f = v_i^2$ and $i \neq n$ (see Fig. 3 (b)), $C = (H[v_1^2, v_{i-1}^2|C(1, i-1)], v_i^1, H[v_{i+1}^2, v_{n-1}^2|C(i+1, n-1)], v_n^2)$. When $v_f = v_n^2$, $C = H[v_1^1, v_1^2|C(1, n-1)] + (v_1^1, v_1^2)$. Now, we consider the case that $v_f \in W$. When $v_f = v_1^1$ (see Fig. 3 (c)), $C = (v_1^2, H[v_2^2, v_{i-1}^2|C(2, i-1)], v_i^2, H[v_{i+1}^2, v_{n-2}^2|C(i+1, n-2)], v_{n-1}^2, v_n^2)$. When $v_f = v_i^2$ and $i \neq 1, n-1$ (see Fig. 3 (d)), $C = (v_1^2, H[v_2^2, v_{i-1}^1|C(2, i-1)], v_i^1, H[v_{i+1}^1, v_{n-2}^2|C(i+1, n-2)], v_{n-1}^2, v_n^2)$. When $v_f = v_1^2$, $C = H[v_2^1, v_2^2|C(2, n-1)] + (v_2^1, v_2^2)$. When $v_f = v_{n-1}^2$, $C = H[v_1^1, v_1^2|C(1, n-2)] + (v_1^1, v_1^2)$.   □

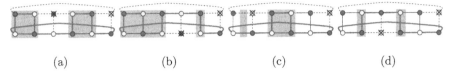

Fig. 3. Illustration of the proof of Lemma 6

Theorem 2. $P_m \times C_n$, $n \geq 4$ even, is 2-vertex-fault L^{opt}-cyclic.

Proof. The proof is by induction on m. We are sufficient to construct an L^{opt}-cycle C for the case that $m \geq 3$ and $f = f_v = 2$. We assume w.l.o.g. that at most one faulty vertex is contained in $R(1)$ due to the similarity of $P_m \times C_n$.

Case 1   There is one faulty vertex in $R(1)$. We assume w.l.o.g. that v_n^1 is faulty, and let v_f be the faulty vertex other than v_n^1. When $v_f \in B$ (see Fig. 4 (a)), $C = (v_1^1, v_2^1, \cdots, v_{n-1}^1, P')$, where P' is an L^{opt}-path between v_{n-1}^2 and v_1^2 in $G\langle R(2, m)\rangle$. The existence of P' is due to Theorem 1. When $v_f \in W$ and $v_f \neq v_1^2$ (see Fig. 4 (b)), $C = (v_1^1, v_2^1, \cdots, v_{n-2}^1, P')$, where P' is an L^{opt}-path between v_{n-2}^2 and v_1^2 in $G\langle R(2, m)\rangle$. When $v_f = v_1^2$ (see Fig. 4 (c)), $C = (v_2^1, v_3^1, \cdots, v_{n-1}^1, P')$, where P' is an L^{opt}-path between v_{n-1}^2 and v_2^2 in $G\langle R(2, m)\rangle$.

Case 2   There is no faulty vertex in $R(1)$. We let C' be an L^{opt}-cycle in $G\langle R(2, m)\rangle$. If C' passes through an edge (x, y) in $G\langle R(2)\rangle$, a merge of C' and $G\langle R(1)\rangle - (x', y')$ w.r.t. (x, x') and (y, y') results in an L^{opt}-cycle, where x' and y' are the vertices in $R(1)$ adjacent to x and y, respectively. No such an edge (x, y) exists only when $n = 4$ and a pair of vertices with the same color in $R(2)$ are faulty. We consider the case that $n = 4$ and two white vertices are faulty. For $m \geq 4$ (see Fig. 4 (d)), $C = (v_2^2, v_2^1, v_3^1, v_4^1, v_4^2, P')$, where P' is an L^{opt}-path in $G\langle R(3, m)\rangle$ between v_4^3 and v_2^3. For $m = 3$, $C = (v_2^2, v_2^1, v_3^1, v_4^1, v_4^2, v_4^3, v_3^3, v_2^3)$. Similarly, we can construct an L^{opt}-cycle for the case that two black vertices are faulty. $\square$

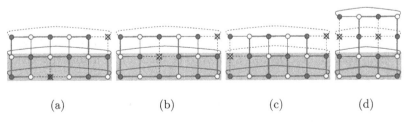

Fig. 4. Illustration of the proof of Theorem 2

A bipartite graph is called *1-vertex and 1-edge-fault L^{opt}-cyclic* if it has an L^{opt}-cycle when $f_v \leq 1$ and $f_e \leq 1$.

Theorem 3. $P_m \times C_n$, $n \geq 4$ even, is 1-vertex and 1-edge-fault L^{opt}-cyclic.

Proof. We consider the case that $f_v = 1$ and $f_e = 1$ by Theorem 1. We let v_f and (x, y) be the faulty vertex and edge, respectively. We assume w.l.o.g. that x has a color different from v_f. We find an L^{opt}-cycle C regarding x (as well as v_f) as a faulty vertex by using Theorem 2. C does not pass through (x, y) and v_f, and the length of C is $mn - 2$. Thus, C is a desired L^{opt}-cycle. □

Contrary to Theorem 2 and 3, $P_m \times C_n$, $n \geq 4$ even, with two faulty edges does not always have an L^{opt}-cycle since both faulty edges may be incident to a common vertex of degree 3. Note that, when there are no faulty vertices, an L^{opt}-cycle means a hamiltonian cycle. There are some other fault patterns which prevent $P_m \times C_n$ from having a hamiltonian cycle. For example, $P_2 \times C_n$ has no hamiltonian cycle if (v_n^1, v_1^1) and (v_2^2, v_3^2) are faulty (see Fig. 5 (a)): supposing that there is a hamiltonian cycle, a faulty edge (v_n^1, v_1^1) (resp. (v_2^2, v_3^2)) forces (v_1^1, v_2^1) and (v_1^1, v_1^2) (resp. (v_2^2, v_1^2) and (v_2^2, v_2^1)) to be included in the hamiltonian cycle, which is impossible. Similarly, we can see that $P_m \times C_4$ has no hamiltonian cycle if (v_2^1, v_2^2) and (v_4^1, v_4^2) are faulty (see Fig. 5 (b)).

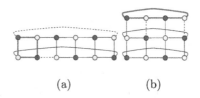

(a)   (b)

Fig. 5. Nonhamiltonian $P_m \times C_n$ with $f = f_e = 2$

Theorem 4. *$P_m \times C_n$, $n \geq 4$ even, with two faulty edges has a fault-free cycle of length at least $mn - 2$.*

Proof. We let (x, y) and (x', y') be the faulty edges, and assume w.l.o.g. that x and x' are black. We find an L^{opt}-cycle C regarding x and y' as faulty vertices by using Theorem 2. C does not pass through (x, y) and (x', y') as well as x and y', and the length of C is $mn - 2$, as required by the theorem. □

Theorem 5. *$P_m \times C_n$, $m \geq 3$ and even $n \geq 6$, with two faulty edges has a hamiltonian cycle if both faulty edges are not incident to a common vertex of degree 3.*

Proof. The proof is by induction on m. We let e_f and e'_f be the faulty edges, and will construct a hamiltonian cycle C.

  Case 1 $e_f, e'_f \in G\langle R(1) \rangle$. We assume w.l.o.g. that $e_f = (v_n^1, v_1^1)$ and $e'_f = (v_i^1, v_{i+1}^1)$. By assumption, we have $i \neq 1, n - 1$. When i is even, $C = (H[v_1^m, v_i^m | C(1, i)], H[v_{i+1}^1, v_n^m | C(i + 1, n)])$. When both i and m are odd, $C = (H[v_1^m, v_i^m | C(1, i)], H[v_{i+1}^m, v_n^m | C(i + 1, n)])$. The existence of hamiltonian paths in $G\langle C(1, i) \rangle$ and in $G\langle C(i + 1, n) \rangle$ is due to Lemma 1 (b). When i is odd and m is even, $C = (H[v_1^m, v_i^{m-1} | C(1, i)], H[v_{i+1}^{m-1}, v_n^m | C(i + 1, n)])$.

Case 2 $e_f \in G\langle R(1)\rangle$ and $e'_f \notin G\langle R(1)\rangle$. W.l.o.g., we let $e_f = (v_n^1, v_1^1)$. By assumption, both (v_1^1, v_1^2) and (v_n^1, v_n^2) are fault-free. $C = (v_1^1, v_2^1, \cdots, v_n^1, P')$, where P' is a hamiltonian path in $G\langle R(2, m)\rangle$ between v_n^2 and v_1^2 due to Theorem 1.

Case 3 $e_f, e'_f \notin G\langle R(1)\rangle$.

Case 3.1   There is a faulty column edge joining a vertex in $R(1)$ and a vertex in $R(2)$. There exists i such that both (v_i^1, v_i^2) and (v_{i+1}^1, v_{i+1}^2) are fault-free since $f_e = 2$ and $n \geq 6$. $C = (H[v_i^1, v_{i+1}^1|R(1)], P')$, where P' is a hamiltonian path in $G\langle R(2, m)\rangle$ between v_{i+1}^2 and v_i^2.

Case 3.2   There is no faulty column edge joining a vertex in $R(1)$ and a vertex in $R(2)$. First, we consider the case that there is a faulty edge in $G\langle R(2)\rangle$. We assume w.l.o.g. that $e_f = (v_n^2, v_1^2)$. We find a hamiltonian path P' in $G\langle R(2, m)\rangle$ between v_n^2 and v_1^2 regarding e_f as a fault-free edge by using Theorem 1. Obviously, P' does not pass through e_f. Thus, we have a hamiltonian cycle $C = (H[v_1^1, v_n^1|R(1)], P')$. The construction of a hamiltonian cycle for the base case $m = 3$ is completed since the case that there is a faulty edge in $G\langle R(3)\rangle$ is reduced to Case 1 and 2, and the case that there is a faulty edge joining a vertex in $R(2)$ and a vertex in $R(3)$ is reduced to Case 3.1.

The remaining case is that $m \geq 4$ and there is no faulty edge in $G\langle R(1, 2)\rangle$. The faulty edges are contained in $G\langle R(2, m)\rangle$, and both of them are not incident to a common vertex in $R(2)$. That is, we have $G\langle R(2, m)\rangle$ isomorphic to $P_{m-1} \times C_n$ such that both faulty edges are not incident to a common vertex of degree 3. Thus, we have a hamiltonian cycle C' in $G\langle R(2, m)\rangle$ by the induction hypothesis. A merge of C' and $G\langle R(1)\rangle - (x', y')$ w.r.t. (x, x') and (y, y') results in a hamiltonian cycle in $P_m \times C_n$, where x and y are the vertices in $R(2)$ such that (x, y) is an edge on C', and x' and y' are the vertices in $R(1)$ adjacent to x and y, respectively. This completes the proof.   □

4 $C_m \times C_n$ with Even m and n

Let us consider fault-hamiltonicity of a bipartite $C_m \times C_n$ with $m, n \geq 4$. $C_m \times C_n$ is bipartite if and only if both m and n are even.

Theorem 6. *(a)* $C_m \times C_n$, m *and* n *even, is 1-fault strongly hamiltonian-laceable.* *(b)* $C_m \times C_n$, m *and* n *even, is 2-fault* L^{opt}*-cyclic.*

Proof. The statement (a) is due to Theorem 1. It is sufficient to show that $C_m \times C_n$ with two faulty edges has a hamiltonian cycle by Theorem 2 and Theorem 3. $C_m \times C_n$ has a spanning subgraph isomorphic to $P_m \times C_n$ or $P_n \times C_m$ which has at most one faulty edge. Thus, $C_m \times C_n$ has a hamiltonian cycle by Corollary 1.   □

5   Concluding Remarks

We proved that $P_m \times C_n$, $n \geq 4$ even, is 1-fault strongly hamiltonian-laceable, 2-vertex-fault L^{opt}-cyclic, 1-vertex and 1-edge-fault L^{opt}-cyclic. If there are two faulty edges, $P_m \times C_n$ has a fault-free cycle of length at least $mn - 2$. It was also proved that $P_m \times C_n$, $m \geq 3$ and even $n \geq 6$, is hamiltonian if both faulty edges are not incident to a common vertex of degree 3. By employing fault-hamiltonicity of $P_m \times C_n$, we found that a bipartite $C_m \times C_n$ is 1-fault strongly hamiltonian-laceable and 2-fault L^{opt}-cyclic.

References

1. C.C. Chen and N.F. Quimpo, "On strongly hamiltonian abelian group graphs," in. *Australian Conference on Combinatorial Mathematics (Lecture Notes in Mathematics #884)*, pp. 23–34, 1980.
2. S.-Y. Hsieh, G.-H. Chen, and C.-W. Ho, "Hamiltonian-laceability of star graphs," *Networks* **36(4)**, pp. 225–232, 2000.
3. A. Itai, C.H. Papadimitriou, and J.L. Czwarcfiter, "Hamiltonian paths in grid graphs," *SIAM J. Comput.* **11(4)**, pp. 676–686, 1982.
4. H.-C. Kim and J.-H. Park, "Fault hamiltonicity of two-dimensional torus networks," in *Proc. of Workshop on Algorithms and Computation WAAC'00*, Tokyo, Japan, pp. 110–117, 2000.
5. J.S. Kim, S.R. Maeng, and H. Yoon, "Embedding of rings in 2-D meshes and tori with faulty nodes," *Journal of Systems Architecture* **43**, pp. 643–654, 1997.
6. M. Lewinter and W. Widulski, "Hyper-hamilton laceable and caterpillar-spannable product graphs," *Computers Math. Applic.* **34(11)**, pp. 99–104, 1997.
7. J.-H. Park and K.Y. Chwa, "Recursive circulants and their embeddings among hypercubes," *Theoretical Computer Science* **244**, pp. 35–62, 2000.
8. C.-H. Tsai, J.M. Tan, Y.-C. Chuang, and L.-H. Hsu, "Fault-free cycles and links in faulty recursive circulant graphs," in *Proc. of Workshop on Algorithms and Theory of Computation ICS2000*, pp. 74–77, 2000.

How to Obtain the Complete List of Caterpillars
(Extended Abstract)

Yosuke Kikuchi, Hiroyuki Tanaka, Shin-ichi Nakano, and Yukio Shibata

Department of Computer Science, Gunma University,
1-5-1 Tenjin-cho, Kiryu, Gunma, 376-8515 Japan
{kikuchi,hiroyuki,nakano,shibata}@msc.cs.gunma-u.ac.jp

Abstract. We propose a simple algorithm to generate all caterpillars without repetition. Also some other generation algorithms are presented. Our algorithm generates each caterpillar in a constant time per caterpillar.

1 Introduction

There are several studies for enumerating graphs with some properties [2,3,4,5,6, 8,10]. Such researches have been accomplished by combinatorial method. On the other hand, it was hard to actually generate all graphs having some properties, since the number of such graphs is very huge in general. However, due to the improvement of the performance of computers in recent years, now it is possible to generate all such graphs, and various algorithms have been invented for such generation problems [1,7,9].

In this paper, we enumerate all caterpillars without repetition. The caterpillars form an important subclass of trees. The structure of a caterpillar is simple. Whereas caterpillars have some interesting properties. It is known that a tree is an interval graph if and only if the tree is a caterpillar[11].

Generating algorithms for trees have been studied in many researches. For example, the algorithm in [1] generates all rooted trees, the algorithm in [12] generates all (un-rooted) trees, and the algorithm in [9] generates all plane trees.

It is known that the number of caterpillars with $n + 4$ vertices is $2^n + 2^{\lfloor n/2 \rfloor}$ [5].

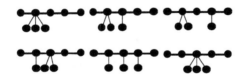

Fig. 1. Caterpillars with 8 vertices and diameter 4.

In this paper, we give an algorithm to generate all caterpillars with n vertices and diameter d. Figure 1 shows all caterpillars with eight vertices and diameter

T. Warnow and B. Zhu (Eds.): COCOON 2003, LNCS 2697, pp. 329–338, 2003.
© Springer-Verlag Berlin Heidelberg 2003

four. Then, using the algorithm as a subroutine we design an algorithm to generate all caterpillars with n vertices. We also design an algorithm to generate all "plane" caterpillars.

Then, based on the same idea, we design two more generating algorithms. We generate all spiders and all scorpions, those are subclasses of trees.

The rest of the paper is organized as follows. Section 2 gives some definition. In Section 3, we give an algorithm to generate all caterpillars with n vertices and diameter d. In Section 4, we improve the running time of the algorithm. In Section 5, we give an algorithm to generate all "plane" caterpillars. In Section 6 we give two algorithms to generate all spiders and all scorpions. Finally Section 7 concludes the paper.

2   Definition

Let $G = (V, E)$ be a simple graph with vertex set V and edge set E. The *degree* of a vertex $v \in V$, denoted by $d(v)$, is the number of neighbors of v in G. A *path* is an ordered list of distinct vertices $v_1, v_2, \ldots, v_n$ such that (v_{i-1}, v_i) is an edge for all $2 \leq i \leq n$. We denote a path with n vertices P_n. The *length* of a path means the number of edges in the path. For two vertices u and v in G, $\ell(u, v)$ is the shortest length of paths between u and v. For a vertex u in G, the *eccentricity* of u, denoted by $\mathrm{ecc}(u)$, is $\max_{v \in V} \ell(u, v)$. The *diameter* of G is $\max_{v \in V} \mathrm{ecc}(v)$.

A *tree* is a connected graph without cycles. A vertex v in a tree is a *leaf* if $d(v) = 1$. A tree is a *caterpillar* if removing all leaves remains a path. In this paper, we assume the number of vertices in a caterpillars is three or more. Figure 1 shows some examples of caterpillars. A leaf of a caterpillar is called a *foot*, and an edge incident to a foot is called a *leg*. Given a caterpillar C with diameter d, we designate one path with length d, and call it the *backbone* of C. We denote the backbone by a sequence of vertices, say $P_d = (u_0, u_1, \ldots, u_d)$.

A *spider* is a tree having at most one vertex having degree more than two[10]. See Figure 2(b). The vertex having degree more than two is called the *body* of the spider. A path connecting the body and a leaf is called a *leg*. If all legs but possible exception of one have length at most two, then the spider is called a *scorpion*[10]. See Figure 2(c).

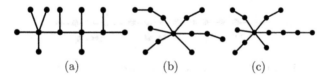

(a)   (b)   (c)

Fig. 2. Illustrations of (a)a caterpillar (b)a spider and (c)a scorpion.

3   Generating All Caterpillars with Diameter d

In this section, we give an algorithm to generate all caterpillars with n vertices and diameter d. Here, note that, the number of foots in these caterpillars is $n - d + 1$.

First, we define a unique label sequence for each caterpillar as follows.

For a caterpillar C with n vertices and diameter d, let $P_d = \{u_0, u_1, \ldots, u_d\}$ be an arbitrary backbone in C. We call a vertex in the caterpillar a *cocoon* if the vertex is not in the backbone. For each vertex u_i on P_d, let $l(u_i)$ be the number of cocoons adjacent with u_i. For each i we have $l(u_i) \leq n - d - 1$ since the total number of cocoons is $n - d - 1$. Thus $l(u_0) + l(u_1) + \cdots + l(u_d)$ is also $n - d - 1$. For C we define its label sequence $L = (\, l(u_0), l(u_1), \ldots, l(u_d)\,)$. For example, for the caterpillar in Figure 2 (a), if we choose the horizontal segment chain from left to right as P_d then we have $L = (0, 3, 1, 2, 1, 0)$, otherwise if we choose P_d in the other direction then we have $L = (0, 1, 2, 1, 3, 0)$. By the structure of a caterpillar $l(u_0) = l(u_d) = 0$ always holds. So we ignore these two labels.

It is easy to see that each caterpillar has at most two label sequences, because P_d has to include the path consisting of all vertices with degree two or more, and we have only choice for its direction. If a caterpillar has two label sequences, then they are mirror-copy each other, so one sequence can be derived from the other by reversing. If a caterpillar has only one label sequence, then the label sequence is symmetric. For example, we obtain only $L = (0, 3, 0)$ for the caterpillar at the lower right side of Figure 1.

For a label sequence $L = (l(u_1), l(u_2), \ldots, l(u_{d-1}))$, if $l(u_i) = l(u_{d-i})$ holds for each $i = 1, 2, \ldots, d-1$, then L is called a *symmetric* label sequence. Otherwise let j be the minimum integer such that $l(u_j) \neq l(u_{d-j})$. If $l(u_j) > l(u_{d-j})$ then L is called a *forward* label sequence, otherwise called a *backward* label sequence.

If a caterpillar has two label sequences then we choose the forward one for "the" label sequence of the caterpillar. For example, the caterpillar at the lower left side of Figure 1 has two label sequences $(3, 0, 0)$ and $(0, 0, 3)$, then we choose $(3, 0, 0)$ as its label sequence.

If two label sequences defined above are distinct, then corresponding two caterpillars are not isomorphic. So we have defined a unique label sequence for each caterpillar. One can also observe that any label sequence corresponds to a caterpillar with n vertices and diameter d if

(i)   it consists of $d - 1$ labels,

(ii) each label is an integer less than $n - d$,

(iii) the sum of labels is $n - d - 1$, and

(iv) it is symmetric or forward.

Given d and n, we are going to design an algorithm to generate all label sequences corresponding to caterpillars with n vertices and diameter d. Our algorithm regards these label sequences as $(n - d)$-ary $(d - 1)$ digit numbers, and generates these label sequences in decreasing order.

For example, caterpillars with eight vertices and diameter four correspond to label sequences which are 4-ary 3 digit numbers. We generate such 4-ary 3 digit numbers in decreasing order as follows:

$$(3,0,0) \rightarrow (2,1,0) \rightarrow (2,0,1) \rightarrow (1,2,0) \rightarrow (1,1,1)$$
$$\rightarrow (1,0,2) \rightarrow (0,3,0) \rightarrow (0,2,1) \rightarrow (0,1,2) \rightarrow (0,0,3).$$

These 4-ary 3 digit numbers include four forward label sequences, two symmetric label sequences and four backward label sequences, and these four forward label sequences and two symmetric label sequences correspond to six caterpillars in Figure 1.

We can easily count the number of those label sequences above. That equals to the total number of division of $(n - d - 1)$ identical cocoons into $(d - 1)$ labeled points. So the number is $\binom{(n - d - 1) + (d - 1) - 1}{(d - 1) - 1} = \binom{3 + 3 - 1}{3 - 1} =$ 10. However how can we actually generate all such label sequences?

We can define the successor of a label sequence $(c_1, c_2, \cdots, c_{d-1})$ as follows.
Case 1: For some i $(1 \leq i \leq d - 2)$, there exists $c_i \neq 0$.

Let $i = s$ be the maximum index with $c_i \neq 0$ and $1 \leq i \leq d-2$. The successor of $(c_1, c_2, \cdots, c_{d-1})$ is $(c_1, c_2, \cdots, c_s - 1, 1 + c_{d-1}, 0, 0, \cdots, 0)$.
Case 2: $c_i = 0$ for all i $(1 \leq i \leq d - 2)$.

$(c_1, c_2, \cdots, c_{d-1})$ does not have the successor and the generation is completed.

At first, we choose the lowest digit except for the last digit with nonzero value and let it be s-th digit.

If $s = d - 2$, that means the digit c_s is the 2nd lowest digit, then decrease the value of c_s by one, and increase the value of c_{d-1} by one.

If $s < d - 2$, then decrease the value of c_s by one, set $c_{s+1} = 1 + c_{d-1}$ and set $c_{d-1} = 0$. Note that the values of $c_{s+1}, c_{s+2}, \ldots, c_{d-2}$ are all zero in the given label sequence $(c_1, c_2, \cdots, c_{d-1})$.

For example, the successor of $(1, 2, 3, 4, 5)$ is $(1, 2, 3, 3, 6)$, the successor of $(1, 2, 3, 0, 2)$ is $(1, 2, 2, 3, 0)$.

If we output only forward label sequences and symmetric label sequences among generated label sequences, then we can generate all caterpillars with n vertices and diameter d. We have the following algorithm.

Algorithm generate-all-caterpillars(d, n)
{Generate all $(n-d)$-ary $(d-1)$digit numbers such that the sum of each digit is $n - d - 1$.}
1   initialize $(c_1, c_2, \cdots, c_{d-1}) = (n - d - 1, 0, 0, \cdots, 0)$ and output this label sequence;
2   **while** $(c_1, c_2, \cdots, c_{d-2})$ is not all zero;
3   **do** {The label sequence has the successor.}
4   **begin**
5   Let $i = s$ be the maximum index with $c_i \neq 0$ and $1 \leq i \leq d - 2$;
6   **if** $s = d - 2$
7   **then** decrease c_{d-2} by one and increase c_{d-1} by one;
8   **else** {$s > 2$ }
9   **then** decrease c_s by one, set $c_{s+1} = 1 + c_{d-1}$, set $c_{d-1} = 0$;
10   **if** the obtained label sequence is forward or symmetric

11   **then** output the label sequence;
12   **end**

Lemma 1. *The algorithm generates all caterpillars with n vertices and diameter d. The algorithm uses $O(d)$ space in total and runs in $O(d \cdot g(n))$ time, where $g(n)$ is the total number of caterpillars with n vertices and diameter d.*

Proof. This algorithm updates the label sequence consisting of $(d-1)$ labels. We need at most a constant amount of memory except for the "current" label sequence, so it is clear that the algorithm needs $O(d)$ space in total.

Computation in line 2 and 5 in the algorithm can be executed efficiently as follows. Here we need to choose the maximum index i with $c_i \neq 0$. By connecting the nonzero elements among $c_1, c_2 \cdots, c_s$ by a doubly linked list, (we can update such a doubly linked list in $O(1)$ time per label sequence), line 2 and 5 can be accomplished in $O(1)$ time by checking the last element of the doubly linked list.

Line 7 and 9 take $O(1)$ time. Note that the number of labels to be updated is at most three.

Line 10 and 11 take $O(d)$ time.

The algorithm outputs more than half of all generated label sequences, since the number of forward label sequences is equal to the number of backward label sequences, and the algorithm outputs all forward label sequences and symmetric label sequences.

Thus the algorithm runs in $O(d \cdot g(n))$ time.   □

4   Improvement

The algorithm in Section 3 generates all label sequences, those correspond to $(n-d)$-ary $(d-1)$ digit numbers such that the sum of each digit is $n-d-1$. Then the algorithm checks whether each generated label sequence is either forward, symmetric or backward, and outputs only forward or symmetric label sequences. This check takes $O(d)$ time per label sequence.

However if we employ the following data structure, then this check can be accomplished in only $O(1)$ time.

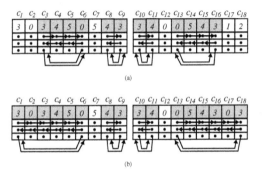

Fig. 3. The data structure for efficient running time.

We memorize all "symmetric pair", that is a pair of lebels (c_i, c_{d-i}) having the same value. See Figure 3, where such labels are hatched. Then we connect consecutive occurrence of such labels by a doubly linked list, and also connect the extremes of those doubly linked lists by a pointer. With the pointers we can decide in $O(1)$ time the current sequence is either forward, symmetric or backward.

A label sequence $L = (c_1, c_2, \cdots, c_{d-1})$ differs from its successor at at most three labels. Two of these are consecutive, those are c_s and c_{s+1}, and c_s is decreased by one, and c_{s+1} is set to $1 + c_{d-1}$. Especially, if $s \neq d - 2$, then c_{d-1} becomes zero. Hence the above data structure can be updated in constant time per a label sequence. We have the following lemma.

Lemma 2. *The algorithm uses $O(d)$ space and runs in $O(g(n))$ time, where $g(n)$ is the total number of caterpillars with n vertices and diameter d. So the algorithm generates each caterpillar in $O(1)$ time per caterpillar on average.*

Then we improve the algorithm more.

How can we generate only symmetric and forward label sequences, without generating any backward label sequences? We need to observe the label sequences generated by the algorithm.

Suppose now the algorithm outputs forward or symmetric label sequence $L = (c_1, c_2, \cdots, c_{d-1})$. Let $L' = (c_1', c_2', \cdots, c_{d-1}')$ be the successor of L. If some backward label sequence are generated after L consecutively, then these label sequences should be skipped without output. Which label sequence should we output after L? Let L_{next} be the next label sequence that should be output after L.

If $c_1 = 0$, then $c_{d-1} = 0$ for all output sequence after L, since we never output backward label sequences. So, by removing the first and the last digit from the current label sequence, we can reduce the generating problem to a generating problem with two fewer labels. If the first digit's value becomes zero then we fix the last digit's value to zero. By applying this idea to the algorithm, we have the following label sequences if $n = 12$ and $d = 8$. All backward label aequences are underlined.

$$(3,0,0,0,0,0,0) \to (2,1,0,0,0,0,0) \to (2,0,1,0,0,0,0) \to (2,0,0,1,0,0,0)$$
$$\to (2,0,0,0,1,0,0) \to (2,0,0,0,0,1,0) \to (2,0,0,0,0,0,1) \to (1,2,0,0,0,0,0)$$
$$\to (1,1,1,0,0,0,0) \to (1,1,0,1,0,0,0) \to (1,1,0,0,1,0,0) \to (1,1,0,0,0,1,0)$$
$$\to (1,1,0,0,0,0,1) \to (1,0,2,0,0,0,0) \to (1,0,1,1,0,0,0) \to (1,0,1,0,1,0,0)$$
$$\to (1,0,1,0,0,1,0) \to (1,0,1,0,0,0,1) \to (1,0,0,2,0,0,0) \to (1,0,0,1,1,0,0)$$
$$\to (1,0,0,1,0,1,0) \to (1,0,0,1,0,0,1) \to (1,0,0,0,2,0,0) \to (1,0,0,0,1,1,0)$$
$$\to \underline{(1,0,0,0,1,0,1)} \to \underline{(1,0,0,0,0,1,1)} \to \underline{(1,0,0,0,0,0,2)} \to (\text{-},3,0,0,0,0,\text{-})$$
$$\to \overline{(\text{-},2,1,0,0,0,\text{-})} \to \overline{(\text{-},2,0,1,0,0,\text{-})} \to \overline{(\text{-},2,0,0,1,0,\text{-})} \to (\text{-},2,0,0,0,1,\text{-})$$
$$\to (\text{-},1,2,0,0,0,\text{-}) \to (\text{-},1,1,1,0,0,\text{-}) \to (\text{-},1,1,0,1,0,\text{-}) \to (\text{-},1,1,0,0,1,\text{-})$$
$$\to (\text{-},1,0,2,0,0,\text{-}) \to (\text{-},1,0,1,1,0,\text{-}) \to (\text{-},1,0,1,0,1,\text{-}) \to \underline{(\text{-},1,0,0,1,1,\text{-})}$$
$$\to (\text{-},1,0,0,0,2,\text{-}) \to (\text{-},\text{-},3,0,0,\text{-},\text{-}) \to (\text{-},\text{-},2,1,0,\text{-},\text{-}) \to \overline{(\text{-},\text{-},2,0,1,\text{-},\text{-})}$$
$$\to \overline{(\text{-},\text{-},1,2,0,\text{-},\text{-})} \to (\text{-},\text{-},1,1,1,\text{-},\text{-}) \to \underline{(\text{-},\text{-},1,0,2,\text{-},\text{-})} \to (\text{-},\text{-},\text{-},3,\text{-},\text{-},\text{-}).$$

We apply the idea above to our algorithm. Let $L' = (c'_1, c'_2, \cdots, c'_{d-1})$ be the successor of L. If L' is a backward label sequence, then we should skip all consecutively generated backward label sequences after L' and we need to find the next symmetric or forward label sequence L_{next}. When L is given, we can determine L_{next} as follows. We have three cases.

Case 1: $c'_1 = c'_{d-1}$.

We have the following three subcases.

Case 1(a): L' is a symmetric label sequence.

If L' consists of at most two labels, then output L' and exit the algorithm. Otherwise, set $L_{next} = L'$.

Otherwise, we may choose the minimum index j such that $c'_j \neq c'_{d-j}$.

Case 1(b): $c'_j < c'_{d-j}$. Now L' is a backward label sequence.

If $j = 2$, then we choose the maximum index s such that $1 \leq s \leq d-3$ and $c'_s > 0$. If $s < d-3$ then we decrease c_s by one, set $c_{s+1} = 1 + c_{d-2} + c_{d-1}$, set c_{d-2} and c_{d-1} to zero. Thus $L_{next} = (c_1, c_2, \ldots, c_s - 1, 1 + c_{d-2} + c_{d-1}, 0, \ldots, 0)$. If $s = 1$ and $c_1 = 1$, then we remove the first digit and the last digit from L, and obtain $L_{next} = (n - d - 1, 0, 0, \ldots, 0)$ having two fewer labels. If $s = d-3$ then $L_{next} = (c_1, c_2, \ldots, c_{d-3} - 1, c_{d-2} + c_{d-1} + 1, 0)$. Note that by assumption, the first digit is not zero but the last digit is always zero. So it means that the derived label sequence is forward.

If $j > 2$ and $c'_{d-j} > 0$, then we choose the maximum index s such that $1 \leq s \leq d-3$ and $c'_s > 0$, and decrease c_s by one. Set $c_{s+1} = 1 + c_{s+1} + c_{s+2} + \cdots + c_{d-1}$, and set $c_{s+2} = c_{s+3} = \cdots = c_{d-1} = 0$. Thus $L_{next} = (c_1, c_2, \ldots, c_s - 1, 1 + c_{s+1} + c_{s+2} + \cdots + c_{d-1}, 0, \ldots, 0)$. Note that by assumption the first digit is not zero, but the last digit always becomes zero.

If $j > 2$ and $c'_{d-j} = 0$, then it contradicts $c'_j < c'_{d-j}$. So this case never occur.

Case 1(c): $c'_j > c'_{d-j}$ (Thus L' is a forward label sequence.).

Set $L_{next} = L'$.

Case 2: $c'_1 > c'_{d-1}$.

Set $L_{next} = L'$.

Case 3: $c'_1 < c'_{d-1}$.

Since L is a symmetric or forward label sequence, $c_1 = c_{d-1} = c'_{d-1} - 1$ and $0 < c_2 = c_{d-2} = c'_{d-2} + 1$.

If there exists an index j such that $c_j \neq c_{d-j}$ then we choose the minimum index j. Then L is not symmetric and so L is forward and $c_j > c_{d-j}$ holds. In this case we construct L_{next} from L as follows.

If $j \geq 3$ then we choose the maximum index k such that $1 \leq k \leq d-j$ and $c_k \neq 0$. Then we decrease the value of c_k by one, and set $1 + c_{k+1} = c_{k+1} + c_{k+2} + \cdots + c_{d-1}$ and set $c_{k+2} = c_{k+3} = \cdots c_{d-1} = 0$. Thus $L_{next} = (c_1, c_2, \ldots, c_k - 1, 1 + c_{k+1} + c_{k+2} + \cdots + c_{d-1}, 0, \ldots, 0)$ Note that by assumption the value of the first digit is not zero but the value of the last digit always becomes zero.

Otherwise, there is no index j such that $c_j \neq c_{d-j}$, so L is symmetric. Then we choose the maximum index k such that $1 \leq k \leq d-3$ and $c_k \neq 0$. We decrease the value of c_k by one, and set $c_{k+1} = c_{d-2} + c_{d-1} + 1$ and $c_{d-2} = c_{d-1} = 0$.

Lemma 3. *Using the above idea, after $O(d)$ preprocessing time, one can generate each caterpillar in $O(1)$ time. The algorithm uses $O(d)$ space in total.*

5   Generating Other Classes of Trees

By executing the algorithm in Section 4 for each value of diameter $k = 2, 3, \cdots, d$, we generate all caterpillars with n vertices. The algorithm generates each caterpillar in $O(1)$ time.

We can also generate plane caterpillars as follows.

A plane graph is a graph with a fixed plane embedding. In this section, two graphs are distinct if their plane embeddings are different. For example, the four plane caterpillars in Figure 4 are distinct, although they are isomorphic as caterpillar.

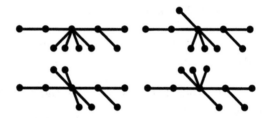

Fig. 4. Plane caterpillars corresponding to the same caterpillars.

Given a plane caterpillar with n vertices and diameter d. We choose a path $P_d = (u_0, u_1, \ldots, u_d)$ with length d as a backbone.

For each $u_i (2 \leq i \leq d - 2)$ on P_d, let $(u_{i-1}, v_1, v_2, \cdots, v_{a(i)}, u_{i+1}, w_1, w_2, \cdots, w_{b(i)})$ be the sequence of vertices given when we list adjacent vertices of u_i clockwise, where $v_1, v_2, \cdots, v_a(i)$ and $w_1, w_2, \cdots, w_{b(i)}$ are adjacent cocoons of u_i. Furthermore, let the number of cocoons adjacent to u_1 and u_{d-1} be $c(1)$ and $c(d - 1)$, respectively.

Hence, for a plane caterpillar with the backbone P_d, we can regard the label sequence to be $(c(1), a(2), a(3), \cdots, a(d-2), c(d-1), b(2), b(3) \cdots, b(d-2))$. There are two choices for P_d and one of which is obtained when we reverse the other one. Then for a plane caterpillar, we can choose at most two label sequences. So these label sequences correspond to $(n - d)$-ary numbers with $2 + (d - 1)2$ digits with summation of values is $n - d - 1$. Therefore we can generate plane caterpillars in $O(1)$ time per caterpillar by similar way of the algorithm before.

6   Generating Spiders and Scorpions

In a similar way to generate caterpillars, we can construct algorithms to generate all spiders with n vertices and all scorpions with n vertices. We sketch these algorithms as follows.

At first we define a label sequence for each spider. Let v be a body of the spider and $d(v)$ the degree of the body. Let $L_1, L_2, \ldots, L_{d(v)}$ be a set of legs. We denote the length of each L_i by $l(L_i)$. Here $(l(L_1) - 1, l(L_2) - 1, \cdots, l(L_{d(v)}) - 1))$ is the label sequence corresponding to a spider. For example, for a spider shown in Figure 2, we get the label sequence by listing six labels $0, 0, 1, 1, 2, 2$ in any order. We can get $(1, 0, 2, 1, 2, 0)$, $(2, 2, 1, 1, 0, 0)$, $(2, 1, 2, 1, 0, 0)$ and so on. These label sequences can be regarded as $(n - d)$-ary d digits numbers with summation of values is $n - d - 1$.

In order to define a unique label sequence for each spider, we choose the maximum label sequence when we regard these label sequences as $(n - d)$-ary d digits numbers. For example, we choose $(2, 2, 1, 1, 0, 0)$ for the spider in Figure 2.

By generating such label sequences in decreasing order, we can generate all spiders with n vertices and the degree of the body is $d(v)$.

Especially, if all legs, with the possible exception of one, have length 2 or less, then the spider is called a scorpion. We can also generate all scorpions by slightly modifying the algorithm.

7   Conclusion

In this paper, we gave an algorithm to generate all caterpillars with n vertices and diameter d without duplications. Our algorithm generates each caterpillars in a constant time per caterpillar. We also proposed some algorithms to generate other subclasses of trees.

How can we generate all trees with a given diameter?

References

1. Beyer, T. and Hedetiniemi, S. M., Constant Time Generation of Rooted Trees, *SIAM J. Comput.*, 9 (1980) 706–712.
2. Biggs, N. L., Lloyd, E. K., Willson, R. J., *Graph Theory 1736-1936*, Clarendon Press, Oxford (1976).
3. Chauve, C., Dulucq, S. and Rechnitzer, A., Enumerating Alternating Trees, *Journal of Combinatorial Theory, Series A*, 94 (2001) 142–151.
4. Chung, K.- L. and Yan, W.- M., On the Number of Spanning Trees of a Multi-Complete/Star Related Graph, *Information Processing Letters*, 76 (2000) 113–119.
5. Harary, F. and Schwenk, A. J., The Number of Caterpillars, *Discrete Mathematics*, 6 (1973) 359–365.
6. Hasunuma, T. and Shibata, Y., Counting Small Cycles in Generalized de Bruijn Digraphs, *Networks*, 29 (1997) 39–47.

7. Li, Z. and Nakano, S., Efficient Generation of Plane Triangulations without Repetitions, *Proc. of ICALP 2001*, Lecture Notes in Comp. Sci., 2079, (2001) 433–443.
8. Lonc, Z., Parol, K. and Wojciechowski, J. M., On the Number of Spanning Trees in Directed Circulant Graphs, *Networks*, 37 (2001) 129–133.
9. Nakano, S., Efficient Generation of Plane Trees, *Information Processing Letters*, 84 (2002) 167–172.
10. Simone, C. D., Lucertini, M., Pallottino, S. and Simeone, B., Fair Dissections of Spiders, Worms, and Caterpillars, *Networks*, 20 (1990) 323–344.
11. West, D. B., *Introduction to Graph Theory, Second Ed.*, Prentice Hall, NJ (2001)
12. Wright, R. A., Richmond, B., Odlyzko, A. and Mckay, B. D., Constant Time Generation of Free Tree, *SIAM J. Comput.*, 15 (1986) 540–548.

Randomized Approximation of the Stable Marriage Problem[*]

Magnús Halldórsson[1], Kazuo Iwama[2], and Shuichi Miyazaki[3], and Hiroki Yanagisawa[2]

[1] Science Institute, University of Iceland
mmh@hi.is
[2] Graduate School of Informatics, Kyoto University
[3] Academic Center for Computing and Media Studies, Kyoto University,
{iwama,shuichi,yanagis}@kuis.kyoto-u.ac.jp

Abstract. While the original stable marriage problem requires all participants to rank all members of the opposite sex in a strict order, two natural variations are to allow for incomplete preference lists and ties in the preferences. Either variation is polynomially solvable, but it has recently been shown to be NP-hard to find a maximum cardinality stable matching when both of the variations are allowed. It is easy to see that the size of any two stable matchings differ by at most a factor of two, and so, an approximation algorithm with a factor two is trivial. In this paper, we give a first nontrivial result for the approximation with factor less than two. Our randomized algorithm achieves a factor of 10/7 for a restricted but still NP-hard case, where ties occur in only men's lists, each man writes at most one tie, and the length of ties is two. Furthermore, we show that these restrictions except for the last one can be removed without increasing the approximation ratio too much.

1   Introduction

An instance of the original *stable marriage problem* (*SM*) [3] consists of N men and N women, with each person having a preference list that totally orders all members of the opposite sex. A man and a woman form a *blocking pair* in a matching if both prefer each other to their current partners. A perfect matching is *stable* if it contains no blocking pair. The stable marriage problem was first studied by Gale and Shapley [1], who showed that every instance contains a stable matching, and gave an $O(N^2)$-time algorithm to find one.

One natural relaxation is to allow for indifference [3,6], in which each person is allowed to include *ties* in his/her preference. This problem is denoted by *SMT* (Stable Marriage with Ties). When ties are allowed, the definition of stability needs to be extended. A man and a woman form a blocking pair if each *strictly* prefers the other to his/her current partner. A matching without such a blocking pair is called *weakly stable* (or simply "stable") and the Gale-Shapley algorithm

---

[*] Supported in part by Scientific Research Grant, Ministry of Japan, 13480081

T. Warnow and B. Zhu (Eds.): COCOON 2003, LNCS 2697, pp. 339–350, 2003.
© Springer-Verlag Berlin Heidelberg 2003

can be modified to always find a weakly stable matching [3]. Another natural variation is to allow participants to declare one or more unacceptable partners. Thus each person's preference list may be incomplete. We refer to this problem as *SMI* (Stable Marriage with Incomplete lists). Again, the definition of a blocking pair is extended, so that each member of the pair prefers the other over the current partner *or* is currently single and acceptable. In this case, a stable matching may not be a perfect matching, but all stable matchings for a fixed SMI instance are of the same size [2]. Hence, finding a maximum cardinality stable matching is trivial.

The problem allowing for both relaxations, denoted by *SMTI* (Stable Marriage with Ties and Incomplete lists), was recently studied in [7,8,4] and it was shown that finding a maximum stable matching (denoted by *MAX SMTI*) is NP-hard. This hardness result was further shown to hold for the restricted case when all ties occur only in one sex, and are of length only two [8].

It is easy to show that stable matchings for any instance differ in size by at most a factor of two, since a stable matching is a maximal matching. Hence approximating MAX SMTI within a factor two is trivial. However, there is no known algorithm whose approximation ratio is less than two. Our goal in this paper is to approximate MAX SMTI with a factor of $2 - \epsilon$. It is interesting to see that MIN SMTI (the problem of finding a stable matching of *minimum* size) is also NP-hard [8], for which no approximation algorithms with a factor of $2 - \epsilon$ were known either. This resembles the situation for approximating a minimum maximal matching in a graph [10,5,9], which is in turn related to approximating a minimum vertex cover in a graph [5,9]. Whether or not approximation with a factor of $2 - \epsilon$ is possible has long been open for both problems.

These resemblances to the famous open problems is apparently a negative factor to our goal. However, SMTI has one good attribute, namely, the difficulty of obtaining (good) solutions can be converted to the difficulty of breaking ties in the SMTI instance. More specifically, suppose that a given instance I of SMTI has stable matchings of size distributing between $s/2$ and s. Then for any stable matching M of size s' ($s/2 \leq s' \leq s$) for I, there is an SMI instance I' which has a stable matching M' (also stable for the original I) of the same size s'. I' can be obtained by breaking ties of I, and M' can be obtained from I' in polynomial time. However, we do not know how to break ties; if we hit a good (bad, resp.) break, then we would obtain a good (bad, resp.) stable matching whose size s' is close to s ($s/2$, resp.). Recall that it appears hard to hit either an extremely good or an extremely bad break. Therefore, a natural conjecture is that it would be easy to hit an "intermediate" break.

The main purpose of this paper is to prove that this conjecture is true. We give a randomized approximation algorithm RandBrk whose expected approximation ratio is at most $10/7 \approx 1.429$. RandBrk is extremely simple; for a given SMTI instance I, we just break its ties uniformly at random and then obtain a stable matching for the resulting SMI instance using the Gale-Shapley algorithm. Unfortunately, however, the ratio $10/7$ is guaranteed only for a restricted class of instances such that (i) ties appear in only men's lists, (ii) each man writes at most one tie, and (iii) the length of each tie is two. (MAX SMTI is still NP-hard under these restrictions [8].) In this paper, we present our analysis for a weaker bound of $5/3 \approx 1.667$. For the analysis for $10/7$ which is almost tight but quite involved, we only show basic ideas.

We also observe in Sec. 4 how results change if we remove each of the conditions (i) through (iii) above. We show that the same approximation ratio is guaranteed without condition (ii). If we remove (i) but restrict the number of ties to at most N (= the number of men in a given instance), the approximation ratio increases to $7/4 = 1.75$, which is still better than two. However, if the third condition is completely removed, then there is a worst-case example for which the approximation ratio of RandBrk becomes as bad as two. For a small constant, say three, RandBrk appears to work well but we cannot prove explicit bounds for its approximation ratio, which is obviously a next target of the research.

Throughout this paper, instances contain equal number N of men and women. A goodness measure of an approximation algorithm T of a maximization problem is defined as usual: the *approximation ratio* of T is the maximum $\max\{opt(x)/T(x)\}$ over all instances x of size N, where $opt(x)$ $(T(x))$ is the size of the optimal (algorithm's) solution, respectively.

2   Algorithm RandBrk and Its Performance Analyses

Recall that our SMTI instances satisfy three conditions (i) through (iii) mentioned in the previous section. Algorithm RandBrk, which receives such an instance $\hat{I}$ and produces a stable matching for $\hat{I}$, consists of the following two steps:

Step 1.   For each man m who writes a tie in $\hat{I}$, break the tie with equal probability, namely, if women w_1 and w_2 are tied in m's list, then w_1 precedes w_2 with probability $1/2$, and vice versa. Let I be the resulting SMI instance.

Step 2.   Find a stable matching M for I by the Gale-Shapley algorithm and output it.

Since Gale-Shapley runs in deterministic polynomial time, RandBrk is a (randomized) polynomial time algorithm. We already know several basic facts about its correctness and performance. Let S denote the set of men who write a tie in $\hat{I}$ and $SMI(\hat{I})$ denote the set of $2^{|S|}$ different SMI instances obtained by breaking ties in $\hat{I}$ (recall that the length of ties is two).

Lemma 1. *[3] For any $I \in SMI(\hat{I})$, any stable matching for I is also stable for $\hat{I}$. (Namely RandBrk outputs a feasible solution.)*

Lemma 2. *[2,3] Let M_1 and M_2 be arbitrary stable matchings for the same SMI instance. Then (i) $|M_1| = |M_2|$ (where $|M|$ denotes the size of the matching M) and (ii) the set of men (women, resp.) matched in M_1 is exactly the same as the set of men (women, resp.) matched in M_2.*

Thus the performance of RandBrk depends only on Step 1. By this lemma, we can define $cost(I)$ for an SMI instance I as the (unique) size of stable matchings for I. Also, let $OPT(\hat{I})$ denote the size of a largest stable matching for SMTI instance $\hat{I}$.

Lemma 3. *[8] (i) There exists $I_1 \in SMI(\hat{I})$ such that $cost(I_1) = OPT(\hat{I})$ and (ii) for any $I_2 \in SMI(\hat{I})$, $cost(I_2) \geq OPT(\hat{I})/2$. (The reason for (ii) is easy: Suppose that m and w are matched in a largest stable matching for $\hat{I}$. Then at least one of them has a partner in any stable matching for I_2. Otherwise they are clearly a blocking pair for that matching.)*

Hence the approximation ratio of RandBrk apparently does not exceed two. Its true value, denoted by $Cost_{RB}(\hat{I})$, is obtained by calculating the expected value for $cost(I)$, namely, by calculating

$$Cost_{RB}(\hat{I}) = \frac{1}{2^{|S|}} \sum_{I \in SMI(\hat{I})} cost(I).$$

Here are some more notations and conventions: Let us fix an arbitrary SMTI instance $\hat{I}$. As defined above, S always denotes the set of men whose preference list includes a tie. Let $I \in SMI(\hat{I})$ and m be a man. If m prefers w_i to w_j in I, we write "$w_i \succ w_j$ in m's list of I." This notation is also used in a woman's list. If the instance I and/or the man m are clear from the context, we often omit them. Let $m \in S$, i.e., m writes a tie in $\hat{I}$. Then we often write "$[w_i \ w_j]$ in m's list of I" to show that women w_i and w_j are tied in m's list of $\hat{I}$ and that the tie is broken into $w_i \succ w_j$ in I. Also, we frequently say that "flip the tie of a man m of I" (although I is an SMI instance), which means that we obtain a new SMI instance I' by changing $[w_i \ w_j]$ in m's list of I into $[w_j \ w_i]$. For SMI instances I_1 and I_2, if I_1 is obtained by flipping the tie of m of I_2, then we write $I_1 = fp(I_2, m)$ (equivalently, $I_2 = fp(I_1, m)$). If a man (woman) has a partner in a stable matching M, then he/she is said to be *matched* in M, otherwise, is said to be *single*. If m and w are matched in M, we write $M(m) = w$ and $M(w) = m$.

2.1   Overview of the Analysis

To evaluate $Cost_{RB}(\hat{I})$, we introduce the following *deterministic* algorithm called TreeGen. TreeGen accepts an SMI instance $I \in SMI(\hat{I})$ and a subset A of S, and produces a binary tree T. Each vertex v of T is associated with some instance I' in $SMI(\hat{I})$ and a subset A' of S. It should be noted that the introduction of TreeGen is only for the purpose of analysis; we are not interested in actually running it or other features, such as its time complexity (which is clearly exponential).

Procedure *TreeGen(I, A)*. (Given an SMI instance I and a subset $A \subseteq S$ of men, construct a binary tree T.)

(1) Create a vertex v whose label is (I, A).

(2) If $A = \emptyset$, return v.

(3) Else, select a man m, denoted by $flip(v)$, in A, and let $TreeGen(I, A - \{m\})$ and $TreeGen(fp(I, m), A - \{m\})$ be the left child and the right child of v, respectively. (How to select $flip(v)$ will be specified later.)

We are interested in the behavior of TreeGen for the special input (I_{opt}, S), where I_{opt} is an SMI instance in $SMI(\hat{I})$ such that $cost(I_{opt}) = OPT(\hat{I})$ (its existence is due to Lemma 3). Then the tree T_{opt} generated by TreeGen from I_{opt} looks as follows: The root is associated with I_{opt}, which can produce an optimal stable matching M_{opt} of the original $\hat{I}$. Let $v = (I, A)$ be a vertex in T_{opt}. Then, if we select a man m in A and if we go to the left child, the associated instance does not change (and of course the associated matching size does not change). However, if we go to the right child, then its associated instance receives a single flip of m and its matching size can decrease. In both cases, m is removed from A as a "touched" man. Now the next lemma is important, which guarantees that the amount of loss in the size of matching when we go to the right child is at most one.

Lemma 4. *[8] Let I_1 and I_2 be in $SMI(\hat{I})$ and m^* be a man in S such that $I_1 = fp(I_2, m^*)$ (equivalently, $I_2 = fp(I_1, m^*)$). Also let M_1 and M_2 be stable matchings for I_1 and I_2, respectively. Then $||M_2| - |M_1|| \leq 1$.*

Our analysis uses TreeGen in the following way: Note that the generated tree T_{opt} has exactly $2^{|S|}$ leaves. Let Γ be the set of instances associated with those leaves. Then one can see that Γ is exactly the same as the set $SMI(\hat{I})$. Therefore, $Cost_{RB}(\hat{I})$ is equal to the average value of $cost(I)$ for all $I \in \Gamma$ since RandBrk produces each instance in $SMI(\hat{I})$ with equal probability. Let $v = (I, A)$ be a vertex of T_{opt}. Then $size(v)$ is defined to be the size of a stable matching associated with v, namely, $size(v) = cost(I)$. Now we define $ave(v)$ as follows: (i) If v is a leaf (i.e., $A = \emptyset$), then $ave(v) = size(v)$. (ii) Otherwise, $ave(v) = \frac{1}{2}(ave(l(v)) + ave(r(v)))$, where $l(v)$ (resp. $r(v)$) is the left child (resp. right child) of v. (We use these notations, $l(v)$ and $r(v)$, throughout this paper.) The following lemma is now immediate:

Lemma 5. *Let v_0 be the root of T_{opt}. Then $ave(v_0) = Cost_{RB}(\hat{I})$.*

Thus all we have to do is to evaluate $ave(v_0)$. Remember that T_{opt} has the property that if we move to the left child, then the size of the stable matching is preserved and if we go to the right child, then the size may decrease by one. Then one might be curious about what kind of result can be obtained for the value of $ave(v_0)$ if we assume this worst case, i.e., if we always lose one when moving to the right. Unfortunately, the result of this analysis is very poor, or we can only guarantee a half of the size of the maximum stable matching, which means that the approximation ratio is as bad as two.

Our basic idea to avoid this worst-case scenario is as follows: (i) If $|S|$ ($=$ the number of ties) is small compared to $\sigma = cost(\hat{I}) = cost(I_{opt})$, say $|S| = \sigma/2$, then even the above simple analysis guarantees a (good) approximation ratio of $4/3$. (Details are omitted, but the following observation might help: If we always traverse left sons, then the path includes zero (losing-one) right edges. If we always traverse right sons, the path includes $\sigma/2$ right edges. So if we traverse "at random", the path would include $\sigma/4$ right edges, which guarantees a cost of $\sigma - \frac{1}{4}\sigma = \frac{3}{4}\sigma$ or an approximation ratio of $4/3$. If $|S|$ is large, say $|S| = \sigma$, then the "average path" can include $\sigma/2$ right edges and we can only guarantee a size of $\sigma - \frac{1}{2}\sigma = \frac{1}{2}\sigma$.) (ii) If $|S|$ is relatively large, then we can select a "good" man m as $flip(v)$ in Step (3) of TreeGen in the following sense: If we flip the tie of m, then either we do not lose the size of matching, or if we do lose the size of matching then we can always select m' in the next round such that flipping his tie does not make the size decrease, due to the lemma below. For a vertex v in T_{opt}, let $height(v)$ be the height of v in T_{opt}. Note that if the label of v is (I, A), then $height(v) = |A|$.

Lemma 6. *Let $v = (I, A)$ be an arbitrary vertex in T_{opt} such that $height(v) > size(v)/2$. Suppose that for any man $m \in A$, selecting m as $flip(v)$ implies that $size(r(v)) = size(v) - 1$. Then there exist two men m_α and m_β in A such that $size(l(v)) = size(v)$, $size(r(v)) = size(v) - 1$, and $size(l(r(v))) = size(r(r(v))) = size(v) - 1$, by choosing $flip(v) = m_\alpha$ and $flip(r(v)) = m_\beta$. (The rightmost figure in Fig. 1 shows how $size(v)$ changes by flipping the ties of m_α and m_β. See Sec. 2.2 for the proof.)*

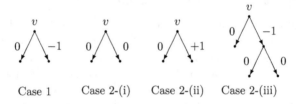

Case 1   Case 2-(i)   Case 2-(ii)   Case 2-(iii)

Fig. 1. Each case of the rule

Now we select m ($=flip(v)$) in TreeGen by the following rule (see Fig. 1):

Case 1. $height(v) \leq size(v)/2$. In this case, set $flip(v)$ to be an arbitrary man in A. (In this case, we assume the worst case, i.e., the size-decrease in every step.)

Case 2. $height(v) > size(v)/2$.

Case 2-(i): If there exists a man $m \in A$ such that letting $flip(v) = m$ makes $size(r(v)) = size(v)$, then set $flip(v) = m$.

Case 2-(ii): Otherwise, if there exists a man $m \in A$ such that letting $flip(v) = m$ makes $size(r(v)) = size(v) + 1$, then set $flip(v) = m$.

Case 2-(iii): Otherwise, set $flip(v) = m_\alpha$ and $flip(r(v)) = m_\beta$ whose existence is guaranteed by Lemma 6.

By the above rule, we can obtain the following lemma whose proof is given in Sec. 2.3.

Lemma 7. *For any node v in T_{opt}, $ave(v) \geq \frac{3}{5}size(v)$.*

By applying Lemma 7 to the root vertex v_0 of T_{opt}, we have that $ave(v_0) \geq \frac{3}{5}size(v_0)$. Since $size(v_0)$ is the optimal cost and $ave(v_0)$ ($= Cost_{RB}(\hat{I})$ by

Lemma 5) is the expected cost of RandBrk's output, we have the following theorem.

Theorem 1. *The approximation ratio of Algorithm RandBrk is at most $\frac{5}{3}$.*

2.2 Proof of Lemma 6

By the assumption of the lemma, $size(r(v)) = size(v) - 1$ no matter how we choose $flip(v)$. Clearly, $size(l(v)) = size(v)$ and $size(l(r(v))) = size(v) - 1$. We only need to show that we can choose m_α and m_β that makes $size(r(r(v))) = size(v) - 1$. We need some preparations.

Lemma 8. *Let $I_1 \in SMI(\hat{I})$, $[w_i \; w_j]$ in m^*'s list of I_1, and M_1 be a stable matching for I_1. Let $I_2 = fp(I_1, m^*)$ and M_2 be a stable matching for I_2. If $|M_1| \neq |M_2|$ then $M_1(m^*) = w_i$.*

Proof. Since $|M_1| \neq |M_2|$, M_1 is not stable in I_2 (otherwise, $|M_1| = |M_2|$ by Lemma 2). So, there is a blocking pair (m, w) for M_1 in I_2. First of all, m must be m^* (otherwise, (m, w) is also a blocking pair for M_1 in I_1 since the preference lists of m and w are same in I_1 and I_2, a contradiction). Secondly, m^* must not be single in M_1 (otherwise, the same contradiction as above). So, suppose that (m^*, w) is a blocking pair for M_1 in I_2. Then since (m^*, w) is a blocking pair in I_2, we must have that $w \succ M_1(m^*)$ in m^*'s list of I_2. However, since (m^*, w) is not a blocking pair in I_1, $M_1(m^*) \succ w$ in m^*'s list of I_1. Since I_1 and I_2 differ only in $[w_i \; w_j]$ or $[w_j \; w_i]$ in m^*'s list, it follows that $M_1(m^*) = w_i$ and $w = w_j$. □

Lemma 9. *Let $I_1, I_2, M_1, M_2, m^*, w_i, w_j$ be the same as Lemma 8. Then $|M_1| = |M_2|$ if (i) m^* is single in M_1 or (ii) $M_1(m^*) \neq w_i$.*

Proof. This is immediate from the previous lemma. □

Now we are ready to prove Lemma 6. Consider a vertex $v = (I, A)$ in T_{opt} satisfying the assumption of Lemma 6. Let M be an arbitrary stable matching for I. For an arbitrary man m_i in A, let $[w_{i_a} \; w_{i_b}]$ in m_i's preference list in I. We claim that (1) $M(m_i) = w_{i_a}$ and (2) w_{i_b} is matched in M.

Proof of claim (1)   By the assumption of Lemma 6, setting $flip(v) = m_i$ makes $size(r(v)) = size(v) - 1$, namely, $size(v) \neq size(r(v))$. Then Lemma 8 implies that $M(m_i) = w_{i_a}$. □

Proof of claim (2)   Suppose that w_{i_b} is single in M. Set $flip(v) = m_i$ and let $I_r(=fp(I, m_i))$ be an SMI instance associated with $r(v)$, a right child of v. Let M_r be a stable matching for I_r. Since we have assumed that $|M_r| = |M| - 1$, $M_r(w_{i_b}) = m_i$ by Lemma 8 (let I and I_r be I_2 and I_1 in Lemma 8, respectively). Thus w_{i_b} is matched in M_r.

Now, construct the bipartite graph G_{M,M_r} as follows (the basic idea of this proof is given in [8]): Each vertex of G_{M,M_r} corresponds to a person in I (or equivalently I_r). There is an edge between m and w if and only if m and w are matched in exactly one of M or M_r. Then the degree of each vertex is at most two. Hence any connected component (including at least one edge) of G_{M,M_r} is a simple path or a cycle.

We first show that a connected component which does not contain m_i is a cycle. Assume that there is a path which does not contain m_i, and suppose that the path starts from a man and ends with a woman. (For other cases, we can do a similar argument.) Now let the path be $m_1, w_1, m_2, w_2, \cdots, m_k, w_k$. Assume that $(m_1, w_1), (m_2, w_2), \cdots (m_k, w_k)$ are couples in M and $(m_2, w_1), (m_3, w_2), \cdots,$ (m_k, w_{k-1}) are couples in M_r. It should be noted that preference lists of these persons are same in I and I_r.

Since m_1 is matched with w_1 in M, m_1's list contains w_1. Then, $m_2 \succ m_1$ in w_1's list, since otherwise, m_1 (who is single in M_r) and w_1 form a blocking pair for M_r. For the same reason, $w_2 \succ w_1$ in m_2's list. Continuing this argument along with the path, we have that $w_k \succ w_{k-1}$ in m_k's list. Also, we can conclude that w_k writes m_k in her list. Then it follows that m_k and w_k form a blocking pair for M_r in I_r, which contradicts the stability of M_r.

Hence, there is only one path in G_{M,M_r} which contains m_i. Recall that w_{i_b} is matched in M_r but single in M. Hence this path starts from w_{i_b}, which means the number of M_r edges is greater than or equal to the number of M edges in this path. Since those numbers are equal in all the other cycles, $|M_r| \geq |M|$, which contradicts the assumption that $|M_r| = |M| - 1$. $\square$

Thus we have shown that for any man m_i in A who has a tie $[w_{i_a} \ w_{i_b}]$, $M(m_i) = w_{i_a}$ and w_{i_b} is matched in M. Now let us take another man m_j in A who has a tie $[w_{j_a} \ w_{j_b}]$. We say that m_i and m_j are disjoint if $\{w_{i_a}, w_{i_b}\} \cap \{w_{j_a}, w_{j_b}\} = \emptyset$. Suppose that all pairs of m_i and m_j are disjoint. Then since none of those w_{i_a}, w_{i_b}, w_{j_a} and w_{j_b} is single as proved above, the matching size $|M|$ $(=size(v))$ is at least $2 \cdot |A|$, which is equal to $2 \cdot height(v)$. This implies $size(v) \geq 2 \cdot height(v)$, which contradicts the assumption of this lemma. Hence there must be a pair of m_i and m_j that are not disjoint. Without loss of generality, we consider the following two cases: (1) $w_{i_b} = w_{j_a}$ or (2) $w_{i_b} = w_{j_b}$. (Note that w_{i_a} and w_{j_a} are matched in M with m_i and m_j, respectively, namely, $w_{i_a} \neq w_{j_a}$.)

Case (1): In this case, set $flip(v) = m_i$ and $flip(r(v)) = m_j$. Let I_r and I_{rr} be SMI instances associated with $r(v)$ and $r(r(v))$, respectively. Let M_r and M_{rr} be stable matchings for I_r and I_{rr}, respectively. By Lemma 8, $M_r(m_i) = w_{i_b}(= w_{j_a})$. This means that $M_r(m_j) \neq w_{j_a}$. Hence by Lemma 9, $|M_{rr}| = |M_r|(= |M| - 1)$ as desired.

Case (2): For clarity, let w_b denote $w_{i_b}(= w_{j_b})$. Without loss of generality, suppose that $m_i \succ m_j$ in w_b's list. Then we set $flip(v) = m_i$ and $flip(r(v)) = m_j$. Let I_r, I_{rr}, M_r and M_{rr} be same as Case (1). Note that, by Lemma 8, $M_r(m_i) = w_b$. Then it turns out that M_r is stable in I_{rr}. (Reason: Assume that M_r is stable in I_r but not stable in I_{rr}. An easy observation shows that the blocking pair must be (m_j, w_b). However, $M_r(m_i) = w_b$ as mentioned above. So it is impossible that this pair blocks M_r in I_{rr} because $m_i \succ m_j$ in w_b's list.) Hence $|M_{rr}| = |M_r|$ by Lemma 2.   □

2.3   Proof of Lemma 7

We introduce the following function $f(s, h)$ for two integers s and h ($s \geq 0, h \geq 0$): For $0 \leq h \leq \frac{s}{2}$, $f(s, h) = s - \frac{h}{2}$, for $\frac{s}{2} < h \leq s$, $f(s, h) = \frac{9}{10}s - \frac{3}{10}h$, and for $h > s$, $f(s, h) = \frac{3}{5}s$.

Lemma 10. $f(s, h) \geq \frac{3}{5}s$.

Proof. If $0 \leq h \leq \frac{s}{2}$, $f(s, h) = s - \frac{h}{2} = \frac{3}{5}s + (\frac{2}{5}s - \frac{h}{2}) \geq \frac{3}{5}s$ since $\frac{2}{5}s \geq \frac{1}{2}h$. If $\frac{s}{2} < h \leq s$, $f(s, h) = \frac{9}{10}s - \frac{3}{10}h = \frac{3}{5}s + \frac{3}{10}(s - h) \geq \frac{3}{5}s$. If $h > s$, $f(s, h) = \frac{3}{5}s$.   □

Lemma 11. *For any vertex v in T_{opt}, $ave(v) \geq f(size(v), height(v))$.*

Proof. By the definition of $ave(v)$, if v is a leaf of T_{opt}, then $ave(v) = size(v)$, and if v is a non-leaf vertex, then $ave(v) = \frac{1}{2}(ave(l(v)) + ave(r(v)))$. We will prove the lemma by induction. First, suppose that v is a leaf of T_{opt}. Then $ave(v) = size(v)$ by definition. By the definition of function f, $f(size(v), height(v)) = f(size(v), 0) = size(v) - \frac{0}{2} = size(v)$. Hence the lemma is true. Next, consider a non-leaf vertex v and assume that the claim is true for all vertices (except for v) in the subtree rooted at v. We will show that the lemma is true for v. We will consider four cases according to the rule in Sec. 2.1, which is used when v is expanded at Step (3) of TreeGen:

Case 1: Suppose that $flip(v)$ is determined using Case 1 of Sec. 2.1. It can be seen that in this case, $size(l(v)) = size(v)$, $size(r(v)) \geq size(v) - 1$ and $height(l(v)) = height(r(v)) = height(v) - 1$. Also by induction hypothesis, $ave(l(v)) \geq f(size(l(v)), height(l(v)))$ and $ave(r(v)) \geq f(size(r(v)), height(r(v)))$. Hence $ave(v) = \frac{1}{2}ave(l(v)) + \frac{1}{2}ave(r(v)) \geq \frac{1}{2}f(size(l(v)), height(l(v))) + \frac{1}{2}f(size(r(v)), height(r(v))) = \frac{1}{2}f(size(v), height(v) - 1) + \frac{1}{2}f(size(r(v)), height(v) - 1)$. Since Case 1 is used, it must be the case that $0 \leq height(v) \leq \frac{size(v)}{2}$. Thus, $\frac{1}{2}f(size(v), height(v) - 1) + \frac{1}{2}f(size(r(v)), height(v) - 1) = \frac{1}{2}(size(v) - \frac{1}{2}(height(v) - 1)) + \frac{1}{2}(size(r(v)) - \frac{1}{2}(height(v) - 1)) \geq size(v) - \frac{1}{2}height(v) = f(size(v), height(v))$.

For Cases 2-(i), 2-(ii) and 2-(iii), we can do the same argument as above, by taking care of the following facts: Since Case 2 applies, $height(v) > size(v)/2$. Furthermore, we can show that if Case 2-(iii) is used to v, then $\frac{1}{2}size(v) < height(v) \leq size(v)$ for the following reason: Suppose that $height(v) > size(v)$. Then, there must be a man in A who is single in a stable matching for I but has an untouched tie in his list, where $v = (I, A)$. By Lemma 9 (i), we can select this man as $flip(v)$, resulting that $size(r(v)) = size(v)$. Thus we must use Case 2-(i) for this vertex v, a contradiction.   □

Now Lemma 7 is immediate from these two lemmas.   □

2.4   Extension to an Upper Bound of 10/7

We can improve the analysis in the previous section to get a better upper bound of 10/7, which is based on the following observation.

Recall the following fact about the binary tree T_{opt}: In the worst case, if we go to the right child, we always lose the size of matching by one. However, in our analysis, we showed that this is not always the case: When $height(v) > size(v)/2$, if we lose the size by one by going to the right child, then we do not lose the size in the next step. We can strengthen Lemma 6 to show that if we lose the size by one when moving to the right child, then we do not lose the size in the *next two steps* or even *next three steps*. To achieve this, however, we need to be much more careful when selecting $flip(v)$ by introducing many different cases, which is a key portion of the proof for this stronger bound.

3   Lower Bound for RandBrk

Here, we give an instance for which Algorithm RandBrk gives approximation ratio of $32/23 \approx 1.391$, which shows that our analysis is almost tight. Consider the following SMTI instance $\hat{I}$.

$$
\begin{array}{ll}
m_1\colon w_1 & w_1\colon m_2 \ m_3 \ m_1 \\
m_2\colon (w_2 \ w_1) & w_2\colon m_2 \\
m_3\colon (w_3 \ w_1) & w_3\colon m_4 \ m_3 \\
m_4\colon (w_4 \ w_3) & w_4\colon m_4
\end{array}
$$

The largest stable matching for this instance is of size 4 (m_i is matched with w_i for $1 \leq i \leq 4$). There are eight SMI instances in $SMI(\hat{I})$. The size of stable matching for each of those eight instances is 4, 3, 3, 3, 3, 3, 2 and 2. Hence the expected size is $(4+3+3+3+3+3+2+2)/8 = 23/8$. Namely, the approximation ratio is 32/23 for this instance.

4   More General Instances

Recall that we imposed the following restrictions to SMTI instances: (i) Ties appear only in men's lists, (ii) each man's list includes at most one tie and (iii) the length of ties is two. In this section, we analyze the performance of RandBrk when some of these conditions are removed.

4.1   Instances with Ties in Both Sides

If ties appear in both men and women's lists, the approximation ratio of Rand-Brk becomes worse. However, if the number of people who write a tie is not too large, then it is still better than two. In this section, we consider the case that the number of such people is N (recall that N is the number of men in a given instance). Given an SMTI instance $\hat{I}$, let S_m and S_w be the set of men and women, respectively, who write a tie in $\hat{I}$. We construct a tree T_{opt} by $TreeGen(I_{opt}, S_m \cup S_w)$, where I_{opt} is the same as before, i.e., an SMI instance corresponding to a largest stable matching for $\hat{I}$. We can prove the following lemma similar to Lemma 6.

Lemma 12. *Let* $v = (I, A)$ *be an arbitrary vertex in* T_{opt} *such that* $height(v) > \frac{2}{3}size(v)$. *Suppose that for any person* $p \in A$, *flipping* p *implies that* $size(r(v)) = size(v) - 1$. *Then there exists a pair of two persons* p_α *and* p_β *in* A *such that* $size(l(v)) = size(v)$, $size(r(v)) = size(v) - 1$, *and* $size(l(r(v))) = size(r(r(v))) = size(v) - 1$, *by choosing* $flip(v) = p_\alpha$ *and* $flip(r(v)) = p_\beta$.

Proof (sketch). Since we assume that flipping p implies that $size(r(v)) = size(v) - 1$, people in A are all matched (see Lemma 9). Partition A into A_m and A_w, where A_m and A_w are sets of men and women in A, respectively. Without loss of generality, assume that $|A_m| \geq |A_w|$. Let W be the multiset of women who appear in A_w, or in the ties of men in A_m. Then $|W| = 2|A_m| + |A_w| \geq \frac{3}{2}|A| = \frac{3}{2}height(v) > size(v)$. As we have discussed in the claims (1) and (2) in Sec. 2.2, all women in W are matched. Hence at least one woman, say w, appears at least twice in W. If w appears in two men's ties, we can do the same argument in Cases (1) and (2) of Sec. 2.2. So, assume that w appears in some man $m(\in A_m)$'s tie and in A_w. Choose m and w as p_α and p_β, respectively. Again, we can use an argument similar to Sec. 2.2 to complete the proof. □

We can obtain $7/4$ upper bound by modifying $f(s, h)$ as follows: For $0 \leq h \leq \frac{2}{3}s$, $f(s, h) = s - \frac{h}{2}$, for $\frac{2}{3}s < h \leq s$, $f(s, h) = \frac{6}{7}s - \frac{2}{7}h$, and for $h > s$, $f(s, h) = \frac{4}{7}s$.

4.2   Multiple Ties for Each Man

The upper bound does not change for this generalization. In the previous analysis, each vertex in T_{opt} is labeled with (I, A), where A is the set of men whose preference list has not been touched yet. Here we generalize A to be the set of *ties* which have not been touched yet. T_{opt} is constructed by $TreeGen(I_{opt}, X)$, where X is the set of all ties in $\hat{I}$. At a vertex $v = (I, A)$, when selecting an element (a tie, this time) in A to create two children $l(v)$ and $r(v)$ of v, we add the following rule which has the highest priority:

> If there are two or more ties in A which belong to the same man m, select a tie which does not include $M(m)$.

We can apply the above rule as long as there are two or more ties belonging to the same man. Suppose that we apply the above rule to $v = (I, A)$. Then, by Lemma 9 (i), we can guarantee that $size(l(v)) = size(r(v)) = size(v)$, namely, we do not lose the size even for the right child. Then it is not hard to see that this analysis gives the same upper bound as before.

4.3 Longer Ties

In this section, we show that the performance of RandBrk becomes poor if we allow arbitrary length of ties. Consider the following instance:

$$m_1: \quad (w_1 \ w_1' \ \cdots \ w_{\ell-1}') \qquad w_1: \quad m_1$$

$$\vdots \qquad \qquad \vdots \qquad \qquad \vdots \qquad \vdots$$

$$m_{\ell-1}: (w_{\ell-1} \ w_1' \ \cdots \ w_{\ell-1}') \qquad w_{\ell-1}: m_{\ell-1}$$
$$m_1': \quad w_1' \qquad \qquad \qquad w_1': \quad m_1 \ \cdots \ m_{\ell-1} \ m_1'$$

$$\vdots \qquad \vdots \qquad \qquad \qquad \vdots \qquad \vdots$$

$$m_{\ell-1}': w_{\ell-1}' \qquad \qquad \qquad w_{\ell-1}': m_1 \ \cdots \ m_{\ell-1} \ m_{\ell-1}'$$

It is not hard to see that there is a stable matching of size $2\ell - 2$ for this example (m_1 to w_1, m_2 to w_2, and so on). If we break ties of men uniformly at random, the expected size of a stable matching we obtain is at most $\ell + \log \ell$ (details are omitted because of the space constraint). The approximation ratio of RandBrk is then at least

$$\frac{2\ell - 2}{\ell + \log \ell} = 2 - O\left(\frac{\log \ell}{\ell}\right).$$

References

1. D. Gale and L. S. Shapley, "College admissions and the stability of marriage," *Amer. Math. Monthly*, Vol.69, pp.9–15, 1962.
2. D. Gale and M. Sotomayor, "Some remarks on the stable matching problem," *Discrete Applied Mathematics*, Vol.11, pp.223–232, 1985.
3. D. Gusfield and R. W. Irving, "The Stable Marriage Problem: Structure and Algorithms," MIT Press, Boston, MA, 1989.
4. M. Halldórsson, K. Iwama, S. Miyazaki and Y. Morita, "Inapproximability Results on Stable Marriage Problems," *Proc. LATIN2002*, LNCS 2286, pp.554–568, 2002.
5. E. Halperin, "Improved approximation algorithms for the vertex cover problem in graphs and hypergraphs," *Proc. 11th Ann. ACM-SIAM Symp. on Discrete Algorithms*, pp. 329–337, 2000.
6. R. W. Irving, "Stable marriage and indifference," *Discrete Applied Mathematics*, Vol.48, pp.261–272, 1994.
7. K. Iwama, D. Manlove, S. Miyazaki, and Y. Morita, "Stable marriage with incomplete lists and ties," In *Proc. ICALP'99*, LNCS 1644, pp. 443–452, 1999.
8. D. Manlove, R. W. Irving, K. Iwama, S. Miyazaki, Y. Morita, "Hard variants of stable marriage," *Theoretical Computer Science*, Vol. 276, Issue 1-2, pp. 261–279, 2002.
9. B. Monien and E. Speckenmeyer, "Ramsey numbers and an approximation algorithm for the vertex cover problem," *Acta Inf.*, Vol. 22, pp. 115–123, 1985.
10. M. Yannakakis and F. Gavril,"Edge dominating sets in graphs," *SIAM J. Appl. Math.*, Vol. 38, pp. 364–372, 1980.

Tetris is Hard, Even to Approximate

Erik D. Demaine, Susan Hohenberger, and David Liben-Nowell

Laboratory for Computer Science, Massachusetts Institute of Technology
200 Technology Square, Cambridge, MA 02139, USA
{edemaine,srhohen,dln}@theory.lcs.mit.edu

Abstract. In the popular computer game of *Tetris*, the player is given a sequence of tetromino pieces and must pack them into a rectangular gameboard initially occupied by a given configuration of filled squares; any completely filled row of the gameboard is cleared and all pieces above it drop by one row. We prove that in the offline version of Tetris, it is NP-complete to maximize the number of cleared rows, maximize the number of tetrises (quadruples of rows simultaneously filled and cleared), minimize the maximum height of an occupied square, or maximize the number of pieces placed before the game ends. We furthermore show the extreme inapproximability of the first and last of these objectives to within a factor of $p^{1-\varepsilon}$, when given a sequence of p pieces, and the inapproximability of the third objective to within a factor of $2 - \varepsilon$, for any $\varepsilon > 0$. Our results hold under several variations on the rules of Tetris, including different models of rotation, limitations on player agility, and restricted piecesets.

1 Introduction

Tetris [13] is a popular computer game that was invented by mathematician Alexey Pazhitnov in the mid-1980s. By 1988, just a few years after its invention, Tetris was already the best-selling game in the United States and England. Over 50 million copies have been sold worldwide. (Incidentally, Sheff [12] gives a fascinating account of the tangled legal debate over the profits, ownership, and licensing of Tetris.)

In this paper, we embark on the study of the computational complexity of playing Tetris. We consider the *offline* version of Tetris, in which the sequence of pieces that will be dropped is specified in advance. Our main result is that playing offline Tetris optimally is NP-complete, and furthermore is highly inapproximable.

The game of Tetris. Concretely, the game of Tetris is as follows. (We give precise definitions in Section 2, and discuss some variants on these definitions in Section 6.) We are given an initial *gameboard*, which is a rectangular grid with some gridsquares filled and some empty. (In typical Tetris implementations, the gameboard is 20-by-10, and "easy" levels have an initially empty gameboard, while "hard" levels have non-empty initial gameboards, usually with the gridsquares below a certain row filled independently at random.)

T. Warnow and B. Zhu (Eds.): COCOON 2003, LNCS 2697, pp. 351–363, 2003.
© Springer-Verlag Berlin Heidelberg 2003

A sequence of *tetrominoes*—see Figure 1—is generated, typically probabilistically; the next piece appears in the middle of the top row of the gameboard. The piece falls, and as it falls the player can rotate the piece and slide it horizontally. It stops falling when it lands on a filled gridsquare, though the player has a final opportunity to slide or rotate it before it stops moving permanently. If, when the

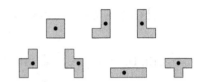

Fig. 1. The tetrominoes Sq ("square"), LG ("left gun"), RG ("right gun"), LS ("left snake"), RS ("right snake"), I, and T, with each piece's center marked.

piece comes to rest, all gridsquares in an entire row h of the gameboard are filled, row h is *cleared*: all rows above h fall one row lower, and the top row of the gameboard is replaced by an entirely unfilled row. As soon as a piece is fixed in place, the next piece appears at the top of the gameboard. To assist the player, typically a one-piece *lookahead* is provided—when the ith piece begins falling, the identity of the $(i+1)$st piece is revealed.

A player *loses* when a new piece is blocked from entirely entering the gameboard by filled gridsquares. Normally, the player can never win a Tetris game, since pieces continue to be generated until the player loses. Thus the player's objective is to maximize his or her score (which increases as pieces are placed and as rows are cleared).

Our results. In this paper, we introduce the natural full-information (offline) version of Tetris: we have a *deterministic, finite* piece sequence, and the player knows the identity and order of all pieces that will be presented. (*Games Magazine* has posed several Tetris puzzles based on the offline game [9].) We study the offline version because its hardness captures much of the difficulty of playing Tetris; intuitively, it is only easier to play Tetris with complete knowledge of the future, so the difficulty of playing the offline version suggests the difficulty of playing the online version. It also naturally generalizes the one-piece lookahead of implemented versions of Tetris.

It is natural to generalize the Tetris gameboard to m-by-n, since a relatively simple dynamic program solves the $m \cdot n = O(1)$ case in time polynomial in the number of pieces. Furthermore, in an attempt to consider the inherent difficulty of the game—and not any accidental difficulty due to the limited reaction time of the player—we initially allow the player an arbitrary number of shifts and rotations before the current piece drops by one row. (We restrict to realistic agility levels later.)

In this paper, we prove that it is NP-complete to optimize any of several natural objective functions for Tetris: (1) maximizing the number of rows cleared while playing the given piece sequence; (2) maximizing the number of pieces placed before a loss occurs; (3) maximizing the number of times a *tetris*—the simultaneous clearing of four rows—occurs; and (4) minimizing the height of the highest filled gridsquare over the course of the sequence. We also prove the extreme inapproximability of the first two (and the most natural) of these objective functions: given an initial gameboard and a sequence of p pieces, for any constant

$\varepsilon > 0$, it is NP-hard to approximate to within a factor of $p^{1-\varepsilon}$ the maximum number of pieces that can be placed without a loss, or the maximum number of rows that can be cleared. We also show that it is NP-hard to approximate the minimum height of the highest filled gridsquare to within a factor of $2 - \varepsilon$.

To prove these results, we first show that the cleared-row maximization problem is NP-hard, and then give extensions of our reduction for the remaining objectives. Our initial proof of hardness proceeds by a reduction from 3-PARTITION, in which we are given a set S of $3s$ integers and a bound T, and asked to partition S into s sets of three numbers each so that the sum of the numbers in each set is exactly T. Intuitively, we define an initial gameboard that forces pieces to be placed into s piles, and give a sequence of pieces so that all of the pieces associated with each integer must be placed into the same pile. The player can clear all rows of the gameboard if and only if all s of these piles have the same height. A key difficulty in our reduction is that there are only a constant number of piece types, so any interesting component of a desired NP-hard problem instance must be encoded by a sequence of multiple pieces. The bulk of our proof of correctness is devoted to showing that, despite the decoupled nature of a sequence of Tetris pieces, the only way to possibly clear the entire gameboard is to place in a single pile all pieces associated with a particular integer.

Our reduction is robust to a wide variety of modifications to the rules of the game. In particular, our results continue to hold in the following settings: (1) with restricted player agility—allowing only two rotation/translation moves before each piece drops in height; (2) under a wide variety of different rotation models—including the somewhat non-intuitive model that we have observed in real Tetris implementations; (3) without any losses—i.e., with an infinitely tall gameboard; and (4) when the pieceset is restricted to $\{LG, LS, I, Sq\}$ or $\{RG, RS, I, Sq\}$, plus at least one other piece.

Related work: Tetris. This paper is, to the best of our knowledge, the first consideration of the complexity of playing Tetris. Kostreva and Hartman [10] consider Tetris from a control-theoretic perspective, using dynamic programming to choose the "optimal" move, using a heuristic measure of configuration quality. Other previous work has concentrated on the possibility of a *perpetual loss-avoiding strategy* in the online, infinite version of the game. In other words, under what circumstances can the player be forced to lose, and how quickly? Brzustowski [2] has characterized all one-piece (and some two-piece) piecesets for which there are perpetual loss-avoiding strategies. He has also shown that, if the machine can adversarially choose the next piece (following the lookahead piece) in reaction to the player's moves, then the machine can force an eventual loss using any pieceset containing $\{LS, RS\}$. Burgiel [3] has strengthened this result, showing that an alternating sequence of LS's and RS's will eventually cause a loss in any gameboard of width $2n$ for odd n, regardless of the player's strategy. This implies that, if pieces are chosen independently at random with non-zero probability mass on both LS and RS, there is a forced eventual loss with probability one.

Recently, Breukelaar, Hoogeboom, and Kosters [1] have given a significant simplification of our reduction and proof of the NP-hardness of maximizing the number of rows cleared in a Tetris game. By using a more restrictive construction to limit piece placement, they are able to give a much shorter proof that all pieces associated with a particular integer must be placed in the same pile. (They have many fewer cases to consider.) The extensions to our reduction that we present in Sections 4, 5, and 6 can also be applied to their reduction to achieve the same results regarding different rules/objectives and inapproximability.

Related work: other games and puzzles. A number of other popular one-player computer games have recently been proven to be NP-hard, most notably the game of Minesweeper [8]—or, more precisely, the Minesweeper "consistency" problem. See the survey of the first author [4] for a summary of other games and puzzles that have been studied from the perspective of computational complexity. These results form the emerging area of *algorithmic combinatorial game theory*, in which many new results have been established in the past few years, e.g., Zwick's positive results on optimal strategies for the two-player block-stacking game *Jenga* [14].

2   Rules of Tetris

Here we rigorously define the game of Tetris, formalizing the intuition of the previous section. For concreteness, we have chosen to give very specific rules, but in fact the remainder of this paper is robust to a variety of modifications to these rules; in Section 6, we will discuss some variations on these rules for which our results still apply.

The *gameboard* is a grid of m rows and n columns, indexed from bottom-to-top and left-to-right. The $\langle i, j \rangle$th *gridsquare* is either *unfilled* (*open, unoccupied*) or *filled* (*occupied*). In a legal gameboard, no row is completely filled, and there are no completely empty rows that lie below any filled gridsquare. When determining the legality of certain moves, we consider all gridsquares outside the gameboard as always-occupied sentinels.

The seven Tetris pieces are exactly those connected rectilinear polygons that can be created by assembling four 1-by-1 gridsquares. The *center* of each piece is shown in Figure 1. A *piece state* $P = \langle t, o, \langle i, j \rangle, f \rangle$ consists of: (1) a *piece type* $t \in \{\mathsf{Sq}, \mathsf{LG}, \mathsf{RG}, \mathsf{LS}, \mathsf{RS}, \mathsf{I}, \mathsf{T}\}$; (2) an *orientation* $o \in \{0°, 90°, 180°, 270°\}$, the number of degrees clockwise from the piece's *base orientation* (shown in Figure 1); (3) a *position* $\langle i, j \rangle \in \{1, \ldots, m\} \times \{1, \ldots n\}$ of the piece's center on the gameboard; and (4) the value $f \in \{\textit{fixed}, \textit{unfixed}\}$, indicating whether the piece can continue to move. (The position of a Sq is the location of the upper-left gridsquare of the Sq, since its center falls on the boundary of four gridsquares rather than in the interior of one.) In an *initial piece state*, the piece is in its base orientation, and the initial position places the highest gridsquares of the piece into row m, and the center into column $\lfloor n/2 \rfloor$, and the piece is unfixed.

For now, rotations will follow the *instantaneous rotation model*. (We discuss other rotation models in Section 6.) For a piece state $P = \langle t, o, \langle i, j \rangle, \textit{unfixed} \rangle$, a gameboard B, and a rotation angle $\theta = \pm 90°$, the rotated piece state $R(P, \theta, B)$ is $\langle t, (o + \theta) \bmod 360°, \langle i, j \rangle, \textit{unfixed} \rangle$ as long as all the gridsquares occupied by the rotated piece are unoccupied in B; if some of these gridsquares are full in B then $R(P, \theta, B) = P$ and the rotation is *illegal*.

Playing the game. No moves are legal for a piece $P = \langle t, o, \langle i, j \rangle, \textit{fixed} \rangle$. The following moves are legal for a piece $P = \langle t, o, \langle i, j \rangle, \textit{unfixed} \rangle$, with current gameboard B: (1) a *rotation*, resulting in the piece state $R(P, \pm 90°, B)$; (2) a *translation*, resulting in the piece state $\langle t, o, \langle i, j \pm 1 \rangle, \textit{unfixed} \rangle$, if the gridsquares adjacent to P are open in B; (3) a *drop*, resulting in the piece state $\langle t, o, \langle i - 1, j \rangle, \textit{unfixed} \rangle$, if all the gridsquares beneath P are open in B; and (4) a *fix*, resulting in $\langle t, o, \langle i, j \rangle, \textit{fixed} \rangle$, if at least one gridsquare below P is occupied in B. A *trajectory* σ of a piece P is a sequence of legal moves starting from an initial state and ending with a fix move. The result of this trajectory on gameboard B is a new gameboard B', as follows:

1. The new gameboard B' is initially B with the gridsquares of P filled.
2. If the piece is fixed so that, for some row r, every gridsquare in row r of B' is full, then row r is *cleared*. For each $r' \geq r$, replace row r' of B' by row $r' + 1$ of B'. Row m of B' is an empty row. Multiple rows may be cleared by the fixing of a single piece.
3. If the next piece's initial state is blocked in B', the game ends and the player *loses*.

For a *game* $\langle B_0, P_1, \ldots, P_p \rangle$, a *trajectory sequence* Σ is a sequence $B_0, \sigma_1, B_1, \ldots, \sigma_p, B_p$ so that, for each i, the trajectory σ_i for piece P_i on gameboard B_{i-1} results in gameboard B_i. However, if there is a losing move σ_q for some $q \leq p$ then the sequence Σ terminates at B_q instead of B_p.

The Tetris problem. For concreteness, we will focus our attention on the following TETRIS problem: given a Tetris game $\mathcal{G} = \langle B, P_1, P_2, \ldots, P_p \rangle$, does there exist a trajectory sequence Σ that clears the entire gameboard of $\mathcal{G}$? (We will consider other Tetris objectives in Section 4.) Membership of TETRIS in NP follows straightforwardly.

3   NP-Completeness of Tetris

We define a mapping from instances of 3-PARTITION [7, p. 224] to instances of TETRIS. Recall the 3-PARTITION problem:

Given: A sequence $a_1, \ldots, a_{3s}$ of non-negative integers and a non-negative integer T, so that $T/4 < a_i < T/2$ for all $1 \leq i \leq 3s$ and so that $\sum_{i=1}^{3s} a_i = sT$.

Output: Can $\{1, \ldots, 3s\}$ be partitioned into s disjoint subsets $A_1, \ldots, A_s$ so that, for all $1 \leq j \leq s$, we have $\sum_{i \in A_j} a_i = T$?

We choose to reduce from 3-PARTITION because it is NP-hard to solve this problem even if the inputs a_i and T are provided in unary:

Theorem 1 (Garey and Johnson [6]). 3-PARTITION *is* NP-*complete in the strong sense.* □

Given an arbitrary instance $\mathcal{P} = \langle a_1, \ldots, a_{3s}, T \rangle$ of 3-PARTITION, we will produce a Tetris game $\mathcal{G}(\mathcal{P})$ whose gameboard can be completely cleared precisely if $\mathcal{P}$ is a "yes" instance. (For brevity, we omit some details; see [5].)

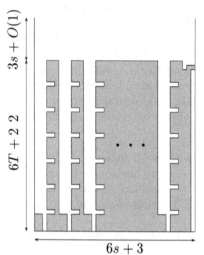

$3s + O(1)$

$6T + 2$ 2

$6s + 3$

Fig. 2. The initial gameboard for a Tetris game mapped from an instance of 3-PARTITION.

The initial gameboard is shown in Figure 2. The topmost $3s + O(1)$ rows form an empty staging area for rotations and translations. Below, there are *s buckets*, each six columns wide, corresponding to the sets $A_1, \ldots, A_s$ for the instance of 3-PARTITION. Each bucket has unfilled *notches* in its fourth and fifth columns in every sixth row, beginning in the fifth row. The first four rows of the first and second columns in each bucket are initially filled, and the sixth column of each bucket is entirely filled. The last three columns of the gameboard form a *lock*, blocking access to the last column, which is unfilled in all rows but the second-highest. Until a piece is placed into the lock to clear the top two rows, no lower rows can be cleared.

The piece sequence consists of the following sequence of pieces for each $a_1, \ldots, a_{3s}$: one *initiator* $\langle \mathsf{I}, \mathsf{LG}, \mathsf{Sq} \rangle$, then a_i repetitions of the *filler* $\langle \mathsf{LG}, \mathsf{LS}, \mathsf{LG}, \mathsf{LG}, \mathsf{Sq} \rangle$, and then one *terminator* $\langle \mathsf{Sq}, \mathsf{Sq} \rangle$. After the pieces associated with $a_1, \ldots, a_{3s}$, we have the following additional pieces: s successive I's, one RG, and $3T/2 + 5$ successive I's. (Without loss of generality, we can assume T is even by multiplying all input numbers by two.)

The TETRIS instance $\mathcal{G}(\mathcal{P})$ has size polynomial in the size of the 3-PARTITION instance $\mathcal{P}$, since $a_1, \ldots, a_{3s}$ and T are represented in unary, and can be constructed in polynomial time.

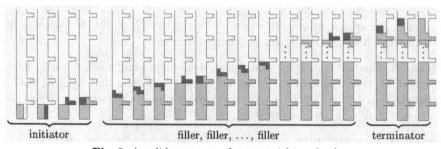

initiator   filler, filler, ..., filler   terminator

Fig. 3. A valid sequence of moves within a bucket.

Lemma 2 (Completeness). *For any "yes" instance $\mathcal{P}$ of* 3-PARTITION, *there is a trajectory sequence Σ that clears the entire gameboard of $\mathcal{G}(\mathcal{P})$ without triggering a loss.*

Proof. In Figure 3, we show how to place all of the pieces associated with the number a_i in a bucket. Since $\mathcal{P}$ is a "yes" instance, there is a partitioning of $\{1, \ldots, 3s\}$ into sets $A_1, \ldots, A_s$ so that $\sum_{i \in A_j} a_i = T$. Place all pieces associated with each $i \in A_j$ into the jth bucket of the gameboard. This yields a configuration in which only the last four rows of the third column of each bucket are unfilled. Next we place one of the s successive I's into each bucket, and the RG into the lock; the first two rows are then cleared. Finally, we place the $3T/2 + 5$ successive I's into the last column. Each of the I's clears four rows; in total, this clears the entire gameboard. $\square$

The proof of soundness is somewhat more involved; here we give a high-level summary and some suggestive details only. Call a trajectory sequence *valid* if it completely clears the gameboard of $\mathcal{G}(\mathcal{P})$, and call a bucket *unfillable* if it is impossible to fill all of the empty gridsquares in it using arbitrarily many pieces from the set $\{LG, LS, Sq, I\}$. Also, we say that a configuration with all buckets as in Figure 4 is *unprepped*.

Fig. 4.
Unprepped buckets.

Proposition 3. *In any valid trajectory sequence:*

1. *no gridsquare above row $6T + 22$ is ever filled;*
2. *all gridsquares of all pieces preceding the RG must all be placed into buckets, filling all empty bucket gridsquares;*
3. *no rows are cleared before the RG in the sequence;*
4. *all gridsquares of all pieces starting with (and including) the RG must be placed into the lock columns, filling all empty lock gridsquares;*
5. *no configuration with an unfillable bucket arises.*
6. *in an unprepped configuration, all pieces in the sequence $I, LG, Sq, r \times \langle LG, LS, LG, LG, Sq \rangle, Sq, Sq$, for any $r \geq 1$, must be placed into a single bucket, yielding an unprepped configuration.*

Proof. For (1), there are only enough pieces to fill and clear the gameboard if every filled gridsquare (from the initial gameboard or from pieces in the sequence) is placed into the lowest $6T + 22$ rows. For (2) and (3), placing any piece other than RG as the first piece to enter the lock columns violates (1), and no row can be cleared until some piece enters the lock. We have (4) from (1,2) and the fact that there are exactly as many gridsquares following the RG as there are empty gridsquares in the lock columns. Finally, (5) follows immediately from (2,3).

The (tedious) details for (6) can be found in [5]; here we give a high-level overview. Call a trajectory sequence *deviating* if it does not place all the pieces into the same bucket to yield an unprepped configuration. We first catalogue ten different classes of buckets that we show to be unfillable. (For example, a bucket with a disconnected region of unfilled gridsquares is unfillable.) We then exhaustively consider the tree of all possible placements of the pieces in the given sequence into the gameboard, and show that an unfillable bucket is produced by every deviating trajectory sequence. The overwhelming majority of deviating trajectory sequences create an unfillable bucket with their first deviating move, while some do not create an unfillable bucket until up to five pieces later. □

Lemma 4 (Soundness). *If there is a valid trajectory sequence for $\mathcal{G}(\mathcal{P})$, then $\mathcal{P}$ is a "yes" instance of* 3-PARTITION.

Proof. If there is a valid strategy for $\mathcal{G}(\mathcal{P})$, then by Proposition 3.2, there is a way of placing all pieces preceding the RG to exactly fill the buckets. By Proposition 3.6, we must place all of the pieces associated with a_i into the same bucket; by Proposition 3.1, we must have exactly the same number of filled gridsquares in each bucket. Define $A_j := \{i : \text{all the pieces associated with } a_i \text{ are}$ placed into bucket $j\}$. The total number of gridsquares placed into each bucket is the same; because every $a_i \in (T/4, T/2)$, we have that each $|A_j| = 3$. Thus the A_j's form a legal 3-partition, and $\mathcal{P}$ is a "yes" instance. □

Theorem 5. *Maximizing the number of rows cleared in a Tetris game is* NP-*complete.* □

4   NP-Hardness for Other Objectives

In this section, we sketch reductions extending that of Section 3 to establish the hardness of optimizing several other natural Tetris objectives. It is easy to confirm that TETRIS remains in NP for all objectives considered below.

Theorem 6. *Maximizing the number of tetrises (the number of times that four rows are cleared simultaneously) in a Tetris game is* NP-*complete.*

Proof. Our gameboard, shown in Figure 5, is augmented with four new bottom rows that are full in all but the sixth column. We append a single I to our previous piece sequence. For a "yes" instance of 3-PARTITION, $(6T + 20)/4 + 1$ tetrises are achievable. For a "no" instance, we cannot clear the top $6T + 22$ rows using the original pieces and thus we clear at most $6T + 22$ total rows. This implies that there were at most $(6T + 20)/4 < (6T + 20)/4 + 1$ tetrises. (Recall T is even.) Therefore we can achieve $(6T + 24)/4$ tetrises exactly when the top $6T + 22$ rows exactly when the 3-PARTITION instance is a "yes" instance. □

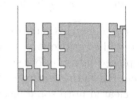

Fig. 5. Gameboard for the hardness of maximizing tetrises.

A different type of objective—considered by Brzustowski [2] and Burgiel [3], for example—is that of *survival*. How many pieces can be placed before a loss must occur? Our original reduction yields some initial intuition on the hardness of maximizing lifetime. In the "yes" case of 3-PARTITION, there is a trajectory sequence that fills no gridsquares above the $(6T + 22)$nd row, while in the "no" case we must fill some gridsquare in the $(6T + 23)$rd row:

Theorem 7. *Minimizing the maximum height of a filled gridsquare in a Tetris game is* NP-*complete.* □

However, this does not imply the hardness of maximizing the number of pieces that can be placed without losing, because Theorem 7 only applies for certain heights—and, in particular, does not apply for height m, because the trajectory sequence from Lemma 2 requires space above the $(6T + 22)$nd row for rotations and translations. To show the hardness of maximizing survival time, we need to do some more work.

Theorem 8. *Maximizing the number of pieces placed without losing is* NP-*complete.*

Proof. We augment our previous reduction as shown in Figure 6. We have created a large *reservoir* of r rows filled only in the first column, and a second *lock* in four new columns, which prevents access to the reservoir until all the top rows are cleared. We append to our piece sequence a single RG (to open the lock) and enough Sq's to

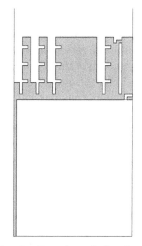

Fig. 6. Gameboard for hardness of maximizing survival time.

completely fill the reservoir. Choose r so that the unfilled area R of the reservoir is more than twice the total area A of the remainder of the gameboard. Observe that the gameboard has odd width, so the unfilled block of the reservoir has even width.

In the "yes" case of 3-PARTITION, we can clear the top part of the gameboard as before, then open the lock using the RG, and then completely fill and clear the reservoir using the Sq's.

In the "no" case, we cannot entirely clear the top part, and thus cannot unlock the reservoir with the RG. No number of the Sq's can ever subsequently clear the lower lock row. We claim that a loss will occur before all of the Sq's are placed. There are an odd number of columns, so only rows that initially contain an odd number of filled gridsquares can be cleared by the Sq's. Thus each row in the top of the gameboard can be cleared at most once; this uses fewer than half of the Sq's. The remainder of the Sq's cover $R/2 > A$ area, and never clear rows. Thus they must cause a loss. □

5   Hardness of Approximation

Theorem 9. *Given a game consisting of p pieces, approximating the maximum number of rows that can be cleared to within a factor of $p^{1-\varepsilon}$ for any constant $\varepsilon > 0$ is* NP-*hard.*

Proof. Our construction is as in Figure 6, with $r > a^{2/\varepsilon}$ rows in the reservoir, where there are a total rows at or above the second lock. As before, we append to the original piece sequence one RG followed by exactly enough Sq's to completely fill the reservoir. As in Theorem 8, in the "yes" case of 3-PARTITION, we can clear the entire gameboard (including the r rows of the reservoir), while in the "no" case case we can clear at most a rows. Thus it is NP-hard to distinguish the case in which at least r rows can be cleared from the case in which at most $a < r^{\varepsilon/2}$ rows can be cleared.

Note that the number of columns c in our gameboard is fixed and independent of r, and that the number of pieces in the sequence is constrained by $r < p < (r+a)c$. We also require that r be large enough that $p < (r+a)c < r^{2/(2-\varepsilon)}$. (Note that r, and thus our game, is still polynomial in the size of the 3-PARTITION instance.) Thus in the "yes" case we clear at least $r > p^{1-\varepsilon/2}$ rows, and in the "no" case we clear at most $a < r^{\varepsilon/2} < p^{\varepsilon/2}$. Thus it is NP-hard to approximate the number of cleared rows to within a factor of $(p^{1-\varepsilon/2})/(p^{\varepsilon/2}) = p^{1-\varepsilon}$. □

Using the construction in Figure 6 (with appropriate choice of r), similar arguments yield the following inapproximability results [5]:

Theorem 10. *Given a game consisting of p pieces, approximating the maximum number of pieces that can be placed without a loss to within a factor of $p^{1-\varepsilon}$ for any constant $\varepsilon > 0$ is NP-hard.* □

Theorem 11. *Given a game consisting of p pieces, approximating the minimum height of the highest filled gridsquare to within a factor of $2 - \varepsilon$ for any constant $\varepsilon > 0$ is NP-hard.* □

6   Varying the Rules of Tetris

Because the completeness of our reduction does not depend on the full set of allowable moves in Tetris—nor soundness on all limitations—our results continue to hold in some modified settings.

In real implementations of Tetris, there is a fixed amount of time (varying with the difficulty level) in which to make manipulations at height h. We consider players with limited dexterity, who can only make a small number of translations and rotations before the piece drops another row.

We defined a loss as the fixing of a piece so that it does not fit entirely within the gameboard. Other models also make sense: e.g., we might define a loss as occurring only *after* rows have been cleared—that is, a piece *can* be fixed so that it extends into the would-be $(m + 1)$st row of the m-row gameboard, so long as this is not the case once all filled rows are cleared.

Finally, we consider a broad class of *reasonable* models for piece rotation. Three models of particular interest are: (1) the instantaneous model of Section 2; (2) the *continuous* (or *Euclidean*) rotation model—the natural model if one pictures pieces physically rotating in space—which extends the instantaneous model by requiring that all gridsquares that a piece *passes through* during its rotation be unoccupied; and (3) the *Tetris rotation model*, illustrated in Figure 7, which we have observed in a number of actual Tetris implementations. In this model, the position of a piece in a particular orientation is determined as follows: within the pictured k-by-k square (fixed independently of orientation), the piece is positioned so that the smallest rectangle bounding it is centered in the square, shifted upwards and leftwards as necessary to align it with the grid. (Incidentally, it took us some time to realize that the "real" rotation in Tetris did not follow the instantaneous model, which is intuitively the most natural one.)

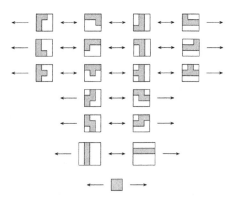

Fig. 7. The Tetris model of rotation. Each piece can be rotated clockwise (respectively, counterclockwise) to yield the configuration on its right (respectively, left).

Theorem 12. *It remains* NP-*hard to optimize (or approximate) the maximum height of a filled gridsquare or the number of rows cleared, tetrises attained, or pieces placed without a loss when any of the following hold:*

1. *the player is restricted to two rotation/translation moves before each piece drops in height.*
2. *pieces are restricted to* {LG, LS, I, Sq} *or* {RG, RS, I, Sq} *plus one other piece.*
3. *losses are not triggered until* after *filled rows are cleared, or no losses occur at all. (In the latter case, the objective of Theorem 8 is irrelevant.)*
4. *rotations follow any reasonable rotation model.*

Proof. For (1), note that completeness (Lemma 2) requires only two translations (to slide a LG into a notch) before the piece falls; soundness (Lemma 4) cannot be falsified by restricting legal moves. For (2), observe that our reduction uses only the pieces {LG, LS, I, Sq, RG}. For the other piecesets, we can take the mirror image of our reduction and/or replace the RG and the lock—our sole requirement on the lock is that that it be opened by a piece that does not appear elsewhere in the sequence. Claim (3) follows since we do not depend on the definition of losses in our proof: the completeness trajectory sequence does not approach the top of the gameboard, and the proof of soundness relies only on unfillability (and not on losses). Finally, for (4), our proof of Lemma 4 assumes an arbitrary reasonable rotation model [5]. Completeness follows even with the reasonability conditions, since the only rotations required for Lemma 2 occur in the upper staging area of the gameboard.   □

7   Future Directions

An essential part of our reduction is a complicated initial gameboard; it is a major open question whether Tetris can be played efficiently with an empty initial configuration.

Our results are largely robust to variations on the rules (see Section 6), but our completeness result relies on the translation of pieces as they fall. At more difficult levels of the game, it may be very hard to make two translations before the piece drops another row in height. Suppose each piece can be translated and rotated as many times as the player pleases, and then falls into place [2]; no manipulations are allowed after the first downward step. Is the game still hard?

It is also interesting to consider Tetris with gameboards of restricted size. What is the complexity of Tetris for an m-by-n gameboard with $m = O(1)$ or $n = O(1)$? Is Tetris fixed-parameter tractable with respect to either m or n? (We have polynomial-time algorithms for the special cases in which $m \cdot n$ is logarithmic in the number of pieces in the sequence, or for the case of $n = 2$.)

We have reduced the pieceset down to five of the seven pieces. For what piecesets is Tetris polynomial-time solvable? (E.g., the pieceset {I} seems polynomially solvable, though non-trivial because of the initial partially filled gameboard.)

Finally, in this paper we have concentrated our efforts on the offline, adversarial version of Tetris. In a real Tetris game, the initial gameboard and piece sequence are generated probabilistically, and the pieces are presented in an online fashion. What can be said about the difficulty of playing online Tetris if pieces are generated independently at random according to the uniform distribution, and the initial gameboard is randomly generated? Some possible directions for this type of question have been considered by Papadimitriou [11].

Acknowledgments. We would like to thank Amos Fiat and Ming-wei Wang for helpful initial discussions, Chris Peikert for comments on an earlier draft, and Josh Tauber for pointing out the puzzles in *Games Magazine* [9]. This work was partially supported by NDSEG and NSF Graduate Research Fellowships.

References

1. R. Breukelaar, H. J. Hoogeboom, and W. A. Kosters. Tetris is hard, made easy. Technical report, Leiden Institute of Advanced Computer Science, 2003.
2. J. Brzustowski. Can you win at Tetris? Master's thesis, U. British Columbia, 1992.
3. H. Burgiel. How to lose at Tetris. *Mathematical Gazette*, July 1997.
4. E. D. Demaine. Playing games with algorithms: Algorithmic combinatorial game theory. In *Proc. MFCS*, pages 18–32, August 2001. cs.CC/0106019.
5. E. D. Demaine, S. Hohenberger, and D. Liben-Nowell. Tetris is hard, even to approximate. Technical Report MIT-LCS-TR-865, 2002. cc.CC/0210020.
6. M. R. Garey and D. S. Johnson. Complexity results for multiprocessor scheduling under resource constraints. *SIAM J. Comput.*, 4:397–411, 1975.
7. M. R. Garey and D. S. Johnson. *Computers and Intractability: A Guide to the Theory of NP-Completeness*. W. H. Freeman and Company, New York, 1979.

8. R. Kaye. Minesweeper is NP-Complete. *Math. Intelligencer*, 22(2):9–15, 2000.
9. S. Kim. Tetris unplugged. *Games Magazine*, pages 66–67, July 2002.
10. M. M. Kostreva and R. Hartman. Multiple objective solution for Tetris. Technical Report 670, Department of Mathematical Sciences, Clemson U., May 1999.
11. C. Papadimitriou. Games against nature. *J. Comp. Sys. Sci.*, 31:288–301, 1985.
12. D. Sheff. *Game Over: Nintendo's Battle to Dominate an Industry.* Hodder and Stoughton, London, 1993.
13. Tetris, Inc. `http://www.tetris.com`.
14. U. Zwick. Jenga. In *Proc. SODA*, pages 243–246, 2002.

Approximate MST for UDG Locally

Xiang-Yang Li

Department of Computer Science, Illinois Institute of Technology
10 W. 31st Street, Chicago, IL 60616, USA
xli@cs.iit.edu

Abstract. We consider a wireless network composed of a set of n wireless nodes distributed in a two dimensional plane. The signal sent by a node can be received by all nodes within its transmission region, which is a unit disk centered at this node. The nodes together define a unit disk graph (UDG) with edge uv iff $\|uv\| \leq 1$. We present the first localized method to construct a bounded degree planar connected structure for UDG whose total edge length is within a constant factor of the minimum spanning tree. The total communication cost of our method is $O(n)$, and every node only uses its two-hop information to construct such structure. We show that some two-hop information is necessary to construct any low-weighted structure. We also study the application of this structure in efficient broadcasting in wireless ad hoc networks. We prove that this structure uses only $O(1/n)$ of the total energy of the previously best-known light weighted structure RNG.

1 Introduction

We consider a wireless ad hoc network composed of n nodes distributed in a two-dimensional plane. Assume that all wireless nodes have distinctive identities and each static wireless node knows its position information either through a low-power Global Position System (GPS) receiver or through some other way. We assume that each wireless node has an omni-directional antenna and a single transmission of a node can be received by *any* node within its vicinity which, we assume, is a unit disk centered at this node. By one-hop broadcasting, each node u can gather the location information of all nodes within the transmission range of u. Consequently, all wireless nodes V together define a unit-disk graph UDG, which has an edge uv iff the Euclidean distance $\|uv\|$ is less than one unit. Throughout this paper, a *local broadcast* by a node means it sends the message to all nodes within its transmission range; a *global broadcast* by a node means it tries to send the message to all nodes in the network by the possible relaying of other nodes. Since the main communication cost in wireless networks is to send out the signal while the receiving cost of a message is neglected here, a protocol's message complexity is only measured by how many messages are sent out by all nodes.

In recent years, there are substantial amount of research on topology control for wireless ad hoc networks [1,2,3,4]. These algorithms are designed for different objectives: minimizing the maximum link length while maintaining the network connectivity [3]; bounding the node degree [2]; bounding the spanning ratio [2,1]; constructing planar spanner locally [1]. Planar structures are used by several localized routing algorithms [5,6]. We [7] recently also proposed the first localized algorithm to construct a bounded

T. Warnow and B. Zhu (Eds.): COCOON 2003, LNCS 2697, pp. 364–373, 2003.
© Springer-Verlag Berlin Heidelberg 2003

degree planar spanner. A structure is called *low weight* if its total edge length is within a constant factor of the minimum spanning tree (MST). However, no localized algorithm is known to construct a low-weighted structure. It was recently shown in [8] that a broadcasting based on MST consumes energy within a constant factor of the optimum.

The best distributed algorithm [9,10] can compute MST in $O(n)$ rounds using $O(m + n \log n)$ communications for a general graph with m edges and n nodes. We can construct the minimum spanning tree of UDG in a distributed manner using $O(n \log n)$ messages. Unfortunately, even for wireless network modeled by a ring, the $O(n \log n)$ number of messages is still necessary for constructing MST of UDG. We present the first localized method to construct a bounded degree planar connected structure H whose total edge length is within a constant factor of MST. The total communication cost of our method is $O(n)$, and every node only uses its two-hop information to construct such structure. We also show that some two-hop information is necessary to construct any low-weighted structure. We also show the application of this structure in efficient broadcasting in wireless ad hoc networks.

Energy conservation is a critical issue in *ad hoc* wireless network for the node and network life, as the nodes are powered by batteries only. In the most common power-attenuation model, the power needed to support a link uv is $\|uv\|^\beta$, where β is a real constant between 2 and 5 dependent on the wireless transmission environment. Minimum-energy broadcast/multicast routing in ad hoc networking environment has been addressed in [11,12]. Wan *et al.* [8] showed that the approximation ratios of MST-based approach is between 6 and 12. Unfortunately, MST cannot be constructed in a localized manner, i.e., each node cannot determine which edge is in the defined structure by purely using the information of the nodes within some constant hops. Relative neighborhood graph has been used for broadcasting in wireless ad hoc networks [13]. It is well-known that $MST \subseteq RNG$. The ratio of the weight of RNG over the weight of MST could be $O(n)$ for n points set [14]. As will show later, the total energy used by this approach could be about $O(n^\beta)$ times optimum.

Notice that a structure with low-weight cannot guarantee that the broadcasting based on it consumes energy within a constant factor of the optimum. We show that the energy consumption using the structure H is within $O(n^{\beta-1})$ of the optimum, i.e., $\omega_\beta(H) = O(n^{\beta-1}) \cdot \omega_\beta(MST)$. This improves the previously known "lightest" structure RNG by $O(n)$ factor since in the worst case $\omega_\beta(RNG) = \Theta(n^\beta) \cdot \omega_\beta(MST)$.

The remainder of the paper is organized as follows. We present our efficient localized method constructing a bounded degree planar structure with low weight in Section 2. In Section 3, we discuss its applications in broadcasting, and find that it saves considerable energy consumption compared with that based on RNG. We conclude our paper in Section 4 with the discussion of possible future works.

2   Low Weight Topology

Let $\|xy\|$ denote the Euclidean distance between two points x and y. A disk centered at a point x with radius r is denoted by $disk(x, r)$. Let $lune(u, v)$ defined by two points u and v be the intersection of two disks with radius $\|uv\|$ and centered at u and v respectively, i.e., $lune(u, v) = disk(u, \|uv\|) \cap disk(v, \|uv\|)$. The *relative neighborhood graph*

[15], denoted by RNG(V), consists of all edges uv such that the *interior* of $lune(u,v)$ contains no point $w \in V$. Notice here if only the boundary of $lune(u,v)$ contains a point from V, edge uv is still included in RNG. Given a geometry graph G over a set of points, let $\omega(G)$ be the total length of the edges in G. More specifically, if the weight of an edge uv is defined as $\|uv\|^b$, then let $\omega_b(G)$ be the total weight of the weighted edges in G, i.e., $\omega_b(G) = \sum_{uv \in G} \|uv\|^b$. When $b = 1$, b is often omitted from the notations. It is known that $MST \subseteq RNG$.

2.1   Modified RNG

Our low-weight structure is based on a modified relative neighborhood graph. Notice that, traditionally, the relative neighborhood graph will always select an edge uv even if there is some node on the boundary of $lune(u,v)$. Thus, RNG may have unbounded node degree, e.g., considering $n - 1$ points equally distributed on the circle centered at the nth point v, the degree of v is $n - 1$.

The modified *relative neighborhood graph* consists of all edges uv such that (1) the *interior* of $lune(u,v)$ contains no point $w \in V$ and, (2) there is no point $w \in V$ with $ID(w) < ID(v)$ on the boundary of $lune(u,v)$ and $\|wv\| < \|uv\|$, and (3) there is no point $w \in V$ with $ID(w) < ID(u)$ on the boundary of $lune(u,v)$ and $\|wu\| < \|uv\|$, and (4) there is no point $w \in V$ on the boundary of $lune(u,v)$ with $ID(w) < ID(u)$, $ID(w) < ID(v)$, and $\|wu\| = \|uv\|$. See Figure 1 for an illustration when an edge uv is *not* included in the modified relative neighborhood graph. We denote such structure by RNG' hereafter. Obviously, RNG' is a subgraph of traditional RNG. We prove that RNG' has a maximum node degree 6 and still contains a MST as a subgraph. The proof of the following lemma is omitted due to space limitation.

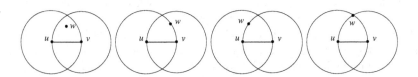

Fig. 1. Four cases when edges are not in the modified RNG.

Lemma 1. *The maximum node degree in graph RNG' is at most 6.*

Lemma 2. *The graph RNG' contains a Euclidean MST as a subgraph.*

PROOF. One way to construct MST is to add edges in the order of their lengths to the MST if it does not create a cycle with previously added edges. We break the tie as follows. We label an edge uv by $(\|uv\|, \max(ID(u), ID(v)), \min(ID(u), ID(v)))$, and an edge uv is ordered before an edge xy if the lexicographic order of the label of uv is less than that of xy. Let T be the MST constructed using the above edge ordering. We can show that $T \subseteq RNG'$. The detail is omitted due to space limitation.   ⌑

Obviously, graph RNG' still can be constructed using n messages. Each node first locally broadcasts its location and ID to its one-hop neighbors. Then every node decides which edge to keep solely based on the one-hop neighbors' location information collected. Since the definition is still symmetric, the edges constructed by different nodes are consistent, i.e., an edge uv is kept by a node u iff it is also kept by node v. The computational cost of a node u is still $O(d \log d)$, where d is its degree in UDG. A simple edge by edge testing method has time complexity $O(d^2)$.

Although graph RNG' has possibly less edges than RNG, its total edge weight could still be arbitrarily large compared with the MST. Figure 2 (a) illustrates an example where $\omega(RNG')/\omega(MST) = O(n)$ for a set of n points. Here $n/2$ points are equally distributed with separation $\epsilon \leq 2/n$ on two parallel vertical segments with distance 1 respectively. Obviously, all edges forming RNG' have total weight $n/2 + (n-2)\epsilon$ and the MST has total weight $1 + (n-2)\epsilon$. On the other hand, we have

Lemma 3. *For any sparse graph G with $O(n)$ edges, containing MST as subgraph, $\omega_b(G) = O(n^b) \cdot \omega_b(MST)$ for $b \geq 1$, and it has length spanning ratio at most $O(n)$.*

PROOF. For any edge $uv \in G$, if $uv \in MST$, then $\|uv\|^b < \omega_b(MST)$. If $uv \notin MST$, then there is a path in MST with edges not longer than uv connecting u and v. Let a_j, $1 \leq j \leq k \leq n$ be the lengths of these edges. Then $\|uv\| < \sum_{1 \leq j \leq k} a_j$. Thus, $\|uv\|^b < (\sum_{1 \leq j \leq k} a_j)^b \leq n^{b-1} \sum_{1 \leq j \leq k} a_j^b \leq n^{b-1} \cdot \omega_b(MST)$. Consequently, $\omega_b(G) = O(n^b) \cdot \omega_b(MST)$ since G has only $O(n)$ edges. Similar proof can show that G has length spanning ratio at most $O(n)$. ☐

2.2 Bound the Weight

In this section, we give a communication efficient method to construct a sparse topology from RNG' whose total edge weight is within a constant factor of $\omega(MST)$. Previously no localized method is known to construct a structure with weight $O(\omega(MST))$.

We first show by example that it is *impossible* to construct a low-weighted structure using only one hop neighbor information. Assume that there is such algorithm. Consider a set of points illustrated by Figure 2 (a). Let's see what this hypothetical algorithm will do to this point set. Since it uses only one-hop information, at every node, the algorithm only knows that there is a sequence of nodes evenly distributed with small separation, and another node which is one-unit away from current node. Since the algorithm has the same (or almost same) information at each node, the algorithm cannot decide whether to keep the long edge. If it keeps the long edge, then the total weight of the final structure is $O(n \cdot \omega(MST))$. If it discards the long edge, however, it may disconnect the graph since the nodes known by the algorithm at one node may be the whole network. See Figure 2 (b) for an illustration. We then give the first localized algorithm that constructs a low-weighted structure using only some two hops information.

Algorithm 1 *Construct Low Weight Structure*

1. All nodes together construct the graph RNG' in a localized manner.

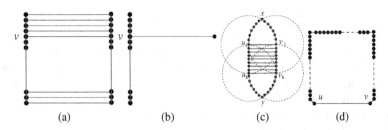

Fig. 2. The hypothetical algorithm cannot distinguish two cases (a) and (b) here. (c): Broadcasting based on RNG is expensive. (d) $\omega_\beta(H) = O(n^{\beta^\square\,1}) \cdot \omega_\beta(MST)$.

2. Each node u locally broadcasts its incident edges in RNG' to its one-hop neighbors. Node u listens to the messages from its one-hop neighbors.

3. Assume node u received a message informing existence of edge xy from its neighbor x. For each edge uv in RNG', if uv is the longest among xy, ux, and vy, node u removes edge uv. Ties are broken by the label of the edges. Here assume that $uvyx$ is the convex hull of u, v, x, and y.

Let H be the final structure formed by all remaining edges in RNG', and we call it low weighted modified relative neighborhood graph. Obviously, if an edge uv is kept by node u, then it is also kept by node v. To study the total weight of this structure, we will show that the edges in H satisfies the *isolation property* [16], which is defined as follows. Let $c > 0$ be a constant. Let E be a set of edges in d-dimensional space, and let $e \in E$ be an edge of weight l. If it is possible to place a hyper-cylinder B of radius and height $c \cdot l$ each, such that the axis of B is a subedge of e and B does not intersect with any other edge, i.e., $B \cap (E - \{e\}) = \phi$, then edge e is said to be *isolated* [16]. If all the edges in E are isolated, then E is said to satisfy the *isolation property*. The following theorem is proved by Das *et al.* [16].

Theorem 1. *[16] If a set of line segments E in d-dimensional space satisfies the isolation property, then $\omega(E) = O(1) \cdot \omega(SMT)$.*

Here SMT is the Steiner minimum tree over the end points of E. Obviously, total edge weight of SMT is no more than that of the minimum spanning tree. Generally, $\omega(MST) = O(\omega(SMT))$ for a set of points in Euclidean space. It is also known [16] that, in the definition of the isolation property, we can replace the hyper-cylinder by a hypersphere, a hypercube etc., without affecting the correctness of the above theorem. We will use a disk and call it *protecting disk*. Specifically, the protecting disk of a segment uv is $disk(p, \frac{\sqrt{3}}{4}\|uv\|)$, where p is the midpoint of segment uv. Obviously, we need all such disks do not intersect any edge except the one that defines it.

We first partition the edges of H into at most 8 groups such that the edges in each group satisfy the isolation property. Notice, given any node u, any cone apexed at u with angle less than $\pi/3$ will contain at most one edge of H incident on u since $H \subseteq RNG'$. Thus, we partition the region surrounded the origin by 8 equal-sized cones, say $\mathbb{C}_1$, $\mathbb{C}_2$, $\cdots$, $\mathbb{C}_8$ (the cone is half-open and half-close). The cones at different nodes are just a simple shifting of cones from the origin. Let E_i be the set of edges at cone C_i (one end-point is the apex of the cone and the other end-point is inside the cone).

Lemma 4. *No two edges in E_i share an end-point.*

PROOF. Assume that there are two edges xu and yu share a common node u, then obviously, these two edges cannot be from the cone apexed at node u; Clearly, angle $\angle xuy \leq 2\pi/8$. However, we already showed that there are no two edges incident on u form an angle less than $\pi/3$. This finishes the proof.   ⌷

Theorem 2. *The total edge weight of H is $O(\omega(MST))$.*

PROOF. We basically just show that the edges in E_i satisfy the isolation property, for $1 \leq i \leq 8$. For the sake of contradiction, assume that E_i does not satisfy the isolation property. Consider any edge uv from E_i and assume that it is not isolated. Thus, there is an edge, say xy, intersects the protecting disk of uv. There are four different cases: Case (a): $x \in disk(u, \|uv\|)$ and $y \in disk(v, \|uv\|)$; Case (b): $x \in disk(u, \|uv\|)$ and $y \notin disk(v, \|uv\|)$; Case (c): $x \notin disk(u, \|uv\|)$ and $y \in disk(v, \|uv\|)$; Case (d): $x \notin disk(u, \|uv\|)$ and $y \notin disk(v, \|uv\|)$. These four cases are illustrated by Figure 3. We will show that none of these four cases is possible.

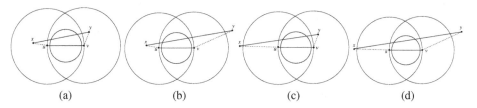

(a)   (b)   (c)   (d)

Fig. 3. Four cases that an edge uv is not isolated. Assume edge xy intersects the protecting disk.

For the first case, since x is in $disk(u, \|uv\|)$ and y is in $disk(v, \|uv\|)$, we know that xu and yv are both shorter than uv. Here, xu and yv need not be in the structure E_i. Thus, either uv or xy is the longest edge among uv, xy, xu and yv. Consequently, our algorithm will remove either uv or xy (whichever is longer).

For the remaining three cases, we will show that edge xy is the longest of these four edges. First of all, nodes x and y cannot be on the different side of the line passing through nodes u and v. Assume that they do, and x is below the line uv. Assume that x is outside of the disk centered at u with radius $\|uv\|$ since one of the x and y is outside of the corresponding disk. See Figure 4 for an illustration. We first show that $\angle yxu < \pi/3$. Let q be the intersection point of segment xy with line uv. Let p be the corner point of the lune $lune(u, v)$ that is on the same side of uv as y. Obviously, $\angle yxu < \angle yqu < \angle pqu < \angle puv = \pi/3$. We then show that $\|xy\| > \|xu\|$. Let z be the intersection point of xy with the boundary of $lune(u, v)$ and closer to u than v. Obviously, $\angle xuz > \pi/2$, thus, $\|xy\| > \|xz\| > \|xu\|$. Consequently, point u is inside the lune defined by points x and y, which is a contradiction to the fact that $xy \in RNG'$.

We then prove that the Case (b) is impossible. Assume that y is outside of disk centered at v with radius $\|uv\|$. See Figure 4 (2) for an illustration of the proof that

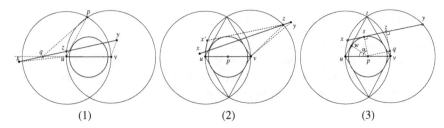

Fig. 4. (1) Node x is below uv and y is above. (2) Case (b) is impossible. (3) Edge xy is longest.

follows. Let z be the intersection point of xy with $disk(v, \|uv\|)$ that is closer to y. Let x' be the point on the disk $disk(v, \|uv\|)$ such that segment zx' is tangent to the protecting disk of segment uv. Obviously, $\angle ux'z > \pi/2$. Then $\|zx\| > \|zx'\|$. We can show that $\|zx'\|$ is at least $\|zv\|$ (the proof is presented in next paragraph). Then

$$\|yx\| > \|yz\| + \|zx'\| > \|yv\| - \|zv\| + \|zx'\| > \|yv\|$$

Then xy is the longest segment of the convex hull $xyvu$ since $\|xu\| \le \|uv\| \le \|vy\|$. This is a contradiction since our algorithm will remove edge xy. Notice here edge xy is the longest edge implies that node u is a neighbor of x and node v is a neighbor of y. Thus both node x and node y will know the existence of edge uv, and thus will remove edge xy according to our algorithm.

Figure 4 (3) illustrates the proof of $\|zx'\| \ge \|zv\|$ that follows. Consider any chord xy tangent on the protecting disk for uv. We then prove $\|xy\| \ge \|yv\| = \|uv\|$. Let z be the midpoint of xy, i.e., vz is perpendicular to xy. To make xy shorter, segment vz must be as long as possible. Let p be the midpoint of uv and s be the point on xy such that segment ps is perpendicular to segment xy. Then clearly, $\|vz\| = \|ps\| + \|pv\| \cdot \cos(\angle ups)$. Thus, xy is minimized when angle $\angle ups$ is minimized. However, $\angle ups > \angle upw$ since x and y are all above the line uv. Here w is the only intersecting point of chord ut with the protecting disk $disk(p, \frac{\sqrt{3}}{4}\|uv\|)$. It is easy to show $\|ut\| = \|tv\| = \|uv\|$. Thus, the minimum length of segment xy is $\|uv\|$ when $\angle ups = \angle upw$.

The proof of Case (c) is exactly the same as that for Case (b). For Case (d), same to the proof of Case (b), we know that $\|xu\| < \|xy\|$ and $\|vy\| < \|xy\|$. Then edge xy is also the longest edge of the convex hull $xyvu$. This is a contradiction since our algorithm will remove edge xy (nodes x and y will be informed by u and v respectively of the existence of edge uv since $\|xu\| < 1$ and $\|vy\| < 1$). This finishes the proof. $\boxdot$

Lemma 5. *The constructed topology H still contains a MST as a subgraph.*

PROOF. Consider the minimum spanning tree T constructed in the proof of Lemma 2. We will prove that $T \subseteq H$ by induction on the order of the edges added to the minimum spanning tree T. For the edge with the smallest order, it is clearly still in H. Assume that the first $k - 1$ edges added to T are still in H. Consider the kth edge, say uv, added to T. If uv is not in H, there must have two points x and y such that edge uv has the largest lexicographical label among edges on the convex hull $uvyx$.

Notice that for RNG', it is easy to show by induction that, for any two points p and q, there is a path in RNG' connecting p and q, whose edges have label less than that of pq. For any edge in this path, if it is not in T, then by definition of T, we know that there is another path with edges in T to connect the two endpoints of this edge. Thus, for any two points p and q, there is a path in T connecting p and q, whose edges have label less than that of pq.

Consequently, for points u and v, there is a path in T connecting them using edges with label lexicographically less than uv. This is a contradiction to the fact that uv is also in the minimum spanning tree T. This finishes the proof. ⊡

3 Application in Broadcasting

Minimum-energy broadcast/multicast routing in a simple ad hoc networking environment has been addressed in [11,12]. Any broadcast routing is viewed as an arborescence (a directed tree) T, rooted at the source node of the broadcasting, that spans all nodes. Let $f_T(p)$ denote the transmission power of the node p required by T. For any leaf node p of T, $f_T(p) = 0$. For any internal node p of T, $f_T(p) = \max_{pq \in T} \|pq\|^\beta$, i.e., the β-th power of the longest distance between p and its children in T. The total energy required by T is $\sum_{p \in V} f_T(p)$. It is known [17] that the minimum-energy broadcast routing problem cannot be solved in polynomial time if $P \neq NP$. Wan $et\ al.$ [8] showed that the approximation ratio of MST based approach is between 6 and 12.

Lemma 6. *For any point set V in the plane, the total energy required by any broadcasting among V is at least $\omega_\beta(MST)/C_{mst}$, where $6 \leq C_{mst} \leq 12$ is a constant related to the geometry minimum spanning tree.*

RNG has been used for broadcasting in wireless ad hoc networks [13]. Obviously, the ratio of the weight in RNG over the weight of MST could be $O(n)$ for n points set [14]. Figure 2 (c) illustrates an example that the total energy used by broadcasting on RNG could be about $O(n^\beta)$ times of the minimum-energy used by an optimum method. Here the n nodes are evenly distributed on the arc xu_1, segment u_1u_k, arc u_ky, arc yv_k, segment v_kv_1, and arc v_1x. Here four nodes u_1, u_k, v_1, and v_k form a unit square. It is not difficult to show that $\omega(MST) = \Theta(1)$ and $\omega_\beta(MST) = \Theta(1/n^{\beta-1})$, while $\omega(RNG) = \Theta(n)$ and $\omega_\beta(RNG) = \Theta(n)$. Together with Lemma 3, we know that in the worst case, $\omega_\beta(RNG') = \Theta(n^\beta) \cdot \omega_\beta(MST)$.

Notice that although the structure H has total edge weight $\omega(H) \leq c \cdot \omega(MST)$ for some constant c, it does not guarantee that $\omega_\beta(H)$ is within a constant factor of $\omega_\beta(MST)$ for $\beta > 1$. Figure 2 (d) illustrates such an example. Here the segment uv has length 1. The other $n - 2$ nodes are evenly distributed along the three segments of a square (with side length $1 + \epsilon$) such that the lines drawn in Figure 2 is indeed the graph RNG'. It is not difficult to show that $H = RNG'$. Obviously, $\omega_\beta(H) = O(1)$ and $\omega_\beta(MST) = O(1/n^{\beta-1})$ for any $\beta > 1$. On the other hand, we have

Lemma 7. $\omega_\beta(H) \leq O(n^{\beta-1}) \cdot \omega_\beta(MST)$.

PROOF. Assume that $\omega(H) \leq c \cdot \omega(MST)$ for a constant c. Let a_i, $1 \leq i \leq k$ be the edge lengths of H, and b_i, $1 \leq i \leq n-1$ be the edge lengths of MST. Here $k = O(n)$ is the number of edges in H. Then $\sum_{1 \leq i \leq k} a_i^{\beta} \leq (\sum_{1 \leq i \leq k} a_i)^{\beta} \leq c^{\beta} \cdot (\sum_{1 \leq i \leq n-1} b_i)^{\beta} \leq c^{\beta} \cdot n^{\beta-1} \cdot \sum_{1 \leq i \leq n-1} b_i^{\beta}$. This finishes the proof. $\square$

Consequently, we know that in the worst case, $\omega_{\beta}(H) = \Theta(n^{\beta-1}) \cdot \omega_{\beta}(MST)$. Figure 2 (d) shows that, to get a structure with weight $O(\omega_{\beta}(MST))$, we have to construct a MST for that example. Notice that the worst case communication cost to build a MST is $O(n \log n)$ under the broadcast communication model. It seems that the we may have to spend $O(n \log n)$ communications to build a structure with weight $O(\omega_{\beta}(MST))$. However, this worst case may not happen at all: for the configurations of nodes that need $O(n \log n)$ communications to build MST, the structure built by our method may be good enough; on the other hand, for the example that our algorithm does not perform well, we may find an efficient way to build a MST using $o(n \log n)$ messages. We leave it as an open problem whether we can construct a structure G whose weight $\omega_{\beta}(G)$ is $O(\omega_{\beta}(MST))$ using only $o(n \log n)$ messages, or even $O(n)$ messages. Here each message has $O(\log n)$ bits always.

4   Conclusion

We consider a wireless network composed of n a set of wireless nodes distributed in a two dimensional plane. We presented the first localized method to construct a bounded degree planar connected structure H with $\omega(H) = O(1) \cdot \omega(MST)$. The total communication cost of our method is $O(n)$, and every node only uses its two-hop information to construct such structure. We showed that some two-hop information is necessary to construct any low-weighted structure. We also studied the application of this structure in efficient broadcasting in wireless ad hoc networks. We showed that the energy consumption using this structure is within $O(n^{\beta-1})$ of the optimum, i.e., $\omega_{\beta}(H) = O(n^{\beta-1}) \cdot \omega_{\beta}(MST)$ for any $\beta \geq 1$. This improves the previously known "lightest" structure RNG by $O(n)$ factor since $\omega(RNG) = \Theta(n) \cdot \omega(MST)$ and $\omega_{\beta}(RNG) = O(n^{\beta}) \cdot \omega_{\beta}(MST)$.

On one aspect, a structure with low-weight does not guarantee that it approximates the optimum broadcasting structure in terms of the total energy consumption. On the other hand, a structure for broadcasting whose total energy consumption is within a constant factor of optimum does not guarantee that it is low-weight. We can show that its total edge length is within $O(\sqrt{n})$ of $\omega(MST)$ for a n-nodes network. Considering this "non-relevance" of the low-weight structure and the optimum broadcasting structure, it remains open how to construct a topology good for broadcasting.

The constructed structure is planar, and has bounded degree, low-weight. We [18] recently gave an $O(n \log n)$-time centralized algorithm constructing a bounded degree, planar, and low-weighted spanner. However, we cannot make that a distributed algorithm using $O(n)$ communications without sacrificing the spanner property. On the other hand, we [7] showed how to construct a planar spanner with bounded degree in a localized manner (using $O(n)$ messages) for unit disk graph. However, the constructed structure does not seem to have low-weight. It remains open how to construct a bounded degree, planar, and *low-weighted spanner* in a distributed manner using only $O(n)$ communications under the local broadcasting communication model.

References

1. Li, X.Y., Calinescu, G., Wan, P.J.: Distributed construction of planar spanner and routing for ad hoc wireless networks. In: 21st Annual Joint Conference of the IEEE Computer and Communications Societies (INFOCOM). Volume 3. (2002)
2. Li, X.Y., Wan, P.J., Wang, Y., Frieder, O.: Sparse power efficient topology for wireless networks. Journal of Parallel and Distributed Computing (2002) To appear. Preliminary version appeared in ICCCN 2001.
3. Ramanathan, R., Rosales-Hain, R.: Topology control of multihop wireless networks using transmit power adjustment. In: IEEE INFOCOM. (2000)
4. Wang, Y., Li, X.Y.: Geometric spanners for wireless ad hoc networks. In: Proc. of 22nd IEEE International Conference on Distributed Computing Systems (ICDCS). (2002)
5. Bose, P., Morin, P., Stojmenovic, I., Urrutia, J.: Routing with guaranteed delivery in ad hoc wireless networks. ACM/Kluwer Wireless Networks 7 (2001) 609–616 3rd int. Workshop on Discrete Algorithms and methods for mobile computing and communications, 1999, 48–55.
6. Karp, B., Kung, H.T.: GPSR: Greedy perimeter stateless routing for wireless networks. In: ACM/IEEE International Conference on Mobile Computing and Networking. (2000)
7. Li, X.Y., Wang, Y.: Localized construction of bounded degree planar spanner for wireless networks (2003) Submitted for publication.
8. Wan, P.J., Calinescu, G., Li, X.Y., Frieder, O.: Minimum-energy broadcast routing in static ad hoc wireless networks. ACM Wireless Networks (2002) Preliminary version appeared in IEEE INFOCOM 2000.
9. Faloutsos, M., Molle, M.: Creating optimal distributed algorithms for minimum spanning trees. Technical Report Technical Report CSRI-327 (also submitted in WDAG '95) (1995)
10. Gallager, R., Humblet, P., Spira, P.: A distributed algorithm for minimumweight spanning trees. ACM Transactions on Programming Languages and Systems 5 (1983) 66–77
11. Clementi, A., Penna, P., Silvestri, R.: On the power assignment problem in radio networks. Electronic Colloquium on Computational Complexity (2001) To approach. Preliminary results in APPROX'99 and STACS'2000.
12. Wieselthier, J., Nguyen, G., Ephremides, A.: On the construction of energy-efficient broadcast and multicast trees in wireless networks. In: Proc. IEEE INFOCOM 2000. (2000) 586–594
13. Seddigh, M., Gonzalez, J.S., Stojmenovic, I.: Rng and internal node based broadcasting algorithms for wireless one-to-one networks. ACM Mobile Computing and Communications Review 5 (2002) 37–44
14. Li, X.Y., Wan, P.J., Wang, Y., Frieder, O.: Sparse power efficient topology for wireless networks. In: IEEE Hawaii Int. Conf. on System Sciences (HICSS). (2002)
15. Toussaint, G.T.: The relative neighborhood graph of a finite planar set. Pattern Recognition 12 (1980) 261–268
16. Das, G., Narasimhan, G., Salowe, J.: A new way to weigh malnourished euclidean graphs. In: ACM Symposium of Discrete Algorithms. (1995) 215–222
17. Clementi, A., Crescenzi, P., Penna, P., Rossi, G., Vocca, P.: On the complexity of computing minimum energy consumption broadcast subgraphs. In: 18th Annual Symposium on Theoretical Aspects of Computer Science, LNCS 2010. (2001) 121–131
18. Li, X.Y., Wang, Y.: Efficient construction of low weight bounded degree planar spanner (2003). COCOON 2003.

Efficient Construction of Low Weight Bounded Degree Planar Spanner

Xiang-Yang Li and Yu Wang

Department of Computer Science, Illinois Institute of Technology
10 W. 31st Street, Chicago, IL 60616, USA
xli@cs.iit.edu, wangyu1@iit.edu

Abstract. Given a set V of n points in a two-dimensional plane, we give an $O(n \log n)$-time centralized algorithm that constructs a planar t-spanner for V, for $t \leq \max\{\frac{\pi}{2}, \pi \sin \frac{\alpha}{2} + 1\} \cdot C_{del}$, such that the degree of each node is bounded from above by $19 + \lceil \frac{2\pi}{\alpha} \rceil$, and the total edge length is proportional to the weight of the minimum spanning tree of V, where $0 < \alpha < \pi/2$ is an adjustable parameter. Here C_{del} is the spanning ratio of the Delaunay triangulation, which is at most $\frac{4\sqrt{3}}{9}\pi$. Moreover, we show that our method can be extended to construct a planar bounded degree spanner for unit disk graphs with the adjustable parameter α satisfying $0 < \alpha < \pi/3$. This method can be converted to a localized algorithm where the total number of messages sent by all nodes is at most $O(n)$ (under broadcasting communication model). These constants are all worst case constants due to our proofs. Previously, only centralized method [1] of constructing bounded degree planar spanner is known, with degree bound 27 and spanning ratio $t \simeq 10.02$. The distributed implementation of this centralized method takes $O(n^2)$ communications in the worst case.

1 Introduction

Let $d_G(u, v)$ be the length of the shortest path in graph G connecting two vertices u and v. Given a set of points V in a two-dimensional plane, a graph $G = (V, E)$ is a t-spanner of another graph H if for any two nodes u and v $d_G(u, v) \leq t \cdot d_H(u, v)$. Here the length of an edge is the Euclidean distance between its two endpoints. When H is the complete graph, we simply say that G is a t-spanner. If graph G has only $O(n)$ edges, then G is called *sparse* spanner. If the total edge length of G is within a constant factor of the Euclidean minimum spanning tree of V, then G is called *low weight* spanner. Many algorithms are known that compute sparse t-spanners with some additional properties such as bounded node degree, small spanner diameter (i.e., any two points are connected by a t-spanner path consisting of only a small number of edges), low weight, and fault-tolerance, see, e.g., [2,3,4,5,6,7,8]. All these algorithms compute t-spanners for any given constant $t > 1$ and thus, the hidden constants all depend on t.

We consider how to construct planar spanners for a set of two-dimensional points or a unit disk graph. Several planar geometry structures are studied before. It is known that the relative neighborhood graph [9,10] and Gabriel graph [9,11,12] are not spanners, while the Delaunay triangulation [13,14,15] is a t-spanner for $t \leq \frac{4\sqrt{3}}{9}\pi$. Hereafter, we

T. Warnow and B. Zhu (Eds.): COCOON 2003, LNCS 2697, pp. 374–384, 2003.
© Springer-Verlag Berlin Heidelberg 2003

use C_{del} to denote the spanning ratio of the Delaunay triangulation. Das and Joseph [16] showed that the minimum weighted triangulation and the greedy triangulation are t-spanners for some constant t. Levcopoulos and Lingas [17] showed, for any real number $r > 0$, how to construct a planar t-spanner from the Delaunay triangulation, whose total edge length is at most $2r + 1$ times the weight of a minimum spanning tree of V, where $t = (1 + 1/r)C_{del}$. Notice that all these structures could have unbounded node degree.

Recently Bose *et al.* [1] proposed a centralized $O(n \log n)$-time algorithm that constructs a planar t-spanner for a given nodes set V, for $t = (1 + \pi) \cdot C_{del} \simeq 10.02$, such that the node degree is bounded from above by 27. As we knew, this algorithm is the first method to compute a planar spanner of bounded degree.

In this paper, we give a simpler method to construct bounded degree planar t-spanner with low weight. In addition, degree bound and spanning ratio of our method are better than those in [1]. The main result of this paper is the following theorem.

Theorem 1. *There is an $O(n \log n)$-time algorithm that, given a set V of n points in a two-dimensional plane, constructs a graph*

1. *that is planar,*
2. *that is a t-spanner, for $t = \max\{\frac{\pi}{2}, \pi \sin \frac{\alpha}{2} + 1\} \cdot C_{del}(1 + \epsilon)$,*
3. *in which each point of V has degree at most $19 + \lceil \frac{2\pi}{\alpha} \rceil$,*
4. *and whose total edge weight is bounded from above by a constant factor of the weight of the Euclidean minimum spanning tree of V. Here the constant factor depends on ϵ.*

Here $0 < \alpha < \pi/2$ is an adjustable parameter.

The rest of the paper is organized as follows. In Section 2, we propose our method constructing bounded degree planar t-spanner with low weight for a two-dimensional point set. In Section 3, we extend our method to construct bounded degree planar t-spanner for any unit disk graph defined over a two-dimensional point set. Moreover, we show this centralized method can be converted to a localized algorithm, which can be used for wireless networks. We conclude our paper in Section 4.

2   Bounded Degree and Planar Spanner on Point Set

Our algorithms borrow some idea from the algorithm by Bose *et al.* [1]. They show that the length stretch factor of the final graph is $\frac{(\pi+1)2\pi}{(3\cos\pi/6)(1+\epsilon)}$ and node degree is at most 27. The running time of their algorithm is $O(n \log n)$. However, their method is impossible to have a localized even distributed version, since they use BFS and many operations on polygons (such as degree-3 partitions). Notice that breadth-first-search may take $O(n^2)$ communications. In this section, we will give a new method for constructing a planar spanner with bounded node degree for a point set V. The basic idea of our methods is to combine Delaunay triangulation and the ordered Yao structure [18].

2.1   Construction Algorithm

Algorithm: Constructing Bounded Degree Planar Spanner with Low Weight

1. First, it computes the Delaunay triangulation of a set V of n nodes, $Del(V)$. Let $N_{Del}(u)$ be the neighbors of node u in the Delaunay triangulation $Del(V)$, and d_u be the degree of node u in $Del(V)$. By proper data structure, $N_{Del}(u)$ and d_u can be achieved in time $O(n)$.

2. Find an order π of V as follows. Let $G_1 = Del(V)$ and $d_{G,u}$ be the node degree of u in graph G. Remove the node u with the smallest value of $(d_{G_i,u}, ID(u))$ from G_i, let $\pi_u = n - i + 1$, and call the remaining graph G_{i+1}. Repeat this procedure for $1 \leq i \leq n$. Let $\pi_{u_n} = 1$. Let P_v denote the predecessors of v in π, i.e., $P_v = \{u \in V : \pi_u < \pi_v\}$. Notice since G_i is always a planar graph, we know that the smallest value of $d_{G_i,u}$ is at most 5. Then, in ordering π, node u at most have 5 edges to its predecessors P_u in $Del(V)$.

3. Let E be the edge set of $Del(V)$, E' be the edge set of the desired spanner. Initialize E' to be empty set and all nodes in V are unprocessed. Then, for each node u in V, following the increasing order π, run the following steps to add some edges from E to E' (we only consider the Delaunay neighbors $N_{Del}(u)$ of u):

   a) We use $v_1, v_2, \cdots, v_k$ to denote the predecessors of node u (see Figure 1). Notice that u can have at most 5 edges to its predecessors (processed Delaunay neighbors) in E, i.e., $k \leq 5$. Then there are $k \leq 5$ *open* sectors at node u whose boundaries are rays emanated from u to the processed neighbors v_i of u in $Del(V)$. For each such sector at u, we divide it into a minimum number of *open* cones of degree at most α, where $\alpha \leq \pi/2$ is a parameter.

   b) For each such cone, let $s_1, s_2, \cdots, s_m$ be the geometrically ordered neighborhood $N_{Del}(u)$ of u in this cone. That is, $s_1, s_2, \cdots, s_m$ are all *unprocessed* nodes that are connected by some edges of E to u in this cone. For this cone, we first add the shortest edge in E that is connected to u to the edge set E', then add to E' all the edges (s_j, s_{j+1}), $1 \leq j < m$.

   c) Mark node u processed.

   Repeat this procedure in the increasing order of π, until all nodes are processed. The final graph formed by edges E' is denoted by $BPS(V)$.

4. Run the greedy spanner algorithm by [7] to bound the weight of the graph.

Notice that in the algorithm we use *open* sectors, which means that in the algorithm we do not consider adding the edges on the boundaries (any edge involved previously processed neighbors). For example, in Figure 1, the cones do not include any edges uv_i. This guarantee the algorithm does not add any edges to node v_i after v_i has been processed. This approach, as we will show it later, bounds the node degree.

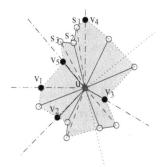

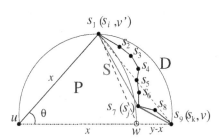

Fig. 1. Constructing planar spanner with bounded degree point set: process node u.

Fig. 2. The shortest path in polygon P.

2.2 Analysis of Algorithm

To show degree of $BPS(V)$ is bounded by a constant, we prove following theorem.

Theorem 2. *The maximum node degree of the graph $BPS(V)$ is at most $19 + \lceil \frac{2\pi}{\alpha} \rceil$.*

PROOF. Notice that for a node u there are 2 cases that an edge uv is added to the $BPS(V)$, let us discuss them one by one.

Case 1: When we process node u, some edges uv have already been added by some processed nodes w before. There are two subcases for this case.

Subcase 1.1: The edge uv has been added by a processed node v ($w = v$). For example, in Figure 1 , node u has edges from v_2, v_3 and v_5 before it is processed. For each predecessor v, it only adds one edge to node u.

Subcase 1.2: The edge uv has been added by processed node w (w is not v), node v is also an unprocessed node when processing w. For example, in Figure 1 , node s_2 have edges from s_1 and s_3 added by processing node u before node s_2 is processed. Notice that both v and u are neighbors of this processed node w. For each predecessor w, it at most adds two edges to node u.

Because for each u, it can only have at most 5 predecessor neighbors (processed neighbors), and each of predecessor can at most add 3 edges to it (either Subcase 1.1 or Subcase 1.2, or both). Thus, the number of this kind of edges (edges added by its predecessors before u is processed) is bounded by 15.

Case 2: When node u is processed, we can add one edge uv for each cones. Since we have at most 5 sectors emanated from u and each cone must have angle at most α, it is easy to show that we can at most have $4 + \lceil \frac{2\pi}{\alpha} \rceil$ cones at u. So the number of this kind of edges is also bounded by $4 + \lceil \frac{2\pi}{\alpha} \rceil$.

Notice that after node u is processed, no edges will be added to it. Consequently, the degree of each node u is bounded by $19 + \lceil \frac{2\pi}{\alpha} \rceil$ in the final structure.   ⊡

For example, when $\alpha = \pi/2$, then the maximum node degree is at most 23; when $\alpha = \pi/3$, then the maximum node degree is at most 25. Either case improves the previous bound 27 on the maximum node degree by Bose *et al.* [1].

It is trivial that $BPS(V)$ is a planar graph. Since $Del(V)$ is a planar graph and the algorithm only adds the Delaunay edges to $BPS(V)$. Notice that all edges $s_i s_{i+1}$ are also in $Del(V)$ since s_i and s_{i+1} are consecutive Delaunay neighbors of node u.

Finally, we prove that the graph $BPS(V)$ is a spanner.

Theorem 3. *The graph $BPS(V)$ is a t-spanner, where* $t = \max\{\frac{\pi}{2}, \pi \sin \frac{\alpha}{2} + 1\} \cdot C_{del}$.

PROOF. First, remember that $Del(V)$ is a spanner with a constant length stretch factor $C_{del} = \frac{4\sqrt{3}}{9}\pi \approx 2.42$. Keil and Gutwin [15] proved it using induction on the order of the lengths of all pair of nodes (from the shortest to the longest). We can show that the path connecting nodes u and v constructed by the method given in [15] also satisfies that all edges of that path is shorter than $\|uv\|$. So if we can prove this claim: *for any edge $uv \in Del(V)$, there exists a path in $BPS(V)$ connecting u and v whose length is at most a constant ℓ times $\|uv\|$*, then we know $BPS(V)$ is a $\ell \cdot C_{del}$-spanner.

Then we prove the above claim. Consider an edge uv in $Del(V)$. If $uv \in BGP(V)$, the claim holds. So assume that $uv \notin BGP(V)$.

Assume w.l.o.g. that $\pi_u < \pi_v$. It follows from the algorithm that, when we process node u, there must exist a node v' in the same cone with v such that $\|uv\| > \|uv'\|$, $uv' \in BPS(V)$, and $\angle v'uv < \alpha \leq \pi/2$. Let $v' = s_1, s_2, \cdots, s_k = v$ be this sequence of nodes in the ordered unprocessed neighborhood of u from v' to v.

Same with the proof in [1], consider the polygon P, consisting of nodes $u, s_1, \cdots, s_k$. We will show that the path $s_1 s_2 \cdots s_k$ has length that is at most a small constant factor of the length $\|uv\|$. Let us consider the shortest path from s_1 to s_k that is *totally inside* the polygon P. Let $S(s_1, s_k)$ denote such path. This path consists of diagonals of P. For example, in Figure 2, $S(s_1, s_k) = s_1 s_7 s_9$.

Assume that $\|uv'\| = x$. Let w be the point on segment uv such that $\|uw\| = \|uv'\|$. Assume that $\|uv\| = y$, then $\|wv\| = y - x$. Notice that node v' is the closest Delaunay neighbors in such cone. Obviously, all Delaunay neighbors s_i in this cone is outside of the sector defined by segments uw and uv'. We will show that such path $S(s_1, s_k)$ is contained inside the triangle $\triangle w s_1 s_k$. First, if no Delaunay neighbors is inside $\triangle w s_1 s_k$, then $S(s_1, s_k) = s_1 s_k$. Thus, the claim trivially holds. If there is some Delaunay neighbors inside $\triangle w s_1 s_k$, then s_1 will connect to the one S_i forming the smallest angle $\angle u s_1 s_j$. Similarly, node s_k will connect to the one s_j forming the smallest angle $\angle u s_k s_j$. Obviously s_i and s_j are inside $\triangle w s_1 s_k$, thus, the shortest path connecting them is also inside $\triangle w s_1 s_k$. Since path $S(s_1, s_k)$ is the shortest path inside the polygon P to connect s_1 and s_k, by convexity, the length of $S(s_1, s_k)$ is at most $\|v'w\| + \|wv\| = 2x \sin \frac{\theta}{2} + y - x$. Here $\theta = \angle v'uv < \alpha$.

An edge $s_i s_j$ of $S(s_1, s_k)$ has endpoints s_i and s_j in the neighborhood of u. Let $D(s_i, s_j)$ be the sequence of edges between s_i and s_j in the ordered neighborhood of u, which are added by processing u. For example, in Figure 2, $D(s_1, s_7) = s_1 s_2 s_3 s_4 s_5 s_6 s_7$. This path is in $BPS(V)$. We can bound the length of $D(s_i, s_j)$ by $\pi/2 \|s_i s_j\|$ by the argument in [1,19]. In [19], it is shown that the length of $D(s_i, s_j)$ is at most $\pi/2$ times $\|s_i s_j\|$, provided that (1) the straight-line segment between s_i and s_j lies outside the Voronoi region induced by u, and (2) that the path lies on one side of the line through s_i and s_j. In other words, we need $D(s_i, s_j)$ to be *one-sided Direct*

Delaunay path [1] [13]. In [1], they showed that both these two conditions hold when $\angle s_i u s_j < \pi/2$. This is trivially satisfied since $\angle s_i u s_j < \alpha \leq \pi/2$. Thus, we have a path $u s_1 s_2 \cdots s_k$ to connect u and v with length at most

$$x + (2x\sin\frac{\theta}{2} + y - x)\frac{\pi}{2} \leq y(\frac{\pi}{2} + \frac{x}{y}(\pi\sin\frac{\alpha}{2} - \frac{\pi}{2} + 1)) \leq y\cdot\max\{\frac{\pi}{2}, \pi\sin\frac{\alpha}{2} + 1\}$$

Putting it all together, we know $BPS(V)$ is a spanner with length stretch factor at most $\max\{\frac{\pi}{2}, \pi\sin\frac{\alpha}{2} + 1\}\cdot C_{del}$.   ▣

For example, when $\alpha = \pi/2$, then the spanning ratio is at most $(\frac{\sqrt{2}\pi}{2} + 1)\cdot C_{del}$; when $\alpha = \pi/3$, then the spanning ratio is at most $(\frac{\pi}{2} + 1)\cdot C_{del}$; when $\alpha = 2\arcsin(\frac{1}{2} - \frac{1}{\pi}) \simeq 20.9°$, then the spanning ratio is at most $\frac{\pi}{2}\cdot C_{del}$. We expect to further improve the bound on the spanning ratio by using the following property: all such Delaunay neighbors s_i is inside the circumcircle of the triangle uvv'; see Figure 2. Notice that, the method by Bose *et al.* [1] actually achieves the same spanning ratio as this one, although they did not prove this. However, the node degree of the graph generated by our method is smaller than that by [1].

Notice that the time complexity of our centralized algorithm is $O(n\log n)$ too. We can build Delaunay triangulation in $O(n\log n)$, and do ordering in time $O(n\log n)$ (using heap for the ordering based on degrees), and Yao structure in $O(n)$ (each edge is processed at most a constant times and there are $O(n)$ edges to be processed). When using heap for the ordering, initially building a heap needs $O(n\log n)$, then we remove one node and it has at most 5 adjacent edges, it needs at most 5 times updating the heap based on degree (each of which can be done in time $O(\log n)$). So the ordering can be done in $O(n\log n)$. Consequently, the time complexity is $O(n\log n)$, same with the method by Bose *et al.* [1]. However, our algorithm has smaller bounded node degree, and (more importantly) our algorithm has potential to become a localized version for wireless ad hoc networks application as we will describe later.

3   Bounded Degree and Planar Spanner on Unit Disk Graph

We consider a wireless ad hoc network (or sensor network) with all nodes distributed in a two-dimensional plane. Assume that all wireless nodes have distinctive identities and each static wireless node knows its position information either through a low-power Global Position System (GPS) receiver or through some other way. For simplicity, we also assume that all wireless nodes have the same maximum transmission range and we normalize it to one unit. By one-hop broadcasting, each node u can gather the location information of all nodes within the transmission range of u. Consequently, all wireless nodes V together define a unit-disk graph $UDG(S)$, which has an edge uv if and only

---

[1] For any pair of nodes u and v, let $u = w_1, w_2, \cdots, w_k = v$ be the sequence of nodes whose Voronoi region intersect segment uv and the Voronoi regions at w_i and w_j share a common boundary segment. Then the Direct Delaunay path $DT'(u, v)$ is $w_1 w_2 \cdots w_k$.

if the Euclidean distance $\|uv\|$ between u and v is less than one unit. In this section we give two centralized algorithms to construct planar spanner with bounded degree for $UDG(V)$. Then, we show the first centralized method can be converted to a localized algorithm using $O(n)$ messages, which can be used for wireless ad hoc networks.

3.1   Construction Algorithms

Algorithm 1: Constructing Planar Spanner with Bounded Degree for $UDG(V)$

1. Same with the algorithm for point set, first, compute Delaunay triangulation $Del(V)$.
2. Removing the edges whose length is longer than 1 in $Del(V)$. Call the remaining graph unit Delaunay triangulation $UDel(V)$. For every node u, we know its unit Delaunay neighbors $N_{UDel}(u)$ and its node degree d_u in $UDel(V)$.
3. Then, same with the algorithm for point set, find an order π of V as follows: Let $G_1 = UDel(V)$ and $d_{G,u}$ is the node degree of u in graph G. Remove the node u with the smallest value of $(d_{G_i,u}, ID(u))$ from G_i, let $\pi_u = n - i + 1$, and call the remaining graph G_{i+1}. Repeat this procedure for $1 \le i \le n$. Obviously, in ordering π, node u at most have 5 edges to its predecessors P_u in $UDel(V)$.
4. Let E and E' be the edge sets of $UDel(V)$ and the desired spanner. Initialize $E' = \emptyset$ and all nodes in V are unprocessed. Then, for each node u in V, following the increasing order π, run the following steps to add some edges to E':

   a) Node u uses its predecessors (processed Unit Delaunay neighbors) in E to define at most 5 *open* sectors at node u (see Figure 3). For each sector, we divide it into a minimum number of *open* cones of degree α, where $\alpha \le \pi/3$.
   b) For each cone, first add the shortest edge in E that is adjacent to u to the edge set E', then add to E' all the edges $s_j s_{j+1}$ between its geometrically ordered unprocessed neighbors in this cone, $1 \le j < m$. Notice that, here such edges $s_j s_{j+1}$ are not necessarily in $UDel(V)$. For example, when node u has a Delaunay neighbor x such that ux intersects edge $s_i s_{i+1}$ and $\|ux\| > 1$.
   c) Mark node u processed.

   Repeat this procedure in order of π, until all nodes are processed. Let $BPS_1(UDG(V))$ denote the final graph formed by edge set E'.

Algorithm 2: Constructing Planar Spanner with Bounded Degree for $UDG(V)$

1. Run the algorithm for point set to build $BPS(V)$ with parameter $\alpha \le \pi/3$.
2. Removing the edges whose length is longer than 1 in $BPS(V)$. The final graph is denoted by $BPS_2(UDG(V))$.

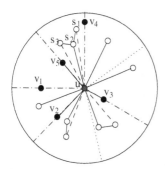
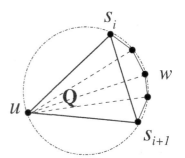

Fig. 3. Constructing planar spanner with bounded degree for $UDG(V)$: process node u. $v_1, \cdots, v_5$ are the processed neighbors of node u in $UDel(V)$.

Fig. 4. No new edges can be added by other nodes to intersect $s_i s_{i+1}$, where $s_i s_{i+1}$ is added by node u and not in $UDel(V)$.

Notice that in both these algorithms for $UDG(V)$, we change the cone angle bound from $\pi/2$ to $\pi/3$. The reason is in the proof of spanner property we need to guarantee the edge $s_i s_j$ and vv' must be in $UDG(V)$, i.e., $\|s_i s_j\| \le 1$ and $\|vv'\| \le 1$.

Notice that the constructed graphs $BPS_1(UDG(V))$ and $BPS_2(UDG(V))$ could be different since (1) the ordering of nodes could be different; (2) $BPS_1(UDG(V))$ could add some edges (some $s_i s_{i+1}$ type edges) that do not belong to $UDel(V) = Del(V) \cap UDG(V)$, while $BPS_2(UDG(V))$ always uses the edges from $UDel(V)$.

3.2   Analysis of Algorithms

The bounded node degree properties of these two final structures are trivial. The proof is similar to the one for point set. Only difference is that the angle of open cone is $\alpha \le \pi/3$ instead of $\alpha \le \pi/2$. Notice that node degree is bounded by 25 if $\alpha = \pi/3$.

Since $BPS_2(UDG(V))$ is a subgraph of planar graph $BPS(V)$, it must be a planar graph. So we only need to prove that the graph $BPS_1(UDG(V))$ is a planar graph.

Theorem 4. $BPS_1(UDG(V))$ *is a planar graph.*

PROOF. Observe that $UDel(V)$ is a planar graph. When each node u is being processed, we add two kinds of edges: (1) edge us_i, where s_i is the nearest unprocessed node in some cone divided by u; (2) some edges $s_i s_{i+1}$, when s_i and s_{i+1} are consecutive unprocessed neighbors of u in graph $UDel(V)$. See Figure 3 for illustration. These edges us_i belong to $UDel(V)$, so they will not intersect each other. If edge $s_i s_{i+1}$ is in $UDel(V)$, then it will not break the planar property of the graph also. Otherwise, the only possible reason which makes $s_i s_{i+1} \notin UDel(V)$ is that there are some edges (such as uw in Figure 4) in $Del(V)$ between us_i and us_{i+1} with length longer than 1. Then all such endpoints w of these long edges and s_i, s_j, u will form a polygon, denoted by Q, in $UDel(V)$. We will show that after $s_i s_{i+1}$ is added no intersecting edges can be added in $BPS_1(UDG(V))$. Notice that all the edges which are possible to add in $BPS_1(UDG(V))$ must be diagonals of some polygons in $UDel(V)$. However, all the diagonals of polygon Q intersecting $s_i s_{i+1}$ are longer than 1, as uw is, i.e., they will never be considered by our algorithm. Consequently, adding edge $s_i s_{i+1}$ will not break the planar property. This finishes our proof.   ⌘

Finally, we prove $BPS_1(UDG(V))$ and $BPS_2(UDG(V))$ are spanners.

Theorem 5. $BPS_1(UDG(V))$ *is a* $\ell \cdot C_{del}$-*spanner, where* $\ell = \max\{\frac{\pi}{2}, \pi \sin \frac{\alpha}{2} + 1\}$.

PROOF. Keil and Gutwin [15] showed that the Delaunay triangulation is a t-spanner for a constant $C_{del} = \frac{4\sqrt{3}}{9}\pi$ using induction on the increasing order of the lengths of all pair of nodes. We can show that the path connecting nodes u and v constructed in [15] also satisfies that all edges of that path is shorter than $\|uv\|$. Consequently, for any edge $uv \in UDG(V)$ we can find a path in $UDel(V)$ with length at most a $t = \frac{4\sqrt{3}}{9}\pi$ times $\|uv\|$, and all edges of the path is shorter than $\|uv\|$. So we only need to show that for any edge $uv \in UDel(V)$, there exists a path in $BPS_1(UDG(V))$ between u and v whose length is at most a constant ℓ times $\|uv\|$. Then $BPS_1(UDG(V))$ is a $\ell \cdot C_{del}$-spanner.

Consider an edge uv in $UDel(V)$. If edge uv is in $BPS_1(UDG(V))$, the claim trivially holds.

Then consider the case $uv \notin BPS_1(UDG(V))$. The rest of the proof is similar to the proof of Theorem 3. There must exist a node v' in the same cone with v such that $\|uv\| > \|uv'\|$, $uv' \in BPS(V)$, and $\angle v'uv < \alpha \leq \pi/3$. Let $v' = s_1, s_2, \cdots, s_k = v$ be the sequence of nodes in the ordered unprocessed neighborhood of u in $UDel(V)$ from v' to v. Let $v' = w_1, w_2, \cdots, w_k = v$ be the sequence of nodes in the ordered unprocessed neighborhood of u in $Del(V)$ from v' to v. Obviously, the set $\{s_1, s_2, \cdots, s_k\}$ is a subset of $\{w_1, w_2, \cdots, w_k\}$. Similar to Theorem 3, we know that the length of the path $uw_1w_2 \cdots w_k$ to connect u and v with length at most $\max\{\frac{\pi}{2}, \pi \sin \frac{\alpha}{2} + 1\} \cdot \|uv\|$, where $w_1 = s_1$ is the nearest neighbor of u in the cone, and $w_k = v$. Since any such node w_i is not inside the polygon Q (defined in the Figure 4 of proof for Theorem 4), the path $us_1s_2 \cdots s_k$ is not longer than the length of path $uw_1w_2 \cdots w_k$. This finishes the proof.   ⌘

Theorem 6. $BPS_2(UDG(V))$ *is a* $\ell \cdot C_{del}$-*spanner, where* $\ell = \max\{\frac{\pi}{2}, \pi \sin \frac{\alpha}{2} + 1\}$.

PROOF. Since $BPS_2(UDG(V))$ is a subgraph of $BPS(V)$, by removing edges longer than one, and $BPS(V)$ is a spanner, we only need to prove the spanner path $D(v', v)$ constructed in $BPS_2(V)$ (in our spanner proof) does not have edges longer than one for each u and v if $uv \in UDG(V)$.

This is trivial. Since the angle of cone is $\pi/3$ here, $\|s_i s_j\| < \|uv\| \leq 1$. From the proof given by Keil and Gutwin [15], we know all the edges in the spanner path $D(s_i, s_j)$ constructed in $BPS_2(V)$ are bounded by $\|s_i s_j\|$. Consequently, they all have length at most one. So the spanner path $D(v', v)$ survives after removing long edges. This finishes the proof.   ⌘

Notice that the computation costs of both algorithms are $O(n \log n)$. The first centralized algorithm can be extended to a localized algorithm [20]. The basic idea is as follows: first construct a planar spanner, localized Delaunay triangulation (LDel), for UDG using method in [21]; then build a local order based on node degree in LDel; finally apply the same technique in previous algorithms to bound the node degree following the local order. The total communication cost of the algorithm is bounded by $O(n)$. We prove in [20] that the constructed final topology is still planar, has bounded node degree, and has bounded spanning ratio. (The proof is surprisingly much more complicated than the centralized counterpart because the distributed method adds some extra edges, and removes some edges compared with the centralized method.)

4   Conclusion

In this paper, we first proposed a new structure which is a planar spanner with bounded node degree for any point set V. Then we show two centralized algorithms to construct this structure for $UDG(V)$. We can further bound the total weight of the structure by applying the method by Gudmundsson et al. [7]. The centralized algorithms can be implemented in time $O(n \log n)$. A localized algorithm [20] can be implemented using $O(n)$ messages under the broadcast communication model for wireless networks. The basic idea of this new method is to use (localized) Delaunay triangulation to make planar spanner graph, then apply some ordered Yao graph to bound the node degree. It is carefully designed to not lose all good properties when combining them.

Acknowledgment. The authors would like to thank Prosenjit Bose and Peng-Jun Wan for valuable discussions on paper [1].

References

1. Bose, P., Gudmundsson, J., Smid, M.: Constructing plane spanners of bounded degree and low weight. In: Proceedings of European Symposium of Algorithms. (2002)
2. Arya, S., Smid, M.: Efficient construction of a bounded degree spanner with low weight. Algorithmica **17** (1997) 33–54
3. Arya, S., Das, G., Mount, D., Salowe, J., Smid, M.: Euclidean spanners: short, thin, and lanky. In: Proc. 27th ACM STOC. (1995) 489–498
4. Levcopoulos, C., Narasimhan, G., Smid, M.: Improved algorithms for constructing fault tolerant geometric spanners. Algorithmica (2000)
5. Chandra, B., Das, G., Narasimhan, G., Soares, J.: New sparseness results on graph spanners. In: Proc. 8th Annual ACM Symposium on Computational Geometry. (1992) 192–201
6. Das, G., Narasimhan, G.: A fast algorithm for constructing sparse euclidean spanners. International Journal on Computational Geometry and Applications **7** (1997) 297–315
7. Gudmundsson, J., Levcopoulos, C., Narasimhan, G.: Improved greedy algorithms for constructing sparse geometric spanners. In: Scandinavian Workshop on Algorithm Theory. (2000) 314–327
8. Lukovszki, T.: New results on fault tolerant geometric spanners. Proceedings of the 6th Workshop on Algorithms an Data Structures (WADS'99), LNCS (1999) 193–204
9. Bose, P., Devroye, L., Evans, W., Kirkpatrick, D.: On the spanning ratio of Gabriel graphs and β-skeletons. In: Proc. of the Latin American Theoretical Infocomatics (LATIN). (2002)

10. Jaromczyk, J., Toussaint, G.: Relative neighborhood graphs and their relatives. Proceedings of IEEE **80** (1992) 1502–1517
11. Gabriel, K., Sokal, R.: A new statistical approach to geographic variation analysis. Systematic Zoology **18** (1969) 259–278
12. Eppstein, D.: β-skeletons have unbounded dilation. Technical Report ICS-TR-96-15, University of California, Irvine (1996)
13. Dobkin, D., Friedman, S., Supowit, K.: Delaunay graphs are almost as good as complete graphs. Discr. Comp. Geom. (1990) 399–407
14. Keil, J., Gutwin, C.: The Delaunay triangulation closely approximates the complete euclidean graph. In: Proc. 1st Workshop Algorithms Data Structure (LNCS 382). (1989)
15. Keil, J.M., Gutwin, C.A.: Classes of graphs which approximate the complete euclidean graph. Discr. Comp. Geom. **7** (1992) 13–28
16. Das, G., Joseph, D.: Which triangulations approximate the complete graph? In: Proceedings of International Symposium on Optimal Algorithms (LNCS 401). (1989) 168–192
17. Levcopoulos, C., Lingas, A.: There are planar graphs almost as good as the complete graphs and almost as cheap as minimum spanning trees. Algorithmica **8** (1992) 251–256
18. Bose, P., Gudmundsson, J., Morin, P.: Ordered θ graphs. In: Proc. of the Canadian Conf. on Computational Geometry (CCCG). (2002)
19. Bose, P., Morin, P.: Online routing in triangulations. In: Proc. of the 10 th Annual Int. Symp. on Algorithms and Computation ISAAC. (1999)
20. Li, X.Y., Wang, Y.: Localized construction of bounded degree planar spanner for wireless networks (2003) Submitted for publication.
21. Li, X.Y., Calinescu, G., Wan, P.J.: Distributed construction of planar spanner and routing for ad hoc wireless networks. In: 21st IEEE INFOCOM. Volume 3. (2002)

Isoperimetric Inequalities and the Width Parameters of Graphs[*]

L. Sunil Chandran[1], T. Kavitha[1], and C.R. Subramanian[2]

[1] Max-Planck-Institute für Informatik,
Stuhlsatzenhausweg 85, 66123 Saarbrücken, Germany.
{sunil,kavitha}@mpi-sb.mpg.de
[2] The Institute of Mathematical Sciences, Chennai, 600113, India.
crs@imsc.res.in

Abstract. We relate the isoperimetric inequalities with many width parameters of graphs: treewidth, pathwidth and the carving width. Using these relations, we deduce

1. A lower bound for the treewidth in terms of girth and average degree
2. The exact values of the pathwidth and carving width of the d-dimensional hypercube, H_d
3. That treewidth $(H_d) = \Theta\left(\frac{2^d}{\sqrt{d}}\right)$.

Moreover we study these parameters in the case of a generalization of hypercubes, namely the Hamming graphs.

1 Introduction

Discrete isoperimetric inequalities have recently become important in combinatorics. Usually two kinds of isoperimetric problems are considered: the vertex isoperimetric problem and the edge isoperimetric problem. Let $G = (V, E)$ be a graph. For $S \subseteq V$, the vertex boundary $\Phi(S)$ can be defined as follows.

$$\Phi(S) = \{w \in V - S : \exists v \in S \text{ such that } \{w, v\} \in E\} \tag{1}$$

The vertex isoperimetric problem is to minimize $|\Phi(S)|$ over all subsets S of V with $|S| = \ell$ for a given integer ℓ. We denote this minimum value by $b_v(\ell, G)$, i.e., $b_v(\ell, G) = \min_{S \subseteq V(G), |S| = \ell} |\Phi(S)|$. (We denote by $V(G)$ and $E(G)$ the vertex and edge sets of G, respectively.)

The edge isoperimetric problem can be defined in a similar way. Given a subset S of V, its edge boundary $\delta(S)$ can be defined as

$$\delta(S) = \{e \in E : \text{exactly one end point of } e \text{ is in } S\} \tag{2}$$

---

[*] This is a combined announcement of the results in two different papers. The results explained in Section 3 is from *Girth and Treewidth* (*L. S. Chandran, C. R. Subramanian*) [8]. The rest of the results are from *Lower bounds for width parameters of graphs using Isoperimetric Inequalities* (*L.S. Chandran, T. Kavitha*) [7].

T. Warnow and B. Zhu (Eds.): COCOON 2003, LNCS 2697, pp. 385–393, 2003.
© Springer-Verlag Berlin Heidelberg 2003

The edge isoperimetric problem is to minimize $|\delta(S)|$ over all subsets S of V with $|S| = \ell$ for a given integer ℓ. Let $b_e(\ell, G) = \min_{S \subseteq V(G), |S| = \ell} |\delta(S)|$.

Many methods have been discovered to attack the isoperimetric inequalities – including martingale techniques [22], eigen value analysis [1] and purely combinatorial methods. (See [19] for a survey.) The discrete isoperimetric inequalities are of interest, not only because they answer extremely natural and basic questions about graphs, but also because they have numerous applications, most notably to random graphs and geometric functional analysis. In this paper, we explore yet another application of isoperimetric inequalities, namely to provide lower bounds for certain "width" parameters of graphs.

The notions of treewidth and pathwidth were introduced by Robertson and Seymour [20,21] in their series of fundamental papers on graph minors.

Definition 1. *A tree decomposition of* $G = (V, E)$ *is defined to be a pair* (X, T) *where* $X = \{X_i : i \in I\}$ *is a collection of subsets of* V *(we call these subsets the nodes of the decomposition) and* $T = (I, F)$ *is a tree having the index set* I *as the set of vertices such that the following conditions are satisfied:*

1. $\bigcup_{i \in I} X_i = V$.
2. $\forall (u, v) \in E, \exists i \in I : u, v \in X_i$.
3. $\forall i, j, k \in I$: *if* j *is on a path in* T *from* i *to* k, *then* $X_i \cap X_k \subseteq X_j$.

The width of a tree decomposition $(\{X_i : i \in I\}, T)$ *is defined to be equal to* $\max_{i \in I} |X_i| - 1$.

The treewidth of G *is defined to be the minimum width over all tree decompositions of* G *and is denoted by* $tw(G)$.

A path decomposition of $G = (V, E)$ *is a tree decomposition* (X, T) *in which* T *is required to be a path. The pathwidth of* G *is defined to be the minimum width over all path decompositions of* G *and is denoted by* $pw(G)$.

Recent research has shown that many NP-complete problems become polynomial or even linear time solvable, or belong to NC, when restricted to graphs with small treewidth (See [2,3]). The decision problem of checking, given G and k, whether $tw(G) \leq k$ is known to be NP-complete. Hence the problem of determining the treewidth of an arbitrary graph is NP-hard.

Another interesting "width" parameter is the carving width. This notion is used by Seymour and Thomas in [23] in their algorithm to decide whether there exists a routing tree with congestion less than k, when the underlying graph is planar. We present below the definition of carving width.

Let V be a finite set with $|V| \geq 2$. Two subsets $A, B \subseteq V$ cross if $A \cap B, A - B, B - A, V - (A \cup B)$ are all non–empty. A carving in V is a set $\mathcal{C}$ of subsets of V such that

1. $\emptyset, V \notin \mathcal{C}$
2. no two members of C cross
3. $\mathcal{C}$ is maximal subject to the two conditions above.

Let G be a graph. For $A \subset V(G)$, let $\delta(A)$ denote the set of all edges with exactly one end point in A. For each $e \in E(G)$, let $p(e) \geq 0$ be an integer. For

$X \subseteq E(G)$, we denote $\sum_{e \in X} p(e)$ by $p(X)$. If $|V(G)| \geq 2$, we define the p–carving width of G to be the minimum over all carvings in $V(G)$, of the maximum over all $A \in \mathcal{C}$, of $p(\delta(A))$. The carving width of G is the p–carving width of G where $p(e) = 1$ for every edge e and is denoted by $cw(G)$. Seymour and Thomas show that computing $cw(G)$ for general graphs is NP-hard.

The d–dimensional hypercube, H_d, is the graph on 2^d vertices, which correspond to the 2^d d–vectors whose components are either 0 or 1, two of the vertices being adjacent when they differ in just one coordinate. Hypercubes are a well-studied class of graphs, which arise in the context of parallel computing, coding theory, algebraic graph theory and many other areas. They are popular because of their symmetry, small diameter and many interesting graph–theoretic properties.

1.1   Our Results

In this paper we develop lower bounds for treewidth, pathwidth and carving width of a graph in terms of isoperimetric inequalities. As applications of these lower bounds we have the following results.

Result 1 The length of the shortest cycle in a graph G is called the girth of G. We derive a lower bound for the treewidth of G in terms of its girth and average degree. In particular, we show that if G has girth g and average degree d, then $tw(G) \geq \max\left(\lceil \frac{d}{2} \rceil, \frac{1}{4e(g+1)}(d-1)^{\lfloor (g-1)/2 \rfloor} - 2\right)$, where e is the Napier number. In view of a well–known conjecture regarding the existence of graphs with girth g, minimum degree δ and having at most $c(\delta - 1)^{\lfloor (g-1)/2 \rfloor}$ vertices (for some constant c), this lower bound seems to be almost tight (but for a multiplicative factor of $g + 1$). (See section 2 for details.)

Result 2 We study the parameters treewidth, pathwidth and carving width in the context of d–dimensional hyper cubes. The question of estimating the treewidth of hyper cube was proposed by Chlebikova [10] who states that $2^{\lfloor \frac{d}{2} \rfloor} \leq tw(H_d) \leq \binom{d}{\lfloor \frac{d}{2} \rfloor} + \binom{d}{\lceil \frac{d}{2} \rceil}$. In [9], Chandran and Subramanian show that $tw(H_d) \geq \left\lfloor \frac{3 \cdot 2^d}{2(d+4)} \right\rfloor - 1$. In this paper we obtain improved bounds for the treewidth of H_d and determine the exact values for the pathwidth and carving width of H_d.

1. $pw(H_d) = \sum_{m=0}^{d-1} \binom{m}{\frac{m}{2}}$.
2. $c_1 \frac{2^d}{\sqrt{d}} \geq tw(H_d) \geq c_2 \frac{2^d}{\sqrt{d}}$, for some constants c_1, c_2, where $c_1 < 1.2$ and $c_2 > 0.48$, for large d.
3. $cw(H_d) = 2^{d-1}$.

Result 3 In fact, we study these parameters in the context of a more general class of graphs, namely the Hamming graphs. The *Hamming Graph* K_q^d is the graph on q^d vertices, which correspond to the q^d d–vectors with coordinates coming from a set of size q, two of the vertices being adjacent if they differ in just one coordinate. (Clearly, the d–dimensional hypercubes are a special case of the Hamming graphs, K_q^d, namely, when $q = 2$).

1. $pw(K_q^d) \geq c_1 \frac{q^d}{\sqrt{d}}$ for some constant $c_1 > 0$, where $c_1 \approx \sqrt{\frac{2}{\pi}} e^{-2}$ for large d.

2. $pw(K_q^d) \leq c_2 \frac{q^d}{\sqrt{d}}$ for some constant c_2, where $c_2 < 1.2$ for large d.

2   A Lower Bound for Treewidth and an Application

Definition 2. *Let $G = (V, E)$ be a simple connected graph on n vertices with no self loops, and s be a number such that $1 \leq s \leq n$. We define $N_{min}(s) = \min |N(X)|$ over all (non empty) X with $\frac{s}{2} \leq |X| \leq s$.*

It can be shown that $tw(G) \geq N_{min}(s) - 1$, for any s, $1 \leq s \leq |V(G)|$. The details of the proof of this result is is available in the full paper [8]. Noting that $N_{min}(s) = \min_{\frac{s}{2} \leq x \leq s} b_v(x, G)$, we can state the following theorem.

Theorem 1. *Let $G = (V, E)$ be a graph and let s be an integer with $1 \leq s \leq n$. Then $tw(G) \geq \min_{\frac{s}{2} \leq x \leq s} b_v(x, G) - 1$.*

We make use of this lower bound for treewidth to derive the following result. We omit the proof. (It is available in the full paper [8]).

Theorem 2. *Let G be a graph with girth at least g, average degree at least d. Then*

$$tw(G) \geq \max \left(\left\lceil \frac{d}{2} \right\rceil, \ \frac{1}{4e(g+1)} (d-1)^{\lfloor (g-1)/2 \rfloor} - 2 \right)$$

where e is the Napier number.

Perhaps, a natural approach to the question of getting a lower bound for the treewidth of G would be to look for the largest r such that G has a K_r minor (K_r stands for the complete graph on r vertices), since the treewidth of a graph is at least the treewidth of any of its minors and $tw(K_r) = r - 1$. (H is called a minor of G iff H can be obtained from a subgraph of G by a sequence of edge contractions). This problem has received some attention in the literature. It can be inferred from a result of Thomassen [25] in conjunction with a result[1] in [17, 24] that if $girth(G)$ is at least $cr\sqrt{\log r}$ and the minimum degree of G is at least 3 then G has a K_r minor. This result was improved recently by Diestel and Rempel [12] who proved that if $girth(G) \geq 6 \log r + 3 \log \log r + c$, and $\delta(G) \geq 3$, then G has a K_r minor. One can infer from the above result that if $\delta(G) \geq 3$, then $tw(G) \geq r' - 1 \geq 2^{g/6} - 1$, where r' is the largest integer such that G has a $K_{r'}$ minor. Very recently Kühn and Osthus [18] have improved his result, and have shown using probabilistic methods that if $girth(G) \geq g$ for some odd

---

[1] The result of [17,24] states that if the average degree of a graph G is at least $cr\sqrt{\log r}$ then G has a K_r minor.

g, then for some $c > 0$, G has a K_r minor for some $r \geq \frac{c.(\delta)^{\frac{g+1}{4}}}{\sqrt{\log \delta}}$. Clearly, the lower bound we provide for treewidth in terms of girth and minimum degree is much better than (in fact about the square of) what is derivable from the minor theoretic result of Kühn and Osthus. Moreover, our lower bound is expressed in terms of average degree rather than the minimum degree and thus is stronger.

More fundamentally, Kühn and Osthus argue that it is unlikely that their lower bound for the clique minor size may be improved significantly. Thus, extending their point further, it is unlikely that an approach based on clique minors may give a lower bound for the treewidth (in terms of girth and minimum degree or average degree) comparable to ours.

For a brief exposition of the early developments on the existence of dense minors in graphs of large girth, see Chapter 8 of [11]. For an introductory account of the role of treewidth in minor theory, see Chapter 12 of [11].

Remark on the tightness of the above lower bound: The following is a well known conjecture on the existence of high girth graphs. (See for example, [5], page 164).

Conjecture: *There exists a constant c such that for all integers $g, \delta \geq 3$, there is a graph $G(g, \delta)$ of minimum degree at least δ and girth at least g whose order (number of vertices) is at most $c(\delta - 1)^{\lfloor \frac{g-1}{2} \rfloor}$.*

In view of this conjecture, we see that the lower bound given in Theorem 2 is very close to what is best possible (but for a multiplicative factor of $g + 1$) since the treewidth of $G(g, \delta)$ can be at most $c(\delta-1)^{\lfloor \frac{g-1}{2} \rfloor}$, the total number of vertices, whereas the lower bound proven by us is at least $\frac{1}{4e(g+1)}(\delta - 1)^{\lfloor (g-1)/2 \rfloor} - 2$ since the average degree $d \geq \delta$. It may be noted that, by using $G(g, \delta)$ as components it is easy to construct graphs with minimum degree at least δ, girth at least g, and having an arbitrarily large number of vertices but its treewidth (or pathwidth) still at most $c(\delta - 1)^{\lfloor \frac{g-1}{2} \rfloor}$.

3   A Lower Bound for Pathwidth and Carving Width

Since the $tw(G) \leq pw(G)$, we can, of course, use Theorem 1 to provide a lower bound for the pathwidth of a graph also. But we notice that in the case of pathwidth one can state a much stronger result. Below we develop the necessary concepts.

A graph $G = (V, E)$ is defined to be an interval graph iff its vertices $V = \{v_1, v_2, \cdots, v_n\}$ can be put in one to one correspondence with a set of intervals $\{I_1, I_2, \cdots, I_n\}$ on the real line in such a way that $\{v_i, v_j\} \in E$ if and only if the corresponding intervals I_i and I_j have a non empty intersection. Without loss of generality one can assume that all the intervals are closed intervals, see for example [14], page 13. The reader can easily convince himself that given any graph G there exist super graphs G' of G such that G' is an interval graph. For example, the complete graph on n nodes is an interval graph. The clique number $\omega(G)$ is defined to be the number of vertices in a maximum sized clique in G.

Definition 3. *A graph G has interval–width, $IW(G) = k$ iff k is the smallest non–negative integer such that G is a subgraph of some interval graph H, with $\omega(H) = k + 1$.*

The following characterization of pathwidth in terms of interval graphs is well–known. The omitted proofs (from this section onwards) are available in the full paper [7].

Lemma 1. *Let G be a graph. Then $pw(G) = IW(G)$.*

Lemma 2. *Let $G = (V, E)$ be a connected interval graph and $1 \le s \le n$. Then $\omega(G) \ge b_v(s, G) + 1$.*

Theorem 3. *Let $G = (V, E)$ be any graph on n nodes, and let $1 \le s \le n$. Then $pw(G) \ge b_v(s, G)$.*

Proof. Clearly in any interval super graph G' of G (on the same number of vertices) $b_v(s, G') \ge b_v(s, G)$. The theorem follows from Lemma 1 and Lemma 2. ∎

A lower bound similar to that of Theorem 1 can be proved for the carving width of a graph G in terms of its edge isoperimetry. The following lemma is from [23].

Lemma 3 ([23]). *Let V be a finite set with $|V| \ge 2$, let T be a tree in which every vertex has valency 1 or 3, and let τ be a bijection from V onto the set of leaves of T. For each edge e of T let $T_1(e)$ and $T_2(e)$ be the two components of $T - e$; and let*

$$\mathcal{C} = \{\{v \in V : \tau(v) \in V(T_i(e))\} : e \in E(T), i = 1, 2\}$$

Then $\mathcal{C}$ is a carving in V. Conversely, every carving in V arises from some tree T and bijection τ in this way.

Lemma 4. *Let T be a tree with each of its nodes having valency either 3 or 1. For any edge e, let $T_1(e)$ and $T_2(e)$ be the components of $T \backslash e$. For any $1 \le x \le n, \exists$ an edge e_x such that the number of leaves of one of the trees $T_1(e_x)$ and $T_2(e_x)$ lies between $\frac{x}{2}$ and x.*

Theorem 4. *Let $G = (V, E)$ be a graph on n nodes and let $1 \le x \le n$. Then $cw(G) \ge \min_{\frac{x}{2} \le s \le x} b_e(s, G)$.*

4   Applications to Hypercubes and Hamming Graphs

4.1   Hypercube

A prime example of a graph of combinatorial interest is the hypercube H_d. The reader may be reminded that H_d can be considered as a graph on the power set

$\mathcal{P}(X)$ of a d-point set X in which a vertex corresponding to a set x is adjacent to a vertex corresponding to a set y if and only if $|x \Delta y| = 1$. (Here $x \Delta y$ stands for the symmetric difference of the sets x and y.)

Before we deal with the pathwidth, treewidth and carving width of the hypercube, we introduce another well-known graph theoretic parameter, namely the bandwidth. Let $G = (V, E)$ be a graph. Let a bijection $\phi : V \to \{1, \cdots, n\}$ be called an ordering of the vertices of G. Then for any edge $e = \{u, v\} \in E$, let $\Delta(e, \phi) = |\phi(u) - \phi(v)|$. The bandwidth of G is defined as the minimum over all possible orderings ϕ of $V(G)$, the maximum value of $\Delta(e, \phi)$ over all edges $e \in E$. The following is easy to prove.

$$\text{bandwidth}(G) \geq \max_{1 \leq s \leq n} b_v(s, G) \tag{3}$$

Also, it is not difficult to see from the definition of pathwidth and bandwidth that $bandwidth(G) \geq pw(G)$. Note that this inequality can be strict and the value of $|bandwidth(G) - pw(G)|$ can be arbitrarily large. Below we will show that in the case of H_d, pathwidth=bandwidth.

In [15], Harper addressed the question of finding the bandwidth of H_d. To minimize the value of $\max_{e \in E} \Delta(e, \phi)$ he proposed an optimal ordering ϕ' of the vertices of H_d. He indeed showed that $\max_e \Delta(e, \phi') = max_s b_v(s, H_d)$. Harper's result on the bandwidth of H_d can be summarised as below.

Lemma 5. $bandwidth(H_d) = \max_s b_v(s, H_d) = \sum_{m=0}^{d-1} \binom{m}{\frac{m}{2}}$

The last equality can be proved by induction and is mentioned in the last page of Harper's article [15]. Putting together all the pieces, we get our result on the pathwidth of H_d.

Theorem 5. $pw(H_d) = bandwidth(H_d) = \max_s b_v(s, H_d) = \sum_{m=0}^{d-1} \binom{m}{\frac{m}{2}}$

Proof. By Theorem 3, we know that $pw(H_d) \geq \max_s b_v(s, H_d)$. It follows by Lemma 5, that $pw(H_d) \geq bandwidth(H_d)$. But we know that bandwidth of any graph is at least as much as its pathwidth. The required result follows. ∎

The following result was first proved by Harper [15]. Simpler proofs were later given by Katona [16], Frankl and Füredi [13] etc. See [4], Chapter 16 for an exposition.

Lemma 6. *Let s be an integer such that $s = \sum_{i=0}^{r-1} \binom{d}{i} + m$, where $0 < m \leq \binom{d}{r}$ ($1 \leq r \leq d$.) Then, $b_v(s, H_d) = \binom{d}{r} - m + \partial_u(m)$ where $\partial_u(m)$ is the minimum possible cardinality of the upper shadow of a set of m subsets in X^r (see [4], chapter 5 for details).*

By setting $m = \binom{d}{r}$ in the above lemma, we have the following simple corollary.

Corollary 1. $b_v(\sum_{i=0}^{r} \binom{d}{r}, H_d) = \binom{d}{r+1}$

Now, making use of Lemma 6, Corollary 1 and Theorem 1 and using sterling's approximation for factorials, we can present our bounds for the treewidth of H_d.

Theorem 6. $c_1 \frac{2^d}{\sqrt{d}} \geq tw(H_d) \geq c_2 \frac{2^d}{\sqrt{d}}$, *for some constants c_1, c_2, where $c_1 < 1.2$ and $c_2 > 0.48$, for large d.*

Similarly, using Theorem 4 we can infer the following result.

Theorem 7. $cw(H_d) = 2^{d-1}$.

Proof. It follows from Bollobás [4] and Bollobás and Leader [6] that $\min_{2^{d-2} \leq s \leq 2^{d-1}} b_e(s, H_d) \geq 2^{d-1}$. So, by Theorem 4, we get that $cw(H_d) \geq 2^{d-1}$. In fact we can construct a carving $\mathcal{C}$ on H_d with width $= 2^{d-1}$, thus proving the theorem. (See the full paper [7] for details.) ∎

4.2   Hamming Graphs

Another graph of combinatorial interest is K_q^d, the product of d copies of a complete graph of order q. Equivalently, the vertex set of K_q^d is the set of all strings of size d on the alphabet $\{1, 2, ..., q\}$ and two strings are adjacent if they differ in exactly one place.

Theorem 8. $pw(K_q^d) \geq c_1 \frac{q^d}{\sqrt{d}}$ *for some constant $c_1 > 0$, where $c_1 \approx \sqrt{\frac{2}{\pi}} e^{-2}$ for large d.*

Theorem 9. $pw(K_q^d) \leq c_2 \frac{q^d}{\sqrt{d}}$ *for some constant c_2, where $c_2 < 1.2$ for large d.*

Combining Theorems 8 and 9, we see that the $pw(K_q^d) = \Theta(\frac{q^d}{\sqrt{d}})$.

References

1. N. Alon and V. D. Millman, λ_1, *isoperimetric inequalities for graphs and super concentrators*, Journal of Combinatorial Theory, Series. B, 38 (1985), pp. 73–88.
2. S. Arnborg and A. Proskurowski, *Linear time algorithms for NP–hard problems on graphs embedded in k–trees*, Discrete Applied Mathematics, 23 (1989), pp. 11–24.
3. H. L. Bodlaender, *A tourist guide through treewidth*, Acta Cybernetica, 11 (1993), pp. 1–21.
4. B. Bollabás, *Combinatorics*, Cambridge University Press, 1986.
5. B. Bollobás, *Extremal Graph Theory*, Academic Press, 1978.
6. B. Bollobás and I. Leader, *Edge-isoperimetric inequalities in the grid*, Combinatorica, 11 (1991), pp. 299–314.
7. L. S. Chandran and T. Kavitha, *Lower bounds for width parameters of graphs using isoperimetric inequalities*. Manuscript, 2003.
8. L. S. Chandran and C. R. Subramanian, *Girth and treewidth*, Tech. Rep. MPI-I-2003-NWG2-01, Max-Planck-Institut für Informatik, Saarbrücken, Germany, 2003.
9. ———, *A spectral lower bound for the treewidth of a graph and its consequences*. To appear in Information Processing Letters, 2003.
10. J. Chlebikova, *On the tree-width of a graph*, Acta Mathematica Universitatis Comenianae, 61 (1992), pp. 225–236.
11. R. Diestel, *Graph Theory*, vol. 173, Springer Verlag, New York, 2 ed., 2000.

12. R. Diestel and C. Rempel, *Dense minors in graphs of large girth*. To Appear in Combinatorica, 2003.
13. P. Frankl and Z. Füredi, *A short proof for a theorem of Harper about Hamming spheres*, Discrete Math., 34 (1981), pp. 311–313.
14. M. C. Golumbic, *Algorithmic Graph Theory And Perfect Graphs*, Academic Press, New York, 1980.
15. L. Harper, *Optimal numberings and isoperimetric problems on graphs*, Journal of Combinatorial Theory, 1 (1966), pp. 385–393.
16. G. O. H. Katona, *The hamming-sphere has minimum boundary*, Studia Sci. Math. Hungar., 10 (1975), pp. 131–140.
17. A. V. Kostochka, *Lower bound of the hadwiger number of graphs by their average degree*, Combinatorica, 4 (1984), pp. 307–316.
18. D. Kühn and D. Osthus, *Minors in graphs of large girth*. To appear in Random Structures and Algorithms, 2003.
19. I. Leader, *Discrete isoperimetric inequalities*, Proc. Symp. Appl. Math., 44 (1991), pp. 57–80.
20. N. Robertson and P. D. Seymour, *Graph minors I. excluding a forest*, Journal of Combinatorial Theory, Ser. B, 35 (1983), pp. 39–61.
21. ———, *Graph minors II: algorithmic aspects of tree-width*, Journal of Algorithms, 7 (1986), pp. 309–322.
22. G. Schechtman, *Lévy type inequality for a class of metric spaces*, in Martingale Theory in Harmonic Analysis and Banach Spaces, Lecture Notes in Mathematics, vol. 939, 1982, pp. 211–215.
23. P. Seymour and R. Thomas, *Call routing and the ratcatcher*, Combinatorica, 14 (1994), pp. 217–241.
24. A. G. Thomason, *An extremal function for contractions of graphs*, Math. Proc. Camb. Phil. Soc., 95 (1984), pp. 261–265.
25. C. Thomassen, *Girth in graphs*, Journal of Combinatorial Theory, Ser. B, 35 (1983), pp. 129–141.

Graph Coloring and the Immersion Order*

Faisal N. Abu-Khzam and Michael A. Langston

Department of Computer Science, University of Tennessee, Knoxville, TN
37996–3450, USA

Abstract. The relationship between graph coloring and the immersion
order is considered. Vertex connectivity, edge connectivity and related issues are explored. These lead to the conjecture that, if G requires at least
t colors, then G must have immersed within it K_t, the complete graph
on t vertices. Evidence in support of such a proposition is presented. For
each fixed value of t, there can be only a finite number of minimal counterexamples. These counterexamples are characterized based on Kempe
chains, connectivity, cutsets and degree bounds. It is proved that minimal counterexamples must, if any exist, be both 4-vertex-connected and
t-edge-connected.

1 Introduction

The applications of graph coloring are legion. The usual goal, and the one we
consider here, is to assign colors to vertices so that no two adjacent vertices are
given the same color. Graph coloring has a long and storied history. The study of
four-coloring planar graphs alone has generated interest for over 150 years [21].
Despite all this effort, graph coloring in general remains a notoriously difficult
combinatorial problem.

The chromatic number of G, denoted by $\chi(G)$, is the minimum number of
colors required by G in any proper coloring of its vertices. Of course it is well
known that determining $\chi(G)$ is $\mathcal{NP}$-hard. It is tempting to try to associate
$\chi(G)$ with some sort of clique contained within G. After all, if G contains K_t as
a subgraph, then it is easy to show that G can be colored with no fewer than t
colors. To see that the presence of a K_t subgraph is not necessary, however, one
needs only to observe that C_5, the cycle of order five, requires three colors yet
does not contain K_3 as a subgraph.

Nevertheless, perhaps some weaker form of K_t is present. One possibility is
topological containment, in which taking subgraphs is augmented with removing
subdivisions. An edge is subdivided when it is replaced by a path formed from
two edges and an internal vertex of degree two; subdivision removal reverses this
operation. For example, C_5 contains K_3 topologically. Sometime in the 1940s

* This research is supported in part by the National Science Foundation under
grants EIA–9972889 and CCR–0075792, by the Office of Naval Research under
grant N00014–01–1–0608, by the Department of Energy under contract DE–AC05–
00OR22725, and by the Tennessee Center for Information Technology Research under award E01–0178–081.

Hajós conjectured that if $\chi(G) \geq t$, then G must contain a topological K_t [11]. The conjecture is trivially true for $t \leq 3$. In 1952 Dirac proved it true for $t = 4$ [4]. It was not until Catlin's work in 1979 that Hajós' conjecture was finally settled, and negatively, with a family of counterexamples for $t \geq 7$ [3]. Ironically, one such counterexample is the 15-vertex graph defined by the crossproduct of C_5 and K_3. It requires eight colors but contains no topological K_8. Subsequently, Erdős and Fajtlowicz were able to prove the rather surprising result that almost all graphs are counterexamples [6]. Thus Hajós' conjecture remains open only for $t \in \{5, 6\}$.

Another possibility is the minor order, for which the allowable operations are taking subgraphs and contracting edges. The minor order is a generalization of the topological order, because subdivision removal is just a special case of edge contraction. Hadwiger conjectured in 1943 that, if $\chi(G) \geq t$, then G must contain a K_t minor [10]. This conjecture equates to Hajós' conjecture for $t \leq 4$. Wagner proved in 1964 that, for $t = 5$, it is equivalent to the four color theorem [26]. In 1993 Robertson, Seymour and Thomas proved it true for $t = 6$ [20]. Whether Hadwiger's conjecture holds true in general, however, has thus far not been decided. This is in spite of decades of research, hordes of supporting evidence and a multitude of results on many of its variants and restrictions [1,5,14,23,25, 27]. Even the celebrated Graph Minor Theorem [19] appears to shed no particular light on this question. As of this writing, a resolution of Hadwiger's conjecture seems distant.

In this paper we focus instead on the immersion order. A pair of adjacent edges uv and vw, with $u \neq v \neq w$, is lifted by deleting the edges uv and vw, and adding the edge uw. A graph H is said to be immersed in a graph G if and only if a graph isomorphic to H can be obtained from G by lifting pairs of edges and taking a subgraph. Previous investigations into the immersion order have generally been conducted from a purely algorithmic standpoint. We refer the reader to [2,7,8,9,17] for examples and applications. In contrast, here we mainly consider structural issues. We establish compelling connections between graph coloring and the immersion order, and conjecture that K_t is immersed in any graph requiring t or more colors.

2   Preliminaries

We restrict our attention to finite, simple undirected graphs (multiple edges and loops that may arise from lifting are irrelevant to coloring). G is said to be *t-vertex-connected* if at least t vertex-disjoint paths connect every pair of its vertices. A *vertex cutset* is a set of vertices whose removal breaks G into two or more nonempty connected components. The cardinality of a smallest vertex cutset in G is equal to the largest t for which G is t-vertex-connected (unless G is a complete graph, which can have no vertex cutset). G is said to be *t-edge-connected* if at least t edge-disjoint paths connect every pair of its vertices. An *edge cutset* is a set of edges whose removal breaks G into two or more nonempty connected components. The cardinality of a smallest edge cutset in G is equal to the largest t for which G is t-edge-connected.

If $\chi(G) \leq t$, then G is said to be t-colorable. If $\chi(G) = t$, then G is said to be t-chromatic. If $\chi(G) = t$ and $\chi(H) < t$ for every proper subgraph H of G, then G is said to be t-color-critical. A t-coloring of G is realized by a map c from the vertices of G to the set $\{1, 2, .., t\}$ so that, if G contains the edge uv, then $c(u) \neq c(v)$. Given such a map, c_{ij} is used to denote the subgraph induced by the vertex set $\{u : c(u) \in \{i, j\}\}$. A path contained within c_{ij} is termed a *Kempe chain* [28], so-named in honor of the foundational work done on them by Kempe in [15]. (Ironically, the main result in [15] was a purported proof of the Four Color Theorem that, like so many others, turned out to be fatally flawed.) Of course c_{ij} need not be connected, and so for any $u \in c_{ij}$ we employ $c_{ij}(u)$ to denote the set $\{v : v \text{ resides in the same connected component of } c_{ij} \text{ as does } u\}$. Such sets have useful properties.

Observation 1. *If $\{i, j\} \neq \{k, l\}$, then c_{ij} and c_{kl} are edge disjoint.*

Although the immersion order is traditionally defined in terms of taking subgraphs and lifting pairs of edges, Kempe chains and Observation 1 make it helpful for us to utilize as well the following alternate characterization: H is immersed in G if and only if there exists an injection from the vertices of H to the vertices of G for which the images of adjacent elements of H are connected in G by edge-disjoint paths. Under such an injection, an image vertex is called a *corner* of H in G; all image vertices and their associated paths are collectively called a *model* of H in G.

We use $\delta(G)$ to denote the smallest degree found among the vertices of G. We use $N(u)$ to denote the neighborhood of u. Suppose u has degree $t - 2$ or less in a t-chromatic graph G. Then $G - u$ must also be t-chromatic. Otherwise $G - u$ could be colored with $t - 1$ colors, and u assigned one of the $t - 1$ colors unused within $N(u)$.

Observation 2. *If G is t-color-critical, then $\delta(G) \geq t - 1$.*

It is sometimes advantageous to select, restrict or manipulate colorings. For example, if G is t-chromatic but $G - u$ is only $(t-1)$-chromatic, then it is possible to consider only colorings in which u is assigned a unique color.

Observation 3. *If G is t-color-critical, then for any vertex u there exists a coloring c in which $c(u) = 1$ and $c(v) \neq 1$ for every vertex $v \in G - u$.*

Given the various connections between graph coloring, degrees and connectivity, and in turn the connections between connectivity and the immersion order, we seek to determine just how $\chi(G)$ is related to immersion containment. Our efforts to date prompt us to set the stage for this with the following conjecture. (A superficially similar conjecture has been made by Lescure and Meyniel [22]. Although sometimes called "the immersion conjecture," the notion of containment used there is not the immersion order.)

Conjecture *If $\chi(G) \geq t$, then K_t is immersed in G.*

This speculation motivates our work in the sequel. There we shall present what we believe is compelling preliminary evidence in its support. Our conjecture, like Hadwiger's, is trivially true for $t \leq 4$. This is because the immersion order generalizes the topological order, for which Hajós' conjecture is long known to hold when $t \leq 4$.

Before proceeding, we introduce a notion of immersion-criticality and show how it relates to the possible existence of counterexamples.

Definition *G is* t-immersion-critical *if* $\chi(G) = t$ *and* $\chi(H) < t$ *whenever H is properly immersed in G.*

Because $\chi(K_t) = t$, any counterexample must either be t-immersion-critical or have properly immersed within it another t-immersion-critical counterexample. Similarly, any t-immersion-critical graph distinct from K_t must be a counterexample. Thus our conjecture is equivalent to the statement that K_t is the only t-immersion-critical graph for every t. Although we have thus far fallen short of establishing this one way or the other, we can show that there are at most a finite number of them. To do this, we rely on properties of well-quasi-orders and immersion order obstruction sets. We refer the reader unfamiliar with these concepts to [7,8,16].

Theorem 1. *For each t, there are finitely many t-immersion-critical graphs.*

Proof. Consider the family of graphs $F = \{G : \chi(G) < t$ and $\chi(H) < t$ for every $H \leq_i G\}$. Then, by definition, F is closed in the immersion order. Because graphs are well-quasi-ordered by the immersion relation, it follows that F's obstruction set is finite. This set contains precisely the t-immersion-critical graphs. $\square$

3   Main Results

Graph connectivity has long been a central feature of attempts to settle Hadwiger's conjecture. G is said to be *t-minor-critical* if $\chi(G) = t$ and $\chi(H) < t$ whenever H is a proper minor of G. K_t is of course both $(t-1)$-vertex-connected and $(t-1)$-edge-connected. Thus, if any t-minor-critical graph is not as strongly connected, then Hadwiger's conjecture is false for all $t' \geq t$. So suppose G denotes a t-minor-critical graph other than K_t (in which case the conjecture fails). Some 35 years ago [18], Mader showed that G must be at least 7-vertex-connected whenever $t \geq 7$. This provides evidence in support of the conjecture for $t \in \{7, 8\}$. A few years later [23], Toft proved that G must also be t-edge-connected. This provides additional supporting evidence for all t. Very recently, Kawarabayashi has shown that G must be at least $\lceil \frac{t}{3} \rceil$-vertex-connected as well [13]. Following this approach, we study both the vertex and edge connectivity of t-immersion-critical graphs. We assume $t \geq 5$ unless stated otherwise. Kempe chains play a pivotal role in our investigation.

3.1   Vertex Connectivity

Because they are t-color-critical, it is easy to see that t-immersion-critical graphs are 2-vertex-connected [1]. We now establish that they must in fact be at least 4-vertex-connected. Our work linking coloring to the immersion order begins in earnest with Lemma 4. First, however, we present something of an introduction with three easy but useful lemmas about cutsets, paths and coloring. Lemmas 1 and 2 are probably well known, although they may not be formulated anywhere else in precisely the same way we state them in this treatment. Lemma 2, which we dub *The Patching Lemma*, is especially helpful. Lemma 3 is certainly well known, and mentioned in a variety of sources (see, for example, [12,25,27]).

Lemma 1. *Let S denote a minimum-cardinality vertex cutset in a 2-vertex-connected graph G, and let C denote a connected component of $G \backslash S$. Then any two elements of S must be connected by a path whose interior vertices lie completely within C.*

Two colorings are said to be *equivalent* if the partitions induced by their respective color classes are identical.

Lemma 2. (The Patching Lemma) *Let S denote a vertex cutset of G, and let G_1 and G_2 denote a pair of induced subgraphs for which $G_1 \cup G_2 = G$ and $G_1 \cap G_2 = S$. If G_1 and G_2 admit t-colorings whose restrictions to S are equivalent, then G is t-colorable.*

The Patching Lemma can be used to establish the following well-known fact.

Lemma 3. *No vertex cutset of a t-color-critical graph can be a clique.*

The preceding lemmas tell us a good deal about the make-up of vertex cutsets, and how they relate to coloring. Armed with this information, we are now able to argue more directly about vertex connectivity and the immersion order. To simplify matters, we shall adopt the following conventions for the remainder of this subsection:

- t is at least five,
- G denotes a t-immersion-critical graph,
- S denotes a minimum-cardinality vertex cutset in G,
- C denotes a connected component of $G \backslash S$,
- G_1 denotes the subgraph induced by $C \cup S$, and
- G_2 denotes $G \backslash C$.

Lemma 4. *Every t-immersion-critical graph is 3-vertex-connected.*

Proof. Suppose otherwise, as witnessed by some G with $S = \{a, b\}$. We know from Lemma 3 that the edge ab is not present in G. Let $i \in \{1, 2\}$. By Lemma 1, there must be a path, P_i, with endpoints a and b, whose vertices lie completely

.within G_i. Lifting the edges of P_{3-i} to form the single edge ab, and then taking the subgraph induced by the vertices of G_i, produces a graph H_i properly immersed in G. It follows that H_i is $(t-1)$-colorable. Because ab is present in H_i, any such coloring of H_i assigns different colors to a and b. But G_i is a subgraph of H_i. Thus, there are $(t-1)$-colorings of G_1 and G_2 that each assign different colors to a and b. By the Patching Lemma, this ensures a $(t-1)$-coloring of G, a contradiction. $\square$

Lemma 4 applies to t-topological-critical graphs as well. To see this, note that the two paths defined in the proof are vertex-disjoint. An analog of Lemma 4 does not hold, however, if the graph is only known to be t-color-critical. Such graphs are guaranteed only to be 2-vertex-connected. A t-color-critical graph that is not 3-vertex-connected can be constructed as follows. Begin with a pair of non-adjacent vertices, u and v, a copy of K_{t-1} and a copy of K_{t-2}. Connect u to every vertex but one in the copy of K_{t-1}. Connect v to the remaining vertex in the copy of K_{t-1}. Now connect both u and v to every vertex in the copy of K_{t-2}. Note that these graphs are not t-immersion-critical.

Lemma 5. *If $|S| = 3$, then G_1 and G_2 admit $(t-1)$-colorings that assign more than one color to the elements of S.*

Proof. Let $S = \{u, v, w\}$, and consider the case for G_1. By Lemma 1, there is a path between u and v in G_2. Lifting this path and taking the subgraph induced by the vertices of G_1 produces a graph H properly immersed in G. Because G is t-immersion-critical, and because H contains the edge uv, H must admit a $(t-1)$-coloring that assigns different colors to u and v. As a subgraph of H, G_1 can likewise be colored. A symmetrical argument handles the case for G_2. $\square$

What we have really just shown is that if G is only 3-vertex-connected, then G_1 admits a $(t-1)$-coloring that assigns different colors to any fixed pair of elements of S. This raises the possibility that a single coloring of G_1 may suffice, simultaneously assigning different colors to all three elements of S. We now show that this cannot happen. It follows that the same must then be true for G_2.

Let a and b denote vertices of G, and let c denote a coloring of G in which $c(a) = i \neq j = c(b)$. If a and b belong to the same connected component of c_{ij}, then they are connected by some Kempe chain P_{ij} contained within c_{ij}. In this event, we say that a and b are c-*chained*.

Lemma 6. *If $|S| = 3$, then neither G_1 nor G_2 admits a $(t-1)$-coloring that assigns three different colors to the elements of S.*

Proof Sketch. Suppose otherwise, as witnessed by a $(t-1)$-coloring c of G_1. Let $S = \{u, v, w\}$ and assume, without loss of generality, that $c(u) = 1, c(v) = 2$ and $c(w) = 3$. Let d denote some $(t-1)$-coloring of G_2. By Lemma 5 and the Patching Lemma, it must be that d assigns exactly two colors to the elements of S. So assume, again without loss of generality, that $d(u) = d(v)$. If u and v are not c-chained, then we can exchange colors 1 and 2 in $c_{12}(v)$ to produce a $(t-1)$-coloring c' of G_1 that assigns color 1 to both u and v and leaves the color of w set to 3. This means that the restrictions of c' and d to S are equivalent. But now, by the Patching Lemma, G is $(t-1)$-colorable, which is impossible. Thus it must be that u and v are c-chained by some P_{12} in G_1. The proof proceeds by identifying P_{13} and P_{23} in a similar fashion. These chains are lifted simultaneously, along with one more application of the Patching Lemma. □

Bolstered by the preceding Lemmas, we are now ready to prove that minimum-cardinality vertex cutsets of t-immersion-critical graphs have at least four elements. The use of Kempe chains in Lemma 6 has been especially effective, so much so that we need only paths not chains in what follows.

Theorem 2. *Every t-immersion-critical graph is 4-vertex-connected.*

Proof. Suppose otherwise, as witnessed by some G with $S = \{u, v, w\}$. Let c and d denote $(t-1)$-colorings of G_1 and G_2, respectively. By Lemmas 5 and 6, we restrict our attention to the case in which both c and d assign exactly two colors to elements of S. Without loss of generality, assume $c(u) = c(v)$ and $d(u) = d(w)$. By Lemma 1, there is a path P_1 in G_1 whose endpoints are u and w. Similarly, there is a path P_2 in G_2 whose endpoints are u and v. Lifting P_i and taking the graph induced by the vertices of G_{3-i} produces a graph H_{3-i} properly immersed in G. H_1 contains uv, and so must admit a $(t-1)$-coloring c' that assigns different colors to u and v. G_1 is likewise colored by c'. By Lemma 6, c' cannot assign a third color to w. Lest the restrictions of c' and d to S be equivalent, it must be that $c'(w) = c'(v)$. H_2 contains uw, and so must admit a $(t-1)$-coloring d' that assigns different colors to u and w. G_2 is likewise colored by d'. By Lemma 6, d' cannot assign a third color to v. But if $d'(v) = d'(u)$, then the restrictions of c and d' to S are equivalent. And if $d'(v) = d'(w)$, then the restrictions of c' and d' to S are equivalent. Thus, under some pair of colorings of G_1 and G_2, the Patching Lemma ensures that G is $(t-1)$-colorable, a contradiction. □

3.2   Edge Connectivity

Because the immersion order includes the taking of subgraphs, we know that t-immersion-critical graphs are also t-color-critical. From the work of [24] it follows that they are $(t-1)$-edge-connected. We now show that any t-immersion-critical graph other than K_t is in fact t-edge-connected. We begin a pair of well-known observations (see, for example, [27]).

Observation 4. *A minimum-cardinality edge cutset separates a graph into exactly two connected components.*

Observation 5. *If H is obtained by deleting the edge uv from a t-color-critical graph, then H is $(t-1)$-colorable and, under any $(t-1)$-coloring, u and v are assigned the same color.*

The significance of Observation 5 rests with the next lemma, which plays an essential role in our edge-connectivity arguments. This lemma is probably also well known, although it may not be formulated elsewhere in exactly the same way we state it here.

Lemma 7. *Let H be obtained by deleting the edge uv from a t-color-critical graph. Let c denote a $(t-1)$-coloring of H with $c(u) = c(v) = 1$. Then $v \in c_{1i}(u) \forall i \in \{2, 3, \dots, t-1\}$.*

Proof. Let H and c be defined as stated. Suppose the lemma is false, as witnessed by some i with $v \notin c_{1i}(u)$. Exchanging colors 1 and i in $c_{1i}(u)$ produces c', another $(t-1)$-coloring of H. But then u and v are assigned different colors under c', which is impossible. $\square$

Aided by this information about color-criticality, we are now able to argue more directly about edge connectivity and the immersion order. We shall adopt the following conventions for the remainder of this subsection:

- t is at least 5,
- G denotes a t-immersion-critical graph,
- S denotes a minimum-cardinality edge cutset in G,
- C_1 and C_2 denote the two connected components of $G \backslash S$,
- S_1 and S_2 denote the endpoints of S contained in C_1 and C_2, respectively,
- uv denotes an element of S, with $u \in S_1$ and $v \in S_2$, and
- H denotes $G \backslash \{uv\}$.

Lemma 8. *If G is not t-edge-connected, then every $(t-1)$-coloring of H assigns either one color to S_1 and all $t-1$ colors to S_2 or vice versa.*

Proof Sketch. Suppose G is not t-edge-connected. We know from [24] that S has cardinality $t-1$. Let c denote a $(t-1)$-coloring of H with $c(u) = c(v) = 1$. Lemma 7 ensures that $v \in c_{1i}(u) \forall i \in \{2, 3, .., t-1\}$. Therefore u and v are the endpoints of $t-2$ Kempe chains, where each chain is contained within $c_{1i}(u)$ for some i. By Observation 1, the chains are edge disjoint, and so each contains at least one distinct element of $S' = S \backslash \{uv\}$. Thus there is a one-to-one correspondence between chains and elements of S'. This means that every element of S' has an endpoint assigned color 1 by c. If c assigns only color 1 to S_1, then it must assign all $t-1$ colors to S_2. Similarly, if c assigns all $t-1$ colors to S_1, then it must assign only color 1 to S_2. The only remaining case occurs if c assigns more than one but fewer than $t-1$ colors to S_1. This is handled with a contradiction-based argument and an application of Lemma 7. $\square$

Theorem 3. *Any t-immersion-critical graph other than K_t is t-edge-connected.*

Proof Sketch. Suppose otherwise, as witnessed by some G, not isomorphic to K_t, that is only $(t-1)$-edge-connected. We apply Lemma 8 and, without loss of generality, let c denote a $(t-1)$-coloring of H that assigns color 1 to $S_1 \cup \{v\}$. Thus all $t-1$ colors are assigned to S_2. From here Kempe chains are applied to show that K_{t-1} is immersed in C_2 using a model whose corners are the elements of S_2. With another application of Lemma 8, a K_t is found to be immersed in G using a model whose corners are $u \cup S_2$. $\square$

Corollary 1. *If G is t-immersion-critical and not K_t, then $\delta(G) \geq t$.*

Proof. Immediate from Theorem 3 and the fact that $\delta(G)$ is an upper bound on G's edge connectivity. $\square$

Corollary 2. *If G is t-color-critical with a vertex u of degree $t - 1$, then K_t is immersed in G via a model whose corners are $u \cup N(u)$.*

Proof. Follows from the proof of Theorem 3 by letting S be the set of edges incident on u. $\square$

4   Conclusions

We note that previous work on Hajós conjecture provides additional supporting evidence for both the $t = 5$ and $t = 6$ cases. If our conjecture is true in these cases, then it has no effect on Hajós conjecture. This is because a t-chromatic graph may contain an immersed K_t with or without containing a topological K_t. On the other hand, if our conjecture is false for either case, then it means that Hajós conjecture is also false for that case. This is because a t-chromatic graph without an immersed K_t must also be without a topological K_t. This would be quite a revelation, given that Hajós conjecture for $t \in \{5, 6\}$ has remained open for roughly 60 years.

Settling the general case seems rather foreboding. Perhaps this view is unfairly influenced, however, by knowledge of the long-standing difficulty of settling Hadwiger's conjecture. Observe that Kempe chains are not vertex disjoint. Yet the minor order is inherently dependent on vertex-disjoint paths. In this we sense room for optimism: the immersion order is concerned only with edge-disjoint paths, and Kempe chains are indeed edge disjoint. Given the vast array of applications for coloring and the immersion order, we believe that the nature of their relationship warrants continued study.

References

1. Béla Bollobás. *Extremal Graph Theory*. Academic Press, 1978.
2. H. D. Booth, R. Govindan, M. A. Langston, and S. Ramachandramurthi. Sequential and parallel algorithms for K_4 immersion testing. *Journal of Algorithms*, 30:344–378, 1999.
3. Catlin. Hajös graph-coloring conjecture: variation and counterexamples. *JCTB*, 26:268–274, 1979.
4. G. A. Dirac. A property of 4-chromatic graphs and some remarks on critical graphs. *J. London Math. Soc.*, 27:85–92, 1952.
5. P. Duchet and H. Meyniel. On Hadwiger's number and the stability number. *Annals of Discrete Math.*, 13:71–74, 1982.
6. P. Erdös and S. Fajtlowicz. On the conjecture of Hajös. *Combinatorica*, 1:141–143, 1981.

7. M. R. Fellows and M. A. Langston. Nonconstructive tools for proving polynomial-time decidability. *Journal of the ACM*, 35:727–739, 1988.

8. M. R. Fellows and M. A. Langston. On well-partial-order theory and its application to combinatorial problems of VLSI design. *SIAM Journal on Discrete Mathematics*, 5:117–126, 1992.

9. M. R. Fellows and M. A. Langston. On search, decision and the efficiency of polynomial-time algorithms. *Journal of Computer and Systems Sciences*, 49:769–779, 1994.

10. H. Hadwiger. Über eine klassifikation der streckenkomplexe. *Vierteljahrsschr. Naturforsch. Ges. Zürich*, 88:133–142, 1943.

11. G. Hajös. Über eine konstruktion nicht n-farbbarer graphen. *Wiss. Z. Martin Luther Univ. Halle-Wittenberg Math. Naturwiss. Reihe,*, 10:116–117, 1961.

12. F. Harary. *Graph Theory*. Addison-Wesley, 1969.

13. K. Kawarabayashi. On the connectivity of minimal-counterexamples to Hadwiger's conjecture, to appear.

14. K. Kawarabayashi and B. Toft. Any 7-chromatic graph has K_7 or $K_{4,4}$ as a minor. *Combinatorica*, to appear.

15. A. B. Kempe. How to color a map with four colours without coloring adjacent districts the same color. *Nature*, 20:275, 1879.

16. N. G. Kinnersley. Immersion order obstruction sets. *Congressus Numerantium*, 98:113–123, 1993.

17. M. A. Langston and B. C. Plaut. Algorithmic applications of the immersion order. *Discrete Mathematics*, 182:191–196, 1998.

18. W. Mader. Homomorphiesätze für graphen. *Math. Ann.*, 178:154–168, 1968.

19. N. Robertson and P.D. Seymour. Graph minors XVI: Wagner's conjecture. *Journal of Combinatorial Theory, Series B*, to appear.

20. N. Robertson, P.D. Seymour, and R. Thomas. Hadwiger's conjecture for K_6-free graphs. *Combinatorica*, 13:279–361, 1993.

21. T.L. Saaty and P.C. Kainen. *The Four-Color Problem*. Dover Publications, Inc., 1986.

22. B. Toft. Private communication, 2001.

23. B. Toft. On separating sets of edges in contraction-critical graphs. *Math. Ann.*, 196:129–147, 1972.

24. B. Toft. On critical subgraphs of colour critical graphs. *Discrete Math.*, 7:377–392, 1974.

25. B. Toft. Colouring, stable sets and perfect graphs. In R. Graham, M. Grötschel, and L. Lovász, editors, *Handbook of Combinatorics*, volume 2, chapter 4, pages 233–288. Elsevier Science B.V., 1995.

26. K. Wagner. Beweis einer abschwächung der Hadwiger-vermutung. *Math. Ann.*, 153:139–141, 1964.

27. D. B. West. *Introduction to Graph Theory*. Prentice Hall, 1996.

28. H. Whitney and W.T. Tutte. Kempe chains and the four colour problem. *Utilitas Math.*, 2:241–281, 1972.

Optimal MST Maintenance for Transient Deletion of Every Node in Planar Graphs[*]

Carlo Gaibisso[1], Guido Proietti[1,2], and Richard B. Tan[3]

[1] Istituto di Analisi dei Sistemi ed Informatica "Antonio Ruberti",
CNR, Viale Manzoni 30, 00185 Roma, Italy
gaibisso@iasi.rm.cnr.it.
[2] Dipartimento di Informatica, Università di L'Aquila,
Via Vetoio, 67010 L'Aquila, Italy
proietti@di.univaq.it
[3] Department of Computer Science, Utrecht University,
Padualaan 14, 3584 CH Utrecht, The Netherlands, and
Department of Computer Science, University of Sciences & Arts of Oklahoma,
Chickasha, OK 73018, USA.
rbtan@cs.uu.nl.

Abstract. Given a minimum spanning tree of a 2-node connected, real weighted, planar graph $G = (V, E)$ with n nodes, we study the problem of finding, for every node $v \in V$, a minimum spanning tree of the graph $G - v$ (the graph G deprived of v and all its incident edges). We show that this problem can be solved on a pointer machine in optimal linear time, thus improving the previous known $\mathcal{O}(n \cdot \alpha(n, n))$ time bound holding for general sparse graphs, where α is the functional inverse of Ackermann's function. In this way, we obtain the same runtime as for the edge removal version of the problem, thus filling the previously existing gap. Our algorithm finds application in maintaining wireless networks undergoing transient station failures.

Keywords: Planar Graphs, Minimum Spanning Tree, Transient Node Failures, Radio Networks Survivability.

1 Introduction

Let $G = (V, E)$ be a 2-node connected, undirected and real-weighted planar graph, with n nodes and $m = \mathcal{O}(n)$ edges. Let $T = (V, E_T)$ denote a *minimum spanning tree* (MST) of G, that is, a spanning tree of minimum total edge weight. For any $v \in V$, let $E(v)$ denote the subset of edges of E incident to v, and let $G - v = (V \setminus \{v\}, E \setminus E(v))$. Note that $G - v$ is connected, since G is 2-node connected. In this paper, we consider the problem of finding, for every $v \in V$, an MST of $G - v$.

[*] This work has been partially supported by the CNR-Agenzia 2000 Program, under Grants No. CNRC00CAB8 and CNRG003EF8, and by the Research Project REAL-WINE, partially funded by the Italian Ministry of Education, University and Research.

T. Warnow and B. Zhu (Eds.): COCOON 2003, LNCS 2697, pp. 404–414, 2003.
© Springer-Verlag Berlin Heidelberg 2003

1.1   Motivations

Our problem is motivated by the following application: assume that V is a set of n *sites* that must be interconnected, through a set E of potential *links* between the sites, for which a planar embedding exists. Any of these links $e = \{u, v\}$ has a real weight $c(e)$ associated, representing the inherent cost for communicating between u and v using e. Then, an MST of G serves as a spanning *communication network* of minimum cost. However, apart from its cost, a communication network must also be *reliable*, and therefore one should be ready to maintain the connectivity among the sites as soon as any network component (either a node or an edge) fails. Such a maintenance is generally accomplished by aiming at the same optimization criteria used for building the original network. On this basis, the *replacement* communication network will result in an MST of the graph G now deprived of the failed component. Since the failed component is likely to be repaired soon, the replacement communication network is just temporary, and the old, optimal MST will shortly be restored. Therefore, under these assumptions, it makes sense to study the problem of dealing with the failure of *every* arbitrary component, to precompute all the individual replacement communication networks.

1.2   Related Work

In the last two decades, several results along this direction have been obtained, especially for edge failures. In this context, it is easy to see that a failing edge $e \in E_T$ has to be replaced by a minimum-weight non-tree edge forming a cycle with e in T. Such an edge is named a *replacement edge* for e. The problem of finding all the replacement edges of an MST was originally addressed by Tarjan [19], under the guise of the *sensitivity analysis* of an MST, that is, how much the weight of each individual edge in the MST can be perturbed before the spanning tree is no longer minimal. In his seminal paper, Tarjan considered the case of a general graph G with n nodes and m edges, and solved the problem on a pointer machine in $\mathcal{O}(m \cdot \alpha(m, n))$ time and space, where $\alpha(m, n)$ is the functional inverse of the Ackermann's function defined in [18]. Later on, Dixon *et al.* [9] proposed an optimal deterministic algorithm —for which a tight asymptotic time analysis could not be offered— and a randomized linear time algorithm, both running on a RAM. Then, Booth and Westbrook [1] presented a linear time RAM algorithm for finding all the replacement MSTs for the edge failure case. Finally, a recent result by Buchsbaum *et al.* [2] provides a linear-time MST verification algorithm for a pointer machine. This latter result allows to extend that contained in [1] to a pointer machine model.

In a somewhat related scenario, the problem of finding all the replacement MSTs as a consequence of the failure of each individual node was originally studied by Chin and Houck [4], who gave an $\mathcal{O}(n^2)$ time algorithm. For sparse graphs, more precisely for $m = o\left(\frac{n^2}{\log n}\right)$, the (more general) offline algorithm for the *dynamic* MST problem given by Eppstein [10] can be used to devise a faster $\mathcal{O}(m \log n)$ time algorithm. Subsequently, such a bound has been obtained through a different technique from Das and Loui [8], who also have shown that if the edge weights are sorted in advance, then the problem can be solved in $\mathcal{O}(m \cdot \alpha(m, n))$ time and space. Finally, Nardelli *et al.* [15] have recently proposed an $\mathcal{O}(\min(m \cdot \alpha(n, n), m + n \log n))$ time and linear space pointer machine algorithm. For planar graphs, $m = \mathcal{O}(n)$, and therefore the algorithm in [15] can be used to solve our problem in $\mathcal{O}(n \cdot \alpha(n, n))$ time.

Summarizing, there still exists an efficiency gap between the edge and the node failure case, both for the general and the planar case.

1.3 Summary of Our Results

In this paper, we provide an optimal linear-time algorithm for the node failure case. As a consequence, we get the same time complexity as for the edge failure case, thus filling, for planar graphs, the above mentioned gap. Most remarkably, this is done on a pure pointer machine, without using direct addressing techniques provided by a RAM. From a different viewpoint, our algorithm can be revisited as a pseudo-dynamic algorithm working for any specific sequence of *non-overlapping* transient node deletions. Under this perspective, our algorithm has an $\mathcal{O}(n)$ preprocessing time, and deals in $\mathcal{O}(k)$ time with each MST updating and subsequent recovery induced by a transient node deletion, where k denotes the degree in G of the currently considered node.

Furthermore, a development of our technique makes it possible to optimally solve the problem of finding all the replacement *Euclidean* MSTs (EMST) of a planar set of points of size n, one for each single and temporary point deletion. This yields an application of our result in the context of survivability of *radio networks* in the 2-dimensional plane. Indeed, for radio networks, energy saving is a critical issue, and independently of the communication pattern that has to be established among the stations, the total range allocation should be minimized. Unfortunately, most of these optimization problems are NP-hard, and therefore approximate solutions must be adopted. In particular, it turns out that the EMST provides a constant-ratio approximation of a minimum-energy range assignment to the stations in order to perform both *broadcasting* (1-to-all communication) [7], and *gossiping* (all-to-all communication) [13], i.e., two of the most popular routing schemes. Hence, our technique can be used to re-establish the connectivity of a radio network undergoing non-overlapping transient station failures. As far as we know, this is the first result which is expressly tailored for dealing with fault-tolerance in 2-dimensional radio networks, since previous papers were only focused on either designing of redundant communication protocols [16,14], or on the linear case [11].

The paper is organized as follows: in Section 2 we give some preliminary definitions, while in Section 3 we describe the algorithm for solving the problem, and we provide an analysis of both correctness and complexity. In Section 4, we present the applications to survivability of radio networks. Finally, Section 5 contains conclusions and lists some open problems.

2   Preliminaries

For basic graph terminology, the reader is referred to [12]. Let r denote an arbitrary node in $G = (V, E)$. In the following, the MST $T = (V, E_T)$ of G will be considered as rooted in r. Let $F = E \setminus E_T$ be the set of non-tree edges of G. For any two node-disjoint subtrees T_1 and T_2 of T, let $F(T_1, T_2)$ be the set of non-tree edges having one end-node in T_1 and the other end-node in T_2. Let us denote by $T(v)$ the subtree of T rooted at v, and by $\overline{T}(v)$ the tree T after the removal of $T(v)$. Let v_0 and $v_1, \ldots, v_k$ be the parent and the children of v in T, respectively. Finally, let $\mathcal{H}_v = \{f \in F \mid f \in F(T(v_i), T(v_j)), 1 \leq i, j \leq k, i \neq j\}$, referred to in the following as the set of *horizontal edges* of v, and let $\mathcal{U}_v = \{f \in F \mid f \in F(T(v_i), \overline{T}(v)), 1 \leq i \leq k\}$, referred to in the following as the set of *upwards edges* of v.

A *selected horizontal edge* for $T(v_i)$ with respect to $T(v_j)$ is an edge (if any) $h_{ij} \in F(T(v_i), T(v_j))$ of *minimum* weight. Similarly, a *selected upwards edge* for $T(v_i)$ is defined as an edge (if any) $u_i \in F(T(v_i), \overline{T}(v))$ of *minimum* weight. Let $\mathcal{H}'_v \subseteq \mathcal{H}_v$ and $\mathcal{U}'_v \subseteq \mathcal{U}_v$ be the set of selected edges associated with the subtrees rooted at v. It is easy to see that an MST of $G - v$, say T_{G-v}, can be computed through the computation of an MST $\mathcal{T}_v = (\mathcal{V}_v, \mathcal{R}_v)$ of the graph $\mathcal{G}_v = (\mathcal{V}_v, \mathcal{H}'_v \cup \mathcal{U}'_v)$, where the node set $\mathcal{V}_v = \{\nu_0, \nu_1, \ldots, \nu_k\}$ is obtained by *contracting* each subtree of T created after the removal of v (see Figure 1).

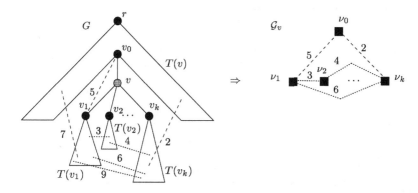

Fig. 1. A node v in G, with the associated set of upwards (dashed) and horizontal (dotted) edges, along with their weights. When node v is removed, subtrees $\overline{T}(v), T(v_1), \ldots, T(v_k)$ are contracted to vertices $\nu_0, \nu_1, \ldots, \nu_k$, respectively, joined by the selected upwards and horizontal edges, to form $\mathcal{G}_v$.

Therefore

$$T_{G-v} = (V \setminus \{v\}, E_T \setminus \{\{v_0, v\}, \{v, v_1\}, \ldots, \{v, v_k\}\} \cup \mathcal{R}_v), \qquad (1)$$

where $\mathcal{R}_v \subseteq \mathcal{H}'_v \cup \mathcal{U}'_v$ is called the set of *replacement edges* for v. The ALL NODES REPLACEMENT (ANR) problem with input G and T is that of finding T_{G-v} (i.e., $\mathcal{R}_v$) for every node $v \in V$.

3   Solving the ANR Problem in Planar Graphs

We first give a high-level description of the algorithm, and then we present it in detail.

3.1   High-Level Description of the Algorithm

A high-level description of our algorithm is the following. We consider all the non-leaf nodes of T in any arbitrary order (if v is a leaf node, then trivially $\mathcal{R}_v = \emptyset$), and we compute an MST of $\mathcal{G}_v$, obtained after the selection of the horizontal and upwards edges of v, as defined above. This selection is the key for the efficiency of our algorithm, since, as we will show, the selected edges for all the nodes in G can be computed in linear time, and moreover the graph $\mathcal{G}_v$ is planar, from which an MST of it can be computed in $\mathcal{O}(|\mathcal{V}_v|)$ time [3]. From this, the linearity of our algorithm follows.

3.2   Computing Efficiently the Selected Edges

As sketched in the previous section, the efficiency problem lies in the computation of the selected edges. We start by proving the following:

Lemma 1. *The selected horizontal edges for all the non-leaf nodes $v \in V$ can be computed in $\mathcal{O}(n)$ time and space.*

Sketch of Proof. By definition, each horizontal edge $f = (x, y)$ is associated with just a *unique* node in V, corresponding to the *least common ancestor* in T of x, y, say $lca(x, y)$. Now, let v_x and v_y denote the children in T of $lca(x, y)$ on the paths (if any) going from $lca(x, y)$ to x and y, respectively. It follows that $f \in F(T(v_x), T(v_y))$, and the horizontal edge h_{xy} can be found by selecting the minimum among all the edges in $F(T(v_x), T(v_y))$. Nodes $lca(x, y), v_x$ and v_y for all the non-tree edges can be obtained in $\mathcal{O}(n)$ time and space by slightly modifying the algorithm proposed in [2], from which the lemma follows.   □

Concerning the upwards edges, observe that each non-tree edge might be a selected upwards edge for several nodes. A different approach is therefore needed. It turns out that for our purpose, we can make use of the technique presented in [15]. Essentially, it consists in transforming the graph G into a multigraph $\mathcal{G} = (V, \mathcal{E})$ such that $\mathcal{E} = E_T \cup F'$, where F' is obtained from F as follows: let $f = \{x, y\} \in F$, and let $lca(x, y), v_x$ and v_y be defined as above. Depending on $lca(x, y), x, y, v_x$ and v_y, edge f is transformed, by maintaining its weight, according to the following rules (notice that x and y are interchangeable):

(i)   if $lca(x, y) = y$, f is replaced by the auxiliary edge $f' = \{x, v_x\}$;
(ii)   if $v_x = x$ and $v_y \neq y$, f is replaced by the auxiliary edge $f' = \{y, v_y\}$;
(iii) if $v_x = x$ and $v_y = y$, f disappears;
(iv)   otherwise, f is replaced by the auxiliary edges $f' = \{x, v_x\}$ and $f'' = \{y, v_y\}$.

The decisive property of the above transformation is that it maintains the planarity, as shown in the following:

Lemma 2. *The multigraph $\mathcal{G} = (V, \mathcal{E})$ is planar.*

Proof. Assume we are given an embedding of G in the plane, say G^*, in which r is the uppermost node of the outer face of G (such an embedding always exists [12]). In the sequel, we specify an order in which edges of F are replaced, and we show that by obeying such an order, G^* can be transformed into a planar embedding of $\mathcal{G}$.

Let R_f be the region of G^* bounded by the fundamental cycle f forms with T, $\forall f \in F$. Let $\sigma = \langle f_1, f_2, \ldots f_{|F|} \rangle$ be any sequence of all the edges in F such that either $R_{f_i} \subset R_{f_k}$, or R_{f_i} and R_{f_k} intersect at most along their boundaries, $\forall\, i < k$. Edges of F are transformed in the same order they appear in σ. We will prove that it is always possible to transform f_i in such a way that the associated auxiliary edges, if any, lie within R_{f_i} without intersecting any other edge.

Trivially, our statement is true for $i = 1$, since no auxiliary edge have still been introduced, and R_{f_1} does not contain any edge of F.

Let us assume we succeeded in transforming $f_1, f_2, \ldots f_{i-1}$. We will show that our claim is true also for $f_i = (x_i, y_i)$. Let us assume f_i is transformed according to rule (iii). Other cases can be similarly treated. By the inductive hypothesis and by our assumptions on σ, any already introduced auxiliary edge either lies within R_{f_i}, or it is external to it. Moreover, recall that auxiliary edges always connect a node to some of its ancestors. Let us now assume it is not possible to connect x_i to v_{x_i} (within R_{f_i}) without affecting the planarity. Then, by the way σ is defined, it follows that it must exist an auxiliary edge $f' = (v, lca(x_i, y_i))$, with v belonging to R_{f_i}.

Let us first analyze the case in which f' has been introduced according to rule (i). Then, f' replaced some edge f_j, $j < i$, whose endpoints are v and the parent of $lca(x_i, y_i)$. By the inductive hypothesis, f_j must lie inside of R_{f_i}. But at the same time, the parent of $lca(x_i, y_i)$ must lie outside of R_{f_i} in G^*, due to the fact that r is the upper-most node of such embedding, thus contradicting the fact G^* was a planar embedding. The cases in which e has been introduced by rules (ii) and (iv) can be treated similarly, from which the thesis follows. $\square$

From the above result, the following can be proved:

Lemma 3. *The selected upwards edges for all the non-leaf nodes $v \in V$ can be computed in $\mathcal{O}(n)$ time and space.*

Proof. Once again, given a non-tree edge $f = \{x, y\}$, nodes $lca(x, y), v_x$ and v_y can be obtained in $\mathcal{O}(1)$ amortized time and space by slightly modifying the algorithm proposed in [2]. Therefore, the multigraph $\mathcal{G}$ can be built in $\mathcal{O}(n)$ time and space. Now let $v \neq r$ be an arbitrary non-leaf node (if $v = r$, then trivially $\mathcal{U}'_v = \emptyset$), having parent v_0 and children $v_1, \ldots, v_k$ in T. By definition, a selected upwards edge $u_i \in F$ for the subtree $T(v_i)$, is an upwards edge of minimum weight among all the non-tree edges forming a fundamental cycle in T

containing $e = \{v_0, v\}$ and $e' = \{v, v_i\}$. From the transformation sketched above, it follows that u_i corresponds to a minimum-weight edge $u'_i \in F'$ forming a cycle with e' (see [15] for the details). Therefore, it follows that u'_i is a replacement edge for e' in $\mathcal{G}$. Hence, given that from Lemma 2 the graph $\mathcal{G}$ is planar, the runtime for finding all the replacement edges (and therefore all the selected upwards edges) is linear [1].   □

3.3   Finding All the Replacement Edges

Once the edges have been selected, to solve the ANR problem we have to compute for every $v \in V$ an MST of $\mathcal{G}_v$, whose set of edges corresponds to $\mathcal{R}_v$. This leads to the main result:

Theorem 1. *The ANR problem for an MST of a 2-node connected, real-weighted, planar graph with n nodes can be solved on a pointer machine in $\mathcal{O}(n)$ time and space.*

Proof. From Lemmas 1 and 3, computing $\mathcal{H}'_v$ and $\mathcal{U}'_v$ for every $v \in V'$ costs $\mathcal{O}(n)$ time and space.

Consider an arbitrary non-leaf node $v \in V$ (if v is a leaf node in T, then trivially $\mathcal{R}_v = \emptyset$). It remains to analyze the total time needed to compute, for every $v \in V'$, an MST of $\mathcal{G}_v$. But $\mathcal{G}_v$ is a minor of G, and therefore it is planar [12]. Hence, an MST of $\mathcal{G}_v$ can be computed in $\mathcal{O}(|\mathcal{V}_v|)$ time [3]. From this and from the fact that $\sum_{v \in V} |\mathcal{V}_v| = \mathcal{O}(n)$, the theorem follows.   □

4   Managing Station Failures in Radio Networks

In this section we describe an application of our results to survivability of radio networks. We start by proving the following:

Theorem 2. *Let $\mathcal{S}$ be a set of n points in the 2-dimensional Euclidean space. Let $T = (\mathcal{S}, E_T)$ be an EMST of $\mathcal{S}$. Then, the problem of finding all the replacement EMSTs, as a consequence of the removal of each individual point in $\mathcal{S}$, can be solved on a pointer machine in $\mathcal{O}(n)$ time and space.*

Sketch of Proof. First of all, we can compute in $\mathcal{O}(n)$ time the Delaunay triangulation of $\mathcal{S}$, say $D(\mathcal{S})$, as constrained by T, by looking at T as to a degenerate simple polygon [5]. $D(\mathcal{S})$ plays the role of the planar graph in which the EMST is embedded. However, differently from what has been previously described, whenever a point p is removed from $\mathcal{S}$, then a set of additional edges, not currently in $D(\mathcal{S})$, must be taken into account for the computation of an EMST of $\mathcal{S} - p$.

More precisely, for a given point $p \in \mathcal{S}$, let $D_p(\mathcal{S})$ be obtained from $D(\mathcal{S})$ by removing p and all its incident edges. Hence, let $\mathcal{S}(p)$ denote the set of adjacent points of p in $D(\mathcal{S})$, and let $D(\mathcal{S}(p))$ be the constrained Delaunay triangulation of $\mathcal{S}(p)$, which can be computed in $\mathcal{O}(|\mathcal{S}(p)|)$ time [5]. The set of edges in $D(\mathcal{S}(p))$ and not in $D(\mathcal{S})$ are called the *additional edges* of p, say $\mathcal{A}_p$.

Then, the problem of finding all the replacement EMSTs can be solved by adopting the same strategy as described in the previous section, with the only further trick of taking into account, whenever horizontal and upwards edges for any given point p are selected, of the associated additional edges $\mathcal{A}_p$ (indeed, an additional edge might be cheaper than the one selected from $D(\mathcal{S})$). This will clearly not affect the overall time and space complexity, from which the result follows. □

The above result finds immediate application to guarantee *survivability* in radio networks undergoing transient station failures. Indeed, assume we are given a set $\mathcal{S} = \{s_1, \dots, s_n\}$ of radio stations (i.e., points) located on the Euclidean plane. A *range assignment* for $\mathcal{S}$ is a real function $R : \mathcal{S} \to \mathbb{R}^+$ assigning to each station s_i a range $R(s_i)$. Since it is commonly assumed that the signal power falls as $1/|s_i - s_j|^\kappa$, where $\kappa \geq 1$ is a constant parameter depending on the transmission environment, and given that in an ideal environment $\kappa = 2$ (see for example [17]), it follows that the *communication power* needed to support a range $R(s_i)$ is $R(s_i)^2$. Hence, the *cost* of R can be defined as

$$Cost(R) = \sum_{s_i \in \mathcal{S}} R(s_i)^2.$$

Given a range assignment R for $\mathcal{S}$, a station s_i *reaches* a station s_j if $d(s_i, s_j) \leq R(s_i)$, where $d(s_i, s_j)$ denotes the Euclidean distance between s_i and s_j. Therefore, the *transmission graph* induced by R over $\mathcal{S}$ is defined as a directed graph $G_R = (\mathcal{S}, A)$ where

$$A = \bigcup_{s_i \in \mathcal{S}} \{e = (s_i, s_j) \mid s_i \text{ reaches } s_j\}.$$

Then, the fundamental trade-off that has to be addressed by any resource allocation algorithm in wireless networks is that of finding a range assignment of *minimum cost* such that the corresponding transmission graph satisfies a given property π. Among others, the following two combinatorial optimization problems play a crucial role in wireless networking:

1. MINIMUM-ENERGY BROADCAST ROUTING (MEBR): Given a source node $s \in \mathcal{S}$, find a minimum-cost range assignment R^* such that the transmission graph G_{R^*} contains a branching (directed spanning tree) rooted at s;
2. MINIMUM-ENERGY COMPLETE ROUTING (MECR): Find a minimum-cost range assignment R^* such that the transmission graph G_{R^*} is strongly connected.

Unfortunately, it turns out that they are both NP-hard (see [7] and [6], respectively). However, both are approximable within a constant factor, through the computation of an MST of the *complete* weighted graph $G_{\mathcal{S}}^{(2)} = (\mathcal{S}, E)$, in which the edge weights are given by the square of the Euclidean distance between any pair of stations. More precisely, ranges are assigned as follows, respectively:

1. For the MEBR-problem, direct the computed MST from the root towards the leaves, and assign to each node a range equal to the Euclidean distance of the farthest children in the directed MST;
2. For the MECR-problem, assign to each node a range equal to the Euclidean distance of the farthest adjacent node in the computed MST.

In this way, it turns out that the MEBR problem can be approximated within 12 [20], while the MECR problem can be approximated within 2 [13].

Additionally, the MST serves as an optimal solution [13] for the MINIMUM-ENERGY ALL-TO-ONE ROUTING-problem, defined as follows: Given a source node $s \in \mathcal{S}$, find a minimum-cost range assignment R^* such that the transmission graph G_{R^*} contains a branching oriented towards s. In this case, it suffices to direct the computed MST towards s, and to assign to each station a range equal to the Euclidean distance from the successor onto the path towards s.

As the usefulness of the MST structure for wireless networks has been established, we can turn our attention towards the problem of managing transient station failures. More formally, the problem we want to solve is the following: given a set of stations $\mathcal{S}$ in the 2-dimensional space, and given a range assignment $R : \mathcal{S} \to \mathbb{R}^+$ induced by an MST of $G_{\mathcal{S}}^{(2)}$, find for every $s \in \mathcal{S}$ an MST of $G_{\mathcal{S}-s}^{(2)}$. We call this the ALL STATIONS REPLACEMENT (ASR) problem.

Theorem 3. *The ASR problem can be solved on a pointer machine in $\mathcal{O}(n)$ time and space.*

Proof. Since the weight of an edge $e = \{s_i, s_j\}$ in $G_{\mathcal{S}}^{(2)}$ is an increasing function of the Euclidean distance of s_i and s_j, then an MST of $G_{\mathcal{S}}^{(2)}$ coincides with an EMST of the point set individuated by $\mathcal{S}$. Therefore, an MST of $G_{\mathcal{S}-s}^{(2)}$ coincides with an EMST of $\mathcal{S} - s$. Hence, the thesis follows from Theorem 2.   □

5   Conclusions and Future Work

In this paper we presented an optimal $\mathcal{O}(n)$ time algorithm for solving the ANR problem in planar graphs, which finds application in managing transient station failures in radio networks.

A natural open problem is to extend the class of graphs for which the linear bound holds. In a broader scenario, for general graphs, we mention the problem of establishing a linear time algorithm for finding all the replacement MSTs for the edge failure case, although this problem seems to be as difficult as computing an MST. Last, we plan to extend our results concerning the EMST to higher dimensional spaces.

Acknowledgements. The authors thank Claudio Gentile and Paolo Ventura for helpful discussions on the topic.

References

1. H. Booth and J. Westbrook, A linear algorithm for analysis of minimum spanning and shortest-path trees of planar graphs, *Algorithmica* **11** (1994) 341–352.
2. A.L. Buchsbaum, H. Kaplan, A. Rogers and J. Westbrook, Linear-time pointer-machine algorithms for least common ancestors, MST verification, and dominators, *Proc. of the 30th Annual ACM Symposium on Theory of Computing (STOC'98)*, (1998) 279–288.
3. D. Cheriton and R.E. Tarjan, Finding minimum spanning trees, *SIAM J. Comput.*, **5**(4) (1976) 724–742.
4. F. Chin and D. Houck, Algorithms for updating minimal spanning trees, *J. Comput. System Sci.*, **16**(3) (1978) 333–344.
5. F.Y.L. Chin, C.A. Wang, Finding the constrained Delaunay triangulation and constrained Voronoi diagram of a simple polygon in linear time, *SIAM J. Comput.*, **28**(2) (1998) 471–486.
6. A.E.F. Clementi, P. Penna and R. Silvestri, Hardness results for the power range assignment problem in packet radio networks, *Proc. of the Third Workshop on Randomization, Approximation and Combinatorial Optimization (RANDOM-APPROX'99)*, Vol. 1671 of Lecture Notes in Computer Science, Springer, 197–208.
7. A.E.F. Clementi, P. Crescenzi, P. Penna, G. Rossi and P. Vocca, On the complexity of computing minimum energy consumption broadcast subgraphs, *Proc. of the 18th Annual Symposium on Theoretical Aspects of Computer Science (STACS'01)*, Vol. 2010 of Lecture Notes in Computer Science, Springer, 121–131.
8. B. Das and M.C. Loui, Reconstructing a minimum spanning tree after deletion of any node, *Algorithmica* **31** (2001) 530–547. Also available as TR UILU-ENG-95-2241 (ACT-136), University of Illinois at Urbana-Champaign, IL, 1995.
9. B. Dixon, M. Rauch and R.E. Tarjan, Verification and sensitivity analysis of minimum spanning trees in linear time, *SIAM J. Comput.*, **21**(6) (1992) 1184–1192.
10. D. Eppstein, Offline algorithms for dynamic minimum spanning tree problems, *J. of Algorithms*, **17**(2) (1994) 237–250.
11. C. Gaibisso, G. Proietti and R. Tan, Efficient management of transient station failures in linear radio communication networks with bases, *2nd International Workshop on Approximation and Randomized Algorithms in Communication Networks (ARACNE'01)*. Vol. 12 of Proceedings in Informatics, Carleton Scientific, 37–54.
12. F. Harary, *Graph theory*, Addison-Wesley, Reading, MA, 1969.
13. L.M. Kirousis, E. Kranakis, D. Krizanc and A. Pelc, Power consumption in packet radio networks, *Proc. of the 14th Annual Symposium on Theoretical Aspects of Computer Science (STACS'97)*, Vol. 1200 of Lecture Notes in Computer Science, Springer, 363–374.
14. E. Kranakis, D. Krizanc and A. Pelc, Fault-tolerant broadcasting in radio networks, *6th European Symposium on Algorithms (ESA'98)*, Vol. 1461 of Lecture Notes in Computer Science, Springer-Verlag, 283–294.
15. E. Nardelli, G. Proietti and P. Widmayer, Maintaining a minimum spanning tree under transient node failures, *8th European Symposium on Algorithms (ESA 2000)*, Vol. 1879 of Lecture Notes in Computer Science, Springer, 346–355.
16. E. Pagani and G.P. Rossi, Reliable broadcast in mobile multihop packet networks, *Third Annual ACM/IEEE Int. Conf. on Mobile Computing and Networking (MO-BICOM'97)*, 34–42.
17. K. Pahvalan and A. Levesque, *Wireless Information Networks*, Wiley-Interscience, New York, 1995.

18. R.E. Tarjan, Efficiency of a good but not linear set union algorithm, *J. of the ACM*, **22**(2) (1975) 215–225.
19. R.E. Tarjan, Applications of path compression on balanced trees, *J. of the ACM*, **26**(4) (1979) 690–715.
20. P.J. Wan, G. Calinescu, X.Y. Li and O. Frieder, Minimum-energy broadcast routing in static ad hoc wireless networks, *Proc. of the 20th Annual Conference on Computer Communications (INFOCOM'01)*, 1162–1171.

Scheduling Broadcasts with Deadlines

Jae-Hoon Kim[1] and Kyung-Yong Chwa[2]

[1] Department of Computer Engineering,
Pusan University of Foreign Studies,
Pusan 608-738, Korea,
jhoon@taejo.pufs.ac.kr
[2] Department of Electrical Engineering & Computer Science,
Korea Advanced Institute of Science and Technology,
Taejon 305-701, Korea,
kychwa@jupiter.kaist.ac.kr

Abstract. We investigate the problem of scheduling broadcasts in data delivering systems via broadcast, where a number of requests from several clients can be simultaneosly satisfied by one broadcast of a server. Most of prior work has focused on minimizing the total flow time of requests. It assumes that once a request arrives, it will be held until satisfied. In this paper we are conserned with the situation that clients may leave the system if their requests are still unsatisfied after waiting for some time, that is, each request has a deadline. The problem of maximizing the throughput, for example, the total number of satisfied requests, is developed, and there are given online algorithms achieving constant competitive ratios.

1 Introduction

Broadcasting is particularly useful for delivering data to a large population. In this environment, a server broadcasts data items to clients and several clients can simultaneously receive an identical item by one broadcast. The use of broadcast technology is inherent in high-bandwidth networks such as cable television, satellite, and wireless network. For example, in Hughes' DirecPC system [11], clients make requests over phone lines and the server satisfies the requests through broadcasts via satellite.

In data broadcasting, a server broadcasts data items over a broadcast channel where the clients making requests for the items "listen to" the channel. All waiting requests for a item are satisfied when the item is transmitted on the broadcast channel. Typically, there have been proposed two models of broadcasting, *push-based* and *pull-based*. In push-based model, the server delivers data using a pre-determined schedule based on estimated access profiles of data items and it is ignorant of actual client requests. On the other hand, in pull-based model, the server is aware of actual client requests and can interactively determine a broadcast schedule adjusting to newly arriving requests. In this paper we concentrate on the pull-based model.

T. Warnow and B. Zhu (Eds.): COCOON 2003, LNCS 2697, pp. 415–424, 2003.
© Springer-Verlag Berlin Heidelberg 2003

To make data delivery systems more efficient, it is needed a careful consideration of how to schedule broadcasts for requests of various data items. In the literature, prior work is limited to the problem of minimizing the total flow time [1,2,4,8], where the flow time of a request is the time elapsing between its arrival time and its completion time. It assumes that once a data request arrives, it will be held until satisfied. But this assumption is not always available in practice. Actually, clients may leave the system if their requests are still unsatisfied after waiting for some time. In this paper we are concerned with such a situation. In particular, we consider the problem of maximizing the throughput in which each request arrives with a weight and a deadline by which it should be satisfied and the sum of weights of satisfied requests is maximized.

We will distinguish between two cases; all data items are of uniform size, that is, each request for a data item is served during the same time, and data items are of variable size. Also we assume that time is *discrete* or *continuous*. In case time is discrete, any request for a data item can be satisfied in one time unit and new requests arrive only at each (integral) time step. This setting is not general because new requests may arrive while a data item is broadcasted by the server. Such a case is called that time is continuous. For the case of continuous time, we consider the model that the server can abort the current broadcast for more valuable requests and later on, it starts the next broadcast of the aborted item from the beginning. It is different from the *preemptive* model, where an aborted item can be resumed later from where it was interrupted. Under our model, clients with buffers would be sufficient to receive only the remaining part of item in the next broadcast of the same item unless it is over the deadline. It may correspond to the *restart* model in the job scheduling literature where the scheduler can abort a currently running job and restart it from scratch later while meeting its deadline [12,13].

The scheduling algorithm used by the server has no knowledge of requests in advance and makes decisions only with information of requests having already arrived. In other words, the setting is *online*, and the performance of an online algorithm is compared with that of the optimal offline algorithm.

1.1   Previous Work

Most of prior work has focused on reducing the flow times of requests. There are a lot of empirical studies [2], etc., and in the theoretical domain, the problem of minimizing the total flow time was investigated in [4,5,6,8]. In [8], time is discrete and the sizes of items are of uniform. The authors showed that no deterministic online algorithm is $O(1)$-competitive and proposed several online algorithms. For the offline version, it was known in [5] that the problem is NP-hard, and the best known offline algorithms guarantee $\frac{1}{\alpha}$-speed $(\frac{1}{1-\alpha})$-approximation and 4-speed 1-approximation, respectively, given in [6]. For the case where requests require variable size items and the schedule is preemptive, it was shown in [4] that any deterministic online algorithm is $\Omega(\sqrt{n})$-competitive, where n is the number of distinct pages, and there is an $O(1)$-speed $O(1)$-approximation online algorithm. Also minimizing the maximum flow time was studied in [3].

For the problem of maximizing the throughput, very little work is known. In [3], they mentioned that there is a polynomial time offline algorithm to determine if a broadcast schedule exists in which all deadlines are met. The work of [7] is closer to ours, where the objective is to maximize the service ratio, i.e., the percentage of satisfied requests. But they assumed a probability distribution of generated requests. Recently, we have found an independent work [9] to study the online version of broadcast scheduling for the throughput maximization like ours.

In job scheduling literature, our work is closely related to the *interval scheduling* [10,14], where each job should be scheduled or rejected as soon as it arrives, that is, each job has a *tight deadline*. In [10], no preemption is allowed and the goal is to maximize the sum of lengths of accepted jobs. The authors showed that no $O(\log \Delta)$-competitive deterministic online algorithm exists and proposed an $O((\log \Delta)^{1+\epsilon})$-competitive online algorithm, where Δ is the ratio of maximum to minimum length of jobs. In [14], preemption is allowed, that is, a running job may be interrupted to be lost, and weights of jobs are given from a function of their lengths satisfying special conditions. The goal is to maximize the sum of weights of completed jobs. They provided an online algorithm guaranteeing the competitive ratio of four and proved that the ratio of four is best possible for all deterministic online algorithms.

1.2   Our Results

First, we study the case that all data items are of uniform size in Section 3 and 4. In case time is discrete, we show that there is a 2-competitive online algorithm and the ratio of two is tight for any deterministic online algorithm. In case time is continuous, we can consider two types of requests; with tight deadlines and with arbitrary deadlines. Tight deadline requests must be immediately served or rejected when they arrive. This is equivalent to the job scheduling problem studied in [14], where a set of requests for a data item arriving at a time step can correspond to a job with a weight equal to the number of requests. It was shown in [14] that there is a 4-competitive online algorithm and it is best possible for any deterministic online algorithm. For the general case of arbitrary deadline requests, we propose a $(3 + 2\sqrt{2})$-competitive online algorithm.

For the case of variable size data items, we present lower bounds of the competitive ratio of any deterministic online algorithm in Section 5.

2   Model and Definitions

There are n possible data items $P_1, P_2, \cdots, P_n$, which are called *pages*. The broadcast time of every page is either uniform, say one, or variable and each request for a page has a deadline by which it should be satisfied by the server and a positive weight. At a point in time, new requests for a page arrive, and whenever the current broadcast is completed, the server starts to supply the next broadcast if there are waiting requests.

For a set A of requests for a page, we denote the total of weights of requests in A as $\|A\|$ and call it the *value* of A. In particular, in case the weights of all requests are equal to one, $\|A\|$ represents the number of requests in A. In this paper, w.l.o.g., we assume that the weight of each request is equal to one. (Note that it is redundant to assume that the weights are given arbitrarily since all analyses through the paper can derive the same results.) Let OPT denote the set of all requests which are satisfied by the optimal offline algorithm OPT. Then $\|OPT\|$ represents the performance of OPT.

For the case of discrete time, any page is broadcasted during one time unit. At each time step t, new requests for various pages arrive and the server (or the scheduler) selects a page and broadcasts it. After the broadcast, at time $t + 1$, other new requests arrive. We will consider this model in Section 3,

For the case of continuous time, new requests for pages may arrive while a page is currently broadcasted. In this model, we will use the term of a *job* to represent the set of requests for a page and some terminologies of job scheduling. This will be investigated for uniform size pages in Section 4 and for variable size pages in Section 5.

3   Discrete Time

In this section we assume that time is discrete and any page can be broadcasted during one time unit. We will begin with an investigation of the well-known algorithm EDF (Earliest Deadline First). At each time, EDF determines the page which is required by the request with the earliest deadline as the page to be broadcasted. If there are several requests for distinct pages having the earliest deadline, then the page with the largest number of requests is chosen.

Let m be the largest possible number of requests for a page arriving at a time step. Then we provide a worst-case instance against EDF as follows: Assume $n \leq m$. At time 1, there arrive m requests with the deadline m to require it per each of distinct $n - 1$ pages and one request for the other one page, say p, with deadline 1. Then EDF serves such one request at time 1. At each time until $m - 1$, the adversary gives one request for the page p which should be immediately served. Then EDF would satisfy $m - 1$ requests until time $m - 1$ and m requests at time m. But OPT could satisfy $m(n - 1)$ requests with large deadline until time $n - 1$ and $m - n$ requests after $n - 1$. Thus the competitive ratio is $\Omega(n)$. In case $n > m$, we can also show that the competitive ratio is $\Omega(m)$.

Theorem 1. *EDF is $\Omega(\min\{n, m\})$-competitive, where n and m are the number of pages and the largest possible number of requests for a page arriving at a time step, respectively.*

We consider an online algorithm in which at each time t, a page with the largest number of requests is chosen, called *Greedy*. Note that Greedy makes a decision regardless of the deadlines of requests. Here time is divided into *busy periods* during which pages are broadcasted by Greedy and *idle periods* during which no page is broadcasted. Then for each busy period T_i, we will compare the performance of Greedy with that of OPT for all requests arriving in T_i. Consider a busy period T and let $\mathcal{J}$ be the set of all requests arriving in T. W.l.o.g., we assume that the whole instance of problem is only $\mathcal{J}$ and Greedy results in the single busy period T. (This has no effect on the gain of Greedy but it is helpful for OPT.) Let $T = [1..\ell]$. All requests served by OPT after ℓ are also satisfied by Greedy within T, because they have deadlines after ℓ and so if not, they could be served by Greedy after ℓ.

Let GD denote the set of all requests satisfied by Greedy and let $OPT' = OPT \setminus GD$. Then from the above argument, each job in OPT' should be served by OPT within T. For each time step t in T, we define a set $OPT'(t)$ of requests in OPT' that are served by OPT at time t. Fix a time step t. Then all requests in $OPT'(t)$ require an equal page and at time t, they are all alive in Greedy's scheduling. Since Greedy chooses to serve a page with the largest number of requests at time t, we can see that $\| OPT'(t) \| \leq \| GD(t) \|$, where $GD(t)$ represents the set of requests served by Greedy at time t. Therefore it follows that $\| OPT' \| = \sum_{t=1}^{\ell} \| OPT'(t) \| \leq \sum_{t=1}^{\ell} \| GD(t) \| = \| GD \|$. It ensures that Greedy is 2-competitive since $\| OPT \| = \| OPT' \| + \| OPT \cap GD \| \leq 2 \cdot \| GD \|$.

Theorem 2. *Greedy is 2-competitive.*

In fact, we will show that the competitive ratio of two is tight for any deterministic online algorithm. We have an instance of requests to give a lower bound in which all requests have the same deadline D, sufficiently large. Consider any deterministic online algorithm $\mathcal{A}$. W.l.o.g., we assume that $\mathcal{A}$ services requests for a page at each time step if there are some waiting requests. Initially, at time 1, there arrive D requests for distinct pages each other and a request for a page p_1 is served by $\mathcal{A}$. Then at time 2, another request for p_1 arrives. Especially, at each time step between 2 and D, the adversary generates one request for the same page as the one selected by $\mathcal{A}$ at the previous time. Thus $\mathcal{A}$ can service D requests, one at each time, but OPT can do $2D - 1$ requests, all given requests, by serving all existing requests for one of the distinct D pages at each time between 1 and D.

Theorem 3. *Any deterministic online algorithm cannot have a competitive ratio of less than two.*

4 Continuous Time

In this section the broadcast time of every page is also equal and time is continuous. So while a page is broadcasted, new requests may arrive. The requests for pages may have arbitrary deadlines. Here we will describe an online algorithm $\mathcal{AC}$. There is a pool of requests that is prepared for the ones which have already arrived but are not satisfied by $\mathcal{AC}$. Initially, the pool is empty. $\mathcal{AC}$ is activated only when new requests arrive or the present broadcast service is completed. Let $\{1, 2, \cdots, n\}$ denote the set of all existing pages. At any time t, assume that new requests for a page i arrive and a page j is being broadcasted. Let $\mathcal{R}$ be the set of requests for the page i which are either in the pool at time t or one of the newly arriving requests and $\mathcal{R}'$ the set of requests for the page j being served at t. For $i \neq j$, if $\|\mathcal{R}\| \geq C \cdot \|\mathcal{R}'\|$, where $C(> 1)$ is determined later, then requests in $\mathcal{R}'$ are *aborted* and those in $\mathcal{R}$ are scheduled to be served. Otherwise, requests in $\mathcal{R}$ are *rejected*. For $i = j$, let $\mathcal{R}''$ be the subset of $\mathcal{R}'$ in which requests can meet their deadlines if they are scheduled at t. If $\|\mathcal{R}'' \cup \mathcal{R}\| \geq C \cdot \|\mathcal{R}'\|$, then requests in $\mathcal{R}'$ are aborted and those in $\mathcal{R}'' \cup \mathcal{R}$ are scheduled. Otherwise, requests in $\mathcal{R}$ are rejected. Also at any time t when a broadcast for a page is completed, $\mathcal{AC}$ determines the page with the largest number of request in the pool as the next page which it will broadcast. (If there is a tie, then $\mathcal{AC}$ chooses any one of such pages.) In particular, when the requests are aborted or rejected, they enter into the pool, and when expiring in the pool, they are taken away from it.

For simplicity, we will term a set of requests for a page a *job*. The jobs are made only at particular times, when new requests for a page arrive or a broadcast for a page is completed. At any time t when new requests for a page arrive, if the page is different from that currently served, then a new job J is defined to be the set containing both the newly arriving requests and all requests for the page remaining in the pool at time t, and otherwise, J is defined to be the set containing both the newly arriving requests and the currently served ones which can be scheduled at t meeting their deadlines. Also its weight and starting time are defined to be the value of the set, that is, the total number of requests in the set, and the time t, respectively. In $\mathcal{AC}$, at time t, the weight of the job J, denoted by $w(J)$, is compared with that of the currently served job. Especially, the job J may be scheduled by aborting the (currently served) job that consists of requests for the same page. Also for any time t when a broadcast for a page is completed, a new job J is defined to be the set of requests for a page having the largest value in the pool, and its weight and starting time are similarly defined. In $\mathcal{AC}$, the job J is scheduled at time t. Note that any job in a feasible schedule of $\mathcal{AC}$ is scheduled at its starting time. In particular, we can regard $\mathcal{AC}$ as an online algorithm with the restart because even if a job was aborted while running, some requests which had been contained in it might be re-scheduled and satisfied later.

We consider a feasible schedule $\mathcal{S}$ of $\mathcal{AC}$. Let $J_1, \cdots, J_m$ be the jobs in $\mathcal{S}$ in non-decreasing order of their starting times s_i, $1 \leq i \leq m$. Then each J_i is served during $[s_i, s_i + 1]$. We will assign to each J_i a time interval $[\alpha_i, \beta_i]$. Fix some job J_i. From the characteristics of $\mathcal{AC}$, there is a chain of jobs $I_0^i, \cdots, I_\ell^i$ such that I_j^i was aborted by I_{j-1}^i, where $I_0^i = J_i$. Then we set α_i to be the starting time of I_ℓ^i. If $\ell = 0$, that is, there is no job aborted by J_i, then set $\alpha_i = s_i$. Also there may be jobs rejected by J_i at their starting times and then we set β_i to be the latest starting time of the rejected jobs. If there is no such a job, then set $\beta_i = s_i$. Then it is trivial that $\beta_i < s_i + 1 \leq \alpha_{i+1}$. Thus we obtain a partition P of time by the intervals $[\alpha_i, \beta_i]$ and (β_i, α_{i+1}).

Lemma 1. *Given the partition $P = \{[\alpha_i, \beta_i] \mid 1 \leq i \leq m\} \cup \{(\beta_i, \alpha_{i+1}) \mid 1 \leq i \leq m-1\}$ from a feasible schedule of $\mathcal{AC}$, no request arrives during (β_i, α_{i+1})*

and there is no request in the pool of $\mathcal{AC}$ during $[s_i + 1, \alpha_{i+1})$ if $s_i + 1 < \alpha_{i+1}$, where s_i is the starting time of the job determining $[\alpha_i, \beta_i]$.

Here we will merge the intervals $[\alpha_i, s_i + 1)$ into one interval if possible. From $i = 1$, we start to merge as follows; with the first $[\alpha_j, s_j + 1)$ satisfying that $s_j + 1 < \alpha_{j+1}$, the intervals $[\alpha_1, s_1 + 1), \cdots, [\alpha_j, s_j + 1)$ are merged into one, and particularly, we define a new half interval $[\gamma_1, \delta_1)$ to be $[\alpha_1, s_j + 1)$. Continuing this process, we obtain the intervals $[\gamma_k, \delta_k)$. From Lemma 1, during $[\delta_k, \gamma_{k+1})$ elapsing between two consecutive intervals, $\mathcal{AC}$ is *idle*, that is, there is no request which $\mathcal{AC}$ can serve. Also for these intervals, the following lemma holds.

Lemma 2. *For any feasible schedule $\mathcal{O}$ of a given instance and any request r starting to be served at a point during $[\gamma_i, \delta_i)$ in $\mathcal{O}$, if r arrived before γ_i, then it should have been scheduled to completion by $\mathcal{AC}$ before δ_{i-1}.*

Proof. If r is such a request which arrived before γ_i, then from Lemma 1, it actually arrived before δ_{i-1}, and it has a deadline greater than δ_{i-1}. So if r were not satisfied by $\mathcal{AC}$ before δ_{i-1}, then it would be alive at δ_{i-1}. It is a contradiction.

Theorem 4. $\mathcal{AC}$ *is $(3 + 2\sqrt{2})$-competitive.*

Proof. We consider a feasible schedule $\mathcal{S}$ of $\mathcal{AC}$. Let $J_1, \cdots, J_m$ be the jobs in $\mathcal{S}$ in non-decreasing order of their starting times s_i, $1 \leq i \leq m$. Then each J_i is served during $[s_i, s_i + 1]$. From the above statement, we assign to each J_i the time interval $[\alpha_i, \beta_i]$, and also we obtain the merged intervals $[\gamma_i, \delta_i), i = 1, \cdots, \mu$. Fix an interval $[\gamma_i, \delta_i)$. Then we also consider a feasible schedule $\mathcal{O}$ of OPT. Let $O_1, \cdots, O_n$ denote the sets of requests scheduled in $\mathcal{O}$ such that all requests in O_j require the same page and are served in $[u_j, u_j + 1]$, where $u_j < u_{j+1}$. Then we will say that O_j overlaps with a time interval $[\alpha, \beta)$ if $\alpha \leq u_j < \beta$. For the interval $[\gamma_i, \delta_i)$, consider the sets $O_p, \cdots, O_q$ which overlap with it. (There may be O_j's which overlap with some $[\delta_i, \gamma_{i+1})$, but it is easy to see that all requests in such O_j's should be scheduled by $\mathcal{AC}$.) For a request in $U^i = \cup_{j=p}^{q} O_j$, if it arrived before γ_i, then by Lemma 2, it would have been completely served by $\mathcal{AC}$. Thus from now on, we concentrate on only the requests in U^i which arrive in $[\gamma_i, \delta_i)$ and are not completely served by $\mathcal{AC}$. Let R^i denote the set of such requests. Then we partition R^i into two subsets R_1^i and R_2^i. The subset R_1^i consists of requests belonging, at least once, to jobs which are scheduled but aborted by $\mathcal{AC}$ in $[\gamma_i, \delta_i)$, and R_2^i requests belonging only to jobs rejected by $\mathcal{AC}$ in $[\gamma_i, \delta_i)$. Then it is easy to see that each request in R^i belongs to either R_1^i or R_2^i, because it is at least contained in the job generated at which it arrives in $[\gamma_i, \delta_i)$.

The interval $[\gamma_i, \delta_i)$ consists of consecutive intervals $[\alpha_u, s_u + 1), \cdots, [\alpha_v, s_v + 1)$ which correspond to jobs $J_u, \cdots, J_v$, respectively. Let $I_0^h, \cdots, I_{\ell_h}^h$ be the chain of jobs such that I_j^h was aborted by I_{j-1}^h, where $I_0^h = J_h$, $j = 1, \cdots, \ell_h$, and $h = u, \cdots, v$. Then for $h = u, \cdots, v$, we can show that $\sum_{j=1}^{\ell_h} w(I_j^h) \leq$

$\frac{1}{C-1}w(J_h)$. Since each request in R_1^i belongs to some I_j^h, it follows that $\|R_1^i\| \le \frac{1}{C-1}\sum_{j=u}^{v}w(J_j)$.

Now, we turn to the requests of R_2^i. Note that each O_k, $p \le k \le q$, overlaps with one of the intervals $[\alpha_j, s_j+1)$, $u \le j \le v$. Fix an interval $[\alpha_j, s_j+1)$. Then there are determined the sets $O_a, \cdots, O_b$ overlapping with it. W.l.o.g., assume that the last set O_b overlaps with the interval $[s_j, s_j+1)$. Then the other O_k's cannot overlap with $[s_j, s_j+1)$. First, we consider the requests in $R_2^i \cap O_b$. Let t be the latest arrival time of requests in $R_2^i \cap O_b$. At time t, $\mathcal{AC}$ checks a job J containing all requests in $R_2^i \cap O_b$, and J is rejected by $\mathcal{AC}$. In case $t \ge \alpha_j$, if J overlaps with $[s_j, s_j+1)$, i.e., $s_j \le t < s_j+1$, then J is rejected by J_j, that is, $w(J) \le C \cdot w(J_j)$. Otherwise, J is rejected by some I_k^j ($k \ge 1$) of the chain $I_0^j, \cdots, I_\ell^j$ ($I_0^j = J_j$). Then, $w(J) \le (\frac{1}{C})^{k-1}w(J_j)$. In case $t < \alpha_j$, all requests in $R_2^i \cap O_b$ exist in the pool of $\mathcal{AC}$ at α_j and so the number of them is less than or equal to $w(I_\ell^j) \le (\frac{1}{C})^\ell w(J_j)$. Thus we can see that $\|R_2^i \cap O_b\| \le C \cdot w(J_j)$.

Next, we will bound the value of $R_2^i \cap (\cup_{k=a}^{b-1}O_k)$. Before proceeding, in the following, we will also say that a set $\mathcal{O}_k$ overlaps with a job J if $s \le u_k < s+1$, where s is the starting time of J. Note that $O_k, a \le k \le b-1$, overlap with jobs I_h^j, $h = 1, \cdots, \ell$, in the chain. For each $O_k, a \le k \le b-1$, we define a job $I_{h(k)}^j$ as the job with the latest starting time among all such jobs overlapped with it. If O_k overlaps with a job I_h^j, then the other O_l's cannot overlap with I_h^j. So the jobs $I_{h(k)}^j, a \le k \le b-1$, are different from each other. In other words, for each $O_k, a \le k \le b-1$, there is uniquely determined a job $I_{h(k)}^j$ overlapped with it so that such jobs are different from each other. Fix any $R_2^i \cap O_k, a \le k \le b-1$. As in the previous, let t be the latest arrival time of requests in $R_2^i \cap O_k$. In case $t \ge \alpha_j$, at time t, a job J containing all requests in $R_2^i \cap O_k$ is checked by $\mathcal{AC}$ and it is rejected by some job $I_h^j, h \ge h(k)$. So we can see that $\|R_2^i \cap O_k\| \le w(J) \le C \cdot (\frac{1}{C})^{h-h(k)}w(I_{h(k)}^j) \le C \cdot w(I_{h(k)}^j)$. In case $t < \alpha_j$, since all requests in $R_2^i \cap O_k$ exist in the pool of $\mathcal{AC}$ at α_j, $\|R_2^i \cap O_k\| \le w(I_\ell^j) \le w(I_{h(k)}^j)$. Thus, $\|R_2^i \cap (\cup_{k=a}^{b-1}O_k)\| \le C \cdot \sum_{k=a}^{b-1}w(I_{h(k)}^j) \le \frac{C}{C-1}w(J_j)$. It implies that $\|R_2^i \cap (\cup_{k=a}^{b}O_k)\| \le (C + \frac{C}{C-1})w(J_j)$. Thus summing up over $[\alpha_j, s_j+1), u \le j \le v$, we can establish that $\|R_2^i\| \le (C + \frac{C}{C-1})\sum_{j=u}^{v}w(J_j)$.

Consequently, $\|OPT\| = \|OPT \cap AC\| + \sum_i \|R^i\| \le \|AC\| + \sum_i(\|R_1^i\| + \|R_2^i\|) \le \|AC\| + (\frac{1}{C-1} + C + \frac{C}{C-1})\sum_i w(J_i) = (1 + C + \frac{C+1}{C-1})\|AC\|$. Let $f(C) = 1 + C + \frac{C+1}{C-1}$. Then putting $f'(C) = 0$, we can see that f is minimized at $C = 1 + \sqrt{2}$. Thus, we obtain the result.

5   Variable Size Pages

In this section, pages have arbitrary sizes. First we show a lower bound of any deterministic online algorithm. Later, even if the online algorithm uses a faster broadcast server, it is shown that any online algorithm with a constant speed broadcast server cannot have a constant competitive ratio. Here the given lower bounds can also be applied to the job scheduling problems in which the online algorithm can abort and restart jobs.

Let P be the largest size of page and R be the maximum number of requests for a page which arrive at a time instant. (Here the smallest size of page is 1.) In the following, all given requests have tight deadlines. If $R \geq \sqrt{P}$, then at time 0, α requests for a page of size P arrive, where $\alpha = \frac{R}{\sqrt{P}}$. Any online algorithm $\mathcal{A}$ has to serve them in order to have a bounded competitive ratio. Immediately after, R requests for a page of size 1 arrive. If $\mathcal{A}$ rejects α requests already served and schedules R requests, then any request no longer arrives. Otherwise, consecutively, other R requests for a page of size 1 arrive. Continue this argument until either $\mathcal{A}$ schedules R requests for a page of size 1 or $\mathcal{A}$ schedules α requests for a page of size P and rejects the total of $R \cdot P$ requests for a page of size 1. Then in the former case, the optimal (offline) algorithm schedules α requests for a page of size P and the competitive ratio is at least $\frac{\alpha P}{R} \geq \sqrt{P}$. In the latter case, the optimal algorithm schedules $R \cdot P$ requests for a page of size 1 and the competitive ratio is at least $\frac{RP}{\alpha P} \geq \sqrt{P}$. If $R < \sqrt{P}$, then similarly, we can show that the competitive ratio is at least $\min\{\frac{P}{R}, R\} \geq \min\{\sqrt{P}, R\}$.

Theorem 5. *Let P be the largest size of page and R be the maximum number of requests for a page which arrive at a time instant. Then any deterministic online algorithm cannot have a better competitive ratio than $\min\{\sqrt{P}, R\}$.*

Theorem 6. *For any constant s and c, any deterministic online algorithm using an s-speed broadcast server cannot be c-competitive.*

Proof. Given any constant s and c, and assume that a deterministic online algorithm $\mathcal{A}$ is c-competitive on an s-speed broadcast server. Then we consider two pages of large size P and small size 1, respectively. At time 0, one request for the page of size P, say r, arrives and it has the tight deadline, equal to P. Then the online algorithm $\mathcal{A}$ must schedule the request r at some time t in $[0, P]$, otherwise, $\mathcal{A}$ would not be competitive. At time $t + \frac{P}{2s}$, the adversary generates w requests for the page of size 1 with the tight deadline. If $\mathcal{A}$ does not schedule the requests, then the adversary consecutively generates w requests for the page of size 1. Thus the adversary can give a huge bundle of requests in $[t + \frac{P}{2s}, t + \frac{P}{s})$ if $\mathcal{A}$ does not abort r. But it leads to a contradiction, because w can be chosen large enough to satisfy that $\frac{P}{2s}w > cP$, i.e., $w > 2sc$. So $\mathcal{A}$ will abort r and schedule some w requests for the page of size 1. Then the adversary gives no request before $\mathcal{A}$ restarts r. If $\mathcal{A}$ does not restart r, then it leads to a contradiction since P can be chosen large enough to satisfy that $P > cw$. Thus $\mathcal{A}$ will schedule r at some time t' after $t + \frac{P}{2s}$. Here note that $t' - t \geq \frac{P}{2s}$. Then we can repeat the above argument. Consequently, either $\mathcal{A}$ rejects r and schedules at most $(2\lceil s \rceil - 1)w$ requests for the page of size 1 or $\mathcal{A}$ schedules r and kw requests for the page of size 1, for some k, $1 \leq k \leq 2\lceil s \rceil - 2$. But in the former case, the adversary schedules the request r for the page of size P, and if P is chosen sufficiently large such that $P > c(2\lceil s \rceil - 1)w$, then it derives a contradiction. In the latter case, the adversary schedules at least $\frac{P}{2s}w$ requests for the page of size 1. Since $P > c(2\lceil s \rceil - 1)w \geq ckw$, if w is chosen large enough to satisfy that $w > 4sc > 2sc(1 + \frac{kw}{P})$, then $\frac{P}{2s}w > c(P + kw)$, and it derives a contradiction.

References

[1] S. Aacharya and S. Muthukrishnan. Scheduling on-demand broadcasts: new metrics and algorithms. In *ACM/IEEE International Conference on Mobile Computing and Networking*, pages 43–54, 1998.

[2] D. Aksoy and M. Franklin. Scheduling for large scale on-demand data broadcast. In *Proc. of IEEE INFOCOM*, pages 651–659, 1998.

[3] Y. Bartal and S. Muthukrishnan. Minimizing maximum response time in scheduling broadcasts. In *Proc. of the Eleventh Annual ACM-SIAM Symposium on Discrete Algorithms*, pages 558–559, 2000.

[4] J. Edmonds and K. Pruhs. Broadcast scheduling: when fairness is fine. In *Proc. of the Thirteenth Annual ACM-SIAM Symposium on Discrete Algorithms*, pages 421–430, 2002.

[5] T. Erlebach and A. Hall. Np-hardness of broadcast scheduling and inapproximability of single-source unsplittable min-cost flow. In *Proc. of the Thirteenth Annual ACM-SIAM Symposium on Discrete Algorithms*, pages 194–202, 2002.

[6] R. Gandhi, S. Khuller, Y. A. Kim, and Y. C. Wan. Algorithms for minimizing response time in broadcast scheduling. In *Proc. of 9th International Integer Programming and Combinatorial Optimization (IPCO) Conference*, volume 2337 of *Lecture Notes in Computer Science (LNCS)*, pages 425–438. Springer-Verlag, 2002.

[7] S. Jiang and N. Vaidya. Scheduling data broadcasts to "impatient" users. In *Proc. of the ACM International Workshop on Data Engineering for Wireless and Mobile Access*, pages 52–59, 1999.

[8] B. Kalyanasundaram, K. Pruhs, and M. Velauthapillai. Scheduling broadcasts in wireless networks. In *Proc. of 8th Annual European Symposium on Algorithms (ESA)*, volume 1879 of *Lecture Notes in Computer Science (LNCS)*, pages 290–301. Springer-Verlag, 2000.

[9] B. Kalyanasundaram and M. Velauthapillai. Broadcast scheduling under deadline. private communication, 2003.

[10] R. Lipton and A. Tomkins. Online interval scheduling. In *Proc. of the Fifth Annual ACM-SIAM Symposium on Discrete Algorithms*, pages 302–311, 1994.

[11] DirecPC Home Page. http://www.direcpc.com/.

[12] J. Sgall. Online scheduling. In *Online Algorithms: The State of the Art, eds. A. Fiat and G. J. Woeginger*, volume 1442 of *Lecture Notes in Computer Science (LNCS)*, pages 196–231. Springer-Verlag, 1998.

[13] M. van den Akker, H. Hoogeven, and N. Vakhania. Restarts can help in the online minimization of the maximum delivery time on a single machine. In *Proc. of 8th Annual European Symposium on Algorithms (ESA)*, volume 1879 of *Lecture Notes in Computer Science (LNCS)*, pages 427–436. Springer-Verlag, 2000.

[14] Gerhard J. Woeginger. On-line scheduling of jobs with fixed start and end times. *Theoretical Computer Science*, 130:5–16, 1994.

Improved Competitive Algorithms for Online Scheduling with Partial Job Values

Francis Y.L. Chin* and Stanley P.Y. Fung

Department of Computer Science and Information Systems,
The University of Hong Kong, Hong Kong.
{chin,pyfung}@csis.hku.hk

Abstract. This paper considers an online scheduling problem arising from Quality-of-Service (QoS) applications. We are required to schedule a set of jobs, each with release time, deadline, processing time and weight. The objective is to maximize the total value obtained for scheduling the jobs. Unlike the traditional model of this scheduling problem, in our model unfinished jobs also get partial values proportional to their amounts processed.

We give a new non-timesharing algorithm GAP for this problem for bounded m, where m is the number of concurrent jobs or the number of weight classes. The competitive ratio is improved from 2 to 1.618 (golden ratio) which is optimal for $m = 2$, and when applied to cases with $m > 2$ it still gives a competitive ratio better than 2, e.g. 1.755 when $m = 3$. We also give a new study of the problem in the multiprocessor setting, giving an upper bound of 2 and a lower bound of 1.25 for the competitiveness. Finally, we consider resource augmentation and show that $O(\log \alpha)$ speedup or extra processors is sufficient to achieve optimality, where α is the importance ratio. We also give a tradeoff result between competitiveness and the amount of extra resources.

1 Introduction

We consider the following online scheduling problem. Given a set of jobs, each job is characterized by a 4-tuple (r, d, p, w) which are the release time, deadline, processing time and weight (value per unit time of processing) respectively. The *span* of a job is the time interval $[r, d]$. A job is *active* at time t if $t \in [r, d]$ and is not completely processed by time t. Preemption is allowed with no penalty, and the goal is to maximize the total value obtained by processing the jobs.

In the traditional model of this problem, only jobs that are completed receive their values, and partially processed jobs receive no value. Recently there is a new model in which jobs that are partially processed receive a partial value proportional to their amounts processed [3,6,4,5]. This model is more relevant in some problem domains, and is first described as a Quality-of-Service (QoS) problem concerning the transmission of large images over a network of low bandwidth [3]. This is also related to a problem called *imprecise computation* in real-time

---

* This work is supported by an RGC research grant.

T. Warnow and B. Zhu (Eds.): COCOON 2003, LNCS 2697, pp. 425–434, 2003.
© Springer-Verlag Berlin Heidelberg 2003

systems [11], and has applications in numerical computation, heuristic search, database query processing, etc.

Jobs arrive online, i.e., no details of a job is known before it is released, and the online scheduling algorithm has to made its decisions based only on the details of jobs already released. We assume all details of a job are known at the time it is released. We judge the performance of online algorithms by their competitive ratios [13]. An online algorithm is *c-competitive* if, for any instance of jobs, the value produced by the online algorithm is at least $1/c$ that of the offline optimal algorithm.

Tight bounds on the competitive ratio are known for the traditional model: both the upper and lower bounds are $(1 + \sqrt{\alpha})^2$ [1,9], where α denotes the *importance ratio*, i.e., the ratio of maximum to minimum job weights. For previous results on the partial value model, Chang and Yap first gave 2-competitive algorithms and a lower bound of 1.17 on the competitive ratio [3]. The upper bound was then improved to $e/(e-1) \approx 1.58$ [5,6]. The lower bound was also improved to 1.236 [6] and most recently to 1.25 [5].

The $e/(e-1)$-competitive algorithm makes use of *timesharing*, i.e., it allows more than one job running on the processor simultaneously, each at reduced speeds so that the sum of processing speeds at any time does not exceed the processor speed. Although timesharing can be simulated in non-timesharing systems by alternating jobs at a very high frequency, this may not be desirable since it incurs a high cost. In fact we proved that timesharing does help: non-timesharing algorithms cannot be better than 1.618-competitive [5]. No non-timesharing algorithms are known to have competitive ratio $2 - \epsilon$ for constant ϵ in general. However in practice there may be additional contraints, e.g. the job weights may not vary too much, or fall into fixed weight classes; or the system would not be too overloaded, i.e., too many jobs released at a short period of time. We can use these information to devise better algorithms. In [4] we give an algorithm which is $(2 - 1/(\lg \alpha + 2))$-competitive[1] . In Sect. 3 we consider the case when there are bounded number of concurrent active jobs, or bounded number of weight classes. Let m be the bound on either one of these. A new online algorithm GAP is proposed which is 1.618-competitive for $m = 2$ and is optimal. This new algorithm, although not optimal for $m > 2$, gives a competitive ratio better than 2.

All the above results are for the single processor setting, and no previous results for this problem are known in the multiprocessor setting. In Sect. 4 we give a 2-competitive algorithm and a lower bound of 1.25 for the competitiveness in this case.

Using *resource augmentation* as a means of analyzing online algorithms first appeared in [12] and [7]. The idea is to give the online algorithm more resources to compensate for its lack of future information, and analyze the tradeoff between the amount of additional resources and improvement in performance. Since then, many problems are analyzed using this approach. We give a new study of applying the resource augmentation analysis to this problem, by using either a

---

[1]   In this paper lg denotes log to base 2.

faster processor or more processors. The only known result is a lower bound of $\Omega(\log \log \alpha)$ speedup to achieve optimality (1-competitiveness) [6], which applies to both the traditional and partial value models. A $4\lceil \lg \alpha \rceil$ upper bound for the traditional model is known [10,8]. In Sect. 5 we give the first upper bound results for the partial value model. We also give a tradeoff result between the amount of extra resources and the improvement to competitive ratio. Such tradeoff results also exist for the traditional model [7,10].

2   Preliminaries

Let $r(q), d(q)$ and $w(q)$ denote the release time, deadline and weight of a job q, respectively. For an algorithm A, let $A(t)$ denote the job running by A at time t, and $done_A(q,t)$ be the amount of work done of job q by A by time t. Without confusion, 'algorithm' and 'schedule' are used interchangeably. If no job is running on A at time t we call $A(t)$ a *null job*. Let OPT denote the offline optimal algorithm. Let $||S||$ denote the value of a schedule S.

A schedule S is *canonical* if for any two times t_1 and t_2 $(t_1 < t_2)$, the following is satisfied: if $q_1 = S(t_1)$, and $q_2 = S(t_2)$ is not null, then either (i) $r(q_2) > t_1$, or (ii) q_1 is not null and $d(q_1) \le d(q_2)$. Intuitively, it means that among the active jobs at any time, S will either schedule the one with the earliest deadline, or discard it forever. We assume ties on deadlines are always broken consistently, for the offline optimal algorithm and the online algorithm, so that we may assume no two deadlines are equal. It can be shown that OPT is canonical [6].

We bound the competitive ratio of online algorithms by employing a *charging scheme* similar to that in [6]. Let A denote an online algorithm. We charge the values of infinitesimally small time periods (i.e. the 'value rates' or weights) from OPT to those in A. Let $F : \Re \to \Re$ be a function mapping each time in OPT to a time in A. For any time t, suppose q is the job currently running in OPT. If $done_{OPT}(q,t) > done_A(q,t)$, $F(t) = t$. Otherwise, find the time $u < t$ when $done_{OPT}(q,t) = done_A(q,u)$, and $F(t) = u$. In both cases the value rate charged is $w(q)$. It can be seen that all job values in OPT are charged under mapping F.

At any time t, there are at most two charges to A (i.e. two times mapped to t by F), one from time t and another from a time later than t. See Fig. 1. Define the *charging ratio* at any time t to be the sum of values of the charges made to t over the value A is getting at time t. If we can bound the charging ratio at time t for all t, this gives a bound on the competitive ratio of A.

3   An Improved Non-timesharing Algorithm

No non-timesharing algorithms are known to be better than 2-competitive for the general case. In this section we give a non-timesharing algorithm that achieves an improved competitive ratio when the number of concurrent jobs is bounded, or the number of weight classes is bounded.

An active job x *dominates* another active job y if $w(x) \ge w(y)$ and $d(x) < d(y)$. An active job is *dominant* if no other active job dominates it. Let w_i denote

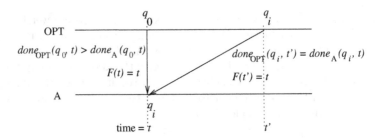

Fig. 1. Charging scheme.

the weight of the i-th heaviest active dominant job at any time. Note that no two active dominant jobs have equal weight.

3.1   Algorithm GAP

Suppose there are at most m active dominant jobs at any moment, where m is known in advance. Later we will see that this assumption generalizes the two conditions mentioned before (bounded number of concurrent jobs or bounded number of weight classes). Intuitively, our algorithm tries to find a sufficiently heavy job which, at the same time, has a weight far away from other lighter jobs. This helps to give a good charging ratio by avoiding jobs with similar weights to charge to one point in time. Formally, algorithm GAP uses a parameter $r > 1$, which depends on m, and is the unique positive real root of the equation $r = 1 + r^{-1/(m-1)}$. The following table shows some values of r and m.

m	2	3	4	5	10	20	∞
r	1.618	1.755	1.819	1.857	1.930	1.965	2

When GAP is invoked, it first finds all active dominant jobs with weights $\geq (1/r)w_1$ (w_1 is the weight of the currently heaviest job). Call this set S. Among jobs in S, find a job q such that, for any other active dominant job q' with $w(q') < w(q)$, we have $w(q)/w(q') \geq r^{1/(m-1)}$. (Note that q' may not be in S). Schedule q. (We will show that such a q always exists. If there are more than one such q then schedule any one of them.) GAP is invoked again when some job is finished, reached its deadline, or a new job arrives.

3.2   Analysis

Theorem 1. *For a system with at most m active dominant jobs at any time, GAP is r-competitive, where r the unique positive real root of $r = 1 + r^{-1/(m-1)}$.*

Proof. We first show that there must be a job that satisfies the above criteria to be scheduled. Let $w_1, w_2, ..., w_p$ be the weights of the active dominant jobs in S. If $w_i/w_{i+1} \geq r^{1/(m-1)}$ for some i in 1, 2, ..., $p - 1$, we are done. Hence suppose $w_1/w_p < r^{(p-1)/(m-1)}$. If there is no other active dominant job outside S, then the job with weight w_p can be scheduled. Otherwise, let w_{p+1} be the weight of the heaviest active dominant job outside S,

$p + 1 \leq m$. We have $w_p > r^{-(p-1)/(m-1)}w_1 \geq r^{-(m-2)/(m-1)}w_1$ and thus $w_p/w_{p+1} > r^{-(m-2)/(m-1)}/(1/r) = r^{1/(m-1)}$. Therefore the job with weight w_p can be scheduled.

By the definition of S and the way the algorithm works, we have the following properties of GAP: for any job y picked by GAP,

(1) $w(y) \geq w_1/r$

(2) no other active dominant job x satisfies $r^{-1/(m-1)}w(y) < w(x) \leq w(y)$.

We use the charging scheme in Sect. 2. Suppose at time t, x, y are the jobs running in OPT, GAP respectively. We consider the charges made to job y. y may receive charges from x and/or charges from y from a later time in OPT. There are three cases:

- Case 1. x does not charge to y. In this case charging ratio $= w(y)/w(y) = 1$.
- Case 2. Only x charges to y. Since x must be active in GAP, and GAP always choose jobs within $1/r$ of the maximum weight (Property (1)), we have charging ratio $= w(x)/w(y) \leq w(x)/(w_1/r) \leq r$.
- Case 3. Both x and y charge to y. By definition of the mapping F, both x and y should be active in GAP at time t. By canonical property of OPT, $d(x) < d(y)$. Therefore $w(x) < w(y)$, or else x dominates y and y would not be scheduled in GAP.

  (i) if x is dominant in GAP, then by Property (2) of GAP, $w(x)/w(y) \leq r^{-1/(m-1)}$.

  (ii) if x is not dominant in GAP, then suppose $z \neq x$ is the 'next' (smaller-weight) active dominant job after y. By Property (2) of GAP, $w(z)/w(y) \leq r^{-1/(m-1)}$, and we must have $w(x) \leq w(z)$ (because y and z are consecutive dominant jobs, there cannot be an active job with weight $> w(z)$ and deadline $< d(y)$), so again we have $w(x)/w(y) \leq r^{-1/(m-1)}$. In both cases, the charging ratio $= (w(x) + w(y))/w(y) = 1 + w(x)/w(y) \leq 1 + r^{-1/(m-1)}$.

Therefore, in any case charging ratio $\leq \max(r, 1 + r^{-1/(m-1)})$. This is minimized by setting r to be the root of $r = 1 + r^{-1/(m-1)}$. In this case, the competitive ratio is r.   □

The above proof only uses the assumption that there are at most m active dominant jobs at any time. Note that whether jobs are active/dominant or not depends on how the algorithm schedules them, not just the instance itself. This is not desirable. However the theorem is still true for the following models, which are more realistic and generalized by the above:

Corollary 1. *GAP is r-competitive if*
(i) at any time t there are at most m jobs with t in their span; or
(ii) there are at most m weight classes, i.e., all jobs are of weight $w_1, w_2, ..., w_m$ for some fixed w_i's.

Proof. (i) automatically implies there are at most m active dominant jobs, while (ii) means there are at most m jobs having different weights at any time, thus at most m active dominant jobs.   □

In [5] it is proved that no non-timesharing algorithms are better than ϕ-competitive, where $\phi \approx 1.618$ is the golden ratio, and the construction is for $m = 2$. When $m = 2$, GAP chooses r to be the root of $r = 1 + 1/r$, i.e. $r = \phi$. Thus we have

Corollary 2. *GAP is an optimal non-timesharing algorithm with competitive ratio 1.618 when m = 2.*

4   The Multiprocessor Case

In this section, we consider the partial job value scheduling problem in a multiprocessor setting. The online algorithm has M processors, and we compare its performance with an offline optimal algorithm also having M processors. Throughout the paper we assume jobs are migratory, i.e., jobs on a processor can be switched to other processors to continue processing.

First consider the upper bound. For the uniprocessor case, FirstFit (i.e., always schedule the heaviest job) is 2-competitive [3]. The same holds true for the multiprocessor case, in which FirstFit always schedules the M heaviest jobs (if there are less than M active jobs, then some processors will idle).

Theorem 2. *In the multiprocessor setting, FirstFit is 2-competitive, and this is tight.*

Proof. We use the charging scheme in Sect. 2. Suppose at time t, $j_1, ..., j_M$ are the M jobs running on the processors of FirstFit, $w(j_1) \geq w(j_2) \geq ... \geq w(j_M)$, and suppose $q_1, ..., q_M$ are jobs running on the processors of the offline optimal algorithm. Some of the q_i's may be the same as some of the j_i's. Without loss of generality assume $j_i = q_i$ for $i \in I \subset \{1, 2, ..., M\}$ (reordering indices of q_i's as necessary). Consider the charges to a certain time t. For those $i \in I$, j_i can charge to time t at most once. For those $i \notin I$, q_i either do not charge to time t, or if they do, then we must have $w(q_i) \leq w(j_M)$ since they are unfinished but not chosen by FirstFit. For these i's, j_i may also charge to t from a later time in OPT. Thus the charging ratio is given by

$$c \leq \frac{\sum_{i \in I} w(q_i) + (M - |I|)w(j_M) + \sum_{i \notin I} w(j_i)}{\sum_{i=1}^{M} w(j_i)}$$

$$= \frac{\sum_{i=1}^{M} w(j_i) + (M - |I|)w(j_M)}{\sum_{i=1}^{M} w(j_i)} \leq \frac{\sum_{i=1}^{M} w(j_i) + Mw(j_M)}{\sum_{i=1}^{M} w(j_i)} \leq 2$$

Consider the following instance of jobs: M copies of $(0, 2, 1, 1+\epsilon)$, and M copies of $(0, 1, 1, 1)$, where $\epsilon > 0$ is very small. FirstFit schedules all weight-$(1+\epsilon)$ jobs and misses all weight-1 jobs, while OPT can schedule all of them. Thus the competitive ratio $\geq (M + M(1 + \epsilon))/(M(1 + \epsilon)) \approx 2$. □

Next we consider the lower bound. In [5] we proved a randomized lower bound of 1.25 for the uniprocessor case. Here we extend the result to the multiprocessor case, with a proof similar to that in [5].

Theorem 3. *No randomized (and hence deterministic) algorithms can be better than 1.25-competitive for any M .*

Proof. (Sketch) We make use of Yao's principle [14]. Basically, it enables us to find a lower bound of randomized algorithms by finding a probability distribution of instances, such that we can bound the ratio of the expected offline optimal value to the expected online value of the best *deterministic* algorithm. This ratio will then be a lower bound of randomized algorithms (see [2]).

Consider a set of $n + 1$ instances:

$J_1 = M$ copies of $\{(0,1,1,1),(0,2,1,2)\}$

$J_i = J_{i-1} \cup M$ copies of $\{(i-1,i,1,2^{i-1}),(i-1,i+1,1,2^i)\}$, for $i = 2,...,n$

$J_{n+1} = J_n \cup M$ copies of $\{(n,n+1,1,2^n)\}$

We form a probability distribution of J_i's with p_i being the probability of picking J_i: $p_i = 1/2^i$ for $i = 1,2,...,n$ and $p_{n+1} = 1/2^n$. Clearly $\sum p_i = 1$.

First consider the offline optimal value. It is easy to see that $||OPT(J_i)|| = (2 + 2^2... + 2^i)M + 2^{i-1}M = [2(2^i - 1) + 2^{i-1}]M$ for $i = 1,2,...,n$, and $||OPT(J_{n+1})|| = (2 + 2^2... + 2^n)M + 2^nM = [2(2^n - 1) + 2^n]M$. Thus

$$E[||OPT||] = \sum_{i=1}^{n} \frac{2(2^i - 1) + 2^{i-1}}{2^i}M + \frac{2(2^n - 1) + 2^n}{2^n}M = (\frac{5n}{2} + 1)M$$

Fix a deterministic online algorithm A. At any time interval $[i-1,i]$ where i is an integer, A is faced with M heavier jobs and M or more lighter jobs. Suppose it spends β_i processor-time (total amount of time available on all processors) on lighter jobs in this time interval (and hence $M - \beta_i$ on heavier jobs). The β_i's completely determine the value obtained by this algorithm (on these instances). We can show that, for $i = 1,2,...,n$,

$$||A(J_i)|| = [\beta_1 + 2(M - \beta_1)] + ... + [2^{i-1}\beta_i + 2^i(M - \beta_i)] + 2^i\beta_i$$

$$||A(J_{n+1})|| = [\beta_1 + 2(M - \beta_1)] + ... + [2^{n-1}\beta_n + 2^n(M - \beta_n)] + 2^n M$$

$$E[||A||] = \frac{1}{2}||A(J_1)|| + \frac{1}{4}||A(J_2)|| + ... + \frac{1}{2^n}||A(J_n)|| + \frac{1}{2^n}||A(J_{n+1})||$$

We can show that all coefficients of β_i's in $E[||A||]$ vanish. Thus $E[||A||]$ only depends on the constant terms, and

$$E[||A||] = \sum_{i=1}^{n} \frac{1}{2^i}(2 + ... + 2^i)M + \frac{1}{2^n}(2 + ... + 2^n + 2^n)M = (2n + 1)M$$

Hence $\dfrac{E[||OPT||]}{E[||A||]} = \dfrac{5n/2 + 1}{2n + 1} \to \dfrac{5}{4}$ as n is very large. Thus no randomized algorithms have competitive ratio better than 1.25. □

5 Resource Augmentation

How much extra resources is required to get 1-competitive algorithms? In this section we give an algorithm that requires $2\lceil \lg \alpha \rceil$ extra resources to achieve 1-competitiveness, in contrast with the $\Omega(\log \log \alpha)$ lower bound [6]. It is in parallel to, but smaller than, the $4\lceil \lg \alpha \rceil$ bound for the traditional model [10,8].

5.1 Importance Ratio = 2

We first consider the case $\alpha = 2$. We use EDF (Earliest Deadline First, i.e., always schedule the job with the earliest deadline) with a speed-2 processor (a speed-s processor is one which has speed s times that of a normal processor).

Theorem 4. *For $\alpha = 2$, EDF with a speed-2 processor is necessary and sufficient to achieve 1-competitiveness.*

Proof. We use a charging scheme almost identical to that stated in Sect. 2. The only difference is that, when $done_{OPT}(q, t) < done_{EDF}(q, t)$ and $done_{OPT}(q, t) = done_{EDF}(q, u)$ for $u < t$ (i.e., charge from t to u), the charges made to time u is $2w(q)$, i.e., twice the weight, to account for the difference in speeds between the offline and online algorithms. Suppose at time t, job q_0 is running in OPT, q_1 is running in EDF. Note that EDF is getting a value of $2w(q_1)$ every unit time since it is running at speed-2. Consider the charges to time t, it consists of a $w(q_0)$ charge from time t and/or a $2w(q_1)$ charge from a later time.

- Case 1. q_0 does not charge to t. Charging ratio $c \leq (2w(q_1))/(2w(q_1)) = 1$.
- Case 2. q_0 charges to t. Thus q_0 is unfinished at time t in EDF, therefore q_1 cannot be null. Suppose all job weights are normalized to be in the range $[1,2]$, then $w(q_0) \leq 2, w(q_1) \geq 1$. If there are no other charges (from later times in OPT), then $c \leq w(q_0)/(2w(q_1)) \leq 1$. Suppose q_1 charges from a later time $u > t$ in OPT. Since OPT is canonical, $d(q_0) \leq d(q_1)$, thus EDF should schedule q_0 instead of q_1. The only possibility is then $q_0 = q_1$, but in this case we still have $c \leq (2w(q_1))/(2w(q_1)) = 1$ since q_0 cannot charge to t at two different times.

Since at any time $c \leq 1$, we showed that EDF is 1-competitive.

Consider using a speed-s processor with $s < 2$. Let $s = 2 - \delta, 0 < \delta < 1$ and $0 < \epsilon < \delta/(1 - \delta)$. Consider the instance consisting of a job $(0, 2 + \epsilon, 2 + \epsilon, 2)$ and 4 copies of $(0, 2, 1, 1)$. OPT gets a value of $4 + 2\epsilon$ by executing the heaviest job. Speed-s EDF gets a value of $2s + 2s\epsilon$. Then $2s + 2s\epsilon = 4 - 2\delta + 4\epsilon - 2\delta\epsilon = (4 + 2\epsilon) + 2\epsilon - 2\delta(1 + \epsilon)$. This is smaller than $4 + 2\epsilon$ if $2\epsilon < 2\delta(1 + \epsilon)$, i.e. $\epsilon < \delta/(1 - \delta)$. □

Due to its sequential nature, a job cannot be running on two processors simultaneously. Thus a speed-2 processor is more powerful than two speed-1 processors, since it can simulate two speed-1 processors by timesharing but not vice versa. However, we still have the following stronger result, using extra processors instead of higher-speed processor to achieve 1-competitiveness. We again use EDF, i.e., the two processors will schedule the earliest-deadline and second-earliest-deadline job respectively. Due to space limitation we omit the proof.

Theorem 5. *EDF with two speed-1 processors is 1-competitive for $\alpha = 2$.*

5.2   General Importance Ratio

In fact, the proof of Theorem 4 can easily be generalized to show that EDF with speed-s processor is α/s-competitive. Thus 1-competitiveness can be achieved for general values of α by using a speed-α processor. However, we can do better: the following algorithm *Grouped-EDF* partitions the weight ranges $[1..\alpha]$ into $\lceil \lg \alpha \rceil$ classes, each having weights in the interval $[1, 2), [2, 4), \ldots, [2^{\lfloor \lg \alpha \rfloor - 1}, 2^{\lfloor \lg \alpha \rfloor})$, $[2^{\lfloor \lg \alpha \rfloor}, \alpha]$. Jobs in each class have importance ratio at most 2. The algorithm assigns two speed-1 processors to process jobs in each class using EDF.

Theorem 6. *Grouped-EDF is 1-competitive using* $2\lceil \lg \alpha \rceil$ *speed-1 processors.*

Proof. Let OPT_i and EDF_i be the 1-processor optimal and 2-processor EDF schedules for scheduling the sub-instance consisting of only the i-th class jobs, respectively. By Theorem 5, $||OPT_i|| \leq ||EDF_i||$ for all i. We also have $||OPT|| \leq \sum ||OPT_i||$ because the processors in each OPT_i can always schedule at least that much obtained by the subset of jobs in OPT restricted to that class. Thus $||OPT|| \leq \sum ||OPT_i|| \leq \sum ||EDF_i|| = ||EDF||$.   □

By using higher-speed processors and timesharing to simulate multiple processors, and by Theorem 4, we have:

Theorem 7. *Grouped-EDF is 1-competitive with a speed-$(2\lceil \lg \alpha \rceil)$ processor, or with $\lceil \lg \alpha \rceil$ speed-2 processors.*

An $O(\log \alpha)$-speed processor may not be practical. If we allow timesharing, we can use a speed-s version of the algorithm MIX in [5] to give a tradeoff between speedup and competitive ratio. The proof is similar to the $e/(e-1)$-competitiveness upper bound proof in [5] and is therefore omitted.

Theorem 8. *The speed-s version of MIX is* $1/(1 - e^{-s})$-*competitive.*

A small amount of additional processing power can give very good competitiveness results, irrespective of α. For example, with $s = 2$ we have $c = 1.16$, and with $s = 5, c = 1.00678$, i.e. just 0.68% fewer than the optimal value.

6   Conclusion

In this paper we consider the online scheduling problem with partial job values, and give new results in the non-timesharing case, the multiprocessor case, and the resource augmentation analysis. Some questions remain open. Most importantly, we do not know whether there are non-timesharing algorithms with competitive ratio better than 2. Another problem is about the exact speedup required for achieving 1-competitiveness: both the traditional and partial value models have bounds $\Omega(\log \log \alpha)$ and $O(\log \alpha)$. Would their true bounds be different? (The partial value model seems 'easier': it has 2-competitive algorithms whereas there is a lower bound of 4 in the traditional model [1].)

References

1. S. Baruah, G. Koren, D. Mao, B. Mishra, A. Raghunathan, L. Rosier, D. Shasha and F. Wang, On the Competitiveness of On-line Real-time Task Scheduling, *Real-Time Systems* 4, 125–144, 1992.
2. A. Borodin and R. El-Yaniv, *Online Computation and Competitive Analysis*, Cambridge University Press, New York, 1998.
3. E. Chang and C. Yap, Competitive Online Scheduling with Level of Service, *Proceedings of 7th Annual International Computing and Combinatorics Conference*, 453–462, 2001.
4. F. Y. L. Chin and S. P. Y. Fung, Online Scheduling with Partial Job Values and Bounded Importance Ratio, *Proceedings of International Computer Symposium*, 787–794, 2002.
5. F. Y. L. Chin and S. P. Y. Fung, Online Scheduling with Partial Job Values: Does Timesharing or Randomization Help? to appear in *Algorithmica*.
6. M. Chrobak, L. Epstein, J. Noga, J. Sgall, R. van Stee, T. Tichý and N. Vakhania, Preemptive Scheduling in Overloaded Systems, preliminary version appeared in *Proceedings of 29th International Colloqium on Automata, Languages and Programming*, 800–811, 2002.
7. B. Kalyanasunaram and K. Pruhs, Speed is as Powerful as Clairvoyance, *Journal of the ACM* 47(4), 617–643, 2000.
8. C.-Y. Koo, T.-W. Lam, T.-W. Ngan and K.-K. To, Extra Processors versus Future Information in Optimal Deadline Scheduling, *Proceedings of 15th ACM Symposium on Parallelism in Algorithms and Architectures*, 133–142, 2002.
9. G. Koren and D. Shasha, D^{over}: An Optimal On-line Scheduling Algorithm for Overloaded Uniprocessor Real-time Systems, *SIAM Journal on Computing* 24, 318–339, 1995.
10. T.-W. Lam and K.-K. To, Performance Guarantee for Online Deadline Scheduling in the Presence of Overload, *Proceedings of 12th Annual ACM-SIAM Symposium on Discrete Algorithms*, 755–764, 2001.
11. J. W. S. Liu, K.-J. Lin, W.-K. Shih, A. C.-S. Yu, J.-Y. Chung and W. Zhao, Algorithms for Scheduling Imprecise Computations, *IEEE Computer* 24(5), 58–68, 1991.
12. C. A. Philips, C. Stein, E. Torng and J. Wein, Optimal Time-critical Scheduling via Resource Augmentation, *Proceedings of 29th Annual ACM Symposium on Theory of Computing*, 140–149, 1997.
13. D. Sleator and R. Tarjan, Amortized Efficiency of List Update and Paging Rules, *Communications of the ACM* 28(2), 202–208, 1985.
14. A. C.-C. Yao, Probabilistic Computations: Toward a Unified Measure of Complexity, *Proceedings of 18th IEEE Symposium on Foundations of Computer Science*, 222–227, 1977.

Majority Equilibrium for Public Facility Allocation
(Preliminary Version)

Lihua Chen[1], Xiaotie Deng[2], Qizhi Fang[3], and Feng Tian[4]

[1] Guanghua School of Management, Peking University,
chenlh@gsm.pku.edu.cn
[2] Department of Computer science, City University of Hong Kong,
csdeng@cityu.edu.hk
[3] Department of Mathematics, Ocean University of China,
fangqizhi@public.qd.sd.cn
[4] Institute of Systems Science, Chinese Academy of Sciences

Abstract. In this work, we consider the public facility allocation problem decided through a voting process uner the majority rule. A locations of the public facility is a majority rule winner if there is no other location in the network where more than half of the voters would have be closer to than the majority rule winner. We develop fast algorithms for interesting cases with nice combinatorial structures. We show that the general problem, where the number of public facilities is more than one and is consider part of the input size, is NP-hard. Finally, we discuss majority rule decision making for related models.

1 Introduction

Majority rule is arguably the most favorite decision making mechanism for public affairs. Informally, a majority equilibrium solution has the property that no other solutions would please more than half of the voters in comparison to it. On the other hand, it is well known that a majority equilibrium may not always exist as shown in the famous Condorcet paradox where three agents have three different orders of preferences, $A > B > C$, $B > C > A$, $C > A > B$ among three alternatives A, B and C. In this work we are interested in computational aspect for the existence of a majority equilibrium in public decision making.

We focus on the public facility location problem. Demange reviewed continuous and discrete spacial models of collective choice, aiming at characterization of the location problem of public services as a result of public voting process [1]. To facilitate a rigorous study of the related problem, Demange proposed four types of majority equilibrium solutions (call Condorcet Winners) and discussed corresponding results [10,8] concerning conditions for their existences. We consider a weighted version of the discrete model of Demange, represented by a network $G = ((V, w), (E, l))$ linking communities together. For each $i \in V$, $w(i)$ represents the number of voters reside at i. For each $e \in E$, $l(e)$ represents the

T. Warnow and B. Zhu (Eds.): COCOON 2003, LNCS 2697, pp. 435–444, 2003.
© Springer-Verlag Berlin Heidelberg 2003

distance between two ends of the road $e = (i, j)$ that connecting the two communities i and j. The location of a public facility such as library, community center, etc., is to be determined by the public via a voting process under the majority rule. We consider a special type of utlity function: each member of the community is interested in minimizing the distance of its location to that of the public facility. While each desires to have the public facility to be close to itself, the desicion has to be agreed upon by a majority of the votes.

Following Demange [1], a locaiton $x \in V$ is a strong (resp. weak) Condorcet winner if, for any $y \in V$, the total weight of vertices that is closer to x than to y is more (resp. no less) than the total weight of vertices that is closer to y than to x. Similarly, it is a quasi-Condorcet winner if we change "closer to x than" to "closer to x than y or of the same distance to x as y". Of the four types of majority winner, strong Condorcet winner is the most restrictive of all, and weak quasi-Condorcet winner is the lest restrictive one and the other two are inbetween them. For discrete models considered by Romero [10], Hansen and Thisse [8], it was known that, the order induced by strict majority relation (the weak Condorcet order) in a tree is transitive. Therefore, a weak Condorcet winner in any tree always exists. In addition, Demange extended the existence condition of a weak Condorcet winner to all single peaked orders on trees [2].

Our study will focus on the algorithmic and complexity aspect of Condorcet Winner, and discuss trees and cycles as well as the cactus graph, that is, a connected graph in which each block is an edge or a cycle. Schummer and R.V. Vohra has recently studied the public facity location following the same network mode [11]. Public decision making process in their discussion is based on strategy-proof *onto* rules. Our study distinguishes from previous work in our focus in algorithmic issues. Recently, there has been a growing research effort in re-examination of concepts in humanity and social sciences via computational complexity approach, e.g., cooperative games [9,5], organizational structure [6], arbitrage [3,4], as well as general equilibrium [7].

In Section 2, we introduce the formal formulation of the public facility location problem with a single facility in a network. Obviously, enumerating through all n locations allows us to have a polynomial time algorithm to find a majority equilibrium location. The issue here is how to improve the time complexity. In particular, we are interested in classify the types of networks for which a Condorcet winner can be found in linear time. As a warm-up example, we present the solution for trees in Section 3. We present a linear algorithm for finding the weak quasi-Condorcet winners of a tree with vertex-weight and edge-length functions (Theorem 1); and prove that in the case, the weak quasi-Condorcet points are the points which minimize the total weight-distance to the individuals' locations (Theorem 2). In Section 4, we give a sufficient and necessary condition for a point to be a weak quasi-Condorcet point for cycles in the case the edge-length function is a constant, and present a much more interesting linear time algorithm. In Section 5, we present an NP-hard proof for finding a majority equilibrium solution when the number of public facilities is taken as the input size, not a constant. We conclude with remarks and discussion on related issues in Section 6.

2   Definition

In [1], Demange surveys and discusses some spatial models of collective choice, and some results concerning the transitivity of the majority rule and the exitence of a majority winner.

Following [1], let $S = \{1, 2, \cdots, n\}$ be a society, that is a set of n individuals, and X a set of alternatives (or choice space). Each individual i has a preference order, denoted $\geq_i$, on X. The n-tuple $(\geq_i)_{i \in S}$ is called the *profile* of the society.

We associate with every profile $(\geq_i)_{i \in S}$ on X three relations on X: for every $x, y \in X$,

(1) xRy if and only if $|\{i \in S : x >_i y\}| \geq |\{i \in S : y >_i x\}|$.

(2) xPy if and only if $|\{i \in S : x >_i y\}| > |\{i \in S : y >_i x\}|$.

(3) xQy if and only if $|\{i \in S : x >_i y\}| > \frac{n}{2}$.

Given S, X and profile $(\geq_i)_{i \in S}$ on X, an alternative x in X is called:

(1)   *Weak quasi-Condorcet winner*   if for every $y \in X$ distinct of x,

$$|\{i \in S : y >_i x\}| \leq \frac{n}{2};$$

i.e.

$$|\{i \in S : x \geq_i y\}| \geq \frac{n}{2}.$$

(2)   *Strong quasi-Condorcet winner*   if for every $y \in X$ distinct of x,

$$|\{i \in S : y >_i x\}| < \frac{n}{2};$$

(3)   *Weak Condorcet winner*   if for every $y \in X$ distinct of x, xRy holds; That is,

$$|\{i \in S : x >_i y\}| \geq |\{i \in S : y >_i x\}|.$$

(4)   *Strong Condorcet winner*   if for every $y \in X$ distinct of x, xPy holds; That is,

$$|\{i \in S : x >_i y\}| > |\{i \in S : y >_i x\}|.$$

In this paper, we consider the public facility location problem with a single facility in a graph. First we introduce some definition and notation. Let $G = (V, E)$ be a undirected graph of order n with a weight function w that assigns to each vertex v of G a positive weight $w(v)$, and a length function l that assigns to each edge e of G a positive length $l(e)$. If P is a chain of G, then we denote by $l(P)$ the sum of lengths of all edges of P. We denote by $d_G(u, v)$ the length of a shortest chain joining two vertices u and v in G, and call it *distance between u and v in G*. For any $R \subseteq V$, we set $w(R) = \sum_{v \in R} w(v)$. In particular, if $R = V$, we write $w(G)$ instead of $w(V)$. A vertex v of G is said to be *pendant* if v has exact one neighbor in G.

Now, for each $v_i \in V$, v_i has a preference order on V induced by the distance on G. We denote the order by $\geq_i$. That is, we have $x \geq_i y$ if and only if

$d_G(v_i, x) \leq d_G(v_i, y)$ for any two vertices x and y of G. The following definition is a extension of that given in [1]

Definition. Given a graph $G = (V, E)$ and profile $(\geq_i)_{v_i \in V}$ on V, a vertex v_0 in V is called:

(1) *Weak quasi-Condorcet winner*, if for every $u \in V$ distinct of v_0,

$$w(\{v_i \in V : u >_i v_0\}) \leq \frac{w(G)}{2};$$

(2) *Strong quasi-Condorcet winner*, if for every $u \in V$ distinct of v_0,

$$w(\{v_i \in V : u >_i v_0\}) < \frac{w(G)}{2};$$

(3) *Weak Condorcet winner*, if for every $u \in V$ distinct of v_0,

$$w(\{v_i \in V : v_0 >_i u\}) \geq w(\{v_i \in V : u >_i v_0\}).$$

(4) *Strong Condorcet winner*, if for every $u \in V$ distinct of v_0,

$$w(\{v_i \in V : v_0 >_i u\}) > w(\{v_i \in V : u >_i v_0\}).$$

In this paper, we will only consider the algorithm for finding weak quasi-Condorcet winner of a tree, a cycle and, more generally, a cactus graph. The properties and algorithms for other three types of Condorcet winners can be discussed by the similar way.

Remark. Let $G = (V, E)$ be a connected graph of order n with a weight function w that assigns to each vertex of G a weight $w(v)$, and a length function l that assigns to each edge of G a length $l(e)$. For each vertex u, set

$$s_G(u) = \sum_{v \in V} w(v) d_G(u, v).$$

Definition. A vertex u_0 is called a *barycenter* if $s_G(u_0) = \min_{v \in V} s_G(v)$.

3   Weak Quasi-Condorcet Winner of a Tree

Romero, Hansen and Thisse pointed out that the family of orders induced by a distance on a tree guarantees the transitivity of P, which implies the existence of a weak Condorcet winner. Furthermore. the weak Condorcet points are the points which minimize the total distance to the individuals' locations (see [1]).

In this section, we propose a linear algorithm for finding the weak quasi-Condorcet winners of a tree with vertex-weight and edge-length functions (Theorem 1); and prove that in the case, the weak quasi-Condorcet points are the points which minimize the total weight-distance to the individuals' locations

(Theorem 2). In fact, the conclusions of the two theorems hold also for the weak Condorcet winner.

Given a vertex $v_0 \in V$, for any vertex u, we define the set of *quasi-friend* vertices of v_0

$$F_G(v_0, u) = \{ v : d_G(v, v_0) \le d_G(v, u) \},$$

and the set of *hostile* vertices of v_0

$$H_G(v_0, u) = \{ v : d_G(v, v_0) > d_G(v, u) \}.$$

By the definition of weak quasi-Condorcet winner, we know that a vertex v_0 of G is a weak quasi-Condorcet winner of G, if for any vertex $u \ne v_0$ of G,

$$w(F_G(v_0, u)) \ge w(H_G(v_0, u)).$$

Theorem 1. *Every tree has one weak quasi-Condorcet winner , or two adjacent weak quasi-Condorcet winners. We can find it or them in linear time.*

Proof. Let $T = (V, E)$ be a tree of order n and let $w(v)$ and $l(e)$ be the weight function and length function on V and E, respectively. We prove the theorem by induction on n.

When $n = 1$, the conclusion is trivial. When $n = 2$, we set $T = (\{u, v\}, uv)$. If $w(u) \ne w(v)$, assuming, without loss of generality, that $w(u) < w(v)$, then it is easy to see that v is a unique weak quasi-Condorcet winner of T. If $w(u) = w(v)$, then both of u and v are weak quasi-Condorcet winners of T.

Now suppose that $n \ge 3$. Let $u_1, u_2, \cdots, u_k$, $k \ge 2$, be all the pendant vertices of T. Assume, without loss of generality, that u_1 is the vertex with minimum weight among these pendant vertices. Hence $w(u_1) < \frac{1}{2}w(T)$. Let u_1v_1 be the corresponding pendant edge. Denote $T^* = T - u_1$, and define the weight function w^* on T^* as follows.

$$w^*(v) = \begin{cases} w(u_1) + w(v_1), & \text{if } v = v_1; \\ w(v), & \text{if } v \ne v_1. \end{cases}$$

Note that $F(u_1, v_1) = \{u_1\}$ and $H(u_1, v_1) = V(G) - \{u_1\}$. Because $w(u_1) < \frac{1}{2}w(T)$, we have $w(F(u_1, v_1)) < w(H(u_1, v_1))$, and hence u_1 is not a weak quasi-Condorcet winner of T.

We show the following

Claim. If v_0 is a weak quasi-Condorcet winner of T^*, then v_0 is also a weak quasi-Condorcet winner of T.

Proof of Claim. Notice that for any vertex $u \ne u_1$, $d_T(u, v_1) = d_{T^*}(u, v_1)$ and $d_T(u, u_1) = d_T(u, v_1) + l(u_1v_1)$. Thus,

(1) $v_1 \in F_T(v_0, u)$ if and only if $v_1 \in F_{T^*}(v_0, u)$; Similarly, $v_1 \in H_T(v_0, u)$ if and only if $v_1 \in H_{T^*}(v_0, u)$.

(2) If $v_1 \in F_{T^*}(v_0, u)$, then $u_1 \in F_T(v_0, u)$; Similarly, if $v_1 \in H_{T^*}(v_0, u)$, then $u_1 \in H_T(v_0, u)$

(3) $w^*(F_{T^*}(v_0, u)) = w(F_T(v_0, u))$; $w^*(H_{T^*}(v_0, u)) = w(H_T(v_0, u))$.

By the definition of weak quasi-Condorcet winner, if v_0 is a weak quasi-Condorcet winner of T^*, then $w^*(F_{T^*}(v_0, u)) \geq w^*(H_{T^*}(v_0, u))$. Thus $w(F_T(v_0, u)) \geq w(H_T(v_0, u))$. That is v_0 is also a weak quasi-Condorcet winner of T.

Theorem 1 follows from Claim. A linear time algorithm follows easily from the proof.

Theorem 2. *Let T be a tree. Then v_0 is a weak quasi-Condorcet winner of T if and only if v_0 is a barycenter of T.*

Proof. Let $T_1, T_2, \cdots, T_k$ be the subtrees of $T - v_0$ and let u_i be the (unique) vertex of T_i adjacent to v_0, $(i = 1, 2, \cdots, k)$.

Obviously, $F_T(v_0, u_i) = V(T) - V(T_i)$, $H_T(v_0, u_i) = V(T_i)$, $(i = 1, 2, \cdots, k)$.

(1) By the definition of weak quasi-Condorcet winner, we get that v_0 is a weak quasi-Condorcet winner of T if and only if for each $i = 1, 2, \cdots, k$, $w(V(T) - V(T_i)) \geq w(T_i)$, i.e., $w(T_i) \leq \frac{1}{2}w(T)$.

For $i = 1, 2, \cdots, k$, set $\overline{T}_i = T - V(T_i)$, It is easy to see that

$$s_T(v_0) = s_{\overline{T}_i}(v_0) + s_{T_i}(u_i) + w(T_i);$$

$$s_T(u_i) = s_{\overline{T}_i}(v_0) + s_{T_i}(u_i) + w(\overline{T}_i).$$

(2) By the definition of barycenter, we get that v_0 is a barycenter of T if and only if for each $i = 1, 2, \cdots, k$, $w(T_i) \leq w(\overline{T}_i)$, i.e., $w(T_i) \leq \frac{1}{2}w(T)$.

From (1) and (2), we get the conclusion of Theorem 2. ∎

4   Weak Quasi-Condorcet Winner of a Cycle

Recall that for a given vertex $v_0 \in V$ and any vertex u, we denote by $F_G(v_0, u)$ the set of quasi-friend vertices of v_0, that is,

$$F_G(v_0, u) = \{ v : d_G(v, v_0) \leq d_G(v, u) \}.$$

Suppose that $C_n = v_1 v_2 \cdots v_n v_1$ is a cycle of order n. We denote by $C[v_i, v_j]$ the sub-chain $v_i v_{i+1} \cdots v_j$ of C_n, and call it $(j - i + 1)$-interval.

Lemma 1. *Let $C_{2r-1} = v_1 v_2 \cdots v_{2r-1} v_1$ be a cycle of order $2r - 1$. Suppose that $l(e) = 1$ for any edge e. Then v_r is a weak quasi-Condorcet winner of C_{2r-1} if and only if the weight of each r-interval containing v_r is at least $\frac{1}{2}w(C_{2r-1})$. That is,*

$$\sum_{k=i}^{r+i-1} w(v_k) \geq \frac{1}{2}w(C_{2r-1}), \qquad i = 1, 2, \cdots, r. \tag{1}$$

Proof. By the definition of weak quasi-Condorcet winner , v_r is a weak quasi-Condorcet winner of C_{2r-1} if and only if for any v_j, $j \neq r$,

$$w(F_C(v_r, v_j)) \geq \frac{1}{2} w(C_{2r-1}). \tag{2}$$

We will show that (1) is equivalent to (2).
For convenience, we set $C = C_{2r-1}$.

Case 1. r is even. Suppose that $r = 2m$.
It is easy to see that for $t = 1, 2, \cdots, m - 1$,

$$F_C(v_r, v_{2t-1}) = F_C(v_r, v_{2t}) = \{v_{m+t}, v_{m+t+1}, \cdots, v_{r+m+t-1}\}; \tag{a_1}$$

$$F_C(v_r, v_{2m-1}) = \{v_r, v_{r+1}, \cdots, v_{2r-1}\}; \tag{b_1}$$

$$F_C(v_r, v_{2m+1}) = \{v_1, v_2, \cdots, v_r\}; \tag{c_1}$$

By the symmetry, for $t = m + 1, m + 2, \cdots, 2m - 1$,

$$F_C(v_r, v_{2t}) = F_C(v_r, v_{2t+1}) = \{v_{t-m+1}, v_{t-m+2}, \cdots, v_{t-m+r}\}. \tag{d_1}$$

The sets of $(a_1) - (d_1)$ are exactly all the r-intervals of C_{2r-1} containing v_r.

Case 2. r is odd. Suppose that $r = 2m - 1$.
Similar to Case 1, we have

$$F_C(v_r, v_1) = F_C(v_r, v_{4m-3}) = \{v_m, v_{m+1}, \cdots, v_{m+r}\}; \tag{a_2}$$

For $t = 1, 2, \cdots, m - 2$,

$$F_C(v_r, v_{2t}) = F_C(v_r, v_{2t+1}) = \{v_{m+t}, v_{m+t+1}, \cdots, v_{r+m+t-1}\}; \tag{b_2}$$

$$F_C(v_r, v_{2m-2}) = \{v_r, v_{r+1}, \cdots, v_{2r-1}\}; \tag{c_2}$$

$$F_C(v_r, v_{2m}) = \{v_1, v_2, \cdots, v_r\}; \tag{d_2}$$

By the symmetry, for $t = m + 1, m + 2, \cdots, 2m - 2$,

$$F_C(v_r, v_{2t-1}) = F_C(v_r, v_{2t}) = \{v_{t-m+1}, v_{t-m+2}, \cdots, v_{t-m+r}\}; \tag{e_2}$$

The sets of $(a_2) - (e_2)$ are exactly all the r-intervals of C_{2r-1} containing v_r. ∎

Similarly, we get the following:

Lemma 2.   Let $C_{2r} = v_1 v_2 \cdots v_{2r} v_1$ be a cycle of order $2r$. Suppose that $l(e) = 1$ for any edge e. Then v_r is a weak quasi-Condorcet winner of C_{2r-1} if and only if the weight of each r-interval containing v_r is at least $\frac{1}{2} w(C_{2r})$. That is,

$$\sum_{k=i}^{r+i-1} w_k \geq \frac{1}{2} w(C_{2r}), \qquad i = 1, 2, \cdots, r. \tag{3}$$

Proof. We will show that (3) is equivalent to (2).
For convenience, we set $C = C_{2r}$.

Case 1. r is even. Suppose that $r = 2m$.
For $t = 1, 2, \cdots, m - 1$,

$$F_C(v_r, v_{2t-1}) = \{v_{m+t}, v_{m+t+1}, \cdots, v_{r+m+t-1}\}; \tag{a_3}$$

$$F_C(v_r, v_{2t}) = \{v_{m+t}, v_{m+t+1}, \cdots, v_{r+m+t-1}, v_{r+m+t}\}; \tag{a_3'}$$

$$F_C(v_r, v_{2m-1}) = \{v_r, v_{r+1}, \cdots, v_{2r-1}\}; \tag{b_3}$$

By the symmetry,

$$F_C(v_r, v_{2m+1}) = \{v_1, v_2, \cdots, v_r\}; \tag{c_3}$$

For $t = m + 1, m + 2, \cdots, 2m - 1$,

$$F_C(v_r, v_{2t}) = \{v_{t-m+1}, v_{t-m+2}, \cdots, v_{t-m+r}, v_{t-m+r+1}\}; \tag{d_3'}$$

$$F_C(v_r, v_{2t+1}) = \{v_{t-m+1}, v_{t-m+2}, \cdots, v_{t-m+r}\}; \tag{d_3}$$

$$F_C(v_r, v_{4m}) = \{v_m, v_{m+1}, \cdots, v_{m+r}\}; \tag{e_3}$$

The sets of $(a_3) - (e_3)$ are exactly all the r-intervals of C_{2r} containing v_r. (If fact, the sets of (a_3') and (d_3') are unnecessary, because the condition (3) will be satisfied for all the sets of (a_3') and (d_3') as long as the same condition (3) is satisfied for all the sets of (a_3) and (d_3)).

Case 2. r is odd. Suppose that $r = 2m - 1$.
For $t = 1, 2, \cdots, m - 1$,

$$F_C(v_r, v_{2t-1}) = \{v_{m+t-1}, v_{m+t}, v_{m+t+1}, \cdots, v_{r+m+t-1}\}; \tag{a_4'}$$

$$F_C(v_r, v_{2t}) = \{v_{m+t}, v_{m+t+1}, \cdots, v_{r+m+t-1}\}; \tag{a_4}$$

By the symmetry, for $t = m, m + 1, \cdots, 2m - 2$,

$$F_C(v_r, v_{2t}) = \{v_{t-m+1}, v_{t-m+2}, \cdots, v_{t-m+r}\}; \tag{b_4}$$

$$F_C(v_r, v_{2t+1}) = \{v_{t-m+1}, v_{t-m+2}, \cdots, v_{t-m+r}, v_{t-m+r+1}\}; \tag{b_4'}$$

$$F_C(v_r, v_{4m-2}) = \{v_m, v_{m+1}, \cdots, v_{m+r-1}\}. \tag{c_4}$$

The sets of $(a_4) - (c_4)$ are exactly all the r-intervals of C_{2r} containing v_r. (If fact, the sets of (a_4') and (b_4') are unnecessary, because the condition (3) will be satisfied for all the sets of (a_4') and (b_4') as long as the same condition (3) is satisfied for all the sets of (a_4) and (b_4)). ∎

By Lemmas 1 and 2, we get the following main result concerning cycles:

Theorem 3. *Let C_n be a cycle of order n. Suppose that $l(e) = 1$ for any edge e. Then $v \in V(C_n)$ is a weak quasi-Condorcet winner of C_n if and only if the weight of each $\lfloor \frac{n+1}{2} \rfloor$-interval containing v is at least $\frac{1}{2}w(C_n)$.*

Corollary 1. *Let C_n be a cycle of order n. Suppose that $C_n[p,\ q]$ is a $\lfloor \frac{n+1}{2} \rfloor$-interval of C. If*

$$\sum_{u \in C_n[p,q]} w(u) < \frac{1}{2} w(C_n),$$

then the interval $C_n[p,\ q]$ contains no weak quasi-Condorcet winner of C_n.

Corollary 2. *Let C_n be a cycle of order n. Suppose that $u_1, u_2 \in V(C_n)$ are the two weak quasi-Condorcet winners of C_n. If $d_{C_n}(u_1, u_2) \le d_{C_n}(u_2, u_1)$, then each vertex of $C_n[u_1, u_2]$ is a weak quasi-Condorcet winner of C_n.*

Proof. Otherwise, assume that $v \in C_n[u_1, u_2]$ is not a weak quasi-Condorcet winner of C_n. Thus by Theorem 3, there exists a $\lfloor \frac{n+1}{2} \rfloor$-interval $C_n[v_1, v_2]$ of C_n containing v such that

$$\sum_{v \in C_n[v_1,v_2]} w(v) < \frac{1}{2} w(C_n).$$

Since $d_{C_n}(u_1, u_2) \le d_{C_n}(u_2, u_1)$, we have $d_{C_n}(u_1, u_2) \le \lfloor \frac{n}{2} \rfloor$, and hence either $u_1 \in C_n[v_1, v_2]$ or $u_2 \in C_n[v_1, v_2]$, which contradicts that u_1 and u_2 are the two weak quasi-Condorcet winners of C_n. ■

As a directed consequence of Corollary 2, we get

Corollary 3. *Let $C_n = v_1 v_2 \cdots v_n v_1$ be a cycle of order n. If v_1 and $v_{\lfloor \frac{n+2}{2} \rfloor}$ are the weak quasi-Condorcet winners of C_n, then each vertex of C_n is a weak quasi-Condorcet winner of C_n.*

Corollary 4. *Let S be the set of all the weak quasi-Condorcet winners of a cycle. If $S \ne \emptyset$ then the subgraph induced by S is connected.*

With Theorem 3, we obtain a linear time algorithm easily.

5 Complexity Issues for Public Location in General Networks

In general, the problem can have various extensions. There may be a number of public facilities to be allocated during one voting process. The community network may be of general graphs. The public facilities may be of the same type, or they may be of different types. The utility functions of the voters may be of different forms. Our discussion in this section will take such variations into consideration. However, in our discrete model, we keep the restriction that the public facilities will be located at nodes of the graph.

We have the following general result:

Theorem 4. *If there are a bounded constant number of public facilities to be located at one voting process under the majority rule, a Condorcet winner (of any of the four tyeps) can be computed in polynomial time. The problem is NP-hard if the number of public facilities to be located is not a constant but considered as the input size.*

6 Remarks and Discussion

In this work, we apply computational complexity approach to the study of public facility location problem decided via a voting process under the majority rule. Our study follows the network model that has been applied to the study of similar problems in economics [10,8,11]. We prove the general problem to be NP-hard and establish efficient algorithms to interesting networks as used in the study of strategy-proof model for public facility location problem [11]. Our mathematical results depend on understanding of combinatorial structures of underlying networks.

Many problems open up from our study. First the decision problem of the majority rule equilibrium. It is not very difficult to problem the solution to be NP-hard in discrete equilibrium problems. The decision problem is harder. The only instance we know of is one recent result for the decision problem for general equilibrium [7]. Secondly, the complexity study for other rules for public facility location is very interesting and deserves further study. Third, it would be interesting to extend our study to other areas and problems of public decision making process.

Acknowledgement. The work described in this paper was supported by a grant from Hong Kong RGC (CityU 1081/02E).

References

1. G. Demange, Spatial Models of Collective Choice, in Locational Analysis of Public Facilities, (eds. J. F.Thisse and H. G. Zoller), North-Holland Publishing Company, 1983.
2. G. Demange, Single Peaked Orders on a Tree, *Math. Soc. Sci.* **3** (1982), pp.389–396.
3. X. Deng, Z. Li and S. Wang, On Computation of Arbitrage for Markets with Friction, *Lecture Notes in Computer Science* **1858** (2000), pp. 310–319.
4. X. Deng, Z.F. Li and S. Wang, Computational Complexity of Arbitrage in Frictional Security Market. International Journal of Foundations of Computer Science, Vol. 13, No.5, (2002), 681–684.
5. X. Deng and C. Papadimitriou, On the Complexity of Cooperative Game Solution Concepts, *Mathematics of Operations Research* **19**(2)(1994), pp. 257–266.
6. X. Deng and C. Papadimitriou, Decision Making by Hierarchies of Discordant Agents, *Mathematical Programming* **86**(2)(1999), pp.417–431.
7. X. Deng, C. Papadimitriou and S. Safra, On Complexity of Equilibrium. STOC 2002, May, 2002, Montreal, Canada, pp.67–71.
8. P. Hansen and J.-F. Thisse, Outcomes of Voting and Planning: Condorcet, Weber and Rawls Locations, *Journal of Public Economics* **16**(1981), pp.1–15.
9. N. Megiddo, "Computational Complexity and the game theory approach to cost allocation for a tree," *Mathematics of Operations Research* 3, pp. 189–196, 1978.
10. D. Romero, Variations sur l'effet Condorcet, Thèse de 3ème cycle, Université de Grenoble, Grenoble, 1978.
11. J. Schummer and R.V. Vohra, Strategy-proof Location on a Network, *Journal of Economic Theory* **104**(2002), pp.405–428.

On Constrained Minimum Pseudotriangulations

Günter Rote[1]*, Cao An Wang[2]**, Lusheng Wang[3]***, and Yinfeng Xu[4]†

[1] Institut für Informatik, Freie Universität Berlin,
Takustraße 9, D-15195 Berlin, Germany,
rote@inf.fu-berlin.de
[2] Department of Computer Science, Memorial University of Newfoundland,
St. John's, NFLD, Canada A1B 3X8,
wang@cs.mun.ca
[3] Department of Computer Science, City University of Hong Kong,
Kowloon, Hong Kong, P.R. China,
lwang@cs.cityu.edu.hk
[4] School of Management, Xi'an Jiaotong University,
Xi'an, 710049, P.R. China,
yfxu@xjtu.edu.cn

Abstract. In this paper, we show some properties of a pseudotriangle and present three combinatorial bounds: the ratio of the size of minimum pseudotriangulation of a point set S and the size of minimal pseudotriangulation contained in a triangulation T, the ratio of the size of the best minimal pseudotriangulation and the worst minimal pseudotriangulation both contained in a given triangulation T, and the maximum number of edges in any settings of S and T. We also present a linear-time algorithm for finding a minimal pseudotriangulation contained in a given triangulation. We finally study the minimum pseudotriangulation containing a given set of non-crossing line segments.

1 Introduction

A *pseudotriangle* is a simple polygon with exactly three vertices where the inner angle is less than π, see Figure 1. These three vertices are called *corners*. The boundary is composed of three pieces of nonconvex chains, where the nonconvex chain has either a reflex inner angle at each inner vertex or is a single edge (the degenerate case). A *pseudotriangulation* of a point set S is a partition of the

* Research by Günter Rote was partly supported by the Deutsche Forschungsgemeinschaft (DFG) under grant RO 2338/2-1.
** The work of Cao An Wang is supported by NSERC grant OPG0041629.
*** Lusheng Wang is fully supported by a grant from the Research Grants Council of the Hong Kong Special Administrative Region, China [Project No. CityU 1087/00E].
† The work of Yinfeng Xu is supported by NSFC(19731001) and NSFC(70121001) for excellent group project. The initial collaboration between Yinfeng Xu and Günter Rote on this paper was sponsored by the graduate program "Graduiertenkolleg *Algorithmische Diskrete Mathematik*" of the Deutsche Forschungsgemeinschaft (DFG), grant GRK219/3, when Yinfeng Xu visited Berlin.

T. Warnow and B. Zhu (Eds.): COCOON 2003, LNCS 2697, pp. 445–454, 2003.
© Springer-Verlag Berlin Heidelberg 2003

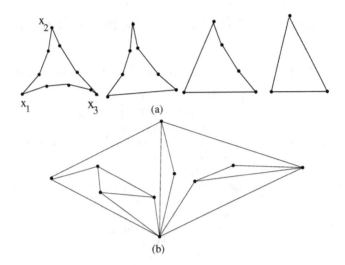

Fig. 1. (a) Some typical pseudotriangles. Vertices x_1, x_2, x_3 are corners, and the three right-hand side cases are degenerate cases of pseudotriangle. (b) A pseudotriangulation of 10 points.

interior of the convex hull of S into a set of pseudotriangles. This geometric structure plays an important role in planning collision-free paths among polyhedral obstacles [4] and in planning non-colliding robot arm motion [2,5]. Previous research on this topic was mainly concentrated on the properties and algorithms for minimum pseudotriangulation of a given point set or a set of convex objects. In those cases, the edges of pseudotriangulations are chosen from the complete edge set of the point set.

It is natural to consider some constraint on the choice of edges. Our work mainly investigates the properties of the minimal pseudotriangulations constrained to be a *subset* of a given triangulation, the minimum pseudotriangulations constrained to be a *superset* of a given set of noncrossing line segments, and on algorithms to find these pseudotriangulations. This investigation is motivated in some applications that one may compromise a minimal pseudotriangulation by a faster construction algorithm, or the environment may be constrained by a set of disjoint obstacles. For example, the paper [1] investigates degree bounds for minimum pseudotriangulations which are constrained by some given subset of edges.

In order to find a minimal pseudotriangulation constrained in a given triangulation, one must be able to identify the edges to be removed. In Section 3, we show a structural property for these edges (Theorem 2). This property allows us to design a linear-time algorithm for finding a minimal pseudotriangulation, which is presented in Section 5.

In contrast to the pseudotriangulation of a set S of n points, where all minimum pseudotriangulations of S have the same cardinality, viz. $2n - 3$ edges [5], the size of the minimum pseudotriangulation constrained in a given triangulation T depends not only on n, but on T.

We investigate the possible sizes of minimal and minimum pseudotriangulations in Section 4. We show that the ratio of the sizes of the best and the worst minimum pseudotriangulation constrained in some T against the size of the minimum pseudotriangulation triangulation of S can vary from 1 to $\frac{2}{3}$. The above bound is optimal asymptotically. Furthermore, the size of a 'minimal' pseudotriangulation constrained in a triangulation depends on the sequence of construction of pseudotriangles. (In a minimal pseudotriangulation, each pseudotriangle has been expanded into its limit, a further expansion will violate the definition of pseudotriangle. A minimal pseudotriangulation may not be minimum with respect to all possible pseudotriangulations constrained in that triangulation.) We show that the ratio of the size of the smallest minimal pseudotriangulation and the size of the largest minimal pseudotriangulation constrained in a same triangulation can vary from 1 to $\frac{2}{3}$. It is known that the size of minimum pseudotriangulation constrained on any setting of S and T is at least $2n - 3$. We show that the *maximum* number of edges in such pseudotriangulations is bounded by $3n - 8$.

In Section 6, we study the pseudotriangulations which *contain* a given set L of noncrossing line segments. Interestingly, we find that the size of a minimum pseudotriangulation for L depends only on the number of reflex vertices of L. The proof uses an algorithm for constructing such a minimum pseudotriangulation.

Finally, we discuss some open questions.

2 Preliminaries

We shall first give some definitions. A *triangulation* T of a planar point set S is a maximal planar straight-line graph with S as vertices. We assume throughout the paper that the points of S lie in general position, i.e., no three points lie on a line, and all angles are different from π.

Let T' be a subgraph of T. For a vertex $p \in S$ define $\alpha(p)$ be the largest angle at p between two neighboring edges incident to p. A vertex p in T' is called a *reflex* point if $\alpha(p) \geq \pi$ in T'.

A *minimum* pseudotriangulation of a point set is one with the smallest number of edges. It is known that the number of edges in any minimum pseudotriangulation of n points is $2n - 3$, see [5].

We now prove some properties for a triangulated pseudotriangle.

Let p be a pseudotriangle, $T(p)$ be a triangulation of p. Let $T(p) - p$ denote the remainder of $T(p)$ after the removal of the edges of p. The *dual graph* of $T(p)$ is defined as usual: Each node in the graph corresponds to a triangle face in $T(p)$, and two nodes determine an edge of the graph if the corresponding triangles share an edge. A *star-chain* consists of three simple chains sharing a common end-node.

Lemma 1. *The dual of any triangulation of a pseudotriangle is a simple chain or a star-chain.*

Proof. See Figure 2a for an illustration. Each interior edge of the triangulation of a pseudotriangle must span on two different chains by the nonconvexity of

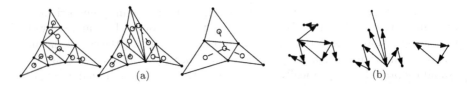

Fig. 2. (a) Different shapes of the dual graph in Lemma 1. (b) The edges of $T(p) - p$ in Lemma 2.

its three chains. This implies that these interior edges can form at most one triangle. The lemma follows.   □

Lemma 2. *Let $T(p)$ be a triangulation of a pseudotriangle p. There is a perfect matching between the edges in $T(p) - p$ and the reflex vertices of p, which matches each edge to one of its vertices.*

Proof. By Lemma 1, the edges of $T(p) - p$ form either a tree which contains exactly one corner of p or a graph with a single cycle, which is formed by a triangle, see Figure 2b. In the first case, we choose the corner as a root and direct all edges of $T(p) - p$ away from the root. Then every reflex vertex will have one edge of the tree pointing towards it, thus establishing the desired one-to-one correspondence between the edges and the reflex vertices. If $T(p) - p$ contains a triangle, we orient the edges of the triangle cyclically, in any direction, and we orient all other edges away from the cycle. Again, every reflex vertex has one edge of the tree pointing towards it. (In fact, the matching between edges and reflex vertices is unique up to reorienting the central triangle.)   □

We can extend the statement of the Lemma 2 from a single pseudotriangle to a pseudotriangulation.

Theorem 1. *Let T be a triangulation of a point set S, and let $P \subseteq T$ be a pseudotriangulation of S. Then there is a perfect matching between the edges in $T - P$ and the reflex vertices of P, which matches each edge to one of its two vertices.*

Proof. Every reflex vertex of P belongs to exactly one pseudotriangle in which it is a reflex vertex. Thus, we can simply apply Lemma 2 to each pseudotriangle of P separately.   □

The following statement is important for our characterization of minimal pseudotriangulations in Theorem 2.

Lemma 3. *Let p be a pseudotriangle, and let E be a nonempty set of edges inside p which partition p into smaller pseudotriangles. Then one of the following two cases holds:*

(a) *E is a triangle.*
(b) *E contains an edge e such that $E - e$ still partitions p into smaller pseudotriangles.*

Proof. Every edge in E connects two different reflex side chains of p. If $|E| \geq 4$, then E contains at least two edges which connect the same pair of reflex side chains of p. We choose among all these edges the edge e which is incident with the pseudotriangle containing the common corner of these chains. Removing e will join two pseudotriangles into a new face which is bounded by portions of two reflex chains and a single edge between these chains. Hence this face is a pseudotriangle, and e is the desired edge for Case (b) of the lemma.

We are left with the case that E contains at most three edges. This case can be treated by an elementary case analysis.   □

3 Minimal Pseudotriangulations

Let T denote a triangulation of S and let P^T denote a pseudotriangulation constrained in T, i.e., $P^T \subseteq T$.

A pseudotriangulation P^T is *minimal* (denoted by P^T_{mal}) if no proper subset of P^T is a pseudotriangulation. P^T is called *minimum* (denoted by P^T_{mum}) if it contains the smallest number of edges over all possible pseudotriangulations constrained in T. For simplicity, we use 'constrained pseudotriangulation P^T' as pseudotriangulation constrained in a given triangulation T.

The definition of a minimal triangulation involves a statement about all subsets of edges. The following theorem shows that is is sufficient to check only a linear number of proper subsets to establish that a pseudotriangulation is minimal.

Theorem 2 (Characterization of minimal pseudotriangulations).
A pseudotriangulation P is minimal if and only if

- *there is no edge $e \in P$ such that $P - e$ is a pseudotriangulation, and*
- *there is no triangular face $\{e_1, e_2, e_3\} \in P$ such that $P - \{e_1, e_2, e_3\}$ is a pseudotriangulation.*

Proof. It is clear that the condition is necessary. Now, suppose that $P' \subset P$ is a pseudotriangulation which is a proper subset of P. We have to show that some edge or triangle of P can be removed. Let p be a pseudotriangle face of P' which contains some edges E of $P - P'$. These edges subdivide p into pseudotriangles, and we can apply Lemma 3 to p. We either get an edge whose removal yields a pseudotriangulation, or E is a triangle, whose removal merges 4 faces of P into p.   □

4 Ratio of the Sizes of Pseudotriangulations

In this section, we show some relationships among the sizes of T, P^T (constrained pseudotriangulation), P^T_{mal} (minimal P^T in T), P^T_{mum} (minimum P^T in T), and $P_{\mathrm{mum}}(S)$ (minimum pseudotriangulation of the point set S).

Theorem 3. *Let S be a set of n points in general position and T be a triangulation of S. The number of edges in P^T_{mum} is at most $3n - 8$, for $n \geq 5$. There are infinitely many values of n for which a triangulation exists where P^T_{mum} has $3n - 12$ edges.*

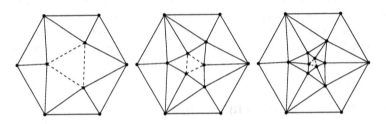

Fig. 3. Three steps of the inductive construction in Theorem 3. The three edges of the dotted central triangle can be removed.

Proof. Suppose that If k vertices lie on the convex hull of S, every triangulation T has $3n - k - 3$ edges, and every pseudotriangulation P (in fact, any noncrossing set of edges) has at most $3n - k - 3$ edges. This follows from Euler's relation. Thus, when $k \geq 5$, the upper bound follows. It is easy to check that when $n \geq 5$ and k is 3 or 4, we can always remove at least $5 - k$ edges and still obtain a pseudotriangulation.

A family of triangulations which show the lower bound is given in Figure 3. The number of vertices is a multiple of 3 and $k = 6$. The instances are constructed inductively, by removing the central triangle and subdividing the resulting pseudotriangle as shown in Figure 3. The new points are slightly twisted about the center in order to obtain a point set in general position, and to ensure that the "direct paths" which lead from the center to the vertices of the outer hexagon make zigzag turns. The only edge set which one can remove is the central triangle. The resulting pseudotriangulation has $3n - 12$ edges. One can check by inspection, using Theorem 2, that it is a minimal pseudotriangulation. Since there was only one way to obtain a pseudotriangulation as a subgraph of T, it is the unique minimal pseudotriangulation. Hence, is is also a minimum pseudotriangulation.   □

We have an example of a minimum pseudotriangulation with $n = 41$ vertices and $3n - 8$ edges. We believe that the upper bound of $3n - 8$ is tight for infinitely many values of n.

Theorem 4. (a) *There are cases of S and T such that the size of T and P^T_{mum}, and all other pseudotriangulations P^T are the same.*

(b) *The ratio between the sizes of two different minimal constrained pseudotriangulations in a given triangulation is between $\frac{2}{3}$ and $\frac{3}{2}$. These bounds are asymptotically tight.*

(c) *The ratio of the size of the minimum pseudotriangulation of S and the minimum pseudotriangulation constrained in T is between 1 and $\frac{2}{3}$, which is* *asymptotically tight. The same bound holds for the size of the minimal constrained pseudotriangulation in T.*

Proof. The bounds on the ratios follow from the fact that a pseudotriangulation of n points has between $2n - 3$ and $3n - 6$ edges. We omit the detailed general proofs that the bounds are tight in this version of the paper, but we show some typical tight instances in Figure 4.

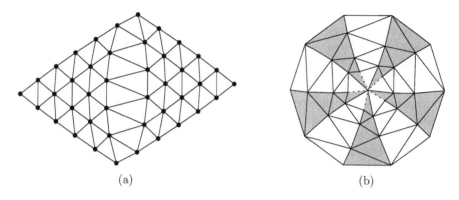

        (a)   (b)

Fig. 4. Examples for the proof of Theorem 4

(a) The triangulation T in Figure 4(a) is obtained by perturbing a triangular grid so that the sides bulge. One can check by inspection, using Theorem 2, that it is a minimal pseudotriangulation, and hence also a minimum pseudotriangulation in T.

(b) In the triangulation of Figure 4(b) we can obtain a minimal triangulation with $3n - 18$ edges by removing the five dotted edges in the center, or we can get another minimal triangulation with $2n - 2$ edges by removing the edges in the shaded funnels.

(c) The example of Figure 3 in Theorem 3 is a minimum and minimal pseudotriangulation $P^T(S)$ with $3n - 12$ edges. A minimum pseudotriangulation of S has always $2n - 3$ edges.   □

5 Constructing a Minimal Pseudotriangulation in a Triangulation

In the following, we shall present a linear-time greedy algorithm to construct a minimal pseudotriangulation in a given triangulation T. By Theorem 2, we just need to check whether we can remove a single edge or a triangle and keep a pseudotriangulation. If this is the case, we remove the edge or triangle and continue with the resulting pseudotriangulation. The following lemma explains how to carry out this test efficiently.

Lemma 4. (a) *Let P be a pseudotriangulation and $e \in P$ be an edge. Then $P - e$ is a pseudotriangulation if and only if the removal of e creates a new* reflex vertex, in other words, if one endpoint of e is not reflex in P and reflex in $P - e$.

(b) Let P be a pseudotriangulation and $\{e_1, e_2, e_3\} \in P$ be a triangular face in P. Then $P - \{e_1, e_2, e_3\}$ is a pseudotriangulation if and only if the removal of the triangle makes all three vertices reflex, or more precisely, if the three vertices of $\{e_1, e_2, e_3\}$ are not reflex in P and reflex in $P - \{e_1, e_2, e_3\}$.

Proof. Removing an edge or a triangle creates a new face from merging two or four pseudotriangles, respectively. We have to check whether this new face contains 3 convex vertices. The proof follows easily by counting the convex angles incident to the affected vertices, before and after removing the edge or the triangle. (It also follows that in case (a), only *one* endpoint of e can be a *new* reflex vertex in $P - e$.)   □

Computationally, the conditions of Lemma 4 can be checked very easily. For example, let $e = ab$ be an edge in a pseudotriangulation P. Let α_1 and α_2 be the two angles incident to e at a, and let β_1 and β_2 be the two corresponding angles at b. Then $P - e$ is a pseudotriangulation if and only if $\alpha_1 < \pi$, $\alpha_2 < \pi$, and $\alpha_1 + \alpha_2 > \pi$, or if $\beta_1 < \pi$, $\beta_2 < \pi$, and $\beta_1 + \beta_2 > \pi$. The condition can be similarly formulated for the removal of a triangle (Lemma 4(b)). Thus, for a given edge or triangle, it can be checked in constant time whether it can be removed.

The algorithm for constructing a minimal pseudotriangulation now works as follows. We call an edge or a triangle *removable* if it satisfies the condition of Lemma 4(a) or (b), respectively. We start with the given triangulation. The algorithm maintains a list of all *removable edges*, which is initialized in linear time by scanning all edges. When a removable edge exists, we simply remove this edge, and update the list of removable edges. The removal of an edge $e = ab$ may affect the removability status of at most four edges of the current pseudotriangulation P (namely, the two neighboring edges at a and at b). These edges can be checked in constant time.

We repeat this procedure until the list of removable edges becomes empty. Now we check if there is any removable *triangle* according to the condition of Lemma 4(b), and we remove it. One can easily show that the removal of a triangle cannot create a new removable edge or a new removable triangle. Thus we can simply scan all faces of P sequentially, in linear time.

In the end we obtain a pseudotriangulation without removable edges or triangles, which is a minimal pseudotriangulation, by Theorem 2.

Theorem 5. *The algorithm produces a minimal pseudotriangulation P^T_{mal} of a given triangulation T in linear time.*   □

6   Constructing a Pseudotriangulation Containing a Given Set of Edges

In this section, we find a minimum pseudotriangulation which *contains* a given set L of non-crossing line segments. The basic idea is to maintain the set of reflex vertices of the given straight-line graph $G(S, L)$ as an invariant when we add extra edges to L to build the pseudotriangulation of L [3].

Theorem 6. *For any noncrossing set of line segments L, there is a pseudotriangulation $T'_L(S) \supseteq L$ which has the same set of reflex vertices as $G(L, S)$.*

Proof. We prove this by gradually adding edges to the set L until we get a pseudotriangulation. First we add all convex hull edges to L. This does not change the set of reflex vertices.

Then the edge set L partitions the interior of the convex hull into faces, which can be considered independently. So let us consider a single face F, see Figure 5. The boundary of F has one component B which is the exterior boundary of F, and possibly several other components inside F. Note that B is a single cycle of edges when we walk along the boundary of F inside B, although this cycle may visit the same edge twice (from two different sides) or it may visit a vertex several times. Nevertheless, we treat B as if it were a simple polygon.

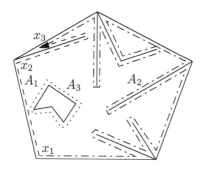

Fig. 5. Illustration for the proof of the Theorem 6.

We will subdivide F into pseudotriangles by repeatedly carrying out the following steps:

- Select a corner x_1 on B and walk clockwise along B until we find the next two corners x_2 and x_3 on B. (B must contain at least 3 corners.) We denote the path from x_1 via x_2 to x_3 along B by A_1, and the remaining part of B by A_2. By A_3, we denote the (possibly empty) set of interior components of the boundary of F, see Figure 5.

- Find the shortest path S from x_1 to x_3 in F which is homotopic to the path A_1 from x_1 to x_3. In other words, we put a string from x_1 along x_2 to x_3 and pull the string taut, regarding B and the components inside F

as obstacles. In other words, we take that shortest path from x_1 to x_3 in F which separates A_1 from $A_2 \cup A_3$.

It is clear that this path S consists of the following pieces:

(a) an initial piece following some part of B from x_1 towards x_2, turning left;
(b) a connecting line segment through the interior of F;
(c) some part of the boundary of the convex hull of $A_2 \cup A_3$, turning right;
(d) a connecting line segment through the interior of F; and
(e) a final piece following some part of B from x_2 to x_3, turning left.

Any of the pieces (a), (c), and (e) may be missing. If (c) is missing, then there is of course only one connecting segment instead of (b) and (d). It follows that the region that is cut off by this path (on the left side of S) is a pseudotriangle that contains no points inside. It may happen that S consists of a single reflex chain from x_1 to x_3 along B. In this case, F was an empty pseudotriangle, and we are done with F. Otherwise, we continue this procedure with the remaining part of F. It is also clear that no edge of S will destroy a reflex vertex. Being a geodesic path, S will only go through reflex vertices (besides the endpoints x_1 and x_3), and it will make left turns when passing around a component that is on the left side, and similarly for right turns.   □

The following immediate consequence of the theorem extends the known results for $L = \emptyset$, where $r = n$.

Corollary 1. *Every minimum pseudotriangulation of a point set S with n points containing a given set L of edges with r reflex vertices has $2n - r - 2$ pseudotriangles and $3n - r - 3$ edges.*   □

7   Conclusion

Several problems remain for further study.

– How to find the minimum pseudotriangulation constrained in T? Is this problem NP-hard?
– Study minimum pseudotriangulations subject to some other constraints.
– How to find the minimum-weight pseudotriangulation?

References

1. Aichholzer O., Hoffmann M., Speckmann B., and Tóth C. D., 'Degree bounds for constrained pseudo-triangulations', in preparation.
2. Kirkpatrick D., Snoeyink J., and Speckmann B., 'Kinetic collision for simple polygons', in *Proc. 16th Ann. Symp. on Computational Geometry*, 2000, pp. 322–329.
3. Pocchiola M. and Vegter G., 'Topologically sweeping visibility complexes via pseudotriangulations', *Discrete and Computational Geometry* **16** (1996), 419–453.
4. Pocchiola M. and Vegter G., 'The visibility complex', *International Journal on Computational Geometry and Applications* **6** (1996), 279–308.
5. Streinu I., 'A combinatorial approach to planar non-colliding robot arm motion planning', *Proc. 41st Ann. Symp. Found. Comput. Sci.* (FOCS), 2000, pp. 443–453.

Pairwise Data Clustering and Applications*

Xiaodong Wu[1]**, Danny Z. Chen[2]***, James J. Mason[3],
and Steven R. Schmid[3]

[1] Department of Computer Science, University of Texas-Pan American,
Edinburg, TX 78539, USA
FAX: 956-384-5099, xwu@cs.panam.edu
[2] Department of Computer Science and Engineering,
University of Notre Dame, Notre Dame, IN 46556, USA
chen@cse.nd.edu
[3] Department of Aerospace and Mechanical Engineering,
University of Notre Dame, Notre Dame, IN 46556, USA
{James.J.Mason.12,Steven.R.Schmid.2}@nd.edu

Abstract. Data clustering is an important theoretical topic and a sharp
tool for various applications. Its main objective is to partition a given
data set into clusters such that the data within the same cluster are
"more" similar to each other with respect to certain measures. In this
paper, we study the pairwise data clustering problem with pairwise simi-
larity/dissimilarity measures that need not satisfy the triangle inequality.
By using a criterion, called the *minimum normalized cut*, we model the
pairwise data clustering problem as a graph partition problem. The graph
partition problem based on minimizing the normalized cut is known to
be NP-hard. We present a $((4 + o(1)) \ln n)$-approximation polynomial
time algorithm for the minimum normalized cut problem. We also give
a more efficient algorithm for this problem by sacrificing the approximation ratio slightly. Further, our scheme achieves a $((2 + o(1)) \ln n)$-
approximation polynomial time algorithm for computing the sparsest
cuts in edge-weighted and vertex-weighted undirected graphs, improving
the previously best known approximation ratio by a constant factor.

1 Introduction

Data clustering is a fundamental problem and a sharp tool for applications
such as information retrieval, data mining, computer vision, pattern recognition, biomedical informatics, and statistics. Its main objective is to partition a
given data set into clusters such that the data within the same cluster are "more"

---

* This research was supported in part by the 21st Century Research and Technology
Fund from the State of Indiana.
** The work of this author was supported in part by the Computing and Information
Technology Center, and by the Faculty Research Council, University of Texas-Pan
American, Edinburg, Texas, USA.
*** The work of this author was supported in part by the National Science Foundation
under Grants CCR-9623585 and CCR-9988468.

T. Warnow and B. Zhu (Eds.): COCOON 2003, LNCS 2697, pp. 455–466, 2003.
© Springer-Verlag Berlin Heidelberg 2003

similar to each other with respect to certain measures. This problem has been intensively studied (e.g., see [11,23]).

The wide variety of applications precludes any single mathematical formulation of the general data cluster problem. Furthermore, many straightforward formulations (e.g., as optimization problems) are NP-hard. The problem of optimally clustering a data set usually occurs in one of two different forms depending on the data representations. One kind of approaches focuses on partitioning data in metric spaces, which recently received a lot of attention. In particular, several combinatorial measures of clustering quality have been investigated. These include *minimum diameter, k-center, k-median, k-means*, and *minimum sum* (e.g., see [1,9,17,22,4,2]). Charikar *et al.* [7] and Guha *et al.* [15] also studied this problem in the dynamic settings. All these algorithms assume that the input data set is in a metric space. However, in a large number of applications arising from computer vision [25], pattern recognition [16], and gene clustering [5], the only available information about the target data set is a similarity/dissimilarity measure between pairwise data. These pairwise measures even need not satisfy the triangle inequality. Hence, there is another kind of approaches, referred to as *pairwise data clustering*, for handling data sets with pairwise measures.

Formally, we define the pairwise data clustering problem as follows: Given $\mathcal{P} = (\mathcal{D}, \mathcal{N}, \mathcal{W}, \mathcal{C})$, where $\mathcal{D}$ is a set of data, $\mathcal{N} \subseteq \mathcal{D} \times \mathcal{D}$ is a set of data pairs, $\mathcal{W} : \mathcal{N} \to \mathcal{R}^+$ measures the similarity/dissimilarity of each data pair in $\mathcal{N}$, and $\mathcal{C}$ is a set of clustering criteria, partition the data set $\mathcal{D}$ into several clusters such that the criteria in $\mathcal{C}$ are optimized. Unlike data clustering in metric spaces, it seems that fewer results on pairwise data clustering are known [16,5,3,17,25]. Aslam *et al.* [3] used the idea that the neighbors of each datum are more reliable in deciding its cluster and some statistical models to cluster data. Kannan *et al.* [17] proposed a bicriteria based on the conductance of a graph to evaluate a clustering.

In this paper, we study pairwise data clustering in a somewhat general setting. We model an input data set as an undirected graph G such that its vertices represent the data items and its edges represent the data pairs in $\mathcal{N}$ whose weights are defined by $\mathcal{W}$. Note that the triangle inequality need not hold on G. We consider a clustering criterion, called the *minimum normalized cut*. This criterion is not only interesting theoretically, but is also capable of delivering impressive clustering results in applications such as image segmentation and pattern recognition [25]. We will focus on partitioning the data set into two clusters (of course, one can recursively apply our partition algorithms to obtain more clusters, if needed), with each cluster corresponding to a subset of vertices in the graph. Thus, we formulate this pairwise data clustering problem as a graph partition problem, as follows.

Minimizing the normalized cut: Given an n-vertex, m-edge undirected graph $G = (V, E)$, each edge $e \in E$ having a non-negative weight $w(e)$, find a *normalized cut* $C \subseteq E$ whose removal partitions V into two disconnected sets S and $\bar{S}$ ($S \cup \bar{S} = V$), such that

$$\alpha(S) = \frac{cut(S, \bar{S})}{assoc(S, V)} + \frac{cut(S, \bar{S})}{assoc(\bar{S}, V)}$$

is minimized, where $cut(S, \bar{S}) = \sum_{e \in C} w(e)$ is the sum of edge weights of the cut C, and $assoc(A, V) = \sum_{(u,v) \in E, u \in A, v \in V} w(u, v)$ for a subset $A \subseteq V$ is the total weight of edges in E between the vertices in A and all vertices of G. $\alpha(S)$ is called the *normalized-cut cost* for the cut C. Note that computing a minimum normalized cut in G is NP-complete [25].

Our main results are summarized as follows.

- For the minimum normalized cut problem, a $((4 + o(1)) \ln n)$-approximation algorithm in polynomial time, and further, for any $\epsilon > 0$, a $((4(1 + \epsilon) + o(1)) \ln n)$-approximation algorithm in $\tilde{O}(\frac{1}{\epsilon^2} m^2)$ time, where $\tilde{O}(\cdot)$ hides a poly-logarithmic factor.

- A $((2 + o(1)) \ln n)$-approximation polynomial time algorithm for computing a sparsest cut in an edge-weighted and vertex-weighted undirected graph, improving Garg *et al.*'s approximation ratio [13] by nearly a factor of 4.

To our best knowledge, no previous provably good approximation polynomial time algorithms for the minimum normalized sum problem were known. Previous work on the minimum normalized cut is mainly concerned with the image segmentation problem. Shi and Malik [25] developed a heuristic approach for the minimum normalized cut problem, but no provable quality is known for their approach. Our algorithms are based on a graph construction that allows the application of the sparsest cut algorithm. However, instead of directly applying the algorithms in [13,21], we judiciously exploit the structures of the region growing technique [13] and make use of the credit scheme in [13,21], thus yielding a better approximation ratio for not only the minimum normalized cut but also the sparsest cut. Further, by using Karakostas's maximum concurrent flow algorithm [18], a more efficient algorithm for the normalized cut is achieved while sacrificing the approximation ratio only slightly.

We should also mention some other closely related work. Given a graph with non-negative edge weights and k commodities, s_i and t_i being the source and sink for commodity i and each commodity being associated with a demand $d(i)$, the *maximum concurrent flow* problem seeks to find the largest λ such that there is a multicommodity flow which routes $\lambda d(i)$ units of commodity i simultaneously in G. Leighton and Rao's seminal work on the *product multicommodity flow* problem [21] (a special case of the maximum concurrent flow problem) shows that the ratio of the capacity of a cut to the demand of the cut is within $\Theta(\log n)$ times the max-flow. This work has led to an array of provably good approximation algorithms for a wide variety of NP-hard problems, including the *sparsest cuts, balanced cuts,* and *separators*. Garg *et al.* [13] used a region growing technique on graphs and obtained an $(8 \ln n)$-approximation algorithm for the sparsest cut problem, improving Leighton and Rao's algorithm [21] by a constant factor. By using the region growing technique in [13] together with the spreading metrics, Even *et al.* [10] achieved $O(\ln n)$-approximation algorithms for *ρ-separators, b-balanced cuts,* and *k-balanced cuts*.

We omit the proofs of some lemmas due to the space limit.

2   Improving Garg *et al.*'s Approximation Algorithm for the Sparsest Cut

Our approximation algorithm for the normalized cut is based on our improved sparsest cut algorithm. In this section, we present our $((2 + o(1)) \ln n)$-approximation polynomial time algorithm for the sparsest cut, improving Garg *et al.*'s approximation ratio [13] by nearly a factor of 4.

Given an n-vertex, m-edge undirected graph $G = (V, E)$, each edge e (resp., vertex v) of G having a non-negative weight $w(e)$ (resp., $w(v)$), the *sparsest cut problem* seeks a cut $C \subset E$ whose removal partitions V into two disconnected sets S and $\bar{S}$ ($S \cup \bar{S} = V$), such that the *sparsity* $\beta(S)$ for C (or S), with

$$\beta(S) = \frac{\sum_{e \in C} w(e)}{\sum_{v \in S} w(v) \cdot \sum_{v \in \bar{S}} w(v)}$$

is minimized. Garg *et al.* [13] gave an $(8 \ln n)$-approximation sparsest cut algorithm. However, we are able to do better. By judiciously choosing the parameters for the region growing technique and using Garg *at el.*'s approach [13], we can obtain a $((2 + o(1)) \ln n)$-approximate sparsest cut in G. The sphere growing procedure presented in this section is similar to the region growing procedure in [13]; the differences are the parameters used in the volume definition and the upper bound on the radii of spheres (to be defined precisely later). Our algorithm achieves a tighter upper bound than that in [13] for the total weight of the cuts produced by our sphere growing procedure, which leads to a better approximation ratio. In addition, unlike [13] (in which an approximate *flux* is computed, and then an approximate sparsest cut is induced from the flux), we compute the desired approximate sparsest cut directly.

Intuitively, the main idea of our algorithm for the sparsest cut, similar to that in [13,21], is as follows. We first assign a *length* to each edge in the graph G by using a linear programming; this defines a lower bound for the sparsest cut. Then, based on the assigned edge lengths, a set B of disjoint "spheres" in G is generated by our *SphereGrowing* procedure. We prove that either there exists a sphere in the set B which gives the desired sparse cut, or the set B enables us to choose a vertex of G from which another sphere can be grown for the desired sparse cut.

2.1   Linear Programming Relaxations

By assigning a non-negative edge length $d(e)$ to each edge $e \in E$, we formulate a linear programming relaxation for the sparsest cut problem on G, as follows:

$$(LP) \min \sum_{e \in E} w(e) \cdot d(e)$$

$$\text{s.t.} \sum_{u,v \in V} w(u) \cdot w(v) \cdot \mathbf{dist}_G(u, v) \geq 1$$

$$\forall e \in E, d(e) \geq 0,$$

where $\mathbf{dist}_H(u,v)$ denotes the distance between vertices u and v in a subgraph H of G with respect to the metric induced by $d(\cdot)$. Actually, (LP) is a dual of the maximum multicommodity concurrent flow problem.

Let τ denote the *value* of the optimal solution $d(\cdot)$ for the linear program (LP) (i.e., $\tau = \sum_{e \in E} w(e) \cdot d(e)$), and β^* denote the sparsity of the sparsest cut in G. Obviously, we have the following lemma.

Lemma 1. *[21,13]* $\tau \leq \beta^*$.

2.2 Sphere Growing

Based on the edge lengths in $\{d(e) \mid e \in E\}$ from the linear programming relaxation, we present in this subsection our approximation algorithm for the sparsest cut problem. The algorithm is based on a *sphere growing* procedure for finding "good" cuts in graphs that have "nearly" constant diameters.

(1) Assigning Volumes

As in [10,13], the volumes assigned to spheres that grow around vertices in the graph define the *credits* that are associated with the spheres in order to pay for the cost of the cut. Our definition of volume as in [10], is a little different from that of [13] in order to improve the approximation bounds.

Definition 1. *A sphere $N(v,r)$ in a subgraph $G' = (V', E')$ of G with the metric $d(\cdot)$ that grows from a vertex $v \in V'$ with a radius r is defined by*

$$N(v,r) \triangleq \{u \in V' \mid \mathbf{dist}_{G'}(u,v) < r\},$$

where $\mathbf{dist}_{G'}(u,v)$ is the u-to-v shortest path length in G' with respect to the edge lengths $d(\cdot)$.

For convenience, denote by $E(v,r)$ the set of edges whose two end vertices are both in $N(v,r)$, and $C(v,r)$ the set of edges cutting $N(v,r)$ from $V - N(v,r)$. Let $w(C(v,r))$ be $\sum_{e \in C(v,r)} w(e)$.

Now, we define the volume of a sphere $N(v,r)$, denoted by $vol(v,r)$.

Definition 2. *[10]*
$$vol(v,r) \triangleq \frac{\tau}{n \ln n} + \sum_{e \in E(v,r)} w(e)d(e) + \sum_{\substack{e = (x,y) \in C(v,r) \\ x \in N(v,r)}} w(e) \left(r - \mathbf{dist}_{G'}(v,x)\right)$$

(2) The Sphere Growing Procedure

Next, we discuss the sphere growing procedure and an important lemma that specifies the upper bound of the sphere radii. This procedure takes as input a subgraph $G' = (V', E')$ of G and a given edge length function $d(\cdot)$. It runs Dijkstra's single-source shortest path algorithm [8] from an arbitrary vertex $v \in V'$ to obtain a sphere $N(v,r)$, such that the ratio R between $w(C(v,r))$ and $vol(v,r)$ (i.e., $R = \frac{w(C(v,r))}{vol(v,r)}$) is logarithmic with respect to n. Figure 1 illustrates the procedure, called *SphereGrowing*. The stopping condition of our SphereGrowing procedure is $R \geq \delta$, where $\delta = \frac{W^2}{\sqrt{2}}\sqrt{\ln(n \ln n + 1)} \cdot (\sqrt{2}\sqrt{\ln(n \ln n + 1)} + 1)$,

Procedure *SphereGrowing()*;
    Input: $G' = (V', E')$ and $\{d(e) \mid e \in E'\}$;
    Output: A sphere $N(v,r)$;
begin
    arbitrarily choose a vertex $v \in V'$;
    $S = \{v\}$;
    $r = 0$;
    y is the vertex such that $\mathbf{dist}_{G'}(v,y)$ is min $_{x \in V' - S}$ $\mathbf{dist}_{G'}(v,x)$;
    while $w(C(v,r)) > \frac{W^2}{\sqrt{2}}\sqrt{\ln(n\ln n + 1)} \cdot (\sqrt{2}\sqrt{\ln(n\ln n + 1)} + 1) \cdot vol(v, \mathbf{dist}_{G'}(v,y))$
    and $S \neq V'$ **do**
        $S = S \cup \{y\}$;
        $r = \mathbf{dist}_{G'}(v,y)$;
        y is the vertex such that $\mathbf{dist}_{G'}(v,y)$ is min $_{x \in V' - S}$ $\mathbf{dist}_{G'}(v,x)$;
    return S;
end

Fig. 1. The sphere growing procedure.

where W denotes the total sum of vertex weights in G. Note that δ is a key parameter to the sphere growing technique that determines the upper bound of the sphere radii produced and hence affects the approximation ratio for the sparsest cut. Our choice of δ is different from [13], which leads to a better approximation ratio.

Lemma 2. SphereGrowing *produces a sphere* $N(v,r)$ *in* $G' = (V', E')$ *with*
$$r < \frac{1}{W^2} \cdot \frac{\sqrt{2}\sqrt{\ln(n\ln n+1)}}{\sqrt{2}\sqrt{\ln(n\ln n+1)}+1}.$$

(3) Finding a "Good" Sparse Cut

Now we are ready to compute a cut whose sparsity $\beta(\cdot)$ is at most $((2+o(1)) \cdot \ln n)$ times the sparsity of the sparsest cut β^*.

We call the *SphereGrowing* procedure on $G_0 = G$ to form a sphere $N(v_1, r_1)$, and repeatedly call it on the remaining subgraph G_i of G to form the next sphere $N(v_{i+1}, r_{i+1})$, until the remaining subgraph is empty. Let $N(v_1, r_1), N(v_2, r_2), \ldots$, and $N(v_l, r_l)$ be all the spheres thus produced, and $G_1, G_2, \ldots$, and G_l $(G_l = \phi)$ be the successive remaining subgraphs. Also, we denote $(V - N(v_i, r_i))$ by $\bar{N}(v_i, r_i)$. Note that $C(v_i, r_i)$ denotes the cut between $N(v_i, r_i)$ and $V - N(v_i, r_i)$. Let $C'(v_i, r_i)$ be the set of edges of the cut $C(v_i, r_i)$ that are in G_{i-1}, for $i = 1, 2, \ldots, l$.

If there is a sphere $N(v_p, r_p)$ such that $\beta(N(v_p, r_p)) \leq \beta^* \cdot (2 + o(1)) \cdot \ln n$, then $C(v_p, r_p)$ is a cut that we seek. Otherwise, we claim that there is at least one sphere, say $N(v_q, r_q)$, such that the weight $w(N(v_q, r_q))$ is "heavy" enough (this is stated precisely in Lemma 4). In the latter case, we can also find a "good" sparse cut by growing spheres centered at v_q. To do that, we first need to estimate the total weight of all the cuts $C(v_i, r_i)$, $i = 1, 2, \ldots, l$, as stated in Lemma 3.

Lemma 3. *The total weight of the cuts* $C(v_1, r_1), C(v_2, r_2), \ldots, C(v_l, r_l)$ *is upper bounded by* $\mathcal{B} = \sqrt{2}W^2 \cdot \sqrt{\ln(n \ln n + 1)} \cdot (\sqrt{2}\sqrt{\ln(n \ln n + 1)} + 1) \cdot (1 + \frac{1}{\ln n}) \cdot \tau$.

This upper bound $\mathcal{B}$ of the total weight of the cuts in $\{C(v_i, r_i) \mid i = 1, 2, \ldots, l\}$ is tighter than that in [13]. Intuitively, a tighter $\mathcal{B}$ implies that a "lighter" sphere can be found (among the desired "heavy" spheres), which can help obtain a better approximate sparsest cut.

Lemma 4. *If for every* $i = 1, 2, \ldots, l$, $\beta(N(v_i, r_i)) > (\sqrt{2}\sqrt{\ln(n \ln n + 1)} + 1)^2 \cdot (1 + \frac{1}{\ln n}) \cdot \tau$, *then there exists a sphere* $N(v_q, r_q)$ *such that* $w(N(v_q, r_q)) \geq \frac{W}{1 + \sqrt{2} \cdot \sqrt{\ln(n \ln n + 1)}}$.

Based on Lemma 4, let $N(v_q, r_q)$ be a sphere with $w(N(v_q, r_q)) \geq \frac{W}{1 + \sqrt{2} \cdot \sqrt{\ln(n \ln n + 1)}}$. Then we grow a sphere centered at the vertex v_q in G. But, we do not check the *while-loop* condition in procedure *SphereGrowing*, and terminate the growing process once all vertices of G are included in the sphere. If we sort the vertices of G in the increasing order of their distances from v_q, say as $v_0, v_1, \ldots, v_{n-1}$ (with $v_0 = v_q$), and let r'_i denote $\mathbf{dist}_G(v_q, v_i)$ for $i = 0, 1, \ldots, n-1$, then $N(v_q, r'_i) = N(v_q, r'_{i-1}) \cup \{v_i\}$. Note that $N(v_q, r'_0) = \{v_q\}$, $N(v_q, r'_{n-1}) = V$, and r_q may not be in the set $\{r'_i \mid 0 \leq i \leq n-1\}$. We now show one of the cuts in $\{C(v_q, r'_i) \mid 0 \leq i \leq n-1\}$ is "good", as stated precisely in Lemma 5.

Lemma 5. *There exists an index* $t \in \{0, 1, \ldots, n-1\}$ *such that* $\beta(N(v_q, r'_t)) \leq (\sqrt{2}\sqrt{\ln(n \ln n + 1)} + 1)^2 \cdot (1 + \frac{1}{\ln n}) \cdot \tau$.

Proof. We prove this by making a contradiction to the key constraint of (LP), i.e., $\sum_{u,v \in V} w(u) \cdot w(v) \cdot \mathbf{dist}_G(u, v) \geq 1$. For this purpose, we assume that for each $i = 0, 1, \ldots, n-1$, $\beta(N(v_q, r'_i)) > \kappa$, where $\kappa = (\sqrt{2}\sqrt{\ln(n \ln n + 1)} + 1)^2 \cdot (1 + \frac{1}{\ln n}) \cdot \tau$.

By observing that $\mathbf{dist}_G(u, v) \leq \mathbf{dist}_G(v_q, u) + \mathbf{dist}_G(v_q, v)$ and each unordered pair of u and v appears in $\sum_{u,v \in V} w(u) \cdot w(v)$ exactly once, and recalling that $W = \sum_{u \in V} w(u)$, we get

$$\sum_{u,v \in V} w(u) \cdot w(v) \cdot \mathbf{dist}_G(u, v) \leq W \cdot \sum_{v \in V} w(v) \cdot \mathbf{dist}_G(v_q, v). \quad (1)$$

Let $N(v_q, r'_h)$ be the smallest sphere in the sphere chain $N(v_q, r'_0) \subset N(v_q, r'_1) \subset \cdots \subset N(v_q, r'_{n-1})$ that contains the sphere $N(v_q, r_q)$. Obviously, $N(v_q, r'_h)$ may be a superset of $N(v_q, r_q)$ because some vertices in $N(v_q, r'_h)$ may have been removed before growing the sphere $N(v_q, r_q)$ by using procedure *SphereGrowing*. Thus, we have $r'_h \leq r_q$. Based on Lemma 2, $r_q < \frac{1}{W^2} \cdot \frac{\sqrt{2}\sqrt{\ln(n \ln n + 1)}}{\sqrt{2}\sqrt{\ln(n \ln n + 1)} + 1}$. Hence,

$$r'_h < \frac{1}{W^2} \cdot \frac{\sqrt{2}\sqrt{\ln(n \ln n + 1)}}{\sqrt{2}\sqrt{\ln(n \ln n + 1)} + 1}. \quad (2)$$

We arrange the vertices of V along a ray originating from $v_q = v_0$ based on their distances from v_q in G. Note that for each v_i, $\mathbf{dist}_G(v_q, v_i) = r'_i$ and $r'_0 = 0$.

Hence, the total weighted distance of all vertices in V from v_q, $\sum_{v \in V} w(v) \cdot \mathbf{dist}_G(v_q, v)$, equals $\sum_{v \in V} w(v) \cdot r'_i$. Note that, for each distance interval (r'_i, r'_{i-1}), it contributes as much as $(r'_i - r'_{i-1}) \cdot (W - \sum_{j=0}^{i-1} w(v_j))$ to the total weighted distance of all vertices in V. Therefore,

$$
\sum_{v \in V} w(v) \cdot \mathbf{dist}_G(v_q, v) = \sum_{i=1}^{n-1} (r'_i - r'_{i-1})(W - \sum_{j=0}^{i-1} w(v_j))
$$
$$
= \sum_{i=1}^{h} (r'_i - r'_{i-1})(W - \sum_{j=0}^{i-1} w(v_j)) + \sum_{i=h+1}^{n-1} (r'_i - r'_{i-1})(W - \sum_{j=0}^{i-1} w(v_j)) \quad (3)
$$

As to the first term in formula (3), by using (2),

$$
\sum_{i=1}^{h} (r'_i - r'_{i-1})(W - \sum_{j=0}^{i-1} w(v_j)) \le W \sum_{i=1}^{h} (r'_i - r'_{i-1}) < \frac{1}{W} \cdot \frac{\sqrt{2}\sqrt{\ln(n \ln n + 1)}}{\sqrt{2}\sqrt{\ln(n \ln n + 1)} + 1}
$$
$$
(4)
$$

We also need to bound the second term in formula (3). Recalling the assumption that every $\beta(N(v_q, r'_i)) > \kappa$, we obtain $w(\bar{N}(v_q, r'_i)) < \frac{1}{\kappa} \cdot w(C(v_q, r'_i)) \cdot \frac{1}{w(N(v_q, r'_i))}$. Note that when $i > h$, $w(N(v_q, r'_i)) \ge w(N(v_q, r'_h)) \ge w(N(v_q, r_q))$. Thus, by Lemma 4, $w(N(v_q, r'_i)) \ge \frac{W}{1 + \sqrt{2}\sqrt{\ln(n \ln n + 1)}}$. Hence, for every i such that $h < i \le n-1$, we have

$$
w(\bar{N}(v_q, r'_i)) < \frac{1}{W} \cdot \frac{1}{(\sqrt{2}\sqrt{\ln(n \ln n + 1)} + 1) \cdot (1 + \frac{1}{\ln n}) \cdot \tau} \cdot w(C(v_q, r'_i)).
$$

we have,

$$
\sum_{i=h+1}^{n-1} (r'_i - r'_{i-1})(W - \sum_{j=0}^{i-1} w(v_j)) = \sum_{i=h+1}^{n-1} (r'_i - r'_{i-1}) \cdot w(\bar{N}(v_q, r'_{i-1}))
$$
$$
\le \frac{1}{W} \cdot \frac{\sum_{e \in E} w(e) \cdot d(e)}{(\sqrt{2}\sqrt{\ln(n \ln n + 1)} + 1) \cdot (1 + \frac{1}{\ln n}) \cdot \tau} < \frac{1}{W} \cdot \frac{1}{\sqrt{2}\sqrt{\ln(n \ln n + 1)} + 1} \quad (5)
$$

From (3), (4), and (5), it follows

$$
\sum_{v \in V} w(v) \cdot \mathbf{dist}_G(v_q, v) < \frac{1}{W} \cdot \frac{\sqrt{2}\sqrt{\ln(n \ln n + 1)}}{\sqrt{2}\sqrt{\ln(n \ln n + 1)} + 1} + \frac{1}{W} \cdot \frac{1}{\sqrt{2}\sqrt{\ln(n \ln n + 1)} + 1} = \frac{1}{W}.
$$
$$
(6)
$$

Therefore, based on (1) and (6), we have

$$
\sum_{u,v \in V} w(u) \cdot w(v) \cdot \mathbf{dist}_G(u, v) < 1,
$$

a contradiction. Thus, the lemma holds.   □

Based on Lemma 5, we can find a subset $S \subseteq V$ whose sparsity

$$
\beta(S) \le (\sqrt{2}\sqrt{\ln(n \ln n + 1)} + 1)^2 \cdot (1 + \frac{1}{\ln n}) \cdot \tau = (2 + o(1)) \ln n \cdot \tau.
$$

Along with Lemma 1, the next lemma holds.

Lemma 6. *For any n-vertex undirected graph with non-negative edge weights and non-negative vertex weights, a sparse cut with a $((2 + o(1)) \ln n)$ - approximation ratio can be found in polynomial time.*

3   Minimizing the Normalized Cut

In this section, we consider the *minimum normalized cut* problem on an undirected graph $G = (V, E)$ whose edges e have non-negative weights $w(e)$. Based on our approximation scheme in Section 2, we give a $((4+o(1)) \ln n)$-approximation polynomial time algorithm for this problem, and further, for any $\epsilon > 0$, a faster $\tilde{O}(\frac{1}{\epsilon^2} m^2)$ time algorithm with a $((4(1 + \epsilon) + o(1)) \ln n)$ approximation ratio.

3.1   Our Algorithm

We first define the weight $w(u)$ of each vertex u in G as $\frac{1}{2} \sum_{e=(u,v) \in E} w(e)$. For a subset $S \subseteq V$, let $w(S)$ denote the total weight of vertices in S, and W denote the total weight of all edges in E. Note that W is also equal to $\sum_{v \in V} w(v)$. Recall that for a vertex subset $S \subseteq V$,

$$\alpha(S) = \frac{w(C(S))}{assoc(S,V)} + \frac{w(C(S))}{assoc(\bar{S},V)}$$

is the normalized-cut cost of the cut $C(S)$ induced by S, and $\beta(S) = w(C(S))/(w(S) \cdot w(\bar{S}))$ is the sparsity of the cut $C(S)$, where $assoc(A, V)$ is the total weight of edges in E between the vertices in A and all vertices of G, and $w(C(S))$ is the total weight of edges in $C(S)$. The following lemma states the immediate connection between the sparse cut and the normalized cut.

Lemma 7. *For any vertex subset $S \subseteq V$ in G, $\frac{1}{2} W \cdot \beta(S) \leq \alpha(S) \leq W \cdot \beta(S)$.*

Based on Lemma 7, one may apply Garg *et al.*'s $(8 \ln n)$-approximation algorithm for the sparsest cut to obtain a $(16 \ln n)$-approximate minimum normalized cut. However, we can do better. Based on Lemma 6, a $((2 + o(1)) \ln n)$-approximate sparsest cut in G can be computed in polynomial time. Together with Lemma 7, the theorem below follows.

Theorem 1. *For any n-vertex undirected graph with non-negative edge weights, a normalized cut with a $((4 + o(1)) \ln n)$-approximation ratio can be obtained in polynomial time.*

3.2   Speeding up the Algorithm

The *SphereGrowing* procedure uses Dijkstra's shortest path algorithm, which takes $O(n \log n + m)$ time [8]. Thus, solving the linear program (LP) dominates the total running time. (LP) can be transformed to a linear program with $O(mn^2)$ variables and $O(n^3)$ constraints [20]. Hence, in practice, this linear programming based approach is not efficient for large-size normalized cut problems.

464   X. Wu et al.

Note that (LP) is a dual of the maximum concurrent flow problem. Developing fast algorithms for the maximum concurrent flow problem which achieve provably close solutions to the optimal ones has attracted a lot of attention (e.g., [24, 19,12,18]). For the minimization (resp., maximization) versions of the problem, an ϵ-*approximate solution* for any error parameter $\epsilon > 0$ is one whose value is at most $(1 + \epsilon)$ (resp., at least $(1 - \epsilon)$) times the minimum (resp., maximum) possible value. Experimental results suggested that the concurrent flow based approximation techniques can yield significantly faster running time. Bienstock [6] reported speedups in the order of 20 times on obtaining approximate solutions by these methods versus the time for obtaining similar approximate solutions by standard linear programming techniques. The speedup over the running time for obtaining exact solutions is in the order of 1000.

We can apply Karakostas's maximum concurrent flow algorithm [18] to obtain an ϵ-approximate solution for the linear program (LP) efficiently, as stated in Lemma 8.

Lemma 8. *An ϵ-approximate solution for the linear program (LP) can be found in $\tilde{O}(\frac{1}{\epsilon^2}m^2)$ time.*

Thus, given any $\epsilon > 0$, we can first obtain a metric $d_\epsilon(\cdot)$ for (LP) with a solution value $\tau_\epsilon \leq (1 + \epsilon)\tau$. Applying our algorithm based on the metric $d_\epsilon(\cdot)$, we can then find a subset $S \subseteq V$ whose sparsity is

$$\beta(S) \leq (2+o(1))\ln n \cdot \tau_\epsilon \leq (2+o(1))\ln n \cdot (1+\epsilon) \cdot \tau \leq (2(1+\epsilon)+o(1)) \cdot \ln n \cdot \beta^*.$$

Based on Lemma 7, we obtain the following corollary.

Corollary 1. *Given any $\epsilon > 0$, a $((4(1+\epsilon)+o(1))\ln n)$-approximate normalized cut can be computed in $\tilde{O}(\frac{1}{\epsilon^2}m^2)$ time.*

4 Applications

In this section, we discuss an image segmentation problem arising in computer vision. Our pairwise data clustering algorithm can be applied to this problem to achieve a new approximation result.

Our basic scheme for solving this problem consists of three key steps: 1) build a weighted undirected graph G from the image; 2) use our normalized cut algorithm to partition G into two subgraphs G_1 and G_2; 3) recursively partition the subgraphs if necessary. Our algorithmic framework is likely to be applicable to other problems, such as gene clustering. We leave the details of applications in our full version of this paper.

Acknowledgments. The authors are very grateful to Dr. L. Fleischer, Operations Research, Carnegie Mellon University, Dr. G. Even, Department of Electrical Engineering, Tel Aviv University, Dr. É. Tardos, Department of Computer Science, Cornell University, and Dr. P. Klein, Department of Computer Science, Brown University, for their helpful discussions on the maximum concurrent flow problem.

References

1. P. Agarwal and C. Procopiuc, Exact and Approximation Algorithms for Clustering, *Proc. of ACM-SIAM SODA*, 1998.
2. V. Arya, N. Garg, R. Khandekar, V. Pandit, A. Meyerson, and K. Munagala, Local Search Heuristics for k-median and Facility Location Problems, *Proc. of ACM STOC*, 2001, 21–29.
3. J. Aslam, A. Leblanc, and C. Stein, A New Approach to Clustering, *Proc. of WAE*, 2000.
4. Y. Bartal, M. Charikar, and D. Raz, Approximating Min-Sum k-clustering in Metric Spaces, *Proc. of ACM STOC*, 2001, 11–22.
5. A. Ben-Dor and Z. Yakhini, Clustering Gene Expression Patterns, *Proc. of ACM RECOMB*, 1999, 33–42.
6. D. Bienstock, January 1999. Talk at Oberwolfach, Germany.
7. M. Charikar, C. Chekuri, T. Feder, and R. Motwani, Incremental Clustering and Dynamic Information Retrieval, *Proc. of ACM STOC*, 1997, 626–635.
8. T.H. Cormen, C. E. Leiserson, and R. L. Rivest, *Introduction to Algorithms*, McGraw-Hill, 1990.
9. P. Drineas, A. Frieze, R. Kannan, S. Vempala, and V. Vinay, Clustering in Large Graphs and Matrices, *Proc. of ACM-SIAM SODA*, 1999.
10. G. Even, J. Naor, S. Rao, and B. Schieber, Fast Approximate Graph Partitioning Algorithms, *SIAM J. Computing*, 28(1999), 2187–2214.
11. B. Everitt, *Cluster Analysis*, Oxford University Press, 1993.
12. N. Garg and J. Könemann, Faster and Simpler Algorithms for Multicommodity Flow and Other Fractional Packing Problems, *Proc. 39th IEEE FOCS*, 1998, 300–309.
13. N. Garg, V. V. Vazirani, and M. Yannakakis, Approximate Max-Flow Min-(Multi)Cut Theorems and Their Applications, *SIAM J. Computing*, 25(1996), 235–251.
14. S. Guattery and G. Miller, On the Performance of Spectral Graph Partitioning Methods, *Proc. of ACM-SIAM SODA*, 1995, 233–242.
15. S. Guha, N. Mishra, R. Motwani, and L. O'Callaghan, Clustering Data Streams, *Proc. of IEEE FOCS*, 2000.
16. T. Hofmann and J. Buhmann, Pairwise Data Clustering by Deterministic Annealing, *IEEE Trans. on Pattern Analysis and Machine Intelligence*, 19(1997), 1–14.
17. R. Kannan, S. Vempala, and A. Vetta, On Clusterings — Good, Bad and Spectral, *Proc. of IEEE FOCS*, 2000.
18. G. Karakostas, Faster Approximation Schemes for Fractional Multicommodity Flow Problems, *Proc. 13th ACM-SIAM SODA*, 2002, 166–173.
19. P. Klein, S. Plotkin, C. Stein, and É. Tardos, Faster Approximation Algorithms for the Unit Capacity Concurrent Flow Problem with Applications to Routing and Finding Sparse Cuts, *SIAM J. on Computing*, 23(1994), 466–487.
20. T. Leighton, F. Makedon, S. Plotkin, C. Stein, É. Tardos, and S. Tragoudas, Fast Approximation Algorithms for Multicommodity Flow Problems, *J. of Computer and System Sciences*, 50(1995), 228–243.
21. T. Leighton and S. Rao, Multicommodity Max-Flow Min-Cut Theorems and Their Use in Designing Approximation Algorithms, *J. of the ACM*, 46(1999), 787–832.
22. J. Matousek, On Approximate Geometric k-clustering, *Discrete and Computational Geometry*, 24(2000), 61–84.

23. B. Mirkin, *Mathematical Classification and Clustering*, Kluwer Academic Publishers, 1996.
24. F. Shahrokhi and D. Matula, The Maximum Concurrent Flow Problem. *J. of the ACM*, 37(1990), 318–334.
25. J. Shi and J. Malik, Normalized Cuts and Image Segmentation, *IEEE Trans. on Pattern Analysis and Machine Intelligence*, 22(8) (2000), 888–905.

Covering a Set of Points with a Minimum Number of Turns

Michael J. Collins

University of New Mexico and Sandia National Laboratories*
Albuquerque NM
mcollins@cs.unm.edu, mjcolli@sandia.gov

Abstract. Given a finite set of points in Euclidean space, we can ask what is the minimum number of times a piecewise-linear path must change direction in order to pass through all of them. We prove some new upper and lower bounds for a restricted version of this problem in which all motion is orthogonal to the coordinate axes.

1 Introduction

There are a variety of situations in which the cost associated with a path (taken, say, by a robot or by some mechanical device) should take into account the number of times this path changes direction. In a very general setting, we are given a finite set $\mathcal{C}$ of points in Euclidean space through which a path $\mathcal{P}$ must travel, and can ask what is the minimum total curvature of $\mathcal{P}$. Here we consider a combinatorial version of the problem, in which all motion of $\mathcal{P}$ is parallel to one of the coordinate axes and all points of $\mathcal{C}$ have integer coordinates. We then seek to minimize the number of ninety-degree turns or "corners" in $\mathcal{P}$.

Even when restricted to two dimensions, many variants of the minimum-turn problem are known to be NP-complete, although efficient constant-factor approximation algorithms exist [1,5]. The three-dimensional version, which is a special case of "discrete milling", has been less studied. In this paper we present optimal paths and lower bounds for various families of sets $\mathcal{C}$ in two and three dimensions.

We use the following notation. The set of integers from a to b inclusive is denoted $[a, b]$. We define the *link length* of a piecewise-linear path $\mathcal{P}$, denoted $s(\mathcal{P})$, as the number of line segments (called "links") which make up the path. A *spanning path* (also called a *covering tour*) of $\mathcal{C}$ is a path which passes through all points of $\mathcal{C}$. The minimum link length of all spanning paths of $\mathcal{C}$ is denoted $s(\mathcal{C})$ and is called the link length of $\mathcal{C}$. We can also think of this as one plus the minimum number of times we must change direction in order to move through all the given points. Note that we count a 180-degree turn as two turns; if a path $\mathcal{P}$ goes from (x, y, z) to (x, y, z') and then to (x, y, z''), we say that there is a

* Sandia is a multiprogram laboratory operated by Sandia Corporation, a Lockheed Martin Company, for the United States Department of Energy's National Nuclear Security Administration under contract DE-AC04-94AL85000.

T. Warnow and B. Zhu (Eds.): COCOON 2003, LNCS 2697, pp. 467–474, 2003.
© Springer-Verlag Berlin Heidelberg 2003

link of length zero in between[1], starting and ending at (x, y, z') and "moving" in the X or Y direction. We categorize links by the direction in which they move; a link from (x, y, z) to (x', y, z) is called an X-link, similarly there are Y- and Z-links. The *length* of a link ℓ is the number of points in C covered by ℓ and not covered by any previous link of $\mathcal{P}$; for our purposes this is a far more useful concept than the geometric length of the line segment. Given a subset of links A in $\mathcal{P}$, we define A^+ (A^-) to be the set of successors (predecessors) of the links in A. Note that we do not prohibit $\mathcal{P}$ from intersecting itself, nor do we prohibit it from touching some grid points not in C.

2   Three-Dimensional Grids

2.1   A Lower Bound for the Cube

We define $\mathcal{G}_{r,s,n}$ to be the three-dimensional regular grid with dimensions $r \times s \times n$, i.e. $\mathcal{G}_{r,s,n} = [1, r] \times [1, s] \times [1, n]$ or any translation thereof. We first consider the case of a cube, $r = s = n$. In spite of its geometric simplicity, and in spite of the fact that the analogous two-dimensional problem is rather easy, determining $s(\mathcal{G}_{n,n,n})$ remains an open problem. A trivial lower bound is n^2, since there are n^3 points to cover and each link has length at most n. To get a better lower bound, we must demonstrate that in any path some links must be shorter; indeed we could think of the problem as maximizing the average length of a link. Our fundamental tool is the following lemma:

Lemma 1. *Suppose that a spanning path $\mathcal{P}$ of $\mathcal{G}_{n,n,n}$ has t pairwise-disjoint sets of links $S_1, \ldots, S_t$ such that, for each i, S_i contains $\alpha_i n^2$ links and each such link has length at most $\beta_i n$. Then we have*

$$s(\mathcal{P}) \geq \left(1 + \sum_{i=1}^{t} \alpha_i(1 - \beta_i)\right)n^2 \ . \tag{1}$$

Proof. For each i, the links in set S_i can cover at most $\alpha_i \beta_i n^3$ points, so the $\sum \alpha_i n^2$ links in these sets cover a total of $\sum \alpha_i \beta_i n^3$ points. The remaining $(1 - \sum \alpha_i \beta_i)n^3$ points need at least $(1 - \sum \alpha_i \beta_i)n^2$ links, for a total of $(1 - \sum \alpha_i \beta_i + \sum \alpha_i)n^2 = (1 + \sum \alpha_i(1 - \beta_i))n^2$ links. □

To find such sets S_i, we make use of the observation that a link which covers points near the center of the cube must be followed by a link of bounded length. Let n be odd (the case of even n is similar) and translate $\mathcal{G}_{n,n,n}$ so that $(0, 0, 0)$ is at the center. Define a nested sequence of cubes Q_i for $1 \leq i \leq (n-1)/2$ by $Q_i = [-i, i] \times [-i, i] \times [-i, i]$. Now we can prove

Lemma 2. *In any spanning path of $\mathcal{G}_{n,n,n}$ there must be pairwise-disjoint sets of links S_i for $1 \leq i \leq (n-1)/2$ of size $8i$, such that each link of S_i intersects Q_i.*

---

[1] Assuming of course that z' is not between z and z''; otherwise we just have one link from (x, y, z) to (x, y, z'').

Proof. By induction. We include a set S_0 containing a single link through $(0,0,0)$. The claim is trivial for $i = 1$; at least 9 links are needed to cover a $3 \times 3 \times 3$ cube. So suppose we have sets $S_1, \cdots, S_{k-1}$; these sets contain

$$1 + \sum_{i=1}^{k-1} 8i = 4k^2 - 4k + 1 \tag{2}$$

links which together cover at most $(4k^2 - 4k + 1)(2k + 1)$ points of Q_k. But Q_k contains $(2k + 1)^3$ points, so it contains at least

$$(2k + 1)^3 - (4k^2 - 4k + 1)(2k + 1) = 8k(2k + 1) \tag{3}$$

points not covered by $\cup_{i=1}^{k-1} S_i$. Since no one link can cover more than $2k + 1$ points of Q_k, there must be at least $8k$ other links intersecting Q_k; these can be the members of S_k. □

Now each link of S_i^+ has length at most $(n + 1)/2 + i$. For suppose a link $\ell \in S_i$ goes from (x, y, z) to (x, y, z'). We must have $|x| \leq i, |y| \leq i$. Then say ℓ^+ goes from (x, y, z') to (x', y, z'). Its length is at most $|x' - x|$ which is at most $(n + 1)/2 + i$.

Applying lemma 1 directly to the sets S_i^+ gives $s(\mathcal{G}_{n,n,n}) \geq \frac{7}{6}n^2 - O(n)$ [2]. To obtain a stronger bound, note that each S_i has $8i$ predecessors as well as $8i$ successors, whose lengths are also bounded by $(n + 1)/2 + i$. Of course we cannot simply replace $\alpha_i n^2 = 8i$ by $\alpha_i n^2 = 16i$ in the application of lemma 1, because we might be counting some links twice; a link ℓ could be the successor of something in S_i and also the predecessor of something in S_j. But we observe that the length of such a link can be bound even more tightly: such a link has length at most $i + j + 1$. To prove this, first note that with no loss of generality we may assume $i \leq j$ and that ℓ starts at (x, y, z) with $\max(|x|, |y|) \leq i$ and moves in the X direction to (x', y, z). If $|x - x'| > i + j + 1$ then $|x'| > j$, but since ℓ^+ cannot move in the X direction it could not pass through Q_j.

So let S_i have σ_i successors that are *not* the predecessors of anything in $\cup_j S_j$, and similarly π_i predecessors that are not the successors of anything $\cup_j S_j$. Furthermore let q_{ij} be the number of links with predecessor in S_i and successor in S_j. Then we have

$$\sigma_i = 8i - \sum_{1 \leq j \leq (n-1)/2} q_{ij}$$
$$\pi_i = 8i - \sum_{1 \leq j \leq (n-1)/2} q_{ji} . \tag{4}$$

The sets counted by σ_i, π_i, and q_{ij} are pairwise disjoint, so we can apply lemma 1; $s(\mathcal{G}_{n,n,n})$ is at least

$$n^2 + \sum (\sigma_i + \pi_i)(1 - \frac{i + (n+1)/2}{n}) + \sum \sum q_{ij}(1 - \frac{i + j + 1}{n}) . \tag{5}$$

To obtain a lower bound on the link length, we seek the smallest possible value for (5). It looks like we have to optimize over all possible choices for q_{ij};

but if we use (4) to eliminate σ_i and π_i from (5), we find that all terms involving q_{ij} cancel, and (5) becomes

$$s(\mathcal{G}_{n,n,n}) \geq 3n^2 - 2\sum 8i\frac{i + (n+1)/2}{n} \ . \tag{6}$$

The sum in (6) is $\frac{5}{6}n^2 + O(n)$, thus we obtain

Theorem 1.

$$s(\mathcal{G}_{n,n,n}) \geq= \frac{4}{3}n^2 - O(n) \tag{7}$$

The best known upper bound (which is conjectured to be optimal) is $\frac{3}{2}n^2 + O(n)$ which comes from a spanning path described in [4] and illustrated in Fig. 1. We fix a direction (say the Z axis), and the path is formed by first spiraling around the outer part of the cube in each XY plane until a box of size $\frac{n}{2} \times \frac{n}{2} \times n$ remains in the center. These remaining points are then covered by moving up and down along the Z axis. To see the motivation for this, first note that in two dimensions we have $s(\mathcal{G}_{n,n}) = 2n-1$ [4]. But there are two entirely different ways to achieve the optimal value, as illustrated in Fig. 2: a spiral path in which the length of successive links gradually decreases from n to 1, and a back-and-forth path that alternates between long and short links. So the path of Fig. 1 spirals $n/4$ times in each XY plane, at which point the length of a link has decreased to $n/2$; then it switches to a back-and-forth motion in which the average length of a link is $n/2$. Therefore we call this path the "hybrid" path.

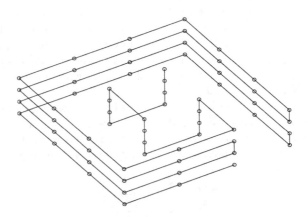

Fig. 1. Conjectured optimal path for $\mathcal{G}_{n,n,n}$

We note that theorem 1 is almost certainly not the best possible. We would not expect it is really possible for every link counted by q_{ij} to have length $i+j+1$; furthermore, it should be possible to choose sets S_i that have size greater than $8i$. We also note that nothing in our proof really hinges on the fact that we have one continuous path; we actually have a lower bound on covering $\mathcal{G}_{n,n,n}$

with a set of loops. In our definition of link-length we do not actually require that $\mathcal{P}$ must loop back and end at the same point where it began; but adding this requirement would increase the link-length by at most three, so it does not matter in this context.

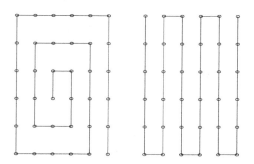

Fig. 2. Two different optimal paths

2.2   Optimal Paths for Non-cubic Grids

We now consider more general $\mathcal{G}_{r,s,n}$. Without loss of generality let $r \leq s \leq n$. It is clear that $s(\mathcal{G}_{r,s,n}) \leq 2rs - 1$, since we can cover all points by moving up and down along the Z axis. We might expect that this is optimal when n is sufficiently large, and this is indeed the case.

Theorem 2. *If $rs \leq n$, then*

$$s(\mathcal{G}_{r,s,n}) = 2rs - 1 \tag{8}$$

Proof. Suppose there exists a spanning path $\mathcal{P}$ with $s(\mathcal{P}) < 2rs - 1$. Then there must be at least one "column" $\{x\} \times \{y\} \times [1, n]$, none of whose points is covered by a Z-link. We need n links in the X and Y directions to cover these points; furthermore no such link could be a successor of another, so there must be at least another $n - 1$ links in $\mathcal{P}$. Therefore we must have $2n - 1 < 2rs - 1$, or simply $n < rs$.   □

Now consider $r = s = \gamma n$ for some fixed $\gamma < 1$. We can generalize theorem 1 to this case. First observe that, given sets S_i which satisfy the conditions of lemma 1, we have

$$s(\mathcal{G}_{\gamma n, \gamma n, n}) \geq \gamma^2 n^2 + n^2 (\sum_{i=1}^{t} \alpha_i (1 - \beta_i)) \ . \tag{9}$$

Now for $1 \leq i \leq \frac{\gamma n + 1}{2}$, we can define S_i precisely as before; the link length is bounded by something very similar to (5), the difference being that the first term is $\gamma^2 n^2$ and the sums run from 1 to $\frac{\gamma n+1}{2}$. We can eliminate the q_{ij} exactly as before to obtain

$$s(\mathcal{G}_{\gamma n, \gamma n, n}) \geq 3\gamma^2 n^2 - 2 \sum_{i=1}^{(\gamma n - 1)/2} 8i \frac{i + (n+1)/2}{n} \tag{10}$$

which gives the desired generalization of theorem 1:

Theorem 3.

$$s(\mathcal{G}_{\gamma n, \gamma n, n}) \geq (2\gamma^2 - \frac{2}{3}\gamma^3)n^2 - O(n) \tag{11}$$

For $\gamma > \frac{1}{2}$, we can apply the idea of the hybrid path, spiraling in the XY-plane until the length of each link is down to $\frac{n}{2}$. This means spiraling $\frac{2\gamma-1}{4}n$ times and yields an upper bound of $(2\gamma - \frac{1}{2})n^2$. We conjecture that these paths are optimal. Note that this conjecture would imply a much stronger result than theorem 2, namely that $s(\mathcal{G}_{r,s,n}) = 2rs - 1$ whenever $s \leq n/2$. This is because in order for the hybrid path to be optimal, its coverage of the central portion $\mathcal{G}_{n/2, n/2, n}$ must be optimal.

2.3   LP-Relaxation Bounds for Small Grids

We have used the AMPL modeling language [3] to develop a linear-programming relaxation of the minimum-turn problem on $\mathcal{G}_{r,s,n}$. There is a variable for each possible link, i.e. for each ordered pair of points differing in only one coordinate. The constraints require that each point be covered at least once and that the number of links entering a point equals the number of links leaving it. We did not include subtour-elimination constraints. For values of n up to 19 we have verified that the link-length of $\mathcal{G}_{n,n,n}$ is at least $\frac{3}{2}n^2$, as expected. At $n = 19$ this LP took about eighteen hours to solve with cplex on a SUN workstation.

3   Optimal Paths for Some Two-Dimensional Configurations

We now consider some two-dimensional configurations, letting $\mathcal{G}_n^m$ denote a rectangular grid $[1, m] \times [1, n]$. In this context we speak of "horizontal" and "vertical" links. Note that $s(\mathcal{G}_n^m) = 2\min(m, n) - 1$ is a special case of theorem 2.

We can generalize this to a disjoint union of two rectangles. Let $\mathcal{C}$ be $\mathcal{G}_1 \cup \mathcal{G}_2$ with $\mathcal{G}_i = \mathcal{G}_{h_i}^{v_i}$. Assume first that the projections of $\mathcal{G}_1, \mathcal{G}_2$ onto the axes are disjoint (say $\mathcal{G}_1 = [1, v_1] \times [1, h_1]$ and $\mathcal{G}_2 = [a, a+v_2] \times [b, b+h_2]$ with $a > v_1, b > h_1$). Note that any spanning path of $\mathcal{C}$ must contain either h_i horizontal links or v_i vertical links intersecting $\mathcal{G}_i$: without h_i such horizontal links, there must be a row containing no horizontal link and thus covered entirely by v_i vertical links. To find $s(\mathcal{C})$ we consider each case separately. Suppose for instance that h_1 horizontal links intersect $\mathcal{G}_1$ and v_2 vertical links intersect $\mathcal{G}_2$. Then we claim the path must have at least $2\max(h_1, v_2) + \min(h_1, v_2) - 1$ links in total. To see this, note that each link has two endpoints at which it is adjacent to other links; note further that each of the v_2 vertical links can be adjacent to at most one of the h_1 horizontal links and vice-versa. So say $h_1 \geq v_2$, then even if each of the v_2 vertical links is adjacent to one of the horizontal links, we still have at least $2h_1 - 2$ endpoints at which new links are needed beyond the ones already counted (the -2 is to account for the start and end points). This requires at least $h_1 - 1$ new links, giving the claimed lower bound. And this bound is in fact achievable as suggested by Fig. 3. So considering all possibilities we have the following theorem:

Theorem 4. *Given $\mathcal{C}$ as defined above,*

$$s(\mathcal{C}) = \min(\quad 2\min(h_1 + h_2, v_1 + v_2) \quad -1,$$
$$2\max(h_1, v_2) + \min(h_1, v_2) -1, \qquad (12)$$
$$2\max(h_2, v_1) + \min(h_2, v_1) -1) \ .$$

The first term corresponds to covering each $\mathcal{G}_i$ in the same direction — in effect covering the entire $(h_1 + h_2) \times (v_1 + v_2)$ rectangle that contains the two smaller rectangles.

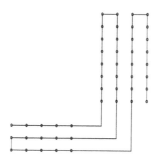

Fig. 3. Covering two disjoint rectangles

An almost identical result holds for two disjoint rectangles that do overlap in one dimension; let us say an overlap of width d along the x-axis, so $\mathcal{G}_1 = [1, v_1] \times [1, h_1]$ and $\mathcal{G}_2 = [v_1 - d + 1, v_1 + v_2 - d] \times [b, b + h_2]$ with $b > h_1$. As before suppose that h_1 horizontal links intersect $\mathcal{G}_1$ and v_2 vertical links intersect $\mathcal{G}_2$. The difference is that now it seems that some horizontal links could be adjacent to vertical links at both endpoints because of the overlap. But in fact we can disregard this case. Consider the subset of $\mathcal{G}_1$ that is not underneath $\mathcal{G}_2$, i.e. the part with x-coordinate less than $v_1 - d + 1$. If we have h_1 horizontal links intersecting this subset, then exactly as before we can say that a link in one direction can adjoin only one link in the other direction. Otherwise we have a row of the subset not touched by any horizontal link, which forces another $v_1 - d$ vertical links and we are back to the case of covering the bounding rectangle. The only change from the previous case is that the first term should be $2\min(h_1 + h_2, v_1 + v_2 - d) - 1$.

Let R_n be the triangular region $\{(x, y) \in \mathcal{G}_n^n : x + y \leq n + 1\}$. We have

Theorem 5. $s(R_n) = \frac{3}{2}n - 1$

Proof. This number of links is achievable by the L-shaped path indicated in Fig. 4. Now for a given spanning path $\mathcal{P}$, let r be the largest value such that the r longest rows of R all contain horizontal links. With no loss of generality we can assume $r \geq \frac{n}{2}$ (otherwise there would be a row of width $r > \frac{n}{2}$ covered entirely by vertical links, and by rotation we get the desired situation). This gives us r horizontal links and $n - r$ vertical links; they are situated so that a vertical link can be adjacent to a horizontal link only at one endpoint. Thus we have $2r$ endpoints at which further links must be added, requiring $r - 1$ further links for a total of $n + r - 1 \geq \frac{3}{2}n - 1$. □

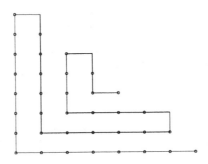

Fig. 4. Optimal path for a triangle

Similar arguments can be used to determine the exact link-length of other simple two-dimensional configurations.

References

1. E. Arkin, M. A. Bender, E. Demaine, S. P. Fekete, J. S. B. Mitchell, and S. Sethia, "Optimal Covering Tours with Turn Costs", *Proceedings of the 12th Annual ACM-SIAM Symposium on Discrete Algorithms (SODA)*, pp. 138–147, 2001
2. M. Collins and B. Moret, "Improved Lower Bounds for the Link Length of Rectilinear Spanning Paths in Grids", *Information Processing Letters* **68** (1998), pp. 317–319
3. R. Fourer, D. M. Gay, and B. W. Kernighan, *AMPL: A Modeling Language for Mathematical Programming*, Boyd and Fraser, Danvers, MA, 1993
4. E. Kranakis, D. Krizanc, and L. Meertens, "Link Length of Rectilinear Hamiltonian Tours on Grids", *Ars Combinatoria* **38** (1994), p. 177.
5. C. Stein and D. P. Wagner, "Approximation Algorithms for the Minimum Bends Traveling Salesman Problem", *Proceedings of IPCO 2001, LNCS 2081*, p. 406

Area-Efficient Order-Preserving Planar Straight-Line Drawings of Ordered Trees*

Ashim Garg and Adrian Rusu

Department of Computer Science and Engineering
University at Buffalo
Buffalo, NY 14260
{agarg,adirusu}@cse.buffalo.edu

Abstract. Ordered trees are generally drawn using order-preserving planar straight-line grid drawings. We therefore investigate the area-requirements of such drawings, and present several results: Let T be an ordered tree with n nodes. Then:

- T admits an order-preserving planar straight-line grid drawing with $O(n \log n)$ area.
- If T is a binary tree, then T admits an order-preserving planar straight-line grid drawing with $O(n \log \log n)$ area.
- If T is a binary tree, then T admits an order-preserving *upward* planar straight-line grid drawing with *optimal* $O(n \log n)$ area.

We also study the problem of drawing binary trees with user-specified arbitrary aspect ratios. We show that an ordered binary tree T with n nodes admits an order-preserving planar straight-line grid drawing Γ with width $O(A + \log n)$, height $O((n/A) \log A)$, and area $O((A + \log n)(n/A) \log A) = O(n \log n)$, where $2 \le A \le n$ is any user-specified number. Also note that all the drawings mentioned above can be constructed in $O(n)$ time.

1 Introduction

An *ordered tree* T is one with a prespecified counterclockwise ordering of the edges incident on each node. Ordered trees arise commonly in practice. Examples of ordered trees include binary search trees, arithmetic expression trees, BSP-trees, B-trees, and range-trees.

An *order-preserving drawing* of T is one in which the counterclockwise ordering of the edges incident on a node is the same as their prespecified ordering in T. A *planar drawing* of T is one with no edge-crossings. An *upward drawing* of T is one, where each node is placed either at the same y-coordinate as, or at a higher y-coordinate than the y-coordinates of its children. A *straight-line drawing* of T is one, where each edge is drawn as a single line-segment. A *grid drawing* of T is one, where each node is assigned integer x- and y-coordinates.

---

* Research supported by NSF CAREER Award IIS-9985136 and NSF CISE Research Infrastructure Award No. 0101244.

T. Warnow and B. Zhu (Eds.): COCOON 2003, LNCS 2697, pp. 475–486, 2003.
© Springer-Verlag Berlin Heidelberg 2003

Ordered trees are generally drawn using order-preserving planar straight-line grid drawings, as any undergraduate textbook on data-structures will show. An upward drawing is desirable because it makes it easier for the user to determine the parent-child relationships between the nodes.

We investigate the area-requirement of the order-preserving planar straight-line grid drawings of ordered trees, and present several results: Let T be an ordered tree with n nodes.

Result 1: We show that T admits an order-preserving planar straight-line grid drawing with $O(n \log n)$ area, $O(n)$ height, and $O(\log n)$ width, which can be constructed in $O(n)$ time.

Result 2: If T is a binary tree, then we show stronger results:

Result 2a: T admits an order-preserving planar straight-line grid drawing with $O(n \log \log n)$ area, $O((n/\log n) \log \log n)$ height, and $O(\log n)$ width, which can be constructed in $O(n)$ time.

Result 2b: T admits an order-preserving *upward* planar straight-line grid drawing with *optimal* $O(n \log n)$ area, $O(n)$ height, and $O(\log n)$ width, which can be constructed in $O(n)$ time.

An important issue is that of the *aspect ratio* of a drawing D. Let E be the smallest rectangle, with sides parallel to x and y-axis, respectively, enclosing D. The *aspect ratio* of D is defined as the ratio of the larger and smaller dimensions of E, i.e., if h and w are the height and width, respectively, of E, then the aspect ratio of D is equal to $\max\{h, w\}/\min\{h, w\}$. It is important to give the user control over the aspect ratio of a drawing because this will allow her to fit the drawing in an arbitrarily-shaped window defined by her application. It also allows the drawing to fit within display-surfaces with predefined aspect ratios, such as a computer-screen and a sheet of paper. We consider the problem of drawing binary trees with arbitrary aspect ratio, and prove the following result:

Result 3: Let T be a binary tree with n nodes. Let $2 \leq A \leq n$ be any user-specified number. T admits an order-preserving planar straight-line grid drawing Γ with width $O(A + \log n)$, height $O((n/A) \log A)$, and area $O((A + \log n)(n/A) \log A) = O(n \log n)$, which can be constructed in $O(n)$ time.

Also note that [3] shows an n-node binary tree that requires $\Omega(n)$ height and $\Omega(\log n)$ width in any order-preserving upward planar grid drawing. Hence, the $O(n)$ height and $O(\log n)$ width achieved by Result *2b* is optimal in the worst case.

Table 1 compares our results with the previously known results.

2   Definitions

We assume a 2-dimensional Cartesian space. We assume that this space is covered by an infinite rectangular grid, consisting of horizontal and vertical channels.

A *left-corner* drawing of an ordered tree T is one, where no node of T is to the left of, or above the root of T. The *mirror-image* of T is the ordered tree

Table 1. Bounds on the areas and aspect ratios of various kinds of planar straight-line grid drawings of an n-node tree. Here, α and ϵ are user-defined constants, such that $0 \leq \alpha < 1$ and $0 < \epsilon < 1$. $[a, b]$ denotes the range $a \ldots b$.

Tree Type	Drawing Type	Area	Aspect Ratio	Reference
Special Balanced Trees such as Red-black	Upward Order-preserving	$O(n(\log \log n)^2)$	$n/\log^2 n$	[6]
Binary	Upward	$O(n \log \log n)$	$(n \log \log n)/\log^2 n$	[6]
	Non-order-preserving	$O(n \log n)$	$[1, n/\log n]$	[1]
	Upward	$O(n^{1+\epsilon})$	$n^{1-\epsilon}$	[2]
	Order-preserving	$O(n \log n)$	$n/\log n$	this paper
	Non-upward Non-order-preserving	$O(n)$	$[1, n^{\alpha}]$	[4]
	Non-upward	$O(n^{1+\epsilon})$	$n^{1-\epsilon}$	[2]
	Order-preserving	$O(n \log n)$	$[1, n/\log n]$	this paper
		$O(n \log \log n)$	$(n \log \log n)/\log^2 n$	this paper
General	Non-upward	$O(n^{1+\epsilon})$	$n^{1-\epsilon}$	[2]
	Order-preserving	$O(n \log n)$	$n/\log n$	this paper

obtained by reversing the counterclockwise order of edges around each node. Let R be a rectangle with sides parallel to the x- and y-axis, respectively. The *height* (*width*) of R is equal to the number of grid-points with the same x-coordinate (y-coordinate) contained within R. The *area* of R is equal to the number of grid-points contained within R. The *enclosing rectangle* E of a drawing D is the smallest rectangle with sides parallel to the x- and y-axis covering the entire drawing. The *height* h, *width* w, and *area* of D is equal to the height, width, and area, respectively, of E. The *aspect ratio* of D is equal to $\max\{h, w\}/\min\{h, w\}$.

A *subtree* rooted at a node v of an ordered tree T is the maximal tree consisting of v and all its descendents. A *partial* tree of T is a connected subgraph of T. A *spine* of T is a path $v_0 v_1 v_2 \ldots v_m$, where $v_0, v_1, v_2, \ldots, v_m$ are nodes of T, that is defined recursively as follows (see Figure 1):

- v_0 is the same as the root of T;
- v_{i+1} is the child of v_i, such that the subtree rooted at v_{i+1} has the maximum number of nodes among all the subtrees that are rooted at the children of v_i.

3   Drawing Binary Trees

We now give our drawing algorithm for constructing order-preserving planar upward straight-line grid drawings of binary trees. In an ordered binary tree, each node has at most two children, called its *left* and *right* children, respectively.

Our drawing algorithm, which we call *Algorithm BT-Ordered-Draw*, uses the divide-and-conquer paradigm to draw an ordered binary tree T. In each recursive step, it breaks T into several subtrees, draws each subtree recursively, and then

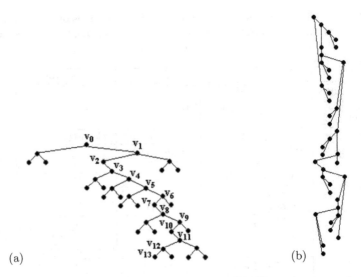

(a)   (b)

Fig. 1. (a) A binary tree T with spine $v_0v_1 \ldots v_{13}$. (b) The order-preserving planar upward straight-line grid drawing of T constructed by *Algorithm BT-Ordered-Draw*.

combines their drawings to obtain an upward left-corner drawing $D(T)$ of T. We now give the details of the actions performed by the algorithm to construct $D(T)$. Note that during its working, the algorithm will designate some nodes of T as either *left-knee, right-knee, ordinary-left, ordinary-right, switch-left* or *switch-right* nodes (for an example, see Figure 2):

1. Let $P = v_0v_1v_2 \ldots v_m$ be a spine of T. Define a *non-spine* node of T to be one that is not in P.
2. Designate v_0 as a *left-knee* node.
3. for $i = 0$ to m do (see Figure 2)
   Depending upon whether v_i is a *left-knee, right-knee, ordinary-left, ordinary-right, switch-left,* or *switch-right* node, do the following:
   a) v_i *is a left-knee node:* If v_{i+1} has a left child, and this child is not v_{i+2}, then designate v_{i+1} as a *switch-right node*, otherwise designate it as an *ordinary-left node*. Recursively construct an upward left-corner drawing of the subtree of T rooted at the non-spine child of v_i.
   b) v_i *is an ordinary-left node:* If v_{i+1} has a left child, and this child is not v_{i+2}, then designate v_{i+1} as a *switch-right node*, otherwise designate it as an *ordinary-left node*. Recursively construct an upward left-corner drawing of the subtree of T rooted at the non-spine child of v_i.
   c) v_i *is a switch-right node:* Designate v_{i+1} as a *right-knee* node. Recursively construct an upward left-corner drawing of the subtree of T rooted at the non-spine child of v_i.
   d) v_i *is a right-knee, ordinary-right, or switch-left node:* Do the same as in the cases, where v_i is a left-knee, ordinary-left, or switch-right node, respectively, with "left" exchanged with "right", and instead

of constructing an upward left-corner drawing of the subtree T_i of T rooted at the non-spine child of v_i, we recursively construct an upward left-corner drawing of the *mirror image* of T_i.

4. Let G be the drawing with the maximum width among the drawings constructed in Step 3. Let W be the width of G.

5. Place v_0 at the origin.

6. for $i = 0$ to m do (see Figures 2 and 3)

Let H_i be the horizontal channel corresponding to the node placed lowest in the drawing of T constructed so far.

Depending upon whether v_i is a *left-knee, right-knee, ordinary-left, ordinary-right, switch-left* or *switch-right* node, do the following:

a) v_i *is a left-knee node:* If v_{i+1} is the only child of v_i, then place v_{i+1} on the horizontal channel $H_i + 1$ and one unit to the right of v_i (see Figure 3(a)). Otherwise, let s be the child of v_i different from v_{i+1}. Let D be the drawing of the subtree rooted at s constructed in Step 3. If s is the right child of v_i, then place D such that its top boundary is at the horizontal channel $H_i + 1$ and its left boundary is one unit to the right of v_i; place v_{i+1} one unit below D and one unit to the right of v_i (see Figure 3(b)). If s is the left child of v_i, then place v_{i+1} one unit below and one unit to the right of v_i (see Figure 3(a)) (the placement of D will be handled by the algorithm when it will consider a switch-right node later on).

b) v_i *is an ordinary-left node:* Since v_i is an ordinary-left node, either v_{i+1} will be the only child of v_i, or v_i will have a right child s, where $s \neq v_{i+1}$. If v_{i+1} is the only child of v_i, then place v_{i+1} one unit below v_i in the same vertical channel as it (see Figure 3(c)). Otherwise, let s be the right child of v_i. Let D be the drawing of the subtree rooted at s constructed in Step 3. Place D one unit below and one unit to the right of v_i; place v_{i+1} on the same horizontal channel as the bottom of D and in the same vertical channel as v_i (see Figure 3(d)).

c) v_i *is a switch-right node:* Note that, since v_i is a switch-right node, it will have a left child s, where $s \neq v_{i+1}$. Let v_j be the left-knee node of P closest to v_i in the subpath $v_0v_1 \ldots v_i$ of P. v_j is called the *closest left-knee ancestor* of v_i. Place v_{i+1} one unit below and $W + 1$ units to the right of v_i.

Let D be the drawing of the subtree rooted at s constructed in Step 3. Place D one unit below v_i such that s is in the same vertical channel as v_i (see Figure 3(e)). If v_j has a left child s', which is different from v_{j+1}, then let D' be the drawing of the subtree rooted at s' constructed in Step 3. Place D' one unit below D such that s' is in the same vertical channel as v_i (see Figure 3(f)).

d) v_i *is a right-knee, ordinary-right, or switch-left node:* These cases are the same as the cases, where v_i is a left-knee, ordinary-left, or switch-right node, respectively, except that "left" is exchanged with "right", and the left-corner drawing of the mirror image of the subtree rooted

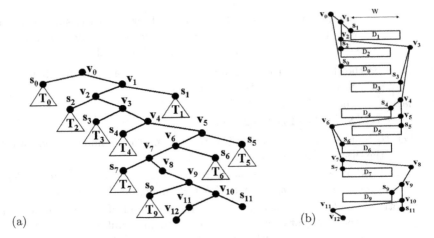

Fig. 2. (a) A binary tree T with spine $v_0v_1 \ldots v_{12}$. (b) A schematic diagram of the drawing $D(T)$ of T constructed by *Algorithm BT-Ordered-Draw*. Here, v_0 is a left-knee, v_1 is an ordinary-left, v_2 is a switch-right, v_3 is a right-knee, v_4 is an ordinary-right, v_5 is a switch-left, v_6 is a left-knee, v_7 is a switch-right, v_8 is a right-knee, v_9 is an ordinary-right, v_{10} is a switch-left, v_{11} is a left-knee, and v_{12} is an ordinary-left node. For simplicity, we have shown $D_0, D_1, \ldots, D_9$ with identically sized boxes but in actuality they may have different sizes.

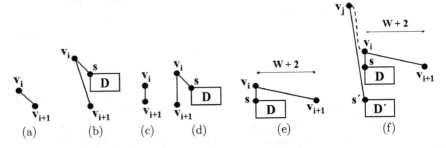

Fig. 3. (a,b) Placement of v_i, v_{i+1}, and D in the case when v_i is a left-knee node: (a) v_{i+1} is the only child of v_i or s is the left child of v_i, (b) s is the right child of v_i. (c,d) Placement of v_i, v_{i+1}, and D in the case when v_i is an ordinary-left node: (c) v_{i+1} is the only child of v_i, (d) s is the right child of v_i. (e,f) Placement of v_i, v_{i+1}, D, and D in the case when v_i is a switch-right node. (e) v_j does not have a left child, (f) v_j has a left child s'. Here, D' is the drawing of the subtree rooted at s'.

at the non-spine child of v_i, constructed in Step 3, is first flipped left-to-right and then is placed in $D(T)$.

To determine the area of $D(T)$, notice that the width of $D(T)$ is equal to $W+3$ (see the definition of W given in Step 3). From the definition of a spine, it follows easily that the number of nodes in each subtree rooted at a non-spine node of T is at most $n/2$, where n is the number of nodes in T. Hence, if we denote by $w(n)$, the width of $D(T)$, then, $W \leq w(n/2)$, and so, $w(n) \leq w(n/2)+3$. Hence,

$w(n) = O(\log n)$. The height of $D(T)$ is trivially at most n. Hence, the area of $D(T)$ is $O(n \log n)$. It is easy to see that the Algorithm can be implemented such that it runs in $O(n)$ time.

[3] has shown a lower bound of $\Omega(n \log n)$ for order-preserving planar upward straight-line grid drawings of binary trees. Hence, the upper bound of $O(n \log n)$ on the area of $D(T)$ is also optimal. We therefore get the following theorem:

Theorem 1. *A binary tree with n nodes admits an order-preserving upward planar straight-line grid drawing with height at most n, width $O(\log n)$, and optimal $O(n \log n)$ area, which can be constructed in $O(n)$ time.*

We can also construct a non-upward left-corner drawing $D'(T)$ of T, such that $D'(T)$ has height $O(\log n)$ and width at most n, by first constructing a left-corner drawing of the mirror image of T using Algorithm *BT-Ordered-Draw*, then rotating it clockwise by $90°$, and then flipping it right-to-left. This gives Corollary 1.

Corollary 1. *Using Algorithm BT-Ordered-Draw, we can construct in $O(n)$ time, a non-upward left-corner order-preserving planar straight-line grid drawing of an n-node binary with area $O(n \log n)$, height $O(\log n)$, and width at most n.*

4 Drawing General Trees

In a general tree, a node may have more than two children. We now briefly describe our algorithm for constructing a (non-upward) order-preserving planar straight-line grid drawing of a general tree T, which we call *Algorithm Ordered-Draw* (for details, see the full paper [5]). This algorithm is similar to the *Algorithm BT-Ordered-Draw* presented in Section 3. Like *Algorithm BT-Ordered-Draw*, it also constructs a left-corner drawing $D(T)$ of its input tree T recursively by splitting T into many subtrees $T_1, T_2, \ldots, T_k$, where each T_i is a tree rooted at a non-spine child of a spine node of T, recursively constructing a drawing of each T_i, and then stacking these drawings one-above-the-other to construct $D(T)$ (see Figure 4). $D(T)$ is constructed as follows: Let $v_0 v_1 v_2 \ldots v_m$ be a spine of T. Each spine node v_i of T is designated as a *left-knee, right-knee, switch-left*, or *switch-right* node as follows: v_0 is designated as a *left-knee* node. For each i, where $0 \le i \le v_{m-1}$, v_{i+1} is designated as a *switch-right (switch-left, left-knee, right-knee*, respectively) node if v_i is a *left-knee (right-knee, switch-left, switch-right*, respectively) node.

For each v_i, where $0 \le i \le m - 1$, it does the following: For simplicity, assume that v_i has at least one non-spine child that precedes v_{i+1}, and at least one non-spine child that follows v_{i+1} in the counterclockwise order of children of v_i (the algorithm can be easily modified to handle the cases when v_{i+1} is the first and/or the last child of v_i in this counterclockwise order). Let $c_1, c_2, \ldots, c_q, v_{i+1}, c_{q+1}, \ldots, c_p$ be the counterclockwise order of the children of v_i. c_p is called the *last* child of v_i.

If v_i is a *left-knee* or *switch-right* node (*right-knee* or *switch-left* node), then it recursively constructs a left-corner drawing D_j of the subtree T_j (image of the subtree T_j) rooted at each c_j (see Figure 4).

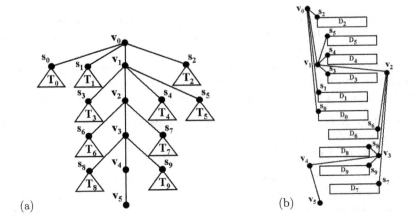

(a)   (b)

Fig. 4. (a) A tree T with spine $v_0v_1 \ldots v_5$. (b) A schematic diagram of the drawing $D(T)$ of T constructed by Algorithm *Ordered-Draw*. Here, v_0 and v_4 are left-knee nodes, v_1 and v_5 are switch-right nodes, v_2 is a right-knee node, and v_3 is a switch-left node. For simplicity, we have shown $D_0, D_1, \ldots, D_9$ with identically sized boxes but in actuality they may have different sizes.

Let W be the maximum width of the drawings of the subtrees rooted at all the non-spine children of all the spine nodes.

If v_i is a *switch-right* node (for example, node v_1 in Figure 4), then v_{i+1} is placed one unit below v_i and at horizontal distance $W+1$ from it. $D_1, D_2, \ldots, D_q$, $D_{q+1}, \ldots, D_p$ are placed one-above-the-other in that order, separated by unit vertical distance, such that their left boundaries are at unit horizontal distance from v_i to its right, and top boundary of D_q is at the same horizontal channel as v_{i+1}.

If v_i is a *right-knee* node (for example, node v_2 in Figure 4), $D_1, \ldots, D_q$ are placed one-below-the-other in that order, separated by unit vertical distance, such that their right boundaries are at unit horizontal distance from v_i to its left, and the top boundary of D_1 is one unit below the subdrawing of $D(T)$ consisting of the subspine $v_0v_1 \ldots v_{i-1}$ and the drawings of the subtrees rooted at the non-spine children of the spine nodes $v_0, v_1, \ldots, v_{i-1}$. v_{i+1} is placed one unit to the left of v_i and at the same horizontal channel as the bottom boundary of D_q. $D_{q+1}, \ldots, D_p$ are placed one-below-the-other in that order, separated by unit vertical distance, such that their right boundaries are at unit horizontal distance from v_i to its left, and the top boundary of D_{q+1} is one unit below the drawing of the image of the subtree rooted at the last child of v_{i+1}.

The placement of $D_1, D_2, \ldots, D_p$ when v_i is a *switch-left* or *left-knee* node is analogous to the case where v_i is a *switch-right* or *right-knee* node, respectively.

Just as for *Algorithm BT-Ordered-Draw*, we can show that the width $w(n)$ of $D(T)$ satisfies the recurrence: $w(n) \leq w(n/2) + 3$. Hence, $w(n) = O(\log n)$. The height of $D(T)$ is trivially at most n. Hence, the area of $D(T)$ is $O(n \log n)$.

Theorem 2. *A tree with n nodes admits an order-preserving planar straight-line grid drawing with $O(n \log n)$ area, $O(\log n)$ width, and height at most n, which can be constructed in $O(n)$ time.*

5 Drawing Binary Trees with Arbitrary Aspect Ratio

Let T be a binary tree. We show that, for any user-defined number A, where $2 \leq A \leq n$, we can construct an order-preserving planar straight-line grid drawing of T with $O((n/A) \log A)$ height and $O(A + \log n)$ width. Thus, by setting the value of A, user can control the aspect ratio of the drawing. This result also implies that we can construct such a drawing with area $O(n \log \log n)$ by setting $A = \log n$.

Our algorithm combines the approach of [1] for constructing non-upward non-order-preserving drawings of binary trees with arbitrary aspect ratio with our approach for constructing order-preserving drawings given in Section 3.

Lemma 1 (Generalization of Lemma 3 of [1]). *Suppose $A > 1$, and f is a function such that:*

- *if $n \leq A$, then $f(n) \leq 1$; and*
- *if $n > A$, then $f(n) \leq f(n^*) + f(n^+) + f(n'') + 1$ for some $n^*, n^+, n'' \leq n - A$ with $n^* + n^+ + n'' \leq n$.*

Then, $f(n) < 6n/A - 2$ for all $n > A$.

An order-preserving planar straight-line grid drawing of a binary tree T is called a *feasible drawing* if the root of T is placed on the left boundary and no node of T is placed between the root and the upper-left corner of the enclosing rectangle of the drawing. Note that a left-corner drawing is also a feasible drawing.

We now describe our algorithm, which we call *Algorithm BDAAR*, for drawing a binary tree T with arbitrary aspect ratio. Let m be the number of nodes in T. Let $2 \leq A \leq m$ be any number given as a parameter to *Algorithm BDAAR*.

Like Algorithm *BT-Ordered-Draw* of Section 3, *Algorithm BDAAR* is also a recursive algorithm. In each recursive step, it also constructs a feasible drawing of a subtree T' of T. If T' has at most A nodes in it, then it constructs a left-corner drawing of T' using Corollary 1 such that the drawing has width at most n and height $O(\log n)$, where n is the number of nodes in T'. Otherwise, i.e., if T' has more than A nodes in it, then it constructs a feasible drawing of T' as follows:

1. Let $P = v_0 v_1 v_2 \ldots v_q$ be a spine of T'.
2. Let n_i be the number of nodes in the subtree of T' rooted at v_i. Let v_k be the vertex of P with the smallest value for k such that $n_k > n - A$ and $n_{k+1} \leq n - A$ (since T' has more than A nodes in it and $n_0, n_1, \ldots, n_q$ is a strictly decreasing sequence of numbers, such a k exists).
3. for each i, where $0 \leq i \leq k - 1$, denote by T_i, the subtree rooted at the non-spine child of v_i (if v_i does not have any non-spine child, then T_i is the empty tree, i.e., the tree with no nodes in it). Denote by T^* and T^+, the subtrees rooted at the non-spine children of v_k and v_{k+1}, respectively, denote by T'', the subtree rooted at v_{k+1}, and denote by T''', the subtree rooted at v_{k+2} (if v_k and v_{k+1} do not have non-spine children, and $k + 1 = q$, then T^*, T^+, and T''' are empty trees). For simplicity, in the rest of the algorithm, we assume that T^*, T^+, T''', and each T_i are non-empty. (The algorithm can be easily modified to handle the cases, when T^*, T^+, T''', or some T_i's are empty).

4. Place v_0 at origin.
5. We have two cases:
   - $k = 0$: Recursively construct a feasible drawing D^* of T^*. Recursively construct a feasible drawing D^+ of the mirror image of T^+. Recursively construct a feasible drawing D''' of the mirror image of T'''. Let s_0 be the root of T^* and s_1 be the root of T^+.

     T' is drawn as shown in Figure 5(a,b,c,d). If s_0 is the left child of v_0, then place D^* one unit below v_0 with its left boundary aligned with v_0 (see Figure 5(a,c)). If s_0 is the right child of v_0, then place D^* one unit above and one unit to the right of v_0 (see Figure 5(b,d)). Let W^*, W^+, and W''' be the widths of D^*, D^+, and D''', respectively. v_1 is placed in the same horizontal channel as v_0 to its right at distance $\max\{W^*+1, W^++1, W'''-1\}$ from it. Let B_0 and C_0 be the lowest and highest horizontal channels, respectively, occupied by the subdrawing consisting of v_0 and D^*. If s_1 is the left child of v_1, then flip D^+ left-to-right and place it one unit below B_0 and one unit to the left of v_1 (see Figure 5(a,b)). If s_1 is the right child of v_1, then flip D^+ left-to-right, and place it one unit above C_0 and one unit to the left of v_1 (see Figure 5(c,d)). Let B_1 be the lowest horizontal channel occupied by the subdrawing consisting of v_0, D^*, v_1 and D^+. Flip D''' left-to-right and place it one unit below B_1 such that its right boundary is aligned with v_1 (see Figure 5(a,b,c,d)).
   - $k > 0$: For each T_i, where $0 \le i \le k-1$, construct a left-corner drawing D_i of T_i using Corollary 1.

     Recursively construct feasible drawings D^* and D'' of the mirror images of T^* and T'', respectively.

     T' is drawn as shown in Figure 6(a,b,c,d). If T_0 is rooted at the left child of v_0, then D_0 is placed one unit below and with the left boundary aligned with v_0. If T_0 is rooted at the right child of v_0, then D_0 is placed one unit above and one unit to the right of v_0. Each D_i and v_i, where $1 \le i \le k-1$, are placed such that:
     - v_i is in the same horizontal channel as v_{i-1}, and is one unit to the right of D_{i-1}, and
     - if T_i is rooted at the left child of v_i, then D_i is placed one unit below v_i with its left boundary aligned with v_i, otherwise (i.e., if T_i is rooted at the right child of v_i) D_i is placed one unit above and one unit to the right of v_i.

     Let B_{k-1} and C_{k-1} be the lowest and highest horizontal channels, respectively, occupied by the subdrawing consisting of $v_0, v_1, v_2, \ldots, v_{k-1}$ and $D_0, D_1, D_2, \ldots, D_{k-1}$. Let d be the horizontal distance between v_0 and the right boundary of the subdrawing consisting of $v_0, v_1, v_2, \ldots, v_{k-1}$ and $D_0, D_1, D_2, \ldots, D_{k-1}$. Let W^* and W'' be the widths of D^* and D'', respectively.

     v_k is placed to the right of v_{k-1} in the same horizontal channel as it, such that the horizontal distance between v_k and v_0 is equal to $\max\{W''-1, W^*+1, d+1\}$. If T^* is rooted at the left-child of v_k, then D^* is flipped left-to-right and placed one unit below B_{k-1} and one unit left of v_k (see

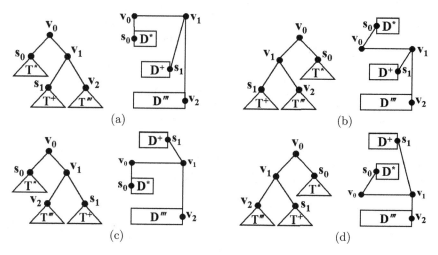

Fig. 5. Case $k = 0$: (a) s_0 is the left child of v_0 and s_1 is the left child of v_1. (b) s_0 is the right child of v_0 and s_1 is the left child of v_1. (c) s_0 is the left child of v_0 and s_1 is the right child of v_1. (d) s_0 is the right child of v_0 and s_1 is the right child of v_1.

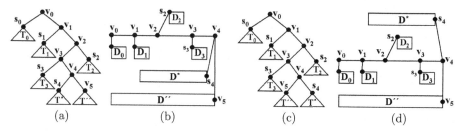

Fig. 6. Case $k > 0$: Here $k = 4$, s_0, s_1, and s_3 are the left children of v_0, v_1, and v_3 respectively, s_2 is the right child of v_2, T_0, T_1, T_2, and T_3 are the subtrees rooted at v_0, v_1, v_2, and v_3 respectively. Let s_4 be the root of T^*. (a) s_4 is left child of v_4. (b) s_4 is the right child of v_4.

Figure 6(a,b)). If T^* is rooted at the right-child of v_k, then D^* is flipped left-to-right and placed one unit above C_{k-1} and one unit to the left of v_k (see Figure 6(c,d)) . Let B_k be the lowest horizontal channel occupied by the subdrawing consisting of $v_1, v_2, \ldots, v_k$, and $D_1, D_2, \ldots, D_{k-1}, D^*$. D'' is flipped left-to-right and placed one unit below B_k, such that its right boundary is aligned with v_k (see Figure 6(b,d)).

Let m_i be the number of nodes in T_i, where $0 \le i \le k - 1$. From Corollary 1, the height of each D_i is $O(\log m_i)$ and width at most m_i. Total number of nodes in the partial tree consisting of $T_0, T_1, \ldots, T_{k-1}$ and $v_0, v_1, \ldots, v_{k-1}$ is at most $A - 1$. Hence, the height of the subdrawing consisting of $D_0, D_1, \ldots, D_{k-1}$ and $v_0, v_1, \ldots, v_{k-1}$ is $O(\log A)$ and width is at most $A - 1$ (see Figure 6).

Suppose T', T^*, T^+, T'', and T'''' have n, n^*, n^+, n'', and n''' nodes, respectively. If we denote by $H(n)$ and $W(n)$, the height and width of the drawing of T' constructed by Algorithm $BDAAR$, then:

$$H(n) = H(n^*) + H(n^+) + H(n''') + 1 \quad \text{if } n > A \text{ and } k = 0$$
$$= H(n^*) + H(n^+) + H(n''') + O(\log A)$$
$$H(n) = H(n^*) + H(n'') + O(\log A) \quad \text{if } n > A \text{ and } k > 0$$
$$H(n) = O(\log A) \quad \text{if } n \le A$$

Since $n^*, n^+, , n'', n''' \le n - A$, from Lemma 1, it follows that $H(n) = O(\log A)(6n/A - 2) = O((n/A)\log A)$. Also we have that:

$$W(n) = \max\{W(n^*) + 2, W(n^+) + 2, W(n''')\} \quad \text{if } n > A \text{ and } k = 0$$
$$W(n) = \max\{A, W(n^*) + 2, W(n'')\} \quad \text{if } n > A \text{ and } k > 0$$
$$W(n) \le A \quad \text{if } n \le A$$

Since, $n^*, n^+, n^* \le n/2$, and $n'', n''' \le n - A < n - 1$, we get that $W(n) \le \max\{A, W(n/2) + 2, W(n - 1)\}$. Therefore, $W(n) = O(A + \log n)$. We therefore get the following theorem:

Theorem 3. *Let T be a binary tree with n nodes. Let $2 \le A \le n$ be any number. T admits an order-preserving planar straight-line grid drawing with width $O(A + \log n)$, height $O((n/A)\log A)$, and area $O((A + \log n)(n/A)\log A) = O(n\log n)$, which can be constructed in $O(n)$ time.*

Setting $A = \log n$, we get that:

Corollary 2. *An n-node binary tree admits an order-preserving planar straight-line grid drawing with area $O(n\log\log n)$, which can be constructed in $O(n)$ time.*

References

1. T. Chan, M. Goodrich, S. R. Kosaraju, and R. Tamassia. Optimizing area and aspect ratio in straight-line orthogonal tree drawings. *Comput. Geom. Theory Appl.*, 23:153–162, 2002.
2. T. M. Chan. A near-linear area bound for drawing binary trees. In *Proc. 10th ACM-SIAM Symposium on Discrete Algorithms (SODA)*, pages 161–168, 1999.
3. A. Garg, M. T. Goodrich, and R. Tamassia. Planar upward tree drawings with optimal area. *Internat. J. Comput. Geom. Appl.*, 6:333–356, 1996.
4. A. Garg and A. Rusu. Straight-line drawings of binary trees with linear area and arbitrary aspect ratio. In *Graph Drawing (GD'02)*, volume 2528 of *Lecture Notes in Computer Science*, pages 320–331. Springer-Verlag, 2002.
5. A. Garg and A. Rusu. Area-efficient order-preserving planar straight-line drawings of ordered trees. Technical Report 2003-05, Department of Computer Science and Engineering, University at Buffalo, Buffalo, NY, 2003.
6. C.-S. Shin, S. Kim, S.-H. Kim, and K.-Y. Chwa. Area-efficient algorithms for straight-line tree drawings. *Comput. Geom. Theory Appl.*, 15:175–202, 2000.

Bounds for Convex Crossing Numbers

Farhad Shahrokhi[1]*, Ondrej Sýkora[2]**, Laszlo A. Székely[3]***,
and Imrich Vrt'o[4]†

[1] Department of Computer Science, University of North Texas
P.O Box 13886, Denton, TX, 76203-3886, USA
[2] Department of Computer Science, Loughborough University
Loughborough, Leicestershire LE11 3TU, The United Kingdom
[3] Department of Mathematics, University of South Carolina
Columbia, SC 29208, USA
[4] Department of Informatics, Institute of Mathematics
Slovak Academy of Sciences
Dúbravská 9, 841 04 Bratislava, Slovak Republic

Abstract. A convex drawing of an n-vertex graph $G = (V, E)$ is a drawing in which the vertices are placed on the corners of a convex $n-$gon in the plane and each edge is drawn using one straight line segment. We derive a general lower bound on the number of crossings in any convex drawings of G, using isoperimetric properties of G. The result implies that convex drawings for many graphs, including the planar 2-dimensional grid on n vertices have at least $\Omega(n \log n)$ crossings. Moreover, for any given arbitrary drawing of G with c crossings in the plane, we construct a convex drawing with at most $O((c + \sum_{v \in V} d_v^2) \log n)$ crossings, where d_v is the degree of v.

1 Introduction

Throughout this paper $G = (V, E)$ denotes an n-vertex graph. Let d_v be the degree of any $v \in V$. A *drawing* of $G = (V, E)$ is a one-to-one placement of the vertices into the plane and representation of edges pq with continuous curves, whose endpoints are the points corresponding to p and q, and do not pass through any other vertex point. A *crossing* is a common interior point for two edges of G. We will consider only the drawings in which any two edges have at most one common point, edges with a common endpoint do not cross, and no three edges cross at the same point. Let $\mathrm{cr}(G)$ denote the *crossing number* of G, i.e. the minimum number of crossings of its edges over all possible drawings of G in the plane. Although the concept of crossing numbers has played a crucial role in

---

* This research was supported by the NSF grant CCR9988525.
** This research was supported by the EPSRC grant GR/R37395/01.
*** This author was visiting the National Center for Biotechnology Information, NLM, NIH, with the support of the Oak Ridge Institute for Science and Education. This research was supported by the NSF contract 007 2187.
† This research was supported by the VEGA grant No. 02/3164/23.

T. Warnow and B. Zhu (Eds.): COCOON 2003, LNCS 2697, pp. 487–495, 2003.
© Springer-Verlag Berlin Heidelberg 2003

settling many problems in combinatorial and computational geometry [10,12,4], many interesting problems involving crossing numbers themselves, remain unresolved or even untouched. An important application area of crossing numbers is graph drawing as number of crossings in visualized graphs influence aesthetics and readability of graphs [5,13]. A *rectilinear drawing* of G is a drawing in which each edge is drawn using a single straight line segment. A *convex drawing* of G is a rectilinear drawing in which the vertices are placed in the corners of a convex n-gon, see Fig.1.

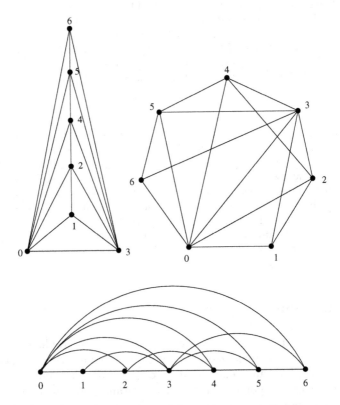

Fig. 1. A graph and its convex and one-page drawings

Let $\overline{cr}(G)$ and $cr^*(G)$ denote the rectilinear crossing number and the convex crossing numbers of G, respectively. Convex crossing numbers or outerplanar crossing numbers were first introduced by Kainen [8] in connection with the book thickness problem. Clearly $cr(G) \leq \overline{cr}(G) \leq cr^*(G)$. In particular, it is well known that $cr(K_8) = 18 < \overline{cr}(K_8) = 19$. In terms of the k-page crossing number ν_k [15,14], it is obvious that $cr^*(G) = \nu_1(G)$ for every graph G.

A main result in this paper is a general lower bound on the convex crossing number. One can observe using Euler's formula that, $cr^*(G) \geq m - 2n$. Consequently, $\mathrm{cr}^*(G) \geq \frac{1}{27} \cdot \frac{m^3}{n^2}$, for $m \geq 3n$, using standard methods such as those in [14]. Moreover, it is well known that $\mathrm{cr}(G) = \Omega(B^2(G) - \sum_{v \in V} d_v^2)$, and hence $\mathrm{cr}^*(G) = \Omega(B^2(G) - \sum_{v \in V} d_v^2)$, where $B(G)$ is the size of $(1/3, 2/3)$ edge separator in G. Our lower bound, presented here, involves isoperimetric properties of G, and is much stronger than the above lower bound in certain cases. Using our lower bound we exhibit classes of graphs for which $\mathrm{cr}^*(G) = \Omega\big((\overline{\mathrm{cr}}(G) + \sum_{v \in V} d_v^2) \log n\big)$. A surprising consequence is that any convex drawing of an n-vertex 2-dimensional grid (which is a planar graph), has $\Omega(n \log n)$ crossings in any convex drawing.

We also derive a general upper bound on $\mathrm{cr}^*(G)$. In particular, given any arbitrary drawing of G with c crossings in the plane, we construct a convex drawing with $O((c + \sum_{v \in V} d_v^2) \log n)$ crossings. Moreover, if the original drawing is represented as a planar graph, where crossings are represented as vertices with degree four, then our construction only takes $O((c+n) \log n)$ time to determine the order in which the vertices of G appear on the convex n-gon. Previously Bienstock and Dean [2] had proved that $\overline{\mathrm{cr}}(G) = O(\Delta \cdot \mathrm{cr}^2(G))$, where Δ denotes the maximum degree of G. We improved their result in [15] by showing that $\mathrm{cr}^*(G) = O((\mathrm{cr}(G) + \sum_{v \in V} d_v^2) \log^2 n)$, where d_v denotes the degree of $v \in V$. Very recently, Even et al. [7], proved that for every degree bounded graph $\mathrm{cr}^*(G) = O((\mathrm{cr}(G) + n) \log n))$. Our new upper bound extends the construction in [7] from degree bounded graphs to arbitrary graphs, and improves our previous bound in [15] by a $\log n$ factor. The upper bound is tight, within a constant multiplicative factor, for many interesting graphs including grids, and hence it can not be improved in general. Our upper bound implies that if $m \geq 4n$, and $\Delta = O((\frac{m}{n})^2)$, then, $\mathrm{cr}^*(G) = O(\mathrm{cr}(G) \log n)$, and therefore, $\overline{\mathrm{cr}}(G) = O(\mathrm{cr}(G) \log n)$. Thus, when G is "semi-regular" and not too sparse, $\mathrm{cr}^*(G)$ is a good approximation for both $\overline{\mathrm{cr}}(G)$ and $\mathrm{cr}(G)$.

2 Lower Bound

We say that G satisfies an $f(x)$-edge isoperimetric inequality if for any k-vertex subset U of V, and $k \leq n/2$, there are at least $f(k)$ edges between U and $V - U$. Define the difference function of f, denoted by Δ_f as

$$\Delta_f(i) = f(i+1) - f(i)$$

for any $i = 0, 1, ..., n - 1$. Next, we derive a general lower bound for the number of crossings in convex drawings of G.

Theorem 1. *Assume that $G = (V, E)$ satisfies an $f(x)$-edge isoperimetric inequality so that Δ_f is positive and decreasing till $n/2$.*

$$\mathrm{cr}^*(G) \geq \frac{n}{8} \sum_{j=1}^{\lfloor \frac{n}{2} \rfloor - 1} f(j)(\Delta_f(j) - \Delta_f(j+1)) - \frac{1}{2}|E(G)|f(3) - \sum_{v \in V} d_v^2.$$

Proof. We prove the theorem for odd n. We extend f by $f(x) = 0$ for $x < 0$. Let D be a convex drawing of G. Without loss of generality we may assume that vertices in D are placed on the perimeter of the unit circle in equidistant positions. Label the vertices by $0, 1, 2, ..., n - 1$. For simplicity we will often identify a vertex with its corresponding integer and all computation will be taken modulo n. Add new edges between i and $i+1$ for $i = 0, 1, ..., n-1$ in D if they are not already there, and note that they produce no crossings. For $u, v \in V$ define the distance between them, denoted by $l(u, v)$, to be $\min\{|u - v|, n - |u - v|\}$, see Fig.2.

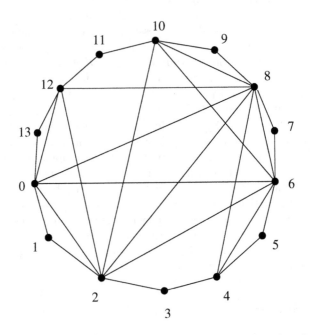

Fig. 2. A convex drawing of a 14-vertex graph. E.g. length of the edge 0,8 is 6.

For any $uv \in E$, let $c(u, v)$ denote the number of crossings of the edge uv with other edges in D, and observe that $c(u, v) \geq f(l(u, v) - 1) - d_u - d_v$. We conclude that

$$c(D) = \frac{1}{2} \sum_{uv \in E} c(u, v) \geq \frac{1}{2} \sum_{uv \in E} (f(l(u, v) - 1) - d_u - d_v)$$

$$= \frac{1}{2} \sum_{uv \in E} f(l(u, v) - 1) - \frac{1}{2} \sum_{v \in V} d_v^2. \tag{1}$$

We say that edge $uv \in E$ in the drawing D covers a vertex i if the unique shortest path between u and v (using only the edges on the boundary of the convex n−gon) contains i. (Note that when uv covers i, we may have $i = u$ or $i = v$.) For any edge $e = uv$ and any vertex i define $load_{u,v}(i)$, (see Fig.3) as

$$load_{u,v}(i) = \begin{cases} \Delta_f \left(\left\lceil \frac{|u-v|}{2} \right\rceil - |i - u| \right) & \text{if } e \text{ covers } i \text{ and} |i - u| < |v - i|, \\ \Delta_f \left(\left\lceil \frac{|u-v|}{2} \right\rceil - |v - i| \right) & \text{if } e \text{ covers } i \text{ and } |i - u| \geq |v - i|, \\ 0 & \text{otherwise.} \end{cases}$$

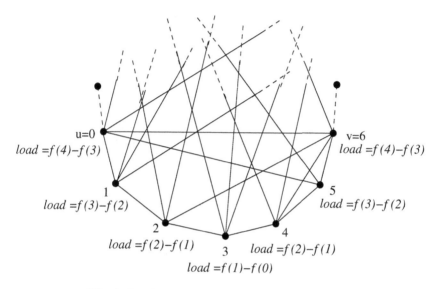

Fig. 3. Loads on the nodes covered by the edge uv

It is easy to see that for any $uv \in E$,

$$\sum_{i \in V} load_{u,v}(i) \leq 2 \sum_{j=0}^{\left\lceil \frac{l(u,v)}{2} \right\rceil} \Delta_f(j) \leq 2f\left(\left\lceil \frac{l(u,v)}{2} \right\rceil + 1 \right). \qquad (2)$$

Note that for $x \geq 4$, $\lfloor \frac{x}{2} \rfloor + 1 \leq x - 1$. We conclude using the above inequality, (2), and (1) that

$$\frac{1}{4} \sum_{uv \in E} \sum_{i \in V} load_{u,v}(i) \leq \frac{1}{2} \sum_{uv \in E} f\left(\left\lceil \frac{l(u,v)}{2} \right\rceil + 1 \right)$$

$$\leq \frac{1}{2}|E(G)|f(3) + \frac{1}{2} \sum_{uv \in E} f(l(u,v) - 1)$$

$$\leq \frac{1}{2}|E(G)|f(3) + c(D) + \sum_{v \in V} d_v^2. \qquad (3)$$

and hence will try to bound from below the sum involving loads.

Let $i \in V$. For $0 \leq s < n/2$, define $E_{i,s}$ to be the set of all edge $uv \in E$ covering vertex i in D such that either $l(i,u) \leq s$ or $l(i,v) \leq s$. Note that for any $i \in V$, and any $uv \in E_{i,s}$, $load_{u,v}(i) \geq \Delta_f(s)$. Let n_s denote $\sum_{i \in V} |E_{i,s}|$. For any $s < n/2$, we have

$$\sum_{i \in V} \sum_{uv \in E_{i,s}} load_{u,v}(i) = \sum_{i \in V} \sum_{j=0}^{s} \sum_{uv \in E_{i,j} - E_{i,j-1}} load_{u,v}(i)$$

$$= \sum_{i \in V} \sum_{uv \in E_{i,0}} load_{u,v}(i) + \sum_{j=1}^{s} \sum_{i \in V} \sum_{uv \in E_{i,j} - E_{i,j-1}} load_{u,v}(i)$$

$$\geq n_0 \Delta_f(0) + \sum_{j=1}^{s} (n_j - n_{j-1}) \Delta_f(j),$$

where the last inequality is obtained by observing that the number of terms in the sum $\sum_{i \in V} \sum_{uv \in E_{i,j} - E_{i,j-1}} load_{u,v}(i)$, is $n_j - n_{j-1}$, and each term is at least $\Delta_f(j)$. It follows that

$$\sum_{i \in V} \sum_{uv \in E_{i,s}} load_{u,v}(i) \geq \sum_{j=1}^{s-1} n_j (\Delta_f(j) - \Delta_f(j+1)),$$

since n_j is a nondecreasing function of j. We also have for all $j < n/2$,

$$n_j \geq \frac{1}{2} n f(j).$$

To see this consider any j consecutive integers $i, i+1, ..., i+j-1$. Then at least $f(j)$ edges leave this j-set, and those edges must cover either i or $i+j-1$. We conclude that

$$\sum_{i \in V} \sum_{uv \in E} load_{u,v}(i) = \sum_{i \in V} \sum_{uv \in E_{i, \lfloor \frac{n}{2} \rfloor}} load_{u,v}(i) \geq \frac{n}{2} \sum_{j=1}^{\lfloor \frac{n}{2} \rfloor - 1} f(j)(\Delta_f(j) - \Delta_f(j+1)).$$

This finishes the proof, using (3).   □

Corollary 1. Let G be an $N \times N$ grid. Then for sufficiently large $n = N^2$

$$cr^*(G) = \Omega(n \log n).$$

Proof. According to Bollobás and Leader [3], $f(x) = \sqrt{2x}$, for $x \leq n/2$, for the $N \times N$ grid. Theorem (1) and some standard calculations give the result.   □

3   Upper Bound

Theorem 2. *If G is drawn in the plane with c crossings, then a convex drawing of G with $O\left((c + \sum_{v \in V} d_v^2) \log n\right)$ crossings can be constructed. Moreover, if the original drawing is properly represented as a plane embedding of a planar graph, then the order in which the vertices of G appear in the in the convex drawing can be determined in $O\left((c + n) \log n\right)$ time.*

Proof. Consider any drawing of G in the plane with c crossings and let the set of crossings be denoted by C. Construct a planar graph denoted by $\hat{G}$, on the vertex set $\hat{V} = V \cup C$ by inserting vertices of degree 4 at the crossings.

Recall Fig.1 that in a *one-page drawing* of $\hat{G}$ all vertices are placed on a straight line l and any edge is drawn using a half circle above the line [15]. The crucial part of the proof is to construct a one-page drawing of $\hat{G}$ with $O((c + \sum_{v \in \hat{V}} d_v^2) \log n)$ crossings. We then modify this drawing to obtain a convex drawing of G by placing the vertices in V on the corners of a convex n-gon in the plane in the order that they appear on the straight line l, and then drawing each edge $ab \in E$ by one straight line segment between a and b. It is easy to see that the number of crossing in this drawing is the same as it was in the one-page drawing.

To obtain the desired drawing of $\hat{G}$ we will first construct a partition tree T [15] of $\hat{G}$. The root of T corresponds to $\hat{G}$, and any non-leaf node in T corresponds to a subgraph of $\hat{G}$ with at least 2 vertices. To describe the tree, it is sufficient to indicate how to construct the left and right children of $\hat{G}$, denoted by $\hat{G}_1$ and $\hat{G}_2$, respectively; the procedure recursively extends itself to the entire tree.

Assign a weight of $w(v) = \frac{d_v^2}{\sum_{y \in \hat{V}} d_y^2}$, to any vertex v of $\hat{V}$, where d_v is the degree of v in $\hat{G}$. Recall a well known theorem of Gazit and Miller [6] that any $\hat{G}$ has a $(1/3, 2/3)$ edge separator of size at most

$$1.6 \sqrt{\sum_{v \in \hat{V}} d_v^2},$$

if for all v, $w(v) \leq 2/3$.

• Case 1. Assume that $w(v) \leq 2/3$ for any vertex $v \in \hat{V}$. Apply the theorem cited above to find an $(1/3, 2/3)$ edge separator of size at most

$$1.6 \sqrt{\sum_{v \in \hat{V}} d_v^2}.$$

Now define $\hat{G}_1$ and $\hat{G}_2$ to be the two components of $\hat{G}$ on the vertex sets $\hat{V}_1$ and $\hat{V}_2$, respectively, that are the obtained by the removal of the $(1/3,2/3)$ separator.

• Case 2. Assume that there is a vertex v in $\hat{G}$ with $w(v) \geq 2/3$. Define $\hat{G}_1$ and $\hat{G}_2$ to be the components of $\hat{G}$ on the vertex sets $\hat{V}_1$ and $\hat{V}_2$, respectively which are obtained by removing all edges incident to v. Thus $\hat{V}_1 = v$ and $\hat{V}_2 = \hat{V} - \{v\}$.

A one-page drawing of $\hat{G}$ is obtained by placing a one-page drawing of $\hat{G}_1$ to the left of a one-page drawing of $\hat{G}_2$, and then drawing the removed edges (between) $\hat{G}_1$ and $\hat{G}_2$ as half circles between the corresponding vertices. Let $b(\hat{G})$ denote the number of edges that have one end point in $\hat{G}_1$ and the other end point in $\hat{G}_2$. Similarly, define $b(\hat{G}_i)$, $i = 1, 2$. It follows from cases 1 and 2 and the recursive definition that

$$b(\hat{G}) \leq 1.6\sqrt{\sum_{v \in \hat{V}} d_v^2},$$

and

$$b(\hat{G}_i) \leq 1.6\sqrt{\sum_{v \in \hat{V}_i} d_{i,v}^2},$$

where $d_{i,v}$ denotes the degree of $v \in \hat{V}_i$ in $\hat{G}_i$, $i = 1, 2$. It follows that

$$b(\hat{G}_i) \leq 1.6\sqrt{\frac{2\sum_{v \in \hat{V}} d_v^2}{3}},$$

$i = 1, 2$. Let $S(\hat{G})$ denote the maximum number of edges that go above any vertex in the obtained one-page drawing of $\hat{G}$. Similarly, define $S(\hat{G}_i)$, $i = 1, 2$. Also note that,

$$S(\hat{G}) \leq b(\hat{G}) + \max\{S(\hat{G}_1), S(\hat{G}_2)\},$$

and therefore

$$S(\hat{G}) = O\left(\sqrt{\sum_{v \in \hat{V}} d_v^2}\right).$$

Now let $c(\hat{G})$ and $c(\hat{G}_i)$ denote the number of crossings in the one-page drawing for $\hat{G}$, and for $\hat{G}_i$, $i = 1, 2$, respectively. Observe that

$$c(\hat{G}) \leq c(\hat{G}_1) + c(\hat{G}_2) + 2b(\hat{G})S(\hat{G}),$$

and thus

$$c(\hat{G}) \leq c(\hat{G}_1) + c(\hat{G}_2) + O(\sum_{v \in \hat{V}} d_v^2).$$

This implies the claimed upper bound, since the depth of the partition tree is logarithmic in $\sum_v d_v^2$, and hence in n, and the sum of the square of degrees is superadditive over the subgraphs. To finish the proof assume that the planar graph $\hat{G}$ is given. Then, the claim regarding the time complexity follows from the fact that computing the edge separators in [6] can be done in the linear time for any planar graph, and hence the partition tree T can be constructed in $O((c + n)\log n)$ time.   □

References

1. S. Bhatt, and F.T. Leighton, A framework for solving VLSI layout problems, *J. Comput. System Sci.*, **28** (1984), 300–331.
2. D. Bienstock, and N. Dean, New results on the rectilinear crossing number and plane embedding, *J. Graph Theory*, **16** (1992), 389–398.
3. B. Bollobás, and I. Leader, Edge-isoperimetric inequalities in the grid, *Combinatorica*, **11** (1991), 299–314.
4. T. K. Dey, Improved bounds for planar *k*-sets and related problems, *Discrete and Computational Geometry*, **19** (1998), 373–382.
5. J. Di Battista, P. Eades, R. Tamassia, and I. G. Tollis, Graph Drawing. Algorithms for the Visualization of Graphs, Prentice Hall, 1999, 432 pp.
6. H. Gazit, and G. Miller, Planar separators and Euclidean norm, *Algorithms, Proc. Int. Symp. SIGAL '90*, LNCS 450, 1990, 338–347.
7. G. Even, S. Guha, and B. Schieber, Improved approximations of crossings in graph drawings and VLSI layout areas, *STOC*, 2000, 296–305. (Full version to appear in SICOMP.)
8. P. C. Kainen, The book thickness of a graph II, *Congressus Numerantium*, **71** (1990), 121–132.
9. F. T. Leighton, *Complexity Issues in VLSI*, MIT Press, 1983.
10. L. A. Székely, Crossing number problems and hard Erdős problems in discrete geometry, *Combinatorics, Probability, and Computing*, **6** (1998), 353–358.
11. J. Pach, and P. K. Agarwal, *Combinatorial Geometry*, Wiley & Sons, NY, 1995.
12. J. Pach, J. Spencer, and G. Tóth, New bounds for crossing numbers, *Discrete and Computational Geometry*, **24** (2000), 623–644.
13. H. Purchase, Which aesthetic has the greatest effect on human understanding?, in: *Proc. Symposium on Graph Drawing, GD'97*, Lecture Notes in Computer Science **1353** (Springer, 1997), 248–261.
14. F. Shahrokhi, O. Sýkora, L. A. Székely, and I. Vrt'o, Crossing numbers: bounds and applications, in: *Intuitive Geometry*, Bolyai Society Mathematical Studies **6**, (I. Bárány and K. Böröczky, eds.), Akadémia Kiadó, Budapest, 1997, 179–206.
15. F. Shahrokhi, O. Sýkora, L. A. Székely, and I. Vrt'o, The book crossing number of graphs, *J. Graph Theory*, **21** (1996), 413–424.

On Spectral Graph Drawing

Yehuda Koren

Dept. of Computer Science and Applied Mathematics
The Weizmann Institute of Science, Rehovot, Israel
yehuda@wisdom.weizmann.ac.il

Abstract. The spectral approach for graph visualization computes the layout of a graph using certain eigenvectors of related matrices. Some important advantages of this approach are an ability to compute optimal layouts (according to specific requirements) and a very rapid computation time. In this paper we explore spectral visualization techniques and study their properties. We present a novel view of the spectral approach, which provides a direct link between eigenvectors and the aesthetic properties of the layout. In addition, we present a new formulation of the spectral drawing method with some aesthetic advantages. This formulation is accompanied by an aesthetically-motivated algorithm, which is much easier to understand and to implement than the standard numerical algorithms for computing eigenvectors.

1 Introduction

A graph $G(V, E)$ is an abstract structure that is used to model a relation E over a set V of entities. Graph drawing is a standard means for visualizing relational information, and its ultimate usefulness depends on the readability of the resulting layout, that is, the drawing algorithm's ability to convey the meaning of the diagram quickly and clearly. To date, many approaches to graph drawing have been developed [4,8]. There are many kinds of graph-drawing problems, such as drawing di-graphs, drawing planar graphs and others. Here we investigate the problem of drawing undirected graphs with straight-line edges. In fact, the methods that we utilize are not limited to traditional graph drawing and are also intended for general low dimensional visualization of a set of objects according to their pair-wise similarities (see, e.g., Fig. 2).

We have focused on spectral graph drawing methods, which construct the layout using eigenvectors of certain matrices associated with the graph. To get some feeling, we provide results for three graphs in Fig. 1. This spectral approach is quite old, originating with the work of Hall [6] in 1970. However, since then it has not been used much. In fact, spectral graph drawing algorithms are almost absent in the graph-drawing literature (e.g., they are not mentioned in the two books [4,8] that deal with graph drawing). It seems that in most visualization research the spectral approach is difficult to grasp in terms of aesthetics. Moreover, the numerical algorithms for computing the eigenvectors do not possess an intuitive aesthetic interpretation.

We believe that the spectral approach has two distinct advantages that make it very attractive. First, it provides us with an exact solution to the layout problem, whereas almost all other formulations result in an NP-hard problem, which can only be approximated. The second advantage is computation speed. Spectral drawings can be computed

T. Warnow and B. Zhu (Eds.): COCOON 2003, LNCS 2697, pp. 496–508, 2003.
© Springer-Verlag Berlin Heidelberg 2003

extremely fast as we have shown in [9]. This is very important because the amount of information to be visualized is constantly growing exponentially.

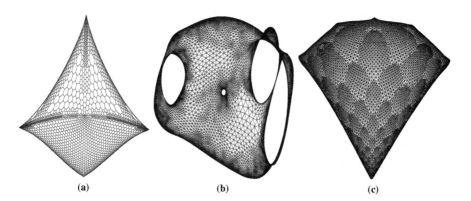

Fig. 1. Drawings obtained from the Laplacian eigenvectors. (a) The 4970 graph. $|V| = 4970$, $|E| = 7400$. (b) The 4elt graph. $|V| = 15606$, $|E| = 45878$. (c) The Crack graph. $|V| = 10240$, $|E| = 30380$.

Spectral methods have become standard techniques in algebraic graph theory; see, e.g., [3]. The most widely used techniques utilize eigenvalues and eigenvectors of the adjacency matrix of the graph. More recently, the interest has shifted somewhat to the spectrum of the closely related Laplacian. In fact, Mohar [11] claims that the Laplacian spectrum is more fundamental than this of the adjacency matrix.

Related areas where the spectral approach has been popularized include clustering [13], partitioning [12], and ordering [7]. However, these areas use discrete quantizations of the eigenvectors, unlike graph drawing, which employs the eigenvectors without any modification. Regarding this aspect, it is more fundamental to explore properties of graph-related eigenvectors in the framework of graph drawing.

In this paper we explore the properties of spectral visualization techniques, and provide different explanations for their ability to draw graphs nicely. Moreover, we have modified the usual spectral approach. The new approach uses what we will call *degree-normalized eigenvectors*, which have aesthetic advantages in certain cases. We provide an aesthetically-motivated algorithm for computing the degree-normalized eigenvectors. Our hope is that this will eliminate the vagueness of spectral methods and will contribute to their recognition as an important tool in the field of graph-drawing and information-visualization.

2 Basic Notions

A graph is usually written $G(V, E)$, where $V = \{1 \ldots n\}$ is the set of n nodes, and E is the set of edges. Each edge $\langle i, j \rangle$ is associated with a non-negative weight w_{ij} that reflects the similarity of nodes i and j. Thus, more similar nodes are connected with "heavier" edges. Henceforth, we will assume $w_{ij} = 0$ for any non-adjacent pair of nodes. Let us denote the neighborhood of i by $N(i) = \{j \mid \langle i, j \rangle \in E\}$. The degree of node i is $\deg(i) \stackrel{\text{def}}{=} \sum_{j \in N(i)} w_{ij}$. Throughout the paper we have assumed, without loss of generality, that G is connected, otherwise the problem we deal with can be solved independently for each connected component.

The *adjacency-matrix* of the graph G is the symmetric $n \times n$ matrix A^G, where

$$A^G_{ij} = \begin{cases} 0 & i = j \\ w_{ij} & i \neq j \end{cases} \qquad i, j = 1, \dots, n.$$

We will often omit the G in A^G.

The *Laplacian* is another symmetric $n \times n$ matrix associated with the graph, denoted by L^G, where

$$L^G_{ij} = \begin{cases} \deg(i) & i = j \\ -w_{ij} & i \neq j \end{cases} \qquad i, j = 1, \dots, n.$$

Again, we will often omit the G in L^G. The Laplacian is positive semi-definite, and its only zero eigenvalue is associated with the eigenvector $1_n \stackrel{\text{def}}{=} (1, 1, \dots, 1)^T \in \mathbb{R}^n$. The usefulness of the Laplacian stems from the fact that the quadratic form associated with it is just a weighted sum of all pairwise squared distances:

Lemma 1. *Let L be an $n \times n$ Laplacian, and let $x \in \mathbb{R}^n$. Then*

$$x^T L x = \sum_{i<j} w_{ij}(x_i - x_j)^2.$$

The proof of this lemma is direct.

Throughout the paper we will use the convention $0 = \lambda_1 < \lambda_2 \leq \dots \leq \lambda_n$ for the eigenvalues of L, and denote the corresponding real orthonormal eigenvectors by $v_1 = (1/\sqrt{n}) \cdot 1_n, v_2, \dots, v_n$.

Let us define the *degrees matrix* as the $n \times n$ diagonal matrix D that satisfies $D_{ii} = \deg(i)$. Given a degrees matrix, D, and a Laplacian, L, then a vector u and a scalar μ are termed *generalized eigen-pairs* of (L, D) if $Lu = \mu Du$. Our convention is to denote the generalized eigenvectors of (L, D) by $\alpha \cdot 1_n = u_1, u_2, \dots, u_n$, with corresponding generalized eigenvalues $0 = \mu_1 < \mu_2 \leq \dots \leq \mu_n$. (Thus, $Lu_i = \mu_i Du_i$, $i = 1, \dots, n$.) To uniquely define $u_1, u_2, \dots, u_n$, we require them to be D-normalized: so $u_i^T D u_i = 1$, $i = 1, \dots n$. We term these generalized eigenvectors *the degree normalized eigenvectors*. It can be shown (see Appendix A) that all the generalized eigenvalues are real non-negative, and that all the degree normalized eigenvectors are D-orthogonal, i.e. $u_i^T D u_j = 0$, $\forall i \neq j$.

3 Spectral Graph Drawing

The earliest spectral graph-drawing algorithm was that of Hall [6]; it uses the low eigenvectors of the Laplacian. Henceforth, we will refer to this method as *the eigen-projection method*. A few other researchers utilize the top eigenvectors of the adjacency matrix instead of those of the Laplacian. E.g., the work of [10], which uses the adjacency matrix eigenvectors to draw molecular graphs. Recently, eigenvectors of a modified Laplacian were used in [1] for the visualization of bibliographic networks.

In fact, for regular graphs of uniform degree deg, the eigenvectors of the Laplacian equal those of the adjacency matrix, but in a reversed order, because $L = deg \cdot I - A$, and adding the identity matrix does not change eigenvectors. However, for non-regular graphs, use of the Laplacian is based on a more solid theoretical basis, and in practice also gives nicer results than those obtained by the adjacency matrix. Hence, we will focus on visualization using eigenvectors of the Laplacian.

3.1 Derivation of the Eigen-Projection Method

We will introduce the eigenprojection method as a solution to a minimization problem. We begin by deriving a 1-D drawing, and then we show how to draw in more dimensions.

Given a weighted graph $G(V, E)$, we denote its 1-D layout by $x \in \mathbb{R}^n$, where $x(i)$ is the location of node i. We take x as the solution of the following constrained minimization problem

$$\min_x \ E(x) \stackrel{\text{def}}{=} \sum_{\langle i,j \rangle \in E} w_{ij}(x(i) - x(j))^2 \qquad (1)$$

$$\text{given: } \text{Var}(x) = 1,$$

where $\text{Var}(x)$ is the *variance* of x, defined as usual by $\text{Var}(x) = \frac{1}{n}\sum_{i=1}^n (x(i) - \bar{x})^2$, and where $\bar{x}$ is the mean of x.

The energy to be minimized, $E(x)$, strives to make edge lengths short. Since the sum is weighted by edge-weights, "heavy" edges have a stronger impact and hence will be typically shorter. The constraint $\text{Var}(x) = 1$ requires that the nodes be scattered in the drawing area, and prevents an overcrowding of the nodes at the same point. Note that the choice of variance 1 is arbitrary, and simply states the scale of the drawing. We could equally have chosen a constraint of the form $\text{Var}(x) = c$. In this way, if x_0 is the optimal solution of variance 1, then $\sqrt{c} \cdot x_0$ is the optimal solution of variance c. Such a representation of the problem reminds the *force-directed graph drawing approach* (see [4,8]), where the energy to be minimized replaces the "attractive forces", and the variance constraint takes the role of the "repulsive forces".

The energy and the constraint are invariant under translation (ensure that for every α: $E(x) = E(x + \alpha \cdot 1_n)$, $\text{Var}(x) = \text{Var}(x + \alpha \cdot 1_n)$). We eliminate this degree of freedom by requiring that the mean of x is 0, i.e. $\sum_{i=1}^n x(i) = x^T \cdot 1_n = 0$. This is very convenient since now the variance can be written in a simple form: $\text{Var}(x) = \frac{1}{n}x^T x$. To simplify the notation we will change the scale, and require the variance to be $\frac{1}{n}$, which is equivalent to $x^T x = \sum_{i=1}^n x(i)^2 = 1$.

Using Lemma 1, we can write the energy in a matrix form: $E(x) = x^T L x = \sum_{\langle i,j \rangle \in E} w_{ij} \cdot (x(i) - x(j))^2$. Now the desired 1-D layout, x, can be described as the solution of the constrained minimization problem

$$\min_x x^T L x \qquad (2)$$

$$\text{given: } x^T x = 1$$

$$\text{in the subspace: } x^T \cdot 1_n = 0.$$

Let us substitute $B = I$ in Claim A (in Appendix A), to obtain the optimal solution $x = v_2$, the second smallest eigenvector of L.

To achieve a 2-D drawing, we need to compute an additional vector of coordinates, y. Our requirements for y are the same as those that we required from x, but in addition there must be no correlation between y and x, so that the additional dimension will provide us with as much new information as possible[1]. Since x and y are centered, we simply have to require that $y^T \cdot x = y^T \cdot v_2 = 0$. Hence y is the solution of the constrained minimization problem

$$\min_y y^T L y \tag{3}$$

$$\text{given: } y^T y = 1$$

$$\text{in the subspace: } y^T \cdot 1_n = 0, y^T \cdot v_2 = 0.$$

Again, use Claim A so that the optimal solution is $y = v_3$, the third smallest eigenvector of L. In order to obtain a k-D drawing of the graph, we take the first coordinate of the nodes to be v_2, the second coordinate to be v_3, and in general, we define the i-th coordinate of the nodes by v_{i+1}.

4   Drawing Using Degree-Normalized Eigenvectors

In this section we introduce a new spectral graph drawing method that associates the coordinates with some generalized eigenvectors of the Laplacian.

Suppose that we weight nodes by their degrees, so the mass of node i is its degree — $\deg(i)$. Now if we take the original constrained minimization problem (2) and weight sums according to node masses, we get the following degree-weighted constrained minimization problem (where D is the degrees matrix)

$$\min_x x^T L x \tag{4}$$

$$\text{given: } x^T D x = 1$$

$$\text{in the subspace: } x^T D 1_n = 0.$$

Substitute $B = D$ in Claim A to obtain the optimal solution $x = u_2$, the second smallest generalized eigenvector of (L, D). Using the same reasoning as in Subsection 3.1, we obtain a k-D drawing of the graph, by taking the first coordinate of the nodes to be u_2, the second coordinate to be u_3, and in general, we define the i-th coordinate of the nodes by u_{i+1}.

We will show by several means that using these degree-normalized eigenvectors is more natural than using the eigenvectors of the Laplacian. In fact Shi and Malik [13] have already shown that the degree-normalized eigenvectors are more suitable for the problem of image segmentation. For the visualization task, the motivation and explanation are very different.

---

[1] The strategy to require no correlation between the axes is used in other visualization techniques like Principal Components Analysis [15] and Classical Multidimensional Scaling [15].

In order to gain some intuition on (4), we shall rewrite it in the equivalent form:

$$\min_x \frac{x^T L x}{x^T D x} \tag{5}$$

in the subspace: $x^T D 1_n = 0$.

It is straightforward to show that a solution of (4) is also a solution of (5).

In problem (5) the denominator moderates the behavior of the numerator, as we are showing now. The numerator strives to place those nodes with high degrees at the center of the drawing, so that they are in proximity to the other nodes. On the other hand, the denominator also emphasizes those nodes with high degrees, but in the reversed way: it strives to enlarge their scatter. The combimation of these two opposing goals, helps in making the drawing more balanced, preventing a situation in which nodes with lower degrees are overly separated from the rest nodes.

Another observation is that degree-normalized eigenvectors unify the two common spectral techniques: the approach that uses the Laplacian and the approach that uses the adjacency matrix. To see this, use the fact that $L = D - A$ and write $Lu = \mu Du$ as $(D - A)u = \mu Du$. By changing sides, we get $Au = (1 - \mu)Du$. Thus, the generalized eigenvectors of (L, D) are also the generalized eigenvectors of (A, D), with a reversed order. In this way, when drawing with degree normalized eigenvectors, we can take either the low generalized eigenvectors of the Laplacian, or the top generalized eigenvectors of the adjacency matrix, without affecting the result.

The degree-normalized eigenvectors are also the (non-generalized) eigenvectors of the matrix $D^{-1}A$. This can be obtained by left-multiplying the generalized eigen-equation $Ax = \mu Dx$ by D^{-1}, obtaining the eigen-equation

$$D^{-1}Ax = \mu x. \tag{6}$$

Note that $D^{-1}A$ is known as the *transition matrix* of a random walk on the graph G. Hence, the degree-normalized eigen-projection uses the top eigenvectors of the transition matrix to draw the graph.

Regarding drawing quality, for graphs that are close to being regular, we have observed not much difference between drawing using eigenvectors and drawing using degree-normalized eigenvectors. However, when there are marked deviations in node degrees, the results are quite different. This can be directly seen by posing the problem as in (5). Here, we provide an alternative explanation based on (6). Consider the two edges e_1 and e_2. Edge e_1 is of weight 1, connecting two nodes, each of which is of degree 10. Edge e_2 is of weight 10, connecting two nodes, each of which is of degree 100. In the Laplacian matrix, the entries corresponding to e_2 are 10 times larger than those corresponding to e_1. Hence we expect the drawing obtained by the eigenvectors of the Laplacian, to make the edge e_2 much shorter than e_1 (here, we do not consider the effect of other nodes that may change the lengths of both edges). However, for the transition matrix in (6), the entries corresponding to these two edges are the same, hence we treat them similarly and expect to get the same length for both edges. This reflects the fact that the *relative* importance of these two edges is the same, i.e. $\frac{1}{10}$.

In many kinds of graphs numerous scales are embedded, which indicates the existence of dense clusters and sparse clusters. In a traditional eigen-projection drawing, dense clusters are drawn extremely densely, while the whole area of the drawing is used to represent sparse clusters or outliers. This is the best way to minimize the weighted sum of square edge lengths, while scattering the nodes as demanded. A better drawing would allocate each cluster an adequate area. Frequently, this is the case with the degree normalized eigenvectors that adjust the edge weights in order to reflect their relative importance in the related local scale.

For example, consider Fig. 2, where we visualize 300 odors as measured by an electronic nose. Computation of the similarities between the odors is given in [2]. The odors are known to be classified into 30 groups, which determine the color of each odor in the figure. Figure 2(a) shows the visualization of the odors by the eigenvectors of the Laplacian. As can be seen, each of the axes shows one outlying odor, and places all the other odors about at the same location. However, the odors are nicely visualized using the degree normalized eigenvectors, as shown in Fig. 2(b).

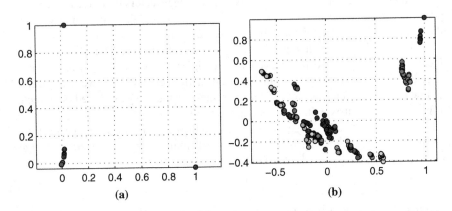

Fig. 2. Visualization of 300 odor patterns as measured by an electronic nose. (a) A drawing using the eigenvectors of the Laplacian. (b) A drawing using the degree-normalized eigenvectors.

5   An Optimization Process

An attractive feature of the degree-normalized eigenvectors is that they can be computed by an intuitive algorithm, which is directly related to their aesthetic properties. This is unlike the (non generalized) eigenvectors, which are computed using methods that are difficult to interpret in aesthetic terms. Now we will derive the algorithm.

Differentiating $E(x)$ with respect to $x(i)$ gives $\frac{\partial E}{\partial x(i)} = 2\sum_{j \in N(i)} w_{ij}(x(i) - x(j))$. Equating this to zero and isolating $x(i)$ we get

$$x(i) = \frac{\sum_{j \in N(i)} w_{ij}x(j)}{\deg(i)} .$$

Hence, when allowing only node i to move, the location of i that minimizes $E(x)$ is the weighted centroid of i's neighbors.

This induces an optimization process that iteratively puts each node at the weighted centroid of its neighbors (simultaneously for all nodes). The aesthetic reasoning is clear. A rather impressive fact is that when initialized with a vector D-orthogonal to 1_n, this algorithm converges *in the direction* of a non-degenerate degree-normalized eigenvector of L. More precisely, it converges either in the direction of u_2 or that of u_n.

We can prove this surprising fact by observing that the action of putting each node at the weighted centroid of its neighbors is equivalent to multiplication by the transition matrix $— D^{-1}A$. Thus, the process we have described can be expressed in a compact form as the sequence

$$\begin{cases} x_0 & = \text{random vector, s.t. } x_0^T D1_n = 0 \\ x_{i+1} & = D^{-1}Ax_i \ . \end{cases}$$

This process is known as the Power-Iteration [5]. In general, it computes the "dominant" eigenvector of $D^{-1}A$, which is the one associated with the largest-in-magnitude eigenvalue. In our case, all the eigenvectors are D-orthogonal to the "dominant" eigenvector $— 1_n$, and also the initial vector, x_0, is D-orthogonal to 1_n. Thus, the series converges in the direction of the next dominant eigenvector, which is either u_2, which has the largest positive eigenvalue, or u_n, which possibly has the largest negative eigenvalue. (We assume that x_0 is not D-orthogonal to u_2 or to u_n, which is nearly always true for a randomly chosen x_0)

In practice, we want to ensure convergence to u_2 (avoiding convergence to u_n). We use the fact that all the eigenvalues of the transition matrix are in the range $[-1, 1]$. Now it is possible to shift the eigenvalues by adding the value 1 to each of them, so that they are all positive, thus preventing convergence to an eigenvector with a large negative eigenvalue. This is done by working on the matrix $I + D^{-1}A$ instead of the matrix $D^{-1}A$. In this way the eigenvalues are in the range $[0, 2]$, while eigenvectors are not changed. In fact, it would be more intuitive to scale the eigenvalues to the range $[0, 1]$, so we will actually work with the matrix $\frac{1}{2}(I + D^{-1}A)$. If we use our initial "intuitive" notions, this means a more careful process. In each iteration, we put each node at the average between its old place and the centroid of its neighbors. Thus, each node absorbs its new location not only from its neighbors, but also from its current location.

The full algorithm for computing a k-D drawing is given in Fig. 3. To compute a degree-normalized eigenvector u_j, we will use the principles of the power-iteration and the D-orthogonality of the eigenvectors. Briefly, we pick some random x, such that x is D-orthogonal to $u_1, \ldots, u_{j-1}$, i.e. $x^T Du_1 = 0, \ldots, x^T Du_{j-1} = 0$. Then, if $x^T Du_j \neq 0$, it can be proved that the series $\frac{1}{2}(I + D^{-1}A)x, (\frac{1}{2}(I + D^{-1}A))^2 x, (\frac{1}{2}(I + D^{-1}A))^3 x, \ldots$ converges in the direction of u_j. Note that in theory, all the vectors in this series are D-orthogonal to $u_1, \ldots, u_{j-1}$. However, to improve numerical stability, our implementation imposes the D-orthogonality to previous eigenvectors in each iteration. The power iteration algorithm produces vectors of diminishing (or exploding) norms. Since we are only interested in convergence *in direction*, it is customary to re-scale the vectors after each iteration. Here, we will re-scale by normalizing the vectors to be of length 1.

Function SpectralDrawing (G – the input graph, k – dimension)
% This function computes $u_2, \ldots, u_k$, the top (non-degenerate) eigenvectors of $D^{-1}A$.
  const $\epsilon \leftarrow 10^{-7}$ % tolerance
  for $i = 2$ **to** k **do**
    $\hat{u}_i \leftarrow$ random   % random initialization
    $\hat{u}_i \leftarrow \frac{\hat{u}_i}{\|\hat{u}_i\|}$
    do
      $u_i \leftarrow \hat{u}_i$
      % D-Orthogonalize against previous eigenvectors:
      for $j = 1$ **to** $i - 1$ **do**
        $u_i \leftarrow u_i - \frac{u_i^T D u_j}{u_j^T D u_j} u_j$
      end for
      % multiply with $\frac{1}{2}(I + D^{-1}A)$:
      for $j = 1$ **to** n **do**
        $\hat{u}_i(j) \leftarrow \frac{1}{2} \cdot \left(u_i(j) + \frac{\sum_{k \in N(j)} w_{jk} u_i(k)}{\deg(j)} \right)$
      end for
      $\hat{u}_i \leftarrow \frac{\hat{u}_i}{\|\hat{u}_i\|}$ % normalization
    while $\hat{u}_i \cdot u_i < 1 - \epsilon$ % halt when direction change is negligible
    $u_i \leftarrow \hat{u}_i$
  end for
  return $u_2, \ldots, u_k$

Fig. 3. The algorithm for computing degree-normalized eigenvectors

The convergence rate of this algorithm when computing u_i is dependent on the ratio μ_i / μ_{i+1}. In practice, we embedded this algorithm in a multi-scale construction, resulting in extremely fast convergence. The multi-scale ascheme is explained in [9].

6   A Direct Characterization of Spectral Layouts

So far, we have derived spectral methods as solutions of optimization problems, or as a limit of a drawing process. In this section we characterize the eigenvectors themselves, in a rather direct manner, to clarify the aesthetic properties of the spectral layout. Once again the degree-normalized eigenvectors will appear as the more natural way for spectral graph drawing.

As we have seen, the quadratic form $E(x) = \sum_{\langle i,j \rangle \in E} w_{ij}(x(i) - x(j))^2$, which motivates spectral methods, is tightly related to the aesthetic criterion that calls for placing each node at the weighted centroid of its neighbors. When the graph is connected, it can be strictly achieved only by the degenerate solution that puts all nodes at the same location. Hence, to incorporate this aesthetic criterion into a graph drawing algorithm, it should be modified appropriately.

Presumably the earliest graph drawing algorithm, formulated by Tutte [14], is based on placing each node on the weighted centroid of its neighbors. To avoid the degenerate solution, Tutte arbitrarily chose a certain number of nodes to be anchors, i.e. he fixed their coordinates in advance. Those nodes are typically drawn on the boundary. This, of course, prevents the collapse; however it raises new problems, such as which nodes should be the anchors, how to determine their coordinates, and why after all such an anchoring mechanism should generate nice drawings. An advantage of Tutte's method is that in certain cases, it can guarantee achieving a planar drawing.

Tutte treats in different ways the anchored nodes and the remaining nodes. Whereas the remaining nodes are located exactly at the centroid of their neighbors, nothing can be said about anchored nodes. In fact, in several experiments we have seen that the anchored nodes are located quite badly.

Alternatively, we do not use different strategies for dealing with two kinds of nodes, but rather, we treat all the nodes similarly. The idea is to gradually increase the deviations from centroids of neighbors as we move away from the origin (that is the center of the drawing). This reflects the fact that central nodes can be placed exactly at their neighbors' centroid, whereas boundary nodes must be shifted outwards.

More specifically, node i, which is located in place $x(i)$, is shifted from the center toward the boundary by the amount of $\mu \cdot |x(i)|$, for some $\mu > 0$. Formally, we request the layout x to satisfy, for every $1 \leqslant i \leqslant n$

$$x(i) - \frac{\sum_{j \in N(i)} w_{ij} x(j)}{\deg(i)} = \mu \cdot x(i) \, .$$

Note that the deviation from the centroid is always toward the boundary, i.e. toward $+\infty$ for positive $x(i)$ and toward $-\infty$ for negative $x(i)$. In this way we prevent a collapse at the origin. We can represent all these n requests compactly in a matrix form, by writing

$$D^{-1} L x = \mu x \, .$$

Left-multiplying both sides by D, we obtain the familiar generalized eigen-equation

$$L x = \mu D x \, .$$

We conclude with the following important property of degree-normalized eigenvectors:

Proposition 1. *Let u be a generalized eigenvector of (L, D), with associated eigenvalue μ. Then, for each i, the exact deviation from the centroid of neighbors is*

$$u(i) - \frac{\sum_{j \in N(i)} w_{ij} u(j)}{\deg(i)} = \mu \cdot u(i) \, .$$

Note that the eigenvalue μ is a scale-independent measure of the amount of deviation from the centroids. This provides us with a fresh new interpretation of the eigenvalues that is very different from the one given in Subsection 3.1, where the eigenvalues were shown as the amount of energy in the drawing.

Thus, we deduce that the second smallest degree-normalized eigenvector produces the non-degenerate drawing with the smallest deviations from centroids, and that the third smallest degree-normalized eigenvector is the next best one and so on.

Similarly, we can obtain a related result for eigenvectors of the Laplacian:

Proposition 2. *Let v be an eigenvector of L, with associated eigenvalue λ. Then, for each i, the exact deviation from the centroid of neighbors is*

$$v(i) - \frac{\sum_{j \in N(i)} w_{ij} v(j)}{\deg(i)} = \lambda \cdot \deg(i)^{-1} \cdot v(i).$$

Hence for eigenvectors of the Laplacian, the deviation between a node and the centroid of its neighbors gets larger as the node's degree decreases.

7 Discussion

In this paper we have presented a spectral approach for graph drawing, and justified it by studying three different viewpoint for the problem. The first viewpoint describes a classical approach for achieving graph layouts by solving a constrained energy minimization problem. This is much like force directed graph drawing algorithms (for a survey refer to [4,8]). Compared with other force-directed methods, the spectral approach has two major advantages: **(1)** Its global optimum can be computed efficiently. **(2)** The energy function contains only $O(|E|)$ terms, unlike the $O(n^2)$ terms appearing in almost all the other force-directed methods.

A second viewpoint shows that the spectral drawing is the limit of an iterative process, in which each node is placed at the centroid of its neighbors. This viewpoint does not only sharpen the nature of spectral drawing, but also provides us with an aesthetically-motivated algorithm. This is unlike other algorithms for computing eigenvectors, which are rather complicated and far from having an aesthetic interpretation.

We have also introduced a third viewpoint, showing that spectral methods place each node at the centroid of its neighbors with some well defined deviation. This new interpretation provides an accurate and simple description of the aesthetic properties of spectral drawing.

Another contribution of our paper is the introduction of a new spectral graph drawing algorithm, using what we have called *degree-normalized eigenvectors*. We have shown that this method is more natural in some aspects, and has aesthetic advantages for certain kinds of data.

References

1. U. Brandes and T. Willhalm, "Visualizing Bibliographic Networks with a Reshaped Landscape Metaphor", Proc. 4th Joint Eurographics – IEEE TCVG Symp. Visualization (VisSym '02), pp. 159–164, ACM Press, 2002.
2. L. Carmel, Y. Koren and D. Harel, "Visualizing and Classifying Odors Using a Similarity Matrix", Proceedings of the ninth International Symposium on Olfaction and Electronic Nose (ISOEN'02), IEEE, to appear, 2003.
3. F.R.K. Chung, *Spectral Graph Theory*, CBMS Reg. Conf. Ser. Math. 92, American Mathematical Society, 1997.
4. G. Di Battista, P. Eades, R. Tamassia and I.G. Tollis, *Graph Drawing: Algorithms for the Visualization of Graphs*, Prentice-Hall, 1999.
5. G.H. Golub and C.F. Van Loan, *Matrix Computations*, Johns Hopkins University Press, 1996.

6. K. M. Hall, "An r-dimensional Quadratic Placement Algorithm", *Management Science* **17** (1970), 219–229.

7. M. Juvan and B. Mohar, "Optimal Linear Labelings and Eigenvalues of Graphs", *Discrete Applied Math.* **36** (1992), 153–168.

8. M. Kaufmann and D. Wagner (Eds.), *Drawing Graphs: Methods and Models*, LNCS 2025, Springer Verlag, 2001.

9. Y. Koren, L. Carmel and D. Harel, "ACE: A Fast Multiscale Eigenvectors Computation for Drawing Huge Graphs", *Proceedings of IEEE Information Visualization 2002 (InfoVis'02)*, IEEE, pp. 137–144, 2002.

10. D.E. Manolopoulos and P.W. Fowler, "Molecular Graphs, Point Groups and Fullerenes", *J. Chem. Phys.* **96** (1992), 7603–7614.

11. B. Mohar, "The Laplacian Spectrum of Graphs", *Graph Theory, Combinatorics, and Applications* **2** (1991), 871–898.

12. A. Pothen, H. Simon and K.-P. Liou, "Partitioning Sparse Matrices with Eigenvectors of Graphs", *SIAM Journal on Matrix Analysis and Applications*, **11** (1990), 430–452.

13. J. Shi and J. Malik, "Normalized Cuts and Image Segmentation", *IEEE Transactions on Pattern Analysis and Machine Intelligence*, **22** (2000), 888–905.

14. W. T. Tutte, "How to Draw a Graph", *Proc. London Math. Society* **13** (1963), 743–768.

15. A. Webb, *Statistical Pattern Recognition*, Arnold, 1999.

A   Solution of Constrained Quadratic Optimization Problems

In this appendix we study a certain kind of constrained optimization problem, whose solution is a generalized eigenvector.

We use two matrices: (1) A — an $n \times n$ real symmetric positive-semidefinite matrix. (2) B — an $n \times n$ diagonal matrix, whose diagonal entries are real-positive. (In fact, it is enough to require that matrix B is positive-definite.) We denote the generalized eigenvectors of (A, B) by $u_1, u_2, \ldots, u_n$, with corresponding eigenvalues $0 \leqslant \lambda_1 \leqslant \lambda_2 \leqslant \cdots \leqslant \lambda_n$. Thus, $Au_i = \lambda_i Bu_i$, $i = 1, \ldots, n$. To uniquely define $u_1, u_2, \ldots, u_n$, we require them to be B-normalized, i.e. $u_i^T Bu_i = 1$, $i = 1, \ldots, n$.

Clearly, for every $1 \leqslant i \leqslant n$, $B^{\frac{1}{2}}u_i$ and λ_i are an eigen-pair of the matrix $B^{-\frac{1}{2}}AB^{-\frac{1}{2}}$. Note that $B^{-\frac{1}{2}}AB^{-\frac{1}{2}}$ is a symmetric positive-semidefinite. Thus, all the eigenvalues are real non-negative, and all the generalized eigenvectors are B-orthogonal, i.e. $u_i^T Bu_j = 0$, $\forall i \neq j$.

Now we define a constrained optimization problem

$$\min_{x} x^T A x \qquad (7)$$

$$\text{given: } x^T B x = 1$$

$$\text{in the subspace: } x^T Bu_1 = 0, \ldots, x^T Bu_{k-1} = 0.$$

Claim. The optimal solution of problem 7 is $x = u_k$, with an associated cost of $x^T A x = \lambda_k$.

Proof. By using the B-orthogonality of $u_1, \ldots, u_n$, we can decompose every $x \in \mathbb{R}^n$ as a linear combination where $x = \sum_{i=1}^{n} \alpha_i u_i$. Moreover, since x is constrained to be B-orthogonal to $u_1, \ldots, u_{k-1}$, we can restrict ourselves to linear combinations of the form $x = \sum_{i=k}^{n} \alpha_i u_i$.

We use the constraint $x^T Bx = 1$ to obtain

$$1 = x^T Bx = \left(\sum_{i=k}^{n} \alpha_i u_i\right)^T B \left(\sum_{i=k}^{n} \alpha_i u_i\right) = \left(\sum_{i=k}^{n} \alpha_i u_i\right)^T \left(\sum_{i=k}^{n} \alpha_i Bu_i\right) =$$
$$= \sum_{i=k}^{n}\sum_{j=k}^{n} \alpha_i u_i \alpha_j Bu_j = \sum_{i=k}^{n}\sum_{j=k}^{n} \alpha_i \alpha_j u_i Bu_j = \sum_{i=k}^{n} \alpha_i^2.$$

The last equation stems from the B-orthogonality of $u_1, u_2, \ldots, u_n$, and from defining these vectors as B-normalized.

Hence, $\sum_{i=k}^{n} \alpha_i^2 = 1$ (a generalization of Pythagoras' Law). Now, we expand the quadratic form $x^T Ax$

$$x^T Ax = \left(\sum_{i=k}^{n} \alpha_i u_i\right)^T A \left(\sum_{i=k}^{n} \alpha_i u_i\right) = \left(\sum_{i=k}^{n} \alpha_i u_i\right)^T \left(\sum_{i=k}^{n} \alpha_i Au_i\right) = \quad (8)$$
$$= \left(\sum_{i=k}^{n} \alpha_i u_i\right)^T \left(\sum_{i=k}^{n} \alpha_i \lambda_i Bu_i\right) = \sum_{i=k}^{n}\sum_{j=k}^{n} \alpha_i u_i \alpha_j \lambda_i Bu_j =$$
$$= \sum_{i=k}^{n}\sum_{j=k}^{n} \alpha_i \alpha_j \lambda_i u_i Bu_j = \sum_{i=k}^{n} \alpha_i^2 \lambda_i \geqslant \sum_{i=k}^{n} \alpha_i^2 \lambda_k = \lambda_k.$$

Thus, for any x that satisfies the constraints, we have $x^T Ax \geqslant \lambda_k$. Since $u_k^T Au_k = \lambda_k$, we can deduce that the minimizer is $x = u_k$. $\qquad\square$

On a Conjecture on Wiener Indices in Combinatorial Chemistry

Yih-En Andrew Ban[1], Sergei Bespamyatnikh[2], and Nabil H. Mustafa[3]

[1] Department of Biochemistry, Duke University, Durham, NC, USA.
[2] Department of Computer Science, University of Texas at Dallas, TX, USA.
[3] Department of Computer Science, Duke University, Durham, NC, USA.
{aban, nabil}@cs.duke.edu, besp@utdallas.edu

Abstract. Drugs and other chemical compounds are often modeled as polygonal shapes called the *molecular graph*, which can be a path, a tree, or in general any graph. An indicator defined over this molecular graph, the *Wiener index*, has been shown to be strongly correlated to various chemical properties of the compound. The Wiener index conjecture for trees states that for any integer n (except for a finite set), one can find a tree with Wiener index n. In this paper, we present progress towards proving this conjecture by presenting a 4-parameter family of trees that we show experimentally to affirm the Wiener index conjecture for very large values of n. Given an integer n, we also present efficient algorithms for finding the tree whose Wiener index is n.

1 Introduction

Drugs and other chemical compounds are often modeled as various polygonal shapes — paths, trees, graphs etc. Each vertex in the polygonal path or tree represents an atom of the molecule, and covalent bonds between atoms are represented by edges between the corresponding vertices. This polygonal shape derived from a chemical compound is often called its *molecular graph*. As the geometry of proteins play an important role in determining the function of the protein, so can the topological properties of the molecular graphs of chemical compounds be correlated to their chemical properties. For some time now, the biochemical community has been using topological indices in an attempt to correlate a compounds molecular graph with experimentally gathered data regarding the compounds characteristics.

Usage of topological indices in biology and chemistry began in 1947 when chemist Harold Wiener developed the most widely known topological descriptor, the Wiener index, and used it to determine physical properties of types of alkanes known as paraffins[13]. In general, topological indices are one of the oldest and most widely used descriptors in quantitative structure activity relationships: Quantitative structure activity relationships (QSAR) is a popular computational biology paradigm in modern drug design [11]. It attempts to encode biological activity such as inhibition or activation into numerical measures by correlation of mass amounts of data from an initial screening of hundreds or thousands

T. Warnow and B. Zhu (Eds.): COCOON 2003, LNCS 2697, pp. 509–518, 2003.
© Springer-Verlag Berlin Heidelberg 2003

of candidate drug compounds. The data is mapped into a structural "activity space" consisting of various descriptors with the hope that these spaces capture or estimate properties of chemical compounds. Constructions of new drugs then proceed by noting the desired activity spaces and creating new compounds which occupy those regions of space.

Amongst the topological indices used as descriptors in QSAR, the Wiener index is by far the most popular index, as it has been shown that the Wiener index has a strong correlation with the chemical properties of the compound [4]. Therefore, to construct a compound with a certain property correlated to some Wiener index, the objective becomes to build substructures in the target chemical compound giving the compound that Wiener index. This in turn leads to the following important problem: given a Wiener index, find a compound with that Wiener index.

An overwhelming majority of the chemical applications of the Wiener index deal with chemical compounds that have acyclic organic molecules. The molecular graphs of these compounds are trees [7]. Therefore most of the prior work on Wiener indices deals with trees, relating the structure of various trees to their Wiener indices (asymptotic bounds on the Wiener indices of certain families of trees, expected Wiener indices of random trees etc.). For these reasons, we concentrate on the Wiener indices of trees as well (see Dobrynin *et al.* [3] for a survey).

1.1 Problem Definition

Definition 1 (Wiener Index). *Given an undirected graph $G = (V, E)$, denote by $d(v_i, v_j)$ the length of the shortest path between two distinct vertices $v_i, v_j \in V$. The Wiener index $W(G)$ is defined as*

$$W(G) = \sum_i \sum_{j>i} d(v_i, v_j) \tag{1}$$

In this paper, we will present results on the Wiener index problems where the graph G is a tree. While the case of graphs has been solved, a major conjecture in this area is whether every positive integer is the Wiener index of some tree.

*Conjecture 1 (***Wiener Index Conjecture** *[6,10]).* Except for some finite set, every positive integer is the Wiener index of some tree.

Towards this goal, we present a family of trees and design fast efficient algorithms for the Inverse Wiener problem, defined as follows [8].

*Problem 1 (***Inverse Wiener Problem,** $P_=$*).* Given an integer n, construct a tree whose Wiener index is n.

An algorithm for solving the problem $P_=$ can be used to verify the above conjecture — the algorithm for problem $P_=$ can be applied n times to verify the conjecture up to any integer n. However, there can be a faster algorithm, and so we introduce the following problem.

Problem 2 **(Inverse Wiener Covering Problem, $P_\leq$).** Given an integer n, for every $i \leq n$ construct a tree with Wiener index i.

1.2 Previous Work

The Wiener index for general graphs can be computed in time $O(|V|^2 \log |V| + |E||V|)$ using any all-pairs shortest paths algorithm. For trees, although the number of edges is $O(n)$, the number of pairs of vertices is still $O(n^2)$. However, the Wiener index of a tree can be computed in $O(n)$ time, through a plethora of computational methods [2,12].

The inverse Wiener problem is easily solvable for graphs: given an integer n, there always exists a graph G such that $W(G) = n$, and it is computable in constant time. This was shown by Goldman *et al.* [6]. However, once the graph is restricted to trees, Goldman *et al.* note that the problem becomes complicated and the complexity is unknown. In fact, the complexity of this question is left unsolved in Gutman and Yeh [8], and conjectured that apart from a finite set, all integers do have corresponding trees with the required Wiener indices. Later Lepović and Gutman [10] presented an exhaustive search algorithm that verifies this conjecture up to integer 1206, still leaving the Inverse Wiener conjecture open. Goldman *et al.* attempt to solve $P_=$ by defining a recurrence relation on Wiener indices of trees by using dynamic programming to construct trees from smaller subtrees. Using their algorithm, they are able to verify the above conjecture for integers up to 10,000. However, the Inverse Wiener conjecture remains open.

1.3 Our Results

We present progress towards proving the Wiener index conjecture by experimentally showing that searching all possible trees is not necessary — a small parameterized subset suffices. More specifically, we present a family of trees, $\mathcal{F}$, and several efficient algorithms for the problem $P_=$ using a decomposition by sorted sequences in time $O(nk)$, where k is a decomposition parameter. We present empirical results on the efficiency of this algorithm, indicating that this algorithm is *orders* of magnitude faster. Finally, we present several efficient algorithms for the problem $P_\leq$, and give empirical results showing that it is faster by a factor of $O(\sqrt{n})$ than the naive implementation.

2 Family of Trees

First, we state that given a tree $T = (V, E)$, one can assign integer weights to the edges of the tree such that the sum of the edge weights is exactly the Wiener index of the tree.

Lemma 1 (Dobrynin *et al.* [3]). *Given a tree $T = (V, E)$, one can assign weights $w(e)$ for each $e \in E$ such that $W(T) = \sum_{e \in E} w(e)$.*

Define $T_k(r_1, \ldots, r_k)$ be a tree of *order k* where

$$V = \{s_1, \ldots, s_k\} \cup \{t_1^1, \ldots, t_{r_1}^1, \ldots, t_1^k, \ldots, t_{r_k}^k\}$$

$$E = \{(s_i, s_{i+1}), \ 1 \le i \le (k-1)\} \ \cup \ \{(t_l^j, s_j), \ 1 \le j \le k, \ 1 \le l \le r_j\}.$$

Note that $m = |V| = \sum_i r_i + k$. Tree T_k is thus defined by k parameters $r_i, i = 1 \ldots k$. It can be shown that the Wiener index of T_k is:

$$W(T_k(r_1, \ldots, r_k)) = \sum_{i=1}^{k-1} \left((\sum_{j=1}^{i}(r_j) + i) \cdot (\sum_{j=i+1}^{k} (r_j) + (k-i)) \right) + (m-k)(m-1)$$

Lemma 2. *The Wiener index of a tree of order k can be computed in $O(k)$ time.*

PROOF. Let $X_i = \sum_{j=1}^{i} r_j$, and $Y_i = \sum_{j=i+1}^{k} r_j$. Numbers X_i and Y_i can be computed in $O(k)$ time, for all $1 \le i \le k$ in an incremental fashion. Hence $W(T_k(r_1, \ldots, r_k))$ can be computed $O(k)$ time.   ☐

We omit details, but we have verified that integers up to 10^8, except for a few numbers, can be represented as $W(T_k)$, for some k, leading us to state the following.

Conjecture 2. Except a set S_1 (Table 1) of 56 numbers ≤ 193, all integers can be represented by $W(T_k)$.

Actually, our experiments indicate that all numbers n, $10^3 \le n \le 10^8$, can be presented by $W(T_k)$, where $k \le 5$. However, it seems that the Wiener indices of the trees T_k, $k \le 4$, do not cover all integers (except a finite set). Based on the above results, we strengthen conjecture 2 by replacing the infinite family of trees T_k by a single family of trees for $k = 5$.

Conjecture 3. Except a set S_2 (Table 1) of 102 numbers ≤ 557, all integers can be represented by $W(T_5)$.

We recall Lagrange's Theorem in number theory, that every integer can be represented as a sum of four squares [9]. For example, the polynomial for the Wiener index (in terms of r_i) for $T_5(r_1, \ldots, r_5)$ consists *only* of quadratic and linear terms. That gives the intuition that a properly chosen four parameter family of trees might be sufficient to represent any Wiener index. One way to remove one parameter would be to impose a constraint on the function $W(T_5)$. After some experimentation, we discovered that the constraint $r_1 = r_5$ still allows for the representation of all integers tested as Wiener indices, except for a certain finite set. Consider the family of trees $T_5(r_1, \ldots, r_5)$ such that $r_1 = r_5$. We denote this family of trees by $F(\cdot)$, where $F(r_1, r_2, r_3, r_4) = T_5(r_1, r_2, r_3, r_4, r_1)$. Then simplifying the Wiener formula for $W(T_5(r_1, \ldots, r_5))$ gives,

$$\begin{aligned} W(F(r_1, r_2, r_3, r_4)) = \ & r_1 \cdot (8r_1 + 8r_2 + 8r_3 + 8r_4 + 28) \\ & + r_2 \cdot (r_2 + 3r_3 + 4r_4 + 11) + r_3 \cdot (r_3 + 3r_4 + 10) \\ & + r_4 \cdot (r_4 + 11) + 20. \end{aligned} \qquad (2)$$

Conjecture 4. Except a set S_3 (Table 1) of 181 numbers ≤ 1177, all integers can be represented by $W(F)$.

3   Algorithm for the Problem $P_=$

In this section we present algorithms for finding trees whose Wiener index is n, given the integer n as input to the algorithm. First, observe that the value of each r_i is bounded in terms of n.

Claim. Given an integer n, any tree T of family $F(r_1, r_2, r_3, r_4)$ with $W(T) = n$ must have $0 \leq r_i \leq \sqrt{n}$.

Recall that each tree in $F(\cdot)$ is defined by four parameters r_1, r_2, r_3 and r_4. Given n, the objective is to find a tree $T \in F(\cdot)$ such that $W(T) = n$. We call the set of all 4-tuples (r_1, r_2, r_3, r_4) the *configuration space* of $W(F(\cdot))$. To find a given integer n, we want to search this configuration space. The straightforward way of computing $F(\cdot)$ is to exhaustively traverse this configuration space, i.e. iterate over all possible r_i's, and compute $W(T)$ for each 4-tuple. By the above Claim, the running time is $O(n^2)$.

However, on examining Equation 2, one finds that the equation is monotone in all parameters r_i. Therefore, a fast algorithm is as follows. Iterate over all values of $r_1, r_2,$ and r_3. So suppose that r_1, r_2 and r_3 are some fixed constants, say c_1, c_2 and c_3 respectively. Perform the binary search over the sequence $W(F(c_1, c_2, c_3, r_4))$, $0 \leq r_4 \leq \sqrt{n}$, to find if $W(T) = n$ for some value of r_4. The running time therefore reduces to $O(n^{3/2} \log n)$. The running time can be further reduced to $O(n^{3/2})$ using Frederickson and Johnson's searching technique [5].

3.1   Decomposition Using Sorted Sequences

We now analyze the structure of Equation 2 more closely and use it to present an even more efficient algorithm for our problem. The first thing to note is the symmetry of the equation between r_2 and r_4, i.e. r_1 "contributes" equally to (the coefficients of) r_2 and r_4, and similarly r_3 "contributes" equally to (the coefficients of) r_2 and r_4. Therefore, instead of fixing r_1 and r_2 as before and trying to find r_3 and r_4 values more efficiently, we fix $r_1 = c_1, r_3 = c_3$. As explained above, this is crucial since r_1 and r_3 contribute symmetrically to r_2 and r_4. Then

$$\begin{aligned}
W(F(c_1, r_2, c_3, r_4)) &= r_2 \cdot (8c_1 + r_2 + 3c_3 + 11) + r_4 \cdot (8c_1 + r_4 + 3c_3 + 11) \\
&\quad + 4r_2 r_4 + K(c_1, c_3) \\
&= (r_2 + r_4) \cdot (8c_1 + 3c_3 + 11) + r_2^2 + r_4^2 + 4r_2 r_4 + K(c_1, c_3) \\
&= (r_2 + r_4) \cdot (8c_1 + 3c_3 + 11) + (r_2 + r_4)^2 + 2r_2 r_4 \\
&\quad + K(c_1, c_3)
\end{aligned}$$

where $K(c_1, c_3) = (c_3^2 + 8c_1^2) + (8c_1 c_3) + (28c_1 + 10c_3) + 20$ is a constant.

Lemma 3. *Given integers* $r_2, r_4, s_2, s_4, c_1, d_1, c_3, d_3,$

$$W(F(c_1, r_2, c_3, r_4)) \geq W(F(c_1, s_2, c_3, s_4)) \implies$$
$$W(F(d_1, r_2, d_3, r_4)) \geq W(F(d_1, s_2, d_3, s_4))$$

if $r_2 + r_4 \geq s_2 + s_4$ *and* $8(d_1 - c_1) + 3(d_3 - c_3) \geq 0.$

PROOF. Assume that $W(F(c_1, r_2, c_3, r_4)) \geq W(F(c_1, s_2, c_3, s_4))$. We will show that the increment in the Wiener index for the configuration (c_1, r_2, c_3, r_4) is larger than the increment for the configuration (c_1, s_2, c_3, s_4), i.e.

$$W(F(d_1, r_2, d_3, r_4)) - W(F(c_1, r_2, c_3, r_4)) \geq$$
$$W(F(d_1, s_2, d_3, s_4)) - W(F(c_1, s_2, c_3, s_4))$$
$$(r_2 + r_4)(8d_1 + 3d_3 + 11 - 8c_1 - 3c_3 - 11) \geq$$
$$(s_2 + s_4)(8d_1 + 3d_3 + 11 - 8c_1 - 3c_3 - 11)$$
$$(r_2 + r_4)(8(d_1 - c_1) + 3(d_3 - c_3)) \geq (s_2 + s_4)(8(d_1 - c_1) + 3(d_3 - c_3))$$

Since $8(d_1 - c_1) + 3(d_3 - c_3) \geq 0$, and $(r_2 + r_4) \geq (s_2 + s_4)$, the proof follows. ⌑

We now use Lemma 3 to identify large subsets of the configuration space that can be searched to find a specific element much more efficiently. Set $r_1 = r_3 = 0$. Define $w((a, b)) = W(F(0, a, 0, b))$ and $c((a, b)) = a + b$. Let $P = \langle p_i = (a_i, b_i) | 0 \leq a_i, b_i \leq \sqrt{n} \rangle$ be sorted by $w(\cdot, \cdot)$. Then the sequence of $|P| = n$ pairs represents the Wiener indices of all possible pairs of parameters r_2 and r_4. Let $P' = \langle p_{i_1}, \ldots, p_{i_k} \rangle$ be the longest subsequence of P such that the sequence $C(P') = \langle c(p_{i_1}), \ldots, c(p_{i_k}) \rangle$ is increasing. P' can be found in time $O(n \log \log n)$ [1]. Note that two conditions hold for any two elements $p_{i_j}, p_{i_{j'}} \in P'$ such that $i_j \leq i_{j'}$, (i) $w(p_{i_j}) \leq w(p_{i_{j'}})$ and (ii) $c(p_{i_j}) \leq c(p_{i_{j'}})$.

Lemma 4. *Given the subsequence* P' *described above, and an integer* n *such that* $n = W(F(r_1, a, r_3, b))$, *where the values of* r_1 *and* r_3 *are known, one can find the values of* a *and* b *in* $O(\log n)$ *time.*

PROOF. Note that in the previous algorithm, we used the binary search over one variable or used Frederickson and Johnson's technique to avoid the binary search, but we had to perform $\Omega(\sqrt{n})$ computations to find $r_2 = a$ and $r_4 = b$. Now we will show that if the subset of configuration space satisfies certain criteria, like P', then we can search in $O(\log n)$ steps.

We need to find a and b such that $W(F(r_1, a, r_3, b)) = n$ where $(a, b) \in P'$. Note that P' is increasing in both $W(F(0, a_i, 0, b_i))$ and in $C(P')$. Take any two pairs $p_{i_j} = (a_{i_j}, b_{i_j})$ and $p_{i_{j'}} = (a_{i_{j'}}, b_{i_{j'}})$ of P', $j \leq j'$. Then $a_{i_j} + b_{i_j} \leq a_{i_{j'}} + b_{i_{j'}}$ from (ii) above, and for any positive r_1 and r_3, $8(r_1 - 0) + 3(r_3 - 0) \geq 0$. Thus the conditions of Lemma 3 are satisfied and

$$W(F(0, a_{i_j}, 0, b_{i_j})) \leq W(F(0, a_{i_{j'}}, 0, b_{i_{j'}})) \implies$$
$$W(F(r_1, a_{i_j}, r_3, b_{i_j})) \leq W(F(r_1, a_{i_{j'}}, r_3, b_{i_{j'}}))$$

$$w(a_{i_j}, b_{i_j}) \leq w(a_{i_{j'}}, b_{i_{j'}}) \implies W(F(r_1, a_{i_j}, r_3, b_{i_j})) \leq W(F(r_1, a_{i_{j'}}, r_3, b_{i_{j'}})) \quad (3)$$

Equation 3 states that the Wiener index of the 4-tuple $(r_1, a_{i_j}, r_3, b_{i_j})$ will always be less than the Wiener index of $(r_1, a_{i_{j'}}, r_3, b_{i_{j'}})$ if $j \leq j'$, *regardless* of the value of r_1 and r_3. Therefore we can do binary search since we have P' sorted by $W(F(0, a_i, 0, b_i))$ already. We only know that the order of the pairs is preserved, although the values of $w(\cdot)$ have changed (since the values of r_1 and r_3 have changed). At each step of the binary search we have to recompute $w(\cdot)$ for each pair, and proceed accordingly. ⌑

Now the algorithm can be completed. Set $r_1 = r_3 = 0$, and compute the set P in time $O(n \log n)$ by sorting all tuples $(0, r_2, 0, r_4)$ by their Wiener indices. Now find the largest increasing subsequence of P in the ordered sequence $C(P)$. This subsequence satisfies the two properties, i.e. increasing with respect to $W(P)$ and $C(P)$. From Lemma 3, the order of this subsequence would remain unchanged with varying values of r_1 and r_3. We store this subsequence as an ordered sequence of pairs P_1. Now iteratively extract largest increasing subsequence P_i in round i till the last round k. Store these ordered sets $\mathcal{P} = \{P_1, \dots, P_k\}$. Now, we vary the values of r_1 and r_3 from 1 to $\sqrt{n}$. Since we don't know which sequence could contain n, we have to search in all k sequences. In each sequence we do binary search as in Lemma 4, achieving the worst case total time $O(nk \log n)$. Note that $k \leq \sqrt{n}$ — we can always define $\mathcal{P} = \{P_1, \dots, P_{\sqrt{n}}\}$, where $P_i = \{(i, 1), \dots, (i, \sqrt{n})\}$. Then, from the monotonicity of the function $F(\cdot)$, each P_i is an increasing sequence in the Wiener index. Again, using Frederickson and Johnson's matrix searching technique, the running time can be improved to $O(nk \log \log n)$ time. We omit details, and conclude the following.

Theorem 1. *Given an integer n, one can find (if they exist) integers c_1, c_2, c_3 and c_4 such that $W(F(c_1, c_2, c_3, c_4)) = n$ in time $O(nk \log \log n)$, where k is a decomposition parameter.*

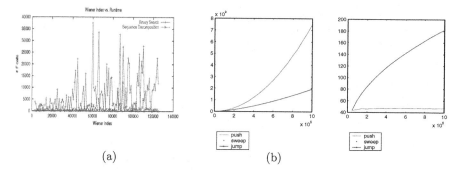

                              (a)   (b)

Fig. 1. (a) Figure showing the running times for problem $P_=$ measured by the number of calls to $F(\cdot)$, (b) The graphs of $T_{push}, T_{sweep}, T_{jump}$ (left) and the ratios T_{naive} versus the functions (right).

We compare our algorithm (without using Frederickson and Johnson technique), denoted SequenceDecomposition, with the binary search algorithm, denoted as Algorithm BinarySearch. We run the above two algorithms for the inverse Wiener problem for Wiener indices $n = 1000, 2000, \dots, 140,000$. For each n, we measure the the running time by counting the number of calls to $W(F(\cdot))$ of both Algorithm BinarySearch, and Algorithm SequenceDecomposition. The results are shown in Figure 1 (a). As expected, the running time for various integers n varies quite a lot (since some searches are lucky to find the number easily) but on the whole it can be easily seen that BinarySearch makes many more calls to $W(F(\cdot))$ than SequenceDecomposition.

4   Algorithm for the Problem $P_\leq$

The problem $P_\leq$ for our class of trees $F(\cdot)$ is as follows: given an integer n, for every integer $0 \leq m \leq n$, compute the configuration (a_1, a_2, a_3, a_4) such that $W(F(a_1, a_2, a_3, a_4)) = m$.

Of course, problem $P_\leq$ can be solved by n calls to problem $P_=$, yielding a $O(n^2 k)$ time algorithm, and with the worst case running time of $O(n^{5/2})$. However, the problem $P_\leq$ can be solved in $O(n^2)$ time by computing $W(F(\cdot))$ for every tuple $a = (a_1, a_2, a_3, a_4)$, where $0 \leq a_i \leq \sqrt{n}$, $i = 1, \dots, 4$. For each tuple, we mark off the integer $W(F(\cdot))$ in an array. After all tuples have been computed, the locations in the array not marked indicate integers not representable. As before, we will measure the running time of the algorithm by the number of calls to the function. Using Equation 2 we can bound $a_1 \leq \sqrt{(n-20)/8}$, $a_i \leq \sqrt{n-20}$, $i = 2, 3, 4$. Therefore the complexity of the naive algorithm is bounded by $\lfloor \sqrt{(n-20)/8} \rfloor \cdot \lfloor \sqrt{n-20} \rfloor^3$. For large n, this is approximately $n^2/(2\sqrt{2})$ and we denote it as $T_{naive}(n)$. Our goal is to design an algorithm that solves $P_\leq$ using a substantially smaller number of computations of $W(F(\cdot))$. The first idea is to make a bound for a that further restricts the search space. We will call our new algorithm as *push algorithm*. The algorithm will sequentially try to cover integers in increasing order. Let

$$s(a) = \lceil \sqrt{2}(2a_1 + a_2 + a_3 + a_4 + 7/2) \rceil$$

and let m be the smallest number whose expression $W(F(\cdot)) = m$ is not computed yet. The value of s can be bounded from below.

Lemma 5. *If $W(F(a)) = m$ then $s(a) > \sqrt{m}$.*

PROOF. The lemma follows from the fact that $W(F(a)) < 2(2a_1 + a_2 + a_3 + a_4 + 7/2)^2$. We prove the inequality

$$- 2(2a_1 + a_2 + a_3 + a_4)^2 - W(F(a)) \geq 0$$

$$- a_1(8a_1 + 8a_2 + 8a_3 + 8a_4 + 28) + a_2(4a_2 + 4a_3 + 4a_4 + 14)$$

$$+ a_3(4a_3 + 4a_4 + 14) + a_4(4a_4 + 14) + 49/2 - W(F(a)) \geq 0$$

$$- a_2(3a_2 + a_3 + 3) + a_3(3a_3 + a_4 + 4) + a_4(3a_4 + 3) + 9/2 > 0$$

□

Table 1. Table presenting the various sets of integers related to the conjectures.

S_1	2, 3, 5, 6, 7, 8, 11, 12, 13, 14, 15, 17, 19, 21, 22, 23, 24, 26, 27, 30, 33, 34, 37, 38, 39, 41, 43, 45, 47, 51, 53, 55, 60, 61, 69, 72, 73, 77, 78, 83, 85, 87, 89, 91, 99, 101, 106, 113, 117, 129, 133, 147, 157, 159, 173, 193
S_2	1, 2, 3, 4, 5, 6, 7, 8, 9, 10, 11, 12, 13, 14, 15, 16, 17, 18, 19, 21, 22, 23, 24, 25, 26, 27, 28, 29, 30, 33, 34, 36, 37, 38, 39, 40, 41, 42, 43, 45, 47, 49, 51, 53, 54, 55, 57, 58, 60, 61, 64, 69, 72, 73, 77, 78, 81, 83, 85, 87, 89, 91, 93, 97, 99, 101, 106, 113, 114, 117, 129, 133, 137, 141, 143, 145, 147, 149, 157, 159, 165, 173, 189, 193, 205, 213, 217, 219, 229, 249, 265, 281, 285, 301, 309, 325, 357, 373, 389, 417, 433, 557
S_3	1, 2, 3, 4, 5, 6, 7, 8, 9, 10, 11, 12, 13, 14, 15, 16, 17, 18, 19, 21, 22, 23, 24, 25, 26, 27, 28, 29, 30, 33, 34, 35, 36, 37, 38, 39, 40, 41, 42, 43, 45, 47, 49, 50, 51, 52, 53, 54, 55, 57, 58, 60, 61, 64, 67, 68, 69, 70, 71, 72, 73, 74, 77, 78, 79, 81, 83, 85, 87, 89, 90, 91, 92, 93, 94, 97, 99, 101, 102, 106, 107, 110, 113, 114, 115, 117, 118, 120, 121, 124, 127, 129, 130, 131, 133, 137, 141, 142, 143, 145, 147, 149, 157, 159, 160, 165, 173, 174, 175, 177, 183, 189, 193, 194, 197, 203, 205, 208, 213, 214, 217, 219, 226, 227, 229, 235, 241, 242, 249, 257, 265, 267, 269, 270, 275, 281, 285, 288, 295, 301, 309, 311, 325, 327, 330, 335, 337, 349, 357, 373, 389, 393, 403, 405, 417, 419, 433, 435, 461, 467, 481, 484, 501, 527, 529, 533, 545, 557, 565, 575, 613, 657, 701, 729, 747, 757, 837, 857, 935, 1061, 1177

The algorithm searches for the solution of $W(F(a)) = m, m \leq n$ in increasing order of $s(a)$. Let M be the current value of $s(a)$. The algorithm enumerates all tuples a so that $s(a) = M$. Let m_0 be the smallest number not representable as $W(F())$[1]. By Lemma 5, if there is solution for $W(F(a)) = m_0$ then M must be greater than $\sqrt{m_0}$. If current value of M is at most $\sqrt{m_0}$, we increase M to $\lfloor \sqrt{m_0} \rfloor$. We implemented the push algorithm and tested it for $n = 10^6$. Let $T_{push}(n)$ be the number of computations of $W(F(\cdot))$. The number of computations T_{push} is essentially quadratic, see Figure 1(b). However, notice that the push algorithm demonstrates a speedup factor of 42 versus T_{naive}.

The second algorithm we implemented, *sweep algorithm*, sweeps tuples a according to the increasing sum $a_1 + a_2 + a_3 + a_4$. The sweep algorithm runs faster than the push algorithm. The ratio $T_{naive}(n)/T_{sweep}(n)$ is approximately 66, as illustrated in Figure 1(b).

We implemented another algorithm that we call *jump algorithm*. The idea is to sweep tuples a lexicographically and skip tuples that do not produce new numbers. We maintain the smallest number m_0 whose representation is not yet found. Suppose that $a_1 = c_1, a_2 = c_2$ and $a_3 = c_3$ are fixed constants and the algorithm starts search for a_4. Let f denote the function $f(a_4) = W(F(c_1, c_2, c_3, a_4))$. f is monotone and the equation $f(a_4) = m_0$ is quadratic. By solving this quadratic equation, we can find the largest value a_4^* so that $f(a_4^*)$ is at most m_0. Then a_4 takes values from a_4^* to $\sqrt{n}$. The experiments show that

1. for $n < 10^5$, the sweep algorithm is faster than others, and
2. for $n \geq 10^5$, the jump algorithm is faster than others, and

---

[1] Note that m_0 must be greater than 1177 since 1177 is not representable as $W(F(\cdot))$.

3. the speedup of the jump algorithm with respect to the naive algorithm grows with n.

We did a non-linear fit of the data using the Levenberg-Marquardt algorithm and the running time of the jump algorithm fits the polynomial equation $1.5841n^{1.51474}$. The asymptotic standard errors of the fit are 0.000655467 for the exponent and 0.0141145 for the coefficient, indicating an accurate fit.

The jump algorithm allows us to verify Conjecture 4 up to 10^8. The running time is 4.6 days on 360Mhz SGI MIPS R12K. We estimate that the sweep and naive algorithms would need 82.1 days and 14.1 years, respectively.

References

1. Bespamyatnikh, S., and Segal, M. Enumerating longest increasing subsequences and patience sorting. *Inform. Process. Lett. 76*, 1-2 (2000), 7–11.
2. Dankelmann, R. Computing the average distance of an interval graph. *Information Processing Letters 48* (1993), 311–314.
3. Dobrynin, A. A., Entringer, R., and Gutman, I. Wiener index of trees: Theory and applications. *Acta Applicandae Mathematicae 66* (2001), 211–249.
4. Estrada, E., and Uriarte, E. Recent advances on the role of topological indices in drug discovery research. *Current Medicinal Chemistry 8* (2001), 1573–1588.
5. Frederickson, G. N., and Johnson, D. B. Generalized selection and ranking: sorted matrices. *SIAM Journal of Computing 13*, 1 (1984), 14–30.
6. Goldman, D., Istrail, S., Lancia, G., and Piccolboni, A. Algorithmic strategies in combinatorial chemistry. In *Proc. 11th ACM-SIAM Sympos. Discrete Algorithms* (2000), pp. 275–284.
7. Gutman, I., and Polansky, O. E. *Mathematical concepts in organic chemistry*. Springer-Verlag, Berlin, 1986.
8. Gutman, I., and Yeh, Y. The sum of all distances in bipartite graphs. *Math. Slovaca 45* (1995), 327–334.
9. Ireland, K. F., and Rosen, M. I. *A classical introduction to modern number theory*. Springer-Verlag, New York, 1990.
10. Lepović, M., and Gutman, I. A collective property of trees and chemical trees. *J. Chem. Inf. Comput. Sci. 38* (1998), 823–826.
11. Martin, Y. C. 3D QSAR. current state, scope, and limitations. *Perspect. Drug Discovery Des. 12* (1998), 3–32.
12. Mohar, B., and Pisanski, T. How to compute the wiener index of a graph. *Journal of Mathematical Chemistry 2* (1988), 267–277.
13. Wiener, H. Structural determination of paraffin boiling points. *J. Amer. Chem. Soc. 69* (1947), 17–20.

Double Digest Revisited: Complexity and Approximability in the Presence of Noisy Data

Mark Cieliebak[1], Stephan Eidenbenz[2], and Gerhard J. Woeginger[3]

[1] Institute of Theoretical Computer Science, ETH Zurich,
cieliebak@inf.ethz.ch
[2] Basic and Applied Simulation Science (CCS-5), Los Alamos National Laboratory[†],
eidenben@lanl.gov
[3] Faculty of Mathematical Sciences, University of Twente and
Department of Mathematics and Computer Science, TU Eindhoven,
g.j.woeginger@math.utwente.nl

Abstract. We revisit the DOUBLE DIGEST problem, which occurs in sequencing of large DNA strings and consists of reconstructing the relative positions of cut sites from two different enzymes: we first show that DOUBLE DIGEST is strongly NP–complete, improving upon previous results that only showed weak NP–completeness. Even the (experimentally more meaningful) variation in which we disallow coincident cut sites turns out to be strongly NP–complete. In a second part, we model errors in data as they occur in real–life experiments: we propose several optimization variations of DOUBLE DIGEST that model partial cleavage errors. We then show APX–completeness for most of these variations. In a third part, we investigate these variations with the additional restriction that conincident cut sites are disallowed, and we show that it is NP–hard to even find feasible solutions in this case, thus making it impossible to guarantee any approximation ratio at all.

1 Introduction

Double digest experiments are a standard approach to construct physical maps of DNA. Given a large DNA molecule, which for our purposes is an unknown string over the alphabet $\{A, C, G, T\}$, the objective is to find the locations of markers, i.e., occurrences of short substrings such as $GAATTC$, on the DNA. Physical maps are required e.g. in DNA sequencing in order to determine the sequence of nucleotides ($A, C, G,$ and T) of large DNA molecules, since current sequencing methods allow only to sequence DNA fragments with tens of thousands of nucleotides, while a DNA molecule can have up to 10^8 nucleotides.

In double digest experiments, two enzymes are used to cleave the DNA molecule. An enzyme is a protein that cuts a DNA molecule at specific patterns, the restriction sites. For instance, the enzyme EcoRI cuts at occurrences of the pattern $GAATTC$. Under appropriate experimental conditions, an enzyme

[†] LA–UR–03:0532; work done while at ETH Zurich

T. Warnow and B. Zhu (Eds.): COCOON 2003, LNCS 2697, pp. 519–527, 2003.
© Springer-Verlag Berlin Heidelberg 2003

cleaves at all restriction sites in the DNA. This process is called *(full) digestion*. Double digest experiments work in three stages: First, clones (copies) of the unknown DNA string are digested by an enzyme A; then a second set of clones is digested by another enzyme B; and finally a third set of clones is digested by a mix of both enzymes A and B, which we will refer to as C. This results in three multisets of DNA fragments. The lengths of these fragments (i.e., their number of nucleotides) are then measured for each multiset by using gel electrophoresis, a standard technique in molecular biology. This leaves us with three multisets of distances (the number of nucleotides) between all adjacent restriction sites, and the objective is to reconstruct the original ordering of the fragments in the DNA molecule, which is the DOUBLE DIGEST problem.

More formally, the DOUBLE DIGEST problem can be defined as follows, where $sum(S)$ denotes the sum of the elements in a set S, and $dist(P)$ is the set of all distances between two neighboring points in a set P of points on a line:

Definition 1 (DOUBLE DIGEST). *Given three multisets A, B and C of positive integers with $sum(A) = sum(B) = sum(C)$, are there three sets P^A, P^B and P^C of points on a line, each starting in 0, such that $dist(P^A) = A$, $dist(P^B) = B$ and $dist(P^C) = C$, and such that $P^A \cup P^B = P^C$?*

For example, given multisets $A = \{5, 15, 30\}$, $B = \{2, 12, 12, 24\}$ and $C = \{2, 5, 6, 6, 7, 24\}$ as an instance of DOUBLE DIGEST, then $P^A = \{0, 5, 20, 50\}$, $P^B = \{12, 14, 26, 50\}$ and $P^C = \{5, 12, 14, 20, 26, 50\}$ is a feasible solution (there may exist more solutions).

Due to its importance in molecular biology, the DOUBLE DIGEST problem has been the subject of intense research since the first successful restriction site mappings in the early 1970's [1,2]. The DOUBLE DIGEST problem is NP–complete [3], and several approaches including exponential algorithms, heuristics, additional experiments or computer–assisted interactive strategies have been proposed (and implemented) in order to tackle the problem [4,5,6,7,8]. The number of feasible maps for a DOUBLE DIGEST instance can be characterized by using alternating Eulerian paths in appropriate graph classes and can be exponential in the number of fragments [3,9,10,11]. For a survey, see [12] and [13].

The double digest experiment is usually carried out with two enzymes that cut at different restriction sites. A majority of all possible enzyme pairings of the more than 3000 known enzymes are pairs with such disjoint cutting behavior. On the other hand, some results in the literature rely on enzymes that cut at the same site in some cases (coincidences) [10]. In particular, NP–hardness of the DOUBLE DIGEST problem has so far only been shown using enzymes that allow for coincidences [3,12,14]. Indeed, such enzyme pairs exist, for example enzymes HaeIII and BalI. However, having two enzymes that are guaranteed to always cut at disjoint sites seems more natural and might lead – at least intuitively – to easier reconstruction problems. For example, such instances always fulfill $|C| = |A| + |B| - 1$ (where $|S|$ denotes the cardinality of set S). To reflect these different types of experiments, we define the DISJOINT DOUBLE DIGEST problem, which is equivalent to the DOUBLE DIGEST problem with the additional requirement that the two enzymes may never cut at the same site, or, equivalently, that P^A

and P^B are disjoint except for the first point (which is 0) and the last point (which is sum(A)).

The NP–hardness results for DOUBLE DIGEST in the literature [3,12,14] rely on reductions from weakly NP–complete problems (namely PARTITION). As a first set of results in this paper, we prove in Section 2 that both DOUBLE DIGEST and DISJOINT DOUBLE DIGEST are actually NP–complete in the strong sense by proposing reductions from 3–PARTITION.

In a second part of the paper, we model reality more closely by taking into account that double digest data usually contains errors. A *partial cleavage* error occurs when an enzyme fails to cut at a restriction site where it is supposed to cut; then one large fragment occurs in the data instead of the two (or even more) smaller fragments. Other error types, such as *fragment length* errors, *missing small fragments*, and *doublets* occur as well (see [5,7,6,14]), but we will focus on partial cleavage errors. They can occur for many reasons, e.g. improper reaction conditions or inaccurate DNA concentration (see e.g. [15] for a list of possible causes). A partial cleavage error occurs e.g. when an enzyme fails to cut at a site where it is supposed to cut in the first (or second) stage of the double digest experiment, but then does cut at this site in the third phase (where it is mixed with the other enzyme). Such an error usually will make it impossible to find a solution for the corresponding DOUBLE DIGEST instance. In fact, only $P^A \cup P^B \subseteq P^C$ can be guaranteed for any solution. Vice–versa, if an enzyme cuts only in the first (or second) phase, but fails to cut in the third phase, then we can only guarantee $P^C \subseteq P^A \cup P^B$.

In the presence of errors, usually the data is such that no exact solutions can be expected. Therefore, optimization criteria are necessary in order to compare and gauge solutions. We will define optimization variations of the DOUBLE DI-GEST problem taking into account different optimization criteria; our objective will be to find good approximation algorithms. An optimal solution for a problem instance with no errors will be a solution for the DOUBLE DIGEST problem itself.[1] Thus, the optimization problem cannot be computationally easier than the original DOUBLE DIGEST problem, and (strong) NP–hardness results for DOUBLE DIGEST carry over to the optimization problem.

A straight-forward optimization criterion for DOUBLE DIGEST is to minimize the absolute number of partial cleavage errors in a solution, i.e., to minimize $e(P^A, P^B, P^C) := |(P^A \cup P^B) - P^C| + |P^C - (P^A \cup P^B)|$ (recall that $|S|$ is the cardinality of set S). Here, points in $(P^A \cup P^B) - P^C$ correspond to errors where enzyme A or B failed to cut in the third phase of the experiment, and points in $P^C - (P^A \cup P^B)$ correspond to errors where either enzyme A or B failed to cut in the first resp. second phase. Unfortunately, the corresponding optimization problem MINIMUM ABSOLUTE ERROR DOUBLE DIGEST in which we try to find point sets P^A, P^B and P^C such that $e(P^A, P^B, P^C)$ is minimum cannot be approximated within any finite approximation ratio (unless P = NP), as a polynomial-time algorithm guaranteeing a finite approximation ratio could be used to solve the NP–complete DOUBLE DIGEST problem in polynomial-time.

---

[1] Of course, this only holds if the optimization criterion is well–designed.

We obtain a more sensible optimization criterion as follows: If we add $|A| + |B| + |C|$ as an offset to the number of errors, we obtain an optimization criterion which turns the absolute number of errors into a measure relative to the input size. The corresponding optimization problem is defined as follows:

Definition 2 (MINIMUM RELATIVE ERROR DOUBLE DIGEST). *Given three multisets A, B and C of positive integers with $sum(A) = sum(B) = sum(C)$, find three sets P^A, P^B and P^C of points on a line, each starting in 0, such that $dist(P^A) = A$, $dist(P^B) = B$ and $dist(P^C) = C$, and such that $r(P^A, P^B, P^C) := |A| + |B| + |C| + e(P^A, P^B, P^C)$ is minimum.*

Instead of counting the number of errors, measuring the total size of a solution is an optimization criterion that seems very natural, even if it does not model cleavage errors exactly. In this case, we want to minimize the total number of points in a solution, i.e., we minimize $|P^A \cup P^B \cup P^C|$. This yields the MINIMUM POINT DOUBLE DIGEST problem, which is defined anologous to MINIMUM RELATIVE ERROR DOUBLE DIGEST except for the minimization criterion.

We show in Section 3 that MINIMUM RELATIVE ERROR DOUBLE DIGEST and MINIMUM POINT DOUBLE DIGEST are APX–hard (i.e., there exists a constant $\varepsilon > 0$ such that no polynomial–time algorithm can guarantee to find approximate solutions that are at most a factor $1 + \varepsilon$ off the optimum solution, unless P = NP) by proposing gap–preserving reductions[2] from MAXIMUM TRIPARTITE MATCHING, using MAXIMUM 4–PARTITION as an intermediary problem. We also analyze a straight–forward approximation algorithm that works for both problems and that achieves an approximation ratio of 2 for MINIMUM RELATIVE ERROR DOUBLE DIGEST and a ratio of 3 for MINIMUM POINT DOUBLE DIGEST.

For each optimization problem, a variation can be defined where the enzymes may only cut at disjoint restriction sites (analogous to DISJOINT DOUBLE DIGEST). The corresponding optimization problems are called MINIMUM DISJOINT RELATIVE ERROR DOUBLE DIGEST and MINIMUM DISJOINT POINT DOUBLE DIGEST. In Section 4, we show that – rather surprisingly – they are even harder to solve than the unrestricted problems: it is NP–hard to even find feasible solutions. We establish this result by showing that the problem of disjointly arranging two given sets of numbers is already NP–hard. This arrangement problem – which we call DISJOINT ORDERING – is a subproblem that every algorithm for any DISJOINT DOUBLE DIGEST variations has to be able to solve; thus, no finite approximation ratio can be achieved for our optimization variations of DISJOINT DOUBLE DIGEST (unless P = NP). Moreover, the same result would also hold for other optimization criteria, since the proof depends only on the disjointness requirement.

In Section 5, we conclude with directions for future research. Due to space limitations, we only give proof sketches in this extended abstract for most of our results; detailed proofs will be given in the full paper.

---

[2] For an introduction to gap–preserving reductions, see [16].

2 Strong NP–Completeness of (DISJOINT) DOUBLE DIGEST

In this section, we show strong NP–completeness for the decision problems DOU-
BLE DIGEST and DISJOINT DOUBLE DIGEST. We present reductions from 3–
PARTITION, which is defined as follows: Given $3n$ integers $q_1, \ldots, q_{3n}$ and integer
h with $\sum_{i=1}^{3n} q_i = nh$ and $\frac{h}{4} < q_i < \frac{h}{2}$ for all $1 \le i \le 3n$, are there n disjoint
triples of q_i's such that each triple sums up to h? The 3–PARTITION problem
is NP–complete in the strong sense [17]. First, we extend the NP–completeness
result from [3] for the DOUBLE DIGEST problem.

Lemma 3. DOUBLE DIGEST *is strongly* NP*–complete.*

Proof. We reduce 3–PARTITION to DOUBLE DIGEST as follows: Given an in-
stance $q_1, \ldots, q_{3n}$ and h of 3–PARTITION, let $a_i = c_i = q_i$ for $1 \le i \le 3n$, and let
$b_j = h$ for $1 \le j \le n$. Then the three (multi-)sets of a_i's, b_j's and c_i's build an
instance of DOUBLE DIGEST. If there is a solution for the 3–PARTITION instance,
then there exist n disjoint triples of q_i's (and a_i's as well) such that each triple
sums up to h. Starting from 0, we arrange the distances a_i on a line such that
each three a_i's that belong to the same triple are adjacent. The same ordering
is used for the c_i's. This yields a solution for the DOUBLE DIGEST instance.
On the other hand, if there is a solution for the DOUBLE DIGEST instance, say
P^A, P^B and P^C, then there exist n subsets of c_i's such that each subset sums up
to h, since each point in P^B must occur in P^C as well, and all adjacent points
in P^B have distance h. These n subsets yield a solution for the 3–PARTITION
instance.   □

Lemma 4. DISJOINT DOUBLE DIGEST *is strongly* NP*–complete.*

Proof (sketch). We show strong NP–hardness by reducing 3–PARTITION to DIS-
JOINT DOUBLE DIGEST. Given an instance of 3–PARTITION, let $s = \sum_{i=1}^{3n} q_i$ and
$t = (n+1) \cdot s$. Let $a_i = q_i$ for $1 \le i \le 3n$, $\hat{a}_j = 2t$ for $1 \le j \le n-1$, $b_j = h + 2t$
for $1 \le j \le n-2$, $\hat{b}_k = h + t$ for $1 \le k \le 2$, $c_i = q_i$ for $1 \le i \le 3n$, and $\hat{c}_j = t$
for $1 \le j \le 2n-2$. Let A consist of the a_i's and $\hat{a}_j$'s, and B and C be defined
accordingly. Then A, B and C are our instance of DISJOINT DOUBLE DIGEST.
    Given a solution for the 3–PARTITION instance, we assume w.l.o.g. that the
q_i's (and thus the a_i's and c_i's) are ordered such that the three elements of
each triple are adjacent. The arrangement shown in the figure below yields a
solution for the DISJOINT DOUBLE DIGEST instance. For the opposite direction,
let P^A, P^B and P^C be a solution for the DISJOINT DOUBLE DIGEST instance.
Each two adjacent points in P^B differ by h (plus t or $2t$), and so do $n+1$ points
in P^C. Hence, there must be n subsets of c_i's that each sum up to h, yielding a
solution for the 3–PARTITION instance.

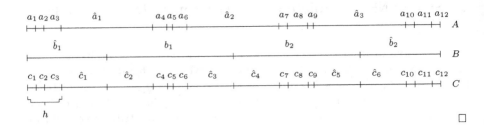

3   Approximability of Minimum Relative Error Double Digest and Minimum Point Double Digest

In this section, we show that Minimum Relative Error Double Digest and Minimum Point Double Digest are both APX–hard, and we propose a straight–forward approximation algorithm that achieves an approximation ratio of 3 respectively 2 for the two problems. For the proof of APX-hardness, we introduce a maximization variation of the well–known 4–Partition problem [17] which is defined as follows:

Definition 5 (Maximum 4–Partition). *Given an integer h and a multiset $Q = \{q_1, \ldots, q_{4n}\}$ of $4n$ integers with $\sum_{i=1}^{4n} q_i = nh$ and $\frac{h}{5} < q_i < \frac{h}{3}$, find a maximum number of disjoint subsets $S_1, \ldots, S_m \subseteq Q$ such that the elements in each set S_i sum up to h.*

While Maximum 4–Partition may be an interesting problem per se, we are mainly interested in it as an intermediary problem on our way to proving APX–hardness for our optimization variations of Double Digest.

Lemma 6. Maximum 4–Partition *is* APX–*hard.*

Proof (sketch). The lemma follows from the original reduction from Maximum Tripartite Matching to 4–Partition given in [17, pages 97–99], if analyzed as a gap-preserving reduction.   □

Lemma 7. Minimum Point Double Digest *is* APX–*hard.*

Proof (sketch). We propose a gap–preserving reduction from Maximum 4–Partition to Minimum Point Double Digest. For a given Maximum 4–Partition instance I, consisting of Q and h, we construct an instance I' of Minimum Relative Error Double Digest as follows: Let $A = C = Q$, and let B contain n times the element h.
    Let OPT denote the size of an optimum solution for I, and let OPT' denote the size of an optimum solution for I'. Then we have: if $OPT \geq n$, then $OPT' \leq 4n + 1$, and if $OPT < (1 - \varepsilon)n$ for a small constant $\varepsilon > 0$, then $OPT' > (4 + \frac{\varepsilon}{2})n + 1$. These two implications describe the reduction as gap-preserving and thus establish the result.   □

Lemma 8. MINIMUM RELATIVE ERROR DOUBLE DIGEST *is* APX–*hard.*

Proof (sketch). The proof uses the same reduction as in Lemma 7 with slightly modified implications.   □

A straight–forward approximation algorithm for our two problems simply arranges all distances from A, B and C on a line in a random fashion, starting at 0. If we analyze this algorithm as an approximation algorithm for MINIMUM POINT DOUBLE DIGEST, we see that this will result in a solution with at most $|A| + |B| + |C| - 1$ points; on the other hand, an optimum solution will always use at least $\max(|A|, |B|, |C|) + 1$ points. Thus, this trivial approximation algorithm achieves an approximation ratio of 3 for MINIMUM POINT DOUBLE DIGEST. The same algorithm yields an approximation ratio of 2 for MINIMUM RELATIVE ERROR DOUBLE DIGEST.

4   NP–hardness of Finding Feasible Solutions for Optimization Variations of DISJOINT DOUBLE DIGEST

In this section, we show for all DOUBLE DIGEST optimization variations in which we disallow coincidences that there cannot be a polynomial–time approximation algorithm with finite approximation ratio, unless P = NP. We achieve this by showing that even finding feasible solutions for these problems is NP–hard. To this end, we introduce the decision problem DISJOINT ORDERING which is defined as follows:

Definition 9 (DISJOINT ORDERING). *Given two multisets A and B of integers with $sum(A) = sum(B)$, find two sets P^A and P^B of points on a line, starting in 0, such that $dist(P^A) = A$, $dist(P^B) = B$, and such that P^A and P^B are disjoint except for the first and the last point.*

Lemma 10. DISJOINT ORDERING *is* NP–*complete.*

Proof (sketch). Obviously, DISJOINT ORDERING is in NP. To show NP–hardness, we reduce 3–PARTITION to it. Given an instance $q_1, \ldots, q_{3n}$ and h of 3–PARTITION, we construct an instance of DISJOINT ORDERING as follows. Let $a_i = q_i$ for $1 \leq i \leq 3n$, $\hat{a}_j = h$ for $1 \leq j \leq n + 1$, $b_i = h + 2$ for $1 \leq i \leq n$, and $\hat{b}_j = 1$ for $1 \leq j \leq (n + 1) \cdot h - 2n$. Let A consist of the a_i's and $\hat{a}_j$'s, and let B consist of the b_i's and $\hat{b}_j$'s. Then $sum(A) = sum(B) = (2n + 1) \cdot h$. In the full proof, we show that the following arrangement makes the reduction work: for A, blocks of three a_i's are separated by one $\hat{a}_j$, and for B, each two b_i's are separated by a block of $h - 2$ distances $\hat{b}_j$ (with the remaining $\hat{b}_j$'s at the beginning and end).   □

We reduce DISJOINT ORDERING to MINIMUM DISJOINT RELATIVE ERROR DOUBLE DIGEST as follows: Let A and B be an instance of DISJOINT ORDERING. We "construct" an instance of MINIMUM DISJOINT RELATIVE ERROR DOUBLE DIGEST by simply letting sets A and B be the same sets, and set C be the empty set. If an approximation algorithm for MINIMUM DISJOINT RELATIVE ERROR DOUBLE DIGEST finds a feasible solution for this instance, this yields immediately a solution for the DISJOINT ORDERING instance, since any solution feasible solution for MINIMUM DISJOINT RELATIVE ERROR DOUBLE DIGEST must arrange the elements from A and B in a disjoint fashion. The same argument applies for MINIMUM DISJOINT POINT DOUBLE DIGEST, and for any other (reasonable) optimization variation of DISJOINT DOUBLE DIGEST since the reduction is totally independent of the optimization criterion. Thus, we have:

Lemma 11. *No polynomial–time approximation algorithm can achieve a finite approximation ratio for* MINIMUM DISJOINT RELATIVE ERROR DOUBLE DIGEST, MINIMUM DISJOINT POINT DOUBLE DIGEST, *or any other (reasonable) optimization variation of* DISJOINT DOUBLE DIGEST, *unless* P = NP.

5   Conclusion

In this paper, we showed that DOUBLE DIGEST and DISJOINT DOUBLE DIGEST are strongly NP–complete; in a second part, we defined several optimization variations of DOUBLE DIGEST that model partial cleavage errors, proved APX–hardness for MINIMUM RELATIVE ERROR DOUBLE DIGEST and MINIMUM POINT DOUBLE DIGEST, and analyzed straight–forward approximation algorithms for these problems that achieve constant approximation ratios. In a last set of results, we showed for DOUBLE DIGEST optimization variations where conincidences are not allowed that even finding feasible solutions is NP–hard.

While our approximability results are tight for all DISJOINT DOUBLE DIGEST variations, our results leave considerable gaps regarding the exact approximability threshold for MINIMUM RELATIVE ERROR DOUBLE DIGEST and MINIMUM POINT DOUBLE DIGEST, which present challenges for future research. In a different direction of future research, optimization variations of DOUBLE DIGEST that model the three other error types (i.e., fragment length, missing small fragments, and doublets) or even combinations of different error types should be defined and studied. On a meta–level of arguing, it seems unlikely that an optimization variation that models partial cleavage errors *and* some of the other error types could be any easier than the problems that model only partial cleavage errors, but there is a possibility that some error types might offset each other in a cleverly defined optimization problem.

References

1. Smith, H.O., Wilcox, K.W.: A restriction enzyme from hemophilus influenza. I. Purification and general properties. Journal of Molecular Biology **51** (1970) 379–391
2. Danna, K.J., Nathans, D.: Specific cleavage of simian virus 40 DNA by restriction endonuclease of hemophilus influenzal. Proc. of the National Academy of Sciences USA **68** (1971) 2913–2917

3. Goldstein, L., Waterman, M.S.: Mapping DNA by stochastic relaxation. Advances in Applied Mathematics **8** (1987) 194–207
4. Bellon, B.: Construction of restriction maps. Computer Applications in the Biosciences (CABIOS) **4** (1988) 111–115
5. Allison, L., Yee, C.N.: Restriction site mapping is in separation theory. Computer Applications in the Biosciences (CABIOS) **4** (1988) 97–101
6. Wright, L.W., Lichter, J.B., Reinitz, J., Shifman, M.A., Kidd, K.K., Miller, P.L.: Computer–assisted restriction mapping: an integrated approach to handling experimental uncertainty. Computer Applications in the Biosciences (CABIOS) **10** (1994) 435–442
7. Inglehart, J., Nelson, P.C.: On the limitations of automated restriction mapping. Computer Applications in the Biosciences (CABIOS) **10** (1994) 249–261
8. Kao, M.Y., Samet, J., Sung, W.K.: The enhanced double digest problem for DNA physical mapping. In: Proc. of the 7th Scandinavian Workshop on Algorithm Theory (SWAT00). (2000) 383–392
9. Schmitt, W., Waterman, M.S.: Multiple solutions of DNA restriction mapping problems. Advances in Applied Mathematics **12** (1991) 412–427
10. Martin, D.R.: Equivalence classes for the double–digest problem with coincident cut sites. Jounal of Computational Biology **1** (1994) 241–253
11. Pevzner, P.A.: DNA physical mapping and alternating Eulerian cycles in colored graphs. Algorithmica **13** (1995) 77–105
12. Waterman, M.S.: Introduction to Computational Biology. Chapman & Hall (1995)
13. Pevzner, P.A.: Computational Molecular Biology. MIT Press (2000)
14. Setubal, J., Meidanis, J.: Introduction to Computational Molecular Biology. PWS Publishing Company (1997)
15. Promega GmbH http://www.promega.com/guides/re_guide/toc.htm: Restriction Enzymes Resource. (2002)
16. Arora, S., Lund, C.: Hardness of approximations. In Hochbaum, D., ed.: Approximation Algorithms for NP–Hard Problems. PWS Publishing Company (1996) 399–446
17. Garey, M.R., Johnson, D.S.: Computers and Intractability: A Guide to the Theory of NP–Completeness. Freeman (1979)

Fast and Space-Efficient Location of Heavy or Dense Segments in Run-Length Encoded Sequences

(Extended Abstract)

Ronald I. Greenberg

Loyola University, 6525 N. Sheridan Rd., Chicago, IL 60626, USA,
rig@cs.luc.edu,
http://www.cs.luc.edu/~rig

Abstract. This paper considers several variations of an optimization problem with potential applications in such areas as biomolecular sequence analysis and image processing. Given a sequence of items, each with a weight and a length, the goal is to find a subsequence of consecutive items of optimal value, where value is either total weight or total weight divided by total length. There may also be a specified lower and/or upper bound on the acceptable length of subsequences. This paper shows that all the variations of the problem are solvable in linear time and space even with non-uniform item lengths and divisible items, implying that run-length encoded sequences can be handled in time and space linear in the number of runs. Furthermore, some problem variations can be solved in constant space. Also, these time and space bounds suffice for certain problem variations in which we call for reporting of many "good" subsequences.

Keywords: maximum consecutive subsequence sum, maximum-density segments, biomolecular sequence analysis, bioinformatics, image processing, data compression

1 Introduction

Let S be a sequence comprised of n *runs*, where the ith run $(1 \leq i \leq n)$ has weight w_i and length $l_i \geq 0$. Initially, we define a segment of S to be a consecutive subsequence of runs, i.e., segment $S(i, j)$ is comprised of runs i through j. The weight of $S(i, j)$ is

$$weight(i, j) = \sum_{k=i}^{j} w_k \ ,$$

the length of $S(i, j)$ is

$$length(i, j) = \sum_{k=i}^{j} l_k \ ,$$

T. Warnow and B. Zhu (Eds.): COCOON 2003, LNCS 2697, pp. 528–536, 2003.
© Springer-Verlag Berlin Heidelberg 2003

and the density of $S(i,j)$ is

$$density(i,j) = weight(i,j)/length(i,j) \ .$$

Prior works on algorithms for biomolecular sequence analysis have considered the problem of finding a segment of S that is *heaviest* (maximizing $weight(i,j)$) or *densest* (maximizing $density(i,j)$), subject to constraints that the segment length must be at least L and/or at most U [1,2,3,4].[1] For brevity, we will refer to these as an L constraint and/or U constraint. In addition, the heaviest segment problem with no constraints on segment length is discussed by Bentley [5]. (This version of the problem was motivated by image processing tasks.) Note that for heaviest segments only we may consider empty segments, which may be represented by choosing $i > j$.

Most of the prior results were for the *uniform* version of the problem in which $l_i = 1$ for all i; exceptions will be noted below. (In the uniform case, the problem is often described as one of finding a subsequence of consecutive items of maximum sum or of maximum average.) We will also introduce below a new variation of the non-uniform version of the problem that we will refer to as the non-uniform case with breakable or non-atomic runs. With non-atomic runs, we will allow each end of a segment to include just a portion of the length of a run. When a run is partially included in a segment, a pro rata portion of its weight will be included in the weight of the segment.

While the non-uniform problem with atomic runs was considered by Goldwasser et al. [1] and an interesting application might be discovered, a particularly interesting use of the non-uniform model would be for working with sequences that have been compressed. That is, given a sequence under the uniform model, a simple compaction would be to replace any set of r consecutive items of weight w with a run of length r and weight wr under the non-uniform model, which corresponds to the standard compression technique of *run-length encoding*. Run-length encoding tends to be particularly useful in monochrome image processing. It could also have potential for such applications as DNA sequence analysis, since each item in a DNA sequence is chosen from just four different nucleotides. (Furthermore, in DNA sequence analysis, researchers may often be interested, on the first pass, in just a binary distinction between C/G and other [6,7,8,9,10].) To work with such a compressed sequence but be able to find a heaviest or densest segment in the uncompressed sequence, we must be able to break runs.

We begin by reviewing prior results for heaviest segments and then for densest segments. When discussing the problems together, we may use the term *optimal* to mean heaviest or densest, and we may refer to the weight or density of a segment as its *value*.

The first result was an unpublished result of Kadane [5] for the unconstrained heaviest segment problem. This solution uses $O(n)$ time and constant space (be-

---

[1] Note that Lin et al. [3] use the term "heaviest" to mean what we define as "optimal" below.

yond the space used to represent the input)[2]. With an L constraint, an algorithm of Huang [4] can be used to find a heaviest segment in $O(n)$ time and $O(n)$ space as noted by Lin et al. [3]. Lin et al. further showed how to obtain the same time and space bounds with both an L constraint and a U constraint [3].

In the case of finding a densest segment, the problem is trivially solvable in $O(n)$ time and constant space if there is no L constraint; just find the single run of maximum density. Lin et al. [3] showed that with an L constraint, a densest segment can be found in time and space $O(n \lg L)$. Goldwasser et al. [2, 1] improved the time and space bounds to $O(n)$. They further showed that the same bounds hold with both an L constraint and a U constraint [1]. In addition, they showed that with only an L constraint, the results could be extended to the non-uniform version of the problem [1]. Finally, they showed that with both constraints, $O(n + n \lg(U - L + 1))$ time suffices when $l_i \geq 1$ for all i. Goldwasser and I have, however, observed that the analysis in [1] can be modified to yield $O(n)$ time and space for any lengths satisfying $l_i \geq 0$ for all i.

This paper explains in Sect. 2 why all the results mentioned so far can be extended to the non-uniform model with atomic runs. In Sect. 3, we show that the space usage for finding a heaviest segment with an L constraint can be reduced to constant space. In Sect. 4, we show that all the results can be maintained even if we allow breakable runs. As indicated above, the use of breakable runs is of particular interest in connection with run-length encoded sequences, but the results of Sect. 4 apply even when l_i values are allowed to be nonintegral and when runs can be broken into any fraction. In Sect. 5, we consider variations on the problem in which we seek not just one optimal segment but all optimal segments or all optimal segments of maximal or minimal length, or even a more general concept as considered by Huang [4].

2   The Non-uniform Model

Most of the prior algorithms for finding a heaviest segment or finding a densest segment may essentially be cast into the following basic framework. (The approach of Huang [4] is somewhat exceptional, and we show in Section 3 that it can be greatly simplified when we seek only a heaviest segment with L constraint.) We sweep left to right across the given sequence considering each position in turn as a possibility for the right endpoint of an optimal segment. At each such step we determine a best choice of the left endpoint, called a "good partner" for the current right endpoint, that is at least as far to the right as the prior good partner. It is relatively easy to see that the good partner should never "back up" when seeking a heaviest segment. For densest segments, the

---

[2] This measure of space usage is analogous to the concept of algorithms that sort in-place (e.g., [11]) by using only a constant amount of storage outside the input array. Interestingly, the algorithm of Kadane has an even stronger property that one need not store the entire input array at one time; rather one may read the input piecemeal and never use more than constant storage in a strict sense.

correctness of this approach is based on the following lemma that is essentially the same as one proven by Goldwasser et al. [2, Lemma 9]:

Lemma 1. *Let $S(i,j)$ be a densest segment among those ending at index j and having length at least L. Similarly, let $S(i',j')$ be a densest segment among those ending at index j' and having length at least L. If $j' > j$ and $i' < i$, then density$(i,j) \geq$ density(i',j').*   □

Except in the unconstrained heaviest segment problem, the existing algorithms make use of a cleverly precomputed data structure of size $O(n)$ to determine how far to move the left endpoint at each step without overshooting the proper location for the good partner of the right endpoint. (In the unconstrained heaviest segment problem, no such data structure is necessary, because a simple check indicates whether the left endpoint should stay at the same position as in the last step or move to the same position as the right endpoint.)

For the most part, the l_i values of the input sequence are irrelevant to the operation of the algorithms that find an optimal segment. The main place they have an effect is in providing an additional constraint (beside the constraint that the good partner does not back up) on the range of indices to consider for the current good partner. In the uniform case, the additional constraint is trivial; a good partner of position j must be in the range $j - U$ to $j - L$. In the non-uniform case, however, these constraints are easily precomputed. In $O(n)$ time and space, a simple scan through the input sequence allows us to calculate U_j and L_j for all j, such that the good partner for j is between U_j and L_j. Instead of precomputing, these constraints can actually be managed on the fly, so that constant space will suffice for finding a heaviest sequence with an L constraint as shown in Section 3.

There is one more complication involved in finding a densest segment with L and U constraints. This problem is actually solved by dividing the input sequence into contiguous blocks of length $U - L$ before proceeding with any other operations. Thus, the problem of finding a good partner breaks down into a problem of comparing a good partner found in a specific block with no explicit U constraint to a good partner found in the next farther block with no explicit L constraint. Whereas Goldwasser et al. [1] proposed dividing the sequence into blocks of $U - L$ runs, Goldwasser and I have observed that dividing into blocks of length at least $U - L$ and as close as possible to $U - L$ yields $O(n)$ time and space for finding a densest segment.

The above observations are encapsulated in the following theorem:

Theorem 2. *$O(n)$ time suffices to find a length-constrained heaviest segment or densest segment even with non-uniform run lengths.*   □

All results given later in this paper will also be applicable to the non-uniform model.

3   Constant-Space Location of a Heaviest Segment with an L Constraint

In this section, we show that a heaviest segment with an L constraint but no U constraint can be found in $O(n)$ time and constant space, improving on the $O(n)$ space result that follows from the approach of Huang [4].

As in the approaches discussed in Sect. 2, we make a scan left to right across the input sequence, considering each position in turn as a possible location for the right endpoint of a heaviest segment. As we do so, we keep track of the good partner (a best left endpoint for the current right endpoint), which also moves rightward. Since $weight(i, j)$ is just $weight(i, j-1) + w_j$, a good partner p of $j-1$ serves as a good partner of j unless a segment of higher weight than $S(p, j)$ is obtained by considering left endpoints p' with $length(p', j) \geq L \geq length(p', j-1)$. By keeping track of the location that is length L away from j, as well as keeping track of the current good partner and the heaviest segment seen so far, we can find a heaviest segment in $O(n)$ time and constant space. We present the algorithm in Fig. 1, but, for simplicity, we find only the $weight$ of a heaviest segment; it should be clear that we could easily keep track of an actual heaviest segment as well. The pseudocode in Fig. 1 also incorporates the use of non-uniform lengths as discussed in Sect. 2.

```
1     Lherestart ← 1
2     maxsofar ← maxendinghere ← Lherewt ← Lherelength ← 0
3     for j ← 1 to n do
4         Lherelength ← Lherelength + l_j
5         Lherewt ← Lherewt + w_j
6         maxendinghere ← maxendinghere + w_j
7         while Lherestart ≤ j and Lherelength − l_Lherestart ≥ L do
8             Lherelength ← Lherelength − l_Lherestart
9             Lherewt ← Lherewt − w_Lherestart
10            Lherestart ← Lherestart + 1
11            maxendinghere ← max{maxendinghere, Lherewt}
12        endwhile
13        if Lherelength ≥ L
14        then maxsofar ← max{maxsofar, maxendinghere}
15        endif
16    endfor
```

Fig. 1. The algorithm to find the weight of a heaviest segment with L constraint and non-uniform lengths in $O(n)$ time and constant space.

We summarize with the following theorem:

Theorem 3. *A heaviest segment with length greater than or equal to L can be found in $O(n)$ time and constant space.*   □

4 Non-atomic Runs

In this section, we show that all the results so far can be extended to work with non-atomic runs. Note that all the running times remain linear in the number of runs and that L, U, and the l_i values may even be nonintegral. We make use of the following two lemmas:

Lemma 4. *To find an optimal (heaviest or densest) segment, we need only consider a segment containing a partial run if its length is exactly L or U.*

Proof. Consider a segment of length strictly between L and U that contains a partial run. Then the length constraints allow us to use more of this run or trim off part of this run. One of these changes must not decrease the value (weight or density) of the segment, since the density of a run is considered to be uniform. $\square$

Lemma 5. *To find an optimal segment, we may limit attention to segments with a partial run on at most one end.*

Proof. Consider a segment with partial runs on each end. Without loss of generality, suppose that that the run truncated on the left has lower density than the run truncated on the right. If the utilized portion of the run on the left is shorter than the unutilized portion of the run on the right, we can completely eliminate the run on the left and add a corresponding portion of the run on the right. On the other hand, if the utilized portion of the run on the left is longer than the unutilized portion of the run on the right, we can completely include the run on the right at the expense of the run on the left. Either way, neither the weight nor density of the segment will decrease. $\square$

With these lemmas in mind, we see that we need not deviate too far from working with atomic runs to obtain an optimal segment when we are allowed to break runs. We need only consider small adjustments in addition to each of the segments considered as a possible optimal segment when working with atomic runs. For example, to update the algorithm of Fig. 1 to work with breakable runs, we can just add the following code after Line 11:

> **if** $Lherelength - L < l_j$
> **then** $Lhereadjwt \leftarrow Lherewt - (Lherelength - L)w_j/l_j$
> $\quad maxendinghere \leftarrow \max\{maxendinghere, Lhereadjwt\}$
> **endif**

and the following code after Line 14:

> $Lhereadjwt \leftarrow Lherewt - (Lherelength - L)w_{Lherestart}/l_{Lherestart}$
> $maxsofar \leftarrow \max\{maxsofar, Lhereadjwt\}$

(Note that this modification is correct under the assumption that $L \geq 0$, which places no restriction on the utility of the algorithm, since $l_i \geq 0$ for all i.)

Incorporating other variations of the problem as well, we can state the following theorem:

Theorem 6. $O(n)$ *time and space suffices to find a length-constrained heaviest segment or densest segment even with non-atomic runs. Furthermore, constant space suffices to find a heaviest segment with no upper bound on segment length.*   □

5   Finding "All" Optimal Segments

Huang's algorithm [4] actually generates not only a heaviest segment (of length at least L) but all heaviest segments that cannot be extended, i.e., that are of maximal length. In fact, Huang's algorithm finds all segments of maximal length that are at least as heavy as any overlapping segment.[3] In this section, we show that if one seeks to report even all optimal segments, the run time may be $\Theta(n^2)$ in the worst case, but reporting all optimal segments of *minimal* length does not change any of the time and space bounds discussed so far. We also show that Huang's goal can be achieved for heavy segments with a simpler algorithm and that the space usage can be reduced from two numeric arrays of length n to a single boolean array of length n.

We begin with the worst-case lower bound for finding all optimal segments. This result is actually quite trivial, since an input sequence with $w_i = 0$ for all i makes every segment optimal with $L = 0$ and no U constraint.

Theorem 7. *Finding all optimal segments requires* $\Theta(n^2)$ *time in the worst case.*   □

(It is also possible to construct more interesting sequences with many optimal segments. Furthermore, while a U constraint will keep the number of optimal segments below n^2, the number of optimal segments may still be $\Theta(nU)$.)

We now note that finding all optimal segments of minimal length can be done with the same time and space bounds as finding one optimal segment:

Theorem 8. *Reporting all optimal segments of minimal length can be done in* $O(n)$ *time and space. For heaviest segments with no U constraint, the space can be reduced to $O(1)$.*

Proof. We can begin by simply finding the optimal value (weight or density of an optimal segment) using the algorithms discussed so far. Then we can essentially rerun the same algorithm but report an optimal segment each time that we find a good partner of a right endpoint for which the value of the corresponding segment equals the optimal value. The only remaining detail is that where there is a tie among candidates for the good partner, we must choose the rightmost good partner. This choice is easy to make; for example in the algorithm for heaviest sequences with an L constraint (Fig. 1), we could maintain good partner information as well as best value information at Lines 11 and 14. When there is a tie in the two values to which we are applying the max operator, we would retain the good partner corresponding to *Lherewt* in preference to *maxendinghere*, and *maxendinghere* in preference to *maxsofar*.   □

---

[3] Note that Huang uses the term "optimal" for a different meaning than in this paper.

```
1     maxbeghereneg[n + 1] ← TRUE
2     maxbeginninghere ← 0
3     for j ← n downto 1 do
4         if maxbeginninghere + w_j < 0 then
5             maxbeginninghere ← 0
6             maxbeghereneg[j] ← TRUE
7         else
8             maxbeginninghere ← maxbeginninghere + w_j
9             maxbeghereneg[j] ← FALSE
10        endif
11    endfor
12    i ← 1
13    maxendinghere ← 0
14    for j ← 1 to n do
15        if maxendinghere + w_j < 0 then
16            maxendinghere ← 0
17            i ← j + 1
18        else
19            maxendinghere ← maxendinghere + w_j
20            if maxbeghereneg[j + 1] then Output (i, j) endif
21        endif
22    endfor
```

Fig. 2. The algorithm to report all segments of maximal length that are at least as heavy as any overlapping segment. In this code, $maxendinghere$ and $maxbeginninghere$ represent weights of possibly empty segments, whereas $maxbeghereneg$ is an boolean array for flagging positions from which the heaviest non-empty interval is negative.

Finally, we give a simpler method than Huang's [4] to find all segments of maximal length (with no U constraint) that are at least as heavy as any overlapping segment, and we reduce the space usage. For simplicity, we stick to the uniform model ($l_i = 1 \forall i$) and $L = 0$. The enhancements of non-uniform lengths, breakable runs, and $L > 0$ can be incorporated by combining ideas presented in earlier sections of this paper. In the case we focus on now, however, our statement about space usage can be particularly strong; it includes even space used to store input data as long as we are allowed to read the input twice.

Theorem 9. *All segments of maximal length that are at least as heavy as any overlapping segment can be reported in $O(n)$ time using just one boolean array of length n plus constant additional space.*

Proof. The key observation is that $S(i, j)$ satisfies the criterion if and only if (1) the heaviest non-empty segment ending with position $i - 1$ and the heaviest non-empty segment beginning with position $j + 1$ each have negative weight (if they exist), and (2) for any position from i to j, the heaviest segment beginning there and the heaviest segment ending there have nonnegative weight. The algorithm in Fig. 2 completes the proof. (For simplicity, we report only non-empty segments.)

□

6 Conclusion

We have seen that finding a length-constrained heaviest segment or densest segment in a sequence composed of either atomic or non-atomic items, each of arbitrary weight and nonnegative length, can be accomplished in time and space linear in the number of items. Furthermore, constant space suffices to find a *heaviest* segment when there is no upper bound on segment length. In addition, the same results apply to finding all optimal segments of minimal length. A remaining open problem is to improve on the linear space requirement for variations of the problem that involve an upper bound on segment length or finding a *densest* segment.

Acknowledgments. Thank you to Michael Goldwasser of Loyola University for helpful discussions.

References

1. Goldwasser, M.H., Kao, M.Y., Lu, H.I.: Linear-time algorithms for computing maximum-density sequence segments with bioinformatics applications. Manuscript available at http://www.cs.luc.edu/~mhg/publications/DensityPreprint.pdf (2002)
2. Goldwasser, M.H., Kao, M.Y., Lu, H.I.: Fast algorithms for finding maximum-density segments of a sequence with applications to bioinformatics. In: Proc. of the Second Annual Workshop on Algorithms in Bioinformatics, Springer-Verlag (2002) 157–171
3. Ling Lin, Y., Jiang, T., Mao Chao, K.: Efficient algorithms for locating the length-constrained heaviest segments with applications to biomolecular sequence analysis. In: 27th International Symposium on Mathematical Foundations of Computer Science. Volume 2420 of Lecture Notes in Computer Science., Springer-Verlag (2002) 459–470 To appear in Journal of Computer and System Sciences.
4. Huang, X.: An algorithm for identifying regions of a DNA sequence that satisfy a content requirement. Computer Applications in the Biosciences **10** (1994) 219–225
5. Bentley, J.: Programming Pearls. Second edn. Addison-Wesley (2000)
6. Hannenhalli, S., Levy, S.: Promoter prediction in the human genome. Bioinformatics **17** (2001) S90–96
7. Nekrutenko, A., Li, W.H.: Assessment of compositional heterogeneity within and between eukaryotic genomes. Genome Research **10** (2000) 1986–1995
8. Larsen, F., Gundersen, R., Lopez, R., Prydz, H.: CpG islands as gene marker in the human genome. Genomics **13** (1992) 1095–1107
9. Hardison, R.C., Krane, D., Vandenbergh, C., Cheng, J.F.F., Mansberger, J., Taddie, J., Schwartz, S., Huang, X., Miller, W.: Sequence and comparative analysis of the rabbit alpha-like globin gene cluster reveals a rapid mode of evolution in a C+G rich region of mammalian genomes. Journal of Molecular Biology **222** (1991) 233–249
10. Gardiner-Garden, M., Frommer, M.: CpG islands in vertebrate genomes. Journal of Molecular Biology **196** (1987) 261–282
11. Cormen, T.H., Leiserson, C.E., Rivest, R.L., Stein, C.: Introduction to Algorithms. Second edn. McGraw-Hill (2001)

Genomic Distances under Deletions and Insertions

Mark Marron, Krister M. Swenson, and Bernard M. E. Moret

Department of Computer Science
University of New Mexico
Albuquerque, NM 87131, USA
ma_luen@eece.unm.edu
{kswenson,moret}@cs.unm.edu

Abstract. As more and more genomes are sequenced, evolutionary biologists are becoming increasingly interested in evolution at the level of whole genomes, in scenarios in which the genome evolves through insertions, deletions, and movements of genes along its chromosomes. In the mathematical model pioneered by Sankoff and others, a unichromosomal genome is represented by a signed permutation of a multi-set of genes; Hannenhalli and Pevzner showed that the edit distance between two signed permutations of the same set can be computed in polynomial time when all operations are inversions. El-Mabrouk extended that result to allow deletions and a limited form of insertions (which forbids duplications). In this paper we extend El-Mabrouk's work to handle duplications as well as insertions and present an alternate framework for computing (near) minimal edit sequences involving insertions, deletions, and inversions. We derive an error bound for our polynomial-time distance computation under various assumptions and present preliminary experimental results that suggest that performance in practice may be excellent, within a few percent of the actual distance.

Keywords: inversion distance, reversal distance, genomic distance, Hannenhalli-Pevzner

1 Introduction

Biologists can infer the ordering and strandedness of genes on a chromosome, and thus represent each chromosome by an ordering of signed genes (where the sign indicates the strand). These gene orders can be rearranged by evolutionary events such as inversions (also called reversals) and transpositions and, because they evolve slowly, give biologists an important new source of data for phylogeny reconstruction (see, e.g., [7,11,12,14]). Appropriate tools for analyzing such data may help resolve some difficult phylogenetic reconstruction problems. Developing such tools is thus an important area of research–indeed, the recent DCAF symposium [6] was devoted to this topic.

A natural optimization problem for phylogeny reconstruction from gene-order data is to reconstruct an evolutionary scenario with a minimum number of the permitted evolutionary events on the tree. This problem is NP-hard for most criteria–even the very simple problem of computing the median[1] of *three* genomes with identical gene

---

[1] The median of k genomes is a genome that minimizes the sum of the pairwise distances between itself and each of the k given genomes.

T. Warnow and B. Zhu (Eds.): COCOON 2003, LNCS 2697, pp. 537–547, 2003.
© Springer-Verlag Berlin Heidelberg 2003

content under such models is NP-hard [4,13]. The problem of computing the edit distance between two genomes is difficult; for instance, even with equal gene content and with only inversions allowed, the problem is NP-hard for unsigned permutations [3].

Hannenhalli and Pevzner [9] made a fundamental breakthrough by developing an elegant theory for signed permutations and providing a polynomial-time algorithm to compute the edit distance (and the corresponding shortest edit sequence) between two signed permutations under inversions; Bader et al. [1] later showed that this edit distance is computable in linear time. El-Mabrouk [8] extended the results of Hannenhalli and Pevzner to the computation of edit distances for inversions and deletions and also for inversions and non-duplicating insertions; she also gave an approximation algorithm with bounded error for computing edit distances in the presence of all three operations (inversions, deletions, and non-duplicating insertion).

In this paper, we extend El-Mabrouk's work by providing a polynomial-time approximation algorithm with bounded error to compute edit distances under unrestricted inversions, deletions, and insertions (including duplications). Our basic approach is based on a new canonical form for edit sequences along with the notion of a *cover* to deal with duplicates. We show that shortest edit sequences can be transformed into equivalent sequences of equal length in which all insertions are performed first, followed by all inversions, and then by all deletions. This canonical form allows us to take advantage of El-Mabrouk's exact algorithm for inversions and deletions, which we then extend by finding the best possible prefix of inversions, producing an approximate solution with bounded error.

Section 2 introduces some notation and definitions. Section 3 gives two key theorems that enable us to reduce edit sequences to a canonical form. Section 4 outlines our method for handling unrestricted insertions. Section 5 outlines our alternate method for analyzing the general case of insertion, deletion, and inversion along with an analysis of our algorithm's error bounds. Finally, Section 6 shows some preliminary empirical results.

2   Notation and Definitions

We denote a particular edit sequence with a Greek letter, π, its operations by subscripted letters, o_i, and its contents enclosed in angle brackets: $\pi = \langle o_1, o_2, \dots, o_n \rangle$. We assume that the desired (optimal) edit sequence is that which uses the fewest operations, with all operations counted equally. As in the standard statement of the equal gene content problem, we move from a perfectly sorted sequence S to the given target sequence T.

We say that substring s_i is *adjacent* to substring s_j whenever they occupy sequential indices in the string under study. Let $\text{sign}_{\min}(s_l)$ be the sign of the element of smallest index in s_l and $\text{sign}_{\max}(s_l)$ be the sign of the element of largest index in s_l; we define the *parity* of a pair of ordered strings (s_i, s_j) as $\text{sign}_{\min}(s_i) \cdot \text{sign}_{\max}(s_j)$.

When two strings s_i and s_j each contain a single character, $s_i = e_i$ and $s_j = e_j$, define their ordering $\zeta = e_i - e_j$. Two substrings of a subject sequence, s_i and s_j, are *correctly oriented* relative to each other if and only if:

1. s_i or s_j is empty.
2. s_i and s_j are both of unit length and, whenever s_i is adjacent to s_j with ordering ζ in the target sequence, then s_i is also adjacent to s_j with ordering ζ in the subject sequence.
3. All substrings in s_i are correctly oriented relative to each other, all substrings in s_j are correctly oriented relative to each other and, whenever s_i is adjacent to s_j with parity ξ in the target sequence, then also s_i is adjacent to s_j with parity ξ in the subject sequence.

We say that an operation *splits* s_i and s_j if the two sequences are correctly oriented before the operation, but not after it.

3 Canonical Forms

In this section, we prove several useful results about shortest edit sequences, results that will enable us to obtain a "canonical form" into which any shortest edit sequence can always be transformed without losing optimality.

We make use throughout our derivation of *reindexing*; this reindexing provides a pliability to the indices that operations act upon which enables us to manipulate the order in which these operations appear. For example, take the string $1, 2, 3, -5, -4, 6, 7, 11, 12$ and suppose that the next operation to perform is an inversion starting at index 4 and going to index 7 (inclusive). The result is the new string $1, 2, 3, -7, -6, 4, 5, 11, 12$. Now, suppose that, in order to achieve some desired form, we need to insert the element 10 at index 4 before the application of this inversion. The goal is to maintain the indices of the inversion so that it continues to act on the substring $-5, -4, 6, 7$. After the application of the insertion associated with index 4, we are left with $1, 2, 3, 10, -5, -4, 6, 7, 11, 12$. In order to maintain the integrity of the inversion, we now adjust the start index of the inversion to be at 5 and the end index to be at 8, upon which application of the inversion correctly yields $1, 2, 3, 10, -7, -6, 4, 5, 11, 12$. The other types of reindexing that we use for inversions and deletions follow a similar pattern.

Our first theorem extends an earlier result of Hannenhalli and Pevzner (who proved that a sorted substring need not be split in an inversion-only edit sequence [9]) by showing that, whenever two substrings are correctly oriented, there is always a minimum edit sequence that does not split them. The idea is to rewrite the optimal edit sequence to keep the substrings together. First define $move(s_x, s_y, \xi)$ to move s_x to the immediate left of s_y with parity ξ. Given an edit sequence $\langle o_1, o_2, \dots, o_k, \dots, o_m, \dots \rangle$ where operation o_k is responsible for splitting the substrings s_i and s_j and operation o_m returns them to their correctly oriented state; we rewrite the operations between o_k and o_m to keep the substrings together. To accomplish this each o_x is expanded into a tuple of operations $\langle f_x, o'_x, t_x \rangle_x$. This tuple is constructed so that the xth tuple is functionally equivalent to o_x and that t_x is the inverse of f_{x+1}. Further, the leading and trailing tuples are designed such that f_k and t_m are identity operations.

For example, suppose we have the sequence $14, 15, 11, 12, 13, 16$ and the operation sequence is $inv(2, 5), inv(1, 4), inv(4, 4), inv(4, 5)$. The first operation in this sequence splits the substring $14, 15$ and the fourth restores it. In this case we relocate 14 and 15 together by constructing the tuples as follows:

1. $inv(2,5) \rightarrow \langle move(14,15,1), inv(1,5), move(-14,-13,-1) \rangle$
2. $inv(1,4) \rightarrow \langle move(-14,16,-1), inv(1,3), move(-14,-15,1) \rangle$
3. $inv(4,4) \rightarrow \langle move(-14,16,-1), inv(\epsilon), move(-14,-15,1) \rangle$
4. $inv(5,5) \rightarrow \langle move(-14,16,-1), inv(4,5), move(14,15,1) \rangle$

The original operation sequence produces:

- $14,15,11,12,13,16 \Rightarrow 14,-13,-12,-11,-15,16 \Rightarrow$
  $11,12,13,-14,-15,16 \Rightarrow 11,12,13,14,-15,16 \Rightarrow 11,12,13,14,15,16$

The new operation sequence produces:
$14,15,11,12,13,16$

1. $\Rightarrow 14,15,11,12,13,16 \Rightarrow -13,-12,-11,-15,-14,16 \Rightarrow$
   $14,-13,-12,-11,-15,16 \Rightarrow$
2. $\Rightarrow -13,-12,-11,-15,-14,16 \Rightarrow 11,12,13,-15,-14,16 \Rightarrow$
   $11,12,13,-14,-15,16$
3. $\Rightarrow 11,12,13,-15,-14,16 \Rightarrow 11,12,13,-15,-14,16 \Rightarrow$
   $11,12,13,-14,-15,16$
4. $\Rightarrow 11,12,13,-15,-14,16 \Rightarrow 11,12,13,14,15,16 \Rightarrow 11,12,13,14,15,16$

We have demonstrated how the construction of the tuples can create an operation sequence where each tuple has the same effect as its corresponding operation in the original sequence and the opposing move operations cancel one another's effect (and can thus be discarded), thereby providing the intuition behind a proof of the following theorem.

Theorem 1. *If subsequences s_i and s_j are correctly oriented relative to each other at some step during the execution of the minimum edit sequence π, say at the kth step, then there is another minimum edit sequence, call it π', that has the same first k steps as π, and never splits s_i and s_j.*

Our next theorem shows that it is always possible to take any minimum edit sequence and transform it into a form where all of the insertions come first, followed by all of the inversions and then all of the deletions. The proof is based on the idea of rewriting each operation preceding the first insert such that, at the beginning and end of each operation rewrite group(tuple), the sequence is the same as at each step in the original sequence, but when the terms are regrouped and cancellation occurs, the insert is pushed to the front of the operator sequence. Since each step produces the same sequence we know that the resulting edit sequence is correct and the cancellation maintains the same number of operations in the new sequence as in the old one.

Theorem 2. *Given a minimal edit sequence $\pi = \langle o_1, o_2 \ldots o_{k-1}, ins_1, o_{k+1} \ldots o_m \rangle$ there is a π' such that $\pi' \equiv \pi$ and $\pi' = \langle ins_1 \ldots ins_p, inv_1 \ldots inv_q, del_1 \ldots del_r \rangle$.*

Proof. Reminiscent of our previous tuple expansion, for each o_j s.t. $j \leq k-1$ we create $\hat{o}_j = (ins'_j, o'_j, del'_j)$, where

$$del'_j = \begin{cases} ins'^{-1}_{j+1} & 1 \leq j \leq k-2 \\ ins_1^{-1} & j = k-1 \end{cases},$$

ins'_j is the inverse of del'_j, and o'_j is o_j reindexed to compensate for the insertion. Thus del'_j deletes whatever was inserted by ins'_j when o'_j is applied and the construction of each tuple ensures $o_j \equiv \hat{o}_j$.

Write $\pi' = \langle \hat{o}_1 \dots, o_{k-1}, ins_1, o_{k+1} \dots, o_m \rangle$; expanding each term $\hat{o}_j$, we get $\pi' = \langle (ins'_1, o'_1, del'_1), (ins'_2, o'_2, del'_2), \dots, (ins'_{k-1}, o'_{k-1}, del'_{k-1}), ins_1, \dots, o_m \rangle$; since del'_j and ins'_{j+1} cancel, the expression reduces to $\langle ins'_1, o'_1, o'_2, \dots, o'_{k-1}, o_{k+1}, \dots, o_m \rangle$. The construction for o'_j ensures that each $\hat{o}_j$ sequence is equivalent to o_j and the cancellation of the ins and del operators in $\hat{o}_j$ results in $|\pi| = |\pi'|$.

This reasoning shows how to move the first insertion to the front of the sequence; further insertion operations can be moved similarly.

These two theorems allow us to define a canonical form for edit sequences. That canonical form includes only inversions and deletions in its second and third parts, which is one of the cases for which El-Mabrouk gave an exact polynomial-time algorithm. We can use her algorithm to find the minimal edit sequence of inversions and deletions, then reconstruct the preceding sequence of insertions. Because this approach fixes the sequence of inversions and deletions without taking insertions into account, then only addresses insertions, it is an approximation, not an exact algorithm. We shall prove that the error is bounded and also give evidence that, in practice, the error is in fact very small.

4   Unrestricted Insertions

4.1   The Problem

The presence of duplicates in the sequence makes the analysis much more difficult; in particular, it prevents a direct application of the method of Hannenhalli and Pevzner's and thus also of that of El-Mabrouk's. We can solve this problem by assigning distinct names to each copy, but this approach begs the question of how to assign such names. Sankoff proposed the exemplar strategy [15], which attempts to identify, for each family of gene copies, the "original" gene (as distinct from its copies) and then discards all copies, thereby reducing a multi-set problem to the simpler set version. However, identifying exemplars is itself NP-hard [2]—and much potentially useful information is lost by discarding copies. Fortunately, we found a simple selection method, based on substring pairing, that retains a constant error bound.

4.2   Sequence Covers

Our job is to pick a group of substrings from the target such that every element in the source appears in one of those substrings. To formalize and use this property, we need a few definitions. Call a substring $e_1 e_2 \dots e_n$ *contiguous* if we have $\forall j,\ e_{j+1} = e_j + 1$. Given a contiguous substring s_i, define the *normalized* version of s_i to be s_i itself if the first element in s_i is positive and $inv(s_i)$ otherwise; thus the normalized version of s_i is a substring of the identity. A maximal subsequence S_{nd} of the source string S is the *non-deleted* portion of S if S_{nd}, viewed as a set, is the largest subset of elements in S that is also contained in the target string T, also viewed as a set. (Note that S_{nd} is not a substring, but a subsequence; that is, it may consist of several disjoint pieces of S.) Given a set C of normalized strings which are maximal in C under the substring relation, define $\uplus C$ to be the string produced as follows; order the strings of C lexicographically and concatenate them in that order, removing any overlap. We will say that a set C of contiguous substrings from T is a *cover* for S if S_{nd} is $\uplus C$. Note that a cover must contain only contiguous strings.

Suppose we have $S = 1, 2, 3, 4, 5, 6, 7$ and $T = 3, 4, 5, -4, -3, 5, 6, 7, 8$; then the set of normalized contiguous strings is $\{(3, 4, 5), (3, 4), (5, 6, 7, 8)\}$, S_{nd} is $(3, 4, 5, 6, 7, 8)$, a possible cover for S is $C_p = \{(3, 4, 5), (5, 6, 7, 8)\}$, and we have $\uplus C_p = (3, 4, 5, 6, 7, 8)$.

Let n be the size of (number of operations in) the minimal edit sequence.

Theorem 3. *There exists a cover for S of size $2n + 1$.*

Proof. By induction on n. For $n = 0$, S itself forms its own cover, since it is a contiguous sequence; hence the cover has size 1, obeying the bound. For the inductive step, note that deletions are irrelevant, since the cover only deals with the non-deleted portion; thus we need only verify that insertions and inversions obey the bound. An insertion between two contiguous sequences simply adds another piece. While one inside a contiguous sequence splits it and adds itself, for an increase of two pieces. Similarly, an inversion within a contiguous sequence cuts it into at most three pieces, for a net increase by two pieces, and an inversion across two or more contiguous sequences at worst cuts each of the two end sequences into two pieces, leaving the intervening sequences contiguous, also for a net increase by two pieces. Since we have $(2(n - 1) + 1) + 2 = 2n + 1$, the bound is obeyed in all cases.

4.3   Building a Minimum Cover

Let $C(T)$ be the set of all (normalized versions of) contiguous substrings of T that are maximal (none is a substring of any other). We will build our cover greedily from left to right with this simple idea: if, at some stage, we have a collection of strings in the current cover that, when run through the $\uplus$ operator, produces a string that is a prefix of length i of our target T, we consider all remaining strings in $C(T)$ that begin at or to the left of position i—that can extend the current cover—and select that which extends farthest to the right of position i. Although this is a simple (and efficient) greedy construction, it actually returns a minimum cover, as we can easily show by contradiction. (The proof follows standard lines: use contradiction, assume a first point of disagreement between an optimal cover and the greedy cover, and verify that we can exchange cover elements to move the disagreement farther down the index chain.) However, it should be noted that, in our sorting algorithm, the best choice of cover need not be a minimum cover—a minimum cover simply allows us to bound the error.

5   Our Algorithm

Now that we have a method to construct a minimal cover, we can assign unique labels to all duplicates which in turn enables the use of El-Mabrouk's approximation method.

However, for greater control of the error and to cast the problem into a more easily analyzed form, we choose to use El-Mabrouk's exact method for deletions only, and then to extend the resulting solution to handle the needed insertions.

To do this we will need to look at the target sequence T with all the elements that do not appear in S removed, call this new sequence T_{ir} to denote that all the inserted elements have been removed.

Theorem 4. *Let π be the minimal edit sequence from S to T, using l insertions and m inversions. Let π' be the minimal edit sequence of just inversions and deletions from S to T_{ir}. If the extension $\hat{\pi}$ is obtained by adding an insertion operation to π' for each of the inserted strings in T then $\hat{\pi}$ has at most $l + m$ insertions.*

Proof. Clearly, our method will do at least as well as looking at each inserted string in T and taking that as an insertion for $\hat{\pi}$. Now, looking at the possible effect of each type of operation on splitting a previous insertion, we have 3 cases (in all cases v is an inserted substring and x another):

1. Inserting another substring cannot split an inserted substring—it just creates a longer string of inserted elements: if x is inserted within v, then $uvw = uv_1v_2w$ becomes $uv_1xv_2w = uv'w$.
2. Deletion of a substring cannot split an inserted substring—it just shortens it, even perhaps to the point of eliminating it and thus potentially merging two neighboring strings: if part of v is deleted, then $uvw = uv_1v_2v_3w$ becomes $uv_1v_3w = uv'w$.
3. An inversion may split an inserted substring into two separate strings, thus increasing the number of inserted substrings by one. It cannot split a pair of inserted substrings because the inversion only rearranges the inserted substrings; it does not create new contiguous substrings. For instance, given $uvw = u_1u_2v_1v_2w$, an inversion that acts on u_2v_1 yields the string $u_1\overline{v_1u_2}v_2w = u_1v'\overline{u_2}v''w$.

Thus, if we have l insertions and m inversions in π, there can be at most $l + m \leq |\pi| = n$ inserted substrings in T and $l + m \leq n$ insertions in $\hat{\pi}$.

Now, the optimal edit sequence defines a corresponding optimal cover C_0 of S; if the cover we obtain, C, is in fact the optimal cover, then our algorithm produces the optimal edit sequence. In order to bound the error of our algorithm, then, we look at the differences between the unknown optimal cover C_0 and our constructed cover C.

The proof is constructive and rather laborious due to the multiple cases, so we restrict ourselves to an illustration of one of the cases. Let s_a, s_x, s_u, s_y, s_z be substrings of S; we use a prime accent ($'$) to denote that a particular substring was marked as a copy by a given cover. Let π_{tail} be the inversion and deletion portion of π and set $S = s_a \ldots s_x s_u s_y \ldots s_z$, $T = s_a \ldots s_x s_u s_y \ldots s_u \ldots s_z$, $T_{opt} = s_a \ldots s_x s'_u s_y \ldots s_u \ldots s_z$ (this T_{opt} is T renamed according to the optimal cover C_0), and $T_{chosen} = s_a \ldots s_x s_u s_y \ldots s'_u \ldots s_z$ (this T_{chosen} is T renamed according to the chosen covering). Further, suppose that s_u is at index I_1 in S and that s'_u (according to T_{opt}) is inserted at index I_2. The construction proceeds by moving s_u to the location of the insertion of s'_u, then inserting s'_u in the location that s_u previously occupied. Thus, for each wrong choice in the cover, we need 3 inversions to move and 1 to insert. In the given example s_u should be moved to I_2 and s'_u should be inserted at index I_1; now π_{tail} can be applied to this modified sequence to produce T_{chosen}.

If the minimal edit sequence contains n operations, then we have $|C| \leq 2n + 1$. Assuming that each of the selections in C is in error and taking the results from above, the worst-case sequence that can be constructed from C is bounded by $4(2n+1)+n = 9n+4$ operations. Finally, the extension of the edit sequence to include the insertions adds at most n insertions. Thus, the edit sequence produced by the proposed method has at most $10n + 4$ operations. While this error bound is large, it is a constant and it is also unrealistically large—the assumptions used are not truly realizable. Furthermore, the bounds can be easily computed on a case-by-case basis in order to provide information on the accuracy of the results for each run. Thus, we expect the error encountered in practice to be much lower and that further refinements in the algorithm and error analysis should bring the bound to a more reasonable level.

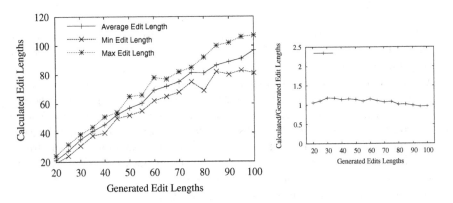

Fig. 1. Experimental results for 200 genes: (a) generated edit length vs. reconstructed length; (b) the ratio of the two.

6   Experimental Results

To test our algorithm and get an estimate of its performance in practice, we ran simulations. We generated pairs of sequences, one the sequence $1, 2, 3, \ldots, n$ (for $n = 200, 400, 800$) and the other derived from the first through an edit sequences. Our edit sequences, of various length, include 80% of randomly generated inversions (the two boundaries of each inversions are uniformly distributed through the array), 10% of deletions (the left end of the deleted string is selected uniformly at random, the length of the deleted string is given by a Gaussian distribution of mean 20 and deviation 7), and 10% insertions (the locus of insertion is uniformly distributed at random and the length of the inserted string is as for deletion), with half of the insertions consisting of new elements and the other half repeating a substring of the current sequence (with the initial position of the substring selected uniformly at random). Thus, in particular, the

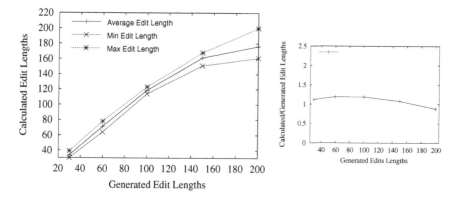

Fig. 2. Experimental results for 400 genes: (a) generated edit length vs. reconstructed length; (b) the ratio of the two.

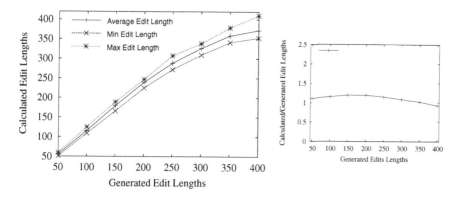

Fig. 3. Experimental results for 800 genes: (a) generated edit length vs. reconstructed length; (b) the ratio of the two.

expected total number of duplicates in the target sequence equals the generated number of edit operations—up to 400 in the case of 800-gene sequences. We ran 10 instances for each combination of parameters (the figures show the average, minimum, and maximum values over the 10 instances).

The results are very gratifying: the error is consistently very low, with the computed edit distance staying below 3% of the length of the generated edit sequence in the linear part of the curve—that is, below saturation. (Of course, when the generated edit sequence gets long, we move into a regime of saturation where the minimum edit sequence becomes arbitrarily shorter than the generated one; our estimated length shows this phenomenon very clearly.) Figures 1, 2, and 3 show our results for sequences of 200, 400, and 800 genes, respectively.

7   Conclusion and Future Directions

An exact polynomial-time algorithm for the computation of genomic distances under insertions, deletions, and inversions remains to be found, but our work takes us a step closer in that direction. More thorough experimental testing will determine how well our algorithm does in practice under different regimes of insertion, deletion, and duplication, but our preliminary results are extremely encouraging. In order to be usable in many reconstruction algorithms, however, a further, and much more complex, computation is required: the median of three genomes. This computation is NP-hard even under inversions only [4,13]—although the algorithms of Caprara [5] and of Siepel and Moret [16] have done well in practice (see, e.g., [10]). Good bounding is the key to such computations; our covering technique may be extendible to median computations.

Acknowledgments. This work is supported by the National Science Foundation under grants ACI 00-81404, DEB 01-20709, EIA 01-13095, EIA 01-21377, and EIA 02-03584.

References

1. D.A. Bader, B.M.E. Moret, and M. Yan. A fast linear-time algorithm for inversion distance with an experimental comparison. *J. Comput. Biol.*, 8(5):483–491, 2001.
2. D. Bryant. The complexity of calculating exemplar distances. In D. Sankoff and J. Nadeau, editors, *Comparative Genomics: Empirical and Analytical Approaches to Gene Order Dynamics, Map Alignment, and the Evolution of Gene Families*, pages 207–212. Kluwer Academic Pubs., Dordrecht, Netherlands, 2000.
3. A. Caprara. Sorting by reversals is difficult. In *Proc. 1st Int'l Conf. on Comput. Mol. Biol. RECOMB97*, pages 75–83. ACM Press, 1997.
4. A. Caprara. Formulations and hardness of multiple sorting by reversals. In *Proc. 3rd Int'l Conf. on Comput. Mol. Biol. RECOMB99*, pages 84–93. ACM Press, 1999.
5. A. Caprara. On the practical solution of the reversal median problem. In *Proc. 1st Workshop on Algs. in Bioinformatics WABI 2001*, volume 2149 of *Lecture Notes in Computer Science*, pages 238–251. Springer-Verlag, 2001.
6. M. Cosner, R. Jansen, B.M.E. Moret, L. Raubeson, L. Wang, T. Warnow, and S. Wyman. An empirical comparison of phylogenetic methods on chloroplast gene order data in Campanulaceae. In D. Sankoff and J. Nadeau, editors, *Comparative Genomics*, pages 99–122. Kluwer Acad. Pubs., 2000.
7. S. Downie and J. Palmer. Use of chloroplast DNA rearrangements in reconstructing plant phylogeny. In P. Soltis, D. Soltis, and J. Doyle, editors, *Plant Molecular Systematics*, pages 14–35. Chapman and Hall, 1992.
8. N. El-Mabrouk. Genome rearrangement by reversals and insertions/deletions of contiguous segments. In *Proc. 11th Ann. Symp. Combin. Pattern Matching CPM 00*, volume 1848 of *Lecture Notes in Computer Science*, pages 222–234. Springer-Verlag, 2000.
9. S. Hannenhalli and P. Pevzner. Transforming cabbage into turnip (polynomial algorithm for sorting signed permutations by reversals). In *Proc. 27th Ann. Symp. Theory of Computing STOC 95*, pages 178–189. ACM Press, 1995.
10. B.M.E. Moret, A.C. Siepel, J. Tang, and T. Liu. Inversion medians outperform breakpoint medians in phylogeny reconstruction from gene-order data. In R. Guigo and D. Gusfield, editors, *Proc. 2nd Int'l Workshop Algorithms in Bioinformatics (WABI'02)*, volume 2452 of *Lecture Notes in Computer Science*, pages 521–536. Springer-Verlag, 2002.

11. R. Olmstead and J. Palmer. Chloroplast DNA systematics: a review of methods and data analysis. *Amer. J. Bot.*, 81:1205–1224, 1994.

12. J. Palmer. Chloroplast and mitochondrial genome evolution in land plants. In R. Herrmann, editor, *Cell Organelles*, pages 99–133. Springer Verlag, 1992.

13. I. Pe'er and R. Shamir. The median problems for breakpoints are NP-complete. *Elec. Colloq. on Comput. Complexity*, 71, 1998.

14. L. Raubeson and R. Jansen. Chloroplast DNA evidence on the ancient evolutionary split in vascular land plants. *Science*, 255:1697–1699, 1992.

15. D. Sankoff. Genome rearrangement with gene families. *Bioinformatics*, 15(11):909–917, 1999.

16. A.C. Siepel and B.M.E. Moret. Finding an optimal inversion median: Experimental results. In O. Gascuel and B.M.E. Moret, editors, *Proc. 1st Int'l Workshop Algorithms in Bioinformatics (WABI'01)*, volume 2149 of *Lecture Notes in Computer Science*, pages 189–203. Springer-Verlag, 2001.

Minimal Unsatisfiable Formulas with Bounded Clause-Variable Difference are Fixed-Parameter Tractable

Stefan Szeider⋆

Department of Computer Science, University of Toronto,
M5S 3G4 Toronto, Ontario, Canada
szeider@cs.toronto.edu

Abstract. Recognition of minimal unsatisfiable CNF formulas (unsatisfiable CNF formulas which become satisfiable by removing any clause) is NP-hard; it was shown recently that minimal unsatisfiable formulas with n variables and $n + k$ clauses can be recognized in time $n^{\mathcal{O}(k)}$. We improve this result and present an algorithm with time complexity $\mathcal{O}(2^k n^4)$ —hence the problem turns out to be fixed-parameter tractable (FTP) in the sense of Downey and Fellows (Parameterized Complexity, Springer Verlag, 1999).
Our algorithm gives rise to an FPT parameterization of SAT ("maximum deficiency") which is incomparable with known FPT parameterizations of SAT like tree-width.

1 Introduction

We consider propositional formulas in conjunctive normal form (CNF) represented as sets of clauses. A formula is *minimal unsatisfiable* if it is unsatisfiable but omitting any of its clauses makes it satisfiable. Recognition of minimal unsatisfiable formulas is computationally hard, shown to be D^P-complete by Papadimitriou and Wolfe [16] (D^P is the class of problems which can be considered as the difference of two NP-problems).

Since for a minimal unsatisfiable formula F the number m of clauses is strictly greater than the number n of variables [1], it is natural to parameterize minimal unsatisfiable formulas with respect to the parameter

$$\delta(F) := m - n,$$

the *deficiency* of F. Following [11] we denote the class of minimal unsatisfiable formulas with deficiency k by MU(k).

It is known that for fixed k, formulas in MU(k) have short resolution refutations and so can be recognized in nondeterministic polynomial time [10]. Moreover, deterministic polynomial time algorithms have been developed for MU(1) and MU(2), based on the very structure of formulas in the respective classes [4,

---

⋆ Supported by the Austrian Science Fund (FWF) Project J2111.

T. Warnow and B. Zhu (Eds.): COCOON 2003, LNCS 2697, pp. 548–558, 2003.
© Springer-Verlag Berlin Heidelberg 2003

11]. Finally it was shown that for any fixed k, formulas in $\mathrm{MU}(k)$ can be recognized in polynomial time [13,7]. The algorithm of [13] relies on the fact that formulas in $\mathrm{MU}(k)$ not only have short resolution refutations, but such refutations can even be found in polynomial time. On the other hand, the algorithm in [7] relies on the fact that the search for satisfying truth assignments can be restricted to certain assignments which correspond to matchings in bipartite graphs (we will describe this approach more detailed in Section 4). Both algorithms have time complexity $n^{\mathcal{O}(k)}$ ([7] provides the more explicit upper bound $\mathcal{O}(n^{k+1/2}l)$ for formulas of length l with n variables).

The degree of the polynomials constituting time bounds of the quoted algorithms [13,7] strongly depends on k, since a "try all subsets of size k"-strategy is embarked. Consequently, even for small k, the algorithms become impracticable for large inputs. The theory of Parameterized Complexity, developed by Downey and Fellows [6], focuses on this issue. A problem is called *fixed-parameter tractable* (*FPT*) if it can be solved in time $\mathcal{O}(f(k) \cdot n^\alpha)$ where n measures the size of the instance and $f(k)$ is any function of the parameter k (the constant α is independent from k).

In this paper we show that $\mathrm{MU}(k)$ is fixed-parameter tractable, stating an algorithm with time complexity $\mathcal{O}(2^k n^4)$. The obtained speedup relies on the interaction of two concepts, *maximum deficiency* and *expansion*, both stemming from graph theory (the graph theoretic concepts carry over to formulas by means of *incidence graphs*, see Section 4). Ultimately, we make use of a characterization of q-expanding bipartite graphs due to Lovász and Plummer [14] (see Theorem 1 below).

Fixed-parameter tractable SAT decision. As a by-product of our algorithm for minimal unsatisfiable formulas, we obtain a general algorithm for SAT decision which runs in time $\mathcal{O}(2^k n^3)$ on instances F if the *maximum deficiency* over all subsets $F' \subseteq F$ is at most k. (For minimal unsatisfiable formulas, deficiency and maximum deficiency agree.) Known parameterizations of SAT which allow FPT decision are based on structural decompositions (tree-width [9], branch-width [2], clique-width [3]; these graph parameters can be applied to CNF formulas via "incidence graphs" or "primal graphs," see [17]). The following items emphasize the significance of our algorithm.

- Maximum deficiency can be computed in polynomial time by matching algorithms [7]; however, computation of tree-width or branch-width is NP-hard, and it is not known whether graphs with fixed clique-width can be recognized in polynomial time.
- Maximum deficiency and the quoted parameters are incomparable: there are formulas with bounded maximum deficiency and arbitrarily large clique-width (resp. tree-width or branch-width); conversely, there are formulas with bounded clique-width (resp. tree-width or branch-width) and arbitrarily large maximum deficiency [17].

The framework of Parameterized Complexity also offers a completeness theory: the hierarchy $W[1] \subseteq W[2] \subseteq \cdots \subseteq W[P]$ contains parameterized problems which are unlikely to be FPT and so can be considered as fixed-parameter *intractable* (see [6] for details). For example, it is $W[2]$-hard to decide whether a formula can be satisfied by instantiating at most k variables: an instance of the $W[2]$-complete problem HITTING SET (see [6]) is nothing but a formula without negative literals. We will mention a $W[P]$-complete satisfiability problem in Section 4.

Because of space constraints, proofs are sketched only; detailed proofs can be found in [17].

2   Notation and Preliminaries

We assume an infinite supply of propositional *variables*. A *literal* is a variable x or a complemented variable $\overline{x}$; if $y = \overline{x}$ is a literal, then we write $\overline{y} = x$; we also use the notation $x^1 = x$ and $x^0 = \overline{x}$. For a set S of literals we put $\overline{S} = \{\,\overline{x} : x \in S\,\}$; S is *tautological* if $S \cap \overline{S} \neq \emptyset$. A *clause* is a finite non-tautological set of literals. A finite set of clauses is a *CNF formula* (or *formula*, for short). The *length* of a formula F is $\sum_{C \in F} |C|$. For a literal x we write $\#_x(F)$ for the number of clauses of F which contain x. A literal x is a *pure literal* if $\#_x(F) \geq 1$ and $\#_{\overline{x}}(F) = 0$; x is a *singular literal* if $\#_x(F) = 1$ and $\#_{\overline{x}}(F) \geq 1$. A literal x *occurs* in a clause C if $x \in C \cup \overline{C}$; $\mathsf{var}(C)$ denotes the set of variables which occur in C. For a formula F we put $\mathsf{var}(F) = \bigcup_{C \in F} \mathsf{var}(C)$. Let F be a formula and $X \subseteq \mathsf{var}(F)$. We denote by F_X the set of clauses of F in which some variable of X occurs; i.e.,

$$F_X := \{\, C \in F : \mathsf{var}(C) \cap X \neq \emptyset\,\}.$$

$F_{(X)}$ denotes the formula obtained from F_X by restricting all clauses to literals over X, i.e.,

$$F_{(X)} := \{\, C \cap (X \cup \overline{X}) : C \in F_X\,\}.$$

A *truth assignment* is a map $\tau : X \rightarrow \{0,1\}$ defined on some set X of variables; we write $\mathsf{var}(\tau) = X$. If $\mathsf{var}(\tau)$ is just a singleton $\{x\}$ with $\tau(x) = \varepsilon$, then we denote τ simply by $x = \varepsilon$. We say that τ is *empty* if $\mathsf{var}(\tau) = \emptyset$. A truth assignment τ is *total* for a formula F if $\mathsf{var}(\tau) = \mathsf{var}(F)$. For $x \in \mathsf{var}(\tau)$ we define $\tau(\overline{x}) = 1 - \tau(x)$. For a truth assignment τ and a formula F, we put

$$F[\tau] = \{\, C \setminus \tau^{-1}(0) : C \in F,\ C \cap \tau^{-1}(1) = \emptyset\,\};$$

i.e., $F[\tau]$ denotes the result of instantiating variables according to τ and applying the usual simplifications. A truth assignment τ *satisfies* a clause if the clause contains some literal x with $\tau(x) = 1$; τ satisfies a formula F if it satisfies all clauses of F (i.e., if $F[\tau] = \emptyset$). A formula is *satisfiable* if it is satisfied by some truth assignment; otherwise it is *unsatisfiable*. A formula is *minimal unsatisfiable* if it is unsatisfiable, and every proper subset of it is satisfiable. We say that formulas F and F' are *equisatisfiable* (in symbols $F \equiv_{sat} F'$) if either both are satisfiable or both are unsatisfiable.

A truth assignment α is *autark* for a formula F if $\mathrm{var}(\alpha) \subseteq \mathrm{var}(F)$ and α satisfies $F_{\mathrm{var}(\alpha)}$; that is, α satisfies all affected clauses. Note that the empty assignment is autark for every formula, and that any total satisfying assignment of a formula is autark. The key feature of autark assignments is the following observation of [15].

Lemma 1 *If α is an autark assignment of a formula F, then $F[\alpha]$ is an equi-satisfiable subset of F.*

Thus, in particular, minimal unsatisfiable formulas have no autark assignments except the empty assignment. If x^ϵ is a pure literal of F, $\epsilon \in \{0,1\}$, then $x = \epsilon$ is an autark assignment (and $F[x = \epsilon]$ can be obtained from F by the "pure literal rule").

If C_1, C_2 are clauses and $C_1 \cap \overline{C_2} = \{x\}$ holds for some literal x, then the clause $(C_1 \cup C_2) \setminus \{x, \overline{x}\}$ is the *resolvent* of C_1 and C_2. A *resolution refutation* of a formula F is a sequence of clauses $C_1, \ldots, C_q$ such that C_q is empty and each clause C_i either belongs to F or is the resolvent of preceding clauses. A resolution refutation is *regular* if literals which are removed from a clause C_i by resolution are not later re-introduced in a clause depending on C_i (see [17] for a more detailed definition).

Consider a formula F and a literal x of F. We obtain a formula F' from F by adding all possible resolvents w.r.t. x, and by removing all clauses in which x occurs. We say that F' is obtained from F by *Davis-Putnam resolution* and we write $\mathrm{DP}_x(F) = F'$. It is well known that $F \equiv_{sat} \mathrm{DP}_x(F)$. Usually, $\mathrm{DP}_x(F)$ contains more clauses than F, however, if $\#_x(F) \leq 1$ or $\#_{\overline{x}}(F) \leq 1$, then clearly $|\mathrm{DP}_x(F)| < |F|$. In the sequel we will focus on $\mathrm{DP}_x(F)$ where x is a singular literal of F.

3   Graph Theoretic Tools

We denote a bipartite graph G by the triple (V_1, V_2, E) where V_1 and V_2 give the bipartition of the vertex set of G, and E denotes the set of edges of G. An edge between $v_1 \in V_1$ and $v_2 \in V_2$ is denoted as ordered pair (v_1, v_2). $N_G(X)$ denotes the set of all vertices y adjacent to some $x \in X$ in G, i.e., $N_G(X)$ is the (open) neighborhood of X. For graph theoretic terminology not defined here, the reader is referred to [5].

A *matching* M of a graph G is a set of independent edges of G; i.e., distinct edges in M have no vertex in common. A vertex of G is called *matched by M*, or *M-matched*, if it is incident with some edge in M; otherwise it is *exposed by M*, or *M-exposed*. A matching M of G is a *maximum matching* if there is no matching M' of G with $|M'| > |M|$; we denote by $\mathcal{M}(G)$ the set of maximum matchings of G. A maximum matching of a bipartite graph on p vertices and q edges can be found in time $\mathcal{O}(p^{1/2}q)$ by the algorithm of Hopcroft and Karp (see, e.g., [14]). Consider a bipartite graph $G = (V_1, V_2, E)$. We say that G is *q-expanding* if $q \geq 0$ is an integer such that $|N_G(X)| \geq |X| + q$ holds for every nonempty set $X \subseteq V_1$. Note that by Hall's Theorem, G is 0-expanding if and only if G has a matching of size $|V_1|$ (see [14]). The *deficiency* of G is defined as $\delta(G) := |V_2| - |N_G(V_2)|$ (if V_1 contains no isolated vertices, then $\delta(G) = |V_2| - |V_1|$). The *maximum deficiency* of G is defined as $\delta^*(G) := \max_{Y \subseteq V_2} |Y| - |N_G(Y)|$. Note that $\delta^*(G) \geq 0$ follows by taking $Y = \emptyset$.

Let M be a matching of a graph G. A path P in G is called M-*alternating* if edges of P are alternately in and out of M; an M-alternating path is M-*augmenting* if both of its ends are M-exposed. If P is an M-augmenting path, then the *symmetric difference* of M and the set of edges which lie on P is a matching of size $|M| + 1$. In this case we say that M' is obtained from M by *augmentation*. Conversely, by a well-known result of Berge (see, e.g., [14, Theorem 1.2.1]) a matching M is a maximum matching if there is no M-augmenting path. Since M-alternating paths of a bipartite graph $G = (V_1, V_2, E)$ are just directed paths in the digraph obtained from G by orienting the edges in M from V_1 to V_2, and orienting the edges in $E \setminus M$ from V_2 to V_1. Hence we can find an M-augmenting path by breadth-first-search starting from the set of M-exposed vertices in V_2 in linear time, $\mathcal{O}(|V_1 \cup V_2| + |E|)$. Consequently, if M exposes s_i vertices of V_i, $i = 1, 2$, then we can find some $M' \in \mathcal{M}(G)$ in time $\mathcal{O}(\min(s_1, s_2) \cdot (|E| + |V_1 \cup V_2|))$ by augmentation (which is often more efficient than to construct a maximum matching from scratch). We will refer to this way of finding a maximum matching the *augmentation procedure*.

Let M be a matching of G. We define $R_{G,M}$ as the set of vertices of G which can be reached from some M-exposed vertex in V_2 by an M-alternating path. By means of the above breadth-first-search approach we can easily obtain the basic graph theoretic results needed for our considerations (for proofs see [17]):

Lemma 2 *Given a bipartite graph* $G = (V_1, V_2, E)$, $V = V_1 \cup V_2$, *and* $M \in \mathcal{M}(G)$, *then the following statements hold true.*

1. $R_{G,M}$ *can be obtained in time* $\mathcal{O}(|V| + |E|)$.
2. *No edge joins vertices in* $V_1 \setminus R_{G,M}$ *with vertices in* $V_2 \cap R_{G,M}$; *no edge in* M *joins vertices in* $V_1 \cap R_{G,M}$ *with vertices in* $V_2 \setminus R_{G,M}$.
3. *All vertices in* $V_1 \cap R_{G,M}$ *and* $V_2 \setminus R_{G,M}$ *are matched vertices.*
4. *If* G *is not 0-expanding, then* $|V_1 \setminus R_{G,M}| > |N_G(V_1 \setminus R_{G,M})|$.
5. $|V_2 \cap R_{G,M}| - |N_G(V_2 \cap R_{G,M})| = |V_2| - |M|$.
6. *If* $R_{G,M} \neq \emptyset$, *then* $R_{G,M}$ *induces a 1-expanding subgraph of* G.
7. M *exposes exactly* $\delta^*(G)$ *vertices of* V_2.
8. $|Y| - |N_G(Y)| \leq \delta^*(G) - 1$ *holds for every proper subset* Y *of* V_2.

We note in passing that we get the same set $R_{G,M}$ for every $M \in \mathcal{M}(G)$; this follows from the fact that every $M' \in \mathcal{M}(G)$ matches the vertices in $V_1 \cap R_{G,M}$ (these vertices belong to every minimum vertex cover, see [1]).

Theorem 1 below is due to Lovász and Plummer [14, Theorem 1.3.6] and provides the basis for an efficient test for q-expansion. We state the theorem using the following construction: From a bipartite graph $G = (V_1, V_2, E)$, $x \in V_1$, and $q \geq 1$, we obtain the bipartite graph G_{qx} by adding new vertices $x_1, \ldots, x_q$ to V_1 and adding edges such that the new vertices have exactly the same neighbors as x; i.e., $G_{qx} = (V_1 \cup \{x_1, \ldots, x_q\}, V_2, E \cup \{x_i y : xy \in E\})$.

Theorem 1 (Lovász and Plummer [14]) *A 0-expanding bipartite graph $G = (V_1, V_2, E)$ is q-expanding if and only if G_{qx} is 0-expanding for every $x \in V_1$.*

Lemma 3 *Given a bipartite graph $G = (V_1, V_2, E)$ and $M \in \mathcal{M}(G)$. For every fixed integer $q \geq 0$, deciding whether G is q-expanding and, if G is not q-expanding, finding a "witness set" $X \subseteq V_1$ with $|N_G(X)| < |X| + q$, can be performed in time $\mathcal{O}(|V_1| \cdot |E| + |V_2|)$.*

Proof. (Sketch) For a vertex $x \in V_1$ we can construct $M_x \in \mathcal{M}(G_{qx})$ by augmenting M at most q times. By Theorem 1, G is q-expanding if and only if $|M_x| = |V_1| + q$ holds for every $x \in V_1$. Assume that $|M_x| < |V_1| + q$ for some $x \in V_1$. We form R_{G_{qx}, M_x} and consider $X' := (V_1 \cup \{x_1, \ldots, x_q\}) \setminus R_{G_{qx}, M_x}$. Lemma 2(4) yields $|N_{G_{qx}}(X')| < |X'|$. By means of Lemma 2(2) one can show that $\{x, x_1, \ldots, x_q\} \subseteq X'$. For $X := X' \setminus \{x_1, \ldots, x_q\}$ we have $|N_G(X)| = |N_{G_{qx}}(X')| < |X'| = |X| - q$; thus X is a witness set. □

4 Matchings and Expansion of Formulas

To every formula F we associate a bipartite graph $I(F)$, the *incidence graph* of F, whose vertices are the clauses and variables of F, and where each clause is adjacent to the variables which occur in it; that is, $I(F) = (\mathsf{var}(F), F, E(F))$ with $(x, C) \in E(F)$ if and only if $x \in \mathsf{var}(C)$; see Fig. 1. for an example. By means

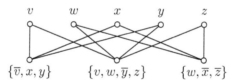

Fig. 1. Incidence graph of the formula $F = \{\{\overline{v}, x, y\}, \{v, w, \overline{y}, z\}, \{w, \overline{x}, \overline{z}\}\}$.

of this construction, concepts for bipartite graphs apply directly to formulas. In particular, we will speak of q-expanding formulas, matchings of formulas, and the (maximum) deficiency of formulas. That is, a formula F is q-*expanding* if and only if $|F_X| \geq |X| + q$ for every nonempty set $X \subseteq \mathsf{var}(F)$. The *deficiency* of a formula F is $\delta(F) = |F| - |\mathsf{var}(F)|$; its *maximum deficiency* is $\delta^*(F) = \max_{F' \subseteq F} \delta(F')$. If $\mathsf{var}(F) = \emptyset$, then F is q-expanding for any q, and we have $\delta^*(F) = |F| \leq 1$. $\mathcal{M}(F)$ denotes the set of maximum matchings of F. Note that 1-expanding formulas are exactly the "matching lean" formulas of [12]. In terms of formulas, parts 7 and 8 of Lemma 2 read as follows (see [12] for an alternate proof of Lemma 5).

Lemma 4 *Every $M \in \mathcal{M}(F)$ exposes exactly $\delta^*(F)$ clauses.*

Lemma 5 *If F is 1-expanding and $F' \subsetneq F$, then $\delta^*(F') \leq \delta^*(F) - 1$.*

A matching M of a formula F gives rise to a partial truth assignment τ_M as follows. For every $(x, C) \in M$ we put $\tau_M(x) = 1$ if $x \in C$, and $\tau_M(x) = 0$ if $\overline{x} \in C$. If $|M| = |F|$, then τ_M evidently satisfies F; thus we have the following (this observation has been made in [18] and [1]).

Lemma 6 *If a formula F has a matching which matches all clauses, i.e., if $\delta^*(F) = 0$, then F is satisfiable.*

Formulas with maximum deficiency 0 are termed *matched formulas* in [8]. For example, the formula F shown in Fig. 1. is matched since the matching $M = \{(v, \{\overline{v}, x, y\}), (w, \{v, w, \overline{y}, z\}), (x, \{w, \overline{x}, \overline{z}\})\}$ matches all clauses; M gives rise to the satisfying truth assignment τ_M with $\tau_M(v) = 0$, $\tau_M(w) = 1$, $\tau_M(x) = 0$.

Lemma 2 yields the following result (which is essentially [7, Lemma 10]).

Lemma 7 *Given a formula F of length l and $M \in \mathcal{M}(F)$, then we can find in time $\mathcal{O}(l)$ an autark assignment α of F such that $F[\alpha]$ is 1-expanding; moreover, $M \cap E(F[\alpha]) \in \mathcal{M}(F[\alpha])$.*

In view of Lemma 1 it follows that minimal unsatisfiable formulas F are 1-expanding and so $\delta^*(F) = \delta(F)$ (see also [1,8]). The next result extends Lemma 6 to formulas with positive maximum deficiency.

Theorem 2 (Fleischner, et al. [7]) *A formula F is satisfiable if and only if $F[\tau]$ is a matched formula for some truth assignment τ with $|\mathsf{var}(\tau)| \leq \delta^*(F)$.*

In particular, for $\delta^*(F) \leq 1$, Theorem 2 yields the following.

Lemma 8 *Let F be a formula of length l on n variables. If $\delta^*(F) \leq 1$, then we can find a satisfying truth assignment of F (if it exists) in time $\mathcal{O}(nl)$.*

Theorem 2 yields an $n^{\mathcal{O}(k)}$ time algorithm for satisfiability of formulas with $\delta^*(F) \leq k$, since for checking satisfiability we just have to consider all instantiations of at most k variables and to check whether the resulting formulas are matched. We note that a similar problem, k-INDUCED 3-CNF SATISFIABILITY, is $W[P]$-complete (see [6]); there, it is asked whether we can instantiate at most k variables such that the resulting formula "unravels" – i.e., if the resulting formula can be satisfied by recursive instantiation of variables which occur in unit clauses.

5   Main Reductions

We call a formula F δ^*-*critical* if $\delta^*(F[x = \epsilon]) \leq \delta^*(F) - 1$ holds for every $(x, \epsilon) \in \mathsf{var}(F) \times \{0, 1\}$. The objective of this section is to reduce a given formula F efficiently to a δ^*-critical formula F' ensuring $\delta^*(F') \leq \delta^*(F)$ and $F \equiv_{sat} F'$. Thus δ^*-critical formulas constitute a "problem kernel" in the sense of [6].

The next lemma (a straightforward proof can be found in [17]) pinpoints a sufficient condition for formulas being δ^*-critical.

Lemma 9 *2-expanding formulas without pure or singular literals are δ^*-critical.*

Consider a sequence $S = (F_0, M_0), \ldots, (F_q, M_q)$ where F_i are formulas and $M_i \in \mathcal{M}(F_i)$, $0 \leq i \leq q$. We call S a *reduction sequence* (starting from (F_0, M_0)) if for each $i \in \{1, \ldots, q\}$ one of the following holds:

- $F_i = F_{i-1}[\alpha_i]$ for some nonempty autark assignment α_i of F_{i-1}.
- $F_i = \mathrm{DP}_{x_i}(F_{i-1})$ for a singular literal x_i of F_{i-1}.

Note that $\mathsf{var}(F_i) \subsetneq \mathsf{var}(F_{i-1})$, hence $q \leq |\mathsf{var}(F_0)|$. By Lemma 1 and since always $\mathrm{DP}_x(F) \equiv_{sat} F$, F_0 and F_q are equisatisfiable. The following can be verified easily.

Lemma 10 *Let $(F_0, M_0), \ldots, (F_q, M_q)$ be a reduction sequence. Any satisfying truth assignment τ_q of F_q can be extended to a satisfying truth assignment τ_0 of F_0; any regular resolution refutation R_q of F_q can be extended to a regular resolution refutation R_0 of F_0.*

Lemma 11 *Let F_0 be a formula on n variables with $\delta^*(F_0) \leq n$, and let $M_0 \in \mathcal{M}(F_0)$. We can construct in time $\mathcal{O}(n^3)$ a reduction sequence $S = (F_0, M_0), \ldots, (F_q, M_q)$, $q \leq n$, such that exactly one of the following holds.*

1. $\delta^*(F_q) \leq \delta^*(F_0) - 1$;
2. $\delta^*(F_q) = \delta^*(F_0)$, F_q *is 1-expanding and has no pure or singular literals.*

Proof. (Sketch.) We construct S inductively; assume that we have already constructed $(F_0, M_0), \ldots, (F_{i-1}, M_{i-1})$ for some $i \geq 1$; we obtain F_i as follows:

1. IF F_{i-1} is not 1-expanding THEN put $F_i := F_{i-1}[\alpha]$ where α is a nonempty autark assignment of F_{i-1} supplied by Lemma 7.
2. ELSEIF F_{i-1} has a pure literal x^ϵ, $\epsilon \in \{0, 1\}$, THEN put $F_i = F_{i-1}[x = \epsilon]$ ($x = \epsilon$ is an autark assignment of F_{i-1}; cf. the discussion in Section 2).
3. ELSEIF F_{i-1} has a singular literal x^ϵ, $\epsilon \in \{0, 1\}$, THEN put $F_i = \mathrm{DP}_x(F_{i-1})$.

As soon as case 2 or case 3 with $|F_i| \leq |F_{i-1}| - 1$ occurs, we have $\delta^*(F_i) \leq \delta^*(F_{i-1}) - 1$; hence we terminate the construction of S. In terminal cases we can use the Hopcroft-Karp algorithm for obtaining M_i; thus the cost of a terminal case is $\mathcal{O}(n^3)$. In non-terminal cases, however, M_i can be constructed directly from M_{i-1} in time $\mathcal{O}(n^2)$. □

By the above results we can efficiently reduce a given formula until we end up with a formula which is 1-expanding and has no pure or singular literals. Next we present further reductions which yield δ^*-critical formulas.

Lemma 12 *Let F be a 1-expanding formula without pure or singular literals and let $X \subseteq \mathsf{var}(F)$ with $|F_X| \leq |X| + 1$. Then $F \setminus F_X \equiv_{sat} F$ and $\delta^*(F \setminus F_X) \leq \delta^*(F) - 1$.*

Proof. (Sketch.) $F_{(X)}$ must be satisfiable, since otherwise it would be minimal unsatisfiable, but this is impossible, since every minimal unsatisfiable formula with deficiency 1 different from $\{\emptyset\}$ has a singular literal (see [4]). Any satisfying total assignment α of $F_{(X)}$ is a nonempty autark assignment of F with $F[\alpha] = F \setminus F_X$. The result now follows by Lemmas 1 and 5. $\square$

Lemma 13 *Let F be a 1-expanding formula without pure or singular literals, $m = |F|$, $n = |\mathrm{var}(F)|$, and let $M \in \mathcal{M}(F)$. We need at most $\mathcal{O}(n^2 m)$ time to decide whether F is 2-expanding, and if it is not, to find an autark assignment α of F with $\delta^*(F[\alpha]) \le \delta^*(F) - 1$ and some $M' \in \mathcal{M}(F[\alpha])$.*

Proof. (Sketch.) In view of Lemma 3, $\mathcal{O}(n^2 m)$ time suffices to decide whether F is 2-expanding, and if it is not, to obtain a set $X \subseteq \mathrm{var}(F)$ with $|F_X| = |X| + 1$ ($\delta^*(F_{(X)}) \le 1$). By the preceding Lemma, $F_{(X)}$ is satisfiable, and by Lemma 8 we can find a total satisfying assignment α in time $\mathcal{O}(n^2 m)$; $\delta^*(F[\alpha]) \le \delta^*(F) - 1$ follows (Lemmas 1 and 5). We need at most one augmentation to obtain M' from $M \cap E(F[\alpha])$. $\square$

We summarize the results of this section:

Theorem 3 *Let F_0 be a formula on n variables with $\delta^*(F_0) \le n$, and let $M_0 \in \mathcal{M}(F_0)$. We can obtain in time $\mathcal{O}(n^3)$ a reduction sequence $(F_0, M_0), \ldots, (F_q, M_q)$, $q \le n$, such that exactly one of the following holds:*

1. $\delta^*(F_q) \le \delta^*(F_0) - 1$;
2. $\delta^*(F_q) = \delta^*(F_0)$ and F_q is δ^*-critical.

6   Proof of Main Results

Theorem 4 *Satisfiability of formulas with n variables and maximum deficiency k can be decided in time $\mathcal{O}(2^k n^3)$. The decision is certified by a satisfying truth assignment or a regular resolution refutation of the input formula.*

Proof. (Sketch.) We may assume $k \le n$, since otherwise the theorem holds by trivial reasons. We form a search tree T where each vertex v is labeled by a reduction sequence S_v as follows (if $S_v = (F_0, M_0), \ldots, (F_r, M_r)$, then we write $\mathrm{first}(v) = F_0$ and $\mathrm{last}(v) = F_r$). We construct some $M \in \mathcal{M}(F)$ by the Hopcroft-Karp algorithm and start with a root vertex v_0; we obtain S_{v_0} applying Theorem 3 to (F, M). Assume now that we have partially constructed T. We halt as soon as $\mathrm{var}(\mathrm{last}(v)) = \emptyset$ for all leaves v. If $\mathrm{var}(\mathrm{last}(v)) \ne \emptyset$ for a leaf v, $S_v = (F_0, M_0), \ldots, (F_r, M_r)$, then by Theorem 3, either (i) $\delta^*(F_r) \le \delta^*(F_0) - 1$, or (ii) $\delta^*(F_r) = \delta^*(F_0)$ and F_r is δ^*-critical. In the first case we add a single child v' to v, and we label v' by a reduction sequence starting from (F_r, M_r); i.e., $\mathrm{first}(v') = \mathrm{last}(v)$. In the second case we add two children v' and v'', labeled by reduction sequences starting from $(F_r[x = 0], M^0)$ and $(F_r[x = 1], M^1)$, respectively (the matchings M^0, M^1 are obtained by the Hopcroft-Karp algorithm).

We observe that the height of T is indeed at most $\delta^*(F) = k$, and so T has at most $2^k - 1$ vertices; thus $\mathcal{O}(2^k n^3)$ time suffices to construct T (Theorem 3). Since F is satisfiable if an only if $\mathsf{last}(v)$ is satisfiable for at least one leaf v of T, the first part of the theorem holds true. By means of Lemma 10 we can read off from T a satisfying assignment or a regular resolution refutation.   □

Theorem 5 *Minimal unsatisfiable formulas with n variables and $n + k$ clauses can be recognized in time $\mathcal{O}(2^k n^4)$.*

Proof. If $k \geq n$, then the theorem holds by trivial reasons, since we can enumerate all total truth assignments of F in time $\mathcal{O}(2^n)$; hence we assume $k < n$. Let $F = \{C_1, \ldots, C_m\}$, $m = n + k < 2n$. If F is minimal unsatisfiable, then it must be 1-expanding and so $\delta^*(F) = \delta(F) = k$; the latter can be checked efficiently (Lemma 7). Furthermore, we have to check whether F is unsatisfiable, and whether $F_i := F \setminus \{C_i\}$ is satisfiable for all $i \in \{1, \ldots, m\}$. This can be accomplished by $m + 1$ applications of Theorem 4 (we have $\delta^*(F_i) \leq k - 1$ by Lemma 5). Thus the time complexity $\mathcal{O}((m + 1)2^k n^3) \leq \mathcal{O}(2^k n^4)$ follows.   □

References

1. R. Aharoni and N. Linial. Minimal non-two-colorable hypergraphs and minimal unsatisfiable formulas. *J. Combin. Theory Ser. A*, 43:196–204, 1986.
2. M. Alekhnovich and A. A. Razborov. Satisfiability, branch-width and Tseitin tautologies. In *Proceedings of the 43rd IEEE FOCS*, pages 593–603, 2002.
3. B. Courcelle, J. A. Makowsky, and U. Rotics. On the fixed parameter complexity of graph enumeration problems definable in monadic second-order logic. *Discr. Appl. Math.*, 108(1-2):23–52, 2001.
4. G. Davydov, I. Davydova, and H. Kleine Büning. An efficient algorithm for the minimal unsatisfiability problem for a subclass of CNF. *Ann. Math. Artif. Intell.*, 23:229–245, 1998.
5. R. Diestel. *Graph Theory*. Springer Verlag, New York, 2nd edition, 2000.
6. R. G. Downey and M. R. Fellows. *Parameterized Complexity*. Springer Verlag, 1999.
7. H. Fleischner, O. Kullmann, and S. Szeider. Polynomial-time recognition of minimal unsatisfiable formulas with fixed clause-variable difference. *Theoret. Comput. Sci.*, 289(1):503–516, 2002.
8. J. Franco and A. Van Gelder. A perspective on certain polynomial time solvable classes of satisfiability. *Discr. Appl. Math.*, 125:177–214, 2003.
9. G. Gottlob, F. Scarcello, and M. Sideri. Fixed-parameter complexity in AI and nonmonotonic reasoning. *Artificial Intelligence*, 138(1-2):55–86, 2002.
10. H. Kleine Büning. An upper bound for minimal resolution refutations. In *Proc. CSL'98*, volume 1584 of *LNCS*, pages 171–178. Springer Verlag, 1999.
11. H. Kleine Büning. On subclasses of minimal unsatisfiable formulas. *Discr. Appl. Math.*, 107(1–3):83–98, 2000.
12. O. Kullmann. Lean clause-sets: Generalizations of minimally unsatisfiable clause-sets. To appear in *Discr. Appl. Math.*
13. O. Kullmann. An application of matroid theory to the SAT problem. In *Fifteenth Annual IEEE Conference on Computational Complexity*, pages 116–124, 2000.

14. L. Lovász and M. D. Plummer. *Matching Theory*. North-Holland Publishing Co., Amsterdam, 1986.
15. B. Monien and E. Speckenmeyer. Solving satisfiability in less than 2^n steps. *Discr. Appl. Math.*, 10:287–295, 1985.
16. C. H. Papadimitriou and D. Wolfe. The complexity of facets resolved. *J. Comput. System Sci.*, 37(1):2–13, 1988.
17. S. Szeider. Minimal unsatisfiable formulas with bounded clause-variable difference are fixed-parameter tractable. Technical Report TR03–002, Revision 1, *Electronic Colloquium on Computational Complexity* (ECCC), 2003.
18. C. A. Tovey. A simplified NP-complete satisfiability problem. *Discr. Appl. Math.*, 8(1):85–89, 1984.

Author Index

Lecture Notes in Computer Science

For information about Vols. 1–2632
please contact your bookseller or Springer-Verlag